Conversion Factors

Multiplying Factors to convert FROM \ TO	I_B (or i_b)	I_C (or i_c)	I_E (or i_e)
I_B (or i_b)	1	β or $\frac{\alpha}{1-\alpha}$	$(1+\beta)$ or $\frac{1}{1-\alpha}$
I_C (or i_c)	$\frac{1}{\beta}$ or $\frac{1-\alpha}{\alpha}$	1	$\frac{1+\beta}{\beta}$ or $\frac{1}{\alpha}$
I_E (or i_e)	$\frac{1}{1+\beta}$ or $(1-\alpha)$	$\frac{\beta}{1+\beta}$ or α	1

Properties of the Basic Differential Amplifier Circuit Configurations

	Unbalanced Input, Balanced Output	Balanced Input, Balanced Output	Balanced Input, Unbalanced Output
Circuit	Fig. 16-5, Fig. 16-8	Fig. 16-6, Fig. 16-9*a*	Fig. 16-7, Fig. 16-9*b*
Voltage gain without external R_E	$\frac{\beta R_L}{(1+\beta)r'_e} \approx \frac{R_L}{r'_e}$	$\frac{\beta R_L}{(1+\beta)r'_e} \approx \frac{R_L}{r'_e}$	$\frac{R_L}{2(1+\beta)r'_e} \approx \frac{R_L}{2r'_e}$
Input resistance without external R_E[a]	$2(1+\beta)r'_e = 2h_{ie}$	$(1+\beta)r'_e = h_{ie}$	$(1+\beta)r'_e = h_{ie}$
Voltage gain with external R_E	$\frac{\beta R_L}{(1+\beta)(r'_e+R_E)} \approx \frac{R_L}{r'_e+R_E}$	$\frac{\beta R_L}{(1+\beta)(r'_e+R_E)} \approx \frac{R_L}{r'_e+R_E}$	$\frac{\beta R_L}{2(1+\beta)(r'_e+R_E)} \approx \frac{R_L}{2(r'_e+R_E)}$
Input resistance with external R_E[a]	$2(1+\beta)(r'_e+R_E)$	$(1+\beta)(r'_e+R_E)$	$(1+\beta)(r'_e+R_E)$

[a]This value is in parallel with any R_B that is used.

Fundamentals of

ELECTRONICS

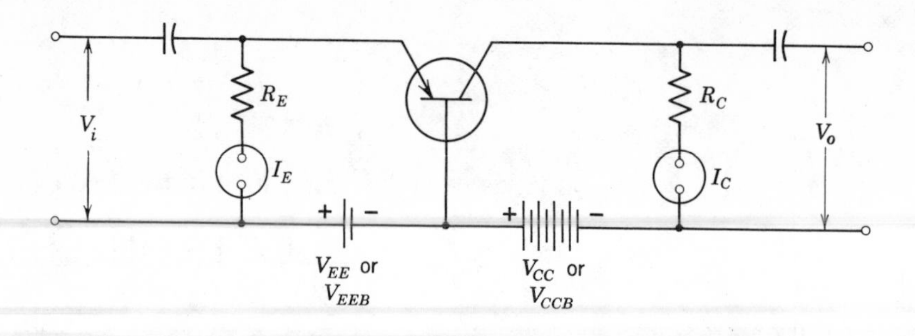

$$p_C = v_C i_C \qquad p_o = v_c i_c \qquad P_o = V_c I_c$$
$$V_i = h_i I_i + h_r V_o = V_1 = h_{11} I_1 + h_{12} V_2$$
$$I_o = h_f I_i + h_o V_o = I_2 = h_{21} I_1 + h_{22} V_2$$

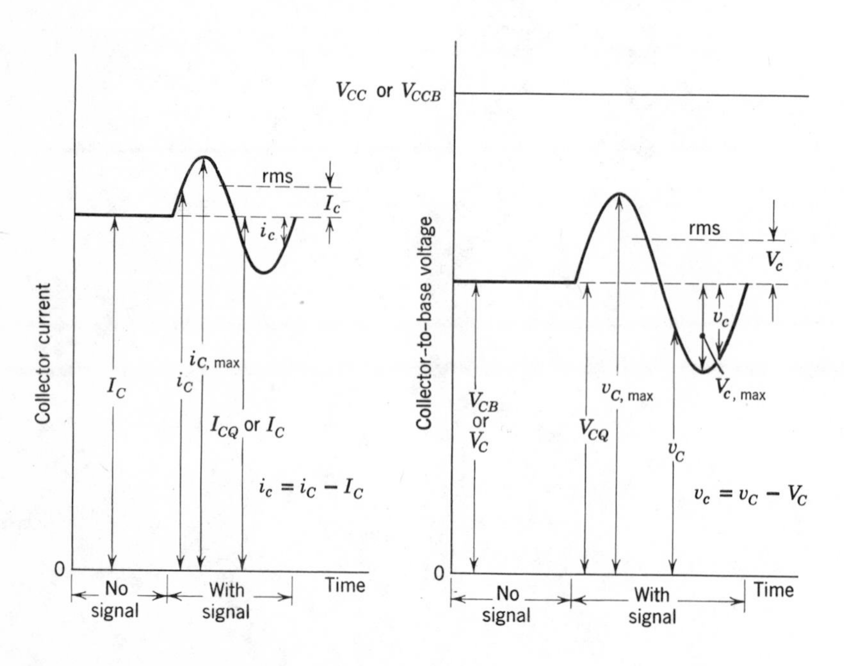

1. Lower-case symbols are instantaneous values.
2. Upper-case subscripts are d-c and total instantaneous values.
3. Lower-case subscripts are for varying component values.
4. If necessary, to separate, use subscripts *max* for peak values and *rms* for effective values.
5. If necessary, use third upper-case subscript on electrode voltage to indicate reference; e.g. V_{EE} or V_{EEB}.
6. Small-signal conditions: lower-case *p*, *h*, and *z*.
7. z_{in} (lower-case) is input impedance when the external impedance is Z_o (upper-case).
8. z_{out} (lower-case) is output impedance when the external impedance is z_s (upper-case).

Fundamentals of ELECTRONICS

SECOND EDITION

E. Norman Lurch

Professor, Electrical Technology
The State University of New York at Farmingdale

JOHN WILEY & SONS, INC.

New York • Chichester • Brisbane • Toronto

Library of Congress Catalogue Card Number: 70-125273

ISBN 0-471-55520-7

Printed in the United States of America

20 19 18 17 16 15 14 13 12 11

To my son,

HENRY S. LURCH, R.I.T.,E.E., '70,

in the hope that this book may be useful in his engineering career.

PREFACE

The following paragraphs, taken from the preface to the original edition of this book, apply to this second edition as well.

"This book is planned to meet the need of the technician who is to work in the field of electronics. It is intended to provide the firm, solid background in fundamentals that is necessary for the study of the more specialized aspects of electronics .... A student using this book should have a working knowledge of d-c fundamentals and, if not prerequisite, he should be studying a-c circuits concurrently with Chapters 1 through 5 which deal with electron devices.

"The level of the text material enables the student to solve such problems as gain calculations, power outputs, graphical solutions, and the determination of modulation applied to a carrier wave. It is not my intention, however, in this book to prepare the user to handle the design calculations and original derivations that are the premise of the electronic engineer. A student with a good working knowledge of algebra and right-angle trigonometry should not have difficulty with the solution of the problems."

In a decade, the electronic art has proliferated in many directions. Most of the text has been rewritten to account for these trends. Solid state takes priority over the electron tube. The integrated circuit and the field effect transistor are treated throughout the book as the need or the application arises. The general approach stresses the model in analyzing a signal circuit.

The first five chapters cover both solid-state and electron-tube devices. The topic of rectifiers (Chapter 6) is extended to include the silicon-controlled rectifier and the triac (Chapter 7). Chapters on decibels (Chapter 8), on auxiliary components (Chapter 9), and on a-c circuits (Chapter 10) have been retained, but now the emphasis is away from the "audio" aspect.

Section 11-8 (Transistor Bias Circuits), because of the importance of the topic, probably will require considerably more class time than its length would indicate. Also the content of Chapter 12 (Models and Small-Signal Analysis) is of major importance. Chapter 13, on hybrid-parameter analysis, is retained as a separate topic that is not a prerequisite for studying the later chapters.

The high-frequency aspect of Chapter 15 (R-C Coupled Amplifiers) is taken as the high-frequency end of the "audio" spectrum. As a result the sophisticated hybrid-π approach is not required and this approach is discussed in Chapter 23 (High-Frequency Amplifiers) where I believe it properly belongs.

Chapter 16 (Special Amplifiers) is particularly important as a general approach to complex, packaged, integrated circuits. The content of Chapter 17 on temperature should be introduced as early as possible in an electronics course in order to understand that a proper heat sink can prevent unnecessary semiconductor failures and burnouts. The emphasis in Chapter 19 (Class-C Amplifiers), in contrast to the rest of the book, is on the electron tube. In the field of high-power RF amplifiers, the solid-state device still cannot compete economically with the electron tube. A generalized approach to feedback (Chapter 20) leads immediately to the operational amplifier, which is treated in detail. The different practical forms of feedback are separately treated.

The sinusoidal oscillator (Chapter 21) is developed from the model. Chapter 22, on nonsinusoidal oscillators, is included in order to introduce a number of circuit concepts that do not readily fit into the flow of the other chapters. Modulation (Chapter 24) and detection (Chapter 25) are included in order to give some description of the communications aspects of electronics. The rapid expansion of electronics in general, during the last decade, has tended to shunt communications to special courses and to specialists. I believe that communications must be maintained in the mainstream of a general electronics program.

In writing this revision, I have tried to retain the approach and flavor that contributed to the success of the original edition. Problems are now found at the end of each section instead of at the end of each chapter. As a result, there are more problems. I have used the problems during this past year in my own classes, and I believe that most of the "bugs" have been eliminated. Also at the conclusion of each chapter there is a set of oral review questions.

When John Lombardo, one of my students with an incisive mind, learned that I was in the process of preparing a second edition of this book, he insisted on reviewing the manuscript on this premise: "A professor who knows the field writes a book and another professor who knows the field reviews the book. The result is that a student who doesn't know the field cannot read the book."

Stony Brook, New York, 1970

E. Norman Lurch

CONTENTS

One Atoms, Electrons, and Current Flow 1

1-1 Composition of Matter 1
1-2 Structure of an Atom 2
1-3 Electron Flow in Solids 6
1-4 Emission 9
1-5 The Structure of a Crystal 12
1-6 *N*-Type Material 15
1-7 *P*-Type Material 17
1-8 Light on Crystals 18
1-9 Mechanism of a Gas Discharge 19
1-10 Generation of Light 22

Two Two-Element Electron Devices 27

2-1 The Semiconductor as a Diode 27
2-2 The Simple Rectifier 34
2-3 The Vacuum Diode 36
2-4 The Zener Diode 41
2-5 Dynamic Diode Characteristics and Diode Models 44
2-6 Cold-Cathode Diodes 51
2-7 Light-Emitting Diodes 52

Three Basic Three-Element Electron Devices 55

3-1 The Action of a Transistor 55
3-2 The Common-Base Circuit 59
3-3 The Common-Emitter Circuit 62
3-4 The Transistor as a Simple Amplifier 67
3-5 Limitations of a Transistor 71
3-6 The Action of a Grid in a Triode Vacuum Tube 73
3-7 Triode Characteristics 75
3-8 The Triode as a Simple Amplifier 78
3-9 Limitations of a Triode 80
3-10 Determination of the Dynamic Coefficients of a Triode 80

Four Other Semiconductor Devices 87

4-1 Transistor Construction 87
4-2 Integrated Circuits 92
4-3 The Field Effect Transistor (FET) 99
4-4 Intrinsic Regions 113
4-5 The Tunnel Diode 115
4-6 The Varactor Diode 117
4-7 The Unijunction Transistor 120
4-8 Thyristor Concepts 123
4-9 The Silicon Controlled Rectifier 128
4-10 The Triac 131

Five Other Electron Tubes 135

5-1 The Tetrode 135
5-2 The Pentode 139
5-3 Remote-Cutoff Tubes 140
5-4 The Beam-Power Tube 143
5-5 Miscellaneous Characteristics of Tubes 145

Six Rectifiers 147

6-1 The Half-Wave Rectifier 147
6-2 The Full-Wave Rectifier 153
6-3 The Full-Wave Bridge Rectifier 157
6-4 Comparison of the Three Basic Rectifier Circuits 159
6-5 Low-Pass Filters 161
6-6 The Capacitor Filter 163
6-7 The Choke Filter 169
6-8 The π Filter 173
6-9 Voltage Multipliers 175
6-10 Shunt Rectifiers 178

Seven Controlled Rectifiers 183

7-1 The Analysis of Load Voltage and Load Current 183
7-2 Phase-Shift Circuits 189
7-3 The Firing of a Silicon-Controlled Rectifier 192
7-4 The Firing of a Triac 202
7-5 The Firing of a Thyratron 205

Eight Decibels and Sound 211

8-1 The Need for a Nonlinear System of Measurement 211
8-2 The Decibel 213
8-3 Logarithms 213
8-4 Decibel Calculations 216
8-5 Loudness 223

Nine Auxiliary Components in Electronic Circuits 227

9-1 Resistors 227
9-2 Potentiometers 231
9-3 Inductors 234
9-4 Transformers 241
9-5 Capacitors 242

Ten A-C Circuits in Electronics 251

10-1 Series Circuits 251
10-2 Parallel Circuits 255
10-3 Bandwidth 262
10-4 Coupled Circuit Theory 265
10-5 Air-Core Transformers – Untuned Primary, Tuned Secondary 268
10-6 Air-Core Transformers – Tuned Primary, Tuned Secondary 269
10-7 Harmonics 273
10-8 Four-Terminal Networks 277

Eleven The Load Line and Bias Circuits 281

11-1 The Load-Line Concept 281
11-2 The Load Line for a Common-Base Amplifier 283
11-3 The Load Line for a Common-Emitter Amplifier 289
11-4 The Load Line for a Field Effect Transistor Amplifier 293
11-5 The Load Line for a Vacuum-Tube Amplifier 296
11-6 Nomenclature 299
11-7 Distortion 303
11-8 Transistor Bias Circuits 310
11-9 Analytic Approach to Load Lines 321

Twelve Models and Small Signal Analysis 325

12-1 Emitter Resistance in a Transistor 325
12-2 The Common-Base Amplifier 328
12-3 Second-Order Considerations in the Common-Base Amplifier 333
12-4 The Common-Emitter Amplifier 334
12-5 The Common-Emitter Amplifier Using Collector-to-Base Feedback 342
12-6 The Common-Emitter Amplifier with Collector Feedback and Emitter Feedback 348
12-7 The Common-Collector or Emitter-Follower Amplifier 352
12-8 The Field Effect Transistor 358
12-9 The Vacuum-Tube 365

Thirteen Hybrid Parameter Analysis 371

13-1 The Common-Base Amplifier 371
13-2 The Common-Emitter Amplifier 382
13-3 The Common-Collector Amplifier 388

Fourteen Transformers and Transformer-Coupled Amplifiers 397

14-1 Theory of the Transformer 397
14-2 Impedance Ratios 399
14-3 The Model for the Transformer 402
14-4 Frequency Response 403
14-5 The Transformer-Coupled Amplifier and the load Line 405
14-6 Vacuum-Tube Circuits 411

Fifteen Resistance-Capacitance Coupled Amplifiers 413

15-1 Graphical Determination of the Load Line 413
15-2 General Considerations of the RC-Coupled Amplifier 418
15-3 Low-Frequency Response 423
15-4 High-Frequency Response 428
15-5 Input Loading 434
15-6 Decibels and Bode Plots 439
15-7 Volume and Tone Controls 446

Sixteen Special Amplifiers 451

16-1 Complementary Symmetry 451
16-2 The Darlington Pair 455
16-3 Differential Amplifier – Single Input, Balanced Output 460
16-4 Differential Amplifier – Dual Input, Balanced Output 464
16-5 Differential Amplifier – Dual Input, Unbalanced Output 468
16-6 Common-Mode Rejection 470
16-7 Constant-Current Stabilization 472
16-8 Clipping Circuits 478
16-9 The Grounded-Grid Amplifier 481

Seventeen Stability, Compensation, and Temperature 485

17-1 General Concepts of Beta Stability 485
17-2 Beta Stability Circuit Analysis 490
17-3 Leakage Currents 499
17-4 Temperature Sensitivity 502
17-5 Graphical Analysis 513
17-6 Diode Biasing and Compensation 518
17-7 Heat Sinks 524

Eighteen Push-Pull and Phase-Inversion 535

18-1 The Basic Circuit 535
18-2 Quantitative Analysis of Harmonics 537
18-3 Qualitative Analysis of Harmonics 539
18-4 Comparison between Push-Pull and Single-Ended Operation 541
18-5 Phase Inverters 543
18-6 Class-AB and Class-B Amplifiers 547
18-7 Complementary Symmetry in Push-Pull 555
18-8 Commercial Audio Amplifiers 559

Nineteen Class-C Amplifiers 565

19-1 The Bias Clamp and Grid-Leak Bias 565
19-2 The Tank Circuit 569
19-3 The Class-C Load Line 571
19-4 Plate-Circuit Analysis 572
19-5 Grid-Circuit Analysis 575
19-6 Summary and Adjustment 578
19-7 Harmonic Operation 580
19-8 An Alternative Approach to Class-C Analysis 581
19-9 Neutralizing Circuits 584

Twenty Feedback 589

20-1 The Fundamental Feedback Equation 589
20-2 Positive Feedback 591

20-3 Negative Feedback 593
20-4 Operational Amplifiers 600
20-5 Negative Voltage Feedback Circuits 607
20-6 Negative Current Feedback 611
20-7 Shunt Feedback 615
20-8 The Emitter Bypass Capacitor 618
20-9 Current Feedback in Vacuum-Tube Circuits 622
20-10 The Cathode Follower 625
20-11 Decoupling 628
20-12 Electronic Regulated Power Supplies 630

Twenty-one Oscillators 637

21-1 Phase Shift Oscillators 637
21-2 Bridge Oscillators 642
21-3 The Armstrong Oscillator 647
21-4 The Hartley Oscillator 651
21-5 The Colpitts Oscillator 655
21-6 Crystal Oscillators 661
21-7 The Electron-Coupled Oscillator (ECO) 671
21-8 Limitations of Oscillators 672

Twenty-two Nonsinusoidal Oscillators 677

22-1 Basic Concepts 677
22-2 The Schmitt Trigger 681
22-3 Bistable Multivibrators – The Flip Flop 685
22-4 The Monostable Multivibrator 689
22-5 Astable Multivibrators 692
22-6 Synchronization 700
22-7 The Blocking Oscillator 703
22-8 Discharge Circuits 706
22-9 Negative Resistance Concepts 707
22-10 Transistor Power Supplies 714

Twenty-three High-Frequency Amplifiers 717

23-1 The Hybrid-π Model for a Transistor 717
23-2 The High-Frequency Model for the FET 725
23-3 The Tuned Amplifier 728

Twenty-four Modulation 739

24-1 The General Problem of Modulation 739
24-2 Amplitude Modulation 740
24-3 AM Circuits 749
24-4 Frequency Modulation 752
24-5 Reactance Circuits 762
24-6 The Balanced Modulator 767
24-7 Pre-emphasis and De-emphasis 772
24-8 Suppressed Carrier and Single Sideband 772

Twenty-five Detection 777

25-1 AM Detectors – Using Diodes 777
25-2 Other AM Detectors 784
25-3 Automatic Gain Control 786
25-4 FM Detectors – The Limiter and Discriminator 790
25-5 FM Detectors – The Ratio Detector 795
25-6 Frequency Conversion 798
25-7 Automatic Frequency Control 802
25-8 Stereophonic Broadcasting – FM Multiplex 804

Answers to Selected Problems 809

Index 819

Chapter One

ATOMS, ELECTRONS, AND CURRENT FLOW

The study of d-c and a-c electric circuits, which is a prerequisite for this book, is usually introduced by an investigation into the basic structure and composition of matter (reviewed in Section 1-1). An understanding of the operation of electronic devices requires a further study of the atom. When semiconductors are encountered, it is necessary to take a number of the advanced theories of modern physics and streamline them to serve the needs of the technician. A discussion of internal atomic-energy levels (Section 1-2) paves the way for an understanding of the mechanism of current flow (Section 1-3) and of electron emission from a surface (Section 1-4). Concepts of the properties and the behavior of a crystalline structure (Section 1-5) and the extension to include N-type and P-type materials (Sections 1-6 to 1-8) form the foundation for the study of semiconductors. The mechanism of current flow in a gas discharge (Section 1-9) has wide application in the field of electron tubes. The generation of light is discussed in Section 1-10.

Section 1-1 Composition of Matter

When a mass of a particular substance is examined, it is found that a division of this mass into small pieces does not affect the properties of the substance. We term the smallest piece that can exist by itself and still retain the original properties the *molecule*. Molecules are so small that an individual molecule is invisible to the eye. A molecule may be divided into parts, but these parts do not have the physical characteristics of the original substance. These new small parts are called *atoms*. An atom is the smallest part of an element that can participate in a chemical change.

Science has shown that all matter consists of combinations of one or more of the 100-odd atomic elements. For instance, a molecule of water is composed of two hydrogen atoms combined with one oxygen atom. A molecule of an organic compound, such as aspirin or gasoline, is a very complex arrangement of many atoms. On the other hand, in many substances, for example, carbon or copper, a single atom constitutes the molecule. Fortunately, in the field of electricity, we are concerned mainly with materials in which the molecule is the single atom.

Section 1-2 Structure of the Atom

A familiar diagram widely used to show the physical form of an atom is the Bohr model (Fig. 1-1). It represents a three-dimensional model in which

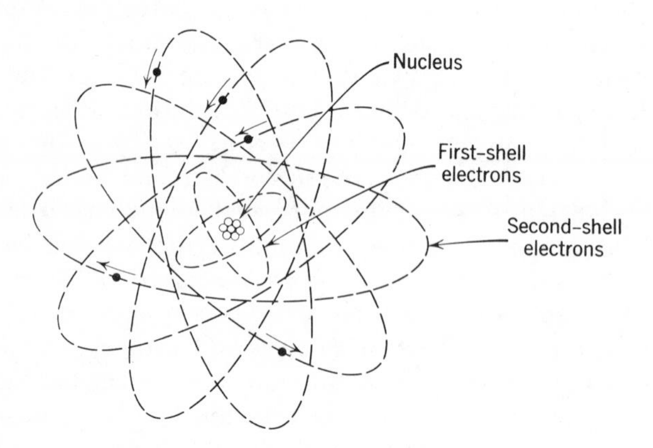

Figure 1-1 The Bohr model of an atom.

the electrons orbit about a central core called the *nucleus*. For purposes of clarity in understanding electronics, the three-dimensional concept is simplified to the two-dimensional forms of Fig. 1-2.

Elemental charged particles called *electrons* orbit around the nucleus in a manner similar to the movement of the planets about the sun. A breakdown of the nucleus shows that its major components are protons and neutrons. The advanced techniques of atomic physics indicate that there are other components in this nucleus, but they do not enter into the theory

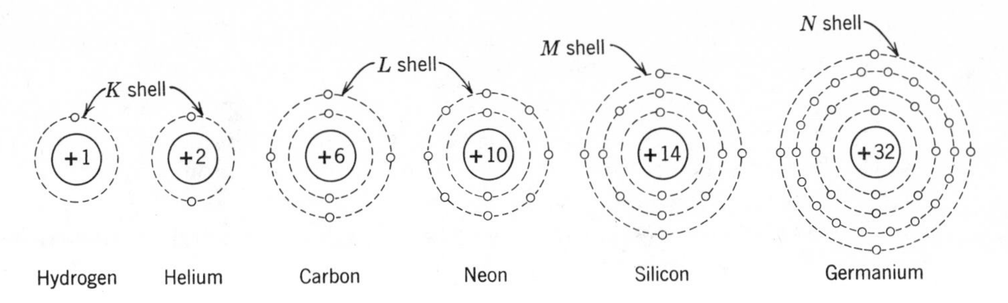

Figure 1-2 Ring models of various atoms.

of electronic devices and may be disregarded. The electron has a fixed electric charge and, from its value, it has been determined that a flow of 6.24×10^{18} electrons per second past a given point constitutes one ampere of electric current. The mass of an electron is incredibly small, 9.1×10^{-31} kg. If the total weight of all the electrons in a steady current flow of 160 A past a fixed point for one year were measured, the result would be 1 oz.

An atom itself in normal state is electrically neutral and without net charge. Since we have considered that each electron represents a certain fixed amount of negative charge, sufficient positive charges are necessary to balance the negative charges of the electrons. These positive particles of matter, located in the central region of the atom, are called *protons*. If a particular atom has 17 electrons in orbit, the nucleus must contain 17 protons to bring the total electric charge to zero. The nucleus also contains neutral components, without a specific charge, which are called *neutrons*. Both the proton and the neutron weigh about 1850 times as much as an electron. Thus the weight of a substance is determined primarily by the total weights of the protons and neutrons in the material.

Physically, an atom consists of a central nucleus with electrons orbiting around the core in much the same manner as the planets travel about the sun. The paths of the planets about the sun lie nearly in a single plane, whereas the orbits of the electrons about the nucleus follow a three-dimensional spherical pattern. The forces in the solar system are gravitational. In the atom, a stability of distance, revolution, and separation is maintained by the action of forces between the elemental charges of the electrons and protons. If the hydrogen atom were enlarged to the size of a 3 ft sphere, the nucleus would be pinhead size at the center. The surface of the sphere would represent the range of possible orbits that the single electron, also pinhead in size, might make around the nucleus. Thus, we observe that the vast relative distance concepts in the solar system and in space are maintained within the atom.

Hydrogen has the simplest atomic structure. It has one proton with a single electron in orbit. We notice that, as electrons are added, the number of protons (and neutrons) increases, making a heavier atom. The atomic table used in chemistry and physics is laid out on principles of orderly increasing numbers of electrons and protons. As examples, neon has 10 electrons, 10 protons, and 10 neutrons within the atom and copper has 29 electrons, 29 protons, and 34 neutrons in its atom.

Additionally, we find that electrons in orbits are confined to specific, finite distances from the atomic center. The innermost orbit, the *K* orbit, may contain up to 2 electrons but no more than 2. The next orbit, the *L* orbit, may contain up to 8 electrons. The third orbit, the *M* orbit, may contain up to 18 electrons. Succeeding orbits have, as maximum numbers of electrons, 18, 12, and 2, in that order. When these orbits are shown on paper for illustration, they are usually represented as concentric rings. As a result, the term *ring* is used quite freely in chemistry to describe many chemical reactions. Since, in physics, it is preferable to think of the atom in a three-dimensional concept, the term *shell* is used to describe the possible paths of orbits. Thus, a hydrogen atom has its electron in the *K* shell. It has been observed that, for the simple atoms which are of interest electrically, an electron may exist in the *N* shell only if the lower shells (*K*, *L*, and *M*) are completely filled. The atomic structures in Fig. 1-2 illustrate this orderly layout.

The outer shell, the *valence* shell, determines the chemical activity of the element. If the outer shell is filled in completely, the substance is inert and does not react chemically. Examples of this are neon, argon, and krypton. If the outer ring is incomplete, it may join in chemical bonds with other atoms to produce the effect of filled outer shells. This action produces molecules of stable chemical compounds, such as water and salt.

The shells are very definitely finite in character and, for an electron to exist within an atom, the electron must exist within one of the specific shells. By this we mean that an electron cannot exist between shells. To permit an electron to move from one shell to another, definite *discrete* amounts of energy, called *quanta*, are required. A quantum of energy is the least unit amount of energy that can be considered in the process. Furthermore, quanta must exist as whole numbers; fractions of a quantum do not exist. If the energy required to shift an electron from one shell to another were three quanta units, an energy level of two quanta would produce no shift. If the energy level were gradually increased, then, suddenly, at a particular instant, the necessary three quanta would be available and the electron would abruptly shift from one shell into the next.

We have stated that the shells are described as finite orbits wherein electrons may travel. If we examine a particular shell carefully, we notice that the *L* shell consists of two very close subshells, the *M* shell has three subshells, and the *N* shell four subshells. Since the subshells are very close to each other within a particular shell, an electron may move between the subshells of a given shell with less energy-level changes required than as if they moved between shells. All this leads to the description of electrons as existing in atoms at definite, *discrete,* or *permissive* energy levels. The permissive energy levels are fixed and invariant. External addition or removal of energy may move an electron from one permissive level to another, but it cannot change the permissive levels. Less than permissive energy amounts can produce heat.

Perhaps, a clearer concept of a discrete energy step can be obtained from the action of heat energy applied to a beaker containing a pound of water. A Btu (British thermal unit) is defined as the heat energy necessary to raise one pound of water one degree Fahrenheit. Successive Btu's raise the temperature of the water from room temperature to 212°F. We find at this point that the water accepts heat without raising its temperature at all. It requires about 970 Btu to evaporate the water into steam which is still 212°F. This large amount of heat is known as the latent heat of vaporization. If there are 10^x molecules of water in one pound, $1/10^x$ is the amount of heat as a fractional part of a Btu that is necessary to raise the molecule one degree in temperature. Let us call this basic amount of energy q. This q is the equivalent energy to what we call a quantum for the electron. Each successive q raises the energy of one molecule one degree to 212°F. At 212°, the addition of $100q$ units of energy does nothing to the water. It takes $970q$ to boil off the molecule. Any amount of energy less than $970q$ does not produce steam. We can consider different water temperatures in degree steps of energy levels comparable to energy levels within a shell, and the large amount of energy, $970q$, as the energy required to move from one shell to the next.

The electrons in the outer shell have a higher energy content than the electrons in the inner shells. External energy then will affect the electrons in the outer shell first. Application of energy in the form of heat will cause the electron to rise to outer permissive levels. Continued increasing heat energy can cause an outer electron to escape from the restraining forces of the atom. This effect is known as *electron emission* and will be discussed in detail in Section 1-4. Electrons in an atom or molecule of a gas may secure enough energy for escape from the kinetic-energy transfer of collision or impact. A detailed discussion of this phenomenon called *ionization* is in Section 1-9. The production of free electrons by radioactivity has a limited application to electronics, particularly for the needs of this text. When atoms are internally bound together to form the solid surface of a metal, electron release obtained by collision is termed *secondary emission* (Section 1-4). Electrons may be literally pulled out or ripped from the restraining forces of the atoms of a cold surface by means of a very high electric field. This is termed *high field emission* (Section 1-9). Light may be considered to consist of definite energy levels or quanta. Light also has the property that, as the frequency of the light increases (moving toward shorter wavelength), the energy content of the light per quantum increases. Certain materials are sensitive to light; that is, electron motion between permissive levels increases with incident light to the point where free electrons actually escape from the surface of the material. This is termed *photoemission.*

Section 1-3 Electron Flow in Solids

The first two subshells of a shell are called *s* and *p*. The inner *s* subshell requires two electrons and the outer *p* subshell requires six electrons for completion. A basic semiconductor material has the *s* subshell filled and the *p* subshell with two electrons. Materials that are suitable for basic semiconductors (see Table A) with this electron structure are silicon and germanium.

Now consider an isolated atom of germanium. The *K*, the *L*, and the *M* shells are completely filled. The inner subshell (*s*) of the *N* shell is filled and

TABLE A Semiconductor Elements

Element	Atomic Number	Electron Content of the Outer *p* Subshell
Boron	5	1
Carbon	6	2
(Nitrogen)[a]	7	3
Aluminum	13	1
Silicon	14	2
Phosphorus	15	3
Gallium	31	1
Germanium	32	2
Arsenic	33	3
Indium	49	1
Tin	50	2
Antimony	51	3
Thallium	81	1
(Lead)	82	2
Bismuth	83	3

[a]The elements listed in parentheses are included to complete the table.

the outer subshell (*p*) contains two electrons. Any activity of electrons in ordinary electronic circuits is concerned with the behavior of the four electrons in the *N* shell. The electrons in the *K*, *L*, and *M* shells are not involved at all.

The *Pauli Exclusion Principle* of modern physics theory requires that no two electrons within a *system* can have exactly the same energy content. Thus the *s* subshell of the *N* shell of the isolated germanium atom contains two different energy states that are filled and the *p* subshell has six possible energy states only two of which are filled. The unfilled energy states are often referred to as *vacancy levels*. Energy in discrete quanta steps can be added to the atom to cause the four outer electrons to shift between the possible eight permissive states. These eight energy levels are shown for the isolated atom at *A* in Fig. 1-3.

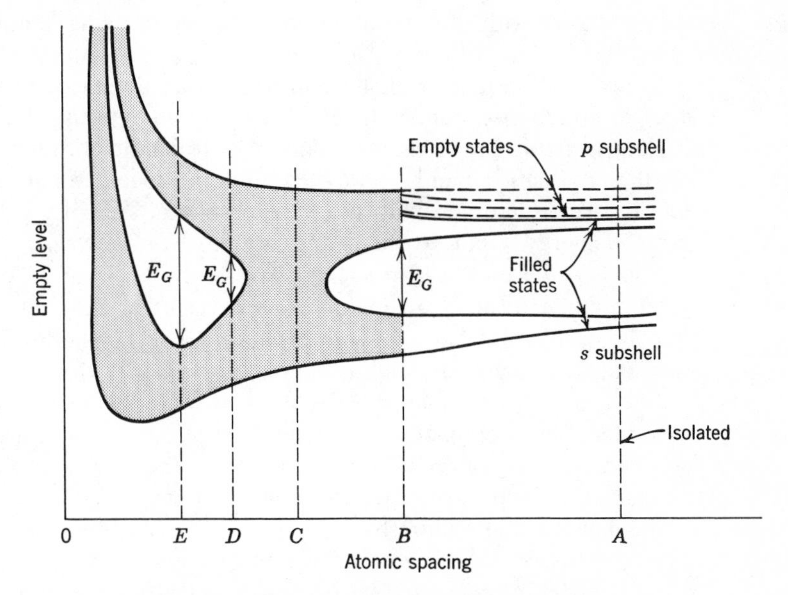

Figure 1-3 Energy levels in the outer shell of an atom.

If we no longer have an isolated atom of germanium but consider a solid, the Pauli Exclusion Principle together with small interatomic distances requires that the number of permissive levels in the *s* and *p* subshells must multiply astronomically. As a result, the multitudinous permissive levels are now described as *bands*, as at *B* in Fig. 1-3, with an energy gap, E_G, between them. This energy gap is a *forbidden* band; that is, an electron may exist in the lower band or in the upper band but *not* in between the two bands.

Energy levels in atoms are measured in units of an *electron-volt*, *eV*, which is the amount of increase in energy that an electron acquires when it is accelerated by the field created by one volt. An electron volt is equivalent to 1.60×10^{-19} joule. A joule is one watt-second. The energy gap, E_G, also is measured in electron-volts.

When the spacing between atoms of the germanium is reduced further, the two permissive bands merge, (*C* in Fig. 1-3). If there are *N* electrons in the outer shells, there are $2N$ possible energy states in the material, and very little energy is required to shift electrons between permissive states.

A further reduction of interatomic spacing separates the merged band into two bands (at *D* and *E* in Fig. 1-3). The upper or higher energy band is the *conduction band* and the lower energy band is the *valence band*. In the case of the materials that originally contained 2+2 electrons in the outer subshells (carbon, silicon, and germanium) all of the electrons are now in the valence band with the conduction band comprising empty permissive states.

In order to have current flow in a material, the electron must be in the upper or conduction band. The application of an external circuit emf causes an electric field, and if the electrons acquire sufficient energy, they move into the conduction band, and a current is produced. When the conduction band is already merged with the valence band (*C* in Fig. 1-3) the required external energy input is small and we have an *electric conductor*. At *E* on Fig. 1-3, the energy gap is large and of the order of 6 eV for carbon in the crystalline form of a diamond. As a result of this high energy gap, a diamond is classified as a good *insulator*. The smaller energy gap at *D* represents the normal crystalline state of a *semiconductor*.

For a broad classification of electrical materials, let us assume that a cubic-inch block of each is available for tests and measurements. If the resistance between opposite faces of each cube is measured, we find that the insulator yields a resistance value of many megohms. The conductor will be measured in millionths of an ohm or micro-ohms. The semiconductor is a cross between the two, and we can expect to find its resistance value in the order of ohms. This method of classification is quite loose and very general because the boundaries between them often tend to be obscure.

The large energy gap at *E* (Fig. 1-3) explains why an insulator will eventually break down with an avalanche of current if the voltage is increased sufficiently. Thyrite, the material used in lightning arresters, at normal voltage levels, acts as an ideal insulator. Under high-voltage surges, the stress of the high potential pulls electrons across the forbidden energy gap into the conduction band, and a protective shunting current is produced. When insulators are operated at high temperatures, the increased energy of heat allows a crossing of electrons into the conduction band with less outside electric stress. Thus, we note that many materials which are good insulators at low temperatures become conductors at high temperatures. The usual representation of the energy levels is shown in Fig. 1-4.

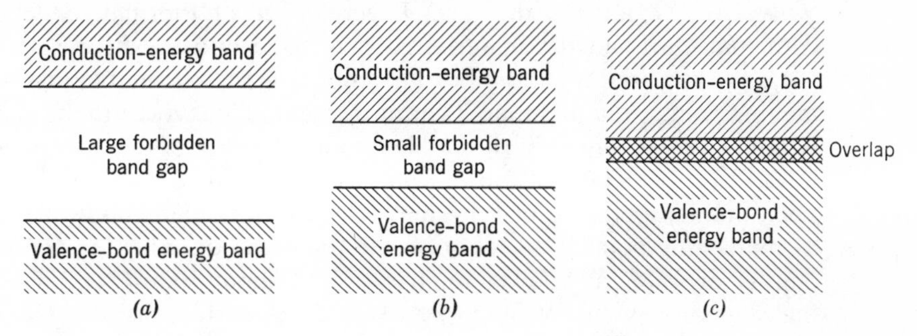

Figure 1-4 Energy levels in the outer shells of different materials. (*a*) Insulator. (*b*) Semiconductor. (*c*) Conductor.

Section 1-4 Emission

Free electrons are obtained to operate vacuum tubes and many forms of gas tubes by heating a surface called the *cathode*. The cathode surface must have the property of being able to liberate the required quantity of electrons while meeting a number of other physical conditions. The emission surface must be able to emit electrons at temperatures below its melting point. It also must be mechanically strong enough to be stable under the conditions of its operation. Also, its efficiency of emission must be high with respect to possible alternative materials.

The energy required to extract an electron from a surface is termed its *work function* and is measured in electron-volts. This energy is required to cause the electron to escape through the surface barrier of a metallic surface, and it is much larger than the amount of energy required to change an electron from one permissive state to another within an energy band. This required energy is secured by heating the material and is likened to the latent heat of vaporization of a liquid.

Pure *tungsten* was one of the first materials to be used as a thermionic surface. Tungsten operates at a temperature of about 4070°F (about 60% of its melting temperature) with an emission rating of 2.7 A/sq in. of surface. The emission efficiency at this operating point is about 4 ma/watt of input heating power. As with all of the other thermionic-emission materials, the characteristics vary greatly with small changes in the operating temperature. A "cold" tungsten filament at 3600°F produces negligible emission. Tungsten has an advantage over other materials in its physical strength and in its ability to have a very long emitting life. It is often used with very large tubes, which are expensive to replace, or with very tiny tubes where the cathode emitting surface is a fine wire.

In the experimental search for other useful emitting materials, *thoriated tungsten* was developed. A small percentage of thorium oxide is mixed with tungsten, and the material is then made into the necessary physical structure required for the tube. The emitter is then heated to just below the melting point. A monatomic layer of thorium oxide works to the surface and is converted to pure thorium. This combination has the advantage of the lower work function of the thorium together with the strength and higher operating temperature of tungsten. Pure thorium alone could not be used, as its melting point is too close to the required operating temperature. The emitter operates at a lower temperature of about 3000°F and has the very high emission value of about 7.6 A/sq in. of surface. It also has the very high efficiency of about 70 mA emission per watt of heating power.

The need for the third type of thermionic surface was brought about by the application of alternating current as a source of heating power in radio receivers. If alternating current is used on a tungsten emitter, electron emission will vary in accordance with the cycles of the alternating current. These fluctuations produce an a-c hum or interference on the electron output of the tube, which in many applications is most undesirable. If it is necessary to develop heat for thermionic emission, an indirect method of heat transfer must be provided. Heat is produced by the alternating current in a resistance wire, and this heat is transferred through a nonmetallic heat conductor to an emitting surface. The thermal time lag through the heat conductor is slow enough to eliminate the effects of any instantaneous variation of heat production in the resistance wire. A device which, when heated by an electric current, produces emission itself is called a *filament*. When we have the indirect heating process, we call the wire that gets hot from the flow of current through itself the *heater* and the separate emitting surface the *cathode*. Technically all emitting surfaces are cathodes, but we must be careful to distinguish a filament from the heater-cathode arrangement by the use of proper terminology.

Usually the heater wire is tungsten, and the cathode surface a combination of barium and strontium oxides. This unit operates at a much lower temperature than the others. At 1560°F, the emission current is 3 A/sq in. of surface with an efficiency of the order of 300 ma emission per watt of heater power. These heater-cathode structures are generally used for the small "receiving-type" tubes. Typical structures are shown in Fig. 1-5.

Another method of obtaining electrons from surfaces is by *photoemission*. Useful light ranges from low frequencies and long wavelengths (infrared) through the visible spectrum, to high frequencies and short wavelengths (ultraviolet). The wavelength of light is measured in angstroms (10^{-8} cm), abbreviated Å, or in micrometers abbreviated μm. One micrometer is 10^{-6} m. The wavelength of visible light is from 4000 to 7000 Å or

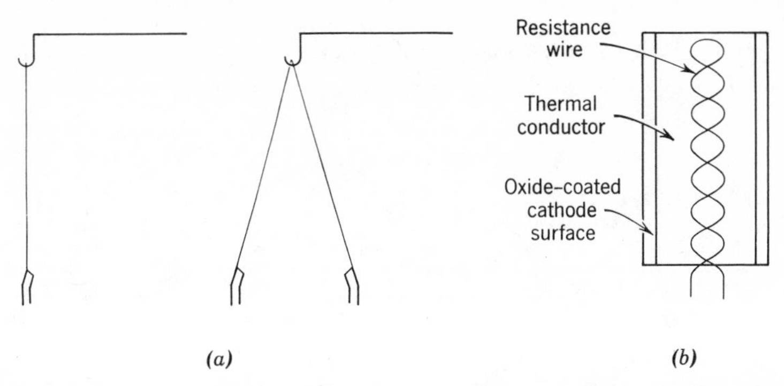

Figure 1-5 The physical construction of typical thermionic emitters. (*a*) Filament structures. (*b*) Heater-cathode structure.

from 0.4 μm to 0.7 μm. The relation between wavelength and color for light is shown in Fig. 1-6. The energy, in quanta, that is available in a ray of light striking the emissive cathode surface, if sufficient, can cause the liberation of an electron. Certain materials have been found and developed that have low enough work functions, so as to be used as light-sensitive surfaces. Some of these substances are indicated on the graph of Fig. 1-7

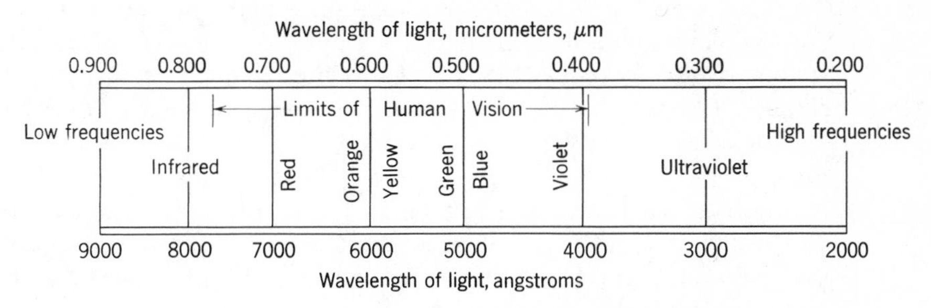

Figure 1-6 The spectral distribution of light energies.

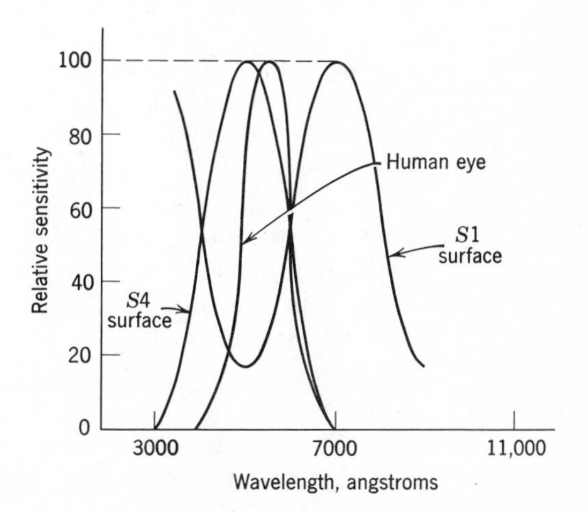

Figure 1-7 Spectral response of different cathode surfaces. (*Courtesy RCA.*)

together with the response of the human eye. It is seen that different materials are sensitive to different color ranges of light.

Section 1-5 The Structure of a Crystal

To understand the operation of a semiconductor, we must discuss atomic-structure theory, but only sofar as it applies to crystals. Fundamentally, a crystalline structure produces special properties that do not exist for the substance in the "ordinary" form. An example may be taken as common carbon, where a special crystalline formation of carbon produces the diamond. Another example of the application of crystalline structures is the field of the metallurgy of steel where different heat-treating processes produce different crystalline formations. The different crystalline structures produce different characteristics, such as a surface hardening, an ability to withstand increased shearing stress, or the ability to be formed or machined. Many metal failures can be explained by the changes in their crystalline structures.

In semiconductor electronics, we are concerned with *single* crystals of silicon and germanium. Natural crystals are usually polycrystalline; that is, they comprise many individual crystals oriented in space in many different directions. The separate single crystals are joined to each other by *grain boundaries*. The primary difficulties faced by the manufacturers are, first, to produce single crystals that must not be polycrystalline and, second, to maintain an extreme degree of purity in the germanium or silicon. If the population of the earth represents the number of molecules in a piece of silicon used for integrated circuits, the tolerance of impurity is represented by three people. The structures of the germanium and silicon atoms are shown in Fig. 1-2.

Semiconductor material has the form of a *face-centered lattice*. In a face-centered crystal, each atom has four neighbors and all are equidistant from each other. The face-centered lattice arrangement is adaptable to both germanium and silicon, as the outer shell has four electrons and eight permissive places for a completely filled shell. Thus, if the atoms in this crystal share electrons in their outer shells with their neighbors, the outer shells of all the atoms will be complete, since its own four electrons plus the four shared from the four neighbors make the complete number of eight electrons in the outer shell. Thus the material is now in a stable form. A three-dimensional model of this crystal is shown in Fig. 1-8. This model can be reduced to the two-dimensional representation of Fig. 1-9 for simplicity.

When we have closely placed atoms, as in the crystal, we may refer again to the permissive energy-band concept we used earlier in discussing insula-

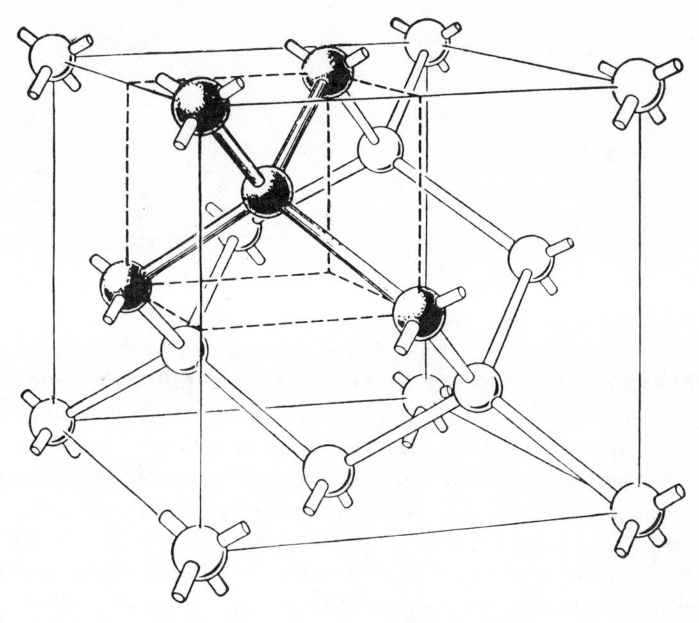

Figure 1-8 Three-dimensional model of a face-centered crystal lattice. (*Courtesy Kittel, Introduction to Solid State Physics.*)

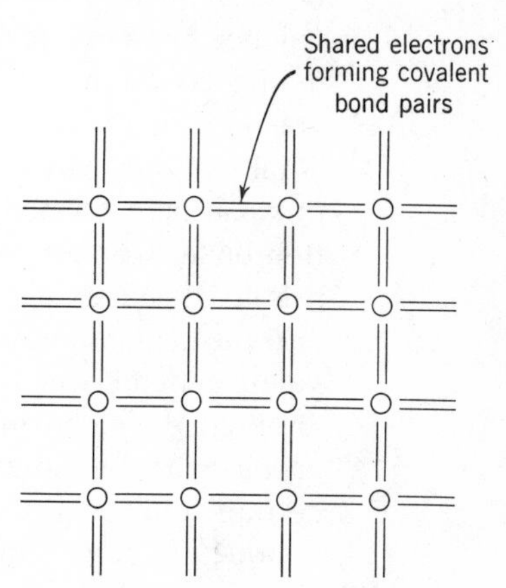

Figure 1-9 Two-dimensional model of a face-centered crystal lattice.

tors and conductors. (Figs. 1-3 and 1-10). The crystalline structure of silicon and germanium have atomic spacing located at D in Fig. 1-3. This corresponds to the valence and conduction bands shown in Fig. 1-10. When the material is at the temperature of absolute zero, all electrons are contained in the valence-bond energy band; none are in the conduction band. Therefore, the material is an ideal insulator. The energy gap, E_G, is 0.785 eV for germanium and 1.21 eV for silicon at absolute zero (0°K). When energy is

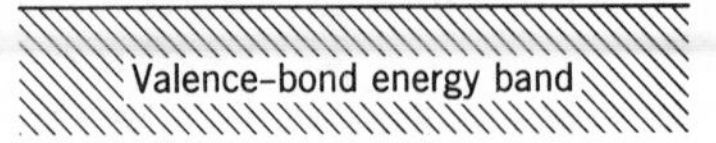

Figure 1-10 Energy bands in the outer shell.

applied to the system in the form of heat, electrons will leave the valence-bond band and jump across to the conduction band. The energy gap becomes smaller as the ambient temperature increases. At room temperature, E_G is 0.72 eV for germanium and 1.10 eV for silicon. Further application of heat causes more and more electrons to cross the energy gap into the conduction band. This means that the resistivity characteristics of germanium and silicon (and carbon which also has four electrons in the outer shell) show a decreasing resistance with an increase in temperature. Thermistors are direct applications of this semiconductor characteristic. Normally, however, semiconductors must operate within definite temperature limits in order to control the resistivity characteristics.

At room temperatures, then, electrons have crossed from the valence-bond band into the conduction band. The absence of the electron in the valence band creates a *hole*. Hence, in pure crystals, the number of electrons in the conduction band is balanced by an equal number of holes in the valence band. A pure crystal above absolute zero must contain these combinations which are called *electron-hole pairs*. The material that has this pairing is termed *intrinsic*. Electrons are free to move within the conduction band. Correspondingly, if an electron in the valence band moves to fill a hole, it leaves a hole where it was before. Thus, we can have not only current in the conduction band but also an independent current within the valence band which is the result of holes "jumping" from one atom to another. An electron in the conduction band has a higher energy level than an electron (or hole) in the valence-bond band. If an electron requires a certain total energy to be moved by an external field, the electron in the conduction band requires less additional energy than the one in the valence-bond band. Accordingly, the usual currents that are encountered in electric circuits are normally conduction-band currents. However, there will be a small value of "hole" current at the same time. Accordingly, in this case; the electrons are the *majority current carriers* and the holes are the *minority current carriers*.

Another definition that must be noted is the term *Fermi level*. At absolute zero, the Fermi level represents the highest possible energy content that an electron may have. In the crystal of pure germanium or silicon, at absolute zero, no electron can exist in the conduction band. Therefore, the Fermi level must be below the conduction band. The Pauli Exclusion Principle requires that no two electrons in a system can be at the same energy level. Consequently, at absolute zero, the electrons in the valence band must have finite energies that are not zero. In fact, the valence-bond band must be filled. Therefore, the Fermi level must be above the valence-bond band. It is shown in the mathematics of modern physics that, for intrinsic semiconductor materials, the Fermi level is in the center of the forbidden energy band, Fig. 1-11.

If a covalent bond is formed, an electron has an energy greater than the Fermi level and a hole has an energy less than the Fermi level.

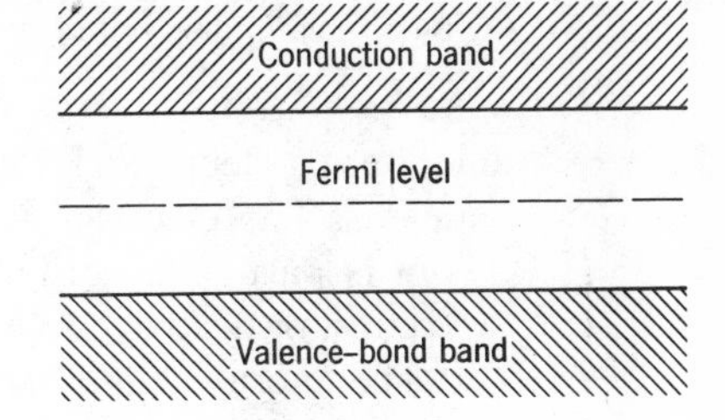

Figure 1-11 Energy-level diagram showing the Fermi level.

Section 1-6 *N*-Type Material

When a desired impurity is introduced into pure germanium or silicon at a carefully controlled rate of the order of one part in ten million by a manufacturing process called *doping,* and when the face-centered lattice structure of the pure germanium or silicon is maintained, the semiconductor is no longer intrinsic but *extrinsic*. Atoms of arsenic or antimony (materials shown in Table A, page 6) have an electron content in the outer *p* shell of three electrons. When they form part of the crystal lattice structure, we now find that electrons are available in excess of the number required for covalent bonds. This new material is called an *N*-type crystal. We may represent this new arrangement by Fig. 1-12. The introduced impurity atoms which add electrons to the crystal system are known as *donor* atoms.

Now this *N* material inherently has more electrons than the pure intrinsic crystal, but it must be remembered that a block of this material is *definitely neutral* as far as overall net charge is concerned. It has no charge any more than a piece of copper could have. The concept of *space-charge neutrality* states that a block of material has an equal number of positive and negative charges. If there is an "extra" free electron, the characteristic of the donor atom is such that it has, in its structure, one more positive charge in its nucleus than the atom of intrinsic material has. This is true because the

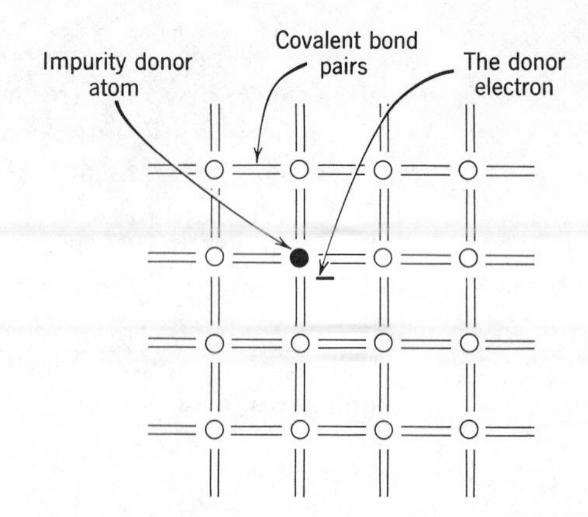

Figure 1-12 Electron affinities within *N* material.

number of electrons in an atom equals the number of positive charges in its nucleus. The electron of the donor atom is only "free" because the donor atom is part of an orderly crystalline structure. If this crystalline pattern were lost or destroyed, this electron would no longer be "free."

The energy required to break covalent bonds in intrinsic semiconductor material is of the order of 0.7 to 1.1 eV. In the formation of *N* material, the energy level of the extra electrons is only about 0.01 eV below the conduction band. Since the average energy supplied to these electrons at room temperature by heat is about 0.025 eV, they shift into the conduction band. Since there is now an added energy level above the levels of intrinsic material, the Fermi level, which represents the highest possible level, is raised (Fig. 1-13).

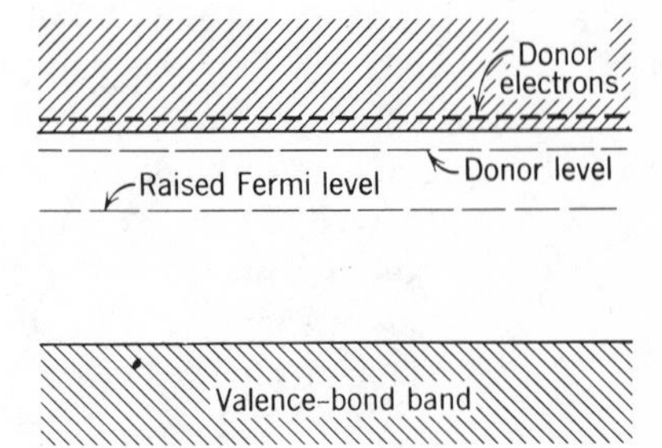

Figure 1-13 Energy levels within *N* material.

An applied external electric field can easily move these donor electrons that are in the conduction band. Accordingly, we can consider that electron flow in *N* material is primarily the flow of these donor electrons in the conduction band. We do have present in the conduction band electron-hole pairs because the material is above absolute zero. Therefore, in addition to the donor electrons, there are also the electrons from the electron-hole pairs in the conduction band and the holes from these pairs in the valence-bond band. These, too, contribute to current flow with the result that we term the electrons the *majority current carriers* and the holes the *minority current carriers* in *N*-type material. At normal temperatures, there are many more electrons than holes. If the temperature of the crystal is raised, more and more electron-hole pairs are formed. At some temperature level, the number of donor electrons becomes negligible with respect to the number of electron-hole paris, and the crystal is then, for all practical purposes, intrinsic. If the heat-generating source is removed, the material returns to its normal extrinsic state. However, there is a limit to the amount of heat that a crystalline substance can take without losing the lattice formation. If lattice pattern is destroyed by excessive temperatures, the device ceases to have the desired properties and must be replaced.

Section 1-7 *P*-Type Material

Like *N* material, *P* material is formed by adding controlled amounts of impurity atoms. *Acceptor* atoms have an outer shell structure of two electrons in the *s* subshell and one electron in the *p* subshell (Table A, page 6). Aluminum, gallium, and indium are in commercial use to provide each of these three electrons in the outer shell. These impurities become an integral part of the crystal-lattice structure but leave certain of the covalent bonds short by one electron. The action of the acceptor atom is to place a vacancy level slightly above (approximately 0.01 eV) the valence-bond band. Thermal energy at room temperature causes valence-band electrons

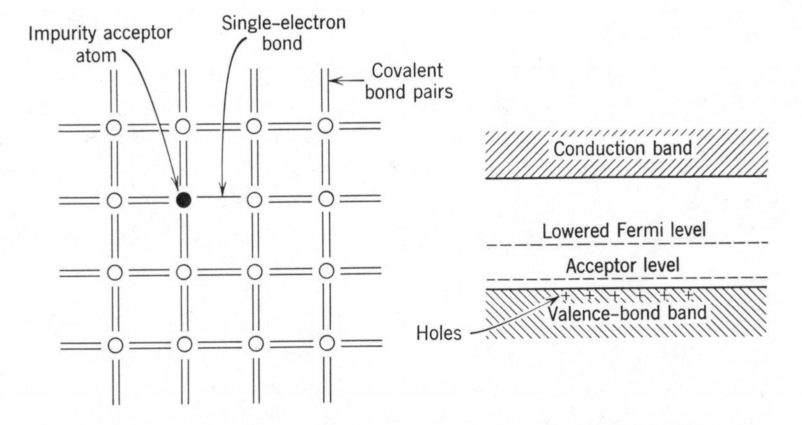

Figure 1-14 Electron affinities and energy levels within *P* material.

to shift into this level, leaving holes in the valence band. There is no change in the conduction band. Accordingly, the Fermi level now shifts downward toward the valence-bond band, as indicated in Fig. 1-14. The few covalent bonds that break into electron-hole pairs establish the minority electron carriers in the conduction band. Current flow in *P* material can be considered as a shift of holes from one atom to another atom within the valence-bond band.

As with *N* material, additional heating causes the number of created electron-hole pairs to overshadow the effect of the hole-producing acceptor atoms. The Fermi level will shift toward the center, and the substance will behave as intrinsic crystal, provided the crystalline structure is not destroyed.

Section 1-8 Light on Crystals

We have stated that added energy on the pure crystal, on *N* material, or on *P* material, can break down the covalent bonds into electron-hole pairs. A larger number of electron-hole pairs in a crystal means that, in it, a larger current will flow under the influence of an external potential source. In other words, we can state that the resistance of the crystal decreases. In this section we shall investigate the photoconductive properties of the lattice.

We have repeated several times, for emphasis, that the breaking of a covalent bond is brought about by energy increments in discrete amounts, and that nothing will happen until the quanta energy requirements of the gap are reached. Earlier it was stated that, for a given light, the energy content per quantum increases with frequency (or, rephrased, increases with a decrease in wavelength). Thus we can experimentally determine that a given light, which at low frequencies will not produce an electron-hole pair, will produce the electron-hole pair if only the frequency of the light is increased. Hence we find that the crystal that is *transparent* to high-frequency light is *opaque* to light at low frequencies. This action is shown in Fig. 1-15.

As the light intensity is raised from zero, the conductivity of the crystal

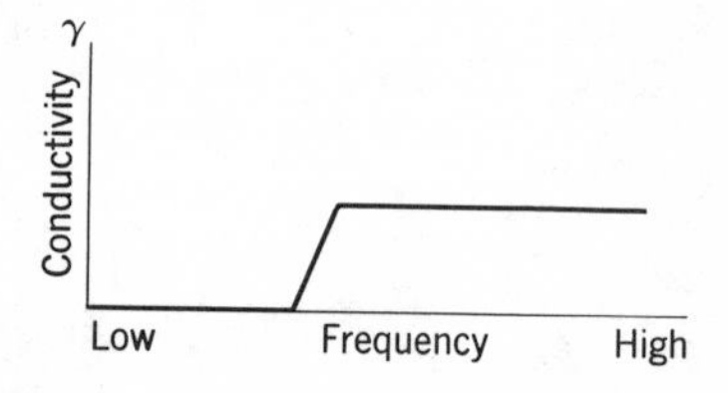

Figure 1-15 The effect of varying light frequency on crystal conductivity.

increases (Fig. 1-16). The residual conductivity γ_0 exists because electron-hole pairs already exist normally at room temperatures that are produced by heat and not by light.

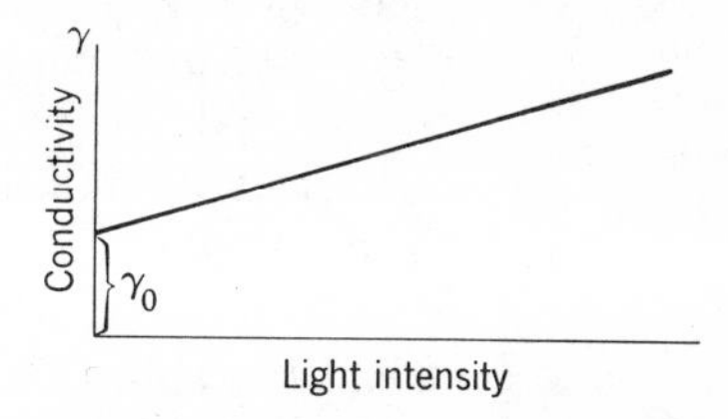

Figure 1-16 The effect of varying light intensity on crystal conductivity.

Section 1-9 Mechanism of a Gas Discharge

In Section 1-2, we stated that free electrons can be produced from collision with gas molecules. In some electron tubes, initial velocities are produced by the presence of a hot-cathode surface. In others, the initial free electron is available because, in any volume of gas, some degree of ionization is naturally present. If two plates have a voltage between them (Fig. 1-17),

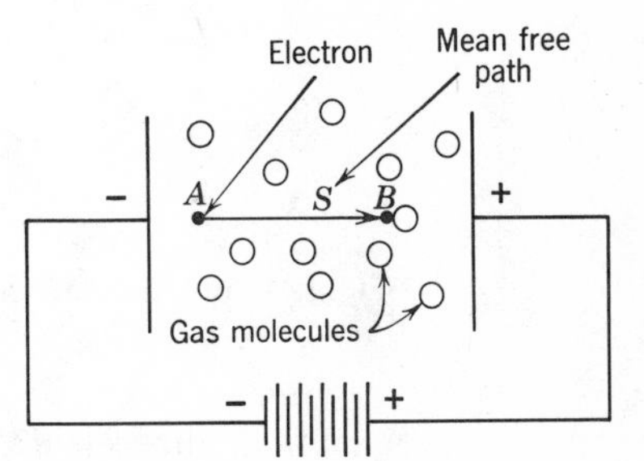

Figure 1-17 The physical condition for ionization of a gas.

a free electron at point A is attracted to the positive plate. The electron travels to the plate along the path S shown by the arrow. At point B it will collide with a gas molecule. If the kinetic energy at point B is sufficient by virtue of having gained sufficient velocity, it liberates an electron from the outer shell of the gas molecule at B. Now two electrons reach the plate, and the newly created positive gas ion at B travels to the negative surface where it picks up an electron on contact and becomes a neutral gas molecule again. This process is cumulative, and the current resulting is limited only by the resistance of the external circuit. The process of gas ionization is accompanied by formation of light energy. We find that different gases and gas combinations produce different colors.

A complete ionization occurs when all gas molecules or particles are ionized. If more input energy (from a higher voltage) is applied between the plates, *double ionization* may occur. Now each gas molecule delivers two electrons, and each gas ion must pick up two electrons at the negative surface. Further stages of *multiple ionization* are possible. A molecule that is subject to multiple ionization has a much greater force when striking the negative surface than has a singly ionized molecule.

The average of all the lengths of the collision paths *S* in Fig. 1-17 is called the *mean free path.* If the mean free path is greater or longer than the separation of the plates, the gas will not ionize. The presence of few gas molecules in a given volume results in a long free-path length. We stated earlier that, unless the energy content of the traveling electron is sufficient, the impacted gas molecules will not ionize. To obtain ionization with a short free-path length, the applied voltage must be increased in order to get the electron up to the required speed within a shorter distance. These results have been summarized as *Paschen's law* and are plotted in Fig. 1-18.

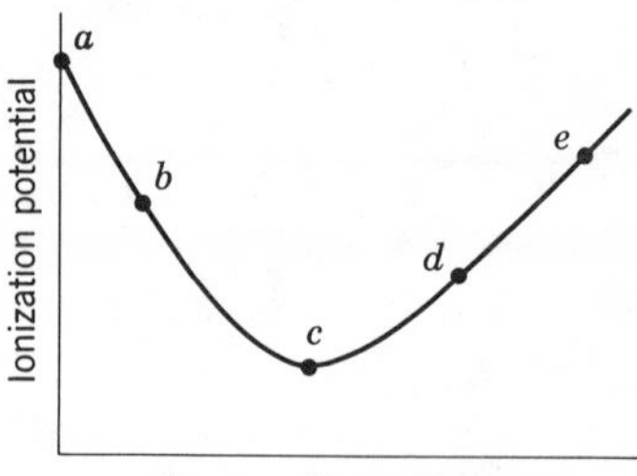

Figure 1-18 Graph illustrating Paschen's law.

Assuming that electrons are not available from a thermionic source, let point *d* represent the operating condition of a sparkplug gap in air. When the plug is subjected to the high-compression pressure of a piston, the actual firing point is at *e* and is under much different conditions than were observed in open air. If, for a fixed gap, the pressure is decreased, we notice that the voltage necessary to produce ionization falls until we reach a minimum at point *c*. Further reduction in pressure increases the length of the mean free path, so that a greater voltage is necessary to make use of the relatively few collisions. Finally, at point *a*, there is a true vacuum. Since the length of the mean free path is infinite, the ionization potential is infinite. A current is produced at point *a* when the voltage is sufficiently high to produce an electron flow by high field emission.

The volt-ampere characteristic of a gas diode that does not have a thermionic source of electrons is shown in Fig. 1-19. Note particularly that the current scale is not linear but logarithmic. A small number of electrons are being continuously produced by energy sources such as

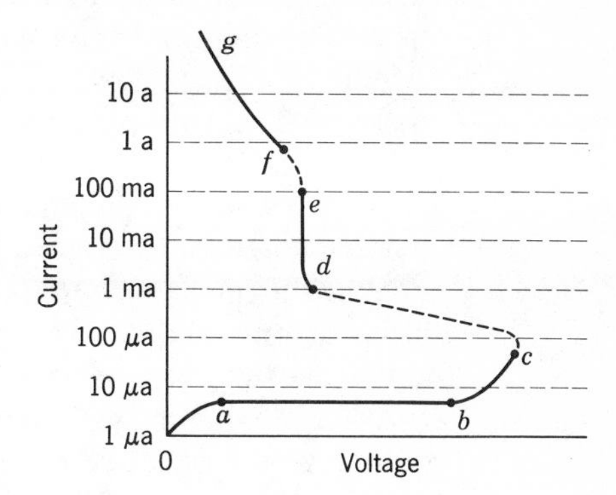

Figure 1-19 The volt-ampere characteristic of a gas discharge.

ambient temperature, light, and even cosmic rays. A small voltage, point *a*, will collect the free electrons that are normally found within the gas. A further increase in voltage to *b* does not increase the current flow. The current consists of *all* the electrons that are being produced naturally, and is termed *saturated.* This saturation current is also called *dark current* because it does not *produce* light.

When the voltage reaches point *b*, ionization occurs and a small increase in voltage increases the current from *b* to *c*. The region *Oabc* is the range of operation of vacuum and gas phototubes.

The characteristic is completely unstable between *c* and *d*. Between points *d* and *e*, we notice that current is independent of voltage. This region is called the *glow-discharge* portion of the curve, and one of its main applications is in the field of light-producing devices such as neon lamps and sodium-vapor lamps. The region between *d* and *e* is also used for voltage-regulator tubes.

Again, the curve is somewhat unstable between *e* and *f*. The portion of the curve between *f* and *g* is the *arc-discharge* region. An arc is generally associated with a very large current, and it has a negative-resistance characteristic wherein current increases with a decrease in voltage. This region is usually accompanied by an abnormal glow.

We summarize this topic of gas discharges by stating a few facts and by making some definitions that were by-passed in order not to detract from the continuity of the discussion. A vacuum tube does not have a perfect vacuum but has a residual gas pressure of the order of 10^{-8} to 10^{-6} mm of mercury. This results in a mean free-path length of the order of 40 m. Thus, in a vacuum tube, ionization of the residual gas is practically impossible. In a gas tube, the pressure is of the order of 10^{-2} or 10^{-1} mm of mercury. Then, the mean free path is of the order of 0.1 mm.

There is a definite and finite *ionization time*. It is possible to reduce this ionization time to microseconds by proper gas combinations and by the use of high ionization potentials, but it is usually of the order of milliseconds. *Deionization time* is also a finite time, and it is necessarily longer than ionization time. A gas is ionized under the force and pressure of an external voltage. Deionization is a natural recombination of the ion and electron without the addition of external voltages. These finite times result in maximum upper-frequency limits in many applications.

When a gas is ionized, the gas exists as a mixture of electrons, ions, and ordinary neutral-gas molecules. This region, called the *plasma,* extends across the tube between the plates. The plasma region acts electrically as if it were a conductor, with the electrons moving to positive plate and the ions moving to the negative plate. Most of the voltage drop occurs right off the negative plate or cathode. This abrupt potential drop is termed *cathode fall.* The light distribution is uniform along the plasma, but it drops off markedly both at the anode and at the cathode.

Section 1-10 Generation of Light

The possible energy-level states for the outer shell electrons for mercury are shown in Fig. 1-20. When an energy of 10.39 eV is supplied, the mercury atom loses an electron and the gas becomes ionized, as explained in the previous section. However, there are a number of possible energy-level states that an electron can take which are less than the one required for full ionization. External energy, for example, raises an electron from the zero or normal state to 4.88 eV. Then the electron drops back to its normal state and, thereby, releases an energy equal to 4.88 eV. This energy appears as light, whose wavelength in Angstroms is given by

$$\lambda = \frac{12{,}400}{E_z - E} \tag{1-1}$$

where E_z and E are energy levels in electron-volts. Using 4.88 eV, λ is 12,400/4.88 or 2537 Å. From Fig. 1-6, we notice that this is an ultraviolet light that can cause a serious sunburn.

The energy state transitions that can produce light are shown by vertical lines on Fig. 1-20. These particular wavelengths of light make up the line-spectrum response of mercury. From these wavelengths, it is observed that a mercury lamp is rich in blue and ultraviolet radiation.

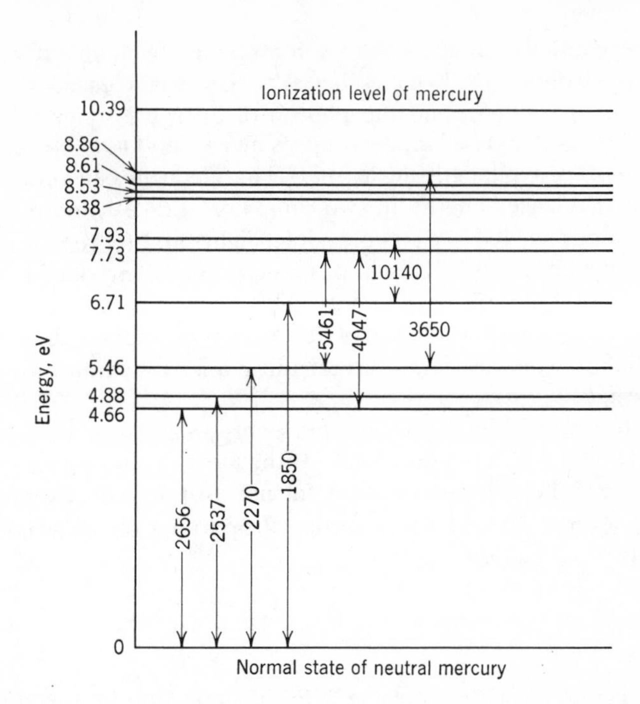

Figure 1-20 The energy levels of a mercury atom. (From *Electronic Devices and Circuits* by Millman and Halkias. Copyright 1967 by McGraw-Hill, Inc. Used with permission of McGraw-Hill Book Company.)

When an electron makes a transition from a high-level state to a lower level state, a specific amount of energy called a *photon* is released. Since an electron volt is 1.60×10^{-19} J, consider the case of a luminous body that gives off 1 W of light energy where the transition is 2.0 eV. The number of photons or *light-packets* given off in one second is

$$\frac{1.0}{1.60 \times 10^{-19} \times 2.0}$$

or 3.12×10^{18}. These photons are radiated out in all directions, but the quantity is so vast that the light is continuous and steady as far as the human eye is concerned.

The energy that is required to raise an electron between permissive states (Fig. 1-20) is specific and, if an incoming radiation is to raise the energy state of the electron, it must have the exact required energy content. When an atom absorbs a photon of energy, the excited atom can return to its normal state in either one jump (the solid vertical lines of Fig. 1-20) or in two or more jumps, which can include the dashed vertical lines of Fig. 1-20.

For example, an absorbed photon coming from an outside source raises the atom from 0 to 7.73 eV. When this energy is released, the electron drops back down 7.73 eV, and the photon of energy is released in the form of light. The energy-level change represents 12,4000/7.73 or 1604 Å, which is very short wave-length ultraviolet. The electron can return to its normal state and releases energy in two jumps, to 4.66 eV and then to zero. This corresponds to 4047 Å (visible violet light) and to 2656 Å which is a low frequency ultraviolet. By this process, a light is produced that is different from the original source of energy. This phenomenon is called *fluorescence*. A direct application is the use of fluorescent powders in the common fluorescent lights. The powders convert ultraviolet radiation into predominately visible light.

Referring to Fig. 1-20, an input energy excitation of 8.38 eV results in the attainment of a possible state. If the atom drops down to an energy level of 7.73 eV, heat is generated. A further drop to 5.46 eV produces a green light photon (5461 Å), and a further drop to the zero normal state produces an ultraviolet photon (2270 Å).

Questions

1. Define or explain each of the following terms: (*a*) neutron, (*b*) valence shell, (*c*) permissive energy level, (*d*) high field emission, (*e*) work function, (*f*) photoemission, (*g*) electron-hole pair, (*h*) intrinsic, (*i*) majority current carrier, (*j*) doping, (*k*) extrinsic, (*l*) opaque, (*m*) double ionization, (*n*) mean free path.
2. Name the component parts of an atom.
3. What are the relative sizes of the particles of an atom?
4. Explain four methods of obtaining free electrons.
5. Distinguish between photoemission and photoconduction.
6. Explain how a conduction band is formed.
7. Explain how one material can be a conductor and another an insulator.
8. What materials are used as sources of electrons in vacuum tubes?
9. What is the difference between a filament and a heater?
10. What is the effect of an increase in temperature on intrinsic semiconductor material?
11. Explain how intrinsic material can have both majority and minority current carriers.
12. What is the effect of doping to produce *N*-type material?
13. Explain the action of minority current carriers in *N*-type material.
14. Why are acceptor atoms so called?
15. How can a vacancy level exist in intrinsic material?

16. How are minority current carriers produced in *P*-type material?
17. What is an opaque crystal? A transparent crystal?
18. Why is the residual conductivity in a photoconductive device other than zero?
19. What is the effect of an increase in temperature on the residual conductivity?
20. Describe what occurs to produce multiple ionization in a gas.
21. Explain by Paschen's law the effect of a high-compression engine on the firing of a spark plug.
22. Why is deionization time longer than ionization time?
23. Why do different gases produce different light color outputs when they are excited by an electrical discharge?
24. Explain how fluorescence occurs.

Chapter Two

TWO-ELEMENT ELECTRON DEVICES

A discussion of the potentials at a *P-N* junction provides the foundation for an understanding of the operation of the semiconductor diode (Section 2-1). A number of the other basic principles of diodes are brought out in the application of the diode to a simple rectifier circuit (Section 2-2). The vacuum diode is approached through a discussion of space charge and saturation (Section 2-3). The properties of a junction under reverse-bias conditions are utilized in the Zener diode (Section 2-4), and equivalent circuits of a diode are developed in Section 2-5. Cold-cathode gas diodes (Section 2-6) have operating characteristics quite similar to the Zener diode. The chapter concludes with a discussion of light-emitting diodes in Section 2-7.

Section 2-1 The Semiconductor as a Diode

A semiconductor is formed into a diode by the creation of a junction between *P* material and *N* material within a crystal during the process of manufacture. The theory of the diode becomes an investigation into the properties of the *P-N* junction. In Fig. 2-1*a* and *b*, we repeat the energy-level diagram for the *N* material with a raised Fermi level and the energy-level diagram for the *P* material with a lowered Fermi level. When the two materials come together at a *P-N* junction (Fig. 2-1*c*), by definition, the Fermi level must be one and the same for the entire crystal. We see that there appears to be an abrupt discontinuity in energy levels from the *N* material to the *P* material. This abrupt discontinuity in level creates electrical forces between the two materials at the junction. In the conduction band, electrons in the *N* material move across the junction to the *P* material. Holes in the valence-bond band move

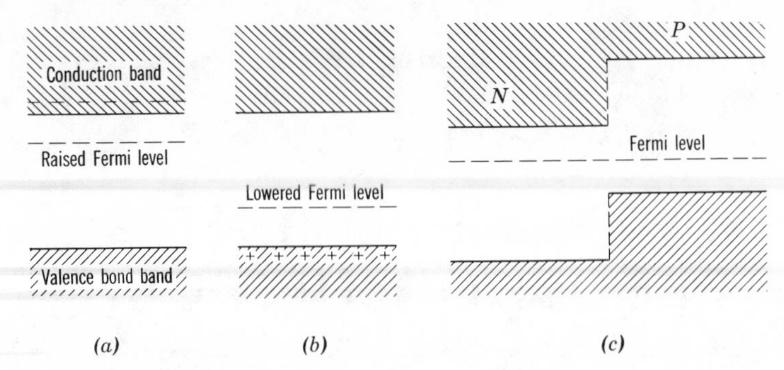

Figure 2-1 Energy-level diagrams in semiconductors. (*a*) *N* material. (*b*) *P* material. (*c*) *N-P* junction.

across the junction from the *P* material to the *N* material. This movement of electrons and holes takes place at the time when the junction is formed and is, then, an inherent characteristic of the boundary. Another approach which may be taken to understand this action is to consider that, in order to have stable conditions at the junction, the *P* material must gain some high-level energy states from the *N* material and lose some of its low-level energy states to the *N* material. Again, we stress that a *P-N* junction is not polycrystalline. It is not made up of an *N* crystal and a *P* crystal, but it is a single crystal, part of which has donor atoms and part of which has acceptor atoms.

The movement of holes into the *N* and the movement of electrons into the *P* are both additive in polarity to develop an emf which becomes a barrier to current flow. The *P* material at the barrier has acquired electrons making it "minus," and the *N* material has acquired holes making it "plus." We represent this emf diagrammatically in Fig. 2-2 where we show it as the junction potential equivalent to the battery voltage E_j. In Fig. 2-3, an external

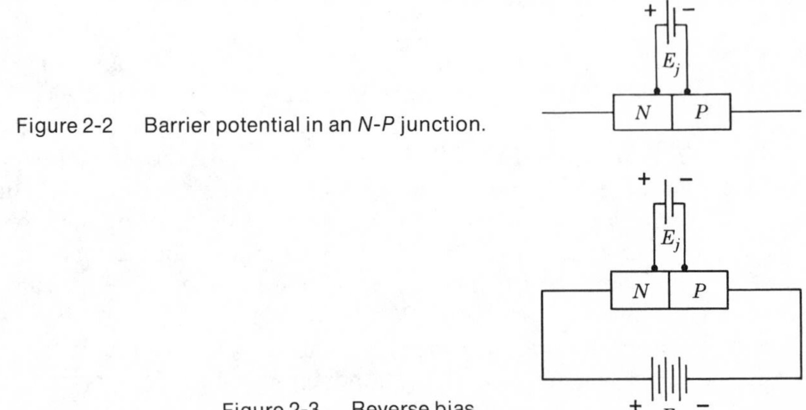

Figure 2-2 Barrier potential in an *N-P* junction.

Figure 2-3 Reverse bias.

battery E is connected with leads to the N material and to the P material. It is obvious that the junction emf is opposing or bucking E. In Fig. 2-4, E is acting in the same direction as or is aiding the junction emf. The connection of Fig. 2-3 will produce less current than the connection of Fig. 2-4. The

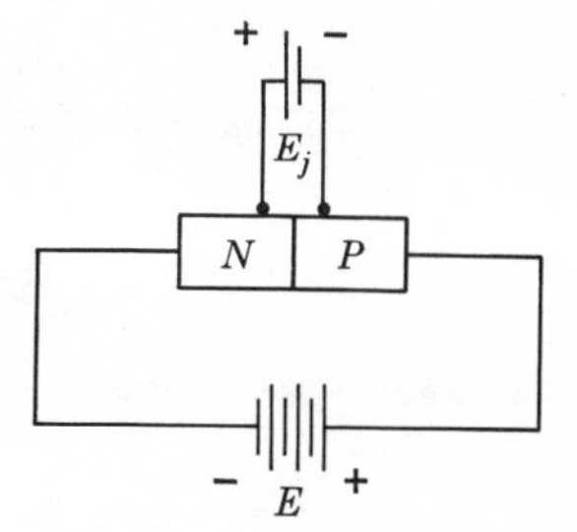

Figure 2-4 Forward bias.

external connection of Fig. 2-3 which discourages current flow is called *reverse bias*. The external connection of Fig. 2-4 which encourages current is called *forward bias*. If we call the positive terminal of the external battery P, we can state that reverse bias is a connection of P to N and of N to P. Forward bias is a connection of N to N and of P to P. This quick method gives a simple approach for determining proper battery polarity connections for semiconductor circuits.

If we represent the N-P junction in Fig. 2-5 as a barrier showing holes

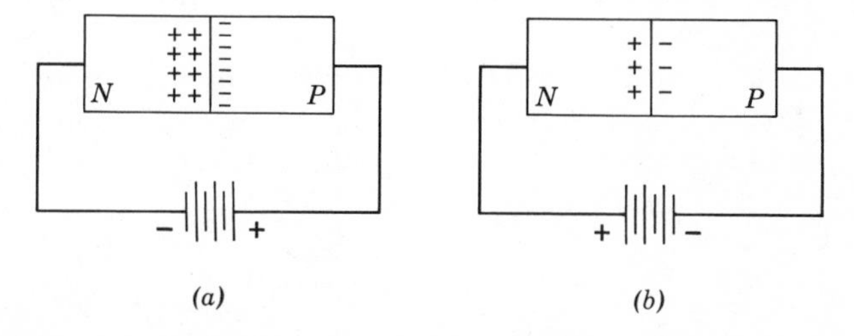

Figure 2-5 Action of holes and electrons at the junction under forward and reverse bias. (*a*) Forward bias. (*b*) Reverse bias.

on the N side of the barrier and electrons on the P side of the barrier, a forward-bias voltage (Fig. 2-5*a*) causes more electrons to cross the barrier from N to P. This action lowers still further the energy-level differences between the N and the P materials. A reverse bias (Fig. 2-5*b*) will *deplete* the junction of these majority current carriers, making the energy-level difference between the N and P greater. We can now sketch energy-level diagrams for the forward- and reverse-bias conditions (Fig. 2-6).

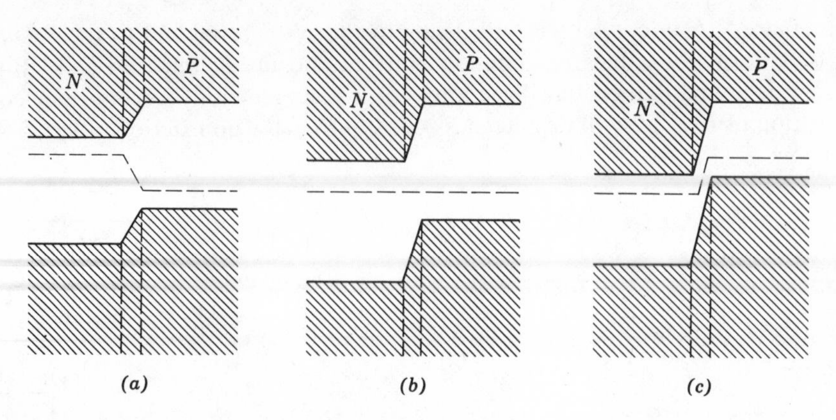

Figure 2-6 Energy levels of an *N-P* junction. (*a*) Forward bias. (*b*) Without bias. (*c*) Reverse bias.

We very explicitly developed the existence of both majority and minority current carriers in Section 1-5. The number of majority current carriers under normal conditions overshadows the number of minority current carriers present. When reverse bias is applied to a *P-N* junction, a current is produced by the minority current carriers in the same manner as forward current in the forward direction. This minority current-carrier flow is called *back current*, *reverse current*, *cutoff current*, or *saturation current* and is designated as I_R.

Figure 2-7 shows the characteristics of a typical silicon diode. When

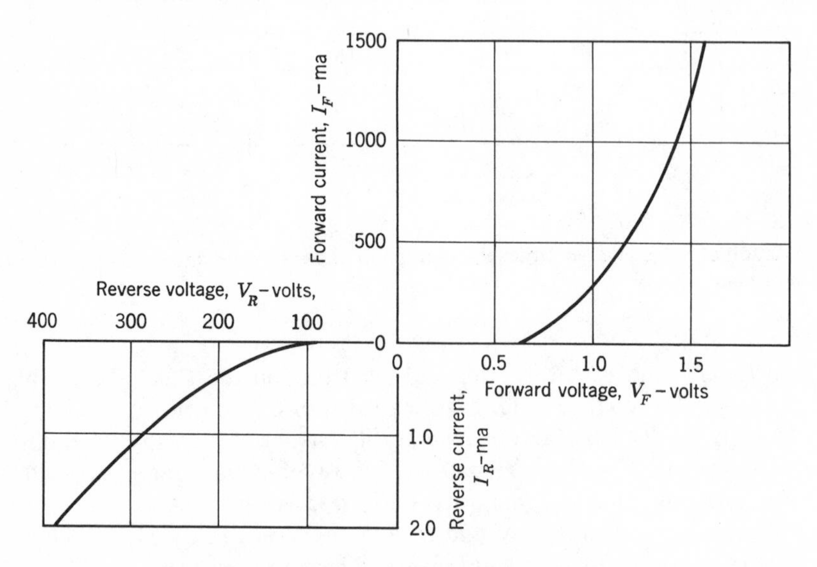

Figure 2-7 Characteristics of a typical diode.

the forward current is 1000 ma, the forward voltage drop is 1.4 V. By using these values, the power dissipation is 1.4 W and the d-c resistance is 1.4 Ω. In the reverse direction 280 V causes a back current of 1.0 ma. This corresponds to a 0.29 W dissipation and a d-c resistance value of 280,000 Ω. If we consider these two points on the diode curve, the front-to-back current ratio is 1000 ma/1.0 ma or 1000/1, and the front-to-back resistance ratio is 1.4/280,000 or 1/200,000. If any other two points on this curve are used, these ratios naturally, will change. We can use an ohmmeter to check out a diode quickly, but the reliable method is to apply a known voltage across the diode and to notice the current flow in each direction.

The symbol representing a semiconductor diode (Fig. 2-8*b*) uses an

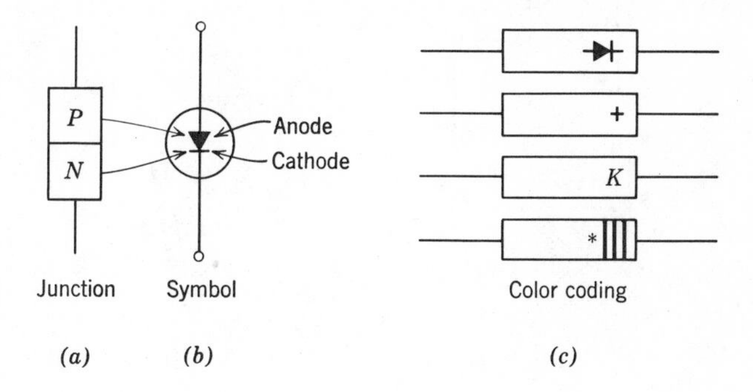

Figure 2-8 Schematic representation of a diode. (*a*) *PN* junction. (*b*) Circuit symbol. (*c*) Diode markings.

arrowhead to show the direction of *conventional* current flow. Notice very carefully that the *electron* flow opposes the arrowhead as in all semiconductor symbols. The color coding of the fourth diode in Fig. 2-8*c* is prefixed by "1N" to indicate its type number. The 1N does not appear on the diode itself.

When voltage is impressed across a diode in the forward direction, we observe from Fig. 2-7 that a small initial voltage is necessary to overcome the effects of the displaced charges on either side of the barrier. The initial forward break-voltage ranges from 0.15 to 0.30 V for germanium units and from 0.60 to 0.70 V for silicon units. Although the voltage drop in the forward direction is small when current flows, there is a heating $V_F I_F$ produced within the diode. If this heat can be dissipated adequately, the semiconductor remains at an equilibrium temperature. If, however, the diode cannot dissipate the heat, the temperature at the junction rises and the increased temperature causes a breaking down of covalent bonds. As a result, the current increases further and the temperature at the junction increases cumulatively to the point where the crystalline structure breaks down. This condition is known as *thermal runaway* or *thermal breakdown.* Thermal runaway can be avoided by limiting the current to a safe value within the capacity of the semiconductor by use of a suitable resistor in series with the diode. Thermal breakdown can occur under conditions of reverse bias as well as in the forward direction.

Because of the many applications of diodes, it is important to examine the reverse characteristics carefully. The barrier layer or junction layer of a *P-N* junction (Fig. 2-9*a*) is created during its manufacture. Electrons

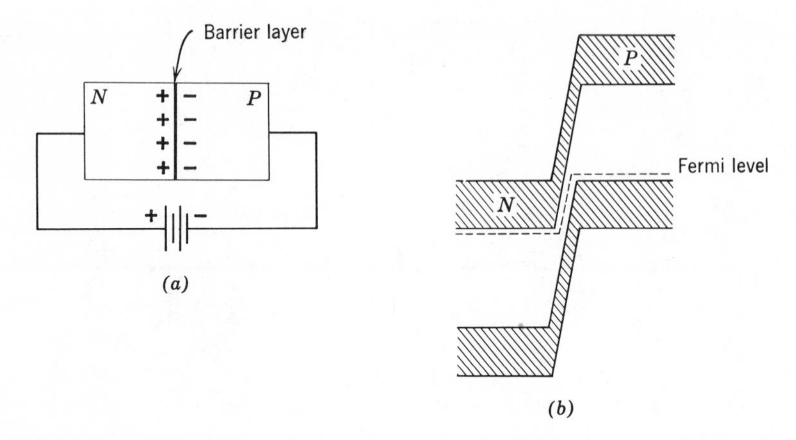

Figure 2-9 Reverse bias conditions in a semiconductor. (*a*) Barrier layer. (*b*) Energy-level diagram.

have crossed the junction from the *N* material to the *P* material and, therefore, the barrier acts to separate the two charges of a capacitor. When a battery is connected to the crystal in reverse bias, there is a flow of electrons from the *N* material into the positive terminal of the battery and a flow of electrons from the negative terminal of the battery into the *P* material. This occurs without a transfer of majority current carriers across the barrier itself. As a result, the barrier is *widened* by a depletion of electrons to the left of the barrier and an increase in concentration of electrons to the right of the barrier.

The barrier itself is very thin of the order of 10^{-3} cm. If the reverse voltage on the diode is 100 V, the voltage gradient across the barrier is 100,000 V/cm. To maintain this high voltage gradient without excessive current flow, it is most important that the material not be contaminated by any impurity atoms that are not explicitly required for the formation of the *P-N* junction.

When the conditions of reverse bias create the energy-level conditions shown in Fig. 2-9*b*, electrons can cross the barrier by moving from the valence band of the *P* material to the conduction band of the *N* material. The potential that causes this current flow is the *Zener potential*, and the process is termed *Zener breakdown*. When an electron under the influence of a high voltage gradient at the barrier collides with an atom, the impact produces an electron-hole pair. This process is cumulative, resulting in large values of current, and is termed an *avalanche*. Although there is a distinction between Zener breakdown and an avalanche breakdown, the term Zener breakdown is often used to describe both cases. If the current is limited by suitable resistance in the external circuit, a Zener breakdown is not necessarily destructive.

In order to examine the mechanism of current flow in detail, consider a battery connected to a diode, as shown in Fig. 2-10*a*. The positive terminal

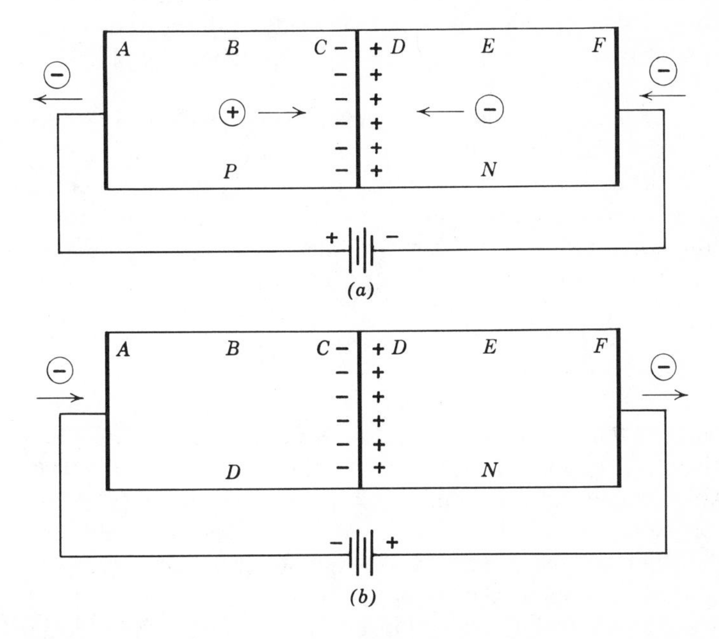

Figure 2-10 The mechanism of current flow. (*a*) Forward bias. (*b*) Reverse bias.

of the battery is connected to the crystal by means of a surface contact plate at *A*, and the negative terminal of the battery is connected to a surface contact plate at *F*. An electron attracted by the positive supply voltage leaves the *P* material and leaves behind a newly created hole. Now there exists in the *P* material near *A* an excess hole. An electron moves in across the barrier from the right to fill a hole in the *P* material. The transfer of an electron across the barrier produces an electron deficiency in the *N* material near the barrier. Electrons shift through the N material to the left, and one electron comes in from the negative battery lead to bring the N material back to its full complement of electrons. By a succession of this process of creating holes at A and of filling in holes at C, a current in the P material is effectively a movement of holes from A to C. Current in the N material is a movement of electrons from F to D. The electrons that are extracted at A are eventually returned to F through the battery circuit by means of the current flow in the battery circuit.

When the diode is reverse biased (Fig. 2-10*b*), holes migrate toward the negative terminal of the battery and electrons migrate toward the positive terminal. This action causes the depletion region to widen. There is no continuous flow of current. The only action is a change in charge at the depletion region, and this gives rise to a capacitive effect at the junction. This transfer mechanism is not a steady state current flow but is similar to the change in the charge on a capacitor when the plates of the capacitor are separated.

If, under the conditions of reverse bias, we consider minority current carriers, we now have the conditions of a forward bias, and a current does flow. The objective in manufacture is to keep the process so controlled that this *reverse current* is very small when compared to the magnitudes of the normal currents of forward bias.

Section 2-2 The Simple Rectifier

If an a-c source is applied to the series circuit formed by a diode and a load resistance, we obtain the simple rectifier circuit of Fig. 2-11. The applied voltage waveform e is, sinusoidal, having a peak value of E_m volts and an effective or rms value of E volts. When m is positive with respect to n, this polarity makes the P material (the anode) of the diode positive with respect to the N material (the cathode). This is the direction of forward bias, and current flows through the entire series circuit. This current develops a voltage drop across the load resistor with the polarity shown on the circuit diagram. When n is positive and m is negative, the N material (the cathode) of the diode is positive with respect to the P material (the anode). This is the reverse-bias condition, and no current can flow in the circuit. Now the voltage appears across the diode. Thus, the only voltage that can exist across the load resistor exists when m is positive. The diode is serving as a *rectifier.* A rectifier is a device or circuit arrangement that makes one-directional (unidirectional) current or voltage from a two-directional or an a-c source.

From the discussion on the semiconductor, in Section 2-1, it is obvious that, when current flows through a diode, there must be an accompanying

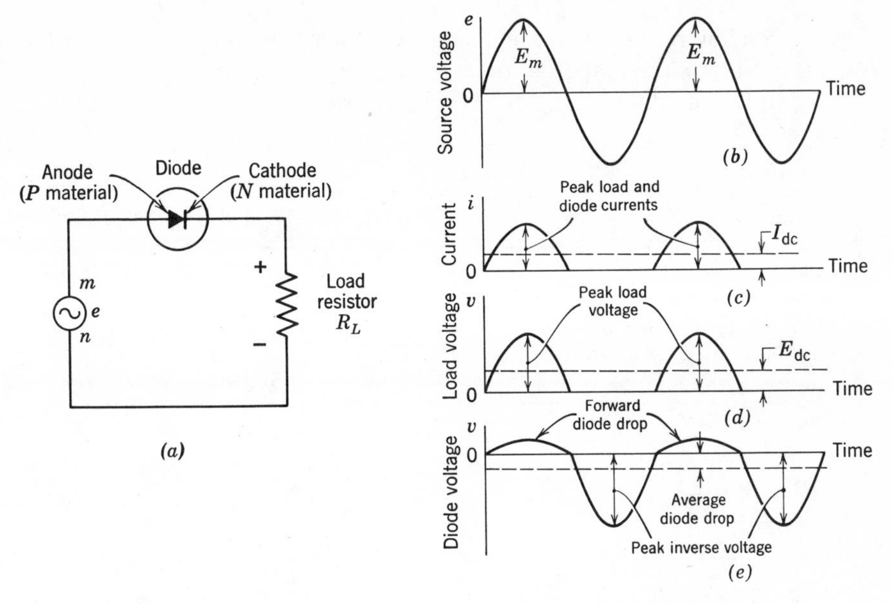

Figure 2-11 Circuit and waveforms for the basic rectifier.

forward voltage drop. This is shown in waveform *e* as the diode drop. The current, waveform *c*, must be a half sine wave, always positive, as is the load-voltage drop, waveform *d*. The whole alternating voltage on the negative half-cycle appears across the diode, since no current can flow in the circuit (waveform *e*). We notice on waveform *e* that the *peak inverse voltage* is, for this circuit, the peak of the alternating line voltage E_m.

If the magnitude of the supply voltage *e* is large, the small forward diode drop may be neglected. Then, effectively, the peak voltage across the load is E_m, and the peak load current is E_m/R_L. By means of calculus, we can show that the average value of a *half* of a sine wave over a *full* a-c cycle is the peak value divided by π. Then, the average load voltage which is the d-c load voltage, is the peak value of the line voltage divided by π. Remembering that the peak values and the effective values are related by $\sqrt{2}$ for sinusoidal waveforms, we have

$$E_{dc} = \frac{E_m}{\pi} = \frac{\sqrt{2}E}{\pi} \tag{2-1a}$$

and

$$I_{dc} = \frac{E_m}{\pi R_L} = \frac{\sqrt{2}E}{\pi R_L} = \frac{E_{dc}}{R_L} \tag{2-1b}$$

If the value of the load resistor R_L is changed, the only change in the waveforms is a change in the amplitude of the current flowing in the load if the diode drop is neglected.

Silicon rectifiers are available commercially in current ratings up to several hundred amperes and in peak reverse voltage ratings up to thousands of volts. The ratings for three typical diodes are given in Table A.

TABLE A

Rating	1*N*270	1*N*1095	1*N*1190
Peak inverse voltage, V_R	100 V	500 V	600 V
Forward d-c current, I_F	200 ma	750 ma	35 A
Forward d-c voltage drop, V_F	1.0 V	1.2 V	1.7 V
Maximum reverse d-c current, I_R	100 μa at 50 V	5 μa at 500 V	10 ma at 600 V

PROBLEMS

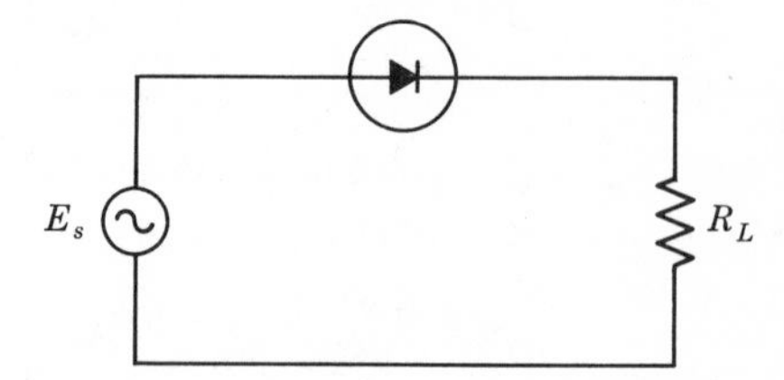

Circuit for Problems 1 and 2.

1. The supply voltage E_s is 120 V rms and the peak current rating of the diode is 1.2 A. Assuming peak rated current flows in the diode, what is the value of R_L and what is the d-c current in R_L?
2. The supply voltage E_s for the rectifier is 1200 V rms and the desired average d-c power in R_L is 100 W. Determine the value of R_L and the d-c current in R_L.
3. Determine the forward d-c resistance and the reverse d-c resistance for each diode listed in Table A.

Section 2-3 The Vacuum Diode

A vacuum diode consists of two elements: the plate or anode and a cathode. The thermionic cathode may be either a filament surface or a heater-cathode combination. For simplicity, the illustrations and discussions are confined to a diode in which the plate and the cathode are parallel planes of finite dimensions (Fig. 2-12). Actually, we do not often encounter this physical arrangement in practice. The plates are usually rectangular in cross section surrounding the cathode or take the shape of a cylinder wall. Also the thermionic emitter is usually one of the forms shown in Fig. 1-5.

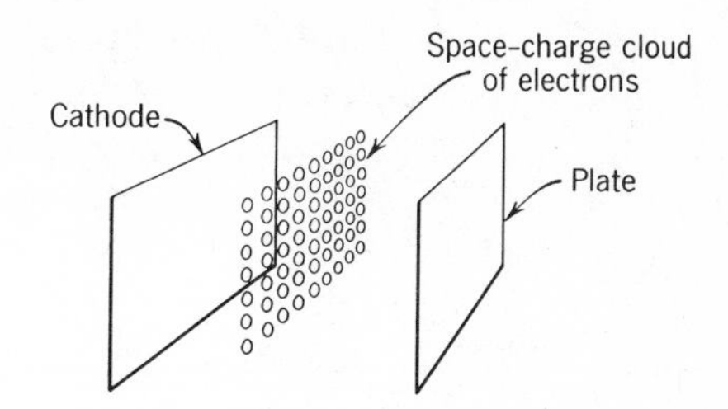

Figure 2-12 The basic structure of a diode.

Figure 2-12 shows that a cloud of electrons exists in space just off the cathode surface. If we assume the plate potential to be zero with respect to the cathode, when electrons are emitted or "boiled off" the cathode, the small velocities obtained in the emitting process carry the electrons away from the cathode in a slow drift which diffuses throughout the tube envelope. If a milliameter is connected from the plate to the cathode forming a complete circuit, the electrons that reach the plate from this small initial velocity return to the cathode through the external circuit giving an indication of a small current flow on the meter. This is known as the *Edison effect.* If electrons are "boiled off" the cathode, the cathode is left with a positive charge. The tendency of the cathode then is to seek back electrons to cancel this positive charge. Accordingly, electrons striking and entering the plate return to the cathode through the external circuit, causing a deflection on the meter.

Another consequence of the resulting positive charge on the cathode is that the cathode draws back unto itself a great many of the electrons that it has emitted. Thus, in the process of emission, there is a heavy "rain" of electrons back on the emitting surface. We noted that, during this process, there is a drift of electrons across the tube. These electrons have a negative charge and tend to repel back into the cathode the next group of electrons that are emitted. The overall result is that there exists in the space just off the cathode a cloud of electrons that is called the *space charge.*

If an external voltage source is connected to a diode so that the plate is negative with respect to the cathode (Fig. 2-13), the meter that reads

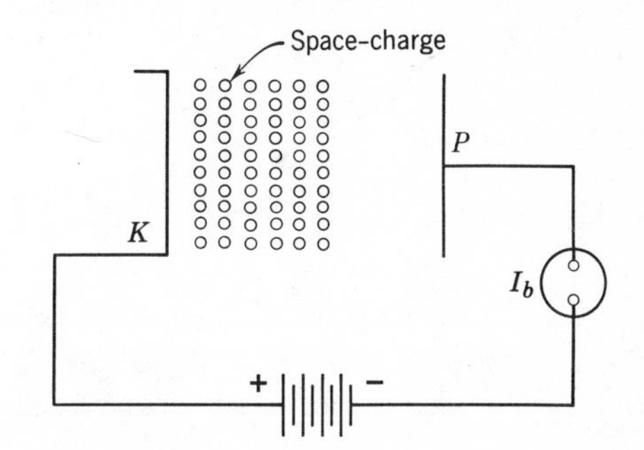

Figure 2-13 Reverse voltage on a diode.

plate current I_b indicates zero. The negative voltage on the plate not only prevents electrons from reaching the plate but also forces the drift electrons back into the space-charge cloud. If the external battery is connected (Fig. 2-14) so that the anode, is positive with respect to the cathode, a plate

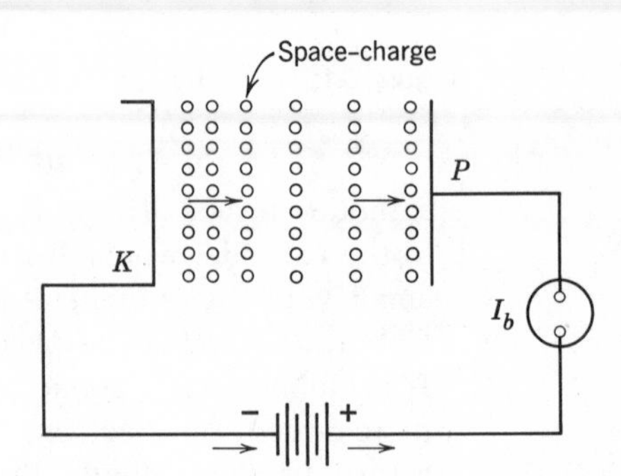

Figure 2-14 Forward voltage on a diode.

current results. The positive anode attracts electrons from the space charge. Those electrons that are removed from the space charge are replaced by further emission from the cathode surface. When the anode potential is increased, the plate current increases. The space charge is reduced as the plate current increases. Eventually, we find that the number of "stored" electrons in the space charge is reduced to zero, and the electrons that are emitted are directly swept to the plate.

When plate current comes from the reservoir of the space charge, the space-charge cloud actually is serving as the cathode of the tube as far as the plate is concerned. The term *virtual cathode* is used to describe this function of the space charge. Also, at those current levels where a space charge exists, the term *space-charge-limited* is used to describe the operation of the tube. When the plate collects all the electrons, a *plate saturation* exists.

If the relation between plate current and plate voltage is plotted, the curve of Fig. 2-15 results. The plate current at point *b* is the Edison-effect

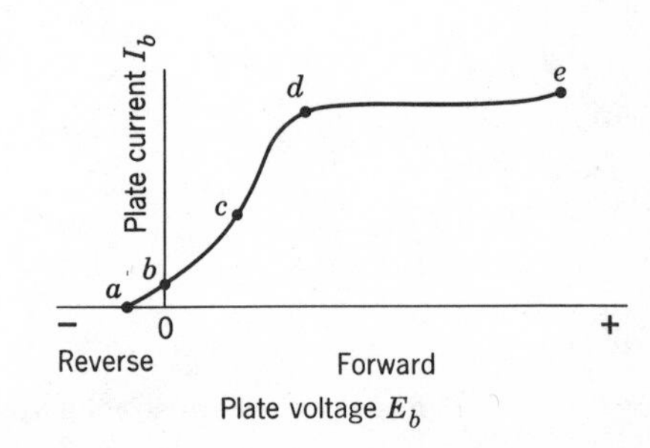

Figure 2-15 Plate characteristic of a diode.

value. A small reverse negative voltage a is required to counteract the initial velocities of emission that produce the Edison-effect current. The current increases from b to c to d with an increasing positive plate potential. At point d, we reach plate saturation. There is very little increase in plate current from d to e.

The magnitude of the Edison effect is so small that, except in certain special applications, it may be neglected and the curve may be redrawn as Fig. 2-16. By several different mathematical methods, an equation can be established for the space-charge-limited region of the curve, *Ocd*:

$$I_b = KE_b^{3/2} \tag{2-2}$$

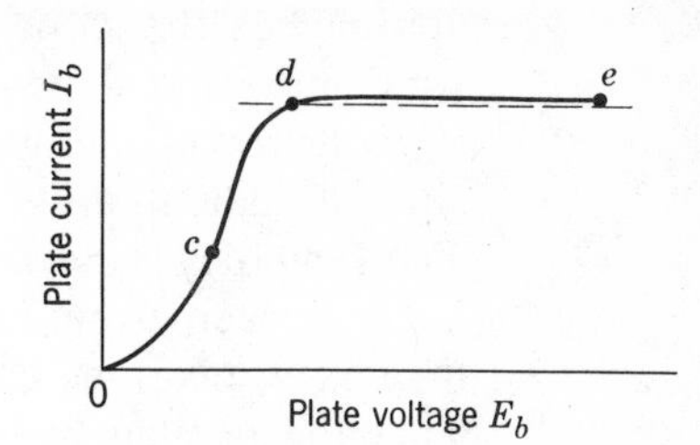

Figure 2-16 Simplified plate characteristic of a diode.

This equation is known as the *three-halves power law*, *Child's law*, the *Langmuir-Child law*, or the *space-charge equation.* If one set of readings is experimentally determined, the value of K can be found by calculation, and then other values of I_b and E_b can be worked out for this value of K. K is proportional to the area of the plate and inversely proportioned to the square of the distance between the cathode and plate. In actual tubes the exponent may not be exactly $\frac{3}{2}$. The $\frac{3}{2}$ value of the exponent is only approximately correct for actual tubes.

Normally, a vacuum tube is operated at levels of plate current that are considerably below the saturation point. However, in diodes, the maximum peak current, which is the saturation value, is important in rectifier circuits and is always listed for the diode in the tube manual. For a typical space-charge-limited diode plate characteristic (Fig. 2-17), point P on the curve

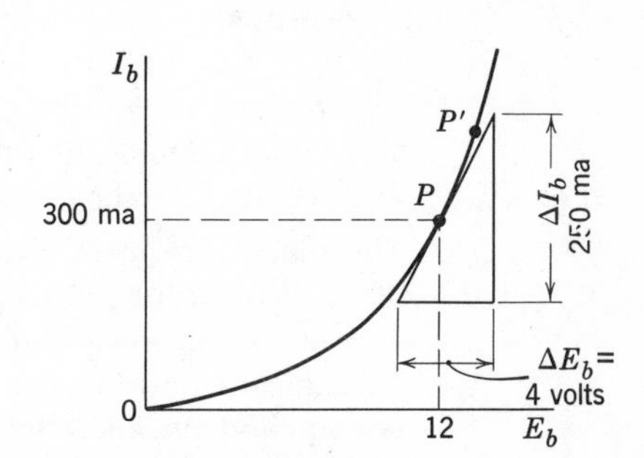

Figure 2-17 Method of determining the static resistance R_p and dynamic resistance r_p for a diode.

is established when the plate voltage E_b causes a 300-ma plate current I_b. If we divide the value of E_b by I_b, we obtain 40 Ω. This is the equivalent value of resistance which the tube presents to I_b and E_b and, since these values are direct current, it is called the *d-c plate resistance* or *static resistance*, using the symbol R_p:

$$R_p \equiv \frac{E_b}{I_b} \tag{2-3a}$$

If the plate voltage increases a small amount, we move from P to P' on the curve. Correspondingly, the plate current increases a small amount. If P' moves closer and closer to P, the direction of change from P to P' approaches the slope of the tangent drawn to the curve at point P. Now we can use the slope of the tangent to determine the ratio of a small change in E_b to the resulting small change in I_b. This ratio value is resistance and, since its definition is based on *small changes*, it must be an a-c resistance. The symbol r_p is used to designate this *a-c plate resistance.*

It is seen from Fig. 2-17 that, if a change in plate voltage along the tangent is taken as ΔE_b, the corresponding change in plate current is ΔI_b. Then the a-c plate resistance r_p is defined as

$$r_p \equiv \frac{dE_b}{dI_b} \equiv \frac{\Delta E_b}{\Delta I_b} \tag{2-3b}$$

The symbols Δ and d are both used in mathematics to be read as "a change in." Other terms that are often used for r_p are *dynamic plate resistance* or *variational plate resistance*.

We must very carefully notice that R_p is a d-c value and that r_p is an a-c value. Also, they are different numerically. By solving for R_p at point P, we obtain 12/0.3 or 40 Ω, whereas the value for r_p is 4/0.250 or 16 Ω. In order for the d-c value of resistance to equal the a-c value, the volt-ampere curve would have to be a straight line going through the origin.

PROBLEMS

In these problems assume that the diodes follow the $\frac{3}{2}$-power law.

1. The forward voltage drop of a type 5*U*4 vacuum-tube diode is 70 V when the current is 300 ma. Determine the diode voltage drop for 0, 50, 100, 150, 200, and 250 ma. Plot the characteristic.
2. The plate current in a particular diode is 60 ma when the plate voltage is 400 V. What is the plate voltage at 40 ma? At 25 ma?
3. The plate current in a diode is 100 ma when the plate voltage is 30 V. The plate dissipation E_bI_b has a maximum permissible value of 6 W. What are the voltage and current ratings for this maximum plate dissipation value?

Section 2-4 The Zener Diode

The property of an avalanche breakdown of a diode in the reverse-bias direction has been adapted to develop the class of semiconductors known as *Zener diodes*, *breakdown diodes*, or *reference diodes*. The symbols in current use for these diodes are shown in Fig. 2-18.

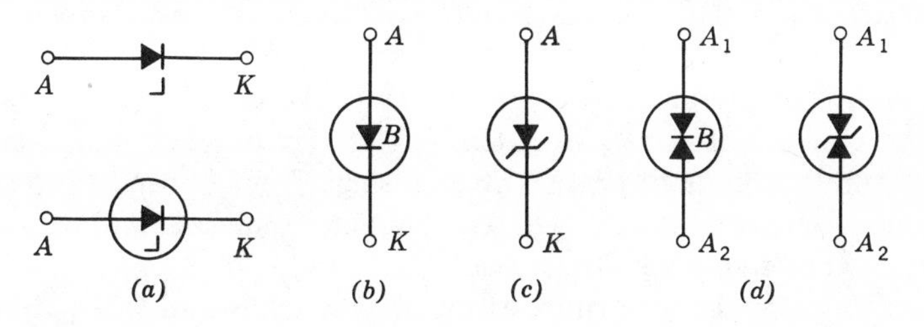

Figure 2-18 Zener diode symbols. (*a*) IEEE standards. (*b*) and (*c*) Alternative symbols used by industry. (*d*) Double breakdown diode.

The typical characteristic of the Zener diode is shown in Fig. 2-19. The forward voltage V_F is naturally much less than the reverse voltage and, since the diode is normally used only in the reverse direction, this portion of the characteristic is drawn only for completeness. The reverse current that

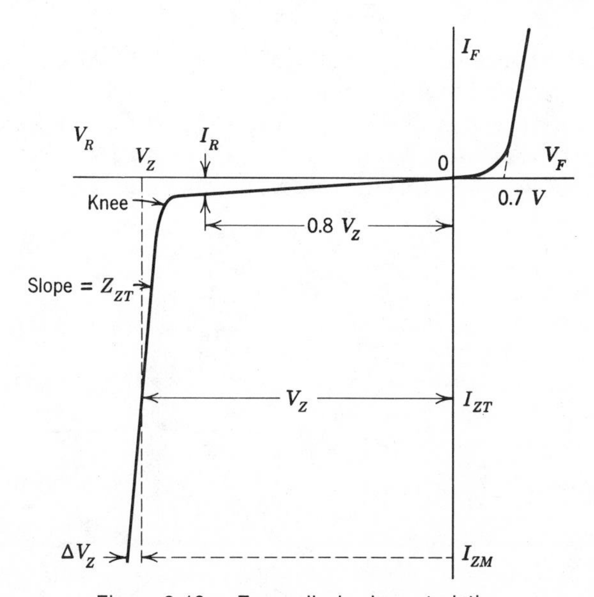

Figure 2-19 Zener diode characteristic.

exists between the origin and the *knee* of the curve is the reverse leakage current of a junction caused by the minority current carriers. This current is specified by giving its value at 80% of the Zener voltage V_Z. When the reverse voltage on the diode is increased from zero, an avalanche takes place at the knee of the curve, and the current rises rapidly with only a very small change in voltage. An external resistance is required to limit this current to the maximum permissible value denoted by I_{ZM}. The Zener voltage V_Z is that voltage which exists across the terminals of the diode, for a current, I_{ZT}, that is the approximate midpoint of its linear range. The structure of the diode and its heat sink can dissipate a maximum value of heat equivalent to P_{ZM} or $V_Z I_{ZM}$ watts. Commercial Zener diodes are presently manufactured in power ratings from $\frac{1}{4}$ to 50 W. Voltage ratings range from 2.4 to 200 V. There are several thousand different combinations of power and voltage ratings available.

The a-c resistance (impedance) Z_{ZT} is measured at I_{ZT}. For example, a half-watt, 20-V Zener diode has a maximum rating for I_{ZM} of 500/20 or 25 ma. This diode has a value for Z_{ZT} of 32 Ω at an I_{ZT} of 5 ma. An increase in Zener current of 5 ma causes the diode voltage to increase by $I_{ZT}Z_{ZT}$ or 0.005×0.160 V. At a current of 25 ma, the voltage drop across the Zener diode is

$$V_Z + Z_{ZT}(I_{ZM} - I_{ZT}) = 20 + 32\ (0.025 - 0.005) = 20.640 \text{ V}$$

At the knee of the curve, the diode voltage is

$$20 - 32 \times 0.005 = 19.840 \text{ V}$$

This corresponds to a total voltage change of 0.800 V or 4% over the whole range of Zener current variation.

Consider the circuit shown in Fig. 2-20. Notice that the Zener is reverse biased for normal operation. The output voltage V_{out} is the Zener potential V_Z. Equations can be formed for this circuit:

$$I_{\text{in}} = I_Z + I_L \tag{2-4a}$$

and

$$R = \frac{E_{\text{in}} - V_Z}{I_{\text{in}}} \tag{2-4b}$$

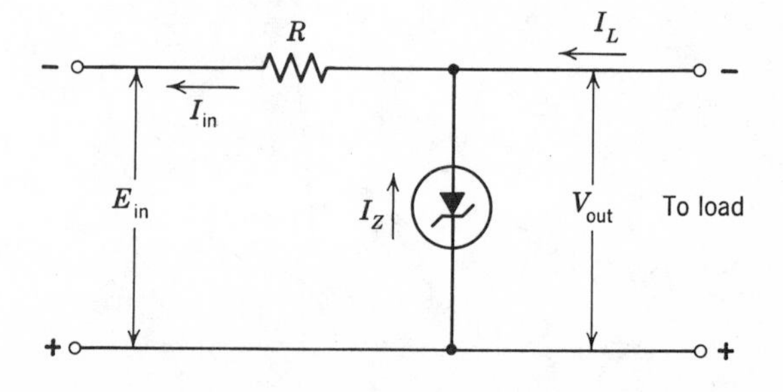

Figure 2-20 Application of a Zener diode in a voltage regulator circuit.

Let us assume that the diode is a 50-V, 50-W unit with a leakage current of 5 ma and that V_{ZT} is 50 V for I_{ZT} equal to 500 ma. Also assume that the Zener is operated at the approximate midpoint of its inverse current rating or 0.500 A. Assume that the load current is 2.000 A and that the supply voltage is 60 V. The supply current from Eq. 2-4*a* is

$$I_{\text{in}} = I_Z + I_L = 0.500 + 2.000 = 2.500 \text{ A}$$

When the load current increases, the current in the Zener diode decreases. Likewise, when the load current decreases, the current in the Zener diode increases. The Zener diode current can either increase 500 ma or decrease 495 ma and still be within its operating range. Accordingly, the load current can vary from 1.500 to 2.495 A and the Zener diode will maintain the load voltage at 50 V.

In an actual circuit, notice that the a-c resistance modifies this calculation slightly. If the diode has a value for Z_{ZT} of 1.2 Ω, the voltage V_Z across the load is

$$V_Z = V_{ZT} + Z_{ZT}(I_Z - I_{ZT}) \tag{2-5}$$

When the load current is 1.500 A, the Zener current is 1.000 A and the load voltage is

$$V_Z = 50 + 1.2(1.000 - 0.500) = 50.60 \ V$$

When the load current is 2.495 A, the Zener current is 5 ma and the load voltage is

$$V_Z = 50 + 1.2\ (0.005 - 0.500) = 49.41 \text{ V}$$

Now, using these numerical values, assume that the diode is operated at 0.500 A with a 2.000-A load. The series resistor R is

$$R = \frac{(60 - 50) \text{ V}}{(2.000 + 0.500) \text{ A}} = 4.0 \ \Omega$$

If the supply voltage E_{in} increases, the Zener diode absorbs the increase in current while maintaining the load voltage at a constant value. At the maximum value of Zener current, the supply voltage is

$$E_{\text{in}} = I_{\text{in}}R + V_{\text{out}} = (1.000 + 2.000)4.0 + 50.60 = 62.60 \text{ V}$$

When the supply voltage decreases, the Zener current can fall to 5 ma and still maintain the load voltage constant:

$$E_{\text{in}} = (0.005 + 2.000)\ 4.00 + 49.41 = 57.43 \text{ V}$$

From these calculations, it is seen that the Zener diode is useful for maintaining a constant load voltage, either for a range of variations in load current or for a supply voltage fluctuation.

Many applications require the use of two Zener diodes connected back-to-back in order to obtain the Zener breakdown characteristic for either direction of the supply polarity. Instead of using two separate diodes, the *double breakdown diode* shown in Fig. 2-18*d* is available. The double breakdown diode is sometimes called a *varistor diode*.

PROBLEMS

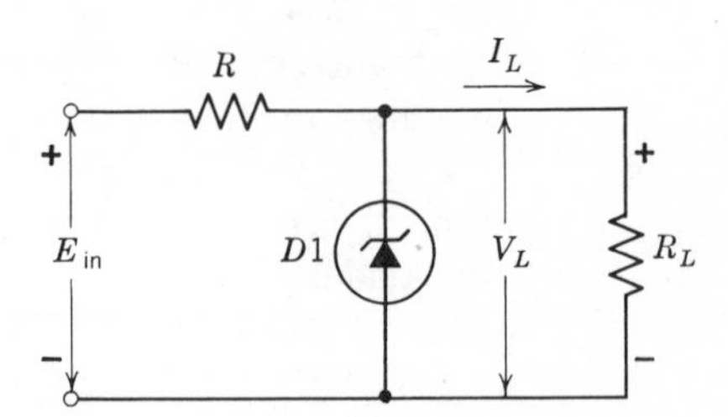

Circuit for Problems 1 to 4.

1. E_{in} is 40 V and R is 50 Ω. The Zener diode $D1$, rated at 3 W, has a value for I_{min} of 5 ma and a value for V_Z of 20 V. Find the range over which R_L can be varied.
2. R is 4 kΩ, R_L is 10 kΩ, and V_Z is 30 V. The input voltage E_{in} varies between 70 and 100 V. What are the maximum and minimum currents through $D1$?
3. R is 3 Ω, R_L is 10 Ω, and V_Z is 10 V. The maximum power dissipation of the Zener diode is 20 W. What is the allowable range of E_{in}, assuming that $I_{Z,min}$ is 50 ma?
4. It is desired to regulate a 15-V load using a Zener diode from a 20-V source. The load current can vary from 0 to 100 ma. What is the power rating for the required Zener diode and what is the value of R? $I_{Z,min}$ for the diode is 10 ma.

Section 2-5 Dynamic Diode Characteristics and Diode Models

The two basic rectifier circuits are shown in Fig. 2-21*a* and *b*. The diodes have the nonlinear characteristic of *c*. Since the reverse current in a diode is effectively zero, we are concerned only with that part of the characteristic that is in the first quadrant. Similarly, the volt-ampere curve for the linear resistor R_L is a straight line shown only in the first quadrant (Fig. 2-21*d*). When the diode is in a circuit with R_L, a composite characteristic can be

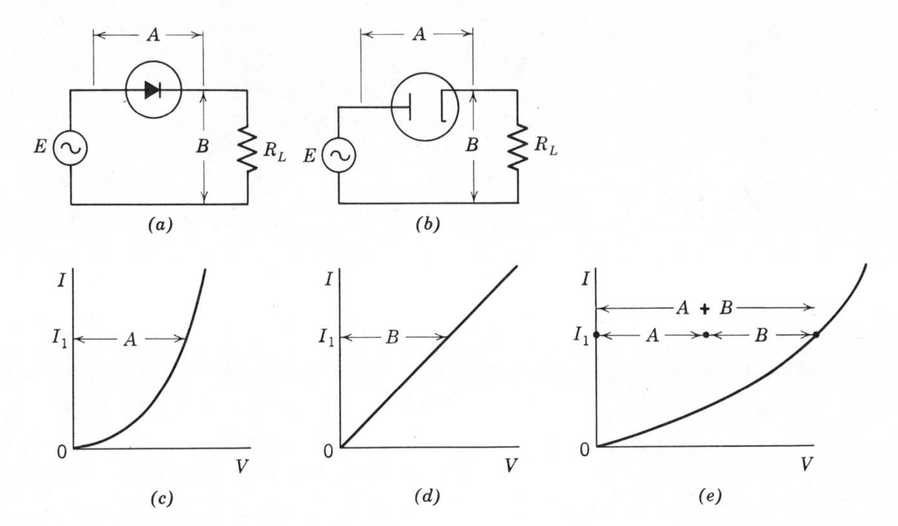

Figure 2-21 Diode load lines. (*a*) Semiconductor and (*b*) vacuum-tube rectifier circuits. (*c*) Diode characteristic. (*d*) Volt-ampere characteristic for R_L. (*e*) Composite characteristic.

formed by adding the individual curves. The new characteristic (Fig. 2-21*e*) is obtained by choosing an arbitrary current value I_1. This current value corresponds to a forward diode voltage drop of A volts (Fig. 2-21*c*) and to an IR drop across R_L of B volts (Fig. 2-21*d*). The sum of these drops is $A+B$ on Fig. 2-21*e*. This process is repeated to secure sufficient data points to yield the smooth composite curve.

The composite characteristic is redrawn in Fig. 2-22. The source voltage E is drawn to a time scale on the vertical axis. A point-by-point extension from the input waveform will yield the actual load current waveform. The points, *a*, *d*, *e*, *f*, and *g*, are the same instants of time and must be laid out to the same time scale. The graphical results of Fig. 2-22 yield the exact waveform in the output, but the method has the disadvantage that the composite characteristic is valid for one particular value of R_L only. When R_L is changed, a new composite characteristic must be drawn.

A method called the model approach has been developed to reduce the nonlinear characteristics of electronic devices to a conventional electric circuit. This approach is particularly suited to semiconductors. The semiconductor diode, Fig. 2-23*a*, has the complete characteristic shown in Fig. 2-23*b*. Straight lines are superimposed on the actual characteristic to produce a characteristic that is made up of several straight-line segments. Then each segment represents a *linear* property over its range. Line *a-b* projects back to the *X*-axis at 0.7 V, which is the effective barrier potential for a typical silicon diode. The line *o-a* is taken as zero current. The reverse

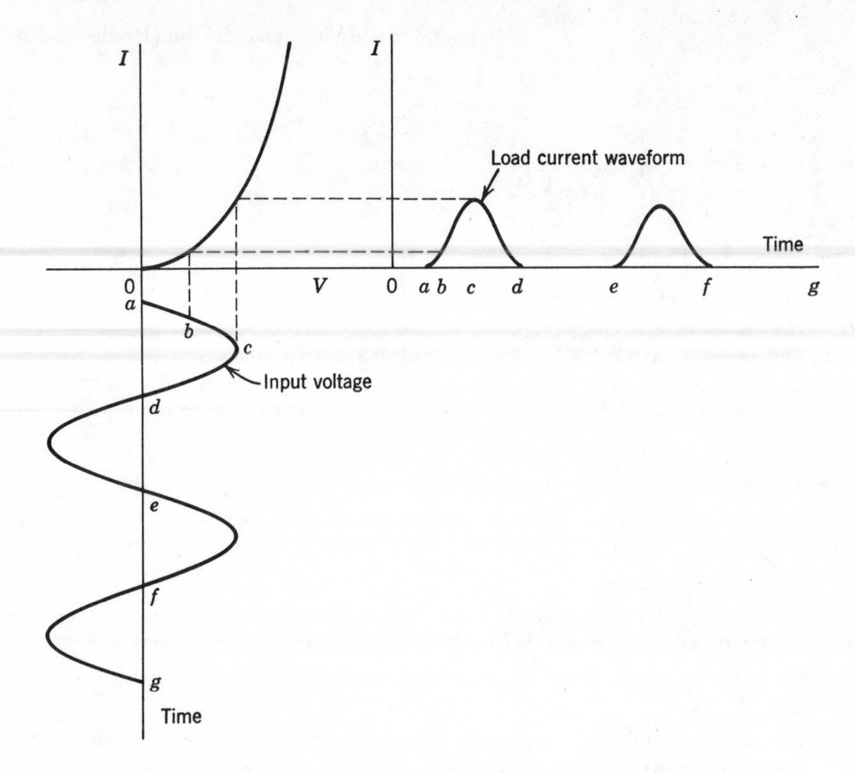

Figure 2-22 Diode dynamic transfer characteristics.

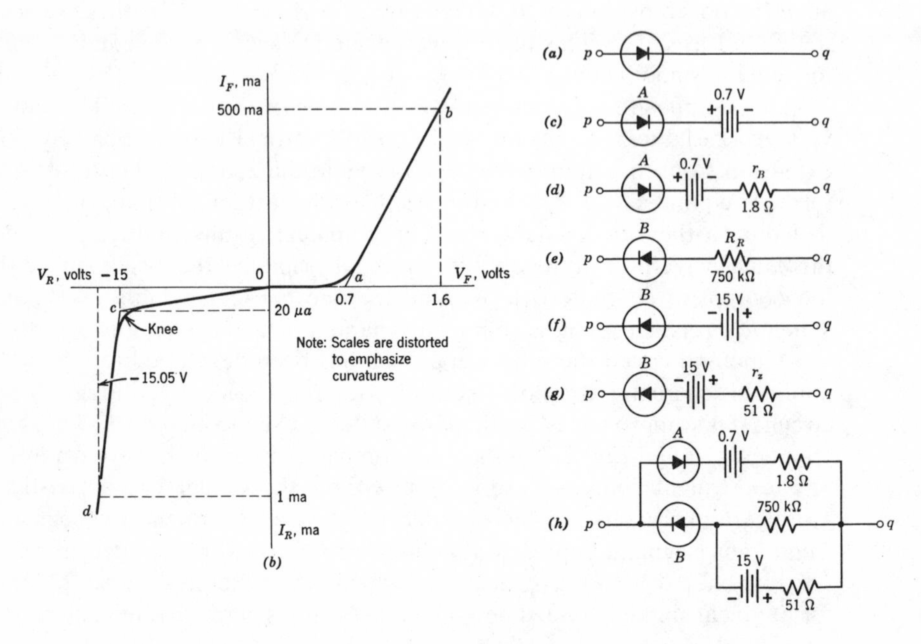

Figure 2-23 Development of a silicon diode model.

characteristic is shown, as usual, to different current and voltage scales. A straight line from *o* to *c* represents the leakage current I_R. The standard method of measuring leakage current for a data sheet is to give the value of leakage current at 80% of the Zener voltage. At the knee of the reverse characteristic *c*, a Zener breakdown occurs, and the current abruptly rises toward *d*. Thus, the reverse characteristic is represented by two straight lines.

The portion of the forward characteristic from *o* to *a* is ideal without current flow up to 0.7 V. This is represented by diode *A* in series with a reversed 0.7 V ideal battery (Fig. 2-23*c*). The region from *a* to *b* has a finite slope other than 90° and indicates that a resistor is in the circuit. This resistance r_B is the bulk resistance of the semiconductor material (the *P* and the *N* sections) and is determined from the slope of the line by $r_B = \Delta V/\Delta I$. The numerical values in this example yield $(1.6-0.7)/(0.500-0)$ or 1.8 Ω for r_B (Fig. 2-23*d*). The addition of the small resistance value of 1.8 Ω will not affect the slope of *o-a* appreciably, since 0.7 V/1.8 Ω is less than $\frac{1}{2}$ ma whereas the value of current at *b* is 500 ma.

To obtain the approximate model for the reverse part of the characteristic, an ideal diode *B* is used to confine any circuit action to this quadrant. The straight line *o-c* represents the reverse resistance R_R, determined from incremental changes as 15.00 V/20 mA or 750,000 Ω and shown by the circuit of Fig. 2-23*e*. The knee voltage at *c* is 15 V and is effectively a battery in the reversed direction (Fig. 2-23*f*). Since there is a finite change in voltage from *c* to *d*, the straight line from *c* to *d* represents a resistance value other than zero. This resistance r_Z is found from incremental changes as (15.05 − 15.00) V/(1 ma-20 μa) or 51 Ω (Fig. 2-23*g*). Now, the whole characteristic is represented by these piecewise circuits put together (Fig. 2-23*h*).

Consider this diode connected to a 75-kΩ load (Fig. 2-24*a*). The complete approximate model together with the load is shown in Fig. 2-24*b*. An incoming voltage waveform is given in Fig. 2-24*c*, and it is required to determine the output voltage waveform. When the incoming signal is positive, only that part of the circuit through ideal diode *A* is considered. Since the bulk resistance, 1.8 Ω, is very small compared to the load resistance, 75 kΩ, the *IR* voltage drop across the bulk resistance is neglected. Thus the barrier potential, 0.7 V, is subtracted from the input voltage to obtain the positive voltage levels shown in Fig. 2-24*d* from *o* to *e*. When the polarity of the input reverses, the circuit is considered through ideal diode *B*. A 1-V signal level causes a reverse current of approximately 1 μA, which does not cause an appreciable voltage drop across the load resistor. When the signal rises to 10 V, the reverse current drop across the load resistor is now 0.9 V, *f* to *g* and *h* to *i*. When the diode breaks down, the reverse resistance (750 kΩ) is bypassed, and the series resistance, 51 Ω, does not produce any appreciable *IR* drop and the load voltage, *g* to *h*, becomes the source voltage less the Zener voltage of the diode or 5 V.

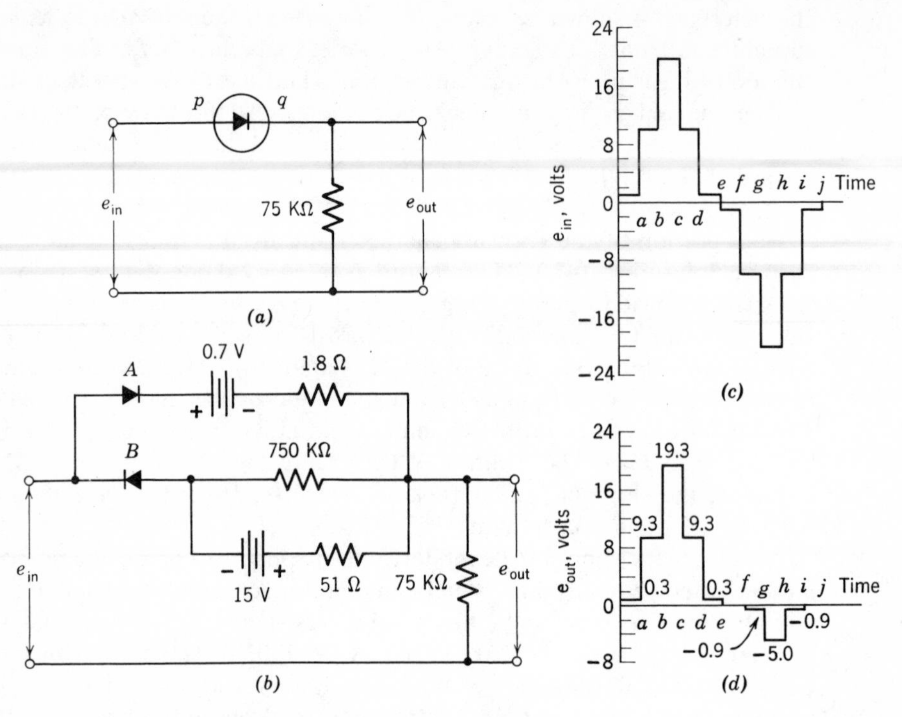

Figure 2-24 The analysis of a large-signal diode circuit. (*a*) Actual circuit. (*b*) Complete approximate model. (*c*) Incoming voltage waveform. (*d*) Output voltage waveform.

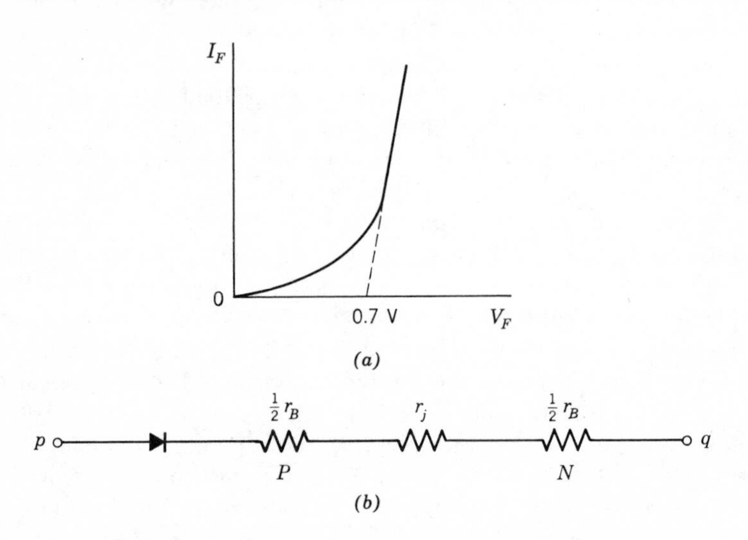

Figure 2-25 The analysis of a small-signal diode circuit. (*a*) Expanded forward characteristic. (*b*) Model.

The expanded characteristic of a diode near the origin in the forward direction is shown in Fig. 2-25*a*. The sharp curvature can be explained in a model (Fig. 2-25*b*) by the introduction of the concept of a junction resistance, r_j. The value of the junction resistance is determined by dividing a numerical value of junction emf by the current in the forward direction. Junction emf's range from 25 to 50 mV. Assume the junction emf is 30 mV. When the current is 1 μA r_j is 30,000 Ω. When the current is 100 μA, r_j is 300 Ω. When the current is 1 ma, r_j is 30 Ω, and when the current is 100 ma, r_j is 0.3 Ω. This value of r_j must be added to r_B. It is only necessary to consider r_j when the currents in the diode are low values that are usually much below the maximum rated forward current of the diode.

PROBLEMS

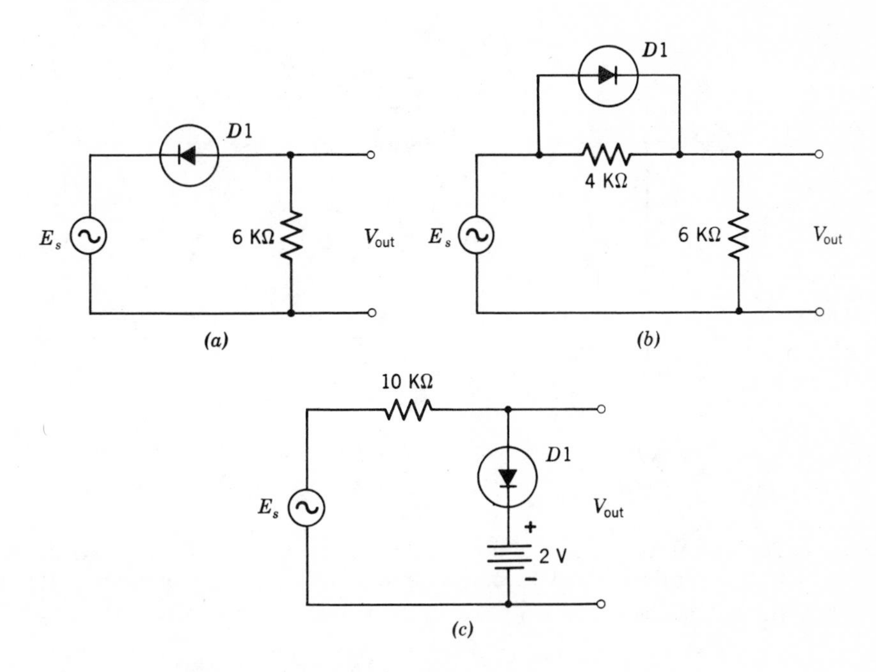

Circuits for Problem 1.

1. An a-c signal E_s is sinusoidal with a peak value of 10 volts is applied to each of the circuits. Draw to scale the output voltage V_{out} for each circuit.
2. The values for the model of a Zener diode are

$$E_j = 0.7\text{ V} \quad r_B = 2\ \Omega \quad R_R = 25\text{ k}\Omega$$
$$V_Z = 50\text{ V} \quad r_Z = 3\ \Omega$$

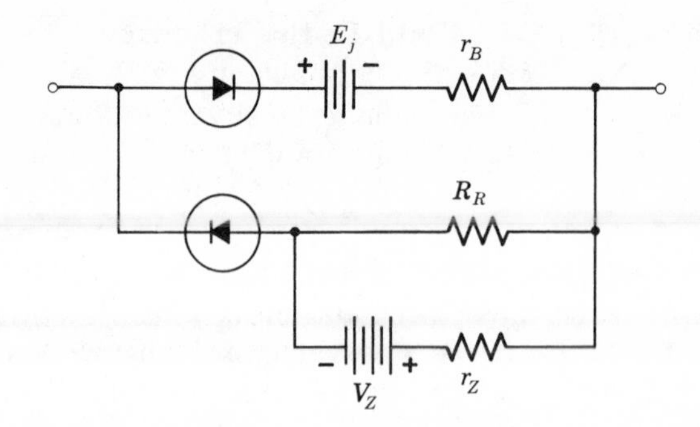

Circuit for Problems 2 and 3.

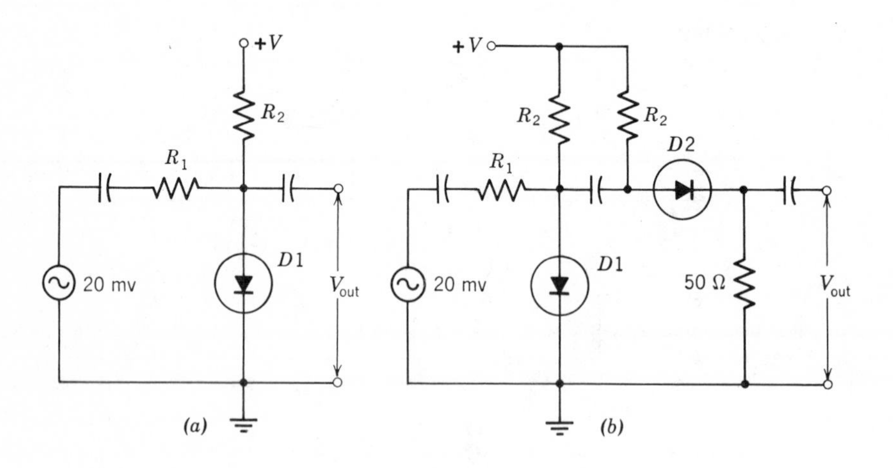

Circuits for Problems 4 to 9. [*Note.* Use 0.7 V for V_F and (50 mv/I) for r_j.]

The Zener diode has a power dissipation limit of 12 W. Draw a detailed forward and a detailed reverse characteristic for the Zener diode.

3. Repeat Problem 2 for a 5-W Zener diode that has the following values:

$$E_j = 0.7\text{ V} \qquad r_B = 3\,\Omega \qquad R_R = 100\text{ k}\Omega$$
$$V_Z = 30\text{ V} \qquad r_Z = 1.2\,\Omega$$

4. R_1 is 50 Ω, R_2 is 20 kΩ, and V is 30 V. Using circuit *a*, determine V_{out}.
5. R_1 is 5000 Ω, R_2 is 5000 Ω, and V_2 is 2 V. Using circuit *a*, determine V_{out}.
6. R_1 is 50 Ω, R_2 is 50 Ω, and V_2 is 5 V. Using circuit *a*, determine V_{out}.
7. Solve Problem 4 for circuit *b*.
8. Solve Problem 5 for circuit *b*.
9. Solve Problem 6 for circuit *b*.

Section 2-6 Cold-Cathode Diodes

Cold cathode diodes are designed to operate in the region of *b* and *c* of Fig. 2-26*c*. Over this part of the characteristic, the voltage drop across the

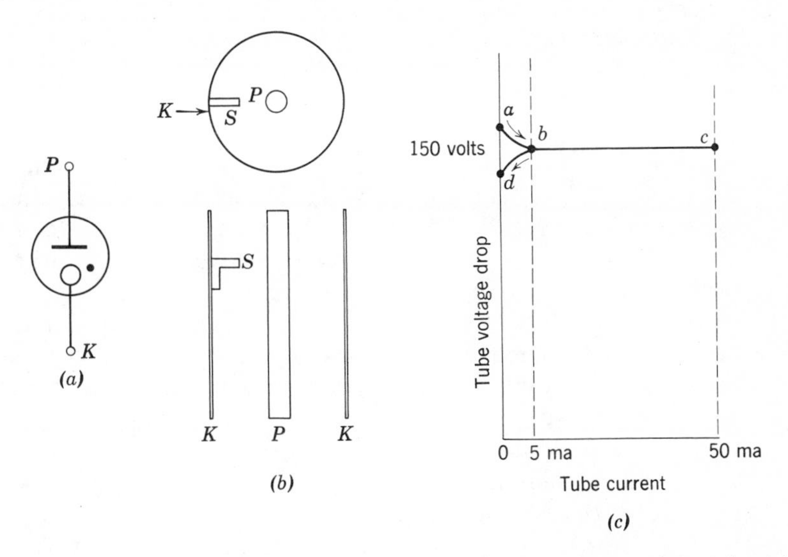

Figure 2-26 Construction and electrical characteristic for an 0D3 voltage-regulator tube. (*a*) Symbol (the solid dot indicates a gas tube). (*b*) Physical structure. (*c*) Electrical characteristics.

tube is constant for different values of current. In the construction of the *OD*3 (Fig. 2-26*b*), a small tip *S* is welded to the cylinder which serves as the cathode. The distance from the end of the starter *S* to the anode is much less than the distance from the cylinder itself to the anode. As a result, ionization starts from *S* to the anode. As more and more current flows through the tube, the area of the glow on the inner cathode surface spreads. When the glow covers the cathode completely, the tube is fully ionized at the single ionization level. For the *OD*3, the tube drop is 150 V. To initiate ionization, a slightly higher voltage, point *a*, is required. This overvoltage causes the current to shift abruptly from point *a* to point *b* on ionization of the gas. Point *c* is the point at which ionization is complete. A further increase of current causes a state of double ionization accompanied by an increasing voltage drop across the tube. When the current is reduced below the value at *b*, the voltage drops a few volts before the tube goes out (the current falls to zero and visible light ceases . . . hence, the generally accepted expression for gas tubes—"goes out".

These electron tubes are used as voltage regulators (Fig. 2-27) in the

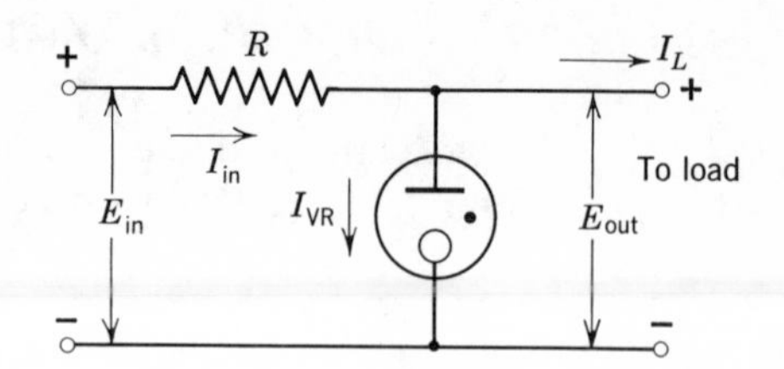

Figure 2-27 The application of the cold-cathode gas diode in a voltage-regulator circuit.

same manner as a Zener diode. Unlike the breakdown diodes, they are only available for a few different voltages, and they are also limited to a 50 ma maximum current. The voltage variation, however, is very small over the useful operating range. A jumper connection is made between two unused pins in the tube base. When the tube is removed from its socket, the power supply is then automatically disconnected.

A neon lamp operates in the same manner, but its characteristic is not as flat as the *OD*3. The voltage across the neon lamp can vary from 55 to 65 V over its operating current range. If a neon lamp is to be used as an indicator on 117 V alternating current, a resistor must be connected in series with it to limit the peak current. Both neon lamps and argon lamps are available with standard lamp bases. The internal leads to the base contacts are made from very high-resistance wire.

PROBLEMS

1. Using the characteristic of the *OD*3 (Fig. 2-26*c*) and the circuit (Fig. 2-27) given in the text, determine the value of *R* when the average load current is 150 ma and the supply voltage is 200 V. Over what range of load-current variation is regulation obtained? Over what range of input voltage variation can regulation be obtained?
2. Solve Problem 3 for a load current of 110 ma and a supply voltage of 175 V.
3. The characteristic of the *OC*3 is the same as that shown in Fig. 2-26*c*, except that the voltage is 105 V. Using the circuit of Fig. 2-27, determine *R* and the range of regulation when the average load current is 75 ma and the supply voltage is 170 V.
4. Repeat Problem 3 for a supply voltage of 145 V and a load current of 110 ma.

Section 2-7 Light-emitting Diodes

When electrons recombine with holes across an *N-P* junction, the actual movement of the electron in relation to the conduction band and valence

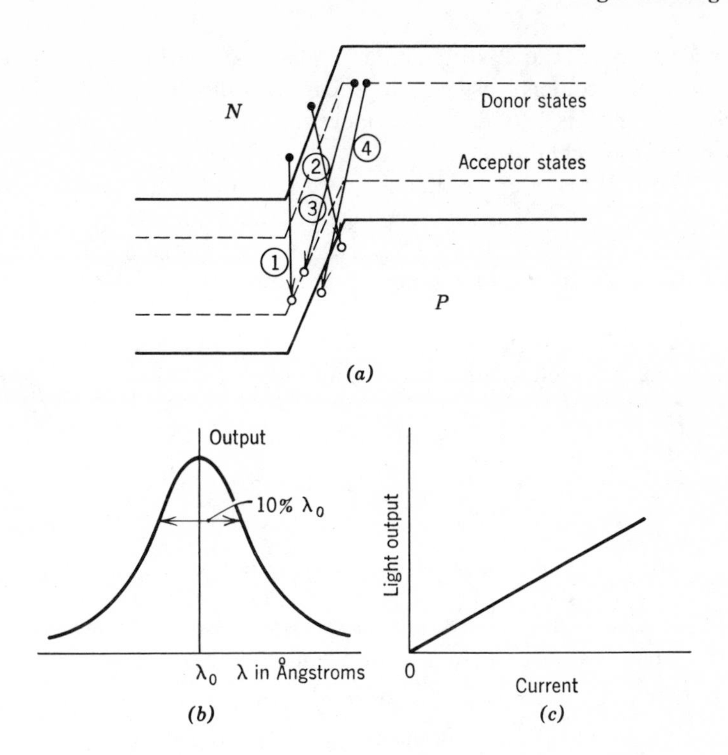

Figure 2-28 Light-emitting diodes. (*a*) Recombination paths. (*b*) Light wavelength. (*c*) Light output.

band energy-level diagram, Fig. 2-28*a*, shows that there are many possible paths for the transition of the electron. When the electron falls from a high to a lower level energy, usually the energy appears as heat, although some energies are converted into light photons. This optical property has been developed to produce devices that have a relatively high efficiency of light conversion (presently to 10%).

The wavelength of the light output is given by the relation

$$\lambda = \frac{12{,}400}{E_G} \text{ Å} \qquad (1\text{-}1)$$

where E_G is the energy gap in electron volts. The first commercial version of the light-emitting diode (LED) used gallium arsenide and has an energy gap of 1.37 V. This diode produces a light having a wavelength of 9100 Å, which is a dark red. A gallium phosphide diode has an energy gap of 2.25 eV that corresponds to 5600 Å, a green light. A number of other materials have been developed to produce other colors.

Since the recombination paths are not exactly the same length (Fig. 2-28*a*) the light is not completely monochromatic but shows a narrow band characteristic (Fig. 2-28*b*). A major advantage of the device, however, is that the light output is very nearly linear with diode current (Fig. 2-28*c*). The covalent bonds broken by an increase in ambient temperature materially reduces the light output.

The LED is operated in a forward-biased direction and the devices have very low values of reverse voltage breakdown. The LED has a relatively high forward voltage drop (0.8 to 1.2 V) in comparison to the diodes designed for use as rectifiers. One commercial application of the LED that has wide acceptance is the use of the device in small, home movie cameras to convert sound into a light signal to produce a sound track on the film. The LED has also been adapted to function as a pilot light.

Questions

1. Define or explain each of the following terms: (*a*) junction potential, (*b*) the arrow on a diode symbol, (*c*) reverse current, (*d*) Zener potential, (*e*) a rectifier, (*f*) Edison effect, (*g*) d-c resistance, (*h*) a-c resistance, (*i*) knee, (*j*) varistor diode, (*k*) bulk resistance, (*l*) ionization potential, (*m*) deionization potential.
2. Explain the differences between thermal runaway and thermal breakdown.
3. How does contamination affect the reverse current in a diode?
4. How does I_F compare numerically with I_R?
5. What is the peak inverse voltage in a rectifier circuit?
6. How is the current in a vacuum-tube diode space-charge limited?
7. Define the d-c plate resistance of a diode. The a-c plate resistance.
8. Under what condition is a static resistance equal to a dynamic resistance?
9. What is Child's law?
10. Why is a Zener diode called a reference diode?
11. In a typical Zener diode, the maximum reverse current is 20 ma and the maximum allowable forward current is 200 ma. Why does this difference in current ratings exist?
12. What is the function of a regulator circuit?
13. Under what conditions can the value of the bulk resistance of a diode be larger than the junction resistance?
14. How is current initiated in a cold-cathode diode?
15. What is the application of a light-emitting diode?
16. What determines the different colors of different light-emitting diodes?

Chapter Three

BASIC THREE-ELEMENT ELECTRON DEVICES

The principles of the operation of a transistor are explained in Section 3-1. An understanding of two basic circuit configurations, the common-base transistor (Section 3-2) and the common-emitter transistor (Section 3-3), form the background for considering the transistor in a basic amplifier (Section 3-4). The common-collector circuit is also discussed in this section. The limitations of a transistor as amplifier are considered in Section 3-5).

A similar procedure is followed to study the operation of a vacuum tube. The action of a grid to control the plate current (Section 3-6) leads to an examination of the standard triode characteristic curves (Section 3-7). The function of the triode as an amplifier (Section 3-8) and its limitations (Section 3-9) are examined. The chapter is concluded (Section 3-10) with a discussion of the dynamic coefficients of the triode.

Section 3-1 The Action of a Transistor

The simplest form of transistor to consider is the single crystal shown in Fig. 3-1. This transistor has the physical form of a rectangular solid containing two separate *P-N* junctions in one crystal. Thus, the crystal may be considered a "sandwich" of thin *P* material between two *N*'s or of thin *N* material between two *P*'s. One end is called the *emitter* and the other end is the *collector*. The *base* is the thin center section. The two different arrangements of the impurity materials are given the self-evident names *NPN transistor* and *PNP transistor*. The standard graphical symbols are shown for the two types. The conventional transistor is often referred to as a *bipolar junction transistor*, *BJT*.

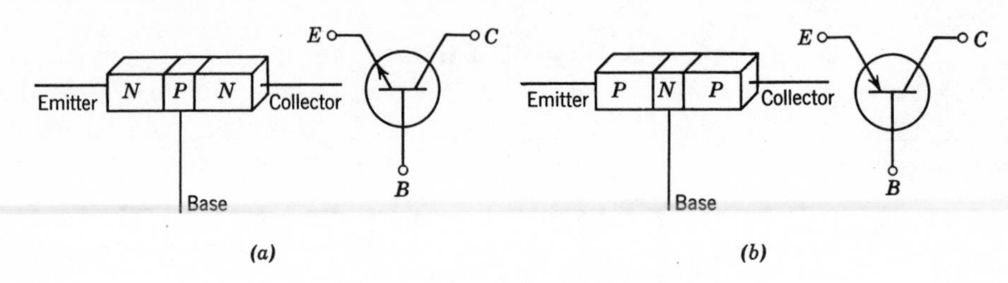

Figure 3-1 Methods of representing transistors. (*a*) *NPN* transistor and symbol. (*b*) *PNP* transistor and symbol.

In the discussion of the semiconductor diode (Section 2-1), we analyzed its operation from the viewpoint of a discussion of the action of the *P-N* junction. In this junction transistor, we have not one but two *P-N* junctions. A transistor is operated with forward bias on the emitter (with respect to the base) and with reverse bias on the collector. In Fig. 3-2, the battery

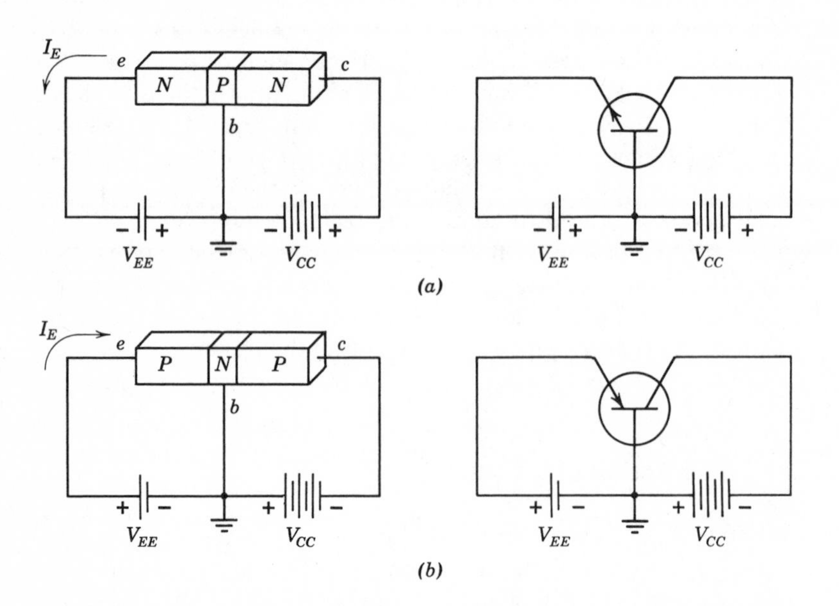

Figure 3-2 Transistor bias connections. (*a*) *NPN* transistor. (*b*) *PNP* transistor.

V_{EE} provides forward emitter bias, and the battery V_{CC} provides reverse collector voltage. The magnitude of the collector supply is often, but not necessarily, greater than the emitter bias supply. In the schematics for the circuits in Fig. 3-2, arrows are drawn showing the direction of *conventional current flow*. These arrows are carried to the transistor symbols and are placed in the emitter leads. The collector leads do not have arrowheads. The arrow points into the symbol when the emitter is *P* material and points out of the symbol when the emitter is *N* material. The direction of the arrowhead on the symbol distinguishes between an *NPN* and a *PNP* transistor.

If we sketch the energy-level diagrams for an *NPN* unit, we obtain Fig. 3-3*a* for the transistor without electrode voltages and Fig. 3-3*b* with the

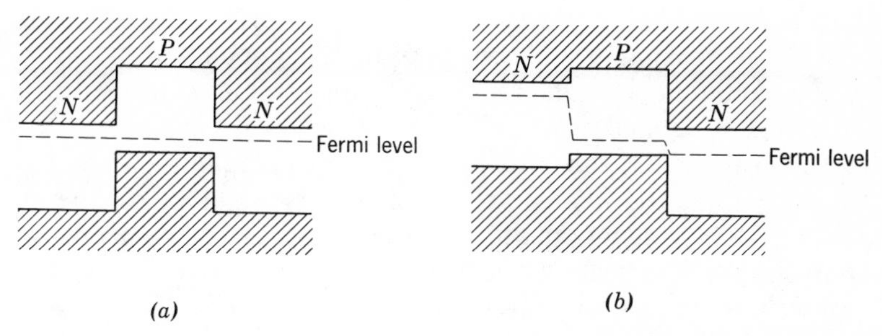

Figure 3-3 Energy levels in the *NPN* transistor. (*a*) Without bias voltages. (*b*) With proper biasing voltages.

biasing voltages in the circuit. We recall from Section 2-1 (Fig. 2-6) that a forward bias lowers the energy-level differences across a *P-N* junction and that a reverse bias increases the energy-level differences across the junction. In this case, for the *NPN* transistor, a forward bias injects electrons from the emitter into the base, reducing the number of holes in the base. The action of the reverse bias on the base-collector junction is to deplete the base of its electrons, thus creating excess holes. If the holes at the emitter-base junction are filled by injected electrons, holes diffuse from the base-collector junction to try to keep the hole density uniform throughout the base material. If the "density" of holes in the *P* material is reduced near the base-collector junction by this drift, the reverse bias will create new and additional holes within the *P* material at the base-collector junction. Thus we find that the current in the collector is caused by the hole drift or hole current in the base. If there is no hole drift or hole current in the base, there can be no current in the collector circuit. We have shown that hole current in the base is produced by forward-bias current in the emitter. Hole drift to the base can only take place if there are injected electrons from an emitter current. In conclusion, a collector current can exist only if there is an emitter current and in an amount that cannot be greater than the emitter current. Also, we may state that the collector current is controlled by the majority current-carrier movement in the base. This discussion suggests that a transistor is a *current-operated* device.

When we considered the depletion of holes in the base caused by the emitter current, we said that the collector, under the conditions of reverse bias, will accept electrons moving across the base-collector junction in order to bring the number of holes in the base back to "normal". It is also possible for the base connection to accept some electrons from the *P* material of the base. As this is the case in an actual transistor, the number of electrons injected into the base from the emitter must be greater than the number of electrons taken from the base by the collector. Thus, in the junction transistor, the collector current I_C is always less than the emitter current I_E.

One of the important constants in semiconductor theory is the ratio of the simultaneously measured collector current I_C to emitter current I_E. This ratio is defined as α_{dc} and, since I_C is always less than I_E, α_{dc} must be less than one. The subscript "*dc*" on α indicates that this ratio is obtained by using I_C and I_E, which are d-c steady-state values. In the hybrid-parameter system of notation, the symbol is $-h_{FB}$. Here, the capital letters in the subscript for h indicate d-c values. The *F* refers to a "forward" sense from emitter to collector, and the *B* refers to the circuit of the transistor in which the base is the common lead between the emitter circuit and the collector circuit. The symbols α_{dc} and h_{FB} are used interchangeably in the literature:

$$\alpha_{dc} \equiv -h_{FB} \equiv \frac{I_C}{I_E} \quad \text{and} \quad \alpha_{dc} < 1 \qquad (3\text{-}1a)$$

Minority current carriers create a collector current without assistance from a forward emitter current. If the emitter voltage is zero and if the emitter current is zero, this current that exists in the collector circuit is the same back current that exists in a crystal diode unit with reverse voltage and is produced by the minority current carriers. As we recall from Section 2-1, a bias that is reverse for the majority current carriers is a forward bias for the minority current carriers. Any current in the collector is the sum of the reverse-bias current plus the normal current produced by forward

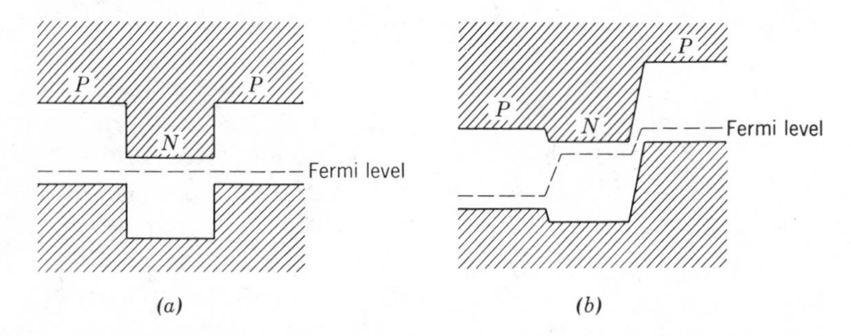

Figure 3-4 Energy levels in the *PNP* transistor. (*a*) Without bias voltage. (*a*) With bias voltages.

bias on the emitter. For example, let us assume that the emitter current in a transistor is 5 ma and the collector current is 4.8 ma. If the transistor is overheated, we note that the collector current is 5.6 ma when the emitter current is 5 ma. Thus a *leakage current* or *cutoff current* I_{CBO} of 0.8 ma is produced by these undesired minority current carriers.

If a *PNP* transistor were used in the preceding discussion, the same approach would be taken, except that the majority current carriers in the base are electrons instead of holes. Then the process would be a depletion of electrons within the base material. Figure 3-4 gives the energy-level diagrams for the *PNP* transistor.

Section 3-2 The Common-Base Circuit

If the circuit of Fig. 3-2*a* is modified so that the emitter and collector potentials may be varied to produce data for characteristic curves, we have the circuit of Fig. 3-5. When the emitter voltage is reduced to zero, the

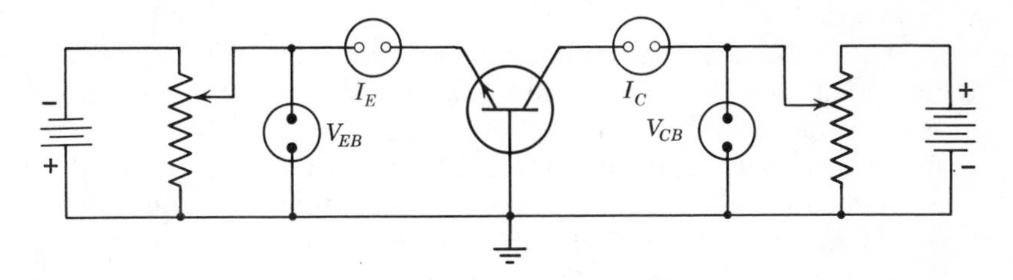

Figure 3-5 Test circuit for common-base *NPN* transistor.

emitter current is zero. Ideally, as the collector voltage is raised from zero to maximum, the collector current remains at zero. However, there must be some leakage current I_{CBO} because of the ever-present minority current carriers in transistor material. The I_{CBO}, the leakage current, is plotted in Fig. 3-6. The notation *CBO* refers to a common-base connection with the emitter open-circuited in order that the emitter current is zero. If the emitter is open-circuited, the emitter current must be zero. This is not true if the emitter is short-circuited to the base when making this measurement.

Now, let us maintain the emitter current at 3 ma. Since the emitter has forward bias, the required voltage to the emitter is less than 1 V. We established that the collector current is produced by the action of emitter current, and it cannot exceed 3 ma, but it is almost 3 ma for a wide range of collector voltage. This is shown on the curve from *d* to *e*. Almost all the rise

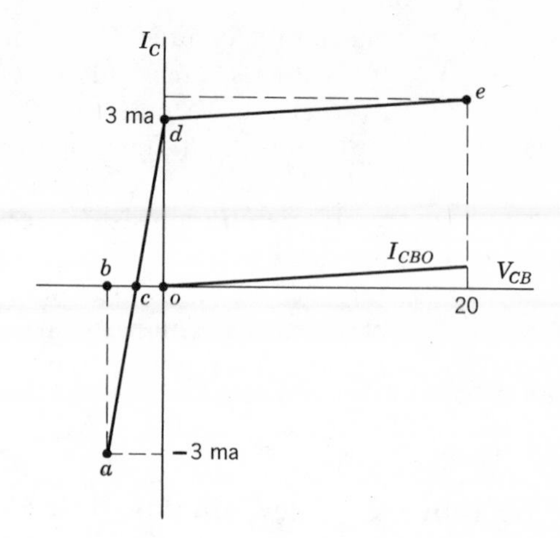

Figure 3-6 The transistor collector characteristic.

in collector current from d to e may be attributed to I_{CBO}. A small "reverse" voltage is necessary on the collector to reduce the collector current to zero, point c on the curve. If the "reverse" voltage on the collector is increased to b, the current will rise to point a, which is 3 ma but is in the direction reverse from normal.

Let us assume that the point e corresponds to 20 V on the collector and that point c is -0.1 V and point b is -0.2 V. The 20 V is a normal reverse bias on the collector, so that the voltages at points b and c are actually forward voltages on the collector. If the voltage from c to b, 0.1 V, produces 3 ma, the full collector voltage applied to the transistor connected backward would produce $3 \times (20/0.1)$ or 600 ma in the collector. This 600 ma current flow would surely destroy the crystalline structure of the transistor. Thus we must be very careful not to make improper battery connections to transistor circuits. With vacuum tubes we often save the tube after an improper connection is made since saturation limits the current, but with the transistor we have no second chance.

We normally do not use any part of the transistor curve except the first quadrant. The collector characteristic is drawn in Fig. 3-7, using the emitter current as the parameter. For an *NPN* transistor, both the collector voltage and the collector current are positive. For a *PNP* transistor, the collector current and the collector voltage are both negative, but the usual practice is to plot the results in the first quadrant in spite of the negative values of I_C and V_{CB}.

We observe in Fig. 3-7 that the emitter current at Point B is 5 ma and that the collector current at Point C is 5 ma. Accordingly, the value of the collector current scale at Point B is less than 5 ma. At Point D, similarly, I_C is less than I_E. Accordingly, α_{dc} or h_{FB} is less than unity throughout the collector characteristic.

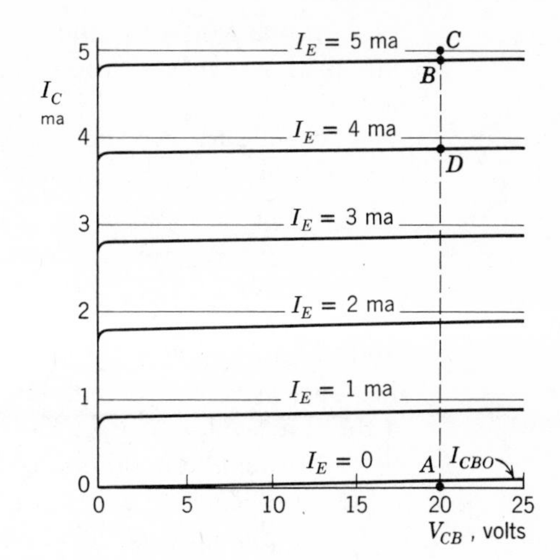

Figure 3-7 Common-base collector characteristic.

The emitter characteristic (Fig. 3-8*b*) indicates that the emitter voltage and current relation is nonlinear. It also indicates the order of the small

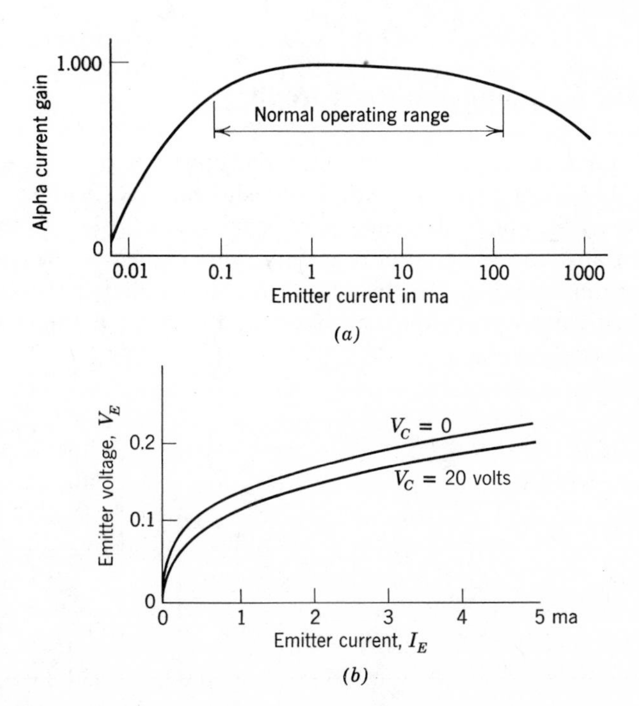

Figure 3-8 Other common-base characteristics. (*a*) The variation of alpha current gain. (*b*) The emitter input characteristic.

forward-bias voltage required to produce emitter current. The slope of these curves shows that we are dealing with resistances of the order of 100 Ω or less.

The a-c current gain, α_{ac} or h_{fb}, is defined on the basis of the ratio of a *change* in collector current to a *change* in emitter current for a constant value of V_{CB}:

$$\alpha_{ac} \equiv -h_{fb} \equiv \left.\frac{\Delta I_C}{\Delta I_E}\right|_{\Delta V_{CB}=0} \tag{3-1b}$$

Here, again, the current gain is less than one. From the curves of Fig. 3-7, the value of α_{ac} can be obtained by taking two points on the curve, *B* and *D*, at the same value of V_{CB}. The fact that points *B* and *D* are both for a V_{CB} of 20 V satisfies the condition of the definition that ΔV_{CB} is zero. h_{fb} is the a-c value of α because the subscripts of h are now lowercase letters.

The alpha current gain of a transistor for normal circuit operating conditions is very close to unity. However, at very low values of I_E, the transfer of current carriers through the transistor is very small and the resulting value of alpha is likewise very small. At very high values of emitter current, alpha tends to fall somewhat, Fig. 3-8*a*.

Section 3-3 The Common-Emitter Circuit

In the previous section, the common return for the bias supplies was the base. The connection of the circuit showing bias supplies that uses the emitter as the common return point is given in Fig. 3-9. In this circuit for an *NPN* transistor, the emitter is negative with respect to the base and provides forward emitter bias. In this circuit the collector is positive, providing the required reverse bias on the collector. For a *PNP* transistor both supply batteries are reversed.

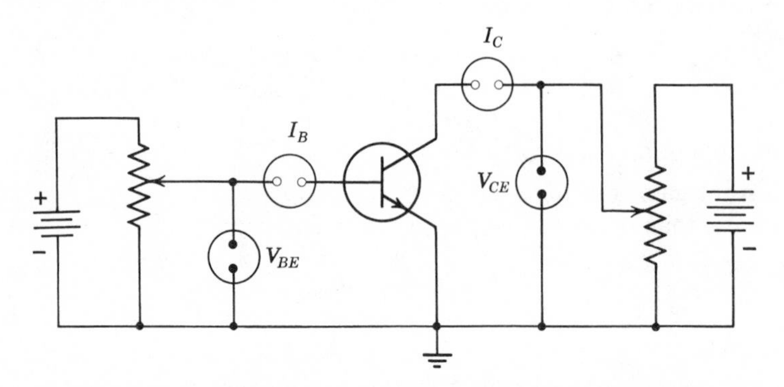

Figure 3-9 Test circuit for common-emitter *NPN* transistor.

In the common-base connection, the base current is the difference between the emitter current and the collector current. In this circuit, the base current is held to a fixed value but the collector voltage is varied to obtain the collector characteristic. Since the base current does not directly control the collector current, we do not find the same degree of linearity that we have in the common-base characteristics. The base currents are measured in microamperes (Fig. 3-10) when the emitter and the collector currents are

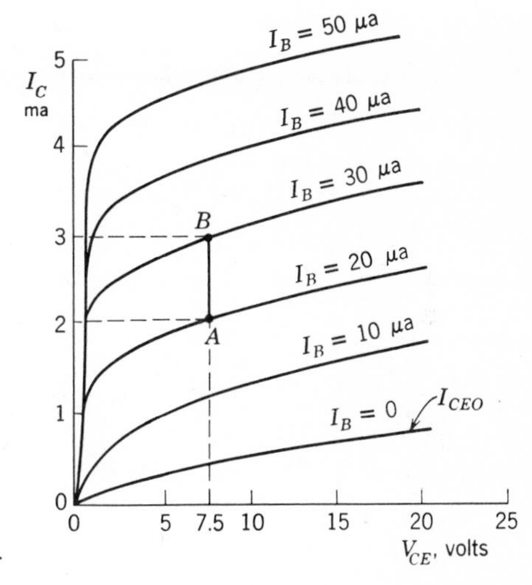

Figure 3-10 Common-emitter collector characteristic.

in milliamperes. The base current is in milliamperes when the order of magnitude of the emitter and the collector currents is in amperes. Here, when the base current is zero, the collector current is the leakage current I_{CEO}.

From Fig. 3-10, the values of the currents at points B and A are

Point B: $I_C = 3$ ma $\quad I_B = 30\ \mu$a $\quad V_{CE} = 7.5$ V
Point A: $I_C = 2$ ma $\quad I_B = 20\ \mu$a $\quad V_{CE} = 7.5$ V

The relation between the collector current and the base current is the current gain of the circuit. Now, in this circuit configuration, the gains are greater than one and both the d-c and the a-c gain factors must be defined:

$$\beta_{\text{dc}} \equiv h_{FE} \equiv \frac{I_C}{I_B} \tag{3-2a}$$

$$\beta_{\text{ac}} \equiv h_{fe} \equiv \left.\frac{\Delta I_C}{\Delta I_B}\right|_{\Delta V_{CE}=0} \tag{3-2b}$$

For the numerical values we have, the d-c beta, β_{dc}, at point B is 3 ma/30 μA or 100, and at point A it is also 2 ma/20 μA or 100. The subscripts on h indicate F is the forward current gain and E is for the common-emitter circuit configuration. Since points B and A are taken for the same value of V_{CE} ($\Delta V_{CE} = 0$), the numerical value of the a-c current gain is

$$\beta_{ac} = h_{fe} = \frac{3\text{ ma} - 2\text{ ma}}{30\ \mu\text{A} - 20\ \mu\text{A}} = 100$$

The variation of the current gain with d-c collector current for a typical transistor is shown in Fig. 3-11*a*.

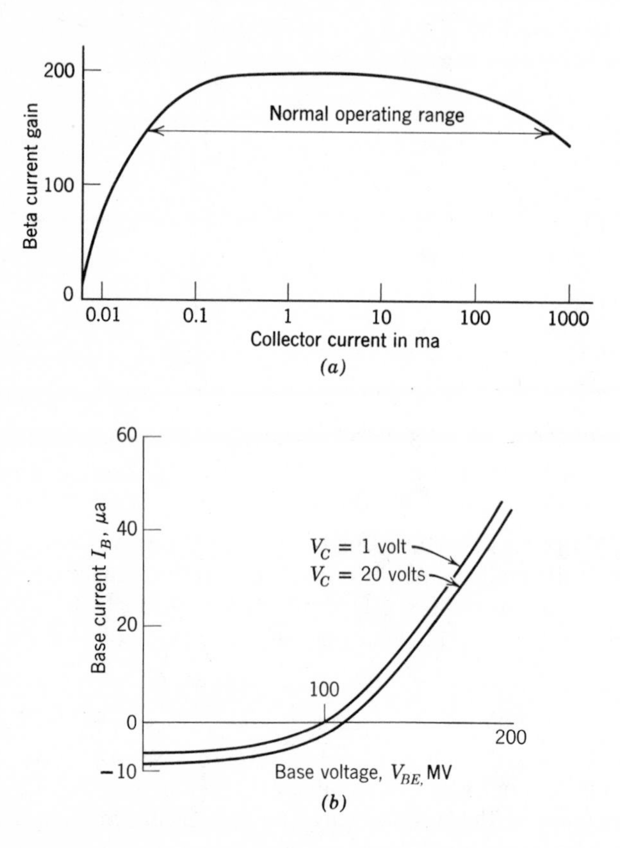

Figure 3-11 Other common-emitter characteristics. (*a*) The variation of beta current gain. (*b*) The base input characteristic.

The leakage current, I_{CBO}, in the common-base circuit is defined as the reverse current that flows in the transistor when the emitter is open-circuited or left disconnected. In a common-emitter circuit I_{CBO} is now a current in the base. Therefore, by the action of beta current gain, this leakage current is amplified by β_{dc} to make a value of βI_{CBO} in both the collector and in the emitter. The total current in the collector is the amplified leakage current plus the leakage current present in the first place:

$$I_{CEO} = I_{CBO} + \beta_{dc} I_{CBO} = I_{CBO} + h_{FE} I_{CBO}$$

or

$$I_{CEO} = I_{CBO}(1 + \beta_{dc}) = I_{CBO}(1 + h_{FE}) \tag{3-3}$$

As a result, the leakage current shown in Fig. 3-10 is much larger than the leakage current shown in Fig. 3-7. I_{CEO} is the collector current measured when the base lead is open-circuited.

We substitute the relation

$$I_B = I_E - I_C$$

Into Eq. 3-2a,

$$\beta_{dc} = \frac{I_C}{I_B} = \frac{I_C}{I_E - I_C}$$

Dividing all terms by I_E,

$$\beta_{dc} = \frac{I_C/I_E}{1 - I_C/I_E}$$

But I_C/I_E defines α_{dc} by Eq. 3-1a:

$$\beta_{dc} = \frac{\alpha_{dc}}{1 - \alpha_{dc}} \tag{3-4a}$$

Solving this equation for α_{dc}:

$$\beta_{dc}(1 - \alpha_{dc}) = \alpha_{dc}$$

$$\beta_{dc} - \beta_{dc}\alpha_{dc} = \alpha_{dc}$$

$$\beta_{dc} = \alpha_{dc} + \beta_{dc}\alpha_{dc}$$

$$\beta_{dc} = (1 + \beta_{dc})\alpha_{dc}$$

$$\alpha_{dc} = \frac{\beta_{dc}}{1 + \beta_{dc}} \tag{3-4b}$$

We have derived the conversion equations from α_{dc} to β_{dc} and from β_{dc} to α_{dc} in terms of the d-c values. The derivation of the conversion equations for the a-c values is very similar with the results

$$\beta_{ac} = \frac{\alpha_{ac}}{1 - \alpha_{ac}} \tag{3-4c}$$

and

$$\alpha_{ac} = \frac{\beta_{ac}}{1 + \beta_{ac}} \tag{3-4d}$$

Consequently, since the forms of the equations are the same, they are usually written without subscripts:

$$\beta = \frac{\alpha}{1-\alpha} \tag{3-4e}$$

and

$$\alpha = \frac{\beta}{1+\beta} \tag{3-4f}$$

The formulation of the conversion equations leads directly to the conversions between I_B, I_C, and I_E, which are d-c current values, and between i_b, i_c, and i_e, which are a-c current values. These conversions are summarized in Table A. These conversion factors must become as automatic to students of electronics as conversions by Ohm's law are to students of electric circuits.

TABLE A Conversion Factors

Multiplying Factors to convert FROM \ TO	I_B (or i_b)	I_C (or i_c)	I_E (or i_e)
I_B (or i_b)	1	β or $\frac{\alpha}{1-\alpha}$	$(1+\beta)$ or $\frac{1}{1-\alpha}$
I_C (or i_c)	$\frac{1}{\beta}$ or $\frac{1-\alpha}{\alpha}$	1	$\frac{1+\beta}{\beta}$ or $\frac{1}{\alpha}$
I_E (or i_e)	$\frac{1}{1+\beta}$ or $(1-\alpha)$	$\frac{\beta}{1+\beta}$ or α	1

The action of the input circuit in the common-emitter transistor circuit (Fig. 3-11*b*) behaves somewhat differently from the input curves for the common-base circuit. Because of the action of I_{CEO}, a reverse current flows through the base lead at the very low base voltages. This curve indicates that the input resistance to the base is about 100 mv/50 μA or 2000 Ω as contrasted to the input resistance of about 100 Ω or less in the common-base circuit.

PROBLEMS

1. If the base current of a transistor is 20 μa when the emitter current is 3.21 ma, what is the value of α_{dc} and β_{dc} for the transistor?

2. The published value of h_{FE} for a transistor states that h_{FE} can vary from 40 to 90. If I_B is fixed at 15 μa, what is the expected variation in I_C?
3. The maximum value of I_{CBO} for a particular transistor is 15 ma. h_{FE} can vary from 150 to 240 for this transistor. What is the corresponding range of I_{CEO}?
4. Convert Equations 3-4*a*, 3-4*b*, 3-4*c*, and 3-4*d* into equations in terms of *h* parameters (h_{FB}, h_{FE}, h_{fb}, and h_{fe}).
5. Find α for each of the following values of β:
   50, 100, 120, 150, and 200
6. Find α for each of the following values of β:
   46, 65, 84, 125, and 165
7. Find β for each of the following values of α:
   0.995, 0.990, 0.9875, and 0.9765
8. Find β for each of the following values of α:
   0.991, 0.962, 0.946, and 0.983

Section 3-4 The Transistor as a Simple Amplifier

In a *common-base* circuit (Fig. 3-12), let us assume that V_{EE} is 0.15 V and that it produces an emitter current I_E of 3 ma. If we assume for the purpose

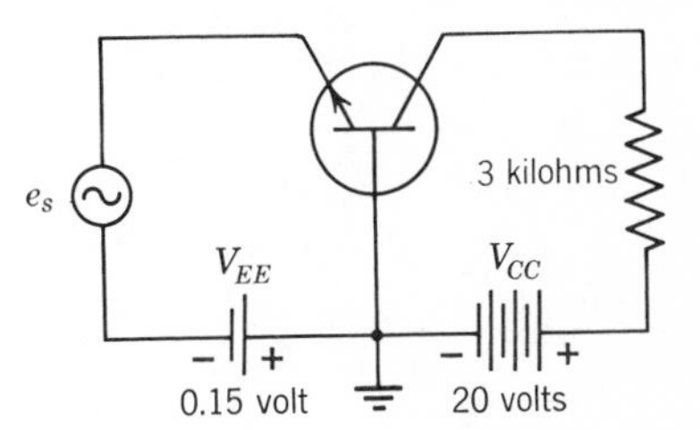

Figure 3-12 The transistor in a simple common-base amplifier.

of a simplified discussion that α is unity, the static or *operating* value of collector current I_C is 3 ma. A 3000-Ω load resistor, at a 3 ma current flow, has a voltage drop of 9 V. Since the collector supply is 20 V, the collector-to-base voltage V_{CB} is 11 V. These values of collector current and voltage are those values that are measured with d-c meters.

A signal, e_s, is placed in series with the emitter bias. Assume that the signal is a sinusoid having a peak amplitude of 0.05 V. When the signal is at its instantaneous positive peak value, the net value of voltage on the emitter is $(-0.15+0.05)$ or -0.10 V. When the signal is at its instantaneous negative peak value, the net value of voltage on the emitter is $(-0.15-0.05)$ or -0.20 V. Assume that the emitter current is linear and proportional to voltage to yield for the emitter:

-0.10 V	2.0 ma
-0.15 V	3.0 ma
-0.20 V	4.0 ma

Accordingly, a total change in the signal of 0.10 V produces a total change in the emitter current of 2.0 ma.

Since α is unity, the total change in collector current is also 2.0 ma and varies in accordance with the emitter current. When the collector current increases to 4 ma, the drop in the load is 12 V, and the collector voltage is 8 V. When I_C decreases to 2 ma, the drop in the load is 6 V, and V_{CE} is 14 V. Thus, a total change of 0.10 V in the input results in a change in the output of 6 V. The voltage gain in the circuit is 6.0/0.10 or 60. Thus this circuit *amplifies* the signal to a higher voltage level. Since the current gain is unity, the power gain, in this example, is 60 also.

We notice in the circuit (Fig. 3-12) that the operating bias voltage V_{EE} on the emitter is negative and that V_{CC} on the collector is positive for an *NPN* transistor. A positive incoming signal makes the emitter *less* negative and *reduces* the emitter current. The collector current accordingly, *decreases*, and the collector voltage, or output signal, *increases* in the positive direction. This circuit, the common-base amplifier, provides voltage and power gain but does not give phase inversion; the input and output signals are *in phase*.

In the *common-emitter* circuit (Fig. 3-13), assume that the operating

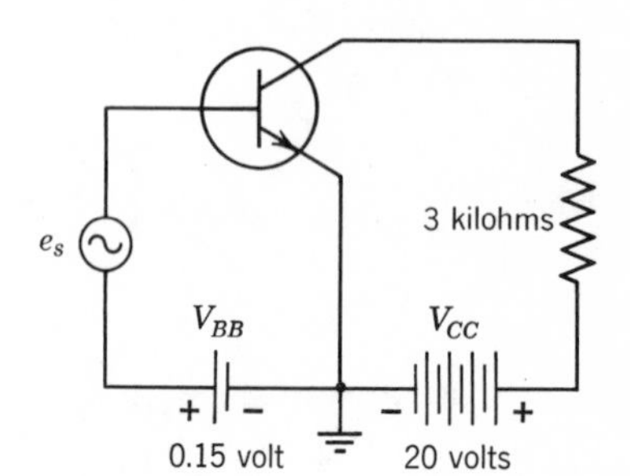

Figure 3-13 The transistor in a simple common-emitter amplifier.

conditions for the circuit are as follows: collector current I_C, 3 ma, base current, I_B, 30 μa, and base voltage, V_{BE}, 0.15 V. The IR drop in the load resistor is 3 (kΩ) $\times$ (ma) or 9 V. Then the collector-to-emitter voltage, V_{CE}, is 11 V. Also assume that an input signal of plus and minus 10 μa is produced by a base voltage change of plus and minus 0.05 V. A 10-μa increase in base current increases the collector current 1 ma from the operating point of 3 ma to 4 ma, lowering the collector voltage from 11 V to 8 V. A decrease in base current of 10 μa decreases the collector current from 3 ma to 2 ma, thus, raising the collector voltage to 14 V. Thus, for this example, the voltage gain is 6 V/0.10 V or 60, which is the same voltage gain as in the previous circuit. However, the current gain is now not unity but is 2 ma/20 μA or 100 and the power gain of the circuit is 60 × 100 or 6000.

We observe from this example that we obtain much greater power gains from the common-emitter circuit than from the common-base base circuit. In practice, we find that most transistor-circuit connections are of the common-emitter form in order to obtain this greater power gain.

In Fig. 3-13, a positive signal increases the base voltage. This increase in base voltage increases the base current which controls the collector current. The collector supply voltage is positive; hence, an increasing collector current causes the collector voltage to be less positive. This less positive condition is negative as far as the signal is concerned. Thus a common emitter circuit produces *phase inversion*.

TABLE B

e_s (Volts)	I_B (Microamperes)	I_C (Milliamperes)	V_{BE} (Volts)	V_{Load} (Volts)	V_{CE} (Volts)
+0.05	40	4	+0.20	+12	+8
0	30	3	+0.15	+9	+11
−0.05	20	2	+0.10	+6	+14

A third method of connecting a transistor in an amplifying circuit is the *common collector* arrangement shown in Fig. 3-14. The numerical values

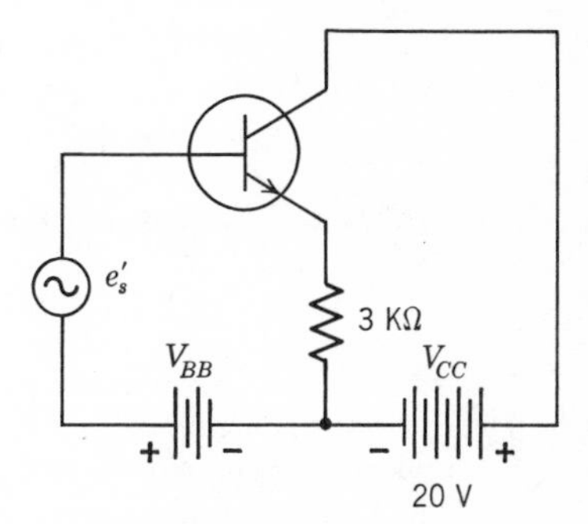

Figure 3-14 The transistor in a simple common-collector circuit.

obtained in the previous common-emitter circuit are shown in Table B. This same set of values is to be applied to the circuit of Fig. 3-14. Now, however, at a no-signal condition ($e_s' = 0$), the emitter is above ground by the voltage drop in the load, +9 V. The base-to-emitter voltage, V_{BE}, is +0.15. Since they are additive, the voltage from base to ground is +9.15 V. Since this is the no-signal condition, V_{BB} must be now 9.15 V. A signal is required to increase the collector current from 3 ma to 4 ma. The load voltage is now +12 V and from the table V_{BE} is +0.20 V. The total voltage from base to ground must be then +12.20 V. Since V_{BB} is 9.15 V, a signal voltage, e_s', of $12.20 - 9.15$ or +3.05 V, is required. When the collector current falls to 2 ma, V_{BE} is +0.10 V and the drop across the 3-kΩ load resistor is +6 V. Therefore, the voltage from base to ground is +6.10 V. Consequently, the value of e_s' required to produce this condition is $+6.10 - 9.15$ or −3.05 V.

The total change in signal voltage is 6.10 V. The total change in output voltage across the load is 6.00 V. Accordingly, the voltage gain of the circuit is 6.00/6.10, which is slightly less than unity. This circuit is particularly characterized by the fact that the voltage gain of the circuit is less than one. On the other hand the current gain of the circuit is the same as before (4.0–2.0 ma)/(40–20 μA) or 100. The power gain of the circuit is (6.00/6.10)100 or approximately the current gain, 100.

From the explanation of the operation of the circuit, we notice that a positive increase in signal produces a positive increase in load voltage. Therefore, the input and output signals are *in-phase*. The two previous circuits handled very small incoming signal levels, ±0.050 V. In constrast this circuit has a large input signal ±3.05 V for the same transistor conditions.

In summary, the common-base circuit has a large voltage gain and a current gain slightly less than unity without producing signal phase inversion. The common-emitter circuit offers the greatest power gain because it has both voltage and current gain with phase inversion. The common-collector circuit or *emitter-follower* circuit, as it is also known, has current gain, but the voltage gain is less than one. It does not invert the phase of the signal, and it can also handle large input signal levels. A summary of the conditions in the three amplifier circuits is given in Table C.

TABLE C Comparison of Amplifier Configurations

	Common-Base Circuit	Common-Emitter Circuit	Common-Collector Circuit
Input voltage change	±0.05 V	±0.05 V	±3.05 V
Input current change	±1.0 ma	±10 μA	±10 μA
Output voltage change	±3.0 V	±3.0 V	±3.0 V
Output current change	±1.0 ma	±1.0 ma	+1.0 ma
Voltage gain	60	60	$\frac{6.00}{6.10} \approx 1.00$
Current gain	1	100	100
Power gain	60	6000	100

PROBLEMS

1. What are α and β for the transistor used in the amplifier shown in Fig. 3-13?
2. How much power is delivered by the collector supply used in the amplifier shown in Fig. 3-13? What is the d-c power dissipated in the load resistor?
3. How much power is used from the emitter supply (V_{EE}) in the amplifier shown in Fig. 3-12? What is the total power supply requirement for the circuit?
4. How much power is used from the base supply (V_{BB}) in the amplifier shown in Fig. 3-13? What is the total power supply requirement for the circuit?
5. Repeat Problem 4 for the amplifier shown in Fig. 3-14.

Section 3-5 Limitations of a Transistor

The limits of operation of a transistor can be established by referring to a collector characteristic curve (Fig. 3-15). In the transistor circuit, we can

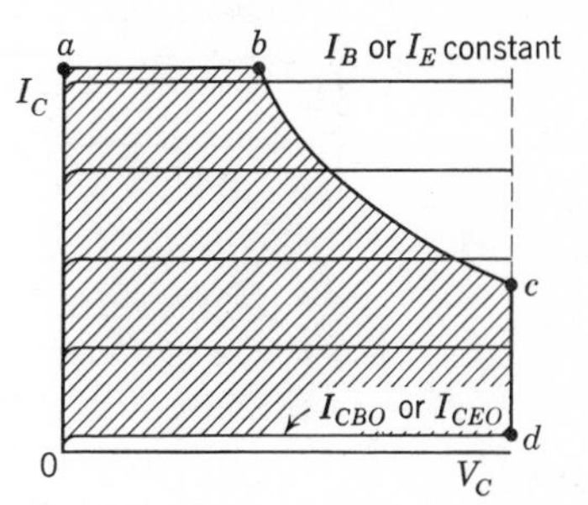

Figure 3-15 The limits of operation of a transistor.

actually operate up to and along the I_C axis, $0a$. A point located along this line is termed *saturated*, since the curves drop off very sharply here. The maximum value of I_C that is permitted is *ab*. By limiting the collector current to the maximum allowable current density in the junction, we automatically establish a limit on the emitter current and on the base current. The curve *bc* is the product of I_C times V_C and represents the maximum allowable collector power dissipation. The product of voltage and current in the input is so small that it may be neglected with respect to the very much greater heating in the collector. If the collector dissipation is exceeded, the excessive heat will permanently damage the crystalline structure. The limit *cd* is the maximum allowable reverse voltage on the collector. A collector voltage exceeding this value may cause a Zener breakdown within the transistor.

An increase in operating temperature breaks down the covalent bonds and produces a larger leakage current I_{CBO} or I_{CEO}. If this leakage current is large, all the curves will rise sharply with increasing collector voltage. The severity of this rise depends on the increase in the operating temperature.

Notice that it is unsafe to exceed power ratings for transistors for even short periods of time because of the small size of the active part and the consequent low ability of the unit to absorb heat without the temperature rise becoming excessive.

These factors, then, limit the operating range of the transistor to the shaded region, $0abcd0$.

In order to be able to interpret the data contained in a transistor specification sheet, the operating region shown in Fig. 3-15 is extended in Fig. 3-16

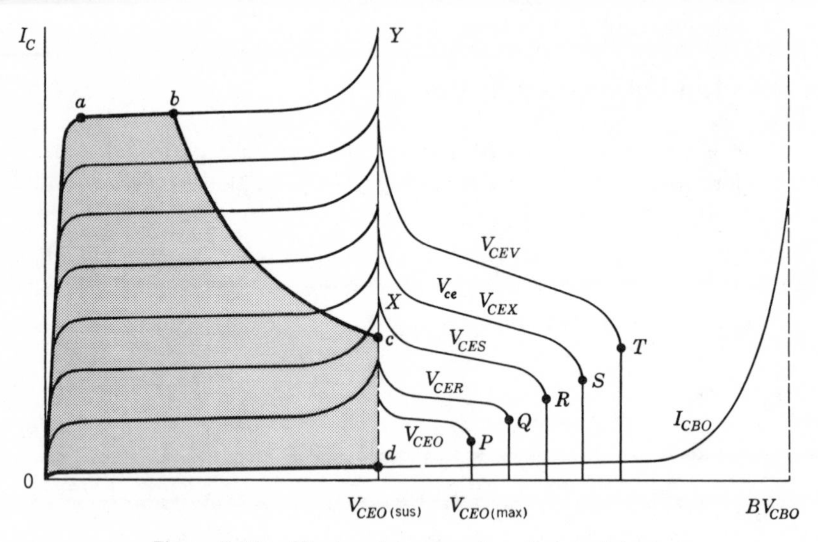

Figure 3-16 The breakdown region of a transistor.

to include the breakdown regions of a transistor. The cutoff current, I_{CBO}, is the current flowing from the collector to base with the emitter open or floating. The reverse voltage applied between the collector and the base with the emitter open is designated V_{CBO}. At a high voltage I_{CBO} rises sharply as an avalanche current without a further increase in voltage. This breakdown voltage is designated by placing a B before V_{CBO} as BV_{CBO}.

When there is a current in the emitter, breakdown voltages occur at lower voltages. The currents within the normal operating characteristic of a transistor within the region $0abcd0$ rise to the values along the line designated X–Y because of thermal runaway. The value of collector-to-emitter voltage that corresponds to X–Y is called the *sustaining* value, V_{CEO} (*sus*).

If, now, the base is floating and a reverse voltage, V_{CEO}, is applied from the collector to the emitter, the current that flows in the circuit is essentially leakage current. However, there is a maximum value, V_{CEO} (*max.*), at which point a large collector current begins to flow and the curve falls back to X–Y. If a resistor is placed between the base and the emitter, this maximum value of voltage is increased as shown by the curve of V_{CER}. When the base is shorted to the emitter, the curve is extended further, as given by V_{CES}. Likewise, when a reverse bias is also placed on the base from a specific external resistive circuit, the curve of V_{CEX} is obtained. When a reverse bias is placed on the base directly from a source of emf, the curve V_{CEV} is obtained. Since all these curves extend over a region of collector voltages, any tabulated breakdown values must be determined for explicit values of collector current. At point P, the breakdown voltage is designated BV_{CEO}; at Q, BV_{CER}; at R, BV_{CES}; at S, BV_{CEX}; and at T, BV_{CEV}.

Two other breakdown voltages are often given for specified current conditions:

BV_{CBO} the collector to base has reverse bias and the emitter is open
BV_{EBO} the emitter to base has reverse bias and the collector is open.

Usually a transistor specification gives some of these values, but not all of them. For example, the breakdown ratings for a Texas Instruments NPN 2N338 transistor are

$BV_{CBO} = 45$ V when $I_C = 50\ \mu$A and $I_E = 0$
$BC_{CEO} = 30$ V when $I_C = 100\ \mu$A and $I_B = 0$
$BV_{EBO} = 1$ V when $I_E = 50\ \mu$A and $I_C = 0$

For a diode the terminology for a reversed-biased breakdown voltage is simply BV_R. When the reverse voltage between the collector and the base of a transistor causes a spreading of the junction region to the point where it extends sufficiently beyond the layer to reach the emitter to cause an electrical short and to cause transistor action to cease, that d-c voltage is the *reach-through* voltage, V_{RT}. This is the more recent term that now is used to supplant the terminology, *punch-through* voltage.

Section 3-6 The Action of a Grid in a Triode Vacuum Tube

The invention of the control grid by De Forest in 1907 initiated the development of today's vast electronics industry. The action of a control grid enables large energies in a plate circuit to be controlled by a very small energy at the control grid. This tube, which consists of a cathode emitting surface, a plate, and a grid, is called a *triode*. The typical physical arrangements for a triode are shown in Fig. 3-17. The considerations given to the design and operation of the cathode and of the plate of the diode in Section 2-3 are also valid for the triode. The grid wires are fine, and the grid assembly is usually placed close to the cathode and far from the plate.

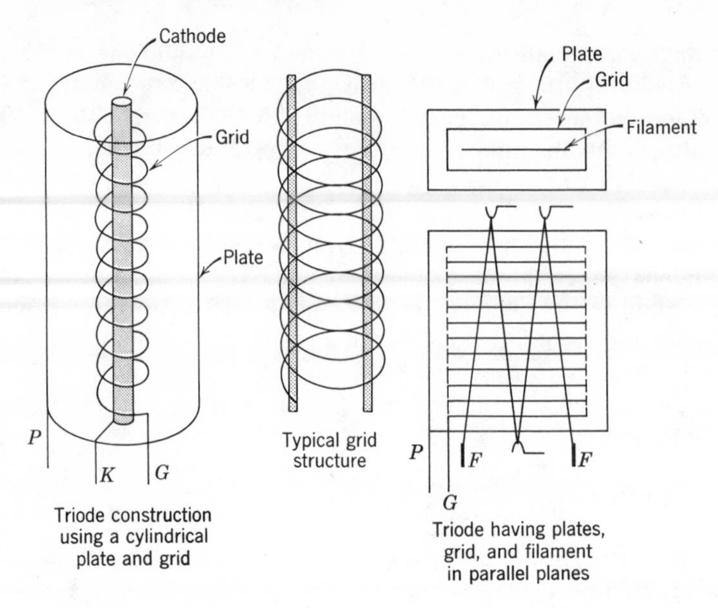

Figure 3-17 The internal mechanical structure of typical triodes.

If the action of the grid is neglected, the plate current is that value which is determined by Child's law. If the grid is maintained at a negative potential (Fig. 3-18), the grid structure is charged negatively. This negative charge creates an electric field that opposes and diminishes the flow of electrons

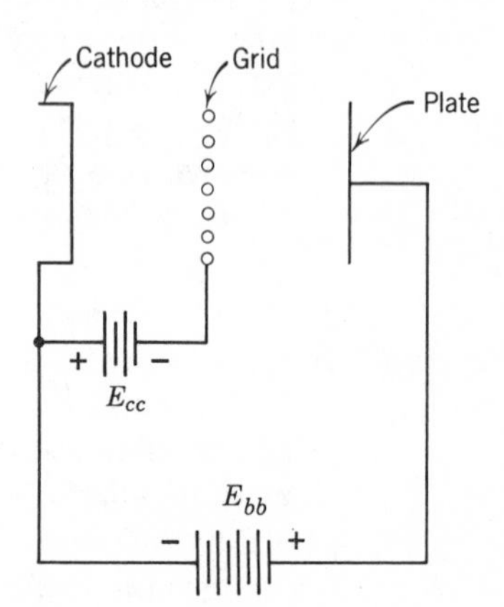

Figure 3-18 Supply voltage connections to the elements of a triode.

from the cathode to the plate. If the negative voltage is sufficiently large, no electrons flow from the cathode to the plate. This condition is called *cutoff*. When the negative grid field is weaker than the cutoff value, current flows to the plate. When the potential on the grid is negative, the grid cannot pick up electrons from the electron stream, and the grid current is zero. Thus, in a vacuum tube, a negative grid potential controls plate current without itself taking power (that is, the product of grid voltage times zero grid current is zero). In many applications, sources that have very low energy levels can be arranged to control a much larger power in the plate circuit of the triode in order to secure an amplification or a gain. When the grid is positive, it collects electrons that produce a grid current. A positive grid also tends to move the space-charge cloud out toward the plate, resulting in a further increase in plate current. If the grid structure forms a fine mesh close to the cathode, it has a much greater control over the plate current than if the mesh were coarse and near the plate.

Section 3-7 Triode Characteristics

In Fig. 3-19, a triode is connected to a variable voltage source for the plate potential E_b with the grid connected to a variable source E_c. The cathode is

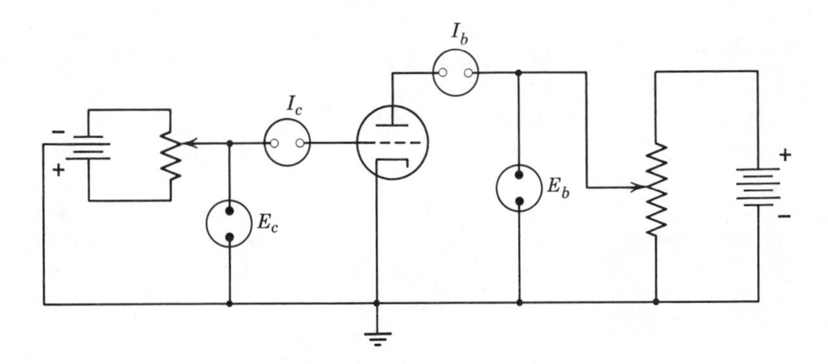

Figure 3-19 A test circuit for triode characteristics.

the common reference point of the two applied potential circuits and, as such, is often called "ground." It is customary to omit the actual heater circuit in schematic diagrams, since the heater only serves the purpose of bringing the cathode up to the proper operating temperature. The cathode symbol is retained in the schematic diagram, as the cathode itself contributes the electron stream for the tube. In this circuit, when the grid is negative (I_c is zero), there are three possible independent variables: E_c, E_b, and I_b. When the grid is positive, there are four independent variables: E_c, E_b, I_c, and I_b.

Several combinations of these variables are conventionally used to show the interdependence of these four quantities because the results must be plotted on graph paper that allows only two variables to be used as the axes. If the plate voltage is varied as the independent variable, and if the grid voltage is held at a constant value, plate current is the dependent variable. This curve is called a *plate family* or a *plate characteristic*. When the plate voltage is held constant, a variation in grid voltage, as the independent variable. This set of curves is called the *transfer characteristic,* or the *mutual characteristic*. When the plate current is held constant, the grid voltage is plotted against the plate voltage to give *constant-current curves*. Since all of these curves are for the same tube, it is possible to develop one curve from the other.

The plate family of a triode is shown in Fig. 3-20. When the grid voltage

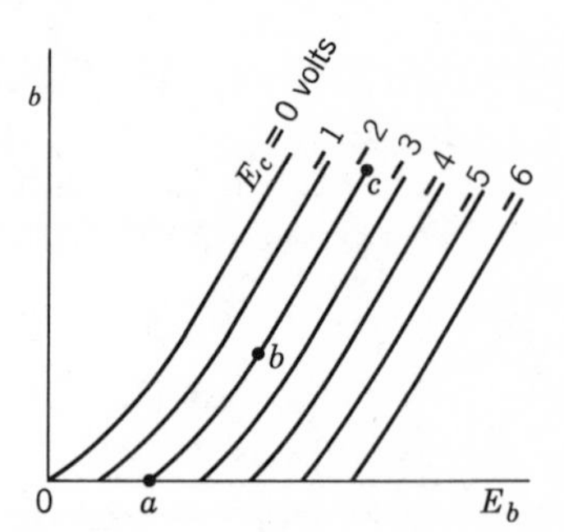

Figure 3-20 The plate characteristics of a triode.

is zero, the current takes the same form as the diode plate characteristic. A negative grid requires a larger plate voltage to overcome the negative repelling effect of the grid field on the electron stream attracted to the plate. Further increasing amounts of negative grid voltage requires that the plate voltage be increased accordingly. Thus the curves are effectively the diode curve shifted in uniform steps to the right with increasing negative grid voltage. Point *a*, for example, on the characteristic is a typical cutoff value. The curve is nonlinear from *a* to *b*, but it is very close to a straight line between *b* and *c*. These curves for constant grid potentials are, for all purposes, parallel.

Figure 3-21 shows the region of the plate characteristic for positive values of grid voltage. When the grid is positive, it extracts electrons from the electron stream that would ordinarily go to the plate. Thus we note a change in the curvature of the characteristics. At the same time a grid current is produced. For a constant positive grid voltage, as the plate voltage increases, fewer electrons go into the grid wires, and more electrons go on to the much greater plate potential. Thus the grid-current curves, which are also plotted on this plate family, decrease with increasing plate voltage.

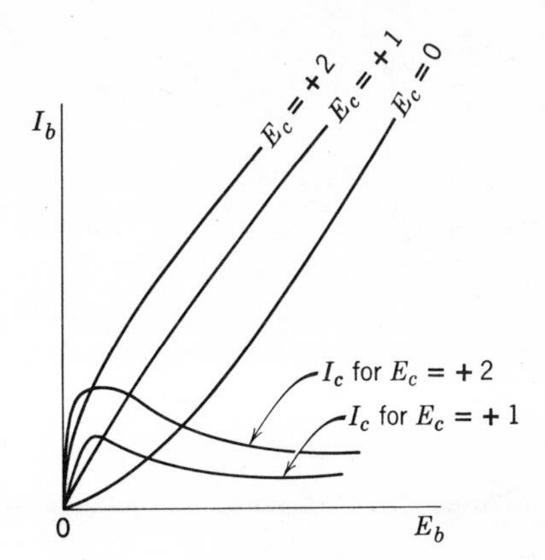

Figure 3-21 The plate characteristics of a triode in the positive grid region.

The grid family or transfer characteristic shows the grid voltage as the independent variable and the plate current as the dependent variable, using constant plate voltages as the parameter (Fig. 3-22). The curve shows that, for a particular plate voltage, a grid voltage that becomes more negative

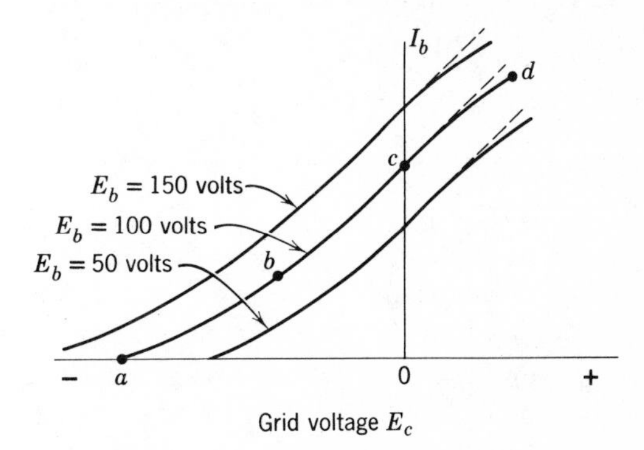

Figure 3-22 Transfer characteristics of a triode.

allows less plate current to flow. The curves in the negative region are again essentially parallel. The curvature is more pronounced between cutoff, point a, and point b. The curve is fairly straight between b and c. When the grid is positive, some of the electron stream transfers to the grid as a grid current, and the plate-current curves tend to drop off from the straight-line extension of the negative grid region.

Constant-current curves are plotted in Fig. 3-23. We show the curves with E_b as the independent variable and E_c as the dependent variable. The axes could very well be inverted, using E_c as the independent variable and E_b as the dependent variable. The practice varies between the different tube manufacturers. The curves are linear in the negative grid region. The line

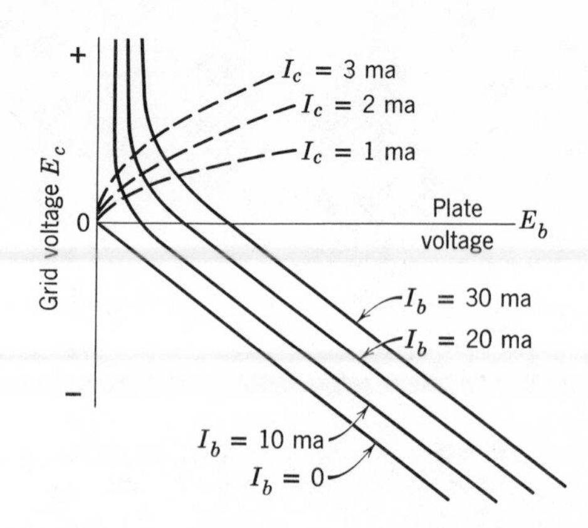

Figure 3-23 Constant-current curves for a triode.

representing zero plate current is the locus of the combinations of E_b and E_c that cut off the tube. This line is called the *cutoff line*.

If any one of the curves is available, it is possible to draw the other two without actually taking further laboratory data. The correspondence between the points of the three characteristic curves is shown in detail in Fig. 3-24. All of the corresponding points are labeled with lowercase letters and all of the corresponding voltages and currents are labeled with capital letters.

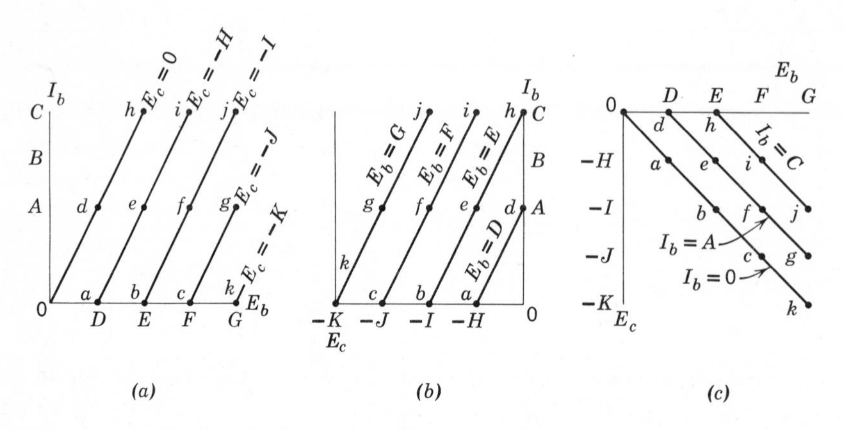

Figure 3-24 Common points to show interrelation of the various tube characteristics. (*a*) Plate family. (*b*) Mutual characteristics. (*c*) Constant-current characteristics.

Section 3-8 The Triode as a Simple Amplifier

The details of waveforms, of the standard nomenclature, and of the process for handling the load resistance R_c of the circuit of Fig. 3-25 are discussed in

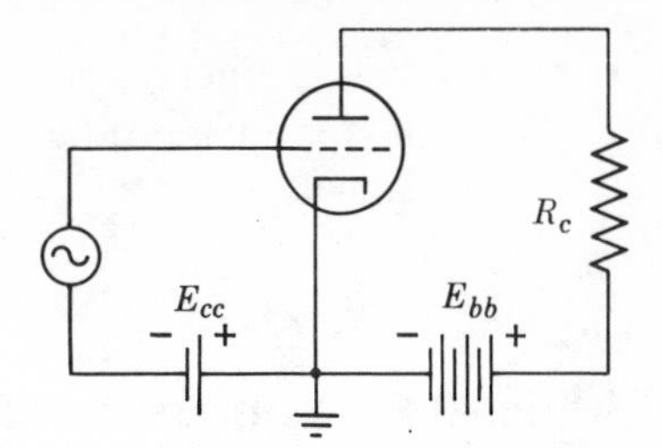

Figure 3-25 Basic triode amplifier circuit.

Chapter 11. Only enough information is given in this section to provide an understanding of the operation of the triode as a simple amplifying device. The battery supplies E_{cc} and E_{bb} establish a particular operating point on the tube characteristic curve. A direct plate current I_b flows, causing an I_bR_c drop in the load resistance. The operating plate potential E_b is less than E_{bb} by the amount of the I_bR_c drop in the load resistance. Assume that the operating plate current of the triode is 10 ma, that R_c is 10 kΩ, and that the plate supply voltage is 250 V. Without signal, the I_bR_c drop is 100 V, and E_b is then 150 V. Also, assume that the grid supply voltage E_{cc} is -10 V.

The total instantaneous grid voltage on the tube varies in accordance with the signal. When the signal is $+1$ V, the total grid signal is $-10+1$ or -9 V. Now the plate current increases, causing an increased drop in R_c. The new plate current is, say, 15 ma. The I_bR_c drop is now 150 V, and the plate voltage is 100 V. When the signal is -1 V, the total grid voltage is -10, -1, or -11 V. Assume that the plate current is now 5 ma. The I_bR_c drop is 50 V, and the plate voltage is 200 V. These results are summarized in Table D.

TABLE D

Signal Voltage	I_b	I_bR_c	E_b	Plate Voltage Relative to the No-Signal Condition
0	10	100	150	0
+1	15	150	100	−50
−1	5	50	200	+50

Now it can be seen that the total change of 2 V in the signal produces a change of 100 V at the plate of the tube. Thus the *voltage amplification* of the vacuum-tube circuit with the load resistance shown is 50. Also, it is very important to note that, when the signal goes positive or increases, the plate voltage decreases or goes negative. When the signal goes negative, the plate voltage increases or goes positive. Thus, like the common-emitter amplifier, this circuit produces, for a resistive load, a *phase inversion* of 180° on the signal.

Section 3-9 Limitations of a Triode

The region of maximum allowable operation of a triode is shown graphically on a plate characteristic (Fig. 3-26). Line $0a$ is the maximum emission that

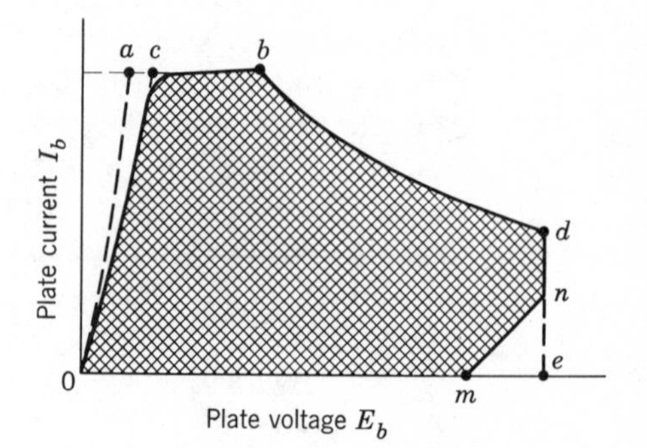

Figure 3-26 Limitations on the operation of a triode.

the tube can have from Child's law. Actually, in practice, the limit is $0c$, which is determined not by Child's law but by the maximum allowable grid current. If the grid current exceeds a certain value, the heat developed in the grid wire will damage or even melt the grid structure. Line cb is the maximum allowable plate current for the tube. The product of E_b times I_b is the *plate dissipation* of the tube. If the rated value of plate heating is exceeded, the tube will be damaged. The maximum allowable plate dissipation is the curved line between b and d. If the plate voltage exceeds de, the plate voltage may arc over to the grid wires. Line mn is the maximum allowable negative grid voltage to prevent arcing or breakdown, either from the grid to the cathode or from the grid to the plate. Thus the allowed region of operation of the tube is within the shaded area, $0cbdnm0$.

Section 3-10 Determination of the Dynamic Coefficients of a Triode

In Fig. 3-27, point P is shown on each of the three basic tube curves. If the plate current is held constant, we move through P from A to B. For the

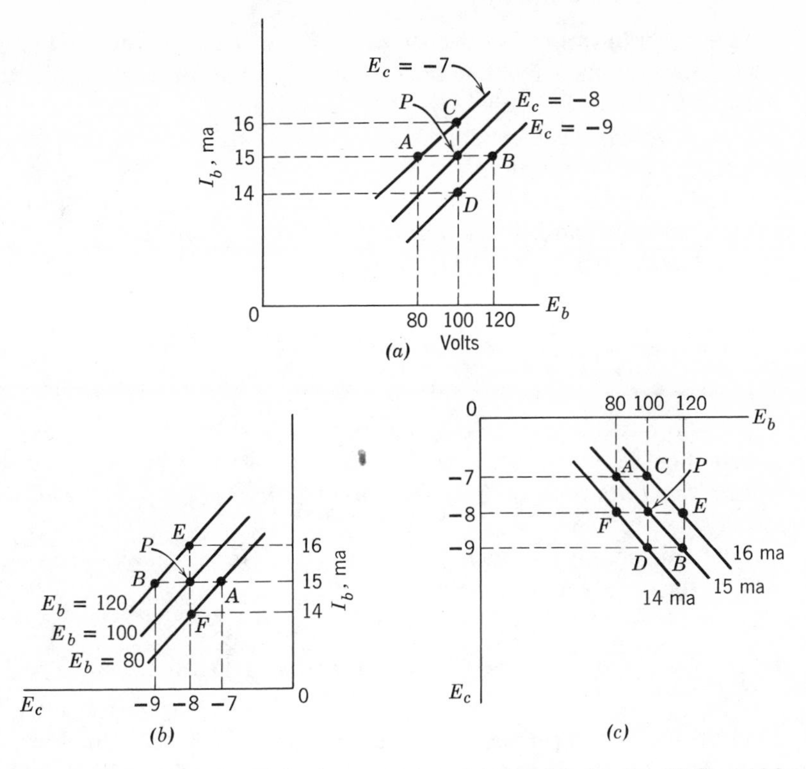

Figure 3-27 Graphical determination of μ, r_p, and g_m. (*a*) On plate family. (*b*) On grid family. (*c*) On constant-current curves.

numerical values we have indicated, this means that a change in plate voltage of 40 V (120 − 80) is obtained from a change of 2 V on the grid. The ratio of this change, 40/2 or 20, is termed the *amplification factor* or μ (Mu) of the tube. A small change on the grid (2 V in this case) produces a large change (40 V) on the plate for a fixed plate current. We may express this definition as an equation:

$$\mu \equiv -\frac{\partial E_b}{\partial E_c} \approx -\frac{\Delta E_b}{\Delta E_c} \qquad (I_b \text{ is constant}) \qquad (3\text{-}5a)$$

The sense of the negative sign associated with the definition of μ is to indicate the phase inversion of the triode amplifier. For the given numerical values:

$$\mu = -\frac{40}{2} = -20$$

This number or ratio μ is dimensionless, since it is a ratio of two voltages.

We can obtain this value of μ from either the plate characteristic or the transfer characteristic directly. If we must find it from the constant plate-current curves, we must take increments *along* the constant current line through P (the 15 ma curve).

If we consider the change from C to D through point P, we obtain a change in plate current for a change in grid voltage with a constant plate voltage. Making this a ratio, we have

$$g_m \equiv \frac{\partial I_b}{\partial E_c} \approx \frac{\Delta I_b}{\Delta E_c} \qquad (E_b \text{ is constant}) \qquad (3\text{-}5b)$$

This ratio has the units of conductance, mhos, but it is usually expressed in micromhos. Since the term relates plate current to grid voltage, it is termed *mutual conductance* or *transfer conductance* or *grid-plate transconductance*.

We can obtain this value of g_m directly from either the plate characteristic or the constant plate-current family as

$$g_m = \frac{2 \text{ ma}}{2 \text{ V}} = 0.001 \text{ mho} = 1000 \, \mu\text{mho}$$

If the value is determined from the transfer characteristic, increments must be taken along the constant-voltage line for a plate voltage E_b of 100 V.

If we consider the change from E to F through point P, the increments obtained are a change in plate voltage and a change in plate current for constant grid voltage. This, as a ratio, is the *dynamic plate resistance* r_p of the triode at point P:

$$r_p \equiv \frac{\partial E_b}{\partial I_b} \approx \frac{\Delta E_b}{\Delta I_b} \qquad (E_c \text{ is constant}) \qquad (3\text{-}5c)$$

r_p may be found directly from the mutual characteristic and from the constant plate-current curves, but the increments must be taken along the constant grid-voltage curve through P in the plate family. From the values we have assumed, the plate resistance is

$$r_p = \frac{40 \text{ V}}{2 \text{ ma}} = 20{,}000 \, \Omega = 20 \text{ k}\Omega$$

If we take the definitions for g_m and r_p and multiply them together, we find that

$$g_m \times r_p = \frac{\Delta I_b}{\Delta E_c} \times \frac{\Delta E_b}{\Delta I_b} = \frac{\Delta E_b}{\Delta E_c}$$

But this is the defined value of μ; so we may write an expression to relate these three dynamic characteristics:

$$-\mu = r_p g_m \tag{3-5d}$$

This equation is very useful in determining tube characteristics from graphs. If two are known, the third may be calculated. This equation is the vacuum-tube counterpart of Eqs. 3-4*e* and 3-4*f*, which relate α and β in a transistor.

Very often it is necessary to determine all three characteristics from one characteristic curve. For instance, the evaluation of r_p from the plate family requires that increments be taken *along* the constant grid-voltage line. If the r_p at point Q (Fig. 3-28) is required, the procedure, as with the deter-

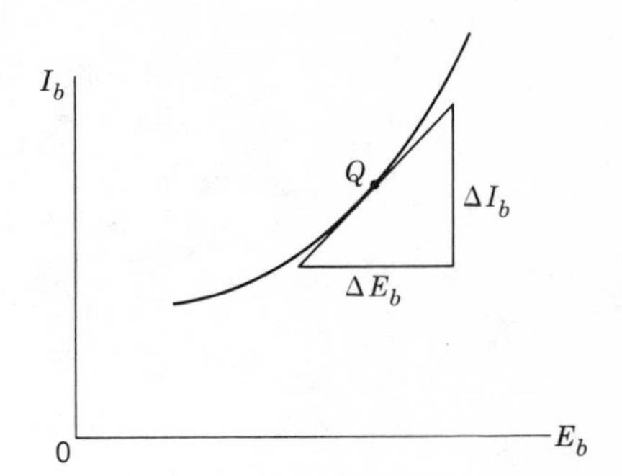

Figure 3-28 Determination of r_p from the plate family by the slope method.

mination of r_p in a diode, is to erect a tangent to the curve at Q and to use the slope of the tangent as the slope of the curve at point Q. In this manner it is easy to determine g_m from a transfer curve and μ from a constant plate-current curve.

The variation of the values of these dynamic characteristics with plate current is shown in Fig. 3-29. The amplification factor is almost constant,

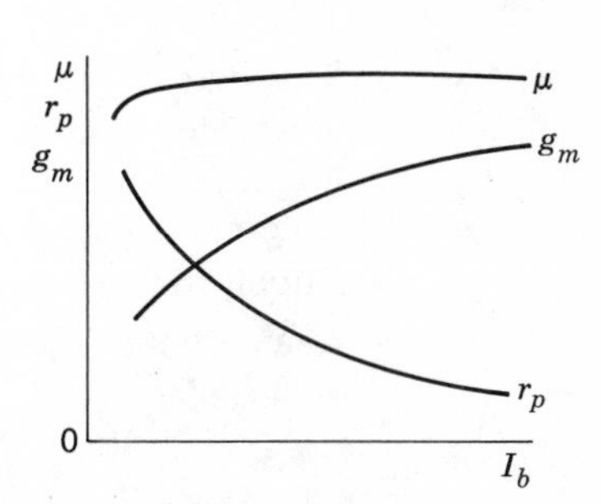

Figure 3-29 Variation of the dynamic characteristics with plate current.

as it is primarily a geometrical factor that involves the shape of the grid structure and its placement with respect to the cathode and to the plate allowing a certain ratio of control over the plate current. The transconductance increases with increasing plate current whereas, naturally, the plate resistance decreases with increasing plate current. In a complex amplifier circuit, the overall amplification is a function of all of these factors.

PROBLEMS

TABLE E

Point	A	B	C	D
I_b ma	9	8	8	2
E_c volts	−4	−4	−6	−6
E_b volts	400	390	460	400

TABLE F

Point	A	B	C	D
I_b ma	17	23	23	?
E_c volts	0	0	−1	−2
E_b volts	100	125	150	175

TABLE G

Vacuum tube	A	B	C	D	E	F
r_p Ω	?	4500	25,000	?	6200	14,000
g_m μmhos	8700	?	2800	5600	?	3430
μ	85	21	?	40	35	?

1. The data in Table E are for four points on a triode plate characteristic. Determine μ, r_p and, g_m.
2. The data in Table F are for four points on a triode plate characteristic. Determine μ, r_p, and g_m. Also determine the missing value of current.
3. Supply the missing items in Table G.
4. At a particular condition of operation, a triode vacuum tube has the following values:

$$g_m = 2000\ \mu\text{mhos}; \quad \mu = 17.5, \quad r_p = 8750\ \Omega,$$
$$E_b = 250\ \text{V}, \quad E_c = -11\ \text{V}, \quad \text{and} \quad I_b = 5.5\ \text{ma}$$

If the grid voltage is changed to -10 V, what is the new value of plate current? If the plate voltage is raised to 260 V, to what value must the grid voltage be set to maintain the plate current at 5.5 ma? If the plate voltage is lowered to 235 V, what is the new value of plate current?

Questions

1. Define or explain each of the following terms: (*a*) emitter, (*b*) collector, (*c*) base, (*d*) *PNP* transistor, (*e*) *NPN* transistor, (*f*) α_{dc}, (*g*) h_{FB}, (*h*) α_{ac}, (*i*) h_{fb}, (*j*) I_{CBO}, (*k*) parameter, (*l*) β_{dc}, (*m*) h_{FE}, (*n*) β_{ac}, (*o*) h_{fe}, (*p*) common-base, (*q*) common-emitter, (*r*) common-collector, (*s*) phase inversion, (*t*) breakdown voltage, (*u*) cutoff, (*v*) plate characteristic, (*w*) transfer characteristic, (*x*) cutoff line.
2. Explain how the transistor is a current-operated device.
3. What is a collector characteristic?
4. If α is given, how is β obtained?
5. If β is given, how is α obtained?
6. Why is I_{CEO} larger than I_{CBO}?
7. Does a common-base amplifier give voltage amplification? Current amplification? Phase inversion?
8. Does a common-emitter amplifier give voltage amplification? Current amplification? Phase inversion?
9. Does a common-collector amplifier give voltage amplification? Current amplification? Phase inversion?
10. What are the limitations of operation of a transistor on a collector characteristic?
11. What happens when a vacuum-tube is operated with a positive grid?
12. How can a grid voltage be applied to reduce plate current to zero?
13. Does a vacuum-tube amplifier give voltage amplification? Current amplification? Phase inversion?
14. What are the limitations of operation of a vacuum tube on a plate characteristic?
15. Relate μ, r_p, and g_m in a vacuum tube. Define each term.

Chapter Four

OTHER SEMICONDUCTOR DEVICES

The field of semiconductor development is continuously extending and expanding into new areas. The purpose of this chapter is to examine some of these developments which have become well established to the point where they are generally available and used as packaged devices. They include various specialized forms of the discrete transistor (Section 4-1), the integrated circuit (Section 4-2), the field-effect transistor (Section 4-3), and transistors that make use of intrinsic materials (Section 4-4). The tunnel or Esaki diode (Section 4-5) and the varactor diode (Section 4-6) are applications of new concepts unknown before the advent of semiconductors. The use of the unijunction transistor (Section 4-7) has been stimulated by its applications in conjunction with a member of the thyristor family (Section 4-8), the silicon controlled rectifier (Section 4-9). The most recent extension of the silicon controlled rectifier concept is the triac (Section 4-10).

Section 4-1 Transistor Construction

Several forms of transistors which are, in a broad sense, called junction transistors have special names that result from the manufacturing process. In Chapter 3, we considered the transistor as being formed of three layers – the emitter, the base, and the collector. We have also stated that the base layer is thin and that it has large surface dimensions compared to its thickness.

To manufacture *rate-grown* junction transistors, a *bar* or ingot of germanium is slowly drawn from the mold and, by a combined process of impurity control called *doping* and of critical temperature control, thin layers

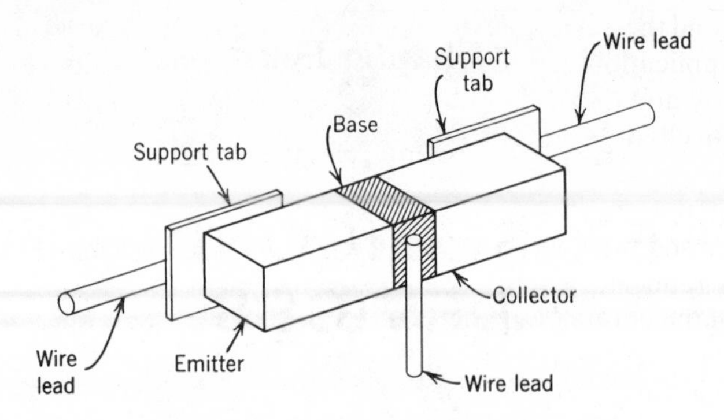

Figure 4-1 The junction transistor.

of P material are formed between slabs of N material. The resulting bar is sawed up into small pieces of NPN material about 0.01 by 0.01 by 0.10 in. for transistor construction (Fig. 4-1). The process is also used to produce small pieces of PNP material. Transistors manufactured by this process are useful up to several megahertz.

The *meltback* process used for special transistor types is similar to the method of manufacturing rate-grown junction transistors. A bar of N-type germanium is cut into small pieces of the size mentioned above. The end of it is melted and refrozen quickly and, by a doping method, a thin layer of P material is produced within the N material.

The capacitance between the emitter and the collector is a limiting factor in the high-frequency use of a transistor. Another limitation is the transit time for majority current carriers to flow through the base. On one hand, if the thickness of the base is reduced, the capacitance between the emitter and the collector increases. On the other hand, an increase in base thickness increases the transit time although it does decrease the capacitance. The objective in the design of high-frequency transistors is to find a way to escape from simultaneous limitations.

The *diffused-alloy* junction transistor is one of the first successful attempts to circumvent these limitations. Holes are drilled into the base material which is N-type material in Fig. 4-2. Small pellets or *dots* of P-type germanium are placed in the holes. A short pulse of high current between the

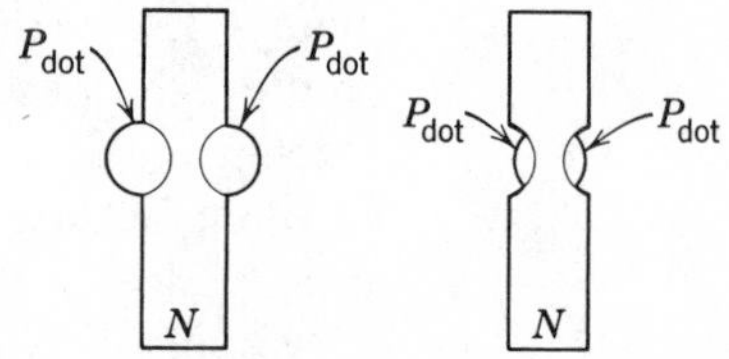

Figure 4-2 Diffused-alloy transistors.

pellet and the base melts the contact surfaces, and they weld or fuse together. The application of heat is not sufficient to destroy the basic characteristic of the *N* and the *P* materials. This process is also applied to the manufacture of diffused-alloy *NPN* transistors.

The separation between the emitter and the collector is reduced, reducing the transit time through the base. The capacitance between the emitter and the collector is actually decreased because, although the separation is reduced, the surface area of the pellets is much less than the surface area of the emitter and the collector in a cut bar transistor.

When we discussed the basic concepts of crystalline structures, we were very careful to state that the conditions concerned the center of the crystal and *not* the surfaces. The action of forces and energy levels at the surface is quite complex but, for the purposes of this discussion, a simplified approach can be taken.

When *N*-type surfaces are exposed to air or to a vacuum (Fig. 4-3*a*), a *surface-barrier layer* of electrons crowds the surface in a state of equilibrium. Just below the barrier layer is a "layer" of holes which has been attracted by the negative charge of the surface-barrier layer, but these holes

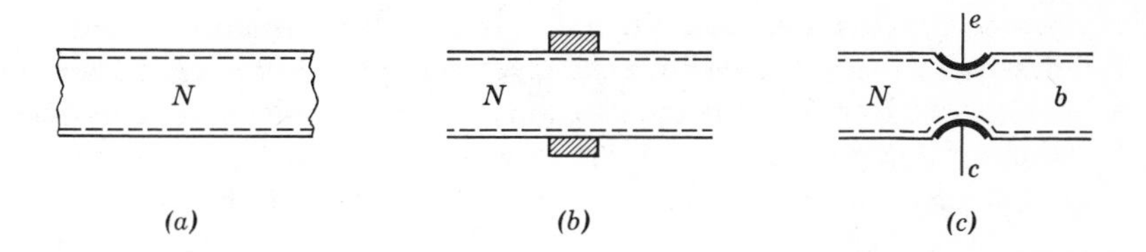

Figure 4-3 Surface-barrier transistors. (*a*) Barrier layers. (*b*) Barrier layer with small electrode plates. (*c*) With etched wells.

do not combine with the surface electrons. A small contact can be made to this surface (Fig. 4-3*b*) without disturbing the barrier layer. In the *surface-barrier transistor*, indentations are made on the base *N* material by an electrochemical etching process. The separation between the bottoms of the indentations is very small. Electrodes serving as the emitter and collector are electroplated within the indentations. A reverse bias on one electrode serving as the collector causes the surface-barrier layer to thicken. The other electrode is forward-biased to serve as the emitter. The emitter injects holes into the base, and these holes cause a neutralization of the surface barrier, allowing a collector current to flow to the extent of the neutralization of the surface barrier. This transistor has a lower capacitance and a shorter transit time than the diffused-alloy junction transistor, and it was developed specifically for high-frequency applications.

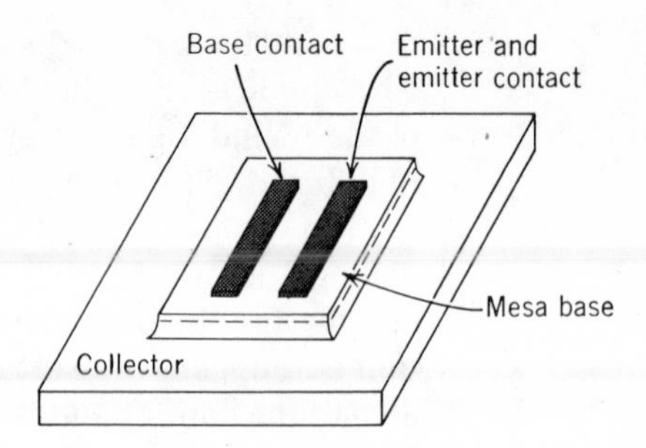

Figure 4-4 The mesa transistor.

In order to obtain better performance at high frequencies together with the capability of higher power, the *mesa transistor* was developed (Fig. 4-4). The base is diffused onto the collector pellet. The strips serving as the base contact and the emitter are evaporated to the base surface. The excess base beyond that required for operation is removed by etching to form the table-like physical form that gives rise to the use of the descriptive term, mesa.

The characteristics of the mesa transistor are improved by using *epitaxial films* to form the *epitaxial mesa transistor.* The epitaxial films are so called because their atoms are aligned to make one continuous structure of the entire transistor. A thin epitaxial layer of high resistance impurity (about 1Ω) is deposited on a low resistance substrate silicon having an impurity of the order of 0.005 Ω-cm to form the collector. The base and emitter are added as in a mesa transistor to form the complete unit (Fig. 4-5).

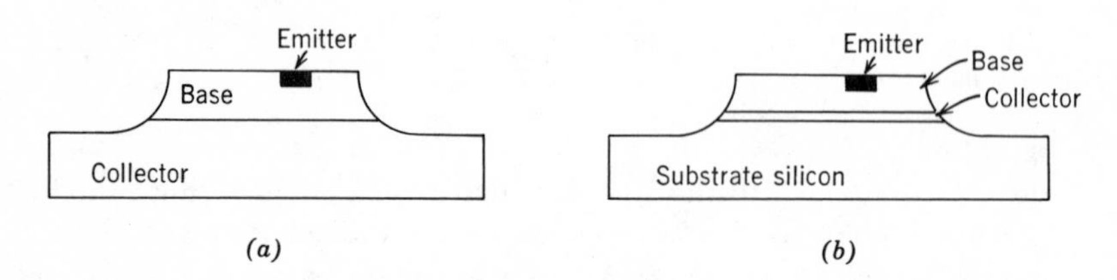

Figure 4-5 Mesa transistors. (*a*) Conventional mesa. (*b*) Epitaxial mesa.

The characteristics of the epitaxial mesa transistor (Fig. 4-6) are more linear than other transistor types. Also this linearity extends over a greater region. Because of its very low capacitance it can be used up to frequencies of the order of hundreds of megahertz.

The *silicon planar transistor* (Fig. 4-7) has characteristics quite similar to the mesa transistor without requiring the mesa forming etching process. The silicon oxide (SiO_2) covering prevents the junctions from absorbing

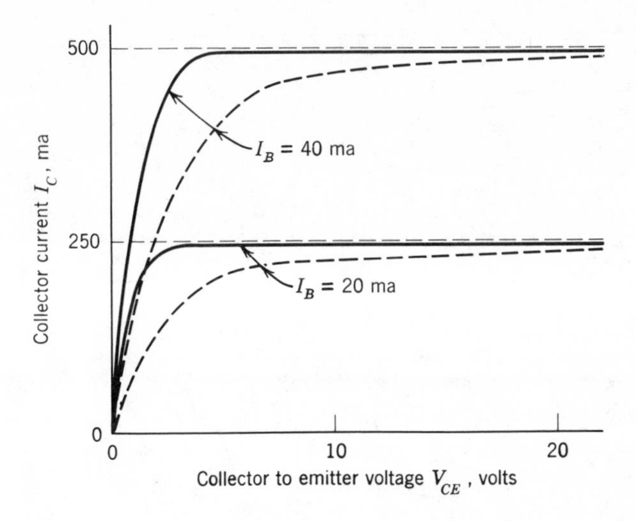

Figure 4-6 Comparison of mesa transistors. Dashed lines—conventional mesa. Solid lines—epitaxial mesa.

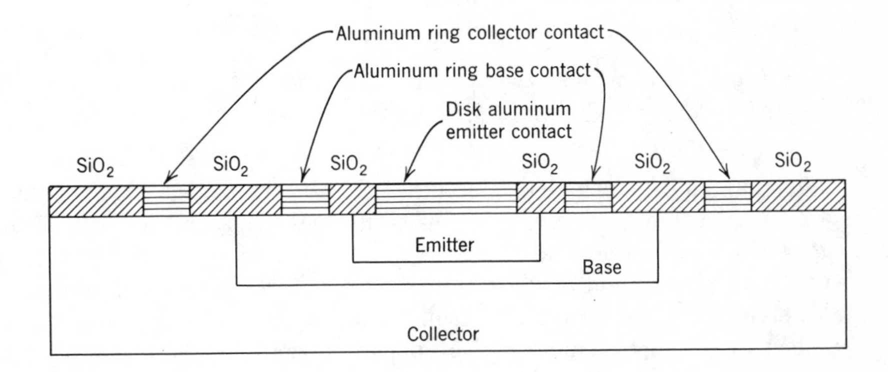

Figure 4-7 A cross section of a silicon planar transistor.

gases. Silicon dioxide is basically what we know as glass. This type of protection produces what is known as a *passivated device* to give better electrical stability to its characteristics.

The *tetrode transistor* is fundamentally an ordinary junction transistor with a second base connection b_2 added to the base portion of the unit (Fig. 4-8). The transistor functions in the normal manner when the "tetrode" base voltage is zero. When a voltage of proper polarity is applied to the base b_2, the majority current carriers are forced down toward b_1. In this manner, the cross-sectional area of the base is materially decreased, reducing the

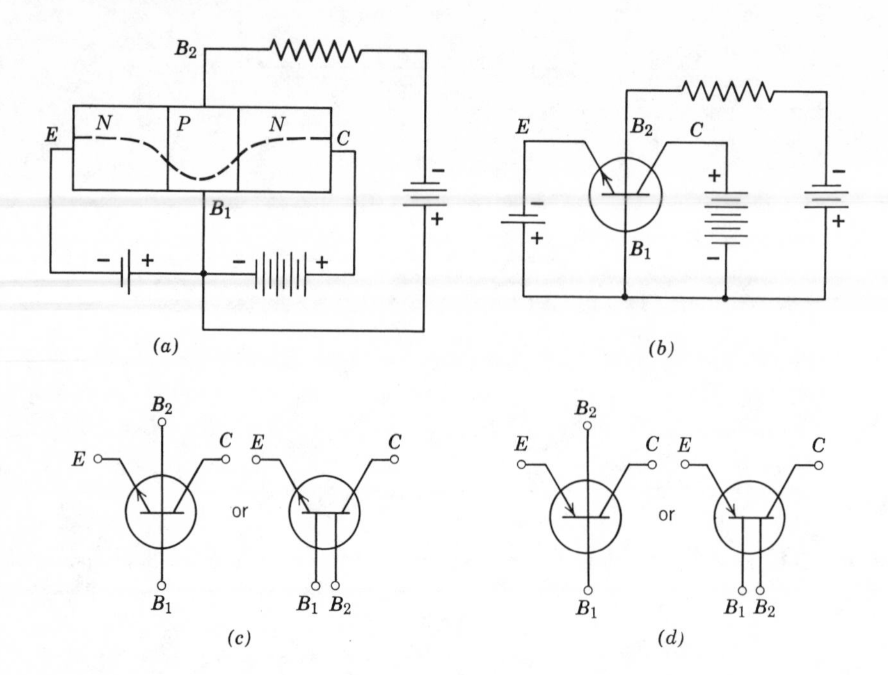

Figure 4-8 The tetrode transistor. (*a*) Physical construction. (*b*) Schematic diagram. (*c*) Symbols for *NPN* tetrode transistor. (*d*) Symbols for *PNP* tetrode transistor.

interelement capacitance between the base and the collector. This reduction in the capacitance enables the transistor to be used at much higher frequencies than would otherwise be possible.

The displacement of the majority current carriers toward b_1 reduces the value of α for the transistor to approximately 0.8. This sacrifice in gain is offset by an improvement in the upper-frequency limits. These transistors are used to about 200 MHz.

Section 4-2 Integrated Circuits

Conventional electronic circuits and conventional electric circuits can be reduced physically in size to the point where 250,000 distinct components can be placed on a surface having an area of 1 sq in. The techniques of circuit design, manufacture, testing, and use comprise what is known as *integrated-circuit (IC) technology* or *microelectronics*. This technology developed because of the importance of reducing the physical size of a complete system. The development of extremely large and complex computers required microelectronic components. A desk-size computer is a reality only because of microelectronics. It has been found also that the production of integrated circuits has enabled the cost of a system to be reduced to the extent that the integrated circuit is usually preferred to the use of discrete components with conventional wiring procedures.

The IC package is used as a "black box" and is connected externally to other IC's (black-boxes) by the use of *interface* wiring and to certain external discrete components that cannot practically be reduced in size to microcircuits. To understand the operation of the IC black box, it is necessary to study and approach electronic circuits and applications by means of conventional discrete components.

Integrated-circuit packages are designed for a specific application by a manufacturer, or they are available as a package that can be purchased the same as a transistor or a vacuum tube. Then they are used as a part of a specific system. Three standard types of available off-the-shelf integrated-circuit packages are shown in Fig. 4-9. These packages are designed to be

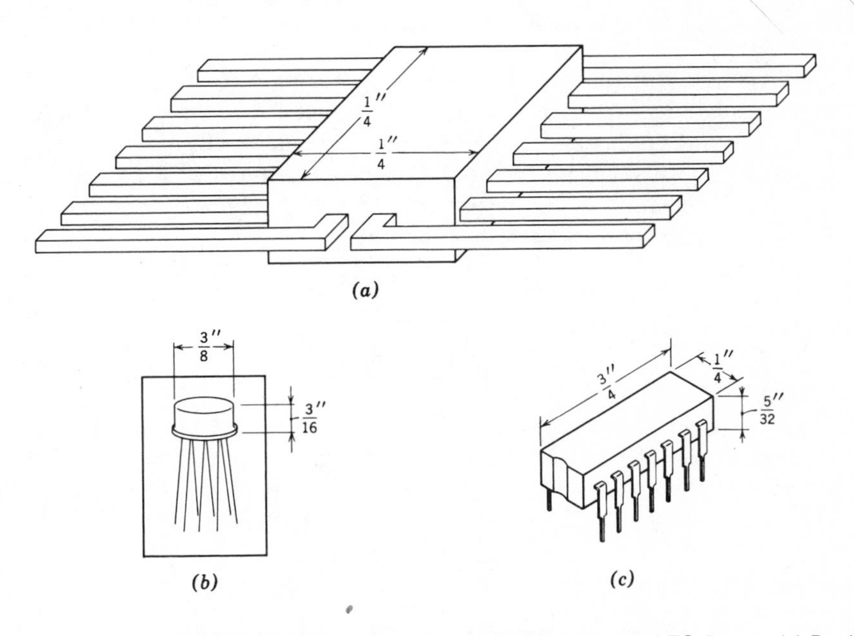

Figure 4-9 Monolithic-circuit packages. (*a*) Flat package. (*b*) Eight-lead TO-5 case. (*c*) Dual-in-line package. (*Dimensions are approximate only.*)

adapted to a series of standard sockets. Some typical examples of IC packages are given in Table A.

Medium-scale integration (*MSI*) refers to a *chip* that contains up to 100

TABLE A Typical Linear Integrated-Circuit Packages

Type	Application	Contents
Dual-in-line	Operational amplifier	18 *NPN* transistor, 3 diodes, 12 resistors
Dual-in-line	General purpose amplifier	18 *NPN* transistors, 12 MOSFET's 20 resistors
TO-5 case	Voltage regulator 100 ma at 34 V	5 *PNP* transistors, 10 *NPN* transistors, 2 Zener diodes, 4 diodes, 5 resistors
TO-5 case	IF amplifier, limiter, discriminator, and preamplifier for FM receiver	20 *NPN* transistors, 3 Zener diodes, 9 diodes, 25 resistors
Dual-in-line	4 general-purpose amplifiers for low-frequency industrial applications	24 *NPN* transistors, 8 diodes, 52 resistors
Flat package	Dual differential amplifier	18 *NPN* transistors, 5 Zener diodes, 14 resistors

integrated components. *Large-scale integration* (*LSI*) refers to a chip that contains more than 100 integrated components. A typical LSI chip is $\frac{1}{16}$ in. by $\frac{1}{8}$ in. and contains 448 components. Obviously, each individual component is extremely small by the standard of the conventional circuit component that is measured in terms of inches. In IC technology one must adapt to measurements suited to this technology:

$$\text{one micrometer } (\mu\text{m}) = 10^{-6} \text{ meter (m)}$$
$$\text{one mil} = 0.001 \text{ inch}$$
$$\text{one mil} = 25 \text{ micrometer } (\mu\text{m})$$
$$\text{one mil} = 250{,}000 \text{ Angstrom units (Å)}$$

The cross-sectional view of an IC transistor is shown in Fig. 4-10. The thickness of the N^+ substrate layer is of the order of 150 μm and the thickness of the epitaxial N layer is approximately 5 μm. The surface layer of

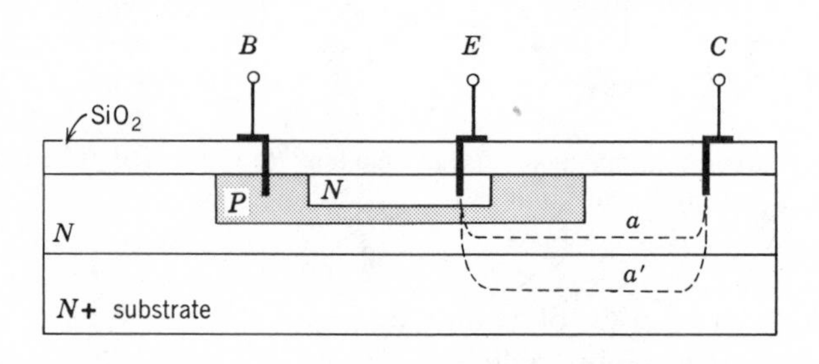

Figure 4-10 Monolithic transistor construction.

silicon dioxide (SiO_2) is approximately $\frac{1}{2}$ μm. Accordingly, the thickness of the base material between the two slabs of N material is only about 1 or 2 μm.

In a conventional transistor, the path of the collector current flow is shown as path a in Fig. 4-10. In a theoretically ideal transistor, the saturation voltage V_{CE} at an I_C of 100 ma is about 0.1 V. In an NPN silicon transistor, the series collector resistance ranges from 6 to 15 Ω. When the voltage drop caused by I_C and this series resistance is added to the ideal 0.1 V, the resulting saturation voltage V_{CE} ranges from 0.7 V to 1.6 V. The substrate material N^+ is more highly doped than the usual N material and, consequently, the resistivity of the N^+ material is much less than the N material. For example, the resistivity of N material is 0.5 Ω-cm and that of N^+ material is 0.01 Ω-cm. Consequently, the parallel path a' in Fig. 4-10 for the collector current offers much less resistance to current flow and by using this N^+ substrate, V_{CE} is reduced to about 0.2 V. Not only does the substrate reduce $V_{CE,SAT}$ but also it reduces the heating loss within the IC itself.

The cross-sectional view of two transistors on a chip is shown in Fig. 4-11. The whole transistor of Fig. 4-10 is contained within a continuous base

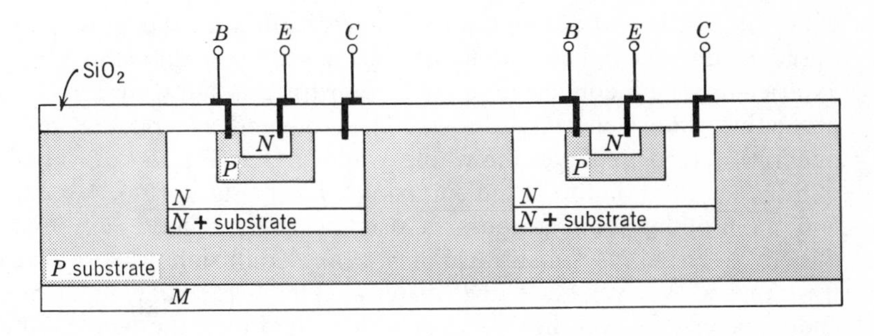

Figure 4-11 Two transistors on one monolithic chip.

of P-substrate material. There is a metallic connection M to the bottom of the P substrate. This diagram shows two NPN transistors. The P substrate is connected to the most negative point in the circuit. Consequently, the N^+ substrates and the N collectors have a large reverse bias with respect to the P substrate. Depletion regions exist in the boundaries between the P substrate and the N^+ substrates and in the boundaries between the P substrate and the N collectors. The effect of this reverse bias is to "float" each transistor in the electrical insulation of a depletion region. Therefore, all transistors (and all of the components in the IC) are insulated from each other.

Silicon crystals of P or N material are grown in rods about $1\frac{1}{4}$ in. in diameter and 6 in. long. The rod is sawed into slices or *wafers* about 10 to 15 mils thick. The slices are lapped, polished, and etched to yield finished wafers about 5 mils thick. The integrated circuit is processed on one side of this wafer; hence, the nomenclature *monolithic* refers to a set of components developed on one smooth side of this wafer. Simultaneously, 100 to 1000 circuits are formed on the side of the wafer. On completion of the processing of the wafer, the wafer is cut up into the individual *chips*. A typical chip size is 0.030 in. by 0.050 in. The chips are then each mounted in a separate holder, appropriate leads are attached, and the final IC packages are completed.

The formation of a single transistor on a wafer is developed in Fig. 4-12. One side of the wafer is coated with SiO_2, Fig. 4-12*a*, and a layer of photosensitive *emulsion* is placed on top of the SiO_2. A *mask*, Fig. 4-12*b*, which has been previously prepared, may contain hundreds or thousands of opaque and transparent areas. The mask is placed over the emulsion on the wafer and is exposed to light. The light that reaches the emulsion through the transparent areas causes the emulsion to harden in the developing process, but the emulsion behind the opaque areas is washed off. Then the SiO_2 that is not protected by the hardened emulsion is etched off. This process of events is called *photolithography*. The P material of the wafer is now exposed below the opaque areas of the mask. By a diffusion process, the exposed areas are converted to an N^+ substrate as shown in Fig. 4-12*c*. The protective SiO_2 material is now etched off and an epitaxial layer of N material is deposited over the whole wafer surface (Fig. 4-12*d*). Then a layer of SiO_2 is placed on top of the epitaxial N layer (Fig. 4-12*e*). A second mask and photolithographic process is used to remove any SiO_2 that is not directly over the N^+ substrate (Fig. 4-12*f*). A diffusion process converts all the exposed N layer back to P material (Fig. 4-12*g*) leaving islands of N^+ and N layers. Silicon dioxide is now deposited over the whole wafer (Fig. 4-12*h*). A third mask and photolithographic process is used to cut windows over the N material (Fig. 4-12*i*). A diffusion process converts the exposed N material to a P layer. The SiO_2 is etched off, and a new SiO_2 layer is placed on the wafer. A fourth mask and photolithographic process is used to create smaller windows over the P layer and a diffusion process converts the exposed surface of the P layer back to N material.

A new layer of SiO_2 is placed over the whole wafer. Holes are formed in the SiO_2 to form wells for leads to the emitter, to the base, and to the collector (Fig. 4-12*j*). *Metalized contacts* are formed to serve as lead terminals and to serve as interconnections between various components on the chip. Now the transistor has the cross-section of Fig. 4-10.

The rather lengthy and involved process of manufacturing steps is both complex and costly. However, in mass production, the cost of producing a wafer can be reduced to, say, $ 20. If there is a *yield* of 400 chips from the

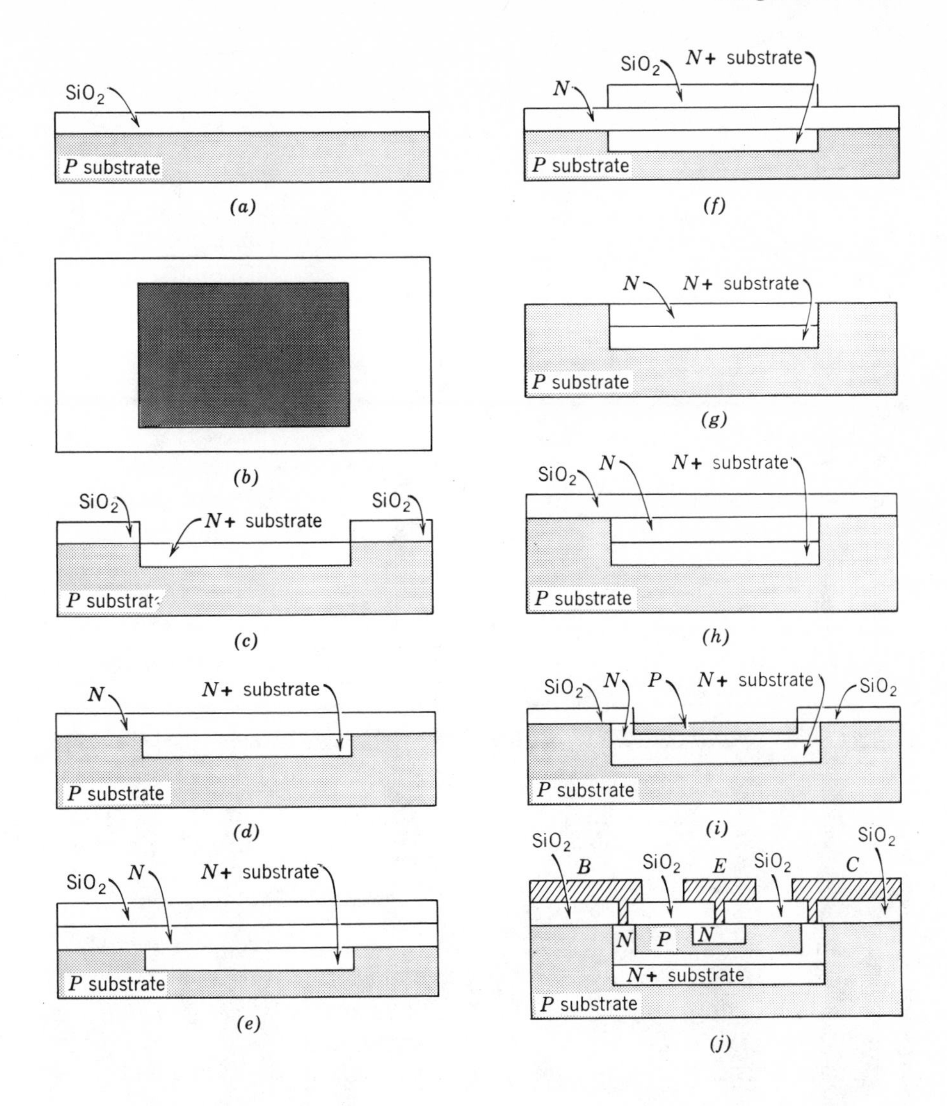

Figure 4-12 Steps in the manufacture of an *NPN* transistor on a chip.

wafer, the cost of each chip is 5 cents. With a cost of packaging and testing a chip at 20 cents, the factory cost of the finished package is then 25 cents. If the chip contains 25 semiconductors and 25 resistors, the cost of each component with its wiring is reduced to $\frac{1}{2}$ cent each. As a result, the integrated circuit must become the predominant tool and building block in the electric art.

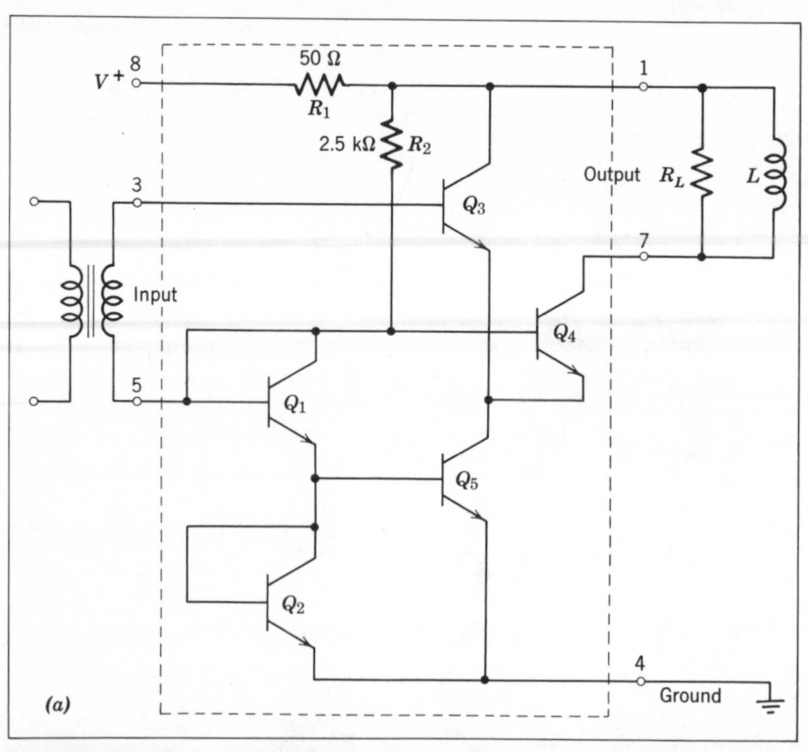

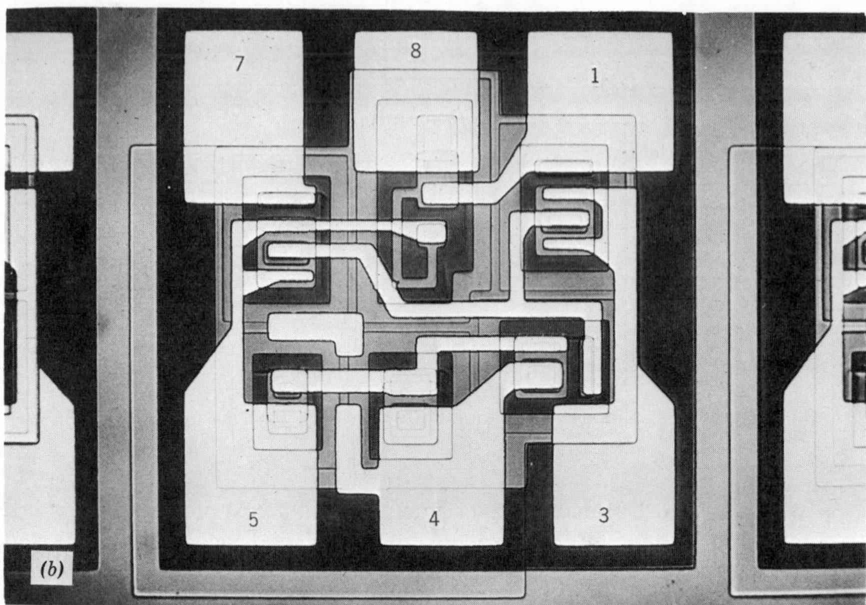

Figure 4-13 A complete IC chip. (*a*) Circuit. (*b*) Microphotograph. (*Courtesy* P. E. Gray and C. L. Searle, *Electronic Principles—Physics, Models, and Circuits*, Wiley, 1969).

A finished IC chip is shown in Fig. 4-13*b*. The chip is 0.020 in. by 0.020 in. The circuit for the chip is given in Fig. 4-13*a* and includes the components that must be added externally to complete the circuit for its application.

Section 4-3 The Field Effect Transistor (FET)

Conventional transistors, *PNP* or *NPN*, function on both hole current and on electron current. Consequently, they are referred to in the literature as *bipolar junction transistors* (BJT). The field effect transistor (FET) operates on *either* electron-current flow *or* on hole-current flow. In contrast to the BJT, the FET is a *unipolar transistor.*

The construction of an *N*-channel junction field effect transistor (JFET) is shown in Fig. 4-14*a*. Metallic contacts, the *source (S)*, and the *drain*

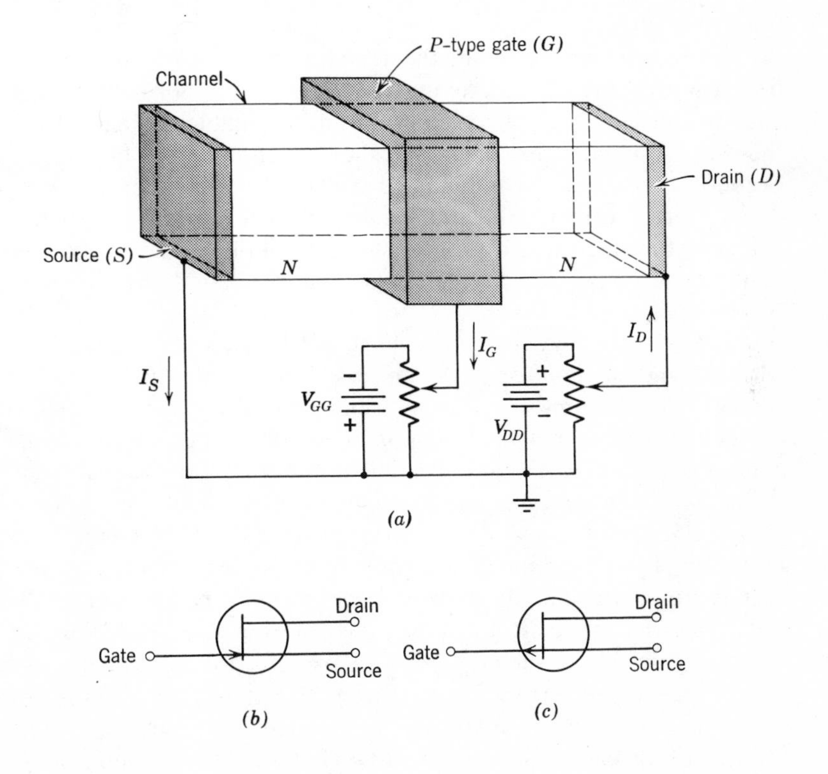

Figure 4-14 The Junction Field Effect Transistor (JFET). (*a*) Construction. (*b*) Symbol for an *N*-channel JFET. (*c*) Symbol for a *P*-channel JFET.

(D), are placed at opposite ends of the *channel*. The contact of the source to the channel and the contact of the drain to the channel are ohmic contacts and are not *P-N* junctions. A *gate (G)* is formed by placing a ring of *P* material around the center of the channel to form a *P-N* junction. Electrode voltages are placed on the JFET as shown in the circuit. The polarities of both V_{GG} and V_{DD} would be reversed if the JFET had a *P* channel. The d-c source current is I_S, the d-c drain current is I_D, and any gate current is I_G. The direct voltage between the gate and the source is V_{GS} and the direct voltage between the drain and the source is V_{DS}. Usually the source is the reference point in the JFET. Most of the literature refers to the JFET simply as an FET with the "J" understood. In this text, we will retain the "J" in order to distinguish the junction type from the other types of field effect transistors.

The circuit symbol for an *N*-channel JFET is shown in Fig. 4-14*b*, and the circuit symbol for a *P*-channel JFET is shown in Fig. 4-14*c*.

The cross-sectional view of the JFET is shown in Fig. 4-15*a*. There is a *P-N* junction between the gate and the channel and, consequently, there is a depletion region in the channel surrounding the gate. Since the action of the JFET does not depend on the depletion region within the gate, the depletion region inside the gate is ignored. The gate is normally reverse biased and, as a result, I_G is zero except for any negligible leakage current. When the gate is made more negative, the depletion region in the *N* channel becomes larger and, eventually, can extend across the full channel (Fig. 4-15*e*) with the result that the cross-sectional area of the channel that is available for electron-current flow to the drain is reduced to zero. The particular gate voltage that reduces the drain current I_D to zero is called the *pinchoff voltage* V_P.

Now assume that the gate voltage is zero and that a drain voltage V_{DS} with the indicated polarity is impressed across the drain and the source (Fig. 4-15*b*). Assume that the interior of the channel has a uniformly distributed resistance between *a* and *b*. The voltage between *a* and *a'* is zero, and the depletion region at *a'* is undisturbed. However, the full drain voltage V_{DS} acts as a reverse voltage between *b* and *b'*. This causes the depletion region to widen and to spread, as shown in Fig. 4-15*c*. Since the drain current in the JFET is a function of the available cross-sectional area of the channel that is not depleted, the drain voltage *must* be positive in an *N*-channel JFET in order to have control over the drain current. When the gate is made negative, Fig. 4-15*d*, the cross-section available for electron flow is reduced, and I_D decreases. At pinchoff (Fig. 4-15*e*) the *N* channel is completely blocked to electron flow by the depletion region. An increase in drain voltage still does not cause any current flow. The increased drain voltage merely extends the depletion region toward the drain contact (Fig. 4-15*f*).

In the *N*-channel JFET, it is possible to operate the gate at a small

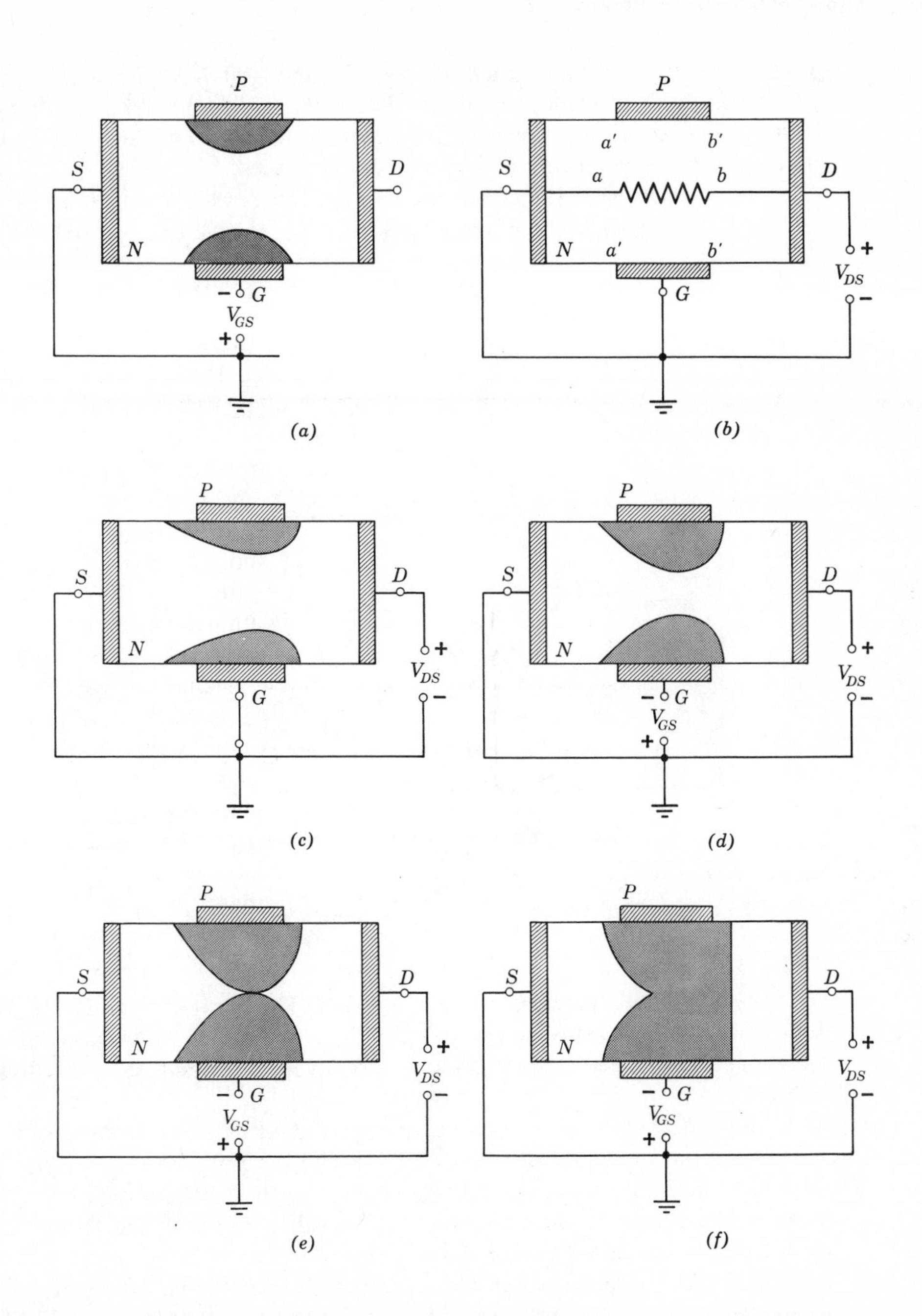

Figure 4-15 The electric field in a FET. (*a*) Depletion region caused by a reverse-biased gate. (*b*) Voltage drop in the channel. (*c*) Effect of drain voltage on depletion region. (*d*) Increased drain voltage with a negative gate. (*e*) Pinch-off at low drain voltage. (*f*) Pinch-off at high drain voltage.

positive voltage of the order of +0.5 V and still have negligible gate current as long as the barrier potential of the *P-N* gate-channel junction is not exceeded. This small positive potential almost eliminates the depletion region at the gate and the drain current increases.

A typical set of curves called the *drain characteristic* obtained by maintaining a fixed gate voltage and by varying V_{DS} is given in Fig. 4-16. The

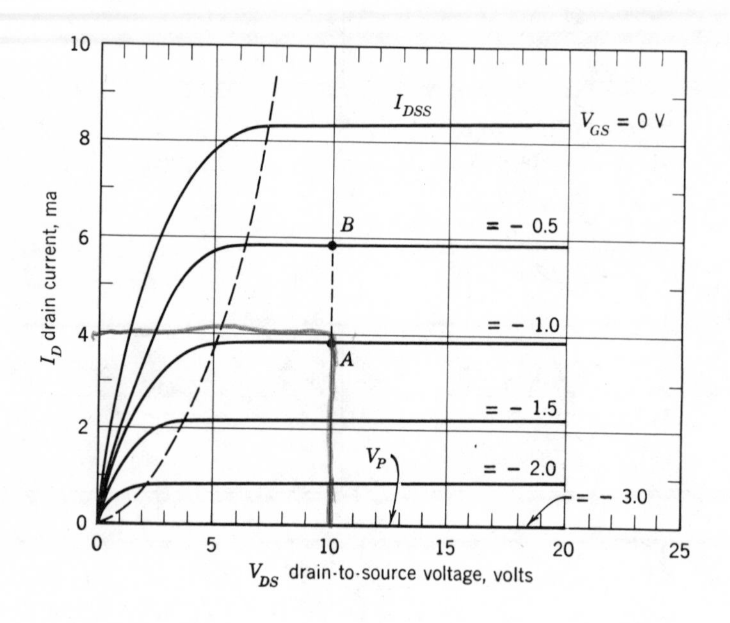

Figure 4-16 Drain characteristic-junction field effect transistor.

value of V_P for this particular JFET is −3 V. The curves of the JFET are quite flat showing that once the value of V_{DS} exceeds the values shown by the dashed curve, the drain current is effectively independent of the drain voltage.

If all of the points on the drain characteristic for a particular value of V_{DS} are plotted on a new set of axes (I_D, V_{GS}), the *transfer characteristic* is obtained. When this is done for V_{DS} equal to 10 V in this example, the transfer characteristic shown in Fig. 4-17 results. Points *A* and *B* are corresponding points on the two characteristics.

The current obtained when V_{GS} is zero and when V_{DS} is sufficiently high to secure a leveling off in current has the specific nomenclature I_{DSS}. The "*SS*" in I_{DSS} indicates the gate is shorted to the source to insure that V_{GS} is zero.

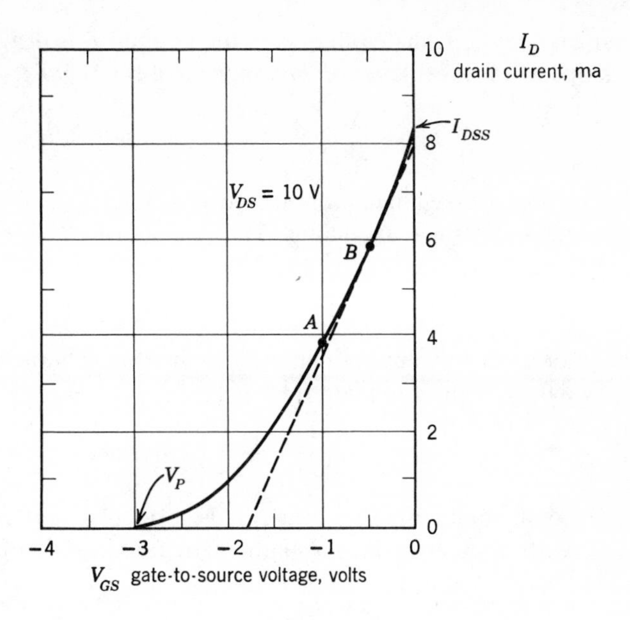

Figure 4-17 Transfer characteristic for JFET.

Methods of an analysis of the JFET, using theoretical physics,* can be used to show that an equation can be developed for I_D in the flat portion of the drain characteristic as

$$I_D = I_{DSS}\left(1 - \frac{V_{GS}}{V_P}\right)^2 \tag{4-1}$$

The JFET used for the characteristics of Fig. 4-16 and Fig. 4-17 has the equation

$$I_D = 8.4\left(1 - \frac{V_{GS}}{-3}\right)^2 \text{ma}$$

in which I_{DSS} is 8.4 ma and V_P is −3 V.

If a slope is drawn to the drain characteristic at point B on Fig. 4-16, the slope of the tangent defines the *a-c drain resistance* r_d as

$$r_d \equiv \frac{\Delta V_{DS}}{\Delta I_{DS}} \qquad \text{for a constant } V_{GS} \tag{4-2}$$

The slope of the characteristic in Fig. 4-16 give a value of 100,000 Ω for r_d.

*For a derivation see Appendix I, E. J. Angelo, *Electronics: BJT's, FET's and Microcircuits,* New York; McGraw-Hill Book Co., 1969.

A tangent to the transfer characteristic at point B is drawn on Fig. 4-17, and its slope defines the *transconductance* g_m of the JFET as

$$g_m \equiv \frac{\Delta I_D}{\Delta V_{GS}} \qquad \text{for a constant } V_{DS} \tag{4-3}$$

Usually, g_m is expressed in micromhos (μmhos).
In this example the transconductance is

$$g_m = \frac{8 \text{ ma}}{1.8 \text{ V}} = 4450\ \mu\text{mhos}$$

The value of g_m can be approximated from the drain characteristic by taking the interval from point A to point B as

$$g_m = \frac{5.8 - 3.7 \text{ ma}}{0.5 \text{ V}} = 4200\ \mu\text{mhos}$$

The transconductance measured at I_{DSS} is termed g_{mo}. If Eq. 4-1 is differentiated* with respect to V_{GS}, a mathematical expression can be obtained for g_m as

$$g_m = -\frac{2I_{DSS}}{V_P}\left(1 - \frac{V_{GS}}{V_P}\right) \tag{4-4}$$

When V_{GS} is zero, g_m is g_{mo}

$$g_{mo} = -\frac{2I_{DSS}}{V_P} \tag{4-5}$$

And, substituting Eq. 4-5 into Eq. 4-4,

$$g_m = g_{mo}\left(1 - \frac{V_{GS}}{V_P}\right) \tag{4-6}$$

Using the values from the characteristics of the example (Figs. 4-16 and 4-17), by Eq. 4-5,

$$g_{mo} = \frac{-2 \times 8.4 \text{ ma}}{-3 \text{ V}} = 5600\ \mu\text{mhos}$$

and at point B, by Eq. 4-6,

$$g_m = 5600\left(1 - \frac{-0.5}{-3.0}\right) = 4650\ \mu\text{mhos}$$

$$^*I_D = I_{DSS}\left(1 - \frac{V_{GS}}{V_P}\right)^2$$

$$\frac{dI_D}{dV_{GS}} = 2I_{DSS}\left(1 - \frac{V_{GS}}{V_P}\right)\left(-\frac{1}{V_P}\right) = -\frac{2I_{DSS}}{V_P}\left(1 - \frac{V_{GS}}{V_P}\right)$$

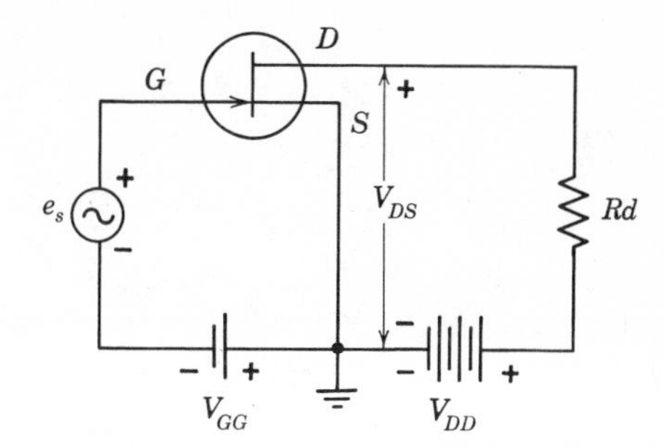

Figure 4-18 Simple JFET amplifier.

A simple amplifier circuit using an N-channel JFET is shown in Fig. 4-18. When e_s has the polarity shown, the sum of e_s and V_{GG} is less negative than V_{GG} alone. The depletion region decreases and I_D increases. The voltage drop $I_D R_d$ increases and the voltage from the drain to the source V_{DS}, which is positive, decreases. When the polarity of e_s reverses, the gate becomes more negative. The drain current decreases and the $I_D R_d$ voltage drop decreases. Consequently, V_{DS} increases in a positive direction. As a result of this action, there is a phase inversion between the input signal at the gate and the output signal at the drain in an FET amplifier. Usually the voltage gain in an FET amplifier is fairly low of the order of 5 to 15.

The advantage of the JFET lies in the fact that I_G is effectively zero. The actual input resistance to the gate is of the order of 10 MΩ.

At this point it is in order to give formal definitions for some of the terms associated with the FET.

BV_{GSS} is the avalanche breakdown voltage from gate to source when the gate is reverse biased. The drain is connected to the source.

BV_{GDS} is the avalanche breakdown voltage from gate to drain when the gate is reverse biased and the source is the common lead.

BV_{DGO} is the avalanche breakdown voltage from drain to gate when the source is open-circuited.

BV_{SGO} is the avalanche breakdown voltage from source to gate when the drain is open-circuited.

BV_{DSS} or BV_{DGS} is the avalanche breakdown voltage when the gate is connected to the source.

I_{GSS} is the leakage current in the gate.

The region to the left of the dashed curve in Fig. 4-16 is replotted to an expanded scale in Fig. 4-19 and the curves are extended into the third quadrant. Equation 4-3 is not valid in this region, and a different equation

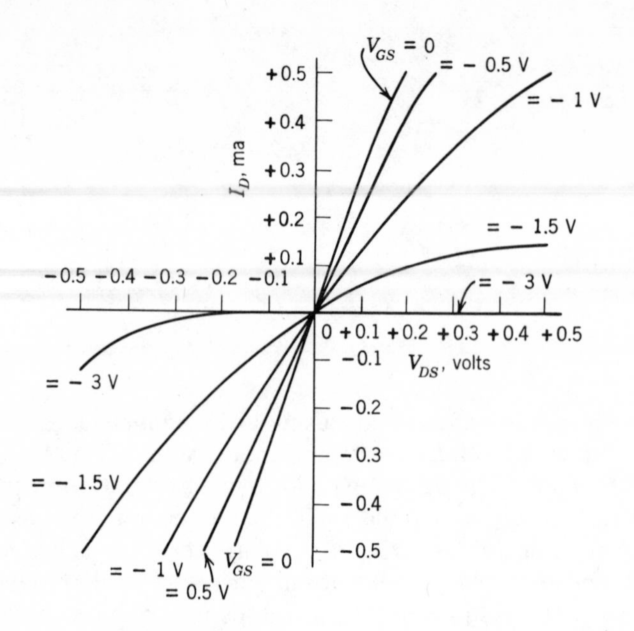

Figure 4-19 Expanded drain characteristic in the region of the origin.

must be used*:

$$I_D = K[2(V_{GS} - V_P)V_{DS} - V_{DS}^2] \tag{4-7}$$

By means of differential calculus†, an equation for the a-c resistance at the origin can be shown to be

$$r_o = \frac{1}{2K(V_{GS} - V_P)} \tag{4-8}$$

r_o is also the reciprocal of the slope of the curves at the origin in Fig. 4-19. The slope of the curve for the gate voltage equal to pinchoff voltage (−3 V) is zero, and the corresponding value for r_o is infinite ohms or an open circuit. The curve for zero gate voltage yields a value for r_o of 350 Ω. In this region,

*For a derivation see Appendix I, E. J. Angelo, *Electronics: BJT's, FET's and Microcircuits,* New York; McGraw-Hill Book Co., 1969.

†The derivative is $\dfrac{dI_D}{dV_{DS}} = K[2(V_{GS} - V_P) - 2V_{DS}]$

At the origin, $V_{DS} = 0$, then

$$\frac{dI_D}{dV_{DS}} = \frac{1}{r_o} = 2K(V_{GS} - V_P)$$

the JFET is useful as a voltage-controlled variable-resistance for many applications.

The construction of the N-channel JFET on an integrated-circuit chip is shown in Fig. 4-20*a*. The *IC* JFET does not use an N^+ substrate because

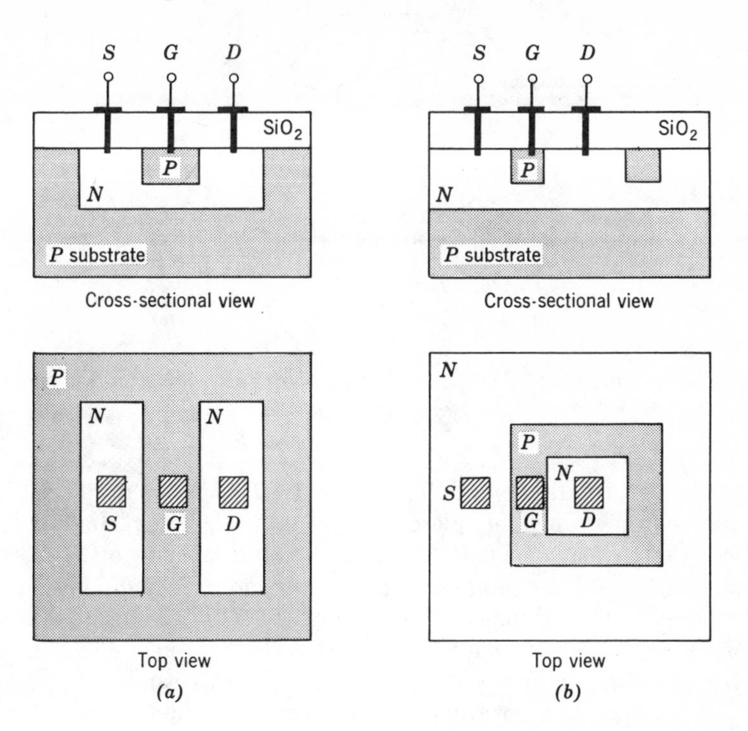

Figure 4-20 FET construction in integrated circuits. (*a*) Linear gate. (*b*) Rectangular gate.

its increased conductivity would defeat the action of a variable depletion region controlled by the gate. The *linear gate* in the JFET shown in Fig. 4-20*a* has a rectangular form. The length of the gate is about 250 μm and its width is of the order of 20 to 30 μm. The depth of the channel under the gate is approximately 1 μm. The *rectangular gate* of the JFET shown in Fig. 4-20*b* surrounds the drain.

When the P substrate is connected to the source, the N channel is "floated" in a reverse-bias depletion region to insulate the JFET electrically from other components in the chip. A disadvantage of the IC JFET is that I_{DSS} and V_P can vary as much as 5 to 1 from wafer to wafer. However, within the same wafer, the spread is much less. The circuit design can be adjusted to compensate for this variation.

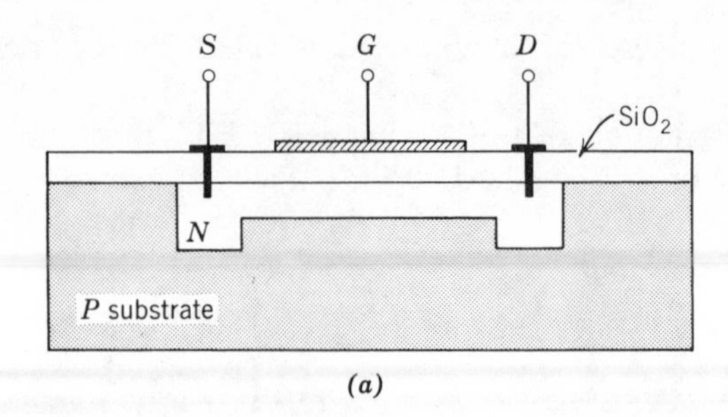

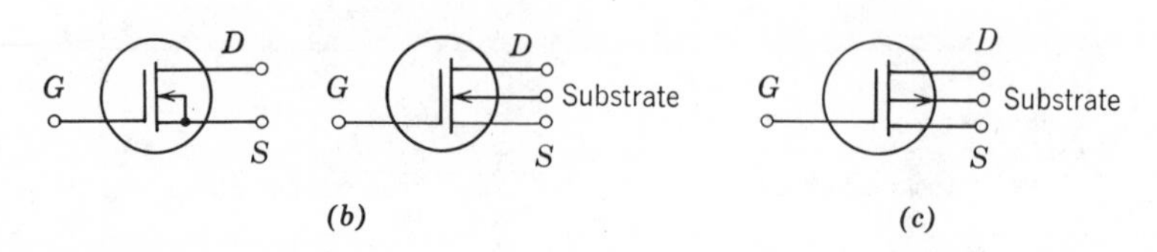

Figure 4-21 Depletion-type MOSFET. (*a*) Cross-sectional view. (*b*) Symbols for *N*-channel MOSFETS. (*c*) Symbol for *P*-channel MOSFET)

The construction of a different type FET is shown in Fig. 4-21. Here the gate is simply a metallic plate that has no *P* or *N* semiconductor property. The gate is insulated from the channel by a layer of SiO_2. This device is called a *depletion-type metal-oxide-semiconductor-field-effect-transistor* or *MOSFET*. Occasionally, the acronym *MOST* is used. An alternative terminology is *depletion-type insulated-gate-field-effect-transistor* or *IGFET*. The symbols used are shown in Fig. 4-21*b* and *c*.

It is imperative not to permit any stray or static voltage to be impressed on the gate or the SiO_2 layer between the gate and the channel will be destroyed. Even picking the MOSFET up can destroy it. Consequently, grounding rings must be used, and they are removed only after the MOSFET is securely wired into the circuit.

The drain characteristic for a typical *N*-channel depletion-type MOSFET is given in Fig. 4-22. The corresponding transfer characteristic is given in Fig. 4-23. The definitions for V_P, I_{DSS}, g_m, and r_d, developed for the JFET, apply to the depletion-type MOSFET without restriction. Similarly, Equations 4-1 to 4-8 are also valid. The equation for I_D for the MOSFET used for the curves of Figs. 4-22 and 4-23 is

$$I_D = 3.1\left(1 - \frac{V_{GS}}{-4.25}\right)^2 \text{ ma}$$

The advantage of the MOSFET lies primarily in the use of the insulated gate. The input resistance to the gate is of the order of 100 MΩ which is the same as the vacuum tube. The leakage current is of the order of 10^{-12} A

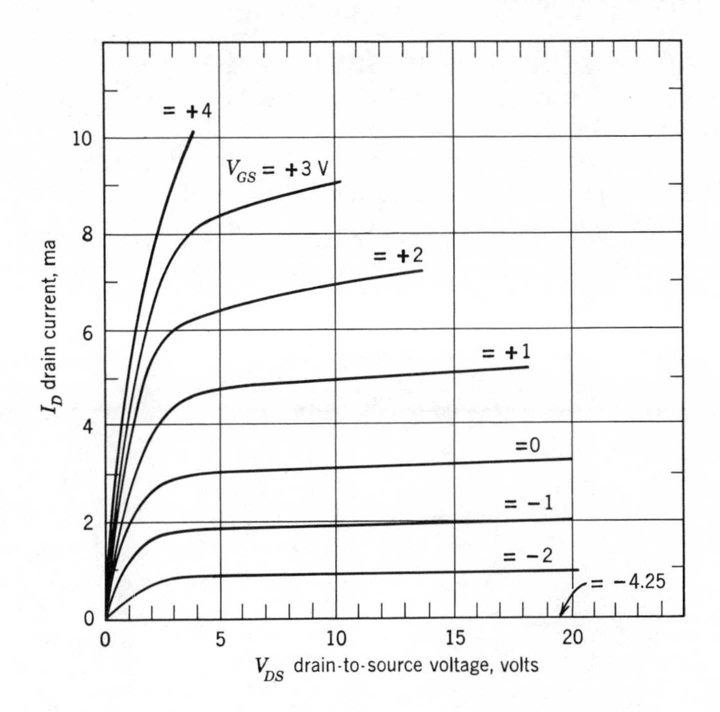

Figure 4-22 Drain characteristic for depletion MOSFET.

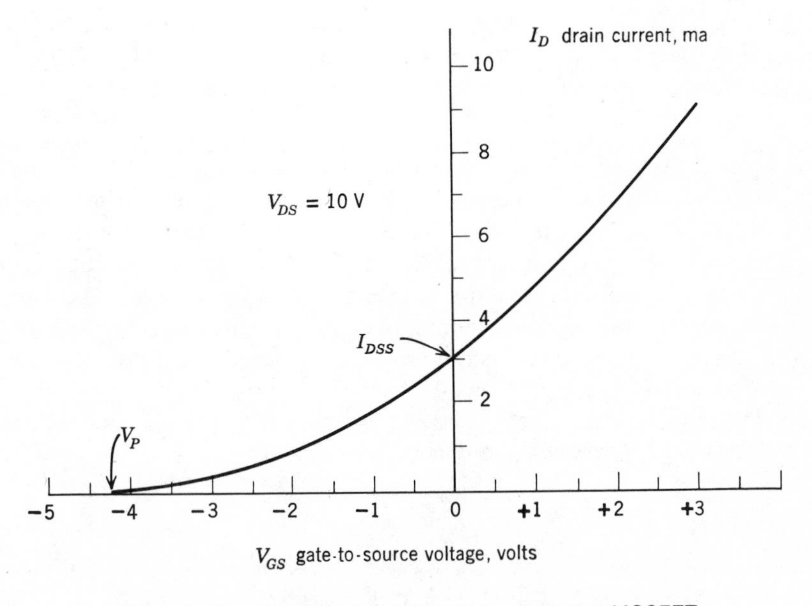

Figure 4-23 Transfer characteristic for depletion MOSFET.

(1pA). The depletion-type MOSFET is usually operated at zero bias ($V_{GG} = 0$), since the gate can normally swing either to a positive voltage or to a negative voltage.

Another integrated-circuit structure is the *enhancement-type MOSFET* shown in Fig. 4-24*a*. When the gate is at a small positive value, a depletion

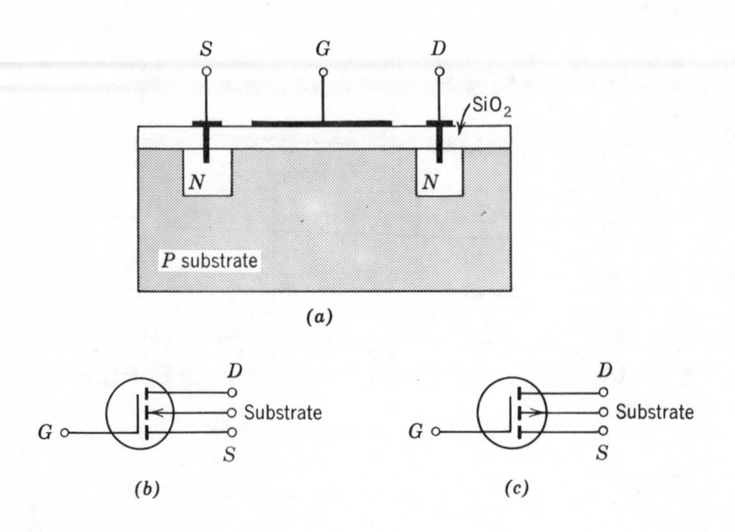

Figure 4-24 Enhancement-type MOSFET. (*a*) Cross-sectional view. (*b*) Symbol for *N*-channel enhancement-type MOSFET. (*c*) Symbol for *P*-channel enhancement-type MOSFET.

layer is formed in the *P* substrate just below the gate. When the positive voltage on the gate is increased, electrons are attracted to the bottom surface of the SiO_2 below the gate. Thus a region of electrons extends from the *N*-type source region to the *N*-type drain region. Consequently, a *virtual* *N*-type channel connects the source and the drain. The least positive voltage that creates this virtual channel is called the *threshold voltage* V_T. Typical drain characteristics are shown in Fig. 4-25 and the corresponding transfer characteristic is given in Fig. 4-26.

Naturally, I_{DSS} and V_P have no meaning for the enhancement-type MOSFET. A new equation* must be used for I_D:

$$I_D = K(V_{GS} - V_T)^2 \tag{4-9}$$

*For a derivation see Appendix I, E. J. Angelo, *Electronics: BJT's, FET's, and Micro-circuits*. New York: McGraw-Hill Book Co., 1969.

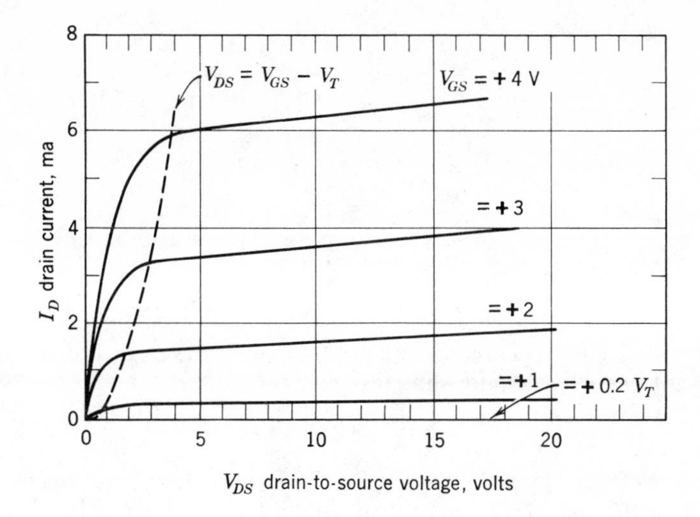

Figure 4-25 Drain characteristic-enhancement MOSFET.

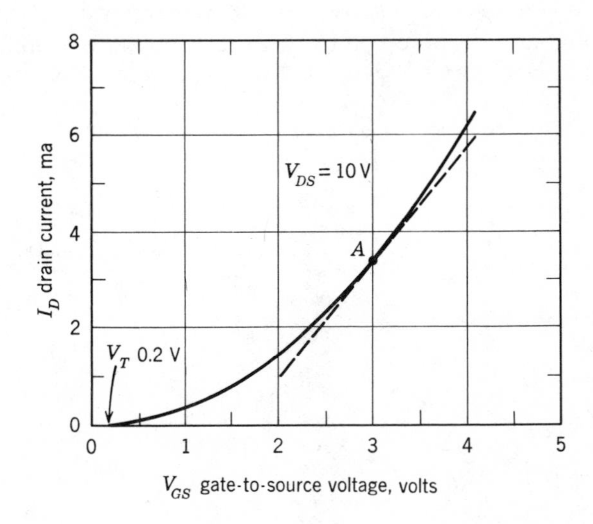

Figure 4-26 Transfer curve for enhancement MOSFET.

The equation for the enhancement-type MOSFET used for Fig. 4-25 and Fig. 4-26 is

$$I_D = 0.445(V_{GS} - 0.2)^2$$

A tangent drawn to the transfer characteristic (Fig. 4-26) at point A gives the value for g_m as

$$g_m = \frac{\Delta I_D}{\Delta V_{GS}} = \frac{6-1 \text{ ma}}{4-2 \text{ V}} = 2.5 \text{ millimhos} = 2500\ \mu\text{mhos}$$

Mathematically, Eq. 4-9 is differentiated to give

$$g_m = 2K(V_{GS} - V_T) \tag{4-10}$$

Substituting values at point A into Eq. 4-7,

$$g_m = 2 \times 0.445(3.0 - 0.2) = 2.5 \text{ millimhos} = 2500\ \mu\text{mhos}$$

The equation* for the drain characteristic curves to the left of the dashed line in Fig. 4-25 is quite similar to Eq. 4-7 as

$$I_D = K[2(V_{GS} - V_T)V_{DS} - V_{DS}^2] \tag{4-11}$$

Another form of the FET is the *N-channel, dual insulated gate MOSFET*, shown in Fig. 14-27*a* as a cross-sectional view of an IC component. The symbols are given in Fig. 14-27*b*. This MOSFET is also available with a *P*-type channel. The drain current is a function of both gate voltages. G1 is usually operated at a negative voltage with respect to the source and G2 is operated at a positive voltage with respect to the source. As an example, if G1 is operated at −1 V and G2 is operated at +4 V, the potential of the

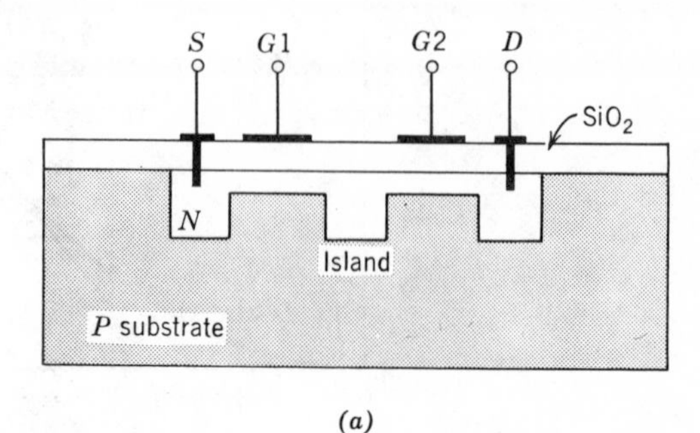

Figure 4-27 *N*-channel, dual insulated gate MOSFET. (*a*) Cross-sectional view. (*b*) Symbols.

*For a derivation see Appendix I, E. J. Angelo, *Electronics: BJT's, FET's, and Microcircuits.* New York: McGraw-Hill Book Co., 1969.

island must be +5 V. The voltage at G2 with respect to the island is −1 V and the voltage at G1 with respect to the source is −1 V. The gate$_2$-to-island voltage must be the same as the gate$_1$-to-source voltage, since I_D is the same in all parts of the channel. These MOSFET's are primarily used in the application of a mixer, Section 25-6.

PROBLEMS

1. The equation for drain current in an N-channel JFET is

$$I_D = 8.4\left(1 - \frac{V_{GS}}{(-3.0)}\right)^2 \text{ ma} \qquad (V_{DS} = 10 \text{ V})$$

Find I_D at each of the following values of V_{GS}:

0, −0.5, −1.0, −1.5, −2.0, and −3.0 V

2. By using the results of Problem 1, plot the drain characteristic and the transfer characteristic.

3. Find g_m at each gate-to-source voltage listed in Problem 1. Plot a curve of g_m against V_{GS}.

4. An N-channel JFET has an I_{DSS} of 10 ma and a pinchoff voltage V_P of −6.0 V. Determine g_{mo} and g_m at $V_{GS} = -2.0$ V.

5. The equation for drain current for an N-channel depletion-type MOSFET is

$$I_D = 3.1\left(1 - \frac{V_{GS}}{(-4.25)}\right)^2 \text{ ma} \qquad (V_{DS} = 10 \text{ V})$$

Find I_D at each of the following values of V_{GS}:

−2.0, −1.0, 0, +1.0, +2.0, and +3.0 V

6. By using the results of Problem 5, plot the drain characteristic and the transfer characteristic.

7. Find g_m at each gate-to-source voltage listed in Problem 6. Plot a curve of g_m against V_{GS}.

8. The equation for the drain current in an N-channel enhancement-type MOSFET is

$$I_D = 0.445\,(V_{GS} - 0.2)^2 \text{ ma} \qquad (V_{DS} = 10 \text{ V})$$

Find I_D at each of the following values of V_{GS}:

1.0, 2.0, 3.0, and 4.0 V

9. Find g_m at each gate-to-source voltage listed in Problem 8. Plot a curve of g_m against V_{GS}.

Section 4-4 Intrinsic Regions

When a *PNP* (or *NPN*) transistor is formed, the effect of the formation of barrier junction potentials *depletes* the adjacent region to the barrier of

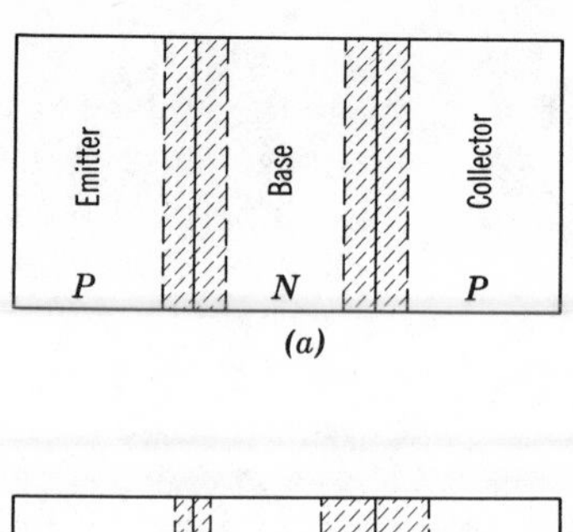

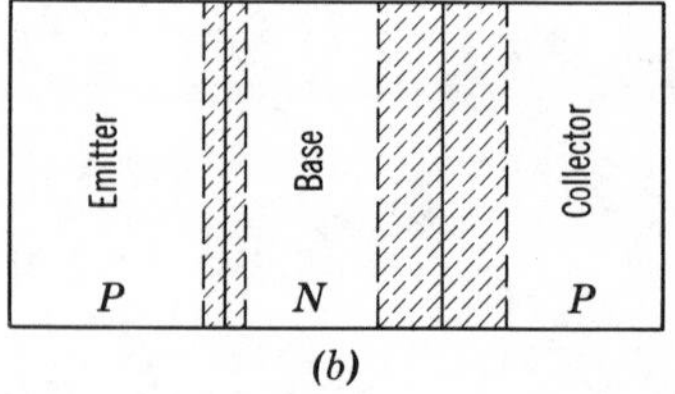

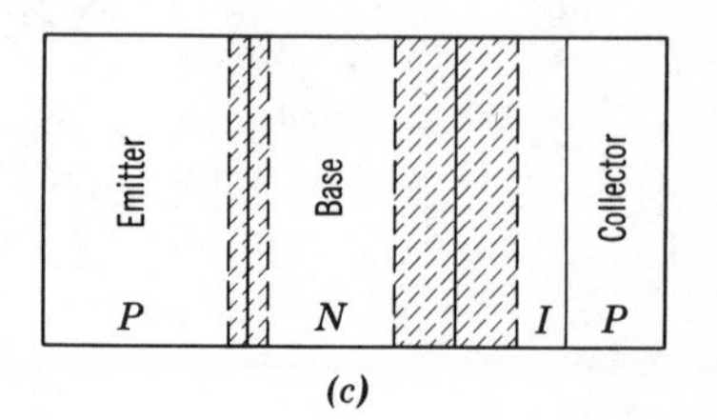

Figure 4-28 Effect of depletion regions. (*a*) Without electrode voltages. (*b*) With normal electrode biasing. (*c*) The *PNIP* transistor.

majority current carriers (Fig. 4-28*a*). A forward bias applied to the emitter decreases its region of depletion and a reverse bias on the collector increases the depletion region (Fig. 4-28*b*). The large depletion region around its base-to-collector barrier extends into the collector and increases the distance that the majority current carriers must travel in order to be effective. This increased distance produces a greater transit time and consequently reduces the upper operating-frequency limit. This limitation is characteristic of the junction transistor.

In order to gain an increased frequency response, an intrinsic region *I* is added between the base and the collector (Fig. 4-28*c*). The transit time of current carriers in the intrinsic material is much shorter than in either *N* or *P* material because there are no retarding forces created by donor or acceptor atoms. These transistors are known as *PNIP* (*or NPIN*) units.

Figure 4-29 shows the symbols used to indicate intrinsic materials in semiconductors. A short slant line (*a*) indicates a transition of material in the base from *N* to *P* or *P* to *N*. The small parallelogram (*b*) defines an intrinsic material *I* placed between two *dissimilar* doped material forming a *PIN* or

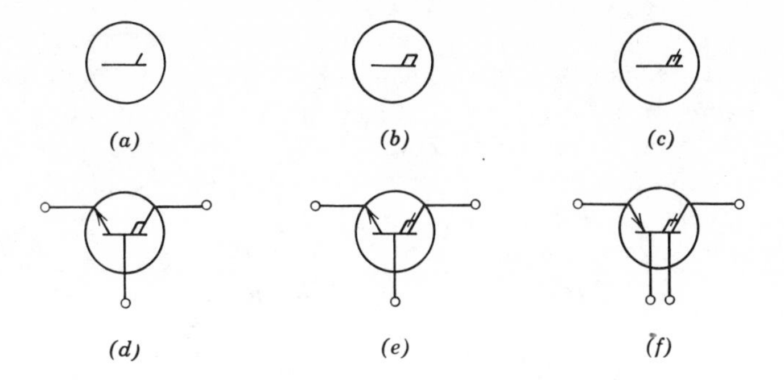

Figure 4-29 Symbols used to indicate intrinsic material. (*a*) *PN* or *NP* base. (*b*) *NIP* or *PIN*. (*c*) *PIP* or *NIN*. (*d*) *NPIN*. (*e*) *NPIP*. (*f*) *PNIN*.

an *NIP* combination. The addition of a small line (*c*) designates an intrinsic layer between two *similar* materials, *NIN* or *PIP*. The direction of the arrow on the emitter in (*d*) indicates that the sequence must be, in this case, *NPIN*. The symbol in (*e*) shows an *NPIP* transistor. When a fourth lead is brought out of the transistor, a direct or *ohmic* connection is made to the intrinsic material. Thus, *PNIN* with ohmic connection to intrinsic region describes (*f*).

The *drift transistor* is a special junction transistor which is equivalent to the *PNIP* transistor in operation. In a *PNP* transistor the base of *N* material is specially processed. At the junction between the emitter and base, the *N* material has a high conductivity, and it gradually changes until it is almost intrinsic material at the base-to-collector junction. The change in voltage gradient across this nonuniform base causes the current carriers to travel faster through the base than they do across a uniform base and thus reduce transit time. The drift transistor is also manufactured in the *NPN* form.

When the emitter and the collector electrodes are manufactured by the process used for the surface-barrier transistor, the resulting transistor is called a *microetched diffused transistor*.

Section 4-5 The Tunnel Diode

In 1958, J. Esaki developed the tunnel or Esaki diode by increasing the amount of doping in a diode from one part in 10^8 to one part in 10^3. The effect of this increasing of donor and acceptor content is to reduce the junction width in the diode from about 5 μm to 0.01 μm or 100Å, which is much shorter than the wavelength of visible light.

The exact mechanism of the tunnel diode is explained by the theory of quantum mechanics, but a concept of its operation can be derived from a much simpler but inexact consideration. When the depletion layer is as thin as the barrier in the tunnel diode, carriers can, at low potentials, break or "tunnel" through the layer instead of having to take the normal energy paths in a conventional *P-N* junction.

The normal energy-level diagram for a *P-N* unbiased junction is shown in Fig. 4-30*a*. The increased doping in the Esaki diode produces the un-

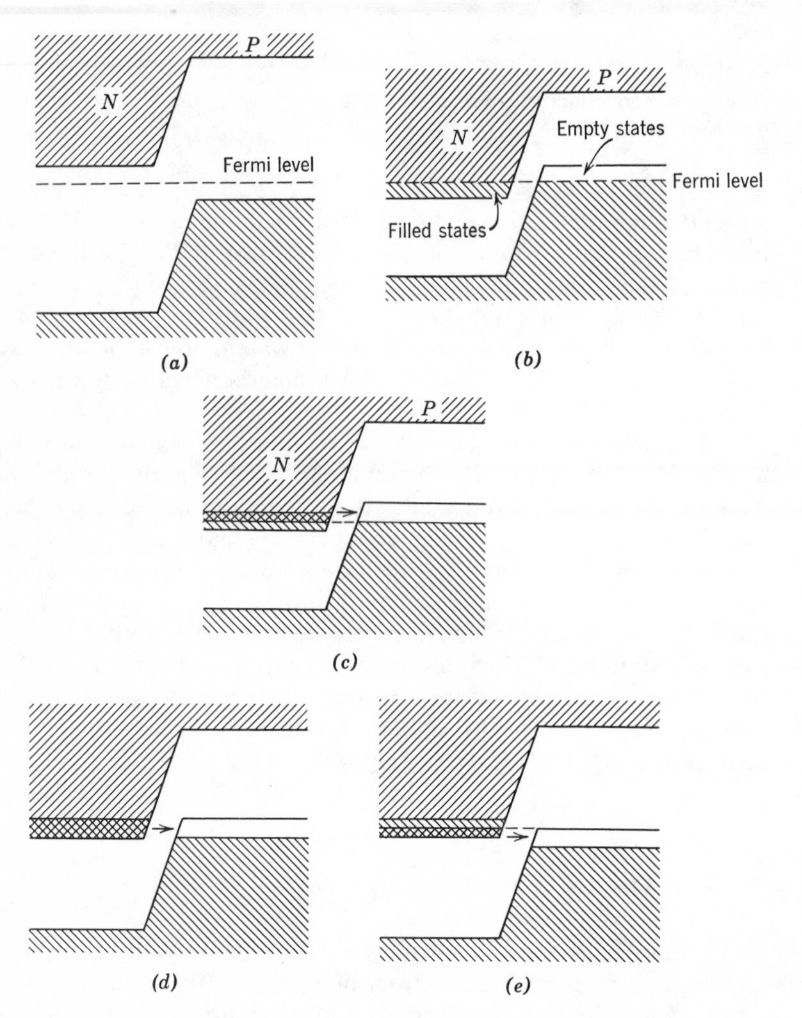

Figure 4-30 The energy level diagrams for tunnel diode. (*a*) Normal unbiased diode. (*b*) Unbiased tunnel diode. (*c*), (*d*), and (*e*) Forward-biased tunnel diode. (Tunneling shown by arrow and dark areas.)

biased energy-level diagram of Fig. 4-30*b*. The fact that the Fermi level cuts through the conduction band of the *N* material and through the valence band of the *P* material causes a shift in electrons (Fig. 4-30*b*) that does not exist in the normal diode. A small increase in forward potential (Fig. 4-30*c*) causes a shift in the relation of the filled states and the empty state with a resultant shift of electrons from left to right to provide a current flow. At this condition, the current on the volt-ampere characteristic is point *c* (Fig. 4-31*b*). Essentially this means that electrons penetrate or *tunnel through* the barrier instead of "climbing over" the barrier as in normal diode current flow. When the forward voltage is increased further. Fig. 4-30*d*, there is a maximum range of possible tunneling and the tunneling current increases, thus, locating the *peak point, d,* on Fig. 4-31*b*. A further increase in forward voltage (Fig. 4-30*e*) decreases the overlap, and tunneling current decreases.

This condition is point *e* (Fig. 4-31*b*). A further increase in forward voltage separates the two bands completely, as in Fig. 4-30*a*, and we now have the action of a normal semiconductor diode. This is the dashed curve of the normal diode in Fig. 4-31*b*.

The *SUM* of the tunneling current and the normal diode current gives the composite characteristic of Fig. 4-31*c*. Typical numerical values are given on the curve. An ordinary diode characteristic drawn to the same scale follows the dashed curve to the valley point and, thence, along the solid curve to the forward point. The tunneling action that creates the peak point is clearly indicated. Because of normal diode action, the valley point (V_V, I_V) has a finite current value. Accordingly, an important tunnel diode definition is the *peak-to-valley ratio*, I_P/I_V, which for this case is 2.0/0.25 or 8.

The transition from the ideal to the actual curve causes a flattened effect termed the *excess current*. The negative resistance of the Esaki diode is specified by determining the slope of a tangent to the curve drawn at the *inflection point* on the curve (V_I, I_I) as a conductance value instead of a resistance value:

$$g_d = \frac{\Delta I}{\Delta V} \tag{4-12}$$

The numerical value from the representative data used for this characteristic is $(-2.0 \text{ ma})/(20 \text{ mv}) = -0.1$ mho.

Section 4-6 The Varactor Diode

When a *P-N* junction is formed, one effect of the formation of barrier junction potential is the depletion of the adjacent region at the barrier of majority current carriers. When forward bias is applied to the diode, the

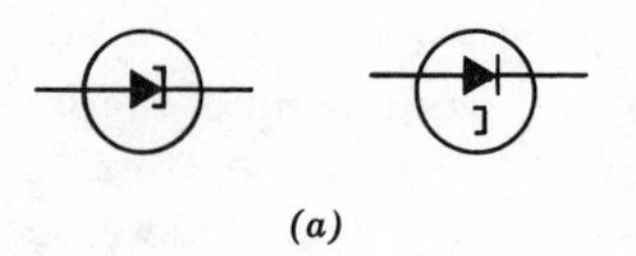
(a)

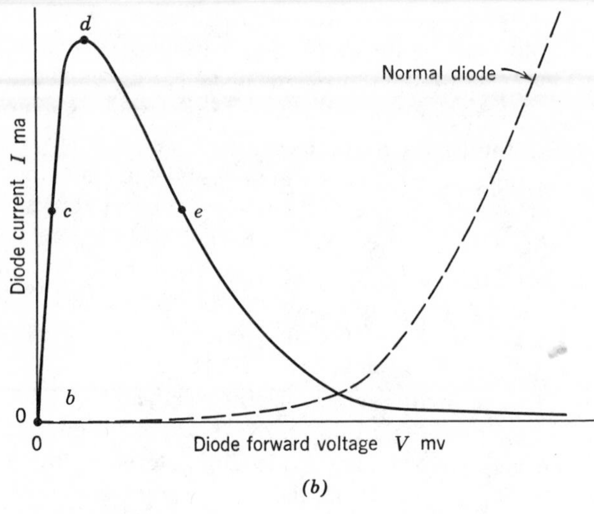

(b)

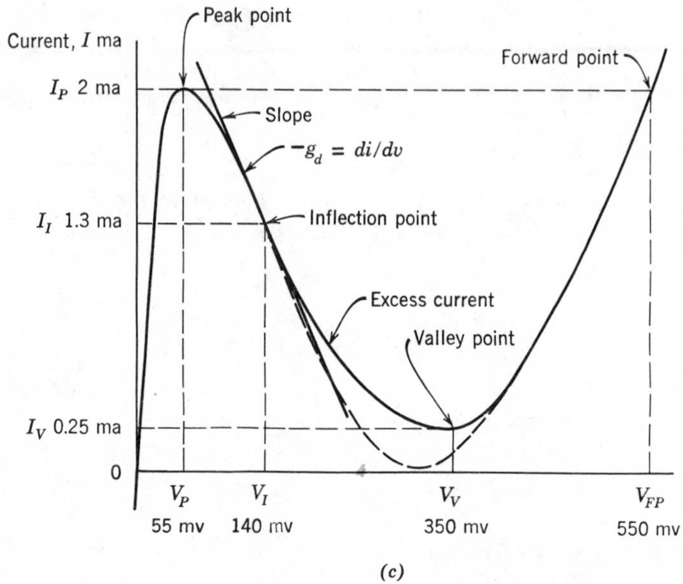

(c)

Figure 4-31 Tunnel diode characteristics. (a) Symbols. (b) Tunnel characteristic. (c) Composite characteristic.

depletion region narrows and, when a reverse bias is applied, the depletion region widens. The depletion region acts as the dielectric of a capacitor, and the regions of the P material and of the N material that are beyond the depletion region serve as the plates of the capacitor. Since the depletion region thickness is a function of voltage, the reverse-biased diode may serve as a variable capacitor. Diodes that are especially constructed for this use are called *variable-voltage capacitors* diodes or *varactor* diodes.

The characteristic and symbols for the varactor diode are shown in Fig. 4-32. As with Zener diodes, these diodes are normally operated with a

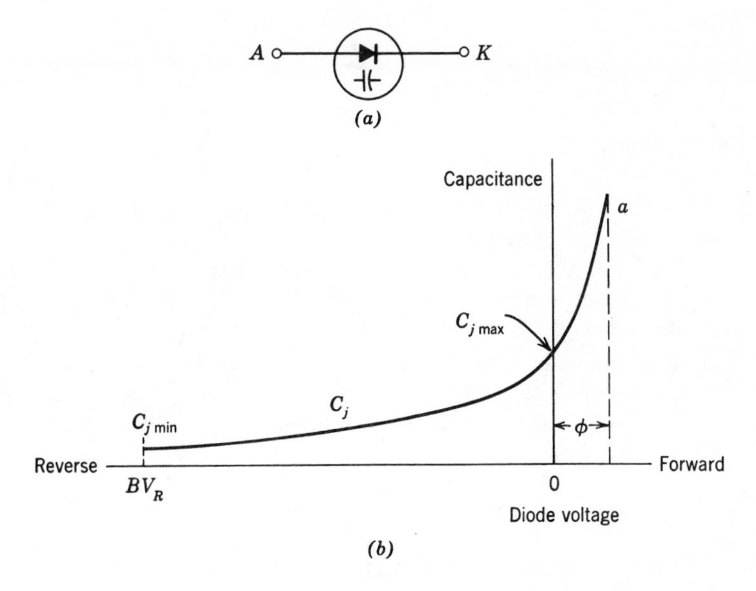

Figure 4-32 The varactor diode. (*a*) Symbol. (*b*) Characteristic.

reverse bias. Only a very small voltage, ϕ, of the order of 0.4 to 1.0 V is required in the forward direction to overcome contact potential. When normal diode current starts to flow at a, the capacitance effect is lost. Therefore, the normal operation of the varactor is contained between C_{jmax} (or C_{jo}), the value of the capacitance at zero voltage and C_{jmin}, the value of the capacitance at the breakdown voltage, BV_R. Any point on the curve, the capacitance, is C_j. Usually, the rating of the capacitance is given for some particular reverse voltage of the order of 4 to 6 V.

The capacitance of the varactor diode is given approximately by

$$C_j = \frac{C_0}{\sqrt[n]{V + V_R}} \text{ pF} \tag{4-13}$$

Where V is the forward barrier potential (0.3 V for germanium and 0.7 V for silicon), V_R is the *magnitude* of the applied reverse voltage, C_0 is a reference value of capacitance in picofarads, and n is a constant (approximately 3 for grown or graded junctions and 2 for alloy or abrupt junctions). The capacitance decreases as the reverse voltage is increased (Fig. 4-32). A typical varactor diode has the following specifications:

C_j at 4 V	47 pF
$C_{j\max}$	120 pF
$C_{j\min}$	14 pF
BV_R	45 V
I_R near BV_R	0.5 μa

PROBLEMS

1. If n is 2 for a silicon varactor diode and if C_0 is 140 pF, find $C_{j\,\max}$, C_j at 4 V, and $C_{j\,\min}$ near BV_R (40 V).
2. If n is 3 for a germanium semiconductor used as a varactor diode, find C_0 when a measured value for C_j yields 60 pF at 7 V. Find $C_{j\,\max}$ and $C_{j\,\min}$ if the breakdown voltage is 30 V.
3. A varactor diode has a value of 3 for n and a value of 140 pF for C_0. The reference value of capacitance in a circuit is taken when the diode voltage is 10 V. What voltages are required to establish twice this capacitance and half this capacitance?

Section 4-7 The Unijunction Transistor

The *double-based diode* or *unijunction transistor* (Fig. 4-33) has a small rod of *P* material extending into the block of *N* material which serves as a *P-N*

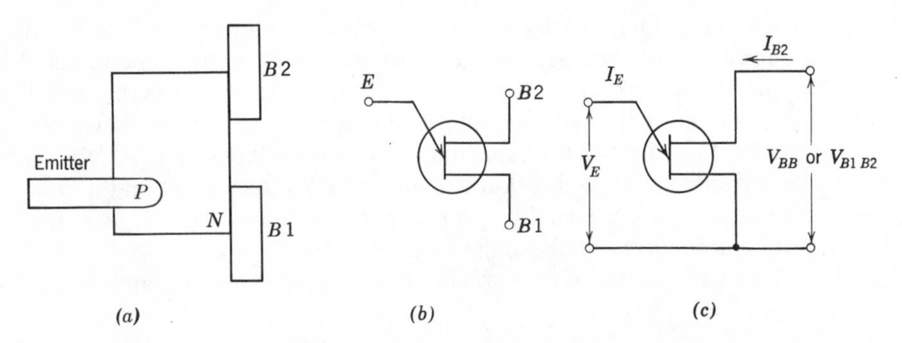

Figure 4-33 The unijunction transistor. (*a*) Construction. (*b*) Symbol. (*c*) Nomenclature.

junction. In one type of construction, two metallic contacts called bases are welded to the N block without creating new junctions. The electrode $B1$ is the common return for the circuit. When a positive voltage is applied to $B2$, there is a uniform potential drop to ground through the N material which has a linear *interbase* resistance of several thousand ohms, measured when the emitter is open-circuited. By a voltage-divider action, the emitter E is located at a point in this potential drop which is ηV_{BB}. The coefficient η is called the *intrinsic standoff ratio*. Typical values of η lie between 0.50 and 0.80.

When the emitter voltage V_E is less than $(\eta V_{BB} + V_D)$, the emitter-to-bar junction is reverse biased. The small contact potential V_D is of the order of 0.70 V for a silicon unit. When V_E is greater than $(\eta V_{BB} + V_D)$, the junction is forward biased. When the junction is reverse biased, the emitter current I_E is negligible. As soon as the junction becomes forward biased, the emitter current becomes large. At this point, holes are injected into the N base material. These holes reduce the resistance of the N material to current flow in the double base circuit. The presence of both holes and electrons in the section of the base between the emitter to ground sharply reduces the voltage drop across this section of the base block. Accordingly, the emitter current rises sharply, showing a negative resistance on the idealized characteristic (Fig. 4-34*a*). The point at which the diode *switches* or *fires* is

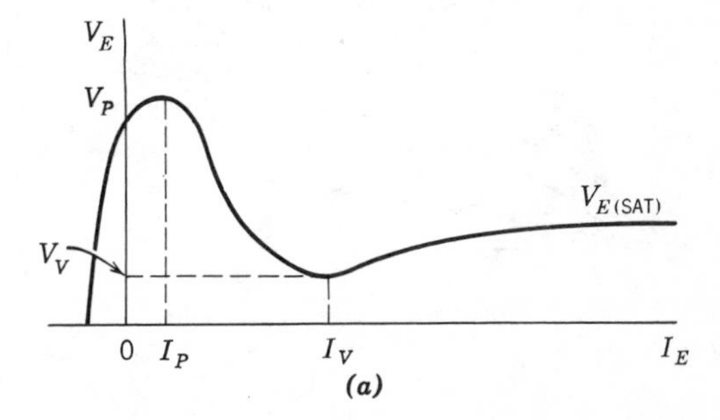

Figure 4-34 Unijunction characteristics. (*a*) Idealized characteristic.

called the *peak point* designated by I_P and V_P. At higher emitter currents, the emitter to base-one voltage drop, V_{EB1}, increases and levels off to the saturation value given by $E_{E(SAT)}$ or $V_{EB1(SAT)}$. The least value on the curve is called the *valley point* (V_V and I_V).

The characteristics of an actual transistor show very sharply defined peak points (Fig. 4-34*b*, see p. 122). The interbase characteristic (Fig. 4-34*c*, see p. 122) shows that the resistance of the interbase is linear only when the emitter current is zero (emitter open).

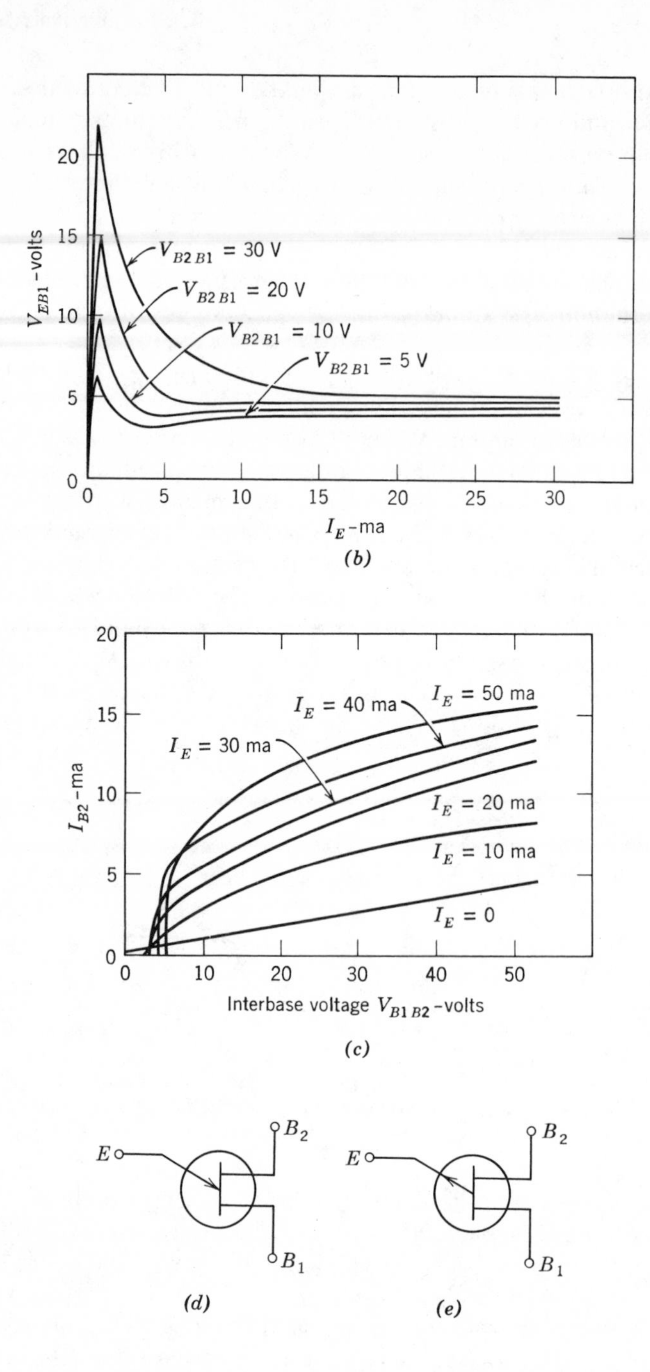

Figure 4-34 (*b*) Emitter characteristic. (*c*) Interbase characteristic. (*d*) Symbol for unijunction transistor with *N*-type base. (*e*) Symbol for unijunction transistor with *P*-type base.

Although this discussion concerns a unijunction transistor with a *P*-type emitter and an *N*-type interbase, they are also commercially available with an *N*-type emitter and a *P*-type interbase (Fig. 4-34*e*). It should be noted that, in accordance with the standardization of semiconductor symbols, the two lines coming into the base line at right angles indicate ohmic connections to the base that do not form junctions. The fact that two separate connections are shown means that the ohmic connections are placed physically at two different points on the base.

Section 4-8 Thyristor Concepts

The *thyristor* is a member of the family of semiconductor devices that have two stable states of operation: one stable state has very low and often negligible current, and the other stable state has a very high current that is usually limited only by the resistance of the external circuit.

The fundamental thyristor is the *NPNP* semiconductor, (Fig. 4-35*a*).

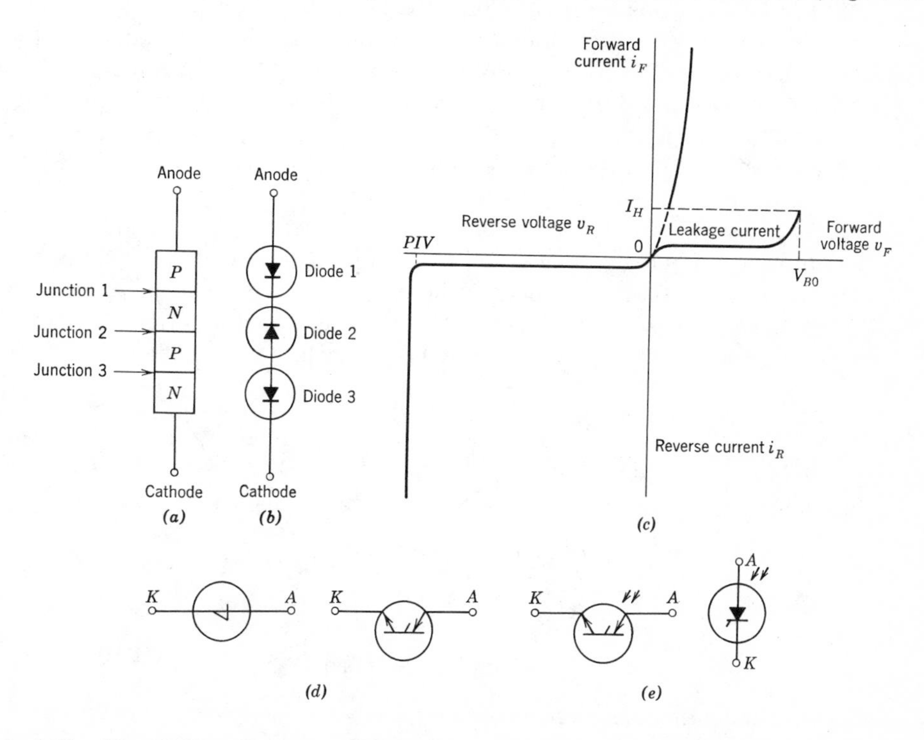

Figure 4-35 The *NPNP* transistor. (*a*) Construction. (*b*) Diode model. (*c*) Breakdown characteristics. (*d*) Graphic symbols. (*e*) Graphic symbols for the light-activated switch.

There are three junctions each of which forms an equivalent diode (Fig. 4-35*b*). When the anode is positive, the *NPNP* semiconductor is termed as being biased in the forward direction. Now diodes 1 and 3 are forward biased and diode 2 is reverse biased. When the *NPNP* is reverse biased (the cathode positive), diodes 1 and 3 block.

This four-layer diode, often referred to as a *Shockley diode*, is formally described as a *reverse blocking diode thyristor*. The characteristic (Fig. 4-35*c*) shows that an avalanche breakdown occurs in the forward direction at V_{BO}, the *break-over voltage*. After breakdown the diode voltage drop falls to a very small value and can permit a large current flow. By having two blocking diodes in series in the reverse direction, the peak-inverse voltage rating is usually much larger in magnitude than V_{BO}. When the reverse voltage exceeds the peak-inverse voltage, a Zener breakdown occurs. Commercial Shockley diodes are available to ratings of 1200 V and 300 A peak.

A modification of the Shockley diode is the *light-activated switch (LAS)*. Light entering an optical window breaks down covalent bonds within the diode allowing break-over voltage to become a function of the intensity of the incident illumination. The symbol for the LAS is shown in Fig. 4-35*e*. The LAS is available in ratings to 200 V and 0.5 A.

The *NPNP* semiconductor can be considered to be the equivalent of two transistors, an *NPN* and a *PNP* in a sort of parallel arrangement (Fig. 4-36). In the forward direction, the two transistors have normal operating bias conditions. In the reverse direction, this is not true. Therefore, this discussion concerns the forward breakdown characteristic and not the reverse region of the properties shown in Fig. 4-35*c*.

From the circuit of Fig. 4-36*c*, it is evident that the anode or cathode current I is the sum of I_{C1} and I_{C2}. Thus, to determine an equation for I in terms of transistor parameters, the expressions for I_{C1} and I_{C2} are required. Both I_{C1} and I_{C2} are collector currents. The collector current of a transistor is the amplified leakage current by Eq. 3-3*c* plus the amplified base current.

$$I = I_{C1} + I_{C2}$$

but $I_{C1} = I_{B2}$ and $I_{C2} = I_{B1}$
Then

$$I_{C1} = (1+\beta_1)I_{CBO1} + \beta_1 I_{B1}$$

By substituting I_{C2} for I_{B1}

$$I_{C1} = (1+\beta_1)I_{CBO1} + \beta_1 I_{C2}$$

In a similar manner, we can obtain I_{C2} as

$$I_{C2} = (1+\beta_2)I_{CBO2} + \beta_2 I_{C1}$$

Cathode $\xleftarrow{I}$ N P N P $\xleftarrow{I}$ Anode

(a)

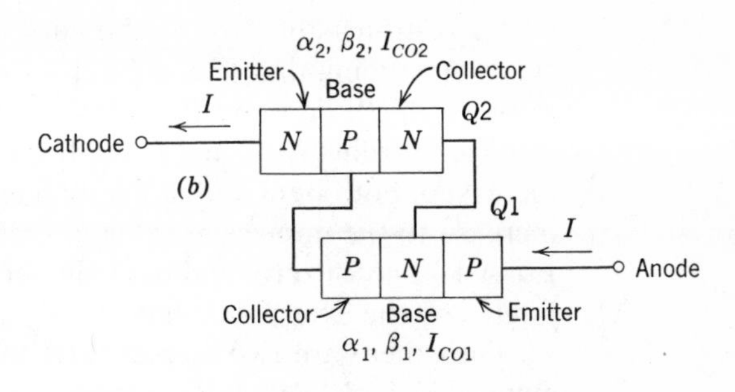

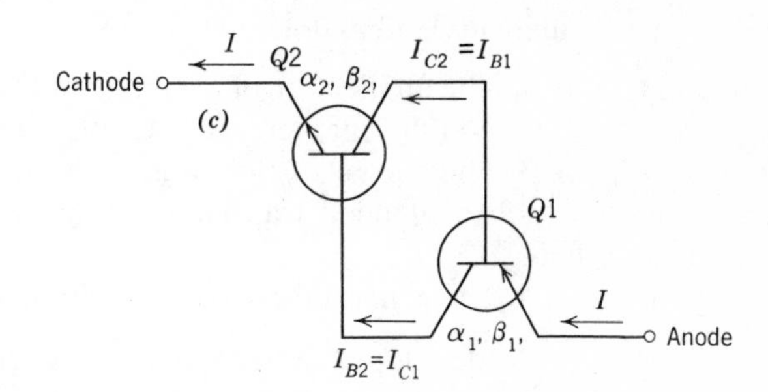

Figure 4-36 The *NPNP* semiconductor as the equivalent of an *NPN* transistor and a *PNP* transistor. (*a*) *NPNP* diode. (*b*) Division into two transistors. (*c*) Circuit.

These two equations can be rearranged to

$$I_{C1} - \beta_1 I_{C2} = (1+\beta_1) I_{CBO1}$$

and

$$-\beta_2 I_{C1} + I_{C2} = (1+\beta_2) I_{CBO2}$$

These equations are solved for I_{C1} and I_{C2} and then are added to yield the external current, I. After simplifying the algebra, the external current becomes

$$I = \frac{(1+\beta_1)(1+\beta_2)(I_{CBO1}+I_{CBO2})}{1-\beta_1\beta_2} \qquad \text{(4-14}a\text{)}$$

The beta current gains can be converted to alpha current gains by substituting:

$$\beta_1 = \frac{\alpha_1}{1-\alpha_1} \qquad \text{and} \qquad \beta_2 = \frac{\alpha_2}{1-\alpha_2} \qquad \text{(3-4}a\text{)}$$

An expression for external current is obtained in terms of alpha current gain:

$$I = \frac{I_{CBO1} + I_{CBO2}}{1 - (\alpha_1 + \alpha_2)} \tag{4-14b}$$

One of the advantages of silicon devices is that the leakage current is very low. Reference to Fig. 3-8*a* and Fig. 3-11*a* shows that for very low values of transistor current, the numerical values of alpha and beta are very small. Assuming that they are zero, we find that the total external current, I, in the *NPNP* diode by Eqs. 4-14*a* and 4-14*b* is effectively the sum of the leakage currents I_{CBO1} and I_{CBO2}. If the current in the diode for any reason is increased, both beta and alpha increase. As soon as either $\beta_1\beta_2$ or $(\alpha_1 + \alpha_2)$ increase to the numerical value of unity, the denominators of Eq. 4-14*a* and Eq. 4-14*b* go to zero, and the value of I goes to infinity. Actually, I must be limited by the external circuit.

Thus there are two stable states in the *NPNP* diode, ON or OFF. In this manner the *NPNP* diode serves as a *solid-state* switch. Any method of creating an increase in current within the layers of the diode will initiate this cumulative breakdown:

1. The application of a voltage sufficiently high to cause breakdown.
2. A sufficient increase in temperature that breaks covalent bonds.
3. The release of electrons by the action of incident light.
4. An induced transistor action by the creation of a forward transistor bias.
5. The generation of current within the diode by capacitive action.

As developed in Section 4-6, a reverse bias on a junction has the effect of a capacitor by depletion. When the voltage across a capacitor is changed, there is a current flow given by fundamental relation:

$$i = C\frac{dv}{dt} \tag{4-15}$$

Thus a sufficient voltage change across the diode can *trigger* a large current flow. This can be a useful method to obtain a breakdown, or it may be necessary that a particular circuit be protected against unwanted voltage transients.

After a thyristor breaks down, the device can be restored to the OFF state by any one or by the combination of several methods.

1. Remove the external source voltage.
2. Reduce the external source voltage to the point where the current falls below the *holding current* value, I_H (Fig. 4-35*c*).
3. Reverse the polarity of the externally applied voltage as in an a-c supply.

There is a finite time required for charges to redistribute within the four-layer diode in order to switch from ON to OFF. In most cases this time is in the order of microseconds.

The four layer diode can be made bidirectional by changing its structure (Fig. 4-37*a*). When Anode No. 1 is positive, the path is *P*1-*N*2-*P*2-*N*3. When

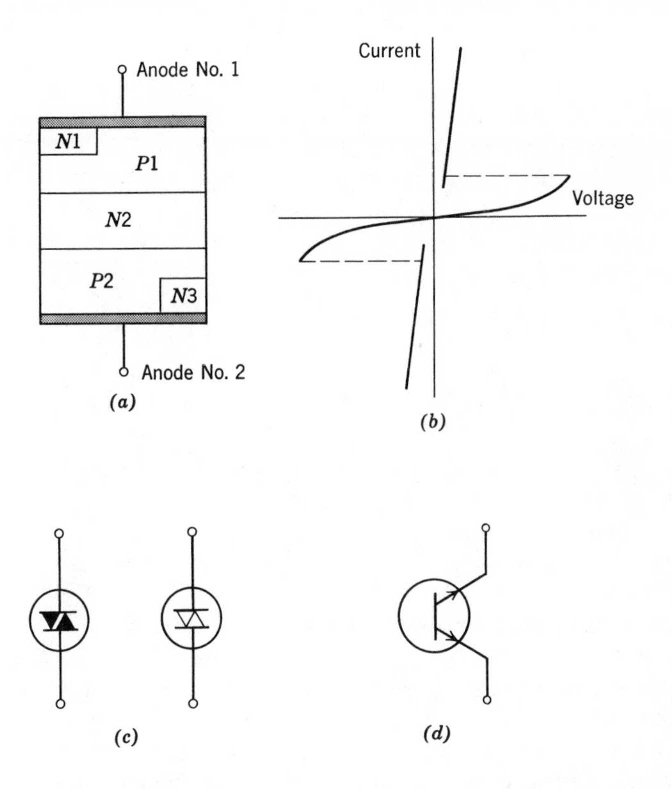

Figure 4-37 Bidirectional thyristors. (*a*) Layer structure. (*b*) Characteristic. (*c*) DIAC symbol. (*d*) Trigger DIAC symbol.

Anode No. 2 is positive, the path is *P*2-*N*2-*P*1-*N*1. The characteristic of the diode, then, is both in the first quadrant and in the third quadrant, yielding symmetrical properties in both forward and reverse directions (Fig. 4-37*b*). This device is termed a *bidirectional dipole thyristor* and is given the acronym, *DIAC*. The symbol is given in Fig. 4-37*c*. A very similar device is the *trigger diac* shown in Fig. 4-37*d*.

PROBLEMS

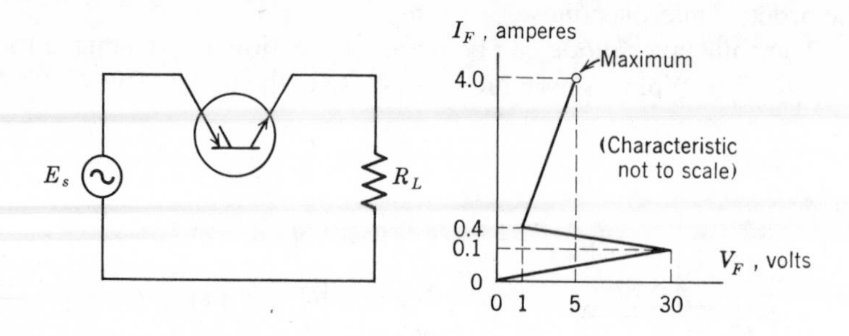

Circuit and characteristic for Problems 1 and 2.

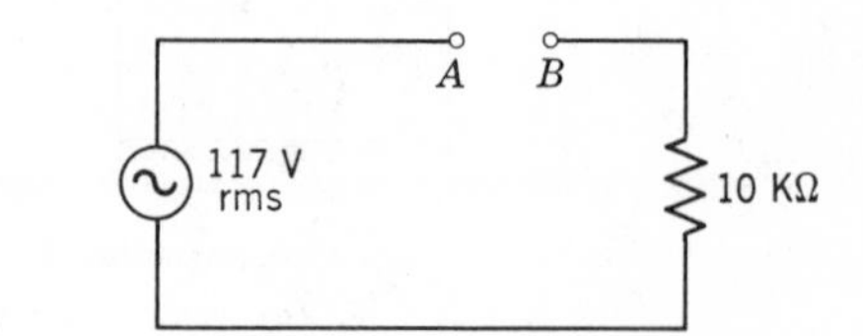

Circuit for Problems 3 to 5.

1. Assume for the purpose of calculation that the characteristic is formed of straight-line segments. The supply voltage is 117 V rms. Determine R_L to limit the current to the maximum value. Determine the value to which e_s must be reduced to turn the diode off. What is the on-state forward resistance of the diode?
2. If R_L is 25 Ω, what is the maximum allowable value of e_s?
3. A Shockley diode is connected between terminals A and B. The diode has a value for V_{BO} of 30 V and for BV_R of 200 V. Sketch the voltage waveform across the load.
4. A light activated switch is connected between terminals A and B. Sketch the voltage waveform across the load for different conditions of incident light.
5. A DIAC is connected between terminals A and B. The DIAC has a value for V_{BO} of 30 V and for BV_R of 200 V. Sketch the voltage waveform across the load.

Section 4-9 The Silicon Controlled Rectifier

The *silicon controlled rectifier (SCR)* is a modification of the Shockley diode. A *gate* connection is formed to the lower *P* layer of the *NPNP* structure (Fig. 4-38*a*). When the gate is forward biased with respect to the

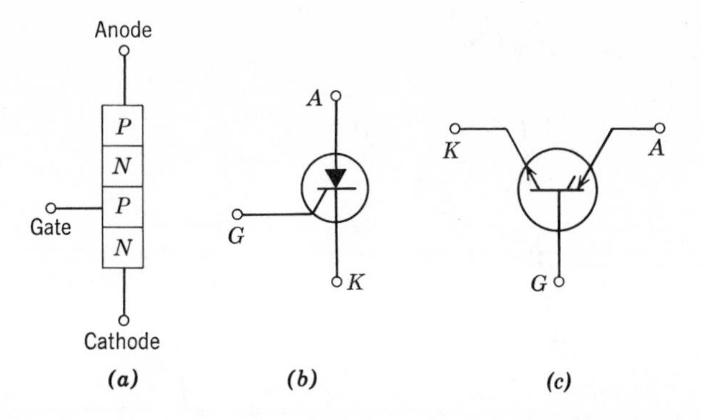

Figure 4-38 The silicon controlled rectifier. (*a*) Construction. (*b*) Symbol. (*c*) Alternative symbol.

cathode, the *PN* junction is biased in the forward direction, and there is a current flow that creates a breakdown between anode and cathode, as explained for the Shockley diode.

When the gate current is zero (Fig. 4-39*a*, see p. 130) the breakdown is that of a Shockley diode. When the gate current is increased (Fig. 4-39*b*, see p. 130) the SCR will *fire* at lower values of forward voltage. In applications, the maximum peak voltage on the anode of the SCR is less than V_{BO}. Accordingly, the SCR will only fire on application of the necessary triggering gate current. In order to trigger the SCR, gate current is required only for a duration of nanoseconds or a few microseconds. When the SCR fires, a minimum anode *holding current*, I_H, is required to maintain the breakdown current. If the minimum current determined by the external circuit is a slightly larger value than I_H, the gate triggering voltage source may be removed or turned off and the SCR will continue to maintain anode current flow. This minimum current is called the *latching current*, I_L.

A circuit giving an application of an SCR that is controlling a load in an automobile circuit is shown in Fig. 4-40, see p. 131. The 600-Ω resistor limits the gate current to 20 ma. A momentary closing of the start contact turns the SCR on. The stop contact bypasses the load current around the anode circuit of the SCR and permits the SCR to regain its blocked or off state. When the stop contact is released, the load current is zero.

The SCR is manufactured in ratings up to several thousand volts and hundreds of amperes average current load. A number of modifications of the SCR are commercially available. One form of the *reverse blocking triode thyristor* has an optical window in addition to the conventional gate contact. Thus, it can be triggered by a combined action of gate current and incident illumination. This device is called a *light-activated SCR (LASCR)* (Fig. 4-41*a*, see p. 131). When a second gate is added to an SCR (Fig. 4-41*b*, see

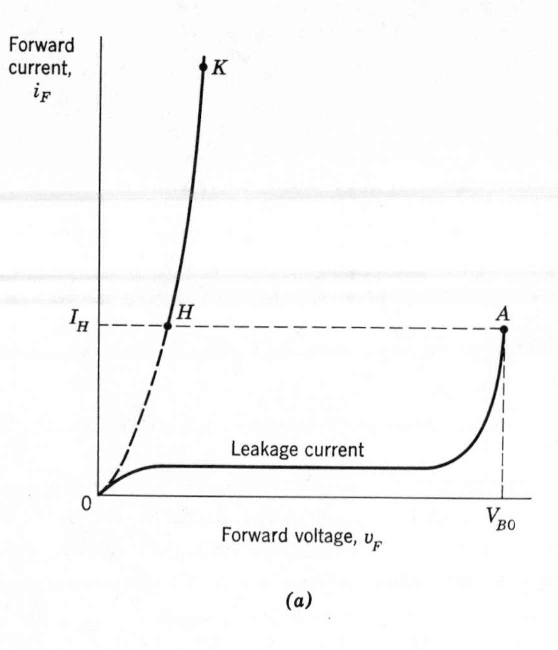

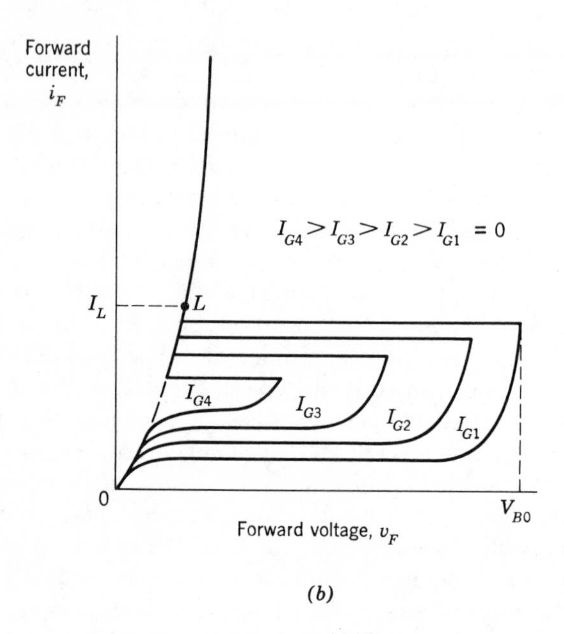

Figure 4-39 SCR characteristics. (*a*) With gate current zero. (*b*) Family of curves.

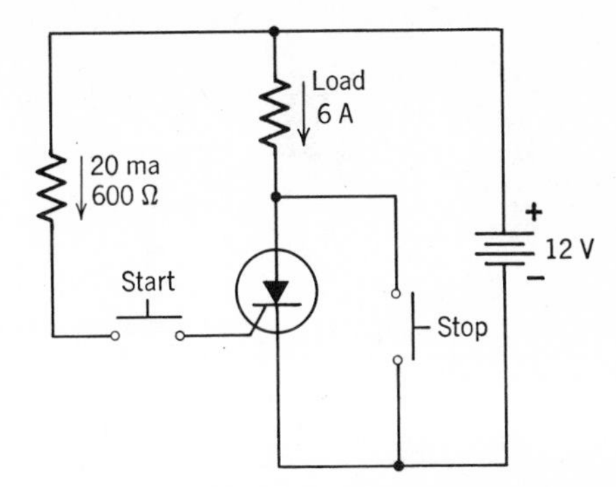

Figure 4-40 SCR application.

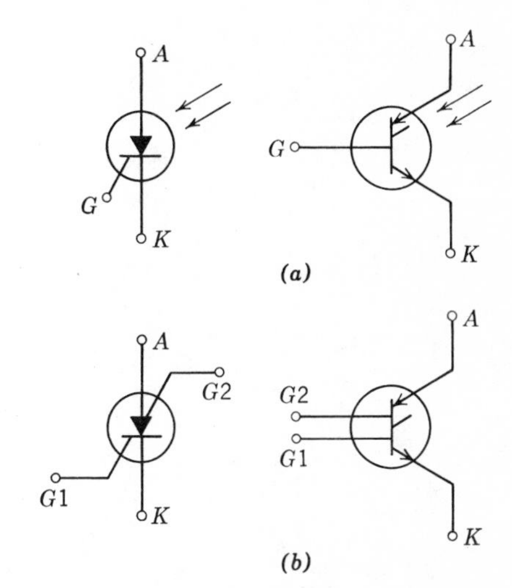

Figure 4-41 Thyristor devices. (*a*) Light-activated SCR (LASCR). (*b*) Tetrode SCR.

p. 131), a *reverse blocking tetrode thyristor* is formed. This device is also known as a *silicon controlled switch (SCS)*. The required triggering pulse on the lower gate is positive. The upper gate is connected to *N* material and, therefore, negative triggering pulses are needed for firing.

Section 4-10 The Triac

The *triac* is a *bidirectional triode thyristor* that has been developed to extend the positive or negative supply of an SCR to allow firing on either polarity with either positive or negative gate current pulses. The cross-

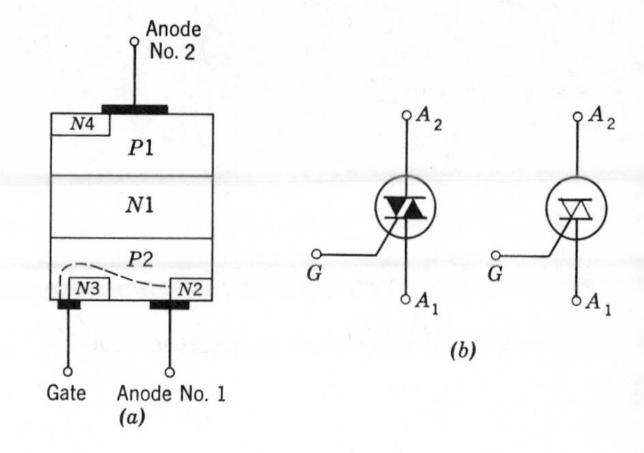

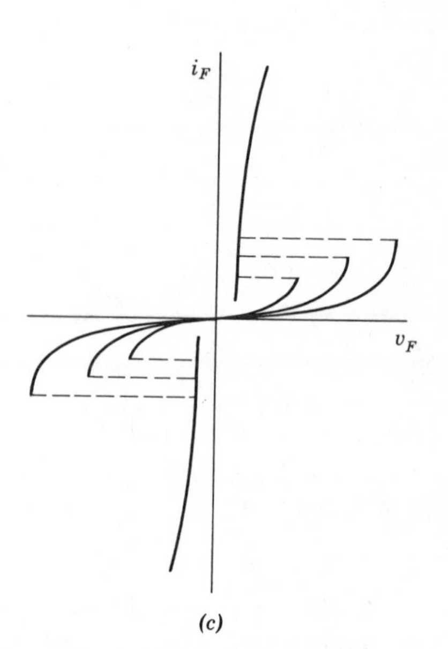

Figure 4-42 The triac. (*a*) Cross-sectional model. (*b*) Symbol. (*c*) Characteristic.

sectional representation is shown in Fig. 4-42*a*. When Anode No. 2 is positive, the path of current flow is *P*1-*N*1-*P*2-*N*2. Junctions *P*1-*N*1 and *P*2-*N*2 are forward biased, and junction *N*1-*P*2 is blocked. A positive gate (with respect to Anode No. 1) biases the junction *P*2-*N*2 in the forward direction, and breakdown occurs as a normal SCR operation. A negative gate (with respect to Anode No. 1) biases the junction *P*2-*N*3 in the forward direction, and the injected current carriers into *P*-2 turn on the four-layer diode. When Anode No. 1 is positive, the path of current flow is *P*2-*N*1-*P*1-

*N*4. Junctions *P*2-*N*1 and *P*1-*N*4 are forward biased and junction *N*1-*P*1 is blocked. A positive gate (with respect to Anode No. 1) injects carriers by forward biasing *P*2-*N*2, and a negative gate injects current carriers by forward biasing *P*2-*N*3. The circuit symbols are shown in Fig. 4-42*b* and typical characteristics are given in Fig. 4-42*c*.

Since the triac can be turned on by any one of the four conditions, it can be used directly on an a-c line to control the amount of current in a load without rectifying. The disadvantage of this device is that a relatively long time is required for it to recover to the off state. Accordingly, it is limited to use on 50, 60, or 400 Hz applications.

Questions

1. Define or explain each of the following terms: (*a*) rate-grown, (*b*) meltback, (*c*) dot, (*d*) bar, (*e*) MSI, (*f*) LSI, (*g*) monolithic chip, (*h*) unipolar, (*i*) bipolar, (*j*) drift transistor, (*k*) micro-etched diffused transistor, (*l*) inflection point, (*m*) peak point, (*n*) valley point, (*o*) LAS, (*p*) LASCR.
2. What is a mesa transistor and what is its advantage?
3. What is the objective of passivation?
4. What is the function of an N^+ substrate? How thick is it?
5. Describe the process of photolithography.
6. Why are metalized contacts used in integrated circuits?
7. How are epitaxial layers formed in an IC?
8. Describe how pinchoff is obtained in a JFET.
9. Relate r_d, g_m, and g_{mo} to FET characteristic curves.
10. What is the input resistance to a gate?
11. Explain how a depletion-type MOSFET functions.
12. Explain how an enhancement-type MOSFET functions.
13. Why is a JFET constructed without a substrate?
14. What is an intrinsic region in a transistor, and why is it used?
15. How does a tunnel diode differ from an ordinary diode?
16. Why is the peak to valley ratio of interest in a tunnel diode?
17. Can a varactor diode be operated with forward bias? Explain.
18. How is the depletion region related to the capacitance in a varactor diode?
19. Theoretically what is the least and what is the maximum capacitance available in a varactor diode?

20. What are the electrodes of unijunction transistor?
21. What is the intrinsic standoff ratio of a UJT?
22. How does a thyristor differ from other semiconductor devices?
23. What is a Shockley diode and what are its characteristics?
24. Compare the action of the Shockley diode with a Zener diode.
25. Explain how a thyristor can be triggered to produce current.
26. Explain the function of the gate in an SCR.
27. Is leakage current required to fire an SCR? Explain.
28. Which is larger, latching current or holding current? Explain.
29. Compare an SCR to a triac. Are the devices interchangeable?
30. Why is a triac more appropriate for use as a light dimmer than an SCR?

Chapter Five

OTHER ELECTRON TUBES

Multigrid tubes, the tetrode (Section 5-1), the pentode (Section 5-2), and the beam power tube (Section 5-4) were developed to overcome the limitations of the triode at high frequencies. The remote-cutoff tube (5-3) is very useful because the gain of an amplifier can be varied by controlling its bias. A short discussion of miscellaneous tube characteristics (Section 5-5) completes the chapter.

Section 5-1 The Tetrode

In the early years of radio communications it was apparent that the range of useful available frequencies for the transmission and reception of signals was limited because of the high interelectrode capacitance between the grid and the plate of the triode (Fig. 5-1). The tetrode was the first successful electron tube developed to reduce this capacitance by several orders of magnitude. Although the tetrode is not used today in receiving equipment, it is extensively used as a high-power amplifier in transmitters. A detailed examination of its properties is useful because it establishes the principles that lead directly to the succeeding development—the pentode.

If a grounded electrostatic shield is placed between the control grid and the plate, capacitance exists between the control grid and ground (the shield), and between the plate and ground (shield), but not between the grid and the plate. In order to have this shielding, the shield would have to be large and solid. It is necessary in vacuum-tube action to have an electron flow from the cathode to the plate. This means that the shield cannot be a solid plate, but it must be an open-wire mesh having the same form as the control grid. When we make this shield or *screen* in the form of a mesh, it loses some of its

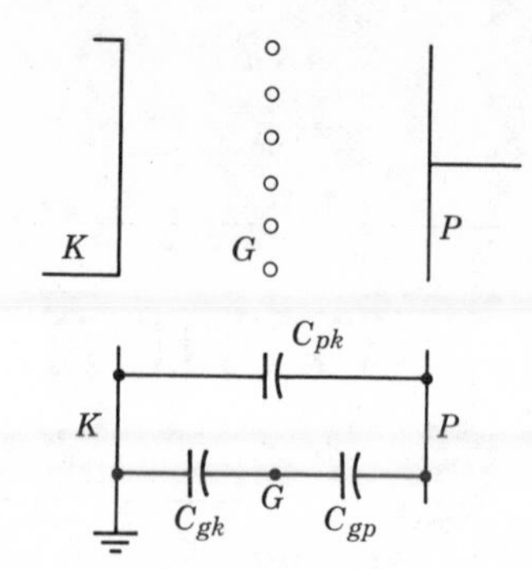

Figure 5-1 Interelectrode capacitance in a triode.

shielding ability. We find that the actual mesh design of this shielding grid is a compromise between the electron flow and satisfactory shielding. As an illustration, with a screen used as an electrostatic shield, C_{gp} is reduced from 3 pF to 0.007 pF.

The tetrode (Fig. 5-2) is a vacuum tube that has four elements: a cathode,

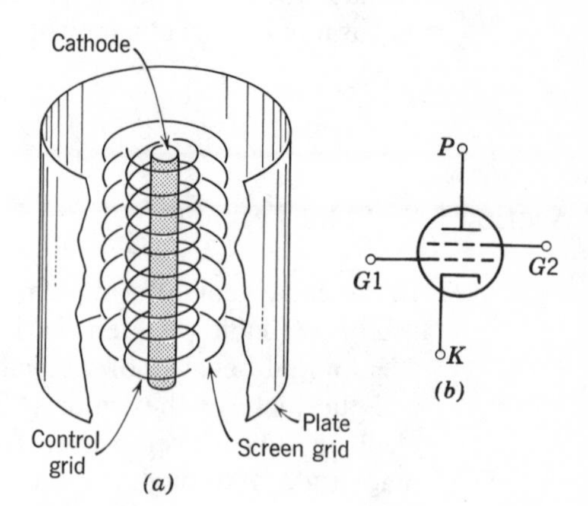

Figure 5-2 The tetrode. (*a*) Mechanical structure. (*b*) Graphic symbol.

a control grid, a screen or shield grid, and a plate. The functions of the cathode, the control grid, and the plate are the same as the three elements of the triode. For the second grid or screen grid to serve as a shield, we found that it must be operated at ground potential. This statement may be modified to include its operation at a fixed d-c potential which is adequately and sufficiently shunted to ground with a bypass capacitor. We find, if the screen were directly connected to the ground or to the cathode, it would be impossible to obtain any plate current at normal tube voltages. Thus, the screen is operated at a fixed positive potential of the order of one third to full plate voltage.

The conventional test circuit used for obtaining the characteristic curves of a tetrode is shown in Fig. 5-3. The screen voltage is maintained at a

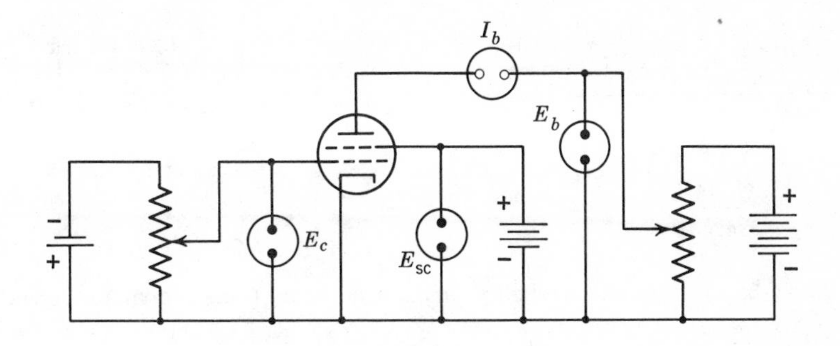

Figure 5-3 Test circuit for a tetrode.

fixed value, and the grid and plate voltages are allowed to vary. For the plate characteristic, the grid voltage is adjusted and held to a specific negative value. When the plate voltage is zero, the positive voltage on the screen attracts electrons from the space-charge cloud. Most of these electrons go to the screen wires, paths *a* and *b* of Fig. 5-4. Some of the electrons

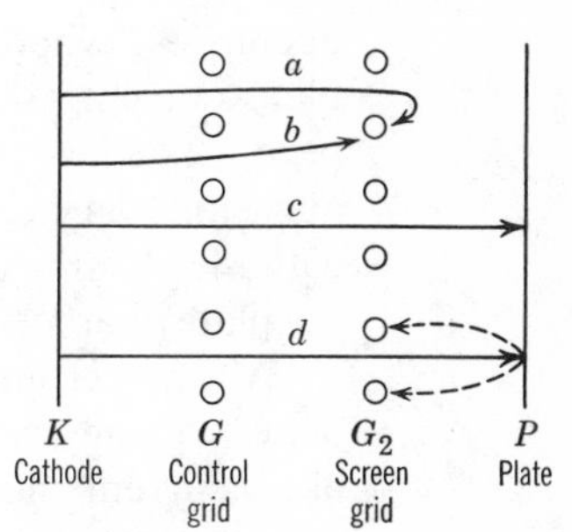

Figure 5-4 Secondary emission in a tetrode.

pass through the screen-wire mesh and continue on until they hit the plate, producing a plate current, path *c* of Fig. 5-4. As the plate voltage is raised from zero, an increasing number of electrons are drawn away from the paths to the screen and go to the plate. This increase in plate current causes a

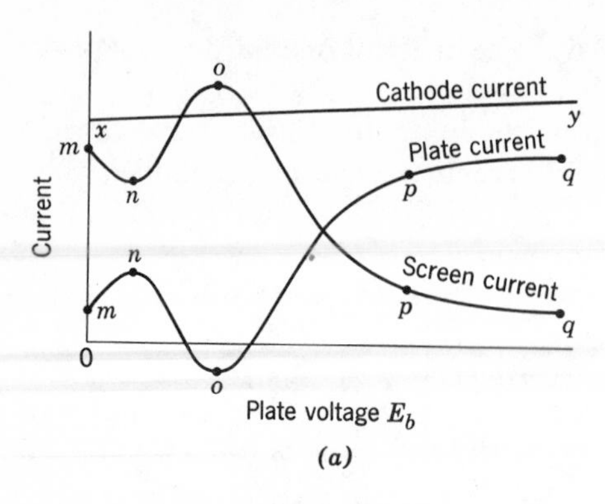

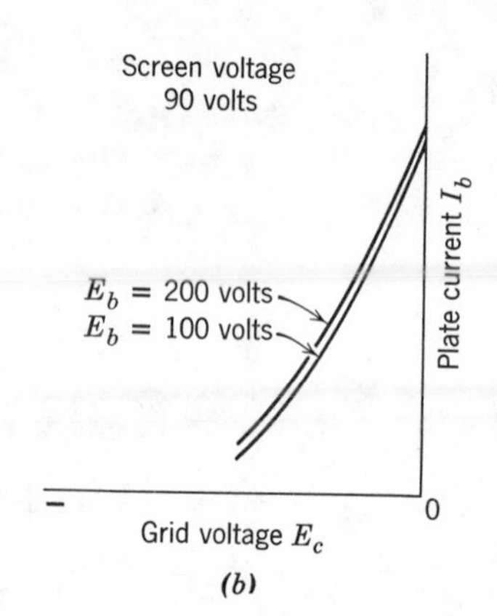

Figure 5-5 Tetrode characteristics. (*a*) Plate family. (*b*) Transfer or mutual family.

corresponding decrease in the screen current. This action is shown in that part of the plate characteristic curves between *m* and *n* (Fig. 5-5*a*).

As the voltage on the plate increases, the kinetic energy of the electrons which reach the plate increases. This increasing kinetic energy now produces a sufficient striking force on impact at the plate to cause *secondary emission*. Electrons are literally knocked off the surface of the plate by the impact to produce the secondary emission, path *d* of Fig. 5-4. The electrons produced by secondary emission are thrown into the space between the screen and the plate. They go to the electrode which is at the higher potential, the screen. This means that, as secondary emission increases, the net plate current, which is the incoming electrons less the secondary emission, decreases, whereas the screen current increases by the amount that the plate current goes down. If the nature of the plate surface is such that it produces a large secondary emission, the quantity of electrons of secondary emission can exceed the number of incoming electrons to the plate from the cathode. In this case, the plate current goes negative. This effect shows on the plate characteristic as the region between *n* and *o* (Fig. 5-5*a*).

After a certain critical value is reached, point *o* of Fig. 5-5*a*, an increasing plate voltage recaptures more and more of the secondary electrons. The plate current increases, and the screen current decreases. When the plate voltage equals and exceeds the screen voltage, point *p* to point *q*, all the secondary emission is drawn back into the plate. The plate current rises somewhat while the screen current continues to decrease.

Over the whole characteristic, the total number of electrons that are involved is determined by the value of the negative grid voltage. The total current, the cathode current (*xy* of Fig. 5-4) is essentially constant. This total current divides between the two positive electrodes, the screen and the plate, in accordance with their relative voltages and the effects of secondary emission, as we have discussed.

The region of the plate characteristic between n and o indicates a negative resistance in which an increasing voltage produces a decreasing current. Normally, the useful range of the tube as an amplifier is limited to the flat portion of the curve between p and q. The discussion of this tube from this point on in the chapter assumes that its operation is confined to this linear region.

The slope of the curve between p and q shows that there is a very small change in plate current for a large change in plate voltage. For a typical small tetrode, a change in plate voltage of 100 V produces a change in plate current of 0.5 ma. Dividing 100 V by 0.5 ma gives a plate resistance r_p, of 200,000 Ω. A high value of plate resistance is characteristic of multigrid vacuum tubes, whereas plate resistances of triodes are relatively low. Since the control grid controls cathode current in the same manner as in a triode, the order of magnitude of the transconductance g_m is the same for tetrodes as for triodes. Then by use of the relation $-\mu = g_m \times r_p$, we notice that tetrodes have much higher amplification factors than triodes. An amplifier stage, using a tetrode, gives a much higher gain than a triode stage. Thus, we find that the tube which was developed to reduce the undesirable effects of a large grid-to-plate capacitance also provides an increased voltage amplification.

Section 5-2 The Pentode

The useful part of the plate characteristic of the tetrode is limited to that region where the plate voltage exceeds the screen voltage. Although the tetrode resolves the problem of grid-to-plate capacitance satisfactorily, the pentode (Fig. 5-6) was developed to extend the useful range of a vacuum tube to include the whole characteristic. A third grid structure, called the *suppressor*, is located between the screen and the plate.

The screen is held at the same positive voltage as in the tetrode. This positive screen attracts electrons in the same fashion detailed in the discussion of the tetrode. The suppressor which is held at cathode potential is less positive than the screen and exerts a repelling effect on the electron flow. When the plate voltage is zero, because of this repelling effect, comparatively few electrons coast on to the plate. When the plate is slightly positive, the attracting force of this plate voltage overcomes the repelling effect of the suppressor, and electrons do get through to the plate. As the plate voltage increases, electrons strike the plate to produce secondary emission as in the tetrode. Now, however, the secondary-emission electrons, which have a low

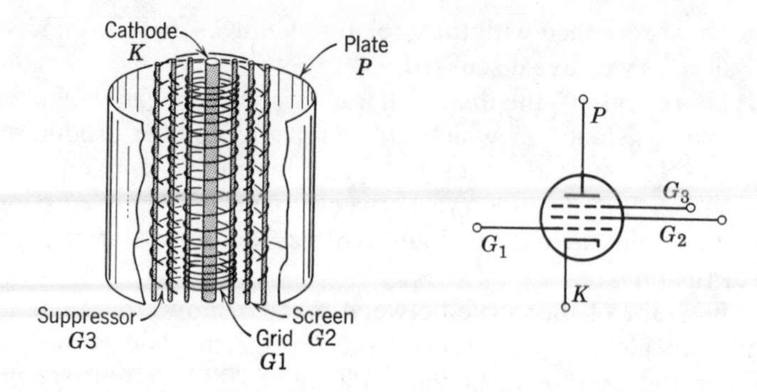

Figure 5-6 Mechanical structure of a pentode.

velocity, do not go to the screen because of the intervening suppressor-grid electric field which tends to repel electrons. The secondary electrons are driven back into the plate because the plate is at a higher potential than the suppressor. The action of the suppressor is then *not* to eliminate secondary emission but to prevent the electrons produced by secondary emission from reaching the screen.

In the pentode, at very low plate voltages, most of the electrons go to the screen. As the plate voltage is increased, more and more of this electron stream is diverted from the screen to the plate. At a relative low plate voltage, the plate current reaches its final value. The plate and the transfer characteristics are shown in Fig. 5-7. Now, in the pentode, we have *two* shield grids between control grid and the plate instead of the one shield grid of the tetrode. These two shield grids reduce the control grid-to-plate capacitance still further. We used 0.007 pF as a typical value for the grid-to-plate capacitance of the tetrode. The additional screening action of the suppressor grid reduces this value to 0.005 pF in a pentode.

The pentode characteristics are distinctly "flatter" than the tetrode curves. This results in still higher values of plate resistance. Since the transconductance is a function of the control grid and screen potentials, the transconductance is substantially the same as the values for a tetrode. Therefore, the values of μ, the amplification factor, are still higher for pentodes than for tetrodes.

Section 5-3 Remote-Cutoff Tubes

Up to this point we have considered vacuum-tube grid structures as being formed of uniformly spaced wires made in the form of a spiral or helix

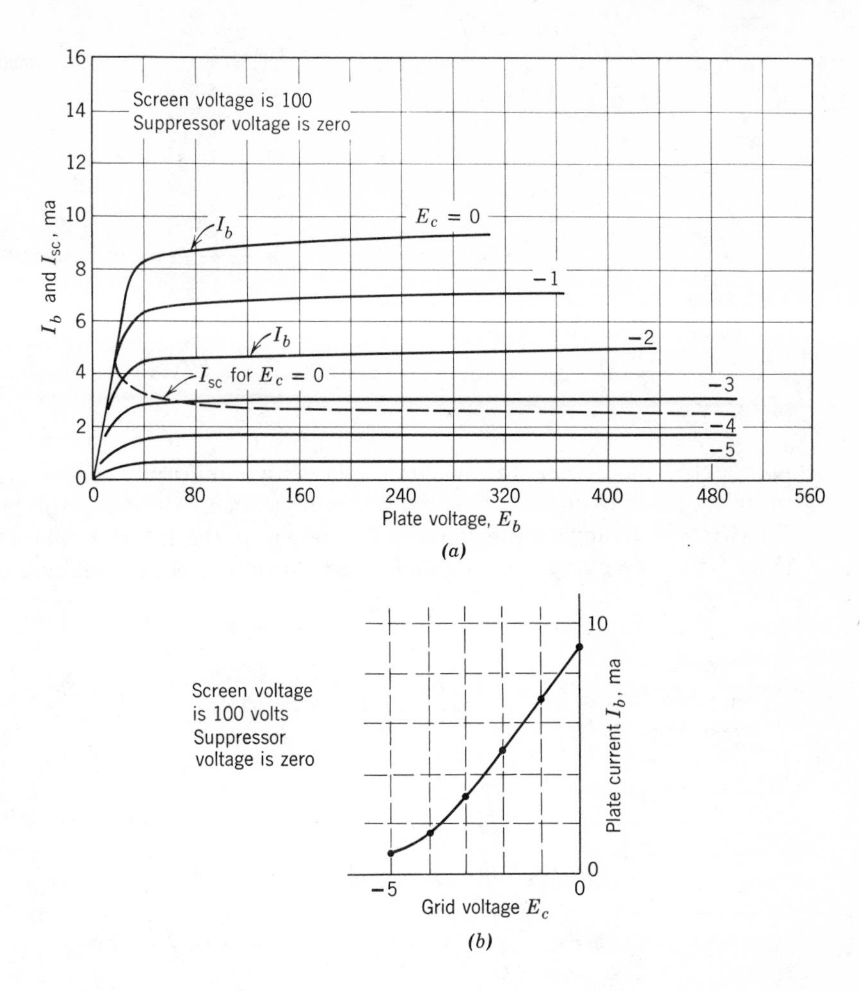

Figure 5-7 Characteristic curves for a pentode. (*a*) Plate characteristics. (*b*) Transfer characteristic. (*Courtesy RCA.*)

(Fig. 5-8*a*). A second structural form (Fig. 5-8*b*) consists of a close spacing at the top and bottom and a gradual spreading toward the center. In the discussion of the operation of the control grid of the triode, we brought out the fact that a fine-wire mesh produces an effective low grid-voltage control over the plate current whereas a wide-grid mesh results in less control of the plate current by the grid voltage.

In a vacuum tube that uses this variable spacing in the control grid construction, a small negative grid voltage cuts off the plate current at the ends of the helix. As the grid is made more negative, the region of cutoff

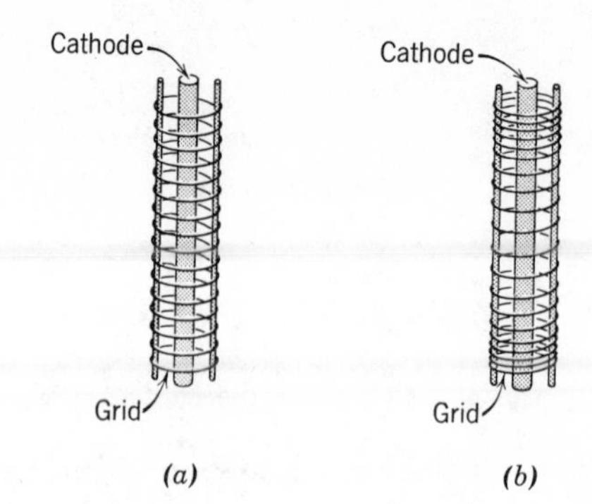

Figure 5-8 Grid structures for sharp and remote-cutoff tubes. (*a*) Sharp-cutoff control grid. (*b*) Remote-cutoff control grid. (*Courtesy RCA.*)

approaches the center. In a vacuum tube with a uniformly spaced grid, cut-off takes place over the whole axial length at once. A comparison between the two grid arrangements can best be shown on the transfer characteristic (Fig. 5-9). The curve for the uniform grid structure is essentially a straight

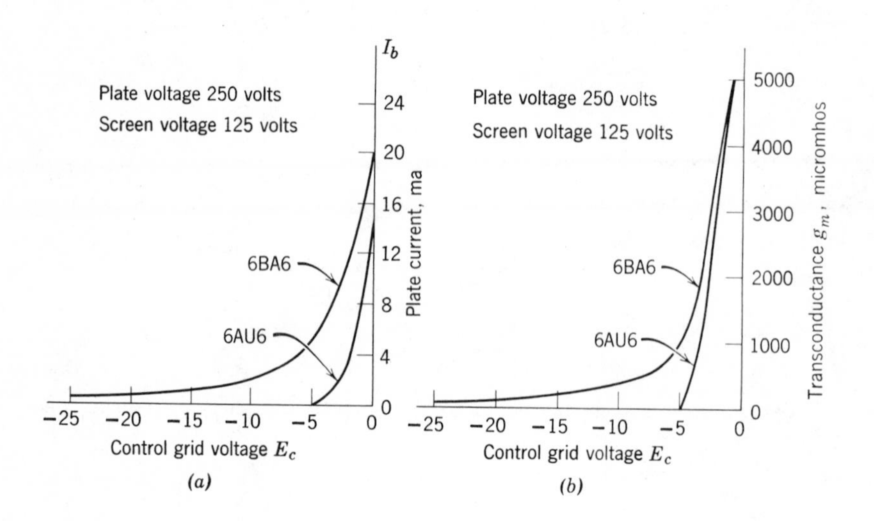

Figure 5-9 Electrical characteristics of a sharp (6AU6) and a remote (6BA6) cutoff tube. (*Courtesy RCA.*)

line, producing a *sharp cutoff*. The variable grid spacing results in a transfer characteristic that is sharply curved with an indefinite cutoff point. The transconductance g_m of the tube is affected in the same manner, since the transconductance is defined as the slope of the transfer curve ($g_m = \Delta I_b/\Delta E_c$

for E_b constant). Because of this, the tube employing this special grid is termed *remote cutoff, supercontrol,* or *variable -mu.*

The use of the variable-control grid spacing has been confined, with only a few exceptions, to multigrid tubes. The supercontrol tube has had its major application in radio-receiving equipment (Section 25-3). However, we do find that the remote cutoff tube is used in special industrial electronic circuits to the extent that the operation and functioning of the tube should be understood by all students of electronics.

Section 5-4 The Beam-Power Tube

The effect of a suppressor grid reduces the power efficiency of a pentode, and the limited range of linear operation reduces the efficiency of the tetrode. The beam-power tube (Fig. 5-10*a*) is the result of developmental efforts that

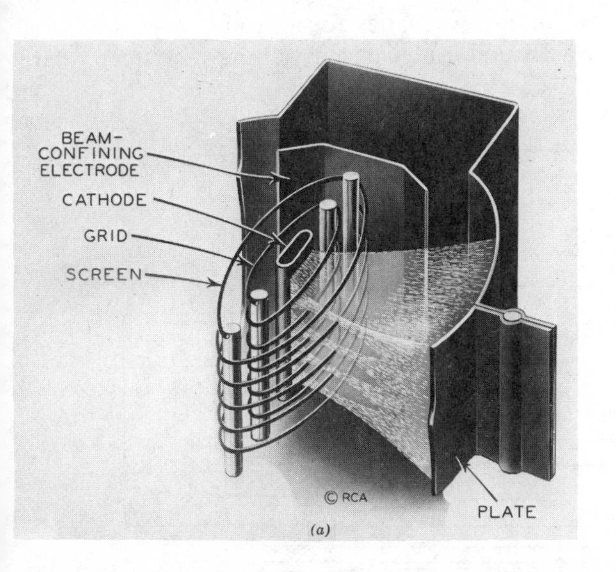

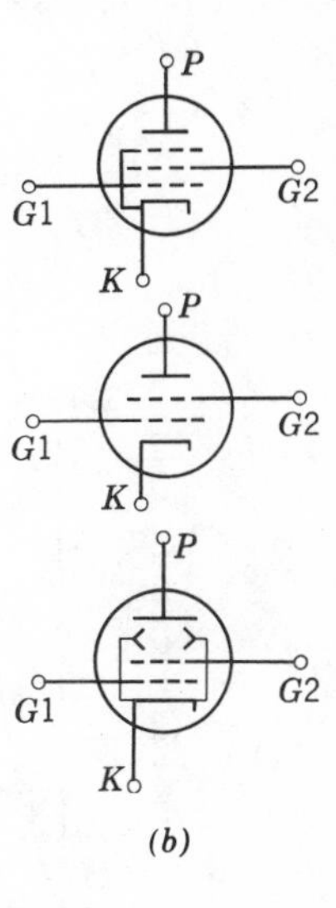

Figure 5-10 The beam-power tube. (*a*) Mechanical structure. (*b*) Symbols in use for beam-power tube. (*Courtesy RCA.*)

seek to combine into one tube structure the best features of each. Because of this similarity, the circuit symbols used for the beam-power tube can be any one of several forms (Fig. 5-10*b*).

In the beam-power tube, the grid wires of the control grid and of the screen grid are carefully aligned so that the electron flow from cathode to plate is in planar bunches or beams. The suppressor grid is omitted, but there are solid beam-directing plates which are electrically tied to the cathode. Since these beam-directing plates are at cathode potential and are located in the area of the tube that is subject to the high screen and plate voltages, these plates repel the electron stream and keep it within tight concentrated paths to the plate. This concentration of electrons in a compact path between the screen and the plate produces a negative field, just as a concentration of electrons in the space charge produces a negative field. The negative-field effect of the beam serves to push the electrons of secondary emission that are produced from impact at the plate back into the plate. The concentration of electrons then acts on secondary emission in the same manner as the suppressor grid in the pentode. The term *virtual suppressor* is used to describe this action of having a suppressor action without actually having the physical structure.

This virtual suppressor action functions only for the high plate-current levels that are found in tubes intended, not as voltage amplifiers, but as power amplifiers. In the plate characteristics for a beam-power tube (Fig. 5-11) the curves are not exactly uniformly spaced, but the very steep, sharp

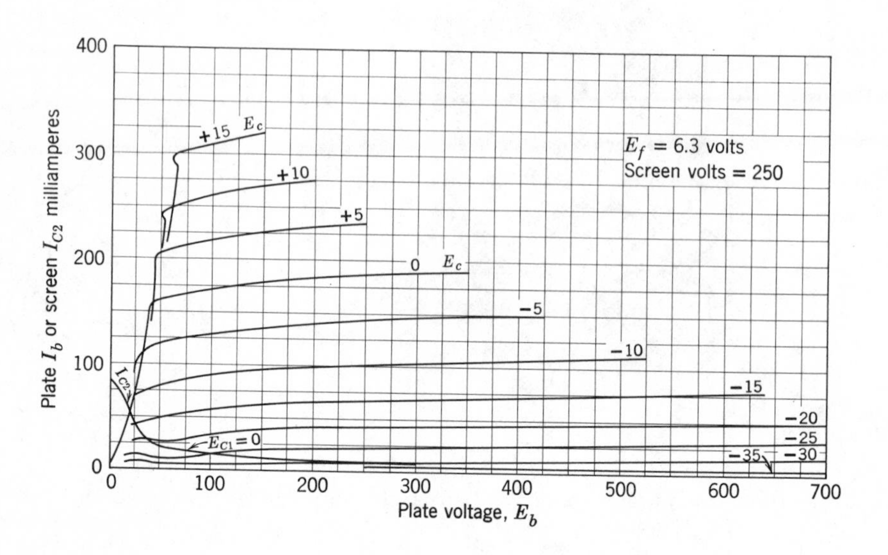

Figure 5-11 Plate characteristic of a beam-power tube. (*Courtesy RCA.*)

rise at low plate voltage before the curves flatten enables a useful operating range to be obtained that encompasses nearly the full quadrant.

Section 5-5 Miscellaneous Characteristics of Tubes

As with semiconductors, when using tubes, it is necessary to have at hand a tube manual. Not only does the manual give details on the electrical characteristics but also it lists the mechanical data for the tube. The physical size, the type of envelope, the socket requirements, the identification of the leads at the base and of special plate or grid caps, if used, are all part of this necessary information. Many of the large tubes have special cooling problems. A certain volume of forced cooling air or water is required for normal operation. All these data are part of the tube manual, and students of electronics should have a copy of one of the several available commercial publications that are issued by the different tube manufacturers.

We have discussed at length the operation of the diode, the triode, the tetrode, and the pentode. Vacuum tubes are also made with four grids (*hexodes*) and five grids (*heptodes* or *pentagrid tubes*). These tubes were developed for special applications in communications. Multipurpose tubes are very common. For instance, a 6AL5 has two plates and two cathodes with a common heater. The 12AX7 and the 12AU7 are examples of tubes that have two separate triodes contained in one envelope with a common heater circuit. The 6AQ5 and 6AV6 are examples of high-μ triodes combined with two diodes in one envelope, using a common heater and a common cathode for all three sections. Very evidently these multi-purpose vacuum tubes satisfy a need of economy where one tube and one base can serve satisfactorily in place of two or more separate tubes.

Many special types of tubes, such as lighthouse tubes, klystrons, magnetrons, and traveling-wave tubes, are used primarily in high-frequency microwave communications systems. A discussion of these tubes does not properly fit in a textbook on basic electronics and electronic circuits. Accordingly, reference should be made for them in any one of the numerous texts in this specialized field.

Questions

1. Define or explain each of the following terms: (*a*) screen voltage, (*b*) secondary emission, (*c*) suppressor, (*d*) sharp cutoff, (*e*) remote cutoff, (*f*) virtual suppressor, (*g*) ionization, (*h*) hexode, (*i*) heptode.
2. Why is a shield grid effective?
3. Explain why the plate current follows *m, n, o, p,* and *q* in Fig. 5-5.
4. What is meant by a negative resistance?
5. Why is the screen of a tetrode held at a lower potential than the plate?

6. Explain the action of secondary emission in a pentode.
7. Why is plate current substantially independent of plate voltage in Fig. 5-7?
8. Describe the construction of a grid structure that produces a remote-cutoff pentode.
9. Explain how the beam-forming plates in a beam-power tube prevent secondary-emission effects.
10. Compare the values of C_{gp} in different types of vacuum tubes.

Chapter Six

RECTIFIERS

The half-wave rectifier (Section 6-1), the full-wave rectifier (Section 6-2), and the bridge rectifier (Section 6-3) are the three basic rectifying circuits in general use in power supplies. Their salient features are compared in Section 6-4. In order to decrease the amount of ripple in the rectifier output, a low-pass filter (Section 6-5) is used. The capacitor filter (Section 6-6), the choke filter (Section 6-7) and the π filter (Section 6-8) are most frequently used. Voltage multipliers (Section 6-9) are used for special applications. The shunt rectifier is treated in Section 6-10.

The electronic-regulated power supply is analyzed in Section 20-12. Inverter circuits used to change direct current into alternating current for the purpose of rectification at a different voltage level are deferred to Section 22-10.

Section 6-1 The Half-Wave Rectifier

The action of the half-wave rectifier with a resistive load was considered in Section 2-2 as an application of the solid-state diode. The basic circuit and the fundamental waveforms for an ideal diode in which the forward voltage drop is zero are redrawn in Fig. 6-1 for convenience. Notice that, for this circuit, the fundamental frequency of the a-c variation in the load-voltage waveform is the same as the frequency of the applied alternating voltage.

If E is the rms or effective value of the transformer secondary voltage, and if E_m is its peak value, we have, from Eq. 2-1,

$$E_{\mathrm{dc}} = \frac{E_m}{\pi} = \frac{\sqrt{2}E}{\pi} = 0.318E_m = 0.450E \tag{6-1a}$$

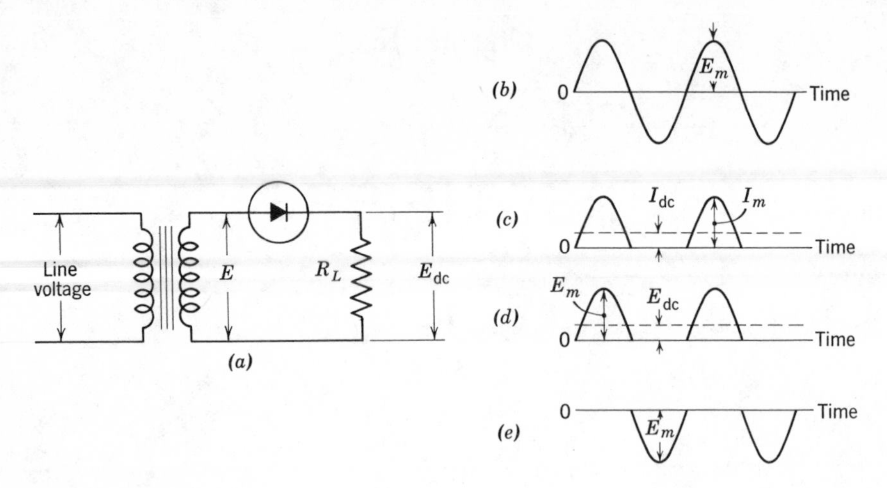

Figure 6-1 The half-wave rectifier. (*a*) Circuit. (*b*) Input voltage. (*c*) Diode and load current. (*d*) Load voltage. (*e*) Diode voltage.

and

$$I_{dc} = \frac{I_m}{\pi} = \frac{E_m}{\pi R_L} = \frac{\sqrt{2}E}{\pi R_L} = \frac{E_{dc}}{R_L} = 0.318 I_m = 0.450 \frac{E}{R_L} \qquad (6\text{-}1b)$$

where

$$E_m = \sqrt{2}E.$$

In electronic circuit analysis there are frequently two or more different sources of emf in series. If one emf is e_1 and the other is e_2, the sum of the two is a *phasor* sum. As long as the frequencies are the same for the two voltages, a direct phasor addition can be made by conventional means. When the frequencies of the two differ, the analysis of the sum is more complex. However, as shown in basic texts on electric circuits, the effective or rms value of the resultant can be determined quite simply from a consideration of power. The total power in a resistor R is the power produced by the effective voltage E_1 at one frequency plus the power produced by the effective voltage E_2 at a second frequency. The total power is given by an equivalent effective voltage E_{eq} across the same resistor:

$$\frac{E_{eq}^2}{R} = \frac{E_1{}^2}{R} + \frac{E_2{}^2}{R}$$

By multiplying through by R and solving for E_{eq},

$$E_{eq} = \sqrt{E_1{}^2 + E_2{}^2}$$

Now let us consider the sum of a d-c voltage, Fig. 6-2*a*, and an a-c

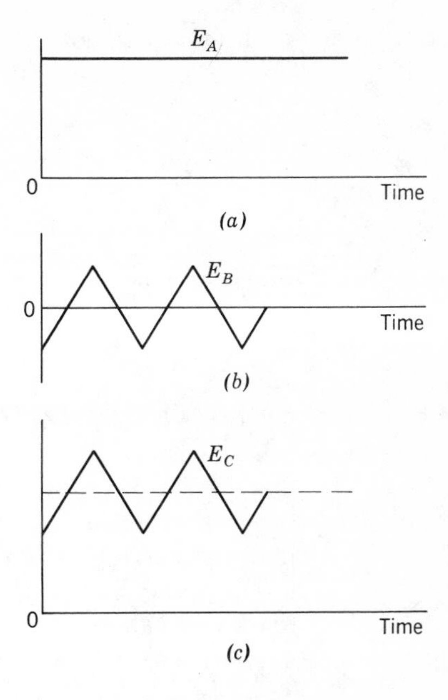

Figure 6-2 Superimposed d-c and a-c waves.

voltage, Fig. 6-2*b*. The a-c waveform is specifically not a sinusoid to show that there is no restriction to a particular waveform in this discussion. When the two waveforms are added point by point, the resultant is shown in Fig. 6-2*c*. If E_A and E_B are the rms values of the components, the rms value of the composite is

$$E_C = \sqrt{E_A^2 + E_B^2}$$

or

$$E_C^2 = E_A^2 + E_B^2$$

In Fig. 6-2 waveform *c* represents the output of a rectifier circuit, waveform *a* is the direct current in the output, and waveform *b* is the alternating current in the output. The a-c component of the output is called the *ripple*. The ratio of the rms value of the ripple to the d-c value is called the *ripple factor* and, when it is expressed in percent, it is called the *percent ripple*. The value of this ripple factor, then, is a measure of the effectiveness of the circuit in rectifying. A d'Arsonval-type d-c instrument reads E_A directly. An a-c meter which reads true effective values, such as an electrodynamometer movement or an iron-vane movement, will read E_C. A blocking capacitor placed in series with the a-c meter keeps out the direct current and allows a reading of E_B to be made directly. Usually the specifications for ripple in a commercial power supply are given as so many millivolts peak-to-peak. This is done because an oscilloscope is a very convenient instrument to use in making ripple measurements. If we know the values for E_A and E_C, then the ripple voltage E_B is

$$E_B^2 = E_C^2 - E_A^2$$

$$E_B = \sqrt{E_C^2 - E_A^2}$$

$$\text{Ripple} = \frac{E_B}{E_A} = \frac{\sqrt{E_C^2 - E_A^2}}{E_A}$$

The rms value of a voltage waveform is defined as

$$V_{\text{rms}} \equiv \sqrt{\frac{1}{T}\int_0^T v^2\,dt}$$

where T is the time for a full period. Since the instantaneous value is squared, and since the square of a negative value is positive, this means that the positive loop and the negative loop have equal contributions to the rms value in a sinusoidal waveform. If the rms value of the rectified load voltage (Fig. 6-1d) is A, the rms value of the part of the waveform which is across the diode (Fig. 6-1e) is also A. We can relate these two rms voltages to the rms value of the applied voltage E by

$$E = \sqrt{A^2 + A^2}$$

Squaring

$$E^2 = 2A^2 \qquad \text{or} \qquad A = \sqrt{\tfrac{1}{2}}\,E$$

Since

$$E \text{ is } E_m/\sqrt{2} \qquad A = \frac{E_m}{2}$$

Therefore the rms value for the half-wave load voltage is $E_m/2$ and the rms value for the half-wave load current is $I_m/2$. If we let the rms value of the ripple voltage be denoted by E_R, we have

$$\left(\frac{E_m}{2}\right)^2 = \left(\frac{E_m}{\pi}\right)^2 + (E_R)^2$$

$$E_R^2 = \frac{E_m^2}{4} - \frac{E_m^2}{\pi^2} = \left(\frac{1}{4} - \frac{1}{\pi^2}\right)E_m^2$$

$$E_R = \sqrt{\frac{1}{4} - \frac{1}{\pi^2}}\,E_m = 0.386E_m = 0.545E \tag{6-2}$$

From Eq. 6-1a, the direct voltage in the output is

$$E_{\text{dc}} = 0.318E_m$$

Then

$$\text{Ripple factor} = \frac{E_R}{E_{\text{dc}}} = \frac{0.386E_m}{0.318E_m} = 1.21 \tag{6-3}$$

or the ripple is 121%.

The conclusion we may draw from this is that, for a half-wave rectifier with resistive load, the amount of alternating current in the output is greater than the amount of direct current in the output.

In the load resistor, the a-c relations are

$$E_{\text{rms}} = \frac{E_m}{2} \qquad \text{and} \qquad I_{\text{rms}} = \frac{E_{\text{rms}}}{R_L} = \frac{E_m}{2R_L}$$

Then the total power P_T in the load is

$$P_T = E_{\text{rms}} I_{\text{rms}} = \frac{E_m^{\,2}}{4R_L}$$

In the same load resistor, the d-c relations are

$$E_{\text{dc}} = \frac{E_m}{\pi} \qquad \text{and} \qquad I_{\text{dc}} = \frac{E_m}{\pi R_L}$$

Then the d-c load power P_{dc} is

$$P_{\text{dc}} = \frac{E_m^{\,2}}{\pi^2 R_L}$$

The *ratio of rectification* or *conversion efficiency* is defined as the ratio of the d-c power, P_{dc}, delivered by the rectifier to the load to the total power, P_T, delivered to the load:

$$\text{Ratio of rectification} = \frac{E_m^{\,2}/\pi^2 R_L}{E_m^{\,2}/4R_L} = \frac{4}{\pi^2} = 0.406 \qquad (6\text{-}4)$$

The ratio of rectification is not an overall efficiency, since it does not include the losses of the diodes and of the transformer. It is an efficiency in the sense that the overall operating efficiency of a half-wave rectifier with a resistive load cannot be greater than 40.6%.

The *rating* of the secondary winding is E or $E_m/\sqrt{2}$ volts and the actual alternating current is $E_m/2R_L$, giving an a-c power equal to $E_m/\sqrt{2} \times E_m/2R_L$ or $E_m^{\,2}/2\sqrt{2}R_L$ watts. The d-c load power has been determined as $E_m^{\,2}/\pi^2 R_L$ watts. The ratio of these two powers is the *transformer utilization factor:*

$$\text{Transformer utilization factor} = \frac{E_m^{\,2}/\pi^2 R_L}{E_m^{\,2}/2\sqrt{2}R_L} = \frac{2\sqrt{2}}{\pi^2} = 0.287 \qquad (6\text{-}5)$$

The meaning of this ratio may best be explained by the use of a numerical example. If a 1-KVA transformer is used in a half-wave rectifier circuit with resistive load, the irregular nonsinusoidal waveforms that occur limit the available d-c power to 287 watts. Since the ratio of rectification is 0.406, the input power from the line is 287/0.406 or 706 watts. In practice, the figure of 287 watts would be too high, since we have assumed that the diode is ideal and does not dissipate power. The transformer utilization ratio yields the volt-ampere rating of the secondary winding of the transformer.

There is a direct current in the diode and in the load. This same direct current must flow in the transformer secondary winding. This direct current may saturate the secondary winding. A condition of saturation materially reduces the transformer output by reducing the amplitude of the secondary voltage wave. In order to prevent an adverse effect caused by saturation, the transformer core size must be increased. This effectively means that the transformer utilization factor of 0.287 must be reduced still further in practice.

In conclusion, we can state that the single-phase rectifiers have four material disadvantages:

1. A very high ripple.
2. A low ratio of rectification (efficiency).
3. A low transformer utilization factor.
4. Definite possibility of d-c saturation of the transformer secondary.

The circuit has the advantage of being the simplest possible arrangement. It is used only for such applications where the advantage of the simple circuit arrangement outweighs the disadvantages. Also, it is commonly used where the load current requirements are very low.

PROBLEMS

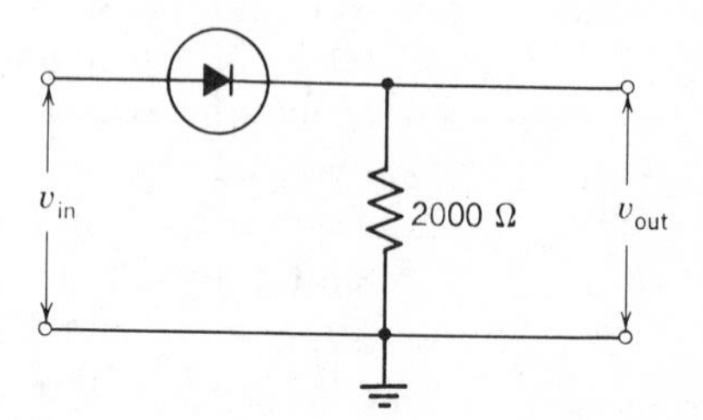

Circuit for Problem 3.

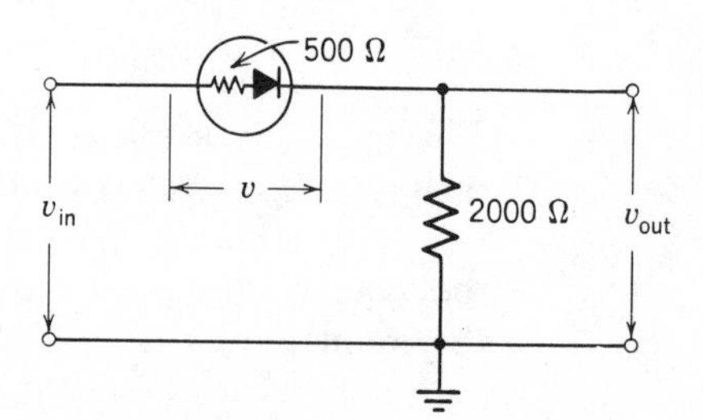

Circuit for Problem 4.

1. A 117-V 60-Hz source is the input to a half-wave rectifier that supplies 100 W to a resistive load at 20 V. Determine the transformer turns ratio, the input power, and the transformer rating in volt-amperes.
2. Solve Problem 1 if the load is 5 W at 3000 V.
3. The input voltage is

$$v_{in} = 100 \cos 377t$$

Sketch and dimension the input-output characteristic that shows v_{out} plotted against v_{in}. The diode is ideal.

4. The input voltage is

$$v_{in} = 100 \cos 377t$$

Sketch and dimension the input-output characteristic that shows v_{out} plotted against v_{in}. Sketch and dimension two full cycles of the waveform of the voltage v.

Section 6-2 The Full-Wave Rectifier

In a full-wave rectifier circuit (Fig. 6-3*a*) the transformer secondary coil has a center tap *b*, which is the common return point of the rectifier circuit. The secondary voltage is considered to be measured from *b* to *c* and from *b* to *a* and not from *c* to *a*. To avoid confusion, the voltage of a secondary winding intended for use in this circuit is specified as, for example, 35-0-35 V. This means that from *b* to *a* we have 35 V rms, and from *b* to *c* the voltage reading is also 35 V rms. Between *c* and *a* the voltage is 70 V rms.

When *a* is positive, diode *A* passes current and *c* is negative, making the anode of diode *B* negative with respect to the cathode. When *c* is positive, there is current in diode *B* and the anode of diode *A* is negative, preventing current in diode *A*. In this manner diode *A* handles the positive half of the a-c cycle and diode *B* handles the negative half of the a-c cycle. Current then flows through the load with each half of the a-c cycle. This full-wave action contrasts with the half-wave rectifier in which a pulse of load current flows only once per a-c cycle. The waveforms in the load are shown in Figs. 6-3*c* and *d*. Each diode by itself carries only half-wave current (Figs. 6-3*e* and *f*). The envelope of the load-current and load-voltage waveforms repeats twice for each full a-c cycle of the supply voltage. The fundamental ripple frequency is then twice the line frequency.

If we consider the instant when *a* is at positive E_m volts, *c* is at negative E_m volts with respect to the common circuit return *b*, diode *A* rectifies and, if it is ideal without a forward diode voltage drop, the load voltage is positive E_m. The cathode of diode *A*, the cathode of diode *B*, and the positive side of the load resistance form a common junction point for the circuit. Now the

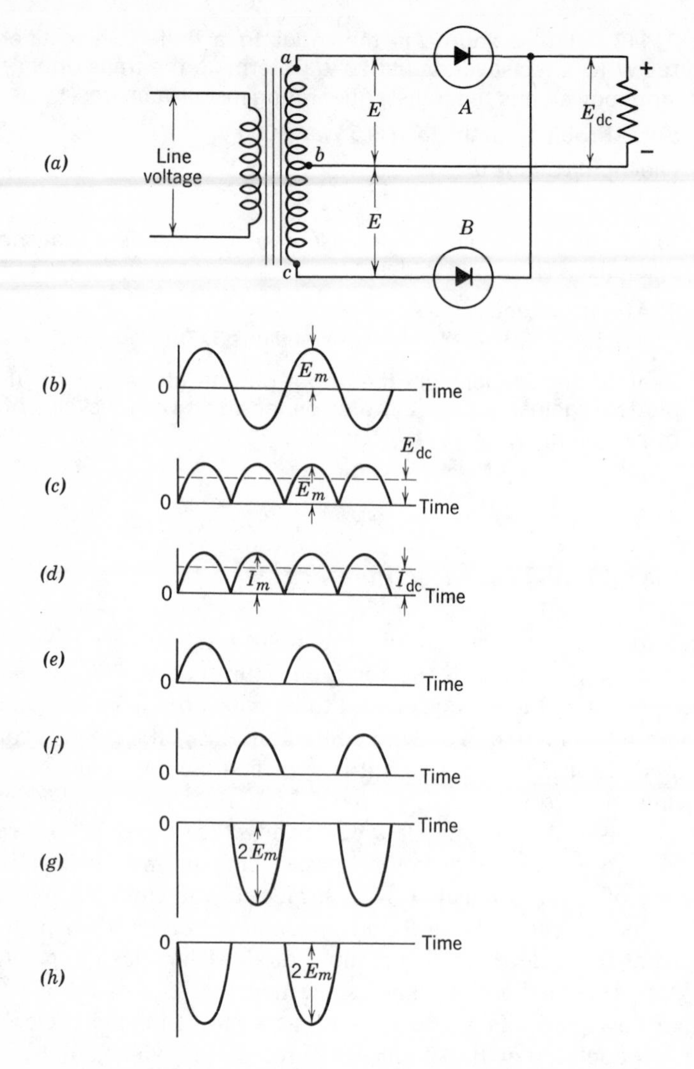

Figure 6-3 The full-wave rectifier. (*a*) Circuit. (*b*) Input voltage. (*c*) Load voltage. (*d*) Load current. (*e*) Diode *A* current. (*f*) Diode *B* current. (*g*) Diode *A* voltage. (*h*) Diode *B* voltage.

cathode of diode *B* is at $+E_m$ volts, and the anode of diode *B* is at $-E_m$ volts. This means that the *peak inverse-voltage* stress on diode *B* is twice the peak of the incoming alternating supply voltage *E*, and is twice the peak of the load voltage. If a diode has a peak inverse-voltage rating of 100 V, the maximum load voltage that can be obtained from a full-wave rectifier circuit using this diode is 50 V. The waveforms showing this inverse voltage are Figs. 6-3*g* and *h*.

Since both halves of the a-c wave are now rectified, many of the values obtained for the half-wave rectifier are changed by a factor of two.

$$E_{\text{dc}} = \frac{2}{\pi}E_m = \frac{2\sqrt{2}E}{\pi} = 0.636E_m = 0.90E \tag{6-6a}$$

and

$$I_{\text{dc}} = \frac{2}{\pi}I_m = \frac{2}{\pi}\frac{E_m}{R_L} = \frac{2\sqrt{2}E}{\pi R_L} = \frac{E_{\text{dc}}}{R_L} = 0.636I_m = \frac{0.90E}{R_L} \tag{6-6b}$$

Using the method developed in the previous section, since the rms load voltage is $E_m/\sqrt{2}$ and the direct load voltage is $2E_m/\pi$, the rms value for the a-c ripple E_R is

$$\left(\frac{E_m}{\sqrt{2}}\right)^2 = \left(\frac{2E_m}{\pi}\right)^2 + E_R^2$$

$$E_R^2 = \frac{E_m^2}{2} - \frac{4E_m^2}{\pi^2} = \left(\frac{1}{2} - \frac{4}{\pi^2}\right)E_m^2$$

$$E_R = \sqrt{\frac{1}{2} - \frac{4}{\pi^2}}E_m = 0.307E_m = 0.434E \tag{6-7}$$

The ripple factor is obtained by dividing E_R by E_{dc}:

$$\text{Ripple factor} = \frac{E_R}{E_{\text{dc}}} = \frac{0.307E_m}{0.636E_m} = 0.482 \tag{6-8}$$

or the ripple is 48.2%.

The a-c load power is

$$\frac{E_m}{\sqrt{2}}\frac{E_m}{\sqrt{2}R_L} = \frac{E_m^2}{2R_L}$$

The d-c load power is

$$\frac{2E_m}{\pi} \times \frac{2E_m}{\pi R_L} = \frac{4}{\pi^2}\frac{E_m^2}{R_L}$$

Taking the ratio of them, we have

$$\text{Ratio of rectification} = \frac{4E_m^2/\pi^2 R_L}{E_m^2/2R_L} = \frac{8}{\pi^2} = 0.812$$

The calculation of the transformer utilization factor for a full-wave rectifier must be done quite carefully to avoid pitfalls. In the secondary winding, we have *two* circuits, each of half-wave rectification. Each half of the winding has within itself a direct current flow. Then, the transformer utilization factor is merely twice that of the half-wave rectifier:

$$\begin{aligned}\text{Transformer utilization factor for secondary} &= 2 \times 0.287 \\ &= 0.574\end{aligned} \tag{6-10a}$$

When we consider the winding as a whole, we are, in effect, considering the primary of the transformer. From Eq. 9-6 we have

$$E_{dc} = \frac{2\sqrt{2}}{\pi}E \qquad \text{and} \qquad I_{dc} = \frac{2\sqrt{2}}{\pi}I$$

Then

$$E = \frac{\pi}{2\sqrt{2}}E_{dc} \qquad \text{and} \qquad I = \frac{\pi}{2\sqrt{2}}I_{dc}$$

Multiplying them together, we have

$$EI = \frac{\pi^2}{8}E_{dc}I_{dc}$$

Then

$$\begin{array}{c}\text{Transformer utilization} \\ \text{factor for primary}\end{array} = \frac{E_{dc}I_{dc}}{EI} = \frac{8}{\pi^2} = 0.812 \qquad (6\text{-}10b)$$

In practice, the average value of these two figures is usually taken:

$$\begin{array}{c}\text{Average transformer} \\ \text{utilization factor}\end{array} = \frac{0.574 + 0.812}{2} = 0.693 \qquad (6\text{-}10c)$$

Using this average transformer utilization factor of 0.693, a transformer that has a rating of 1 KVA can deliver 693 watts direct current to a resistive load in a full-wave rectifier circuit. As the ratio of rectification is 0.812, the required primary demand is 693/0.812 or 854 volt-amperes. Again, it must be remembered that these numerical examples do not consider the power losses of the diodes or the effects of an actual transformer. Since each half of the secondary a-c cycle is used in the full-wave rectifier, the net effect of the d-c flux in the secondary cancels, so that there can be no problem of a d-c saturation.

PROBLEMS

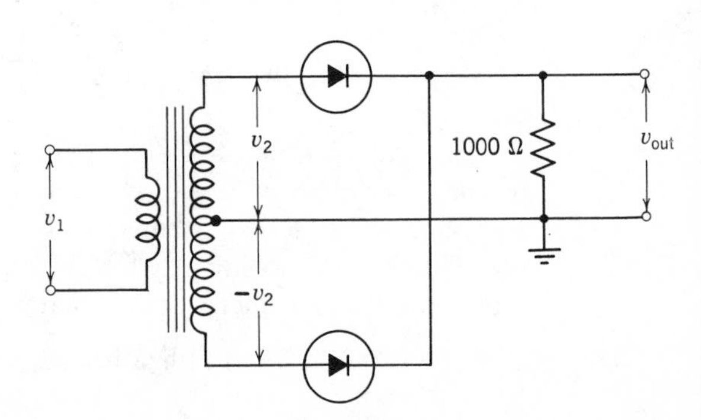

Circuit for Problems 1 and 2.

1. A transformer with a 117-V primary and a 275-0-275 V secondary supplies a 10-KΩ load by a full-wave rectifier. Determine the load voltage, the load current, and the input current, and the input power to the transformer. What is the rating of the transformer in volt-amperes?
2. Solve Problem 1 for a 350-0-350 V secondary and a 2000-Ω load.
3. A full-wave rectifier is used to supply a load in which the d-c load current is 5 A, and the load voltage is 20 V. What are the ratings of the power transformer if the source voltage is 117 V rms?
4. Solve Problem 3 for a 250-ma, 30-V load.
5. The voltages v_1 and v_2 are each 60 cos 377*t*. Sketch and dimension the input-output characteristic that shows v_{out} plotted against v_1. The diodes have a forward resistance of 200Ω.

Section 6-3 The Full-Wave Bridge Rectifier

The full-wave rectifier circuit requires a center tap on the source of the alternating voltage that is to be rectified. In many applications, the advantages of the higher output and efficiency of the full-wave circuit are required, but only a two-terminal source of voltage without a center tap is available. The full-wave bridge rectifier circuit (Fig. 6-4) is used to care for this problem.

From the circuit of the bridge rectifier, it is observed that the cathodes of the diodes are at three different potential levels, *a*, *b*, and *c*. When *b* is positive and *c* is negative, the path of current flow is shown in Fig. 6-4*b*. Diodes *B* and *C* are connected in reverse to this polarity and "block." When *c* is positive and *b* is negative (Fig. 6-4*c*), diodes *A* and *D* block the current flow which is now through diodes *B* and *C*. Thus, a half-cycle of load current and of load voltage occurs for each half-cycle of line voltage. When diode *A* passes current, the full load voltage E_m is across diode *B* as an inverse voltage. The full load-voltage is also across diode *C*, since diode *D* is passing current. In a full-wave bridge rectifier, then, the peak inverse voltage is the peak of the incoming transformer secondary voltage.

The direct load voltage and load current are, as in the full-wave circuit,

$$\begin{aligned} E_{dc} &= \frac{2}{\pi}E_m = \frac{2\sqrt{2}}{\pi}E = 0.636E_m = 0.90E \\ I_{dc} &= \frac{2}{\pi}I_m = \frac{2}{\pi}\frac{E_m}{R_L} = \frac{2\sqrt{2}E}{\pi R_L} = \frac{E_{dc}}{R_L} = 0.636I_m = 0.90I \end{aligned} \qquad (6\text{-}11)$$

Since these equations are the same as Eq. 6-6, the values for the ripple and for the ripple factor must be the same as in Eq. 6-7 and Eq. 6-8:

$$E_R = 0.307E_m = 0.434E \qquad (6\text{-}12)$$

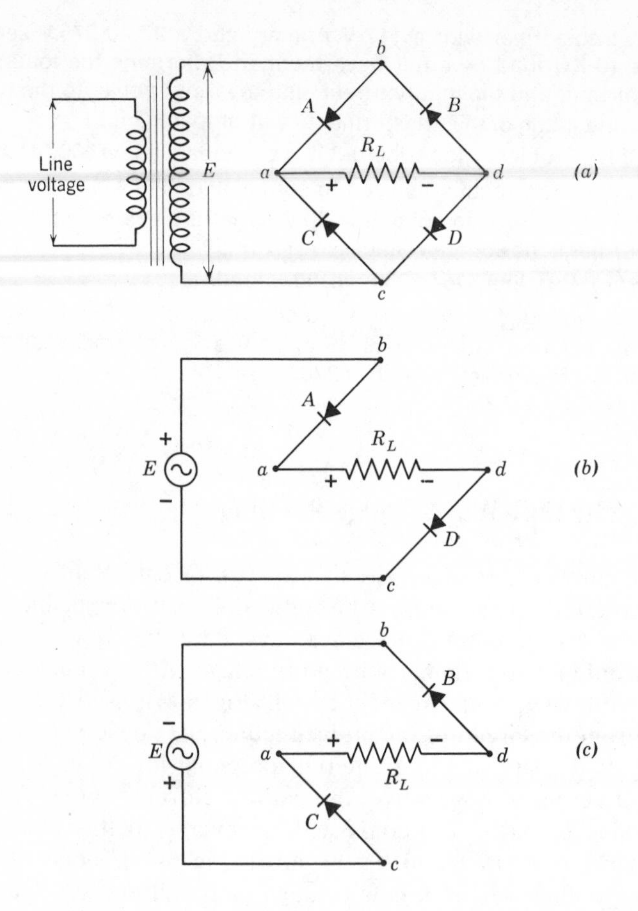

Figure 6-4 The bridge rectifier. (*a*) Circuit. (*b*) Current path when *b* is positive. (*c*) Current path when *c* is positive.

$$\text{Ripple factor} = 0.482 \tag{6-13}$$

or the ripple is 48.2%.

Also, the ratio of rectification must be the same as in Eq. 6-9.

$$\text{Ratio of rectification} = 0.812 \tag{6-14}$$

In this circuit, the transformer secondary current and voltage are purely alternating without any consideration for a d-c flow in the secondary winding. This means that the transformer utilization factor is the same as the ratio of rectification for the full-wave bridge:

$$\text{Transformer utilization factor} = 0.812 \tag{6-15}$$

Now, with a bridge rectifier, a 1-kva transformer can deliver 812 watts of d-c power to a load. The bridge rectifier is the best circuit from the viewpoint of overall performance, but it does have the disadvantage of requiring four diodes instead of one or two. Also, the full secondary voltage of the transformer (or source) is utilized instead of one-half the secondary voltage as in the full-wave center-tapped circuit. When the cost of the transformer is the main consideration in a rectifier assembly, invariably the bridge is used. This is particularly true for large rectifiers which have a high-current rating. There is a small additional power loss in the bridge because two rectifiers are carrying current at all times.

PROBLEMS

1. A transformer with a 117-V primary winding and a 250-V secondary winding is used with a bridge rectifier to supply a 10-kΩ load resistor. Determine the load voltage, the load current, and the input current and input power to the transformer.
2. Solve Problem 1 for a 5000-V secondary winding and a 200,000-Ω load resistor.
3. A bridge rectifier is used to supply a d-c load with 20 A at 20 V from a 117-V source. What are the ratings of the required power transformer?
4. Solve Problem 3 if the d-c load is 100 W at 117 V.

Section 6-4 Comparison of the Three Basic Rectifier Circuits

A comparison among the three rectifier circuits must be made very judiciously so that very serious errors in the reasoning enter and false assumptions will not be made. As an example, consider the full-wave rectifier circuit shown in Fig. 6-5. Assume that the transformer is ideal and is rated at 1-kva. There is a knife switch S in one of the diode leads. When the switch is opened, it is true that the load current and the load voltage decreases by

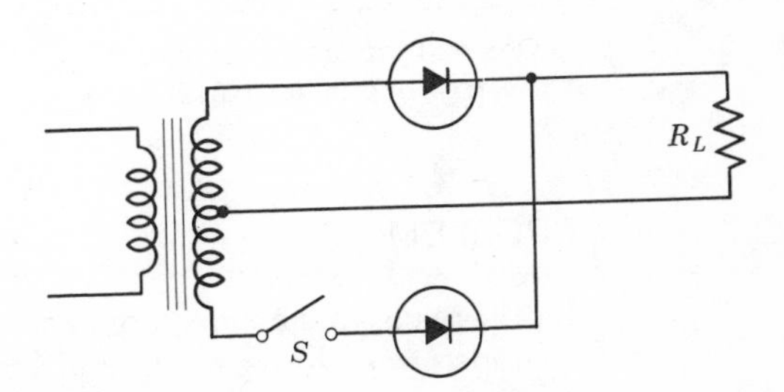

Figure 6-5 A combined half-wave and full-wave rectifier.

one half, but the available loading on the transformer changes. If the power in the load is 693 watts in the full-wave position of the switch, it is 173.25 watts when the switch is opened. Although the transformer utilization factor for a half-wave rectifier is 0.287, it is only 0.1435 when one-half the secondary winding is used. This means that the transformer is fully loaded when delivering only 143.5 watts with the switch opened.

One of the more important considerations in a comparison is the problem of d-c saturation in the half-wave rectifier. The lower ripple factors of the full-wave types are important in applications along with their more efficient transformer utilization factors. The developed relationships and factors are summarized in Table A.

TABLE A Comparison of Rectifier Circuits with Resistive Load

	Rectifier Circuit		
	Half Wave	Full Wave	Bridge
Line voltage ($E_m = \sqrt{2}E$)	E	E^*	E
Number of diodes	1	2	4
Peak inverse voltage	E_m	$2E_m$	E_m
Direct output voltage	$\frac{E_m}{\pi} = 0.318E_m$	$\frac{2E_m}{\pi} = 0.636E_m$	$\frac{2E_m}{\pi} = 0.636E_m$
Ripple factor	1.21	0.482	0.482
Ratio of rectification	0.406	0.812	0.812
Transformer utilization factor	0.287	0.693†	0.812
D-c power available from a 1-kva transformer, watts	287	693	812
Ripple frequency	f	$2f$	$2f$

*One-half secondary voltage.
†Average of primary (0.812) and secondary (0.574).

PROBLEM

1. A service application requires a d-c power of 20 A at 12 V. Determine the transformer turns ratios and the input powers if the equipment designed

to operate from a 117-V, 60-Hz line is (*a*) a half-wave rectifier, (*b*) a full-wave rectifier, and (*c*) a bridge rectifier. What is the rating of the transformer in volt-amperes in each case?

Section 6-5 Low-Pass Filters

A low-pass filter is a three- or four-terminal device (Fig. 6-6) that passes energy at all frequencies below f_c, the *cut-off frequency*, and prevents all energy above the cutoff frequency from appearing in the load R_L. Ideally, the cutoff at f_c is very sharp. Practically, a *roll-off* occurs above f_c (Fig. 6-6*c*).

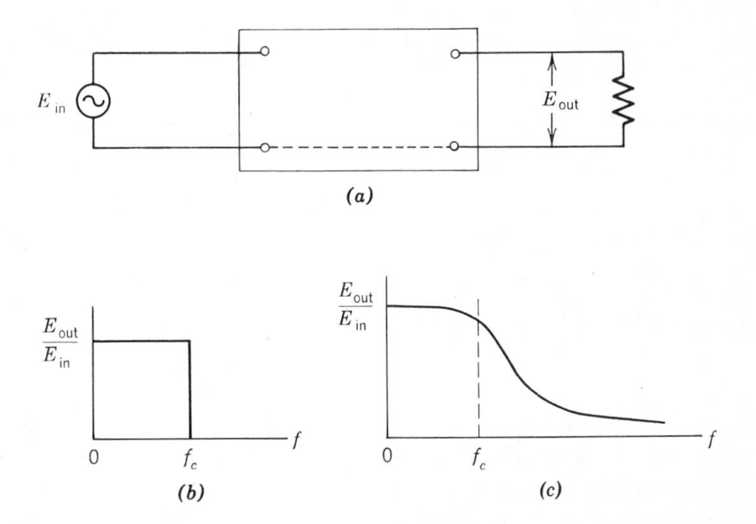

Figure 6-6 Low-pass filter. (*a*) Block diagram. (*b*) Ideal characteristic. (*c*) Actual characteristic.

The steepness of the roll-off is specified by the particular application. Except for more sophisticated designs, the low-pass filter is composed of passive elements, capacitors, inductors, and resistors. For critical applications at high frequencies, quartz plates are used as the elements of *crystal filters*.

The desired output of a power supply is a pure d-c voltage and current in the load without the presence of an a-c ripple content. The frequency of the direct current is zero, since the period of direct current is infinite. The ripple frequency of a power supply is f or $2f$ which for the usual a-c source of power is 60 or 120 Hz dependent on the rectifier circuit used. Therefore the cutoff frequency should lie between zero and 60 or 120 cps.

To understand the action of a low-pass filter, consider the simple circuit shown in Fig. 6-7*a*. The input to the resistance-divider circuit contains a 10-V d-c source plus a 10-V a-c ripple. From the resistance-divider circuit, it is evident that both the d-c and the a-c voltages in the 9000-Ω load

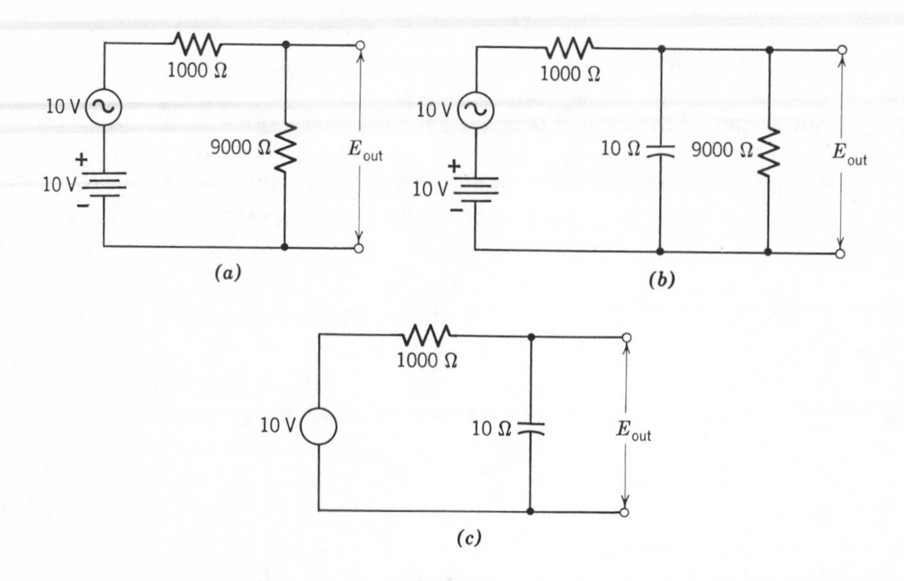

Figure 6-7 R-C filter. (*a*) Without capacitor. (*b*) With capacitor. (*c*) A-C circuit.

resistor are each 9 V. The ripple factor in the load is 9/9 or 1.0 making the ripple 100%. When the R-C filter circuit shown in Fig. 6-7*b* is used, there are two circuits to consider. The value of 10 Ω associated with the capacitor is its reactance at the ripple frequency. As far as direct current is concerned the magnitude of X_C is infinite. The capacitor can be ignored and the circuit without the capacitor for d-c calculations is shown in Fig. 6-7*a*. Thus the d-c output voltage is 9 V. As far as the a-c circuit is concerned, the shunting effect of the 9000-Ω resistor across 10 Ω can be neglected and the a-c circuit reduces to that shown in Fig. 6-7*c*. The current in this series circuit is

$$I = \frac{E}{R - jX_C} = \frac{10}{1000 - j10} \approx \frac{10}{1000} = 0.01 \text{ A}$$

The a-c voltage is

$$E_{\text{out}} = IX_C = 0.01 \times 10 = 0.1 \text{ V} = 100 \text{ mv}$$

Now the ripple factor is 0.1/9 or 0.011 and the ripple is 1.1%. Thus, this simple R-C filter has reduced the ripple in the output from 9 V to 100 mv or by a factor of 90.

A Fourier series development is a mathematical analysis that can be used to determine an equation for a periodic nonsinusoidal waveform. The waveform of the half-wave rectifier shown in Fig. 6-1*d* has the equation:

$$e = \frac{E_m}{\pi}\left[1 + \tfrac{1}{2}\sin \omega t - \tfrac{2}{3}\cos 2\omega t - \tfrac{2}{15}\cos 4\omega t - \tfrac{2}{35}\cos 6\omega t - \cdots\right] \quad (6\text{-}16)$$

This equation shows that the output of the half-wave rectifier has a d-c term (E_m/π) agreeing with Eq. 6-1*a*, a fundamental term which shows that the output contains energy at the line frequency, and a series of higher order terms. The higher order terms are all even multiples of the line frequency and, consequently, are called even harmonics. As the frequency of the harmonic increases, its amplitude decreases. The negative signs merely indicate the phase relationship.

A Fourier development of the full-wave rectifier output shown in Fig. 6-3*c* yields

$$e = \frac{2E_m}{\pi}\left[1 - \tfrac{2}{3}\cos 2\omega t - \tfrac{2}{15}\cos 4\omega t - \tfrac{2}{35}\cos 6\omega t \cdots\right] \quad (6\text{-}17)$$

The d-c term in this equation is $2E_m/\pi$, which agrees with Eq. 6-6*a*. Now there is no fundamental term at line frequency. The lowest frequency content is the second harmonic. The amplitudes of the even harmonics are the same as in Eq. 6-16 and, consequently, the overall harmonic content is less making the ripple in the output less.

Since the amplitudes of the ripple frequencies are given by the Fourier expansions (Eqs. 6-16 and 6-17) an analysis of a filter can be performed. It is only necessary to calculate the harmonic rejection of the filter at the lowest frequency involved. A number of low-pass filters commonly used in power supplies are shown in Fig. 6-8, and their characteristics are listed in Table B. We shall examine in the next sections three of these filter circuits in detail.

Section 6-6 The Capacitor Filter

A single capacitor filter used with a half-wave rectifier circuit is shown in Fig. 6-9. During the positive half of the supply cycle, the capacitor is charged in the time interval between *a* and *b*. When the applied a-c wave falls below the value of the direct voltage on the capacitor, point *b*, the charging current from the diode ceases, and the load current continues to flow by the discharging action of the filter capacitor in the interval from *b* to *c*. Just after point *c*, the increasing supply voltage again exceeds the voltage on the capacitor and the filter capacitor recharges. The load-voltage waveform (Fig. 6-9*c*) is also the capacitor-voltage waveform. The peak-to-

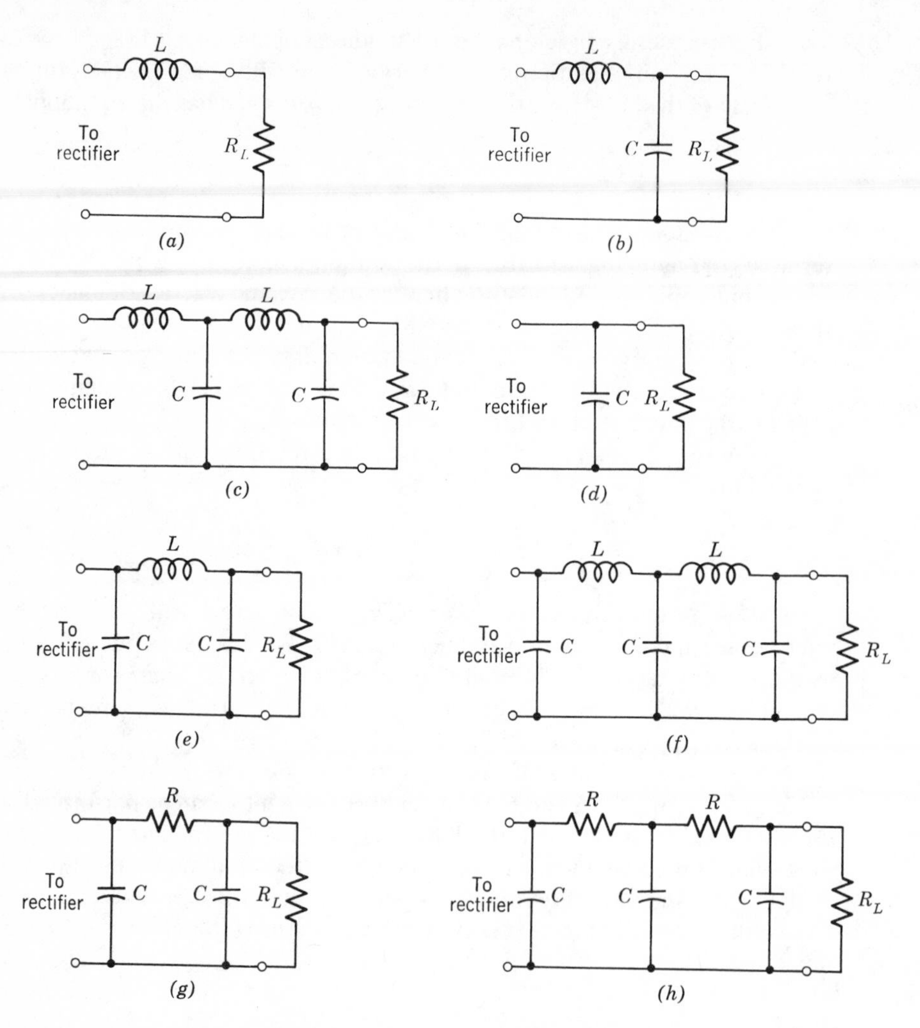

Figure 6-8 Power supply filter configurations.

peak ripple in the load is the voltage measured as the vertical distance from *a* to *b*. The load current (Fig. 6-9*d*) has the same shape as the load-voltage waveform, since the load is resistive. The diode can only pass current during the recharging time of the capacitor, from *a* to *b* and from *c* to *d*. Thus the diode current is in the form of short pulses (Fig. 6-9*e*). The peak inverse voltage is shown in Fig. 6-9*b*. When a filter capacitor is used, the peak inverse voltage can be as high as twice the peak of the alternating line voltage.

The area under the load-current curve (Fig. 6-9*d*) must equal the area

TABLE B Filter Characteristics for the 60-hertz Operation

Filter (Fig. 6-8)	Half Wave		Full Wave	
	Output Voltage	Ripple Factor	Output Voltage	Ripple Factor*
A	–	–	$0.63E_m$	$\frac{R_L}{1600L}$
B	–	–	$0.63E_m$	$\frac{0.83}{CL}$
C	–	–	$0.63E_m$	$\frac{1.455}{C^2L^2}$
D	$E_m - \frac{8350}{C} I_{dc}$	$\frac{4760}{CR_L}$	$E_m - \frac{4170}{C} I_{dc}$	$\frac{2380}{CR_L}$
E	$E_m - \frac{8350}{C} I_{dc}$	$\frac{26000}{C^2LR_L}$	$E_m - \frac{4170}{C} I_{dc}$	$\frac{3300}{C^2LR_L}$
F	$E_m - \frac{8350}{C} I_{dc}$	$\frac{1.82 \times 10^5}{C^3L^2R_L}$	$E_m - \frac{4170}{C} I_{dc}$	$\frac{5810}{C^3L^2R_L}$
G	$E_m - \left[\frac{8350}{C} + R\right] I_{dc}$	$\frac{2.5 \times 10^6}{C^2RR_L}$	$E_m - \left[\frac{4170}{C} + R\right] I_{dc}$	$\frac{10^7}{C^2RR_L}$
H	$E_m - \left[\frac{8350}{C} + 2R\right] I_{dc}$	$\frac{2.6 \times 10^{10}}{C^3R^2R_L}$	$E_m - \left[\frac{4170}{C} + 2R\right] I_{dc}$	$\frac{3.25 \times 10^9}{C^3R^2R_L}$

C is in microfarads and L is in henrys.
*To convert ripple factor into percent ripple multiply by 100.

under the diode-current curve (Fig. 6-9*e*), since the total charge delivered to the capacitor is delivered to the load as load current in the form of discharge. This statement has a slight error since the diode, when it is recharging the capacitor also, at the same time, supplies current into the load. However, the discussion of the operation of many rectifier circuits is greatly simplified by separating the two concepts, and by assuming that the sole function of the diode or diodes is to recharge the filter capacitors, and that the sole function of the filter capacitors is to supply load current by discharge. The diode current takes the form of very sharp, short-duration pulses. If the load current is fixed and if the size of the capacitor is increased, the diode-current pulses become very narrow with a very high amplitude. A vacuum tube has an upper limit of current which is determined by the total emission of the cathode. It is necessary to limit the peak current to a safe value in a semiconductor by placing a resistance between the diode and the line-voltage source. The narrow and sharp diode pulses make the transformer utilization factors still lower than the values obtained with a pure resistive loading.

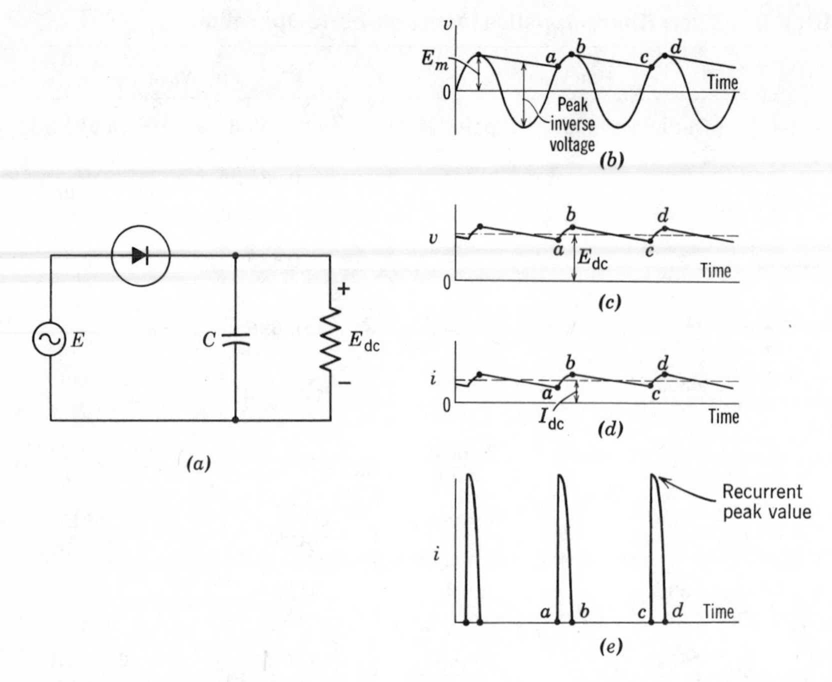

Figure 6-9 Half-wave rectifier with capacitor filter. (*a*) Circuit. (*b*) Action of the capacitor. (*c*) Load voltage. (*d*) Load current. (*e*) Diode current.

When the value of the capacitor is high and the load current is very small, the drop in voltage from b to c is negligible. Then the load voltage is the peak of the line voltage, and the ripple approaches zero. As the load-current demand increases, several things happen. The rate of discharge of the capacitor increases. The recharge action starts at e (Fig. 6-10) instead of at c. Also, the capacitor discharge starts later than d, at f. The average value of the envelope, the direct load voltage, is now lower and the amount of variation of the envelope, the ripple voltage, is greater. The

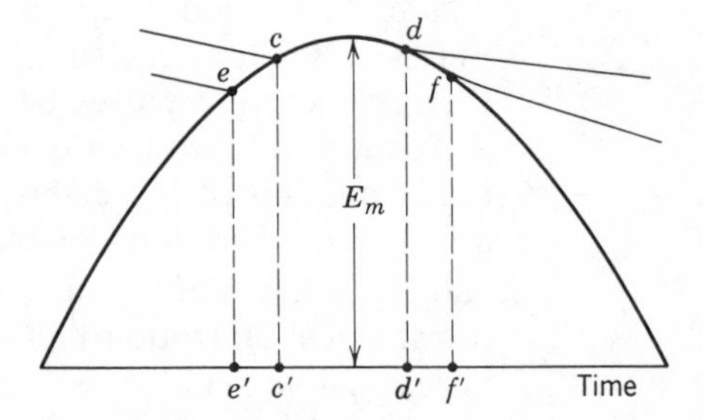

Figure 6-10 Voltage waveform under different loads.

angle or width of diode-current flow increases from $c'd'$ to $e'f'$ in order to handle the extra load current. As we pointed out in the previous paragraph, as long as we are within the peak current limitations of the diode, we can bring the envelope back to points *c* and *d* by adding sufficient capacitance to the filter.

These same considerations apply to the full-wave rectifier that uses a capacitor filter. The waveforms are similar and have two pulses of charging current per cycle, which result in a slightly higher direct load voltage with less ripple.

If we consider a circuit in which the diode drop is not neglected, we must modify Fig. 6-10 and include the voltage drop caused by the circuit resistance (Fig. 6-11). When the capacitor is recharging, the difference

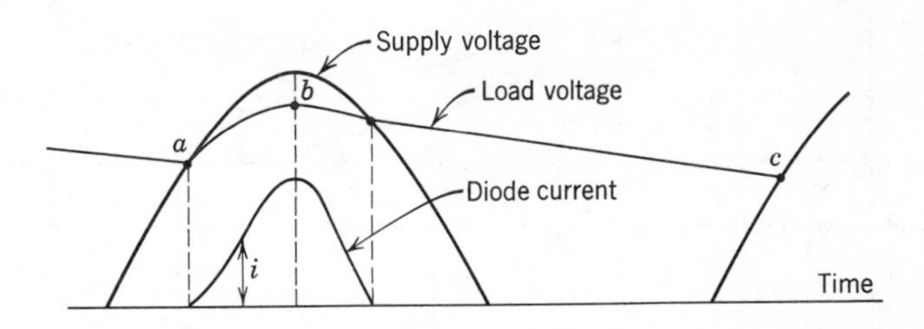

Figure 6-11 Voltage waveforms considering rectifier and source resistance.

between the supply voltage and the load or capacitor voltage is the forward diode voltage drop, *b*. The product of $e \times i$ represents the power dissipation of the diode. When a semiconductor has a possible forward conduction loss in excess of its own dissipation rating, a resistor is usually placed in series with the diode to limit the surge current to a safe value when the circuit is first turned on.

The lNl764 silicon rectifier, for example, has the following typical ratings for use as a half-wave rectifier with a capacitor filter:

RMS supply voltage	150 V
d-c load current	0.5 A
Recurrent peak current	5.0 A
Surge current limit	35.0 A
Maximum input capacitor	250 μF

When the filter capacitor is discharged, it acts as a short circuit when the line voltage is first applied to the rectifier. Accordingly, the magnitude of the required current-limiting resistor is found by dividing the peak a-c voltage by the allowable surge current. For the lNl764 diode the value of the resistor is $\sqrt{2} \times 150/35$ or 6.1 Ω. The largest value that can be used for a filter capacitor is 250 μF in order to keep the recurrent peak value of diode current within the 5-A rating. Typical load curves for this diode as a half-wave rectifier are shown in Fig. 6-12.

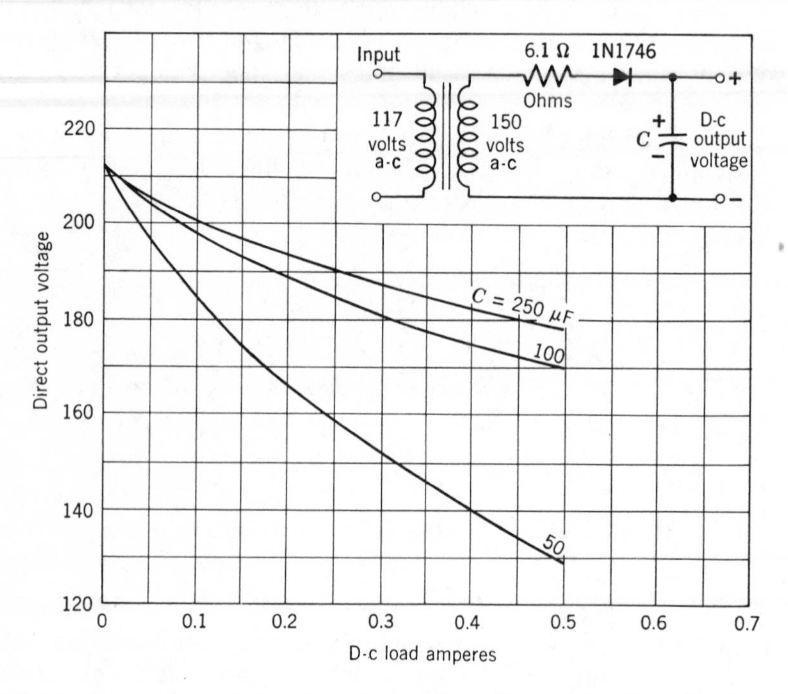

Figure 6-12 Load characteristics of a typical rectifier with capacitor filter. (*Courtesy RCA.*)

A half-wave rectifier used with a capacitor filter, provides a power supply that is used primarily where the load-current requirements are small. It provides a low-cost and lightweight solution for a filtering problem. It has the disadvantage that the direct output voltage decreases with an increase in load and that the percent ripple increases sharply with an increase in load.

PROBLEMS

1. Assume the envelope of the output voltage from a rectifier using a capacitor filter is simplified from that shown in Fig. 6-9*b* to the sawtooth waveform shown. Each time the diode conducts the voltage rises back to the peak

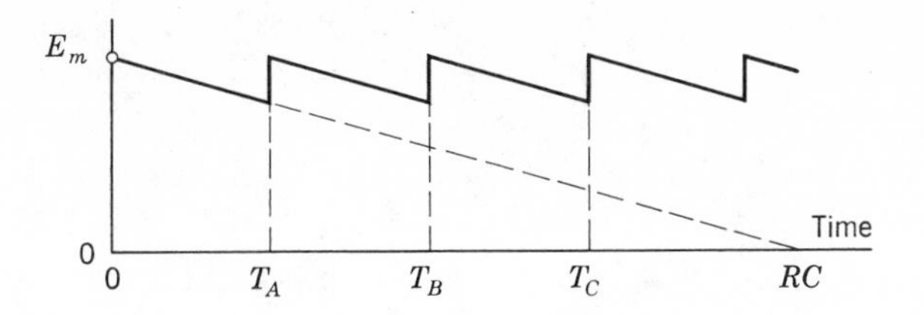

Waveform for Problems 1 and 2.

value. Assuming a 60-Hz source, show that the d-c output voltage is that listed in Table B. The rectifier is half-wave.

2. Solve Problem 1 for a full-wave rectifier.
3. The secondary of a transformer is 100-0-100 V rms. A full-wave rectifier circuit supplies a capacitor filter and a load. The capacitor is 100 μF. The load is 40 ma. What is the no-load output voltage and what is the full-load output voltage? What is the ripple in each case? The frequency is 60 Hz.

Section 6-7 The Choke Filter

A full-wave rectifier that used a *choke* as part of the filter network is shown in Fig. 6-13. The actual filter, the LC combination, is termed either an L *filter* or a *choke-input filter*. The action of the choke is to store up energy in the magnetic field and to release it to the load evenly. Thus the choke increases its energy storage during the time of the peaks of the alternating current and releases it when the rectifier output falls below the load voltage.

When the choke is too small or when the load current is very small, the choke does not deliver current over the full cycle. There are times in the cycle ab and cd (Fig. 6-13) when the choke current is zero. At these times, the overall-filter acts as if it were a simple capacitor filter. The load voltage falls from A to B (Fig. 6-14) with an increase of current from 0 to B'. At B a critical value is reached. Either sufficient inductance is in the filter or the load current has increased, so that the critical value for the inductance is reached. At this critical value, the distances ab and cd (Fig. 6-13a) are just zero. Now current is flowing at all times in the choke. This flow of current in the coil prevents the capacitor from discharging, and the load voltage is maintained at a constant value—from B to C for the ideal rectifier circuit. The voltage at B is ideally 0.63 E_m. The waveforms for this condition are shown in Fig. 6-13b. In an actual circuit, the d-c resistance of the choke and the diode drop cause the voltage to fall from B to D (Fig. 6-14).

A choke that is specifically designed to have a low inductance for the load

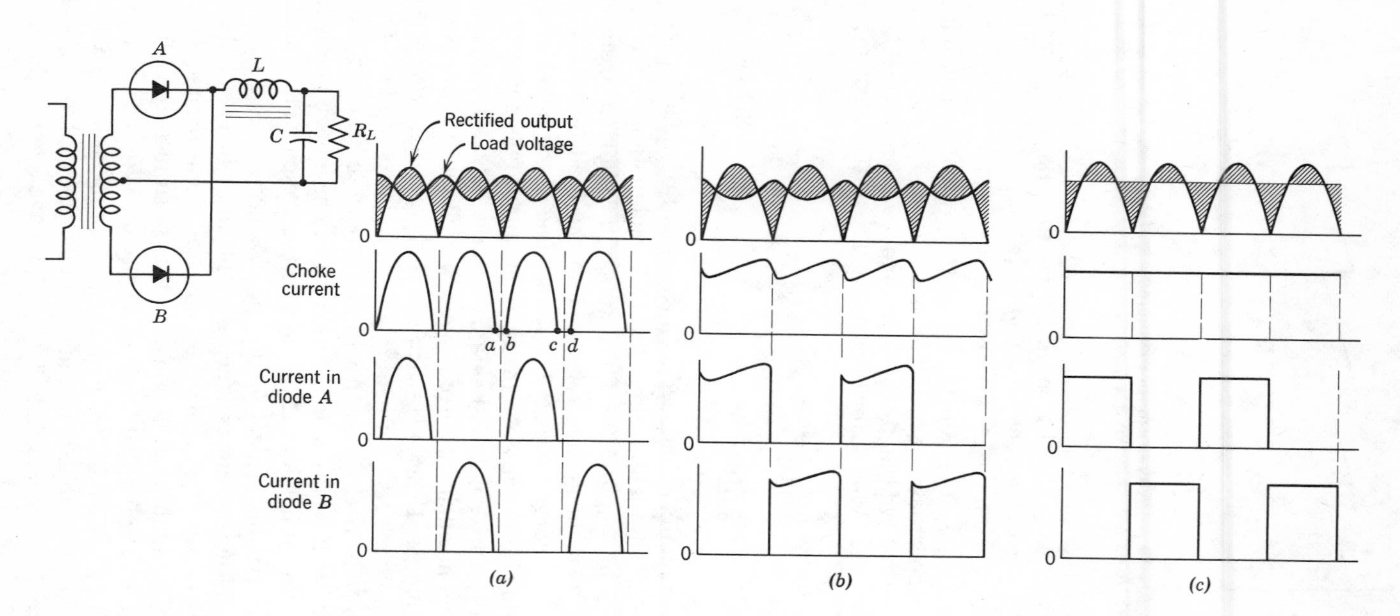

Figure 6-13 Waveforms for the choke filter. (*a*) Small inductance. (*b*) Normal inductance. (*c*) Infinite inductance.

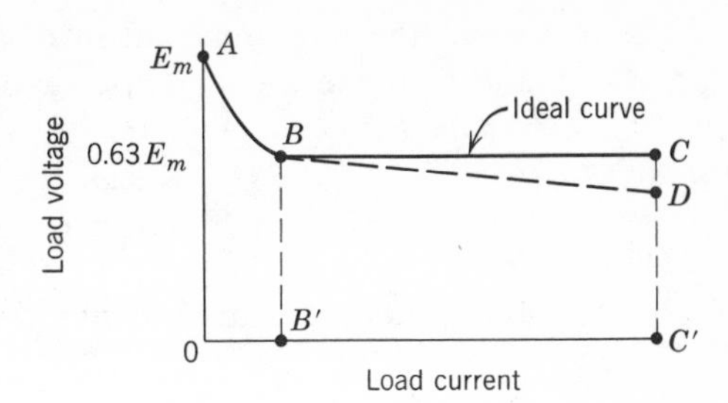

Figure 6-14 Ideal load curve for a choke filter.

current at C' (Fig. 6-14) and a high inductance for low currents is called a *swinging choke*. A properly designed swinging choke will move point B quite close to the voltage axis.

Voltage regulation is a measure of the change of load voltage with load current and is defined as

$$\text{Percent of voltage regulation} \equiv \frac{\text{no load} - \text{full load}}{\text{full load}} \times 100 \qquad (6\text{-}18)$$

A *bleeder resistor* is a resistor that is connected in parallel with the load. A bleeder has a twofold purpose in a rectifier circuit. It discharges the capacitors when the power supply is turned off so that no dangerous residual charge is left on the filter capacitors. Also, the no-load voltage is not point A but point B on Fig. 6-14. In the ideal choke-input rectifier circuit, we can see that the regulation with a bleeder is zero between B and C whereas it is $(E_m - 0.63E_m)/0.63E_m$ or 58.7% without the bleeder.

In the L filter (Fig. 6-13) the reactance of the choke is very much larger than the reactance of the filter capacitor. Likewise, R_L is much larger than X_C. The peak value of the ripple current, I_{2m}, is determined by using the reactance of the choke at the second harmonic of the line frequency and the amplitude of the second harmonic content of the full-wave rectifier output, given by Eq. 6-17. This peak current is

$$I_{2m} = \frac{\dfrac{2Em}{\pi}\left(\dfrac{2}{3}\right)}{2\omega L} = \frac{2E_m}{3\pi\omega L}$$

The d-c current in the load is determined by R_L as

$$I_{\text{dc}} = \left(\frac{2E_m}{\pi}\right) \Big/ R_L = \frac{2E_m}{\pi R}$$

The critical condition of operation of the L filter occurs at point B in Fig. 6-14. This point corresponds to the waveform shown in Fig. 6-13 when the diode current just begins to flow over the whole cycle. That is, the distance *a-b* and the distance *c-d* is just zero. At this point I_{dc} just equals I_{2m}

$$I_{2m} = I_{dc} \qquad \text{and} \qquad \frac{2E_m}{3\pi\omega L} = \frac{2E_m}{\pi R_L}$$

By cancelling terms and inverting,

$$L = \frac{R_L}{3\omega} \tag{6-19a}$$

For a 60 Hz supply, this becomes

$$L = \frac{R_L}{1131} \approx \frac{R_L}{1000} \text{ henries} \tag{6-19b}$$

This value of inductance is the least value of L that can be used to obtain the flat regulation characteristic of the choke filter as shown in Fig. 6-14. The use of 1000 instead of 1131 gives a safety margin in the design. Between points A and B, the choke does not have sufficient inductance to satisfy Eq. 6-19, and the filter operates as a simple capacitor filter.

An improvement in filtering can be made by using two L sections in the filter (Fig. 6-15). The considerations for the operation and for the minimum

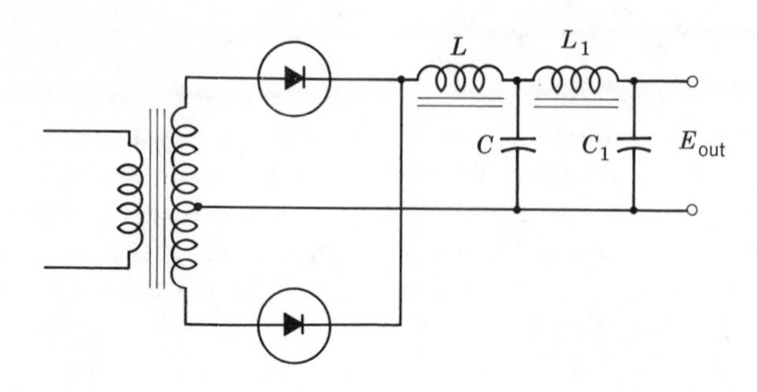

Figure 6-15 Full-wave rectifier with complex filter.

value of inductance remain the same as for the single LC filter. If the circuit load is ideal, and if L_1 has no d-c resistance, the output voltage will not be changed by the addition of the second L section, but the ripple will be reduced:

$$\text{Percent ripple} = \frac{650}{LL_1(C+C_1)^2} \tag{6-20}$$

PROBLEMS

1. The secondary of a transformer using a full-wave rectifier circuit and an *L-C* filter is 275-0-275 V rms. The load resistance is 1500 Ω. A 1% ripple is desired. What are the values of *L*, *C*, and output voltage? The critical-inductance point of the filter as determined by Eq. 6-19*b* is to be at 20% of the designed load current.
2. Solve Problem 1 if the secondary voltage of the transformer is 175-0-175 V and the load resistance is 1000 Ω.
3. The secondary of a transformer is 150-0-150 V. The *L-C* filter comprises a 20 H inductor and a 20 μF capacitor. The rated load is six times the critical-inductance point determined by Eq. 6-19*b*. Determine the full-load current, the full-load voltage, and the full-load ripple.

Section 6-8 The π Filter

A π filter is a modified *L* filter with a capacitor connected across the input to the filter (Fig. 6-16*a*). The π filter has an output characteristic that is higher than the *L* filter (Fig. 6-16*b*). On the other hand, the regulation of the π filter is poorer than that of the *L* filter. Because of the additional filtering effect of the input capacitor, the per cent ripple is lower than the ripple content of the *L*-filter output. A double π-filter circuit (Fig. 6-17) gives a still further reduction in ripple, but the direct voltage drop of the second choke causes the voltage regulation to become still poorer.

The question of which filter should be used for a particular application is resolved by considering a number of factors. The size, the weight, and the cost of the filter components must be balanced against the electrical requirements of the filtering problem, the load current, the regulation, and the permissible ripple. The final filter design is a compromise between these factors. Very often, a simple filter is used with a regulator (Section 20-12). Then a lightweight design is achieved with a very low ripple and an excellent regulation.

PROBLEMS

1. The secondary of a transformer using a full-wave rectifier circuit and a π filter is 275-0-275 V rms. The load resistance is 3000 Ω. A 1% ripple is desired. What are the values of *L*, load voltage, and load current if *C* is 4 μF? Check this result on Fig. 6-16.
2. Repeat Problem 1 for a 500-0-500 V transformer secondary. Determine the operating point on Fig. 6-16 as a check. The allowable ripple is 2%.

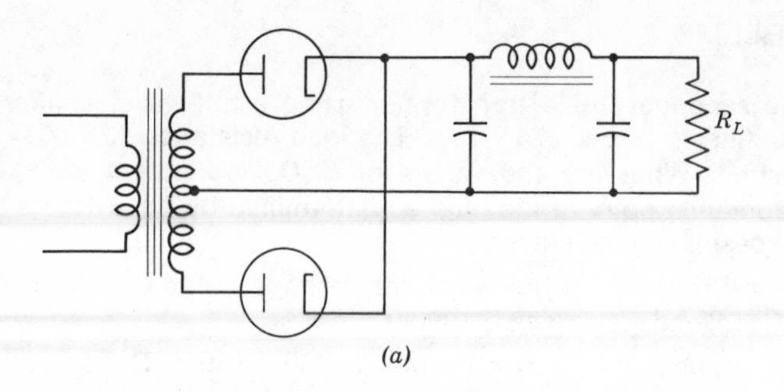

(a)

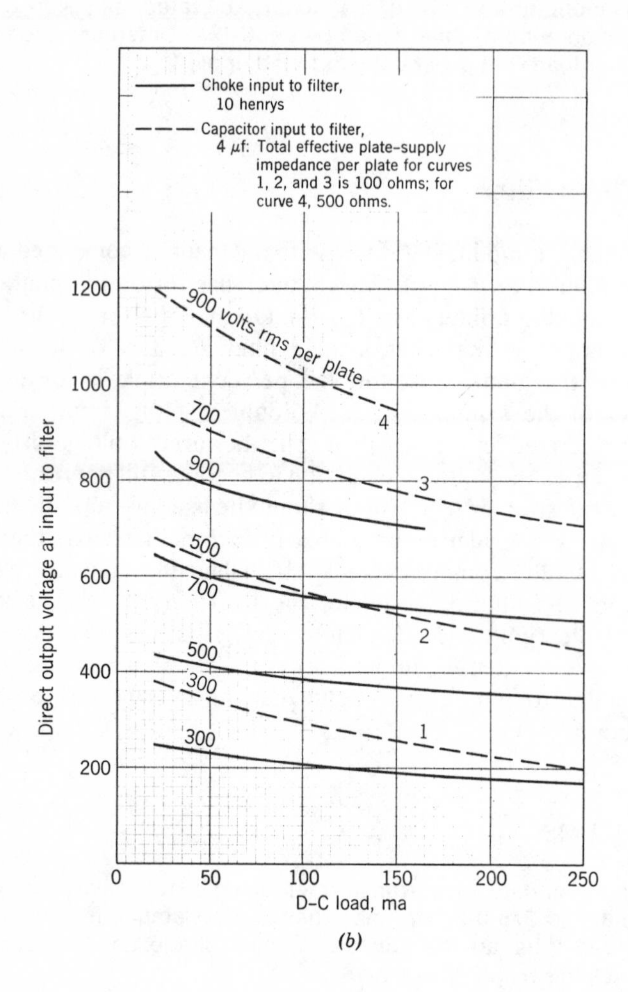

(b)

Figure 6-16 Circuit and load curves for the filter using 5R4-GY. (a) Circuit. (b) Load curves. (*Courtesy RCA.*)

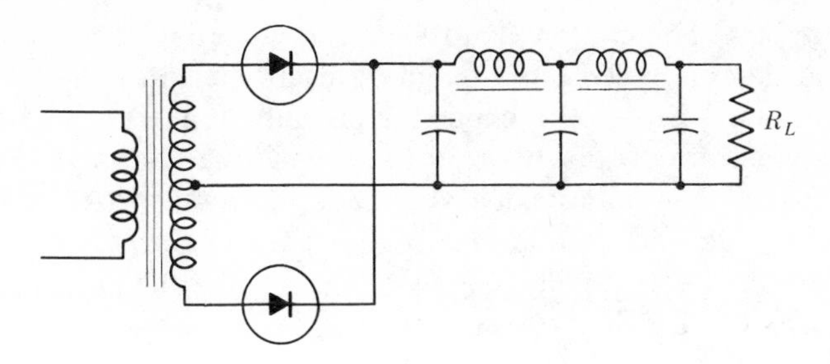

Figure 6-17 Full-wave rectifier using a double-π filter.

Section 6-9 Voltage Multipliers

A *full-wave voltage doubler* is obtained by replacing two diodes in the full-wave bridge rectifier with capacitors, Fig. 6-18*a*. Usually the circuit diagram is given in the form of Fig. 6-18*b*. Diode *A* charges C_A when *m* is positive and *n* is negative. When *n* is positive and *m* is negative, diode *B*

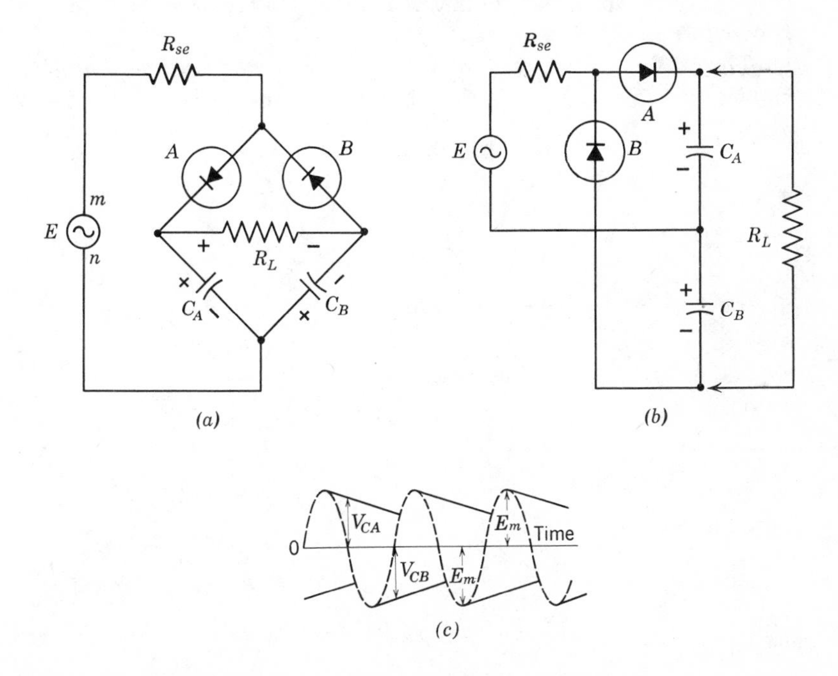

Figure 6-18 The full-wave voltage doubler. (*a*) Circuit. (*b*) Alternate form of circuit layout. (*c*) Waveform.

charges C_B. The two capacitors C_A and C_B are in series; hence, the voltage across them is twice the voltage on each, or is *doubled.* The load resistor R_L is placed across the capacitor combination. The load current through R_L comes from the discharge of the capacitors. The waveforms are shown in Fig. 6-18*c*. The output load voltage is the total spread between the top of the envelope and the bottom of the envelope. When the load current is very small, the load voltage is twice the peak of the line $2E_m$. There are two impulses of charging current into the capacitors per cycle; therefore, the ripple frequency is twice the frequency of the line. The action of the two diodes in the full-wave rectifier charges the whole filter twice each cycle, whereas the charging action of this circuit charges each capacitor once per cycle, but at different times. It is in this sense a full-wave rectifier and not a half-wave rectifier. The ripple in this circuit is greater and the regulation poorer than in the equivalent full-wave rectifier. The peak inverse-voltage ratings of the diodes are twice the peak of the line voltage $2E_m$. Since this circuit is often used on an a-c line without either an isolating transformer or a step-up or step-down transformer, it is important to notice that there is no common connection between the line and the load. When the expense of a line transformer is justified, it is preferable to use the superior circuit of the conventional full-wave rectifier. Typical load curves for the circuit are shown in Fig. 6-20.

The *half-wave voltage doubler* or the *cascade voltage-doubler* circuit is shown in Fig. 6-19*a*. When *n* is positive and *m* is negative, C_A charges

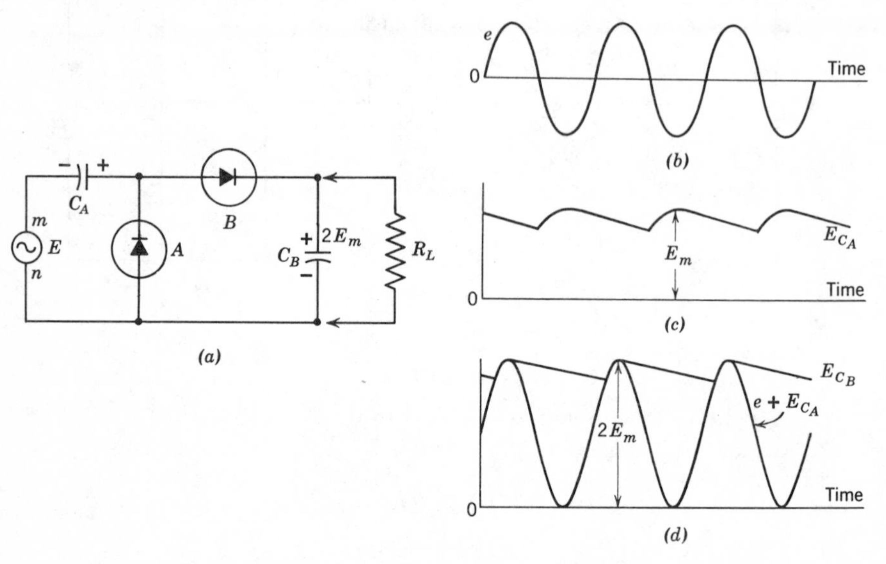

Figure 6-19 The half-wave voltage doubler. (*a*) Circuit. (*b*) Input voltage. (*c*) Waveform across C_A. (*d*) Waveform across C_B.

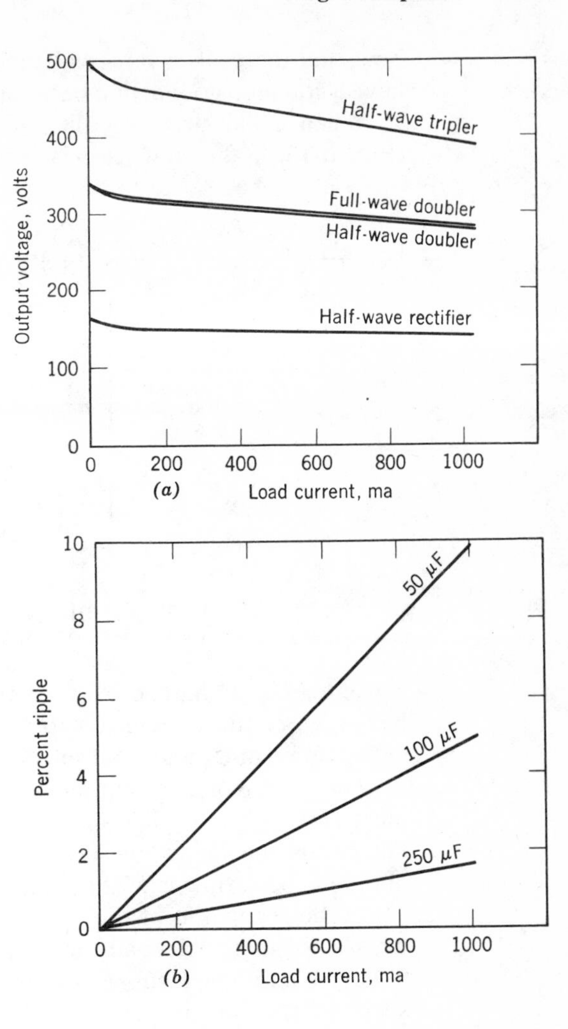

Figure 6-20 Characteristics of voltage multipliers. (*a*) Load voltage. (*b*) Ripple of a full-wave voltage doubler.

through diode A to E_m, the peak of the line voltage. This action is shown in the waveform of Fig. 6-19*c*. When the cycle reverses, n is negative and m is positive. Now, the line voltage e and the voltage across C_A are in series aiding. The maximum value this condition can have is $2E_m$, and C_B charges to $2E_m$ through diode B (Fig. 6-19*d*). The load is connected across C_B. The load receives only one charging pulse per cycle. The ripple frequency is the line frequency giving a basis for the use of the term "half wave." The regulation of this circuit is very poor, and the ripple is very high, even with medium values of load current. The peak inverse voltage on diode A is

$2E_m$, and the peak inverse voltage on diode B is $2E_m$ also. This circuit does have a common connection between the line and the load.

When a half-wave rectifier is added to the half-wave voltage doubler (Fig. 6-19a), the new circuit becomes a voltage tripler, Fig. 6-21. The

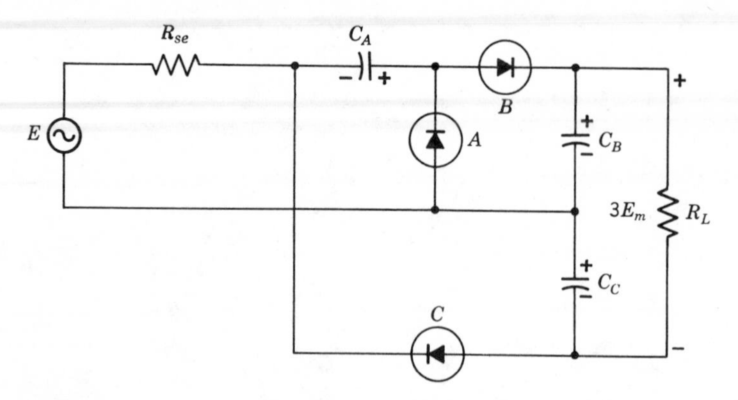

Figure 6-21 Voltage tripler.

capacitor C_B is charged to twice the peak of the supply voltage, $2E_m$. The half-wave circuit charges C_C to the peak of the line voltage E_m. The series combination of C_B and C_C results in a voltage across R_L of $3E_m$.

The basic voltage multiplier can be extended through a ladder arrangement (Fig. 6-22) to yield in N stages a total load voltage of NE_m volts. In this circuit the capacitors have different values, as shown on the circuit for optimum performance. The peak inverse voltage ratings of any diode is $2E_m$. The regulation of this circuit is very poor and its ripple is very high when any appreciable load current is drawn. Accordingly, its use is restricted to very high-voltage supplies where the load-current demands are very low, such as television kinescope supplies and portable Geiger counters.

Section 6-10 Shunt Rectifiers

The shunt rectifier is a form of the half-wave rectifier used in diverse applications in electronics and instrumentation. A half-wave rectifier with a pure capacitive load is shown in Fig. 6-23. The capacitor charges to the peak value of the line voltage and maintains this fixed d-c polarity (Fig. 6-23c). The voltage waveform across the diode is shown in Fig. 6-23d. The sum of the voltage across the capacitor and the voltage across the diode must be the sine-wave input voltage.

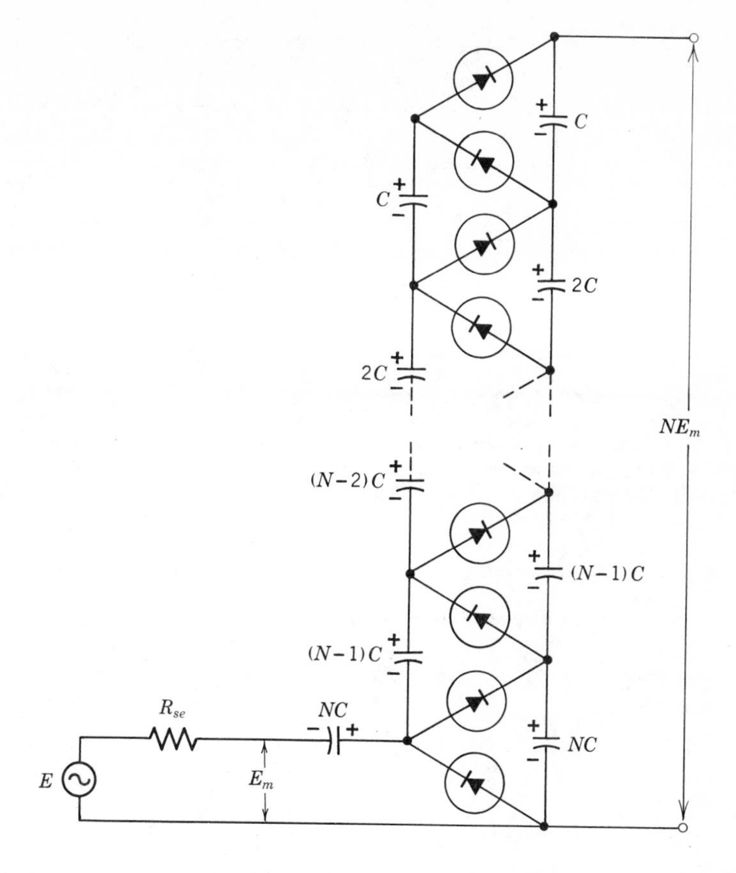

Figure 6-22 General voltage multiplier.

In the shunt rectifier (Fig. 6-24), the conventional positions of the capacitor and the diode rectifier are reversed. The load R_L is now placed parallel with the diode. The peak voltage across R_L is twice the peak of the line voltage, and its average value is the peak voltage of the line. The resistance R_L partially discharges the capacitor between peaks of the line voltage. When R_L is very large, the discharging action is slight. A small current pulse is required from the source to recharge the capacitor back to its peak value. This pulse can be converted to an rms value of current and, when this value of current is divided into the source rms voltage, the effective value of the impedance of the circuit is obtained. Consequently, this circuit presents a very high impedance to the source.

This circuit (Fig. 6-24*a*) can also be used as a *clamp*. A clamping circuit accepts a pure a-c waveform and converts it into a unidirectional waveform.

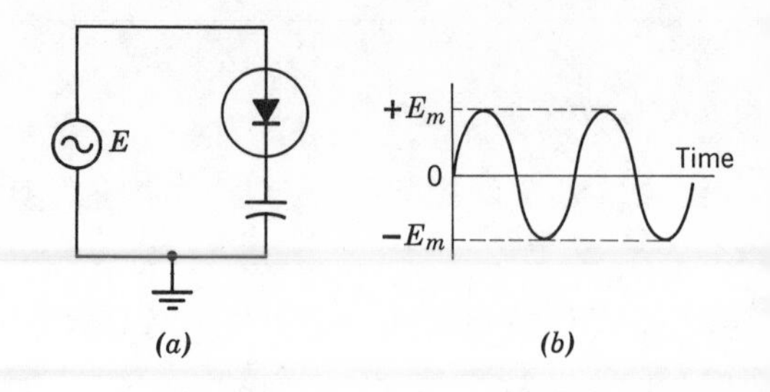

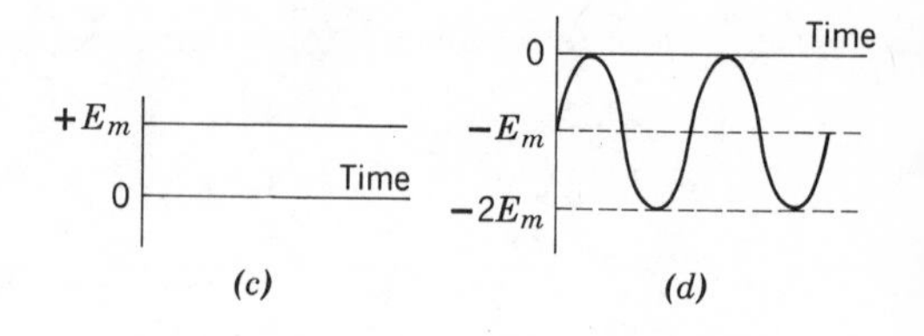

Figure 6-23 The basic half-wave rectifier. (*a*) Circuit. (*b*) Supply voltage. (*c*) Capacitor voltage. (*d*) Inverse voltage across the diode.

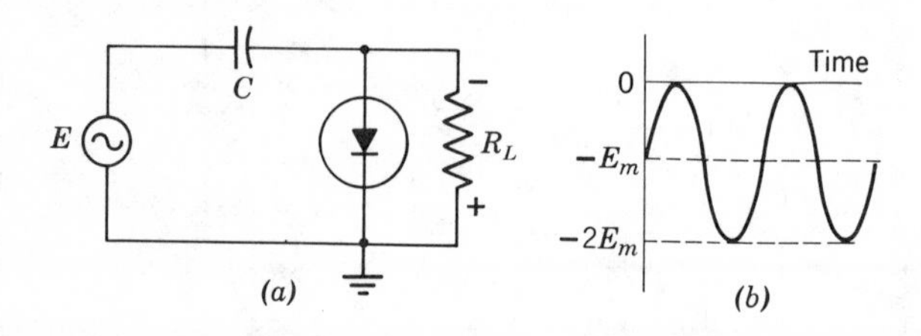

Figure 6-24 The basic shunt rectifier. (*a*) Circuit. (*b*) Output voltage waveform.

When the incoming a-c wave is a train of pulses (Fig. 6-25*a*), the output waveform is negative at all times (Fig. 6-25*b*). When the diode element is reversed, the polarity of the output waveform is reversed (Fig. 6-25*c*).

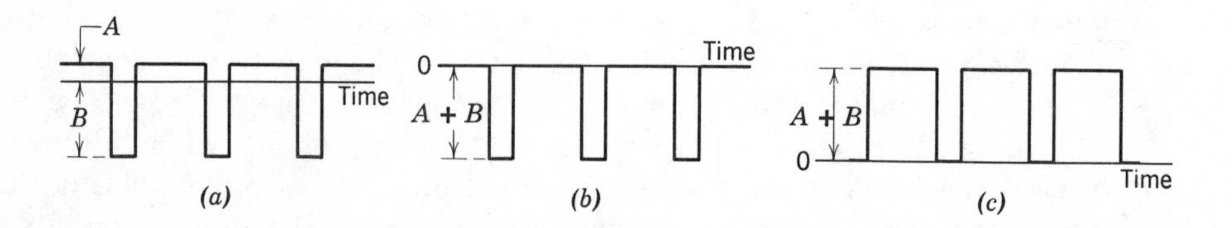

Figure 6-25 Waveforms of a clamp. (*a*) Input. (*b*) Output for circuit of Figure 6-24a. (*c*) Output with diode reversed.

Very often, this basic circuit is referred to in the literature as a clamp and not as a shunt rectifier.

The principles illustrated by the circuits of Figs. 6-24 and 6-25 have very important applications in electronic circuits. A direct voltage which is exactly proportional to the strength of the incoming signal is developed that is used to bias a transistor, an FET, or a vacuum tube. In solid-state electronics the term used to describe this action is *bias clamp*, and in vacuum-tube circuits the action is called *grid-leak bias*.

In Fig. 6-26 an RC filter is added to the basic shunt-rectifier circuit. The filter R_2C_2 establishes a pure direct voltage across R_L which is equal to

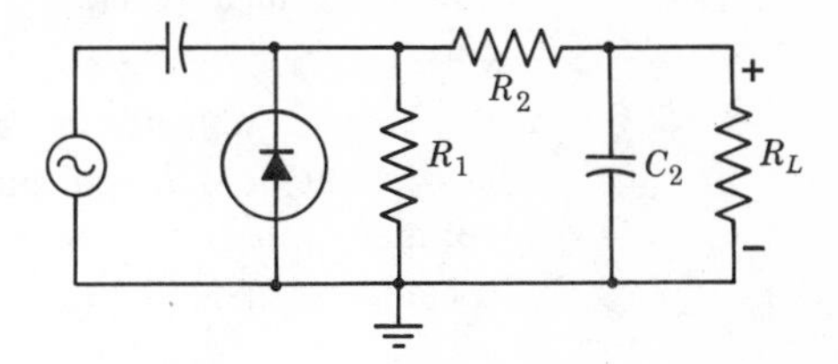

Figure 6-26 Shunt rectifier with output filter.

twice the peak of the source voltage. This circuit is often used in the probes of voltmeters which are designed to measure audio- and radio-frequency voltages without placing a severe shunting load impedance on the circuit where the measurement is taken.

The resistor R_2 in the circuit can be replaced by a filter choke for use as a source of d-c power. This version is used as a means of obtaining a low direct voltage for bias supplies which operate with low current requirements. Also the circuit is used to provide high-voltage low-current d-c sources. When the resistor R_2 in the filter is replaced by a second diode, the voltage-doubler circuit of Fig. 6-19*a* results.

PROBLEMS

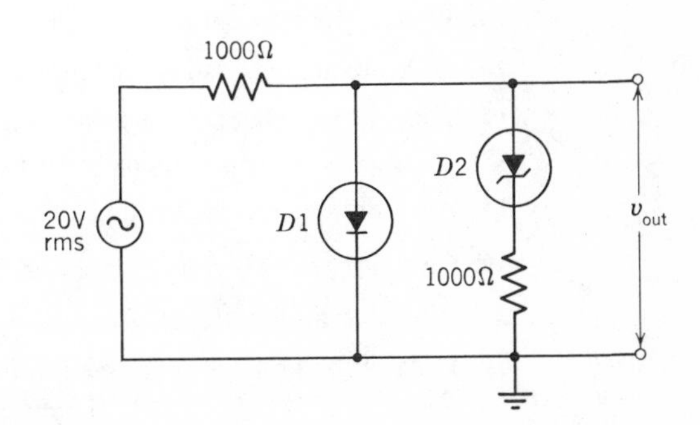

Circuit for Problem 2.

1. In Fig. 6-24 *a* R_L is 100 kΩ and E is $100 \sin 2000\,t$. Sketch the voltage waveform across the load if C is (*a*) 10 μF and (*b*) if C is 10 pF.
2. The Zener diode has a 6-V rating for breakdown. Sketch the voltage waveform across the output terminals (*a*) if the circuit is as shown, (*b*) if the diode *D*1 is reversed, and (*c*) if the diode *D*1 is removed from the circuit.

Questions

1. Define or explain each of the following terms: (*a*) peak inverse voltage, (*b*) ripple factor, (*c*) ratio of rectification, (*d*) transformer utilization factor, (*e*) d-c saturation, (*f*) voltage regulation, (*g*) voltage multiplier, (*h*) bias clamp, (*i*) grid-leak bias.
2. If a d'Arsonval meter movement is used to measure the voltage across a rectifier load, what does it indicate? If an iron-vane voltmeter is used? If an output voltmeter is used?
3. What is the effect on a half-wave rectifier if the diode opens? Short-circuits?
4. Can the numerical value of the ripple factor exceed unity? Explain.
5. What are the advantages and disadvantages of half-wave rectification?
6. Why is the ripple factor for a full-wave rectifier less than half the value for a half-wave rectifier (both with resistive loads)?
7. What are the advantages of a full-wave rectifier?
8. What is the effect on a full-wave rectifier if a diode opens? Short-circuits?
9. What are the advantages and disadvantages of a full-wave bridge?
10. What is the effect on a full-wave bridge if one diode opens? Short-circuits?
11. What is the peak inverse voltage on a diode when a capacitor filter is used?
12. Why is diode current limited to sharp pulses when a capacitor filter is used?
13. What is the significance of a negative voltage regulation?
14. What is meant by a critical value of inductance in a filter?
15. What is the effect of the d-c resistance of chokes in a power supply?
16. Give advantages and disadvantages of using gas tubes as rectifiers.
17. What is the effect on a full-wave voltage doubler if one diode opens? Short-circuits? If the other opens? Short-circuits?
18. Answer question 18 for the half-wave voltage doubler.
19. Compare the two voltage-doubler circuits.
20. Why is a shunt rectifier used in metering circuits?
21. How is the voltage developed in a shunt rectifier related to the rms value of the input?

Chapter Seven

CONTROLLED RECTIFIERS

The general problem of the relations between currents, voltages, and firing angles in a controlled rectifier is considered in Section 7-1. L-R and C-R phase-shift circuits are analyzed (Section 7-2) to show how a 180° phase control can be achieved. Various methods that are used to fire a silicon controlled rectifier (Section 7-3) and a triac (Section 7-4) are examined. The use of the same basic control circuits is extended to include the thyratron (Section 7-5).

Section 7-1 Analysis of Load Voltage and Load Current

In an ordinary diode rectifier circuit, current flows in the diode whenever the instantaneous a-c supply voltage is greater than the voltage across the load at that instant. When the load on a simple diode circuit is a resistive load, load current flows at all times during the half of the a-c cycle that the anode is positive. In a controlled rectifier with a resistive load (Fig. 7-1*a*) the load current is zero at all times unless a control signal is applied to the device to initiate anode current flow. The application of a control signal turns on or fires the rectifier at a specific point, *A*, in the cycle (Fig. 7-1*b*). Point *A* corresponds to an angle, θ_1, which is a point later than the start of the positive half of the a-c cycle.

Once the rectifier has fired, the rectifier remains in conduction until point *B* near the end of the positive half cycle. At this time located at θ_2 degrees, the current falls to zero in the rectifier. In a silicon controlled rectifier, when the anode current falls below the holding current I_H, the conduction current ceases. In a thyratron, when the plate current falls below the least value required to maintain ionization, the tube goes out. The value

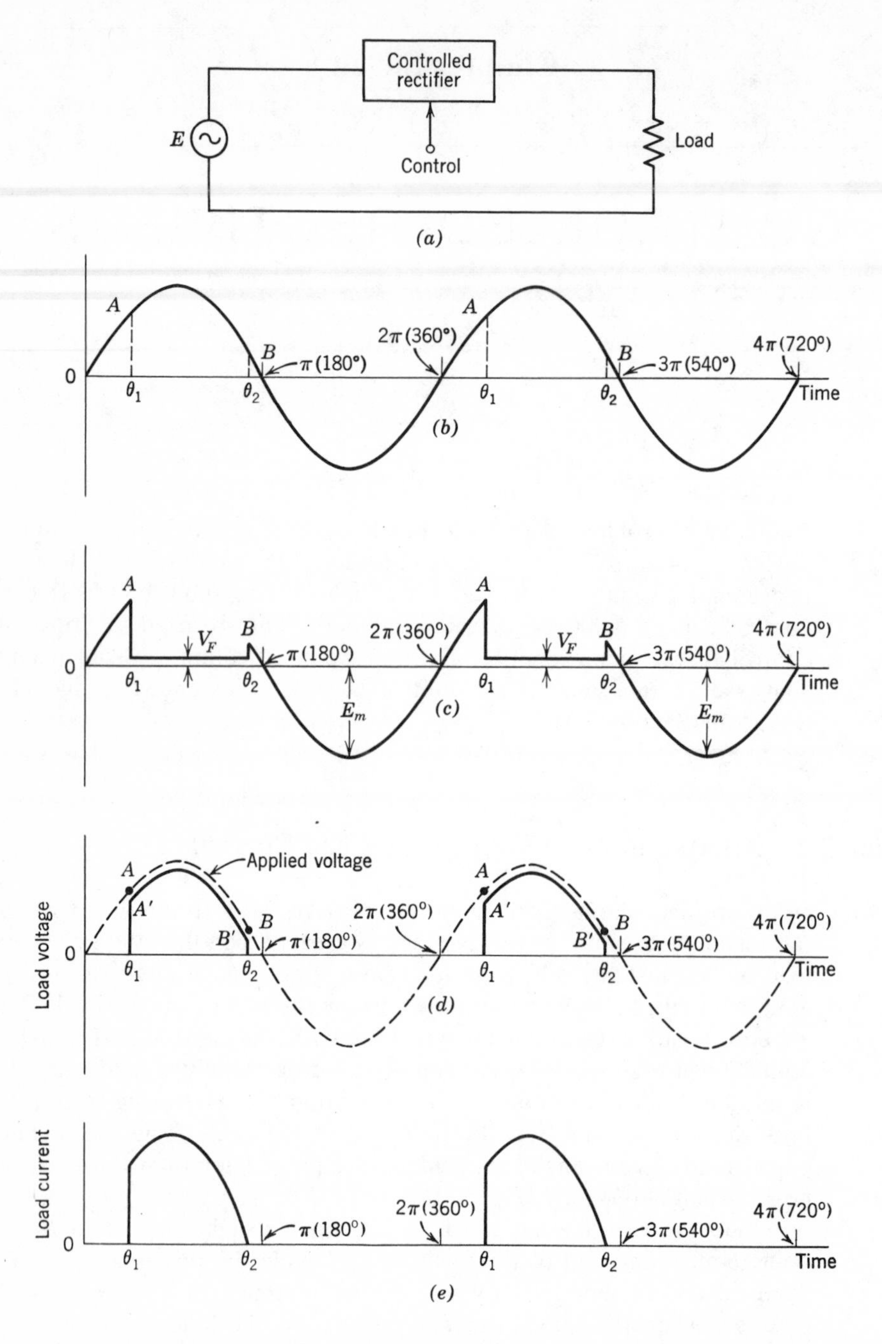

Figure 7-1 The controlled rectifier. (*a*) Block diagram. (*b*) Supply voltage. (*c*) Voltage drop across controlled rectifier. (*d*) Load voltage. (*e*) Load current.

of θ_2 is a function of the rectifier characteristic and is *not* determined by the control signal.

Thus, point A, the firing point, is determined by the angle of delay in the application of the firing signal in the control circuit. As the delay angle θ_1 increases, point A occurs later and later in the cycle and the current in the load decreases.

When the rectifier is turned on at A, the forward voltage drop across the rectifier falls to the level indicated by V_F on Fig. 7-1*c* and remains at that value until the rectifying action ceases at θ_2. During the negative half of the cycle the inverse voltage having a peak value of E_m volts appears across the rectifier in the same manner as in an ordinary diode rectifier circuit. The voltage drop across the load (Fig. 7-1*d*) is the a-c supply voltage between θ_1 and θ_2 less the forward voltage drop, V_F, of the diode. Since the load current (Fig. 7-1*e*) follows Ohm's law, its waveform is proportional to the load-voltage waveform. The d-c voltage across the load and the d-c current in the load is the *average* of the values in the waveform over the *full* cycle from 0 to 2π.

Let us consider the load voltage over one cycle. The load voltage may be expressed by the set of equations for a discontinuous wave:

$$v_L = 0 \qquad \text{when} \qquad 0 \leqslant \theta \leqslant \theta_1 \tag{7-1a}$$

$$v_L = E_m \sin\theta - V_F \qquad \text{when} \qquad \theta_1 \leqslant \theta \leqslant \theta_2 \tag{7-1b}$$

$$v_L = 0 \qquad \text{when} \qquad \theta_2 \leqslant \theta \leqslant 2\pi \tag{7-1c}$$

The definition of an average d-c value in a rectifier is given by

$$V_L \equiv \frac{1}{2\pi}\int_0^{2\pi} v_L d\theta \tag{7-2}$$

Geometrically, this equation states that the average value of load voltage is evaluated by determining the area of the curve shown in Fig. 7-1*d* and by dividing the result by the length of the curve 2π. Since the load voltage is zero at all angles other than the region between θ_1 and θ_2, only this interval need be considered. When Eq. 7-1*b* is substituted, Eq. 7-2 becomes

$$V_L = \frac{1}{2\pi}\int_{\theta_1}^{\theta_2} (E_m \sin\theta - V_F)\, d\theta \tag{7-3}$$

Removing the parenthesis

$$V_L = \frac{1}{2\pi}\int_{\theta_1}^{\theta_2} E_m \sin\theta\, d\theta - \frac{1}{2\pi}\int_{\theta_1}^{\theta_2} V_F d\theta$$

Integrating

$$V_L = -\frac{E_m}{2\pi}\cos\theta\Big]_{\theta_1}^{\theta_2} - \frac{1}{2\pi}V_F\,\theta\Big]_{\theta_1}^{\theta_2}$$

Substituting the limits

$$V_L = -\frac{E_m}{2\pi}(\cos\theta_2 - \cos\theta_1) - \frac{1}{2\pi}V_F(\theta_2 - \theta_1)$$

$$V_L = \frac{E_m}{2\pi}(\cos\theta_1 - \cos\theta_2) - \frac{\theta_2 - \theta_1}{2\pi}V_F$$

In this equation θ_1 and θ_2 are in radians. Usually, these angles are measured in degrees. When this conversion is made and since 2π radians represents 360°, the load voltage is

$$V_L = \frac{E_m}{2\pi}(\cos\theta_1 - \cos\theta_2) - \frac{\theta_2 - \theta_1}{360}V_F \tag{7-4a}$$

and the load current is

$$I_L = \frac{E_m}{2\pi R_L}(\cos\theta_1 - \cos\theta_2) - \frac{\theta_2 - \theta_1}{360}\frac{V_F}{R_L} \tag{7-4b}$$

In most applications of controlled rectifiers, the peak value of the line voltage E_m is very much greater than the forward voltage drop V_F. For example, the peak voltage of a 117-V circuit is 166 V whereas a typical value of V_F for a silicon controlled rectifier is 1 V. Under this condition, the second term may be neglected to simplify the equations to

$$V_L = \frac{E_m}{2\pi}(\cos\theta_1 - \cos\theta_2) \tag{7-5a}$$

and

$$I_L = \frac{E_m}{2\pi R_L}(\cos\theta_1 - \cos\theta_2) \tag{7-5b}$$

Similarly, when E_m is large, θ_2 approaches 180°. When 180° is substituted for θ_2, the error is small. Since the numerical value of cos 180° is -1, the equations become

$$V_L = \frac{E_m}{2\pi}(1 + \cos\theta_1) \tag{7-6a}$$

and

$$I_L = \frac{E_m}{2\pi R_L}(1 + \cos\theta_1) \tag{7-6b}$$

If θ_1 is allowed to go to zero, the equations become

$$V_L = E_m/\pi \qquad \text{and} \qquad I_L = E_m/\pi R_L$$

These equations are the ones of the previous chapter for a half-wave rectifier with a resistive load in which the conduction was assumed for the full positive half of the a-c cycle.

The curve given in Fig. 7-2 is a normalized plot of either Eq. 7-6*a* or Eq. 7-6*b*. From this curve, it is seen that there is a smooth control of the output in the load from a maximum value to zero. Typical waveforms that

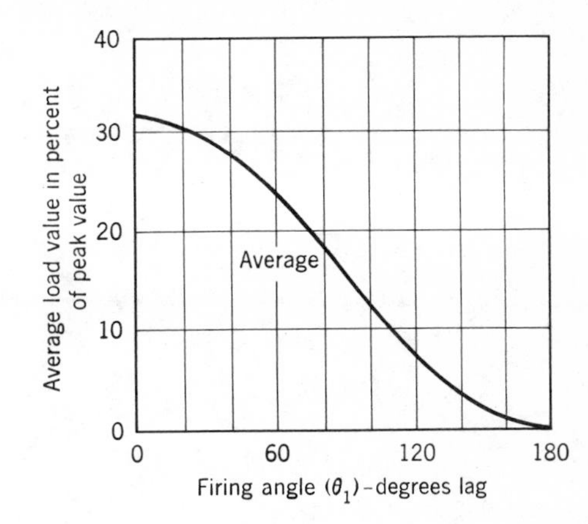

Figure 7-2 Average value of load voltage and load current in a phase-controlled circuit.

may be observed on an oscilloscope for different values of control angle θ_1 are shown in Fig. 7-3.

PROBLEMS

1. The supply voltage is 200-V peak. The forward voltage drop of the rectifier is 2 V. When the load resistance is 1000 Ω, what is the range of d-c load current with firing at 0°, at 45°, at 90°, and at 135°?
2. The supply voltage is 30 V rms, and the forward voltage drop of the rectifier is 2 V. When the load resistance is 10 Ω, what is the range of d-c load current with firing at 0°, at 45°, at 90°, and at 135°?
3. A 10-Ω load is connected to a 117-V rms supply through a controlled rectifier. The load power must be varied between 90 W and 20 W. What is the angular firing control that is required? Assume that V_F is 2 V.
4. A load is connected to a 117-V rms supply through a controlled rectifier. The maximum load power is 50 W and the load power must be controlled down to 20 W. What is the value of the load resistance and over what angular range must firing control be maintained? Assume that V_F is 2 V.
5. A load is connected to a 117-V rms source through a controlled rectifier. The peak current in the load is 4.0 A and the minimum average current is 0.56 A. What is the value of the load resistance, and what is the angular firing control range? What is the range of d-c load current? Assume that V_F is 2 V.
6. Solve Problem 5 if the source voltage is 10 V rms.

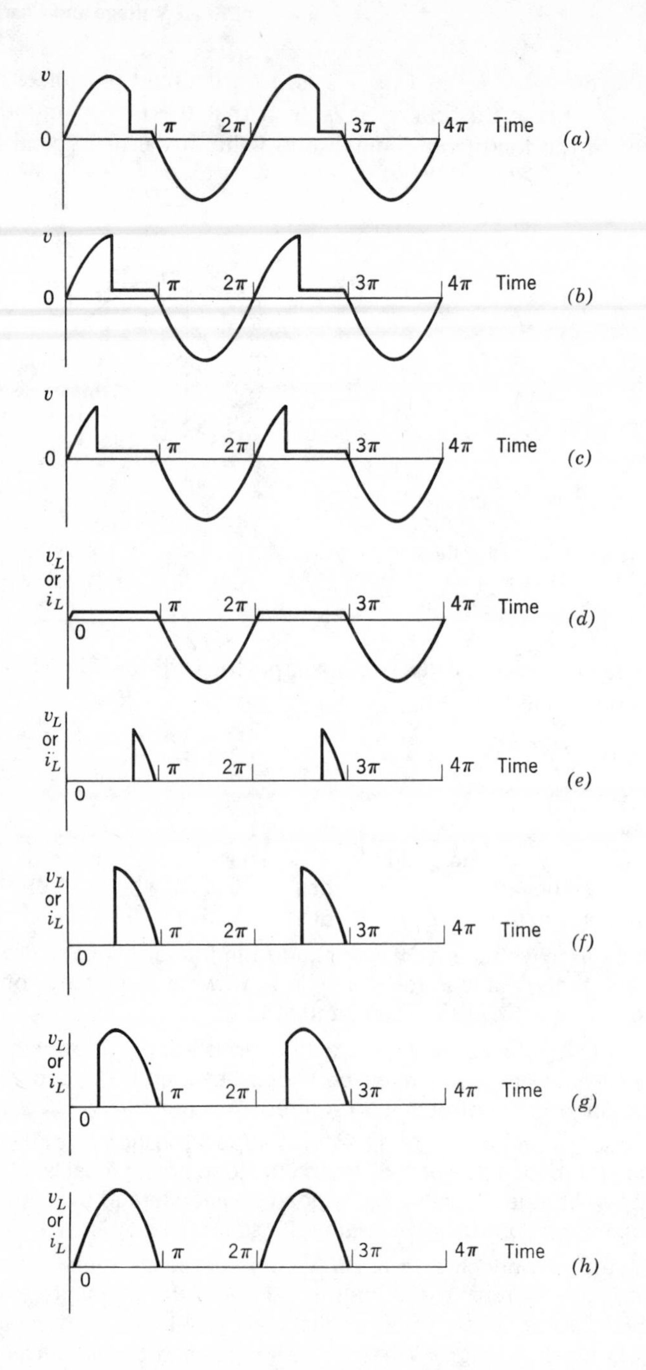

Figure 7-3 Rectifier waveforms. Anode voltage for firing at (*a*) 135°, (*b*) 90°, (*c*) 45°, and (*d*) 0°. Load voltage or load current for firing at (*e*) 135°; (*f*) 90°, (*g*) 45°, and (*h*) 0°.

Section 7-2 Phase-Shift Circuits

A commonly used circuit arrangement to obtain a 180° delay range for the control signal is by the use of a simple *LR* or *CR* series circuit. If a voltage V_{AC} from a center-tapped transformer is applied to a network of R and L (Fig. 7-4), the current I lags V_{AC} by an angle θ. The IR drop is in phase with I,

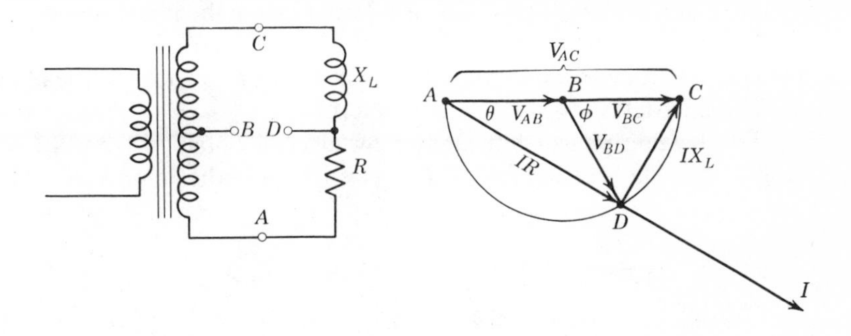

Figure 7-4 *LR* network producing a lagging phase angle.

and the current lags the IX_L drop by 90°. These phasors are sketched in the phasor diagram. B is the midpoint of V_{AC} since it is the center tap on the transformer. Angle ADC is the right angle of a right triangle whose hypotenuse is AC. If either L or R is varied, the lengths of the legs of the right triangle change, but the hypotenuse is fixed. By a theorem in plane geometry, the locus of the point D must follow a semicircle. The voltage V_{BD} represents the voltage from the center tap to the junction of the resistor and the inductance. In the phasor diagram, BC and BD are radii of the semicircle making triangle BCD isosceles. Then, angle BCD equals angle BDC. But, since triangle ACD is a right triangle,

$$\theta + \angle ACD = 90$$

then
$$\angle BCD = 90 - \theta$$

since
$$\angle BCD = \angle BDC$$

and since
$$\angle BCD + \angle BDC + \phi = 180$$

then
$$(90 - \theta) + (90 - \theta) + \phi = 180$$

or
$$\phi = 2\theta \qquad \text{where} \qquad \tan\theta = X_L/R \tag{7-7a}$$

If V_{BC} (or another voltage in phase with V_{BC}) is used as the anode supply, and if V_{BD} is used as the control voltage, a variation in either L or R can produce almost a full range of lagging phase-shift control over 180°. The presence of resistance in the inductance prevents the attainment of a full 180° control. The control voltage V_{BD} is a radius of the semicircle, and will not change its magnitude as the phase angle is varied. When R equals X_L, point D is the midpoint on the arc between A and C, and the phase angle is 90°. If R is the variable element in the phase-shift circuit, an increase in R moves point D from A toward C, decreasing the phase angle. If L is variable, an increase in L moves point D toward A, increasing the phase angle.

If L and R are interchanged, the phase angle is no longer lagging, but leading. The development of the phasor diagram to show this leading phase-shift angle is left to the reader as an exercise.

The lagging phase-shift angle may also be obtained from a CR circuit, (Fig. 7-5). The current in the CR circuit leads the impressed voltage V_{AC}

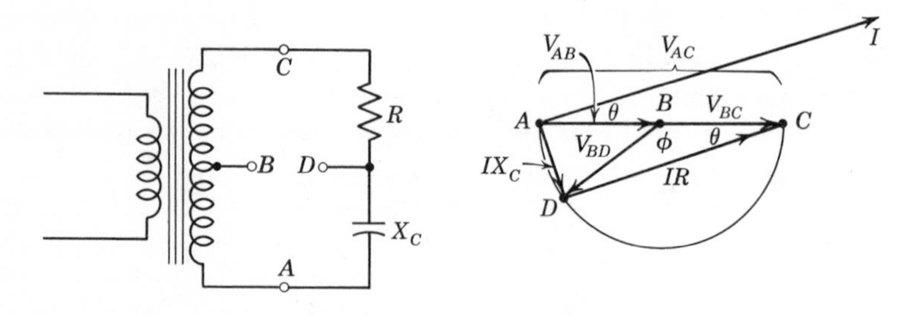

Figure 7-5 *CR* network producing a lagging phase angle.

by θ degrees. The current leads the IX_C voltage phasor by 90° and is in phase with (parallel to) the IR voltage phasor. By the same logic as in the LR circuit, the locus of point D traces out a semicircle. V_{BD} lags V_{BC} by the angle ϕ. ϕ can vary from zero to 180° by changing either R or C. Since $\angle ACD$ is θ, the angle by which V_{BD} lags V_{BC} is

$$\phi = 180 - 2\theta$$

where

$$\tan\theta = \frac{X_C}{R} \tag{7-7b}$$

When the value of the resistor is increased, point D moves toward A, and the phase-shift angle ϕ increases. If the capacitance is increased, X_C decreases, and point D again moves toward A. When R equals X_C, point D is at the midpoint on the semicircle, and the phase shift is 90°. Since capacitors are close to the ideal at power frequencies, a full 180° control can be obtained.

There are many ways of obtaining variable elements for these phase-shift circuits. A direct means is to use a rheostat mounted on a panel as R. The pointer can be calibrated directly in terms of firing angle, conduction angle, load current, or load voltage. In other applications, the variable R is a temperature sensitive element. If the d-c voltage from base to collector is fixed, the current through the transistor is determined by the emitter to base voltage allowing the transistor to serve as the variable resistive element in a phase-shift circuit.

A varactor diode can serve as a voltage-controlled capacitor to be the signal sensing element in the phase-shift circuit. A variable inductance can be obtained from a *saturable-core reactor.* In addition to the normal a-c inductance winding for L on a coil, the saturable-core reactor has a second winding of many turns that carries a d-c control current. When this d-c current is zero, the saturable-core reactor operates below the knee of the magnetic saturation curve and the inductance between the terminals of the a-c winding is high. The d-c current in the control winding can shift the operating point of the coil beyond the knee of the saturation curve. Now with magnetic saturation in the core, the value of the a-c inductance L is very small. Thus, the d-c control current can vary the a-c inductance of the saturable-core reactor from very large values to very small values to yield a large range of output phase angle variations in an R-L phase-shift circuit. A saturation curve is given in Fig. 9-9 with an accompanying explanation.

PROBLEMS

Note: The frequency is 60 Hz for all problems.

1. Interchange R and X_L in Fig. 7-4. Draw the phasor diagram. Determine the relationship between X_L and R to give a 180° phase control for a leading angle.
2. Interchange R and X_C in Fig. 7-5. Draw the phasor diagram. Determine the relationship between X_C and R to give a 180° phase control for a leading angle.
3. The phase-shift control circuit shown in Fig. 7-4 is used to control a controlled rectifier over a 5 to 1 current range. The inductor varies to 10 H. What is the value of R?
4. The phase-shift control circuit shown in Fig. 7-4 is used to control a controlled rectifier over a 3 to 1 current range. The inductor varies to 10 H. What is the value of R?
5. The phase-shift control circuit shown in Fig. 7-5 is used to control a controlled rectifier from 20% to maximum current. The capacitor is 0.05 μF. What size rheostat should be used for R?
6. The phase-shift control circuit shown in Fig. 7-5 is used to control a controlled rectifier from 10% to maximum current. The capacitor is 0.15 μF. What size rheostat should be used for R?

Section 7-3 The Firing of a Silicon-Controlled Rectifier

A summary of the essential characteristics of a typical silicon-controlled rectifier, the Texas Instrument 2N1604, is listed in Table A. The silicon-controlled rectifier is fired or triggered by the initiation of a suitable current

TABLE A Texas Instrument 2N1604 Characteristics

Anode Characteristics		
$V_{F(\mathrm{OFF})}$	Forward voltage in the "off" condition	400 V
V_R	Peak inverse voltage	400 V
I_F	Average rectified forward current	3 A
i_f	Recurrent peak forward current	10 A
$i_{f(\mathrm{surge})}$	Surge current, 1 cycle at 60 cps	25 A
BV_F	Min forward breakover voltage	480 V
BV_R	Min reverse breakdown voltage	480 V
I_R	Max dc reverse current at rated V_R	0.25 ma
V_F	Max forward voltage drop at $I_F = 3$Adc	2 V
I_H	Max holding current	25 ma
$I_{F(\mathrm{OFF})}$	Max dc forward current at V	0.25 ma
Gate Characteristics		
I_G	Forward gate current, maximum	100 ma
V_{GR}	Gate peak inverse voltage	5 V
I_{GT}	Max gate current to trigger[a]	10 ma
BV_G	Min gate breakdown voltage	6 V
V_G	Max forward gate voltage drop at $I_G = 25$ ma	3 V

[a]Positive triggering is assured with a 10 ma gate current.

in the gate. The value of gate current required to guarantee triggering is specified in the data sheet as 10 ma. The gate current must be maintained until the anode current has sufficient time to increase to at least the holding-current value, I_H. The relation between the gate current and the turn-on time is shown in Fig. 7-6.

From the data in Table A, we observe that the voltage required to fire the gate is small, of the order of 3 V. Because of the low values of firing voltage, it is simpler to trigger silicon-controlled rectifiers from some form of pulse instead of by a continuously applied voltage in the control circuit. The simple circuits shown in Fig. 7-7 limit the gate current to safe values by means of current limiting resistors. However, in the d-c applications, the silicon-controlled rectifier is not subjected to reverse voltages, and two separate operations are required. One operation turns the circuit on and the other turns the circuit off.

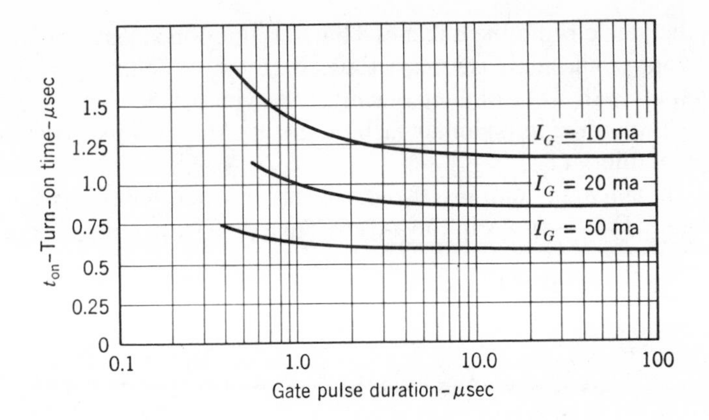

Figure 7-6 Turn-on characteristic for 2N1604.

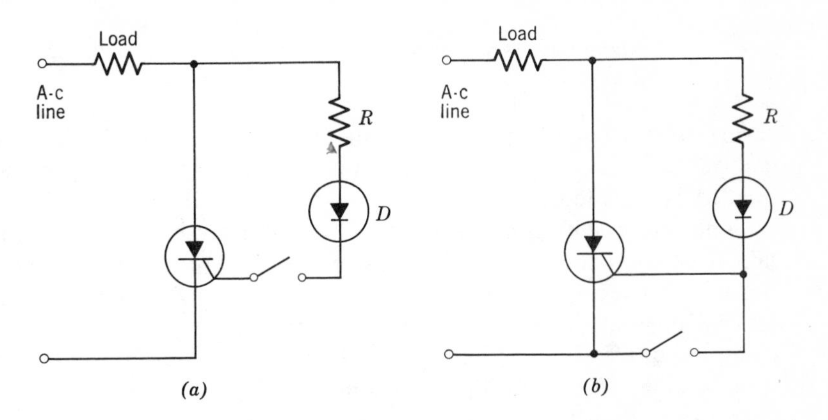

Figure 7-7 A-C switches. (*a*) Series. (*b*) Shunt.

In an a-c application, one switch can be used for both turn-on and turn-off (Fig. 7-7). When the switch is closed in the series switch (Fig. 7-7*a*), current flows in the gate circuit to trigger the anode on the half of the cycle on which the anode is positive. As soon as the SCR fires, the forward voltage drop from anode to cathode V_F is very small, of the order of one volt. Since the entire circuit of R, the diode D, and the gate is a series string in parallel with the anode-cathode path of the rectifier, the gate current falls to a very low value. When the line voltage reverses its polarity, the cathode is positive and the anode is negative. Consequently, the anode current in the silicon-controlled rectifier falls to zero. The diode D, is required to prevent a reverse current flow in the gate. The reverse anode current I_R for the 2N1604 at an inverse voltage of 400 V is only 0.25 ma. If there is any reverse current in the gate circuit, the effect of a barrier junction is lost and the anode reverse current would increase many times 0.25 ma. When the switch is opened, the rectifier will not conduct on any following cycle. The shunt switch (Fig. 7-7*b*) operates in a similar fashion except that it must be opened to turn on the rectifier.

In many applications, the gate circuit is operated from a sinusoidal source of emf (Fig. 7-8). As in the a-c switch, a provision must be made to prevent a

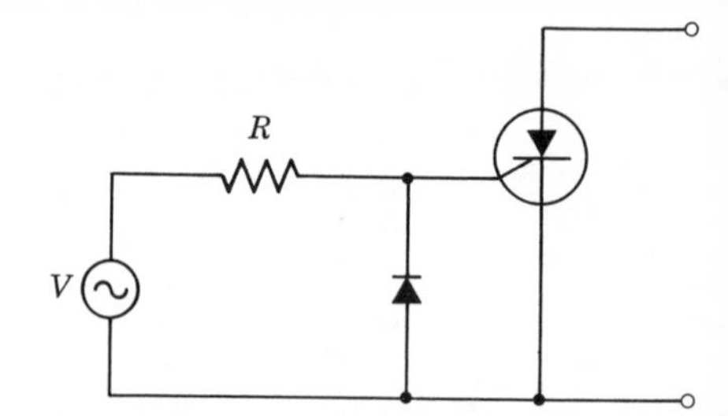

Figure 7-8 A gate clamp.

reverse breakdown current in the gate circuit. A diode is placed between the cathode and gate that effectively shunts or *clamps* the gate circuit when the polarity is reversed.

The phase-shift circuit shown in Fig. 7-9*a* produces a gate voltage that lags the main anode voltage to control the angle of the cycle at which firing occurs. Gate current is limited by a series resistor R. To prevent excessive leakage currents, reverse current flow in the gate is precluded by means of a clamping diode (Fig. 7-9*b*). If $Q1$ is one rectifier in a full-wave circuit and if the voltage from A to B provides the lagging control voltage to trigger $Q1$ (Fig. 7-9*b*), then the voltage from B to A is the proper lagging voltage required to trigger a second SCR, $Q2$ (Fig. 7-9*c*), the other rectifier of the full-wave circuit. This is true because the voltage from B to A is 180° out of phase with the voltage from A to B and the supply voltage on the anode of $Q2$ is

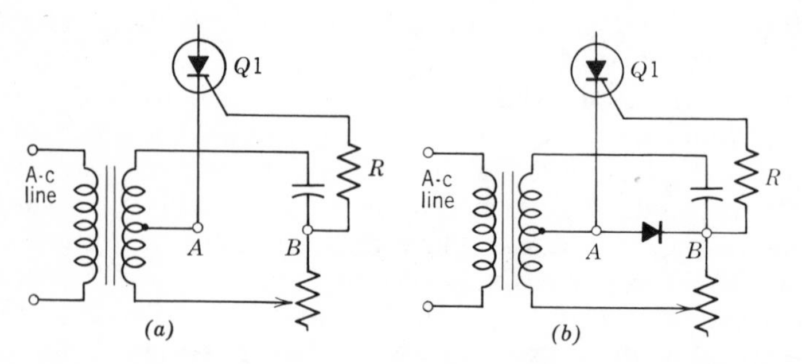

Figure 7-9 Phase-shift control circuits. (*a*) Basic control circuit. (*b*) With clamping diode.

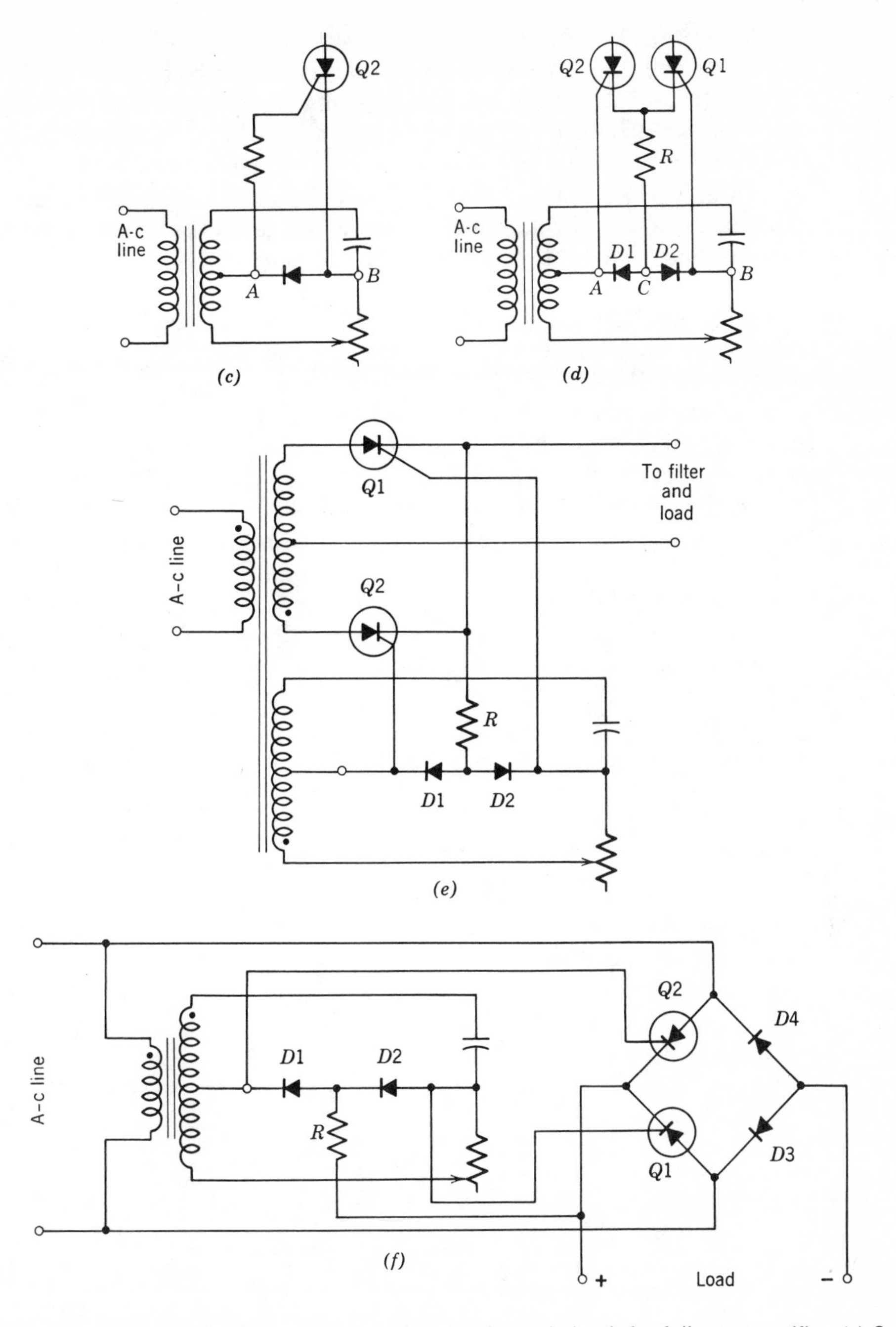

Figure 7-9 (*c*) Control circuit for second rectifier. (*d*) Control circuit for full-wave rectifier. (*e*) Complete full-wave controlled rectifier. (*f*) Controlled bridge rectifier.

180° out of phase with the other anode supply voltage if full-wave rectification is to take place. These two circuits may be combined (Fig. 7-9*d*). The diodes, $D1$ and $D2$, effectively switch point C to A for one half the a-c cycle and then to B for the other half of the a-c cycle. Thus one phase-shift circuit can serve both silicon-controlled rectifiers. The complete rectifier circuit is shown in Fig. 7-9*e*. The addition of two diodes, $D3$ and $D4$, converts this circuit into a phase-controlled full-wave bridge rectifier (Fig. 7-9*f*). The specifications listed for the SCR in Table A state that the gate peak inverse voltage V_{GR} is limited to 5 V. This is a typical value for the SCR, and precautions must be taken to adhere strictly to this specification or immediate destruction of the SCR will occur. All the circuits shown in Fig. 7-7 and in Fig. 7-9 must have protection against this low inverse voltage rating. If a diode is connected between the cathode and the gate so that the anode of the diode is tied to the cathode of the SCR and the cathode of the diode is tied to the gate of the SCR, the reverse voltage cannot exceed the forward voltage of the diode, which is less than 1 V and adequate protection of the SCR is assured. In some circuits a 1000-Ω resistor is connected between the cathode and the gate of the SCR to protect the gate circuit.

When the SCR is used in a full-wave application, the values of V_L and I_L, given by Eqs. 7-4, 7-5, and 7-6, are merely multiplied by two. The curve shown in Fig. 7-2 can be used, provided that the ordinate is multiplied by two also.

Consider the simple series circuit of R_1 and C_1, shown in Fig. 7-10*a*. As long as the switch S is closed, the voltage across the capacitor v_c is zero.

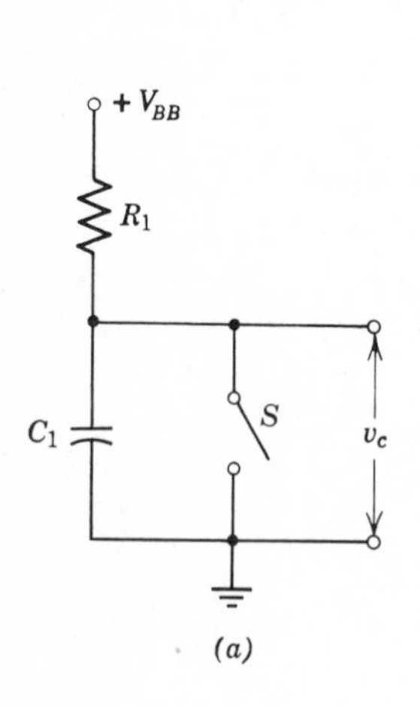

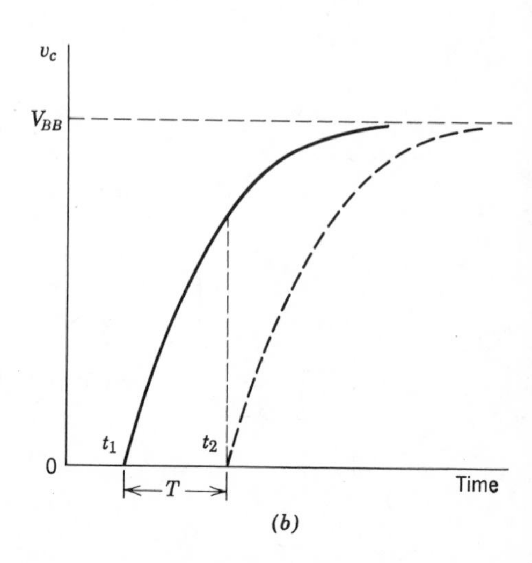

Figure 7-10 The charging of a capacitor. (*a*) Test circuit. (*b*) Waveform.

When the switch is opened at t_1 (Fig. 7-10*b*), the capacitor begins to charge and the voltage rises as an exponential function. If at t_2 the switch is momentarily closed and then reopened, the capacitor is discharged and then the charging of the capacitor begins anew. A unijunction transistor (Section 4-7) is used as the switch across the capacitor (Fig. 7-11). As soon as the voltage

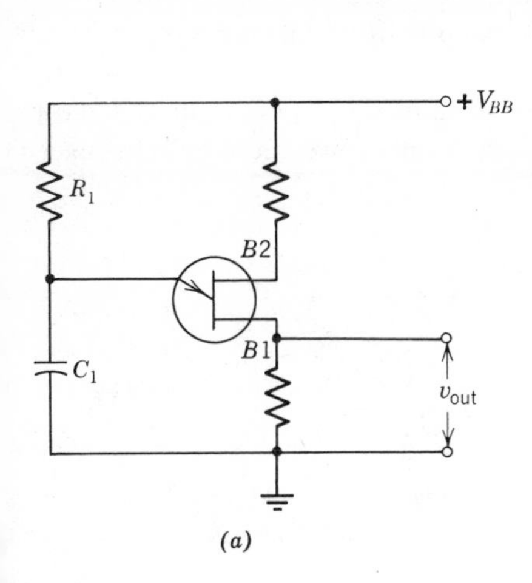

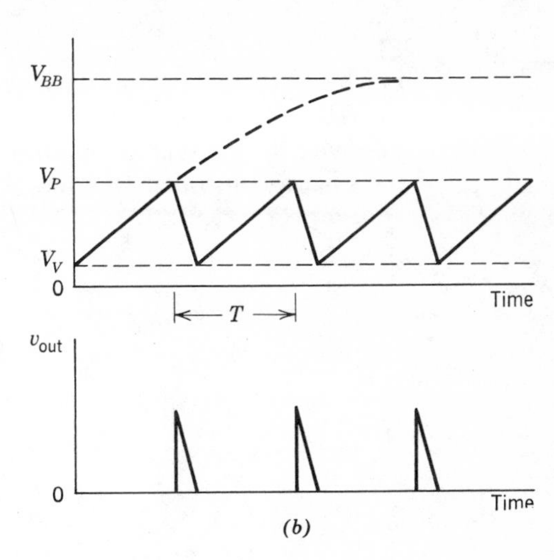

Figure 7-11 Basic unijunction firing circuit.

across the capacitor reaches the peak point V_P, breakdown occurs and the voltage across the capacitor falls to the low value of the valley point V_V. The voltage on the charging capacitor may be expressed by

$$v_c = V_{BB}(1 - \epsilon^{-t/R_1C_1})$$

The peak voltage is related to the intrinsic stand-off ratio η by

$$V_P = \eta V_{BB}$$

At the instant before the unijunction transistor fires, the voltage across the capacitor is

$$v_c = V_P = \eta V_{BB} = V_{BB}(1 - \epsilon^{-T/R_1C_1})$$

Dividing by V_{BB},

$$\eta = 1 - \epsilon^{-T/R_1C_1}$$

transposing

$$\epsilon^{T/R_1C_1} = \frac{1}{1-\eta}$$

Taking logarithms

$$\frac{T}{R_1C_1} = \ln\frac{1}{1-\eta}$$

or

$$T = R_1C_1 \ln\frac{1}{1-\eta} \tag{7-8}$$

in which T is the time for the capacitor to reach V_P from discharge.

When this circuit is used to trigger a silicon-controlled rectifier, the source of power for the unijunction transistor is the same a-c supply that is used for the silicon-controlled rectifiers (Fig. 7-12*a*). A half-wave rectifier $D1$ is used

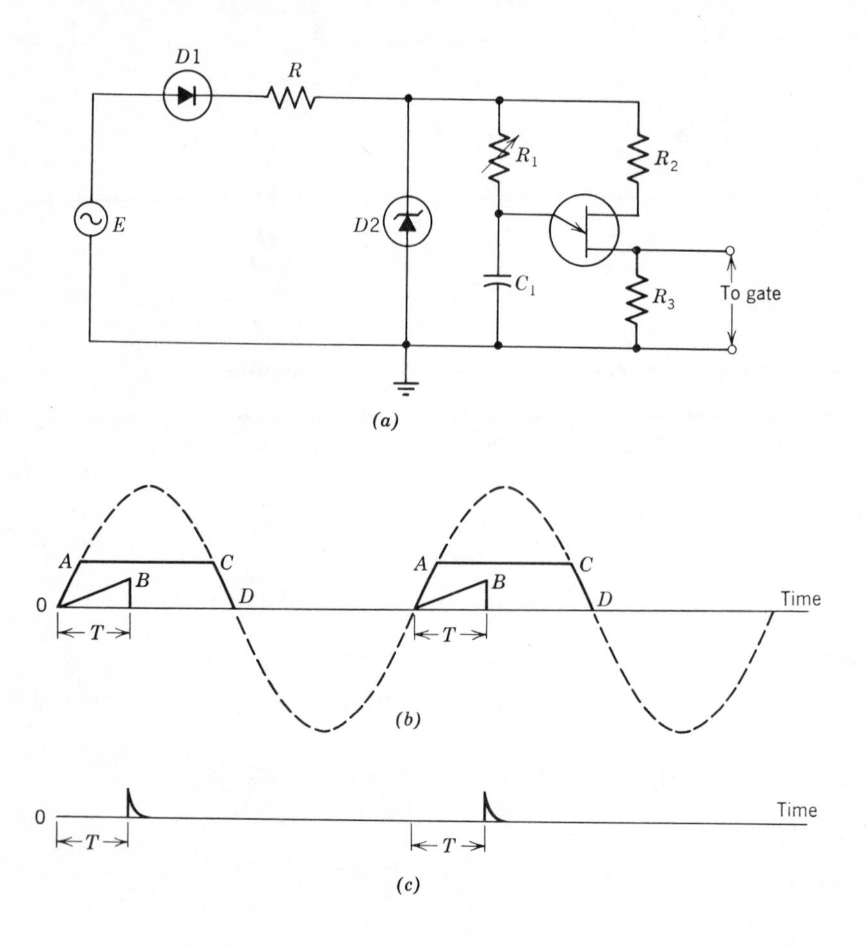

Figure 7-12 Firing of a unijunction transistor from an a-c source. (*a*) Circuit. (*b*) Capacitor voltage. (*c*) Gate input waveform.

in conjunction with a Zener diode $D2$. The waveform of the voltage across the unijunction transistor circuit is the trapezoidal waveform, O-A-B-C-D, shown in Fig. 7-12*b*. On the negative half of the a-c cycle, the diode $D1$ does not conduct. Capacitor C_1 in the emitter circuit begins to charge as soon as the supply voltage goes positive. At point B, the voltage across the capacitor reaches the peak voltage V_P and the unijunction transistor conducts to discharge the capacitor. The time delay T between 0 and B can be converted in terms of degrees lag in the a-c cycle. The angle of lag is controlled by varying the series charging resistor R_1. At B the firing of the unijunction transistor develops the firing pulse as the voltage drop across R_3 (Fig. 7-12*c*) for the silicon-controlled rectifier gate.

In order for the firing pulse to be stable, the capacitor in the unijunction transistor circuit cannot be allowed to recharge once it has discharged in a particular positive half cycle. To accomplish this, the entire unijunction transistor pulse forming circuit is connected between the anode and the cathode of the silicon-controlled rectifier it controls. Before the silicon-controlled rectifier breaks down, the voltage across both the rectifier and the control circuit follows the supply voltage. After the silicon-controlled rectifier fires, the forward voltage drop falls to the value of V_F which is of the order of 1 V. Accordingly, the entire voltage available for the triggering circuit is insufficient to produce a recharge of the capacitor. Such an arrangement is shown in Fig. 7-13. This concept can be readily extended to the full-wave controlled rectifier circuit shown in Fig. 7-14. The waveforms for this circuit are given in Fig. 7-15. By a suitable choice of R_1 and C_1, phase control can be obtained over nearly the full cycle from 0 to 180°.

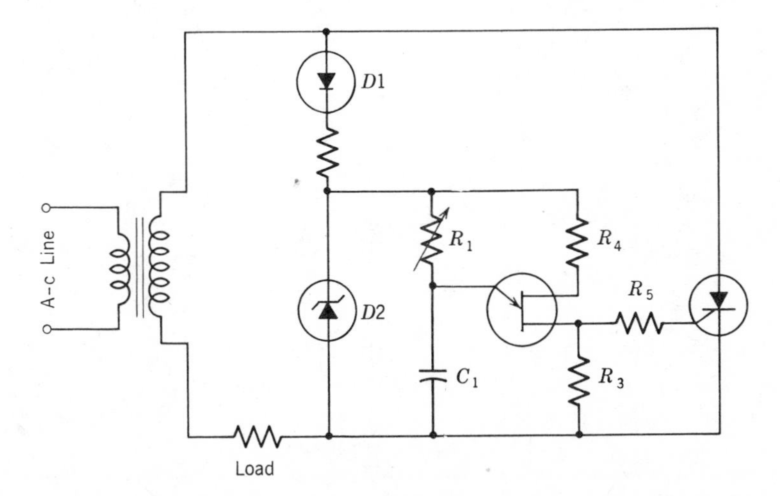

Figure 7-13 Half-wave silicon controlled rectifier circuit with phase-shift control.

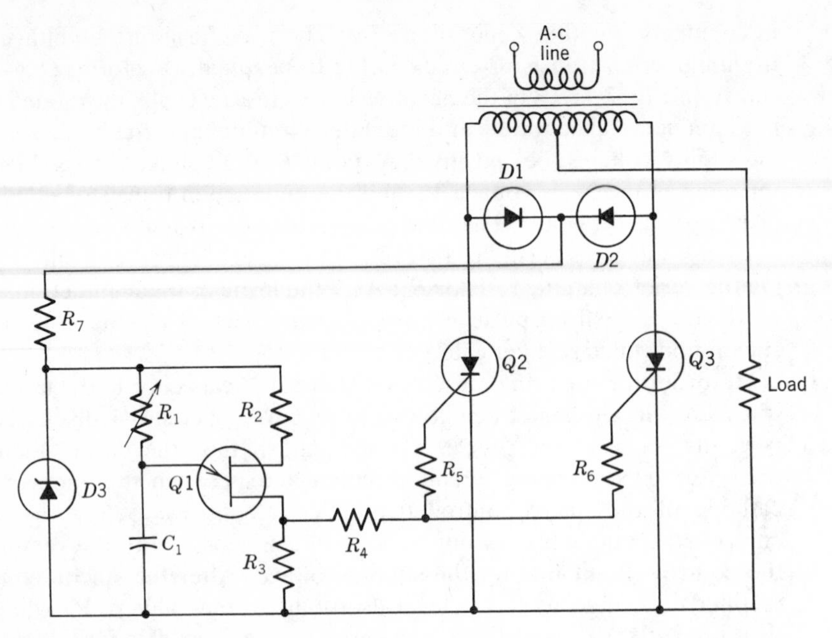

Figure 7-14 Full-wave silicon controlled rectifier circuit with phase-shift control.

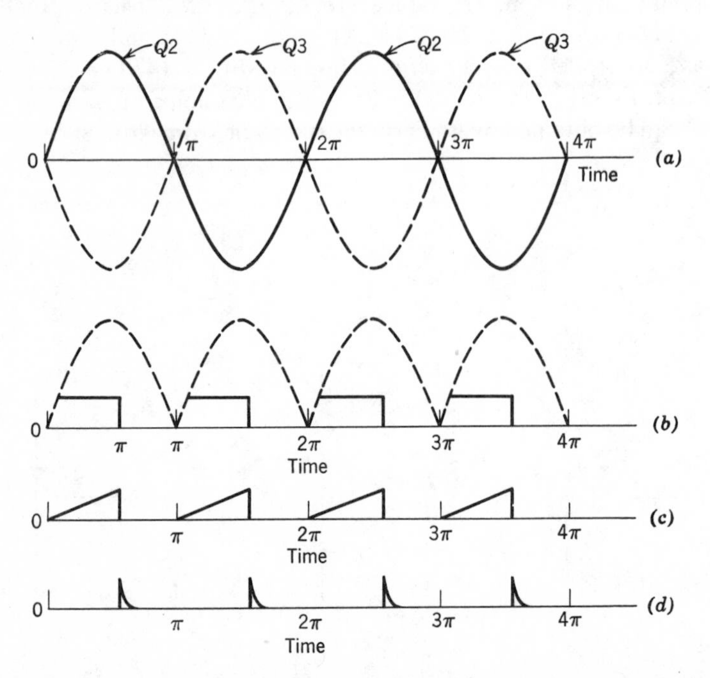

Figure 7-15 Waveforms in a full-wave controlled rectifier. (*a*) Anode supply voltage. (*b*) Voltage waveform across Zener-diode *D*3. (*c*) Voltage waveform across capacitor. (*d*) Gate pulses.

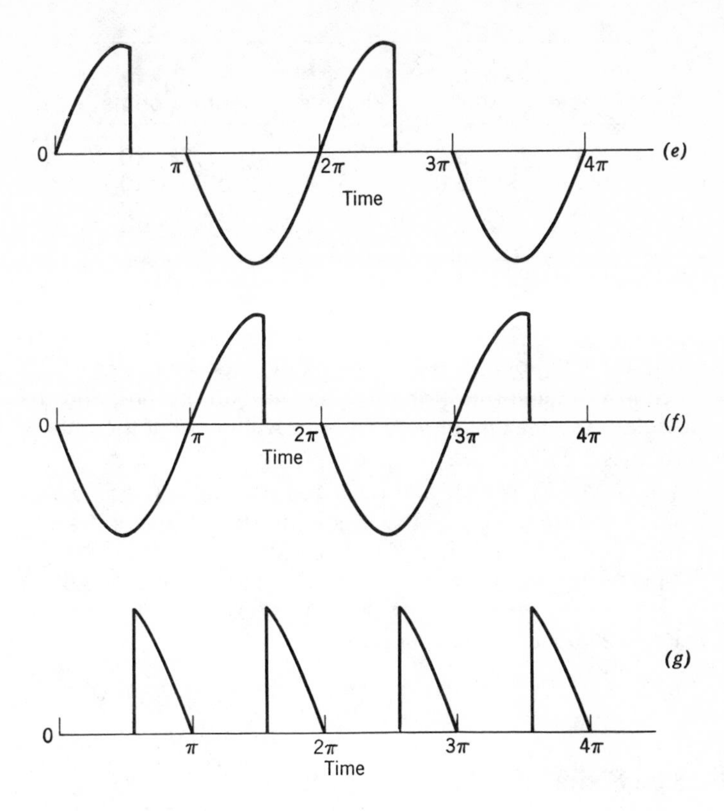

Figure 7-15 (*e*) Waveform voltage waveform across *Q*2. (*f*) Voltage waveform across *Q*3. (*g*) Load voltage or load current waveform.

PROBLEMS

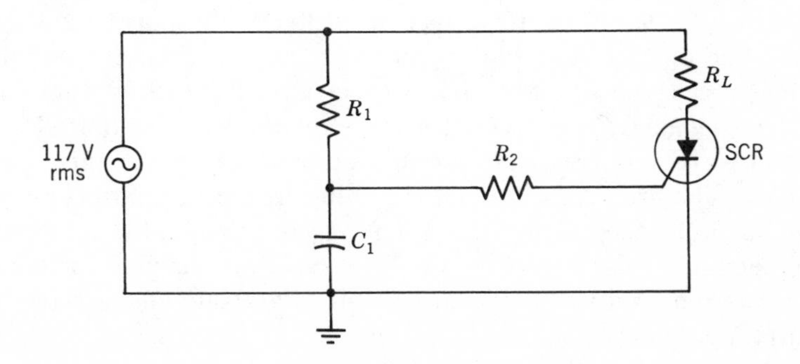

Circuit for Problems 1 to 4.

Problem	R_1 kilohms	X_{C1} kilohms	R_2 kilohms	R_L ohms
1.	2.5	1.2	100	10
2.	5.0	5.0	100	20
3.	5.0	20.0	100	60
4.	2.0	0.5	100	60

For each of the above four problems, draw the phasor diagram. Draw waveforms for line voltage, for v_L, for the voltage across the SCR, and for load current. Calculate the firing angle, the load current, and the load voltage. Calculate the maximum gate current. Assume that the gate firing voltage is zero. V_F is 2.0 V for the SCR.

5. In Fig. 7-11, V_{BB} is 30 V and R_1 is 100 kΩ. The UJT has a value of 0.6 for η. What is the frequency of the triggering pulses when C is 10 μF. Repeat for 1 μF, 0.1 μF, 0.01 μF, and 0.001 μF for C.
6. In Fig. 7-12*a*, *D*2 is a 20-V Zener. R_1 is 1 MΩ, and η is 0.55. What values of C_1 will fire the circuit at 30°, 60°, 90°, 120°, and 150° points on the positive half of a 220-V, 60-Hz a-c supply?

Section 7-4 The Firing of a Triac

A typical circuit used to vary load current in a triac is shown in Fig. 7-16*a*. When the voltage across the trigger diac is sufficiently high, the diac breaks down to provide gate current. The capacitor and resistor combination causes the voltage to the diac to lag the voltage to the triac. The amount of lag adjusted by the rheostat determines the point in the cycle at which the diac and triac turn on and, thus, controls the amount of load current. When the triac turns on, its low forward voltage drop effectively short circuits *a* and *c* and, thus, the diac current falls to zero and the diac recovers its off condition. This process repeats for each half cycle of the line supply (Fig. 7-16*d*).

A single *SCR* produces half-wave rectification. Two *SCR*'s are used in a full-wave rectifier. Both circuits convert a-c power into controlled d-c power. The triac differs by not rectifying. The triac controls a-c power in an a-c load. Consequently, the equations developed for load current, load voltage, and load power for the SCR have no meaning for the triac. Now what is required is the rms or the effective value of the waveform of load current or load voltage of Fig. 7-16*d*.

The average value (the d-c value) of the load current for the triac (Fig. 7-16*d*) is zero. Therefore, this waveform represents an a-c value only. The

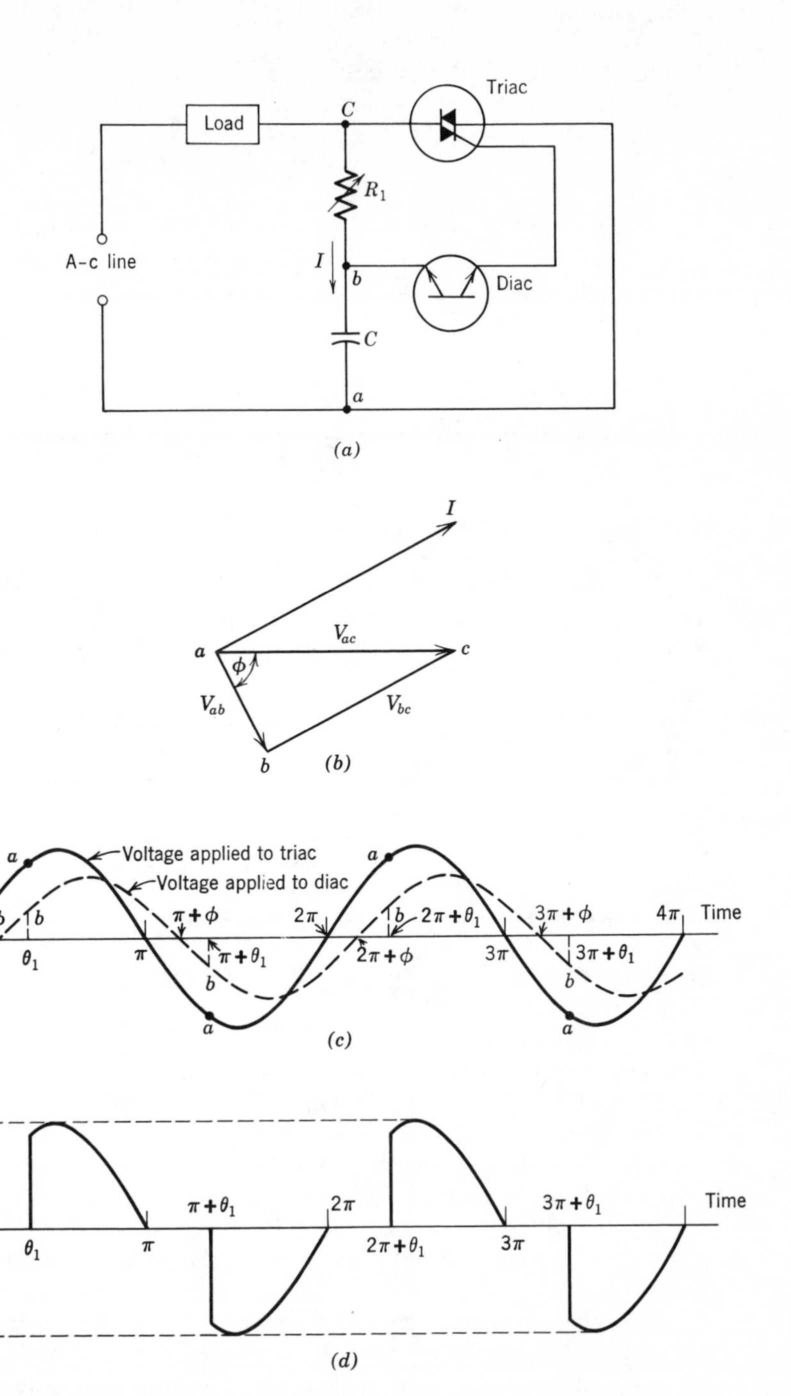

Figure 7-16 Triac application. (*a*) Circuit. (*b*) Phasor diagram. (*c*) Voltage waveforms. (*d*) Load current.

useful result is in terms of an effective or rms value, which is defined by

$$I_{\text{rms}} \equiv \sqrt{\frac{1}{2\pi}\int_0^{2\pi} (I_m \sin\theta)^2\, d\theta}$$

The integration itself is performed over the intervals from θ_1 to π and from $(\pi+\theta_1)$ to 2π. Since these two pulses over the first cycle have symmetry, it is only necessary to perform the integration over one pulse and to double the result:

$$\begin{aligned}
2\int_{\theta_1}^{\pi} (I_m \sin\theta)^2\, d\theta &= 2I_m^2 \int_{\theta_1}^{\pi} \sin^2\theta\, d\theta \\
&= 2I_m^2 \int_{\theta_1}^{\pi} \left(\tfrac{1}{2}-\tfrac{1}{2}\cos 2\theta\right) d\theta \\
&= I_m^2 \int_{\theta_1}^{\pi} d\theta - I_m^2 \int_{\theta_1}^{\pi} \cos 2\theta\, d\theta \\
&= \left[I_m^2\theta + \tfrac{1}{2}I_m^2 \sin 2\theta\right]_{\theta_1}^{\pi} \\
&= I_m^2\left[(\pi-\theta_1) + \tfrac{1}{2}(\sin 2\pi - \sin 2\theta_1)\right] \\
&= I_m^2\left[(\pi-\theta_1) - \tfrac{1}{2}\sin 2\theta_1\right]
\end{aligned}$$

Substituting this evaluation into the definition for the rms value.

$$I_{\text{rms}} = \frac{I_m}{\sqrt{2}}\sqrt{\frac{\pi-\theta_1}{\pi} - \frac{\sin 2\theta_1}{2\pi}} \tag{7-9a}$$

It is usually more convenient to express the first term in degrees

$$I_{\text{rms}} = \frac{I_m}{\sqrt{2}}\sqrt{\frac{180^\circ-\theta_1}{180^\circ} - \frac{\sin 2\theta_1}{2\pi}} \tag{7-9b}$$

where θ_1 is the angle in the cycle at which firing occurs.

An examination of Eq. 7-9 shows that when the firing angle θ_1 is zero, the load current waveform is a complete a-c cycle and Eq. 7-9 reduces to the expected value $I_m/\sqrt{2}$ which is the basic conversion from peak to rms in a-c circuit analysis. When θ_1 is 180°, Eq. 7-9 yields zero, since the value of $\sin 2\theta_1$ is sin 360°, which is zero. Naturally, if the triac fires at 180°, there is zero load current. When θ_1 is 90°, $\sin 2\theta_1$ is sin 180°, which is zero. Now the rms value is $I_m/2$.

Assuming a resistive load, the power in the load is given by evaluating i^2R. When θ_1 is zero, the average load power is $I_m^2R/2$. At a firing angle of 90°, the load power is $I_m^2R/4$. At a firing angle of 180°, the load power is zero.

PROBLEMS

1. In Fig. 7-16, the supply voltage is 117 V rms and the breakdown voltage of the diac is 20 V. What is the least value of θ_1 (Fig. 7-16*d*) that will fire the triac? If C is 1 μF, what value of R_1 will fire the circuit at 20°, at 40°, at 60°, and at 60°?
2. Plot the rms load current and the load power by using a triac with 180° control. Plot these values in percent of maximum obtaining data at 30° intervals.
3. A triac is used to control the input power into a 1000-W heater element that operates from 117 V rms. What firing angles are required for $\frac{3}{4}$, $\frac{1}{2}$, and $\frac{1}{4}$ heat?
4. A triac controls the power to 100-Ω load. A switch reduces the power from maximum to 70% of full power when the device is operated from a 117-V supply. What firing angle should be inserted by the switch when the circuit uses a diac that has a 20-V breakdown voltage?
5. Using Eq. 7-9*b*, plot the power in a 100-Ω resistor controlled by a triac for a 180° variation in phase shift. The supply voltage is 117 V rms. Obtain data at 30° intervals. Assume that the forward voltage drop across the triac is zero after breakdown. Also plot the rms load current.

Section 7-5 The Firing of a Thyratron

A thyratron, when used as a rectifier, may be controlled with either a direct or an alternating voltage on the grid. A circuit which shows the action of a d-c grid control is given in Fig. 7-17. An alternating voltage E is impressed on

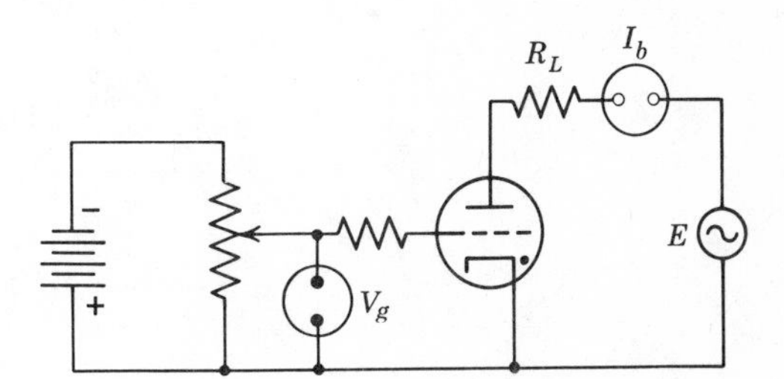

Figure 7-17 Circuit for d-c control of a thyratron.

the plate of the thyratron. The load resistor R_L is in series with the plate and the a-c line. The grid voltage can be varied by adjusting the potentiometer. The firing-control characteristic obtained from the thyratron data sheet is shown in Fig. 7-18*a*. The impressed alternating line voltage is plotted in Fig. 7-18*b*. Let us consider any point Q on the positive half of the cycle. The plate voltage at point Q corresponds to R on the firing characteristic. The grid

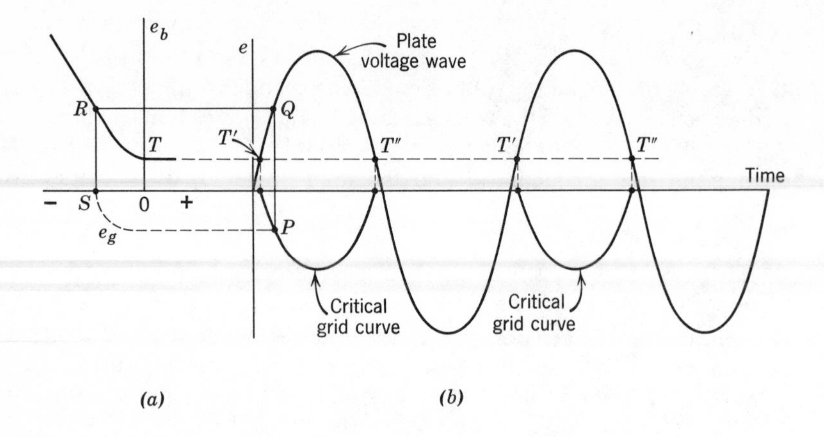

Figure 7-18 D-c control waveforms. (*a*) Control characteristic. (*b*) Development of the critical grid curve.

voltage for point R is S. Now, at that instant of time at which the plate voltage is Q, we plot this firing-control grid voltage S as point P. If different Q's are taken, the corresponding P's plot into a locus which is called the *critical grid curve*.

One positive cycle of the alternating line voltage is redrawn in Fig. 7-19

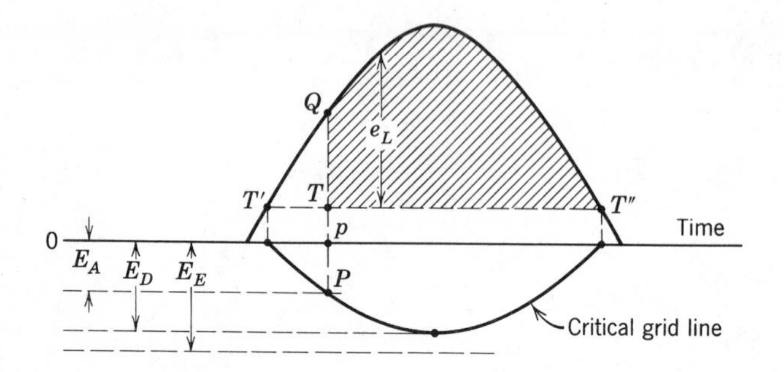

Figure 7-19 Details of the critical grid line.

for greater detail. If the grid bias is set to $-E_A$ volts, it intersects the critical grid line at point P where the plate supply voltage is Q. Up to the time p, the direct grid voltage is more negative than or below the critical grid line. As long as the grid is more negative than the critical grid line, the thyratron will not be fired. At the instant of intersection of the bias line and the critical grid line, which is P at time p, the tube fires. The plate voltage falls from Q to T. The voltage T is the tube drop under conduction. Once the tube is fired, the grid cannot regain control, and the plate voltage remains at this level until T'' is reached in the cycle. At this point, the line voltage e falls below the tube drop, and the thyratron deionizes. If the grid voltage remains at E_A, the thyratron will fire at point Q on each succeeding cycle. If the grid is made less negative, the firing point advances earlier in the cycle. If the grid is made more negative, the firing point occurs later in the cycle. If the grid is made more negative than E_D, for instance E_E, there will be no point of intersection, and the thyratron cannot conduct at all.

When the thyratron fires, the difference between the tube drop, TT'', and the source voltage e is shown as the shaded area in Fig. 7-19. The instantaneous voltage difference between the tube drop, TT'', and the applied voltage e is e_L, the instantaneous load voltage across R_L. Accordingly, the analysis developed in Section 7-1 is valid for this thyratron controlled rectifier circuit with the exception that angular current control cannot extend beyond 90° because the current then falls to zero. Instead of using the values for θ_1 as the variable of control, it is more convenient to use the values of the d-c grid bias. When this is done, the relation shown in Fig. 7-20 results.

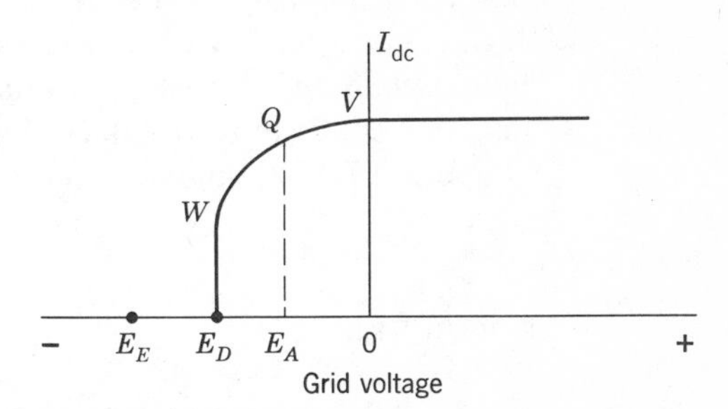

Figure 7-20 The d-c control characteristic.

With d-c control on the thyratron, the current can be adjusted by the grid voltage from a maximum value of a full half-cycle condition continuously and smoothly to one half this maximum value. Also, zero load current may be obtained.

The grid voltage can be provided from the phase-shift circuits discussed in Section 7-2. A line alternating voltage is applied to the plate circuit through the load resistance, and a separate lagging alternating voltage derived from the phase shift circuit is connected between the grid and cathode. In this circuit (Fig. 7-21*a*) the exact magnitude of the grid voltage is not of major importance and, for convenience, we shall assume that e and e_g have the same peak values. However, the phase of the grid voltage with respect to the plate-circuit voltage is controlled by a phase shift circuit.

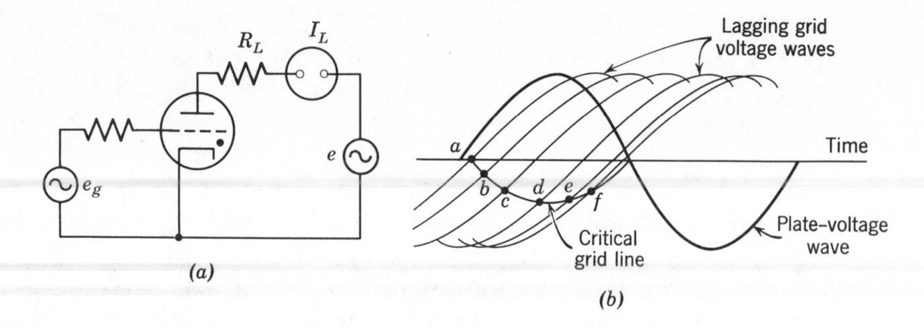

Figure 7-21 A-c control of a thyratron. (*a*) Circuit. (*b*) Plate and grid waveforms for different lagging angles.

The plate-circuit waveform with its critical grid line is shown in Fig. 7-21*b*, along with several different lagging grid voltages. The different grid voltages intersect the critical grid line at different points (a, b, c, d, e, and f), and each causes the thyratron to fire at a different time in the cycle. The greater the angle of the lag of the grid voltage, the later in the cycle the thyratron fires. When the grid voltage lags by an angle of about 160° to 170°, f, the tube fires for only a few degrees before it goes out because of the falling plate-circuit voltage. In this circuit we can obtain full control over the load current from zero to maximum. The equations and waveforms developed in Section 7-1 are valid for this circuit without restriction.

PROBLEMS

In the following problems, assume that the control characteristic of the thyratron (Fig. 7-18*a*) is linear between $E_b = 300$ V, $E_g = -30$ V and $E_b = 20$ V, $E_g = 0$ V.

1. In Fig. 7-17, the alternating plate supply voltage is 117 V rms, and R_L is 2000 Ω. Obtain sufficient data to plot the d-c control characteristic shown in Fig. 7-20.
2. The thyratron shown in Fig. 7-21*a* has a 1500-Ω resistor as the plate load. The plate supply voltage is 230 V rms. The grid voltage is obtained from the phase-shift circuit shown in Fig. 7-5, in which V_{AC} is also 230 V rms. Obtain sufficient data to plot the a-c control characteristic similar to that shown in Fig. 7-2.
3. The peak current in a thyratron rectifier circuit is 450 ma. The average load current is 100 ma. Determine the angle of firing assume the tube drop is 20 V when conducting. The supply voltage is 117 V rms.

4. The peak current in a thyratron rectifier circuit is 900 ma. The average load current is 85 ma. Determine the angle of firing, assuming that the tube drop is 20 V when conducting. The supply voltage is 117 V rms.

Questions

1. Define or explain each of the following terms: (*a*) angle of current flow, (*b*) gate, (*c*) phase shift control, (*d*) gate, (*e*) clamping diode, (*f*) saturable core reactor, (*g*) triggering pulse, (*h*) diac, (*i*) ionization, (*j*) deionization.
2. What determines the angle at which current starts to flow in a controlled rectifier?
3. What determines the angle at which current ceases to flow in a controlled rectifier?
4. Under what conditions can θ_2 be taken as 180° without significant error?
5. Under what conditions can the forward voltage drop across a controlled rectifier be neglected without significant error?
6. Why does a leading phase shift on the control circuit of a controlled rectifier fail to produce current control?
7. What is the advantage of using a UJT to control the firing of an SCR?
8. Why is it necessary to use a clamping diode on the gate of an SCR?
9. Can a UJT be used to control the firing of a triac? Explain.
10. Compare the load current of a triac circuit with the load current produced by a full-wave rectifier using SCR's.
11. What is the value of the direct voltage across the load in a circuit using a triac?
12. Compare the magnitude of the value of forward voltage drop across a thyratron with the value of V_F in an SCR.

Chapter Eight

DECIBELS AND SOUND

The consideration that the ear is not linear in its response (Section 8-1) points out the need of decibels as a logarithmic means of having a physical response and an electrical characteristic represented to the same scale (Section 8-2). A review of logarithms (Section 8-3) preceeds actual calculations using the decibel (Section 8-4). A short discussion of loudness and intensity of sound (Section 8-5) completes the chapter.

Section 8-1 The Need for a Nonlinear System of Measurement

Human sensory response is nonlinear. As an example showing this nonlinearity, a single match, when suddenly ignited in a dark room, produces a lasting glare. In bright sunlight, the same-size match, when struck, does not give off noticeable light. As another example, the noise of an insect can disrupt the calm of a still summer's night. On the other hand, it would take millions of these insects to be heard over the roar of a passing railroad train. In a dark room, two lighted matches give twice the effect of one match on the response of the human eye. In broad daylight, it would take two suns to give twice the effect of one on human vision. These facts would indicate that a true response would be of the order:

Steps of equal response	1	2	3	4	5	6	7	8	9
Quantity of cause	$\frac{1}{16}$	$\frac{1}{8}$	$\frac{1}{4}$	$\frac{1}{2}$	1	2	4	8	16

Each successive step doubles the previous quantity, but the change in response is uniform.

A further indication of the usefulness of such a scheme is given by the system used in music. In music, an increase in one octave doubles the pitch or frequency. The reference frequency used is "A" above "middle C" at 440 Hz. If the relative pitch is plotted on a linear axis, as the keys on a piano, Fig. 8-1 shows that the frequency scale is nonlinear.

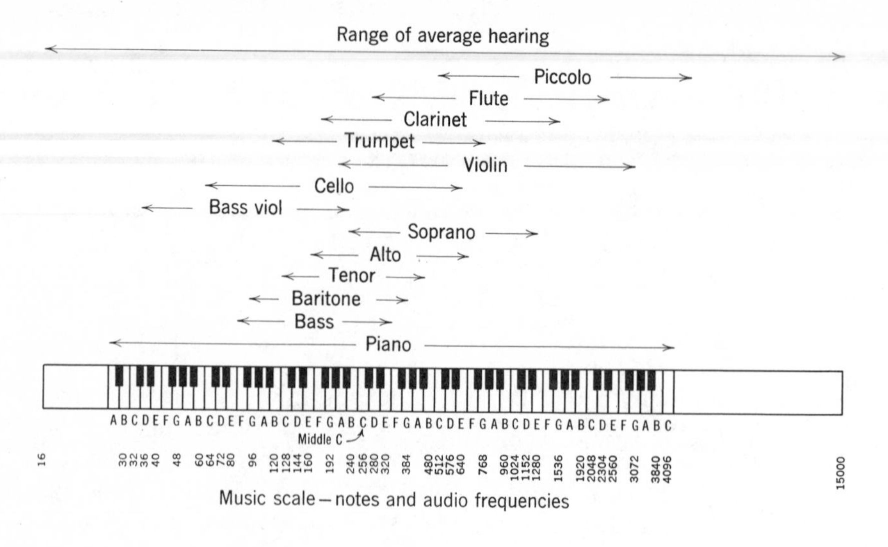

Figure 8-1 Range of frequencies in music.

The nonlinear responses to these and all sensory responses have been generalized under what is called the Weber-Fechner law.

In mathematics the process of taking logarithms of numbers converts a nonlinear scale, such as the musical scale, into a linear scale. Since each octave in the musical scale is a multiplication of the frequency of the preceding octave by two, the spread of one octave on a logarithmic scale is *log 2*, and it is the same number for any one octave.

In showing graphs of frequency response where the independent variable is frequency, the frequency is plotted as the logarithm to the base 10 of the frequency. Since this is the standard conventional practice, graph paper is available called *semilogarithmic* paper in which one axis is logarithmic and the other axis is linear. If we were to use ordinary graph paper for frequency-response curves, it would be necessary to calculate the logarithms of the different frequencies used. In semilogarithmic graph paper, the engraved printing plate is designed so that it is laid out proportionally to the logarithms on one axis scale. When this paper is used, there is no need to calculate logarithms; that work was done in the original design of the graph paper. If it is desired to represent 20 to 20,000 Hz, the required logarithmic axis would be 10 to 100 to 1000 to 10,000 to 100,000 or four-cycle semilogarithmic paper. To represent 20 to 8000 Hz, one would need a logarithmic axis of 10 to 100 to 1000 to 10,000 or three-cycle semilogarithmic paper. Figure 8-3 is drawn on four-cycle semilogarithmic paper.

Section 8-2 The Decibel

In honor of Alexander Graham Bell, the logarithm to the base 10 of the ratio of two powers is defined as a *bel*:

$$\text{Number of bels} \equiv \log_{10} \frac{P_2}{P_1}$$

where P_2 and P_1 represent the two powers that are being compared.

The bel as a unit is cumbersome for general use and for problem solution. In order to have convenient numerical results for problems and applications, we define the *decibel* as one tenth of a bel:

$$\text{Number of decibels} \equiv 10 \log_{10} \frac{P_2}{P_1} \tag{8-1}$$

In audio work, a change in power level of one decibel is barely perceptible to the ear. A change of two decibels is slightly apparent.

Section 8-3 Logarithms

Since the decibel is defined as a logarithm, the technique of the mathematical process of taking logarithms must be studied. In working with decibels, there is a slight variation from the normal mathematical procedure which greatly simplifies the final numerical result. All the work done in taking logarithms is to the base 10. Thus, no subscripts will be used when writing *log*; the 10 is understood.

$$\begin{aligned} \log 10{,}000 &= 4 \\ \log 1000 &= 3 \\ \log 100 &= 2 \\ \log 10 &= 1 \\ \log 1 &= 0 \end{aligned}$$

The numbers, 4, 3, 2, 1, and 0, are known as the *characteristic*. The characteristic numerically is one less than the number of digits in the number to the left of the decimal point. If the number were 834.24, the characteristic is 2. This means that the logarithm of the number lies between 2 and 3. If the number were 8342.4, the logarithm would have the characteristic 3 and lie between 3 and 4. The exact decimal of the logarithm is called the *mantissa*. The mantissa for 834.24 is the same as the mantissa for 8342.4. It is also the same for 8,342,400 or 8.3424. The mantissa is determined by the sequence of the digits and not by the decimal point. The placement of the decimal point in the original number determines the characteristic. In general, the characteristic is determined by visual inspection, and the mantissa is determined by the use of a slide rule. Some examples are:

$$\log 834240 = 5.921$$
$$\log 74 \times 10^5 = 6.870$$
$$\log 231 = 2.364$$
$$\log 3.85 = 0.586$$

A table of logarithms could be used in lieu of a slide rule, but the slide rule has the advantage of saving time while maintaining sufficient accuracy for most problems.

A formal mathematical definition may be given to a logarithm. If a number N is expressed in the form of the x power of 10, the logarithm of N to the base 10 is x.

If $$N = 10^x$$

then $$\log_{10} N = x$$

When a number M is less than 1, it may be written

$$M = \frac{1}{10^y}$$

where y is a positive number greater than zero. Then

$$M = 10^{-y}$$

and by the definition of the logarithm

$$\log_{10} M = -y$$

This approach for a number M, which is less than one but greater than zero, is a little different from the usual method of determining logarithms for calculation of numerical problems in mathematics, but it is the approach that *must* be taken when working with decibels. For example, to find the logarithm of 0.1, convert 0.1 to the fraction $\frac{1}{10}$. Then

$$\log 0.1 = \log \frac{1}{10} = -\log 10 = -1$$

Similar examples are

$$\log 0.001 = \log \frac{1}{1000} = -\log 1000 = -3$$

$$\log 0.20 = \log \frac{1}{5} = -\log 5 = -0.700$$

$$\log 0.375 = \log \frac{1}{2.667} = -\log 2.667 = -0.426$$

$$\log 0.023 = \log \frac{1}{43.5} = -\log 43.5 = -1.638$$

As a general rule, *take logarithms only of numbers that are greater than 1.*

In taking antilogarithms, the method is directly the inverse process used for finding logarithms. Some examples are as follows:

If $\log x = 4$, $\quad x = 10000$

If $\log x = 2$, $\quad x = 100$

If $\log x = 0.254$, $\quad x = 1(\text{antilog } 0.254) = 1(1.795) = 1.795$

If $\log x = 3.621$, $\quad x = 1000(\text{antilog } 0.621) = 1000(4.18) = 4180$

$$\text{If } \log x = -2.00, \ \frac{1}{x} = 100(\text{antilog } 0) = 100(1) = 100$$

$$x = 0.01$$

$$\text{If } \log x = -4.854, \ \frac{1}{x} = 10{,}000(\text{antilog } 0.854) = 10{,}000(7.13)$$

$$= 71{,}300$$

$$x = 0.0000140$$

In working with decibels, we need to evaluate the logarithm of zero.

If
$$0 = \frac{1}{\infty}$$

then
$$\log 0 = \log \frac{1}{\infty} = -\log \infty = -\infty$$

or
$$\log 0 = -\infty$$

Thus, the logarithm of zero is equal to minus infinity.

PROBLEMS

1. Determine the logarithms of the following numbers: (*a*) 2650, (*b*) 132, (*c*) 756,000, (*d*) 1.46, (*e*) 294×10^{16}, (*f*) 0.0023, (*g*) 0.874, (*h*) $\frac{1}{16}$, (*i*) $\frac{3}{64}$, (*j*) 84×10^{-6}.
2. Determine the numbers for which the logarithms are: (*a*) 2.46, (*b*) 6.92, (*c*) 14.20, (*d*) 23.3, (*e*) 0.024, (*f*) −5.78, (*g*) 0, (*h*) −27.4, (*i*) $\frac{1}{16}$, (*j*) 7.23.

Section 8-4 Decibel Calculations

In Section 8-2, we defined the decibel as

$$\text{db} = 10 \log \frac{P_2}{P_1} \tag{8-1}$$

Properly speaking, a decibel is a measure of a power ratio, but very often the measurements are taken in terms of voltage, current, or impedance. In a general case

$$P_1 = \frac{E_1^2}{Z_1 \cos \theta_1} \qquad \text{and} \qquad P_2 = \frac{E_2^2}{Z_2 \cos \theta_2}$$

Then

$$\text{db} = 10 \log \frac{P_2}{P_1} = 10 \log \frac{E_2^2/Z_2 \cos \theta_2}{E_1^2/Z_1 \cos \theta_1}$$

$$\text{db} = 10 \log \frac{E_2^2 Z_1 \cos \theta_1}{E_1^2 Z_2 \cos \theta_2} = 10 \log \left(\frac{E_2}{E_1}\right)^2 \left(\frac{Z_1}{Z_2}\right) \left(\frac{\cos \theta_1}{\cos \theta_2}\right)$$

$$\text{db} = 20 \log \frac{E_2}{E_1} + 10 \log \frac{Z_1}{Z_2} + 10 \log \frac{\cos \theta_1}{\cos \theta_2} \tag{8-2a}$$

In most instances of application, it may be assumed that the two impedances are purely resistive. For pure resistance the power factor $\cos \theta$ is unity, and, as the logarithm of one is zero, the expression simplifies to

$$\text{db} = 20 \log \frac{E_2}{E_1} + 10 \log \frac{R_1}{R_2} \tag{8-2b}$$

When the two resistances are equal or refer to the same resistance, R_1/R_2 becomes unity. The resistance correction term, $10 \log (R_1/R_2)$, is zero, and the decibel relation becomes

$$\text{db} = 20 \log \frac{E_2}{E_1} \tag{8-3}$$

If this decibel relation is evaluated in terms of currents instead of voltages, we have

$$P_1 = I_1^2 R_1 \qquad \text{and} \qquad P_2 = I_2^2 R_2$$

Then

$$\text{db} = 10 \log \frac{I_2^2 R_2}{I_1^2 R_1}$$

$$\text{db} = 20 \log \frac{I_2}{I_1} + 10 \log \frac{R_2}{R_1} \tag{8-4}$$

$$db = 20 \, Log \left| \frac{ES}{Em} \right|$$

As an example, assume that the voltage across a loudspeaker is 2.3 V and, when the volume control is advanced, the speaker voltage becomes 4.8 V. The decibel increase in gain is

$$\text{db} = 20 \log \frac{V_2}{V_1}$$

$$\text{db} = 20 \log \frac{4.8}{2.3} = 20 \log 2.09$$

$$= 20 \times 0.320 = 6.4$$

$$\text{db} = +6.4 \quad \text{(gain)}$$

We did not use the correction factor, $10 \log (R_1/R_2)$ because both measurements are taken across the same loudspeaker.

In another example, the input to a transmission line is 64 V, and the output voltage is 18 V. Since the output is less than the input, we have a loss in gain:

$$\text{db} = 20 \log \frac{E_2}{E_1} = 20 \log \frac{18}{64} = -20 \log \frac{64}{18}$$

$$= -20 \log 3.55 = -20 \times 0.550 = -11.1$$

$$\text{db} = -11.1 \quad \text{(loss)}$$

When the impedance is not specified, it must be assumed that the two values are the same and that the correction term, $10 \log (R_1/R_2)$, accordingly is zero. It is standard practice in decibel calculations to insist that the sign + or − be associated with the numerical value. A +7db means a gain or increase in level of 7 decibels whereas −4db means a decrease in level or a loss of 4 decibels. Sometimes these figures are expressed as "7 db up" and "4 db down." In filter work a +db value ordinarily means loss and a −db means gain.

There are a number of devices used in electronic applications that have a normal inherent loss. Accordingly, it is customary to consider the decibel ratios as positive values. A pad or an attenuator when it is specified as "10 db" infers a loss by its nomenclature. Another example is a filter. When the filter has a rejection or attenuation of 60 db at a given frequency, the output level is −60 db when compared to the output at the "pass" frequency.

Certain decibel values shown in Table A are very convenient to use and should be memorized as they represent whole number ratios. When the voltage ratio is 2, we find db = $20 \log 2 = +6$ db. When the voltage ratio is $\frac{1}{2}$, db = −6. If the power ratio is 2, db = $10 \log 2 = +3$ db. If the power ratio is $\frac{1}{2}$, db = $10 \log \frac{1}{2} = -10 \log 2 = -3$ db. As the power in a resistive circuit varies directly as the square of the voltage, a voltage ratio of 2 or $\frac{1}{2}$ corresponds to a power ratio of $\sqrt{2}$ or $1/\sqrt{2}$.

TABLE A

Decibels	Voltage Ratio	Power Ratio
−6	$\frac{1}{2}$ or 0.500	$\frac{1}{4}$ or 0.250
−3	$1/\sqrt{2}$ or 0.707	$\frac{1}{2}$ or 0.500
0	1	1
3	$\sqrt{2}$ or 1.414	2
6	2	4

As an example of the use of these special values, assume that an initial power level of 60 mw is subject to a +28 db gain. The approximate result is as follows.

0 db is 60 mw.
+6 db gain (power ratio of 4) raises the power level of 240 mw or 0.24 W.
Another +6 db, a total of +12 db, increases the power level by 4 to 0.96 W.
Another +6 db, a total of +18 db, increases the power level by 4 to 3.84 W.
Another +6 db, a total of +24 db, increases the power level by 4 to 15.36 W.
Another +3 db, a total of +27 db, increases the power level by 2 to 30.72 W.
Another +3 db, a total of +30 db, increases the power level by 2 to 61.44 W.

If the last three-decibel steps were linear, the result would be about 41 W. The result by slide-rule calculation is 37.8 W. This method is often very useful for giving a rapidly obtained approximate result without the use of a slide rule or tables.

PROBLEM. Let us assume that a microphone delivers 36 mv at 300 Ω into an amplifier which delivers 15 W into a 16-Ω speaker system at full power. To find the decibel gain of the amplifier, we may use either the power equation or the voltage equation.

Solution 1. The power in the microphone is

$$P_1 = \frac{E_1{}^2}{R_1} = \frac{0.036^2}{300} = 4.32 \times 10^{-6}\ \text{W}$$

$$\text{db} = 10 \log \frac{P_2}{P_1} = 10 \log \frac{15}{4.32 \times 10^{-6}}$$

$$= 10 \log 3.48 \times 10^6 = 10 \times 6.541$$

$$\text{db} = +65.41 \text{ gain}$$

Solution 2. The voltage at the speaker is

$$P_2 = \frac{E_2^2}{R_2}$$

$$15 = \frac{E_2^2}{16}$$

$$E_2 = 15.48$$

$$\text{db} = 20 \log \frac{E_2}{E_1} + 10 \log \frac{R_1}{R_2}$$

$$\text{db} = 20 \log \frac{15.480}{0.036} + 10 \log \frac{300}{16}$$

$$\text{db} = 20 \log 430 + 10 \log 18.75$$

$$\text{db} = 20 \times 2.634 + 10 \times 1.273$$

$$\text{db} = 52.68 + 12.73 = +65.41 \text{ gain}$$

Very often it is very useful to have a meter that is calibrated to read directly in decibels. Since the definition of the term decibel stated that the decibel is a power ratio, a wattmeter with a new scale could be used. However, wattmeters are prohibitively expensive for this use. An a-c voltmeter ordinarily serves as a decibel meter subject to certain restrictions. As 12 V across 30 Ω is not the same power as 12 V across 4000 Ω, the decibel meter needs the additional specification that its scale is accurate only when the meter is used on the specified impedance for which the instrument was calibrated. Several standard references are in common use:

1. The zero decibel reference is a 6-mw power dissipated in a 500-Ω resistive load
2. The zero decibel reference is a 1-mw power dissipated in a 600-Ω resistive load
3. The zero decibel reference is a 1-mw power level without regard to impedance level. This reference is abbreviated *dbm* to distinguish it from the others.

This reference, the *dbm*, has become more popular especially because its use simplifies the calculations and computations involved in complex test equipment and complex operational circuits by keeping all the data in terms of power without regard to impedance values. All the gains and losses can be treated simply by addition and subtraction. In addition, most pieces of test-equipment using the *dbm* scale are standardized at a 50-Ω level.

4. *The Volume Unit (VU).* The volume unit reference, used primarily in the radio broadcasting field, is a 1-mw power dissipated in a 600-Ω resistive load. The volume unit is used only to read power levels in complex waves such as program lines carrying speech or music. A zero volume unit means that a zero-VU complex wave has the same average power content that a 1-mw sinusoidal waveform has at a frequency of 1000 Hz.

The reference for 0 db at 6 mw in 500 Ω is, by calculation, a specific voltage value:

$$P = \frac{E^2}{R}$$

$$0.006 = \frac{E^2}{500}$$

$$E^2 = 3$$

$$E = \sqrt{3} = 1.73 \text{ V}$$

If the 0- to 3-V scale of a meter is calibrated in decibels, then 0 db is located at 1.73 V. The scale marking of −6 db corresponds to a half-voltage ratio, 1.73/2, or 0.865 V. The scale reading of −3 db is at $1.73/\sqrt{2}$ or 1.225 V. Likewise, +3 db corresponds to $1.73\sqrt{2}$ or 2.45 V and +4 db to 2.74 V; +5 db is 3.08 V and is slightly off scale. The zero on the 3-V scale is evaluated as −log ∞ or −∞ db. The two scales are shown in Fig. 8-2.

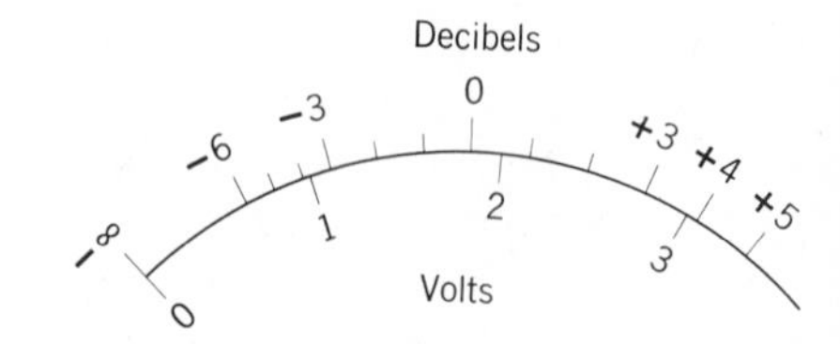

Figure 8-2 Scale of a decibel meter.

All these concepts can be illustrated in the solution of a problem which is commonly encountered in the measurement of audio frequencies. In working with decibels, we often find more than one approach may be taken to determine the final result.

PROBLEM. A multirange a-c meter has a 0- to 3-V scale which is calibrated in decibels (reference: 0 db is 6 mw in a 500-Ω load) as shown in Fig. 8-2. The 0- to 60-V scale is not calibrated in decibels. The meter reads −4.5 db, but the voltage-range selector switch is set on the 60-V range. The load impedance across which the meter is connected is 4500 Ω. Determine the meter reading in decibels, the true decibel value, and the load power.

Solution 1. The scale-range factor, 60/3 or 20, is an increase in voltage and power level. The scale-correction factor introduced by switching to a higher-voltage range is

$$\text{db} = +20 \log 20 = +26 \text{ db}$$

If the 60-V scale were marked in decibels also, it would read for this meter deflection:

$$-4.5 + 26 = +21.5 \text{ db}$$

The meter is calibrated for 500 Ω, but it is used across 4500 Ω. For the same voltage reading, the power dissipated in 4500 Ω must be less than if the meter were connected to the 500 Ω for which it is calibrated. Therefore, the impedance correction factor is negative:

$$\text{db} = -10 \log \frac{4500}{500} = -9.54 \text{ db}$$

The true level is the meter reading plus the impedance correction factor:

$$21.5 - 9.54 = +11.96 \text{ db}$$

The true power is

$$\text{db} = 10 \log \quad \text{(power ratio)}$$
$$11.96 = 10 \log \quad \text{(power ratio)}$$
$$\log \text{(power ratio)} = 1.196$$
$$\text{Power ratio} = 15.70$$

Since the true power is a +db, and since the zero reference level of the meter scale is 6 mw, the true power is 6 × 15.70 or 94.2 mw. (If the true power level were −11.96 db, we would divide 6 by 15.70 instead of multiplying.)

Solution 2. The reference level of 0 db corresponds to 1.73 V. The −4.5-db reading indicates that the meter is reading less than 1.73 V.

$$4.5 = 20 \log \quad \text{(voltage ratio)}$$

$$\log \text{(voltage ratio)} = 0.225$$

$$\text{Voltage ratio} = 1.68$$

Therefore the meter reads:

$$\frac{1.73}{1.68} = 1.03 \text{ V}$$

As the conversion from the 3-V scale to the 60-V scale is a factor of 20, the true load voltage is

$$1.03 \times 20 = 20.6 \text{ V}$$

The true load power is

$$P_L = \frac{E_2^2}{R_L} = \frac{20.6^2}{4500} = 0.0942 \text{ W}$$
$$= 94.2 \text{ mw}$$

The true decibel value is

$$\text{db} = 10 \log \frac{P_2}{P_1} = 10 \log \frac{94.2}{6}$$
$$= 10 \log 15.7 = +11.96 \text{ db}$$

The meter reading is the true decibel value *plus* the impedance correction factor of 9.54 db or +21.5 db.

PROBLEMS

1. The gain of an amplifier is +46 db. The amplifier delivers 3 W into a 4 Ω load. The amplifier input resistance is 150,000 Ω. What input voltage is necessary to produce full output power?
2. The input resistance to an amplifier is 175 Ω, and the output resistance is 3000 Ω. The amplifier gain is +28 db. What is the voltage gain of the amplifier?
3. An amplifier drives a 16 Ω load. The hum-level rating of the amplifier is 90 db below the full power-output rating which is 80 W. What is the hum level in the load, and what voltage does the hum produce across the load?
4. The input resistance of an amplifier is 75 Ω, and the input current is 6 ma. The output resistance is 1500 Ω, and the output voltage is 16 V. What is the amplifier voltage gain, and what is the power gain? Express both in decibels.
5. The input to a 1400-foot 50 Ω transmission line is 64 V. The output is 12 V when the load is matched. What is the loss of the transmission line expressed in decibels per hundred feet?
6. A phonograph pickup develops 15 mv across a 35-Ω input. A 60-W speaker system has an impedance of 16 Ω. What minimum amplifier gain is necessary to produce full power output?

   For the following two problems, the decibel meter is calibrated to a 6-mw 500-Ω standard (0 db). The meter scale is calibrated in decibels for the 0- to 3-V range.
7. The meter is placed across a 2000-Ω load and reads +2.5 db. The scale range switch is set to the 30-V scale. What is the true load power, and what is the true decibel value?
8. The meter is placed across a 75-Ω load and reads −5.5 db. The scale-range switch is set on the 60-V scale. What is the true load power, and what is the true decibel value?

   The general equation that is used for transmission of radio signals in space is

$$P_r = P_t \frac{\lambda^2 G_t G_r}{(4\pi R)^2}$$

where G_t is the gain of the transmitting antenna
G_r is the gain of the receiving antenna

R is the distance in meters of the spacecraft from earth
P_t is the transmitter power
P_r is the receiver power
and λ is the wavelength of the radio signal in meters given by $(300/f)$ where f is the frequency in MHz.

9. The Mariner spacecraft carried a 20-W transmitter operating at 480 MHz. The spacecraft antenna had a gain of +12 db. The receiving antenna used to track the spacecraft had a gain of +80 db and an impedance of 50 Ω. What was the available signal voltage from the tracking antenna from the spacecraft signals? The closest approach distance of Venus was 25,000,000 miles.

10. The moon is approximately 240,000 miles from earth. If a radio wave is sent to the moon, 15% of the signal is reflected back. On earth an antenna having a gain of +80 db is available with a receiver that can respond to a signal at a −110 dbm level. Using an identical antenna for transmission at 100 MHz, what transmitter power is required on earth to obtain a reflected signal from the moon?

Section 8-5 Loudness

H. Fletcher and W. A. Munson presented in the October 1933 issue of the *Journal of the Acoustical Society of America* the results of their lengthy research in sound, "Loudness, its definition, measurement and calculation." In this paper they presented a graph (Fig. 8-3) which relates frequency to

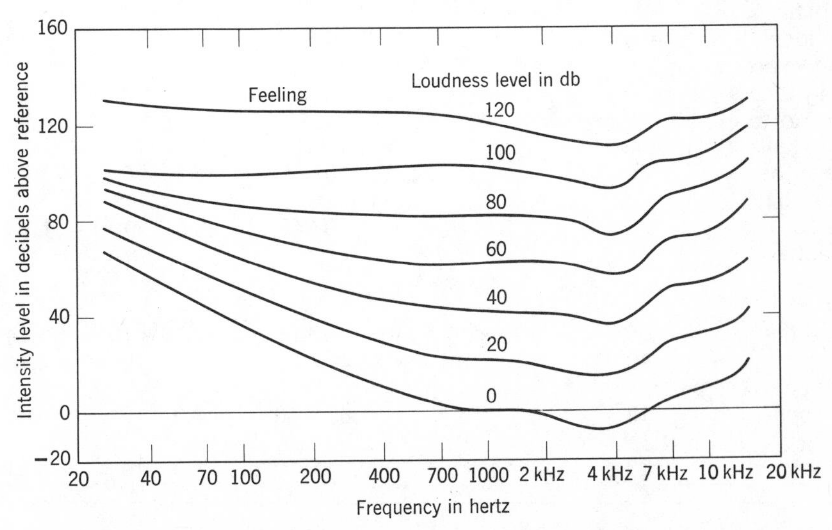

Figure 8-3 Equal-loudness contours. (Reference: Odb = 10^{-16} W/cm^2 at 1000 Hz.)

sound intensity level in decibels above a zero-db reference level. The sound intensity is expressed in watts per centimeter^2 and is compared to a reference level of 10^{-16} W/cm^2 at 1000 Hz. This zero reference corresponds to a root-mean-square sound pressure of 2×10^{-4} dynes/cm^2. This reference sound pressure level is also defined as 2×10^{-4} *microbar*. The microbar unit has gained acceptance for the calibration of transducers and test equipment used in sound measurements in liquids.

From the curves given in Fig. 8-3, we notice that, in order to have threshold audibility at 100 Hz, we must have a +38-db increase in power at 100 cycles over the power required for threshold hearing at 1000 Hz. One of the important results of this research was to show that, as the sound level

TABLE B Intensity Levels.[a]

Type of Sound	Intensity Level (decibels above 10^{-16} watt/centimeter^2)	Intensity (microwatts/centimeter^2)	Root-Mean-Square Sound Pressure (dynes/centimeter^2)	Root-Mean-Square Particle Velocity (centimeters/second)	Peak-to-Peak Particle Displacement for Sinusoidal Tone at 1000 hertz (centimeters)
Threshold of painful sound	130	1000	645	15.5	6.98×10^{-3}
Airplane, 1600 rpm, 18 ft	121	126	228	5.5	2.47×10^{-3}
Subway, local station, express passing	102	1.58	25.5	0.98	4.40×10^{-4}
Noisiest spot at Niagara Falls	92	0.158	8.08	0.31	1.38×10^{-4}
Average automobile, 15 ft	70	10^{-3}	0.645	15.5×10^{-3}	6.98×10^{-6}
Average conversational speech, 3.25 ft	70	10^{-3}	0.645	15.5×10^{-3}	6.98×10^{-6}
Average office	55	3.16×10^{-5}	0.114	2.75×10^{-3}	1.24×10^{-5}
Average residence	40	10^{-6}	20.4×10^{-3}	4.9×10^{-4}	2.21×10^{-7}
Quiet whisper, 5 ft	18	6.3×10^{-9}	1.62×10^{-3}	3.9×10^{-5}	1.75×10^{-8}
Reference level	0	10^{-10}	2.04×10^{-4}	4.9×10^{-6}	2.21×10^{-9}

[a]From *Reference Data for Radio Engineers*, Copyright 1968 by Howard W. Sams & Co., Inc.

increases, the curves tend to flatten out. The curve show that the ear is most sensitive to sounds between 2000 Hz and 4000 Hz in the medium sound intensity range and that the response of the ear falls for both higher and lower frequencies. Some typical sound intensity levels are given in Table B.

An ordinary volume control raises or lowers the gain of an amplifier at all frequencies in accordance with the setting of the control. In high-fidelity equipment the action of a *loudness control* is more complex. At its highest setting the response of the amplifier is flat as in the case of the volume control. However as the volume is reduced by the loudness control, there is more reduction given to the middle frequencies than to the highs and lows. Accordingly, there is an attempt made to reproduce the equal-loudness contours of Fig. 8-3 to yield a more realistic sound.

Questions

1. Define or explain each of the following terms: (*a*) Weber-Fechner law, (*b*) Fletcher-Munson curves, (*c*) semilog graph paper, (*d*) octave, (*e*) decibel, (*f*) mantissa, (*g*) characteristic for log, (*h*) VU, (*i*) dbm.
2. Explain why decibels are used.
3. Why is the plus sign used in decibel terminology?
4. As a voltage ratio, what is 0 db? −3 db? −6 db? +3 db? +6 db?
5. As a power ratio, what is 0 db? −3 db? −6 db? +3 db? +6 db?
6. What is the logarithm of zero?
7. Why are correction factors necessary when decibel meters are used?
8. Distinguish between dbm and VU.
9. Why is it necessary to introduce the term "loudness"?
10. When the volume of a record player is increased, should the bass response be raised or lowered? The treble?

Chapter Nine

AUXILIARY COMPONENTS IN ELECTRONIC CIRCUITS

The study that is prerequisite or corequisite to the study of electronics is usually concerned with the theory of d-c and a-c circuit analysis. It is natural to expect that in these preliminary courses only a small amount of time is devoted to the various features of circuit elements which are designed to meet the specialized needs of the electronic application. It is the intent of this chapter, not to introduce the elements, but to point out certain available types and forms that are useful in transistor and vacuum-tube circuits. Resistors (Section 9-1), potentiometers (Section 9-2), inductances (Section 9-3), transformers (Section 9-4), and capacitors (Section 9-5) are considered in this chapter.

Section 9-1 Resistors

Resistors are used freely in electronic circuits to serve as specific current paths and to serve as circuit elements that either provide a means of reducing a voltage or provide a means of securing a specific voltage drop. The calculation of resistance from the various forms of Ohm's law is very familiar from d-c and a-c circuit analysis. In electronics, the determination of the power rating of a resistor is just as important as the calculation of the resistance value itself. A specification for a resistor is meaningless and incomplete if this power rating in watts is not included. Let us assume that the rating for a resistor is 3.65 W by calculation. The nearest available commercial resistor having the necessary power rating is, say, 5 W. The significance of a 5-W rating is that this unit can safely dissipate 5 W if it is in free space with an unrestricted air circulation. Also, the temperature of the "cooling" air cannot exceed that specified by the manufacturer. When this resistor is wired into the underside of a chassis, the power rating is *derated* so that the surface temperature of the resistor does not exceed its maximum permissible value. Derating factors of five or ten are necessary under extreme conditions.

In most applications that use discrete components, fixed carbon resistors (Fig. 9-1) are suitable. A mixture of carbon in a binder is formed into a short

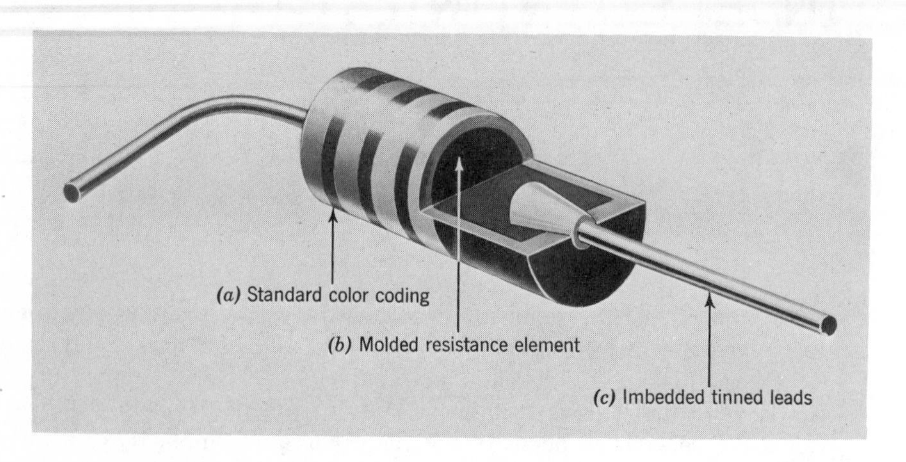

Figure 9-1 Typical carbon resistor. (*Courtesy Allen-Bradley Co.*)

rod. Connection pigtails are placed on the ends of the resistance element. A ceramic form encases the carbon element to keep it protected from humidity and from mechanical damage. The units are made in several sizes capable of dissipating 2 W, 1 W, $\frac{1}{2}$ W, $\frac{1}{4}$ W, or $\frac{1}{8}$ W. The resistance values, which depend on the carbon mixture, range from a fraction of an ohm to about 22 MΩ. Special forms of deposited carbon on glass extend this range to much higher values.

Although carbon has a negative temperature coefficient, the properties of the resistance element have been developed to produce a substantially constant value of resistance over normal temperature variations, extending from about −40°C to +80°C. The carbon resistor has the very important property of being noninductive. This means that, when a carbon resistor is used in a high-frequency circuit, its impedance effectively does not change with frequency.

Carbon resistors are normally not available for handling powers in excess of 2 W. Wire-wound resistors are used for these higher powers. The wire-wound resistor is made from one of several forms of alloy resistance wire such as Nichrome. The wire is coiled on a hollow ceramic tube, and lugs are placed at each end. The whole unit is covered with a heat-resistant ceramic compound and baked. Standoff mounting brackets must be used with the larger units to prevent the contact of the resistor with the chassis or with the other circuit components. These resistors are manufactured with power-handling capacities from 4 or 5 watts to several hundred watts.

Wire-wound resistors are normally inductive and are used in d-c circuits where the inductance of a great many turns has no effect. If they are used in high-frequency circuits, the inductance of the turns and the capacitance between turns makes the calculation of the net impedance of the unit very complex.

Fixed-film resistors are constructed by using a thin layer of resistive material on an insulated core. The thin films are made of a deposited layer of carbon or carbon-boron mixture. These resistors can be made to tolerances as accurately as wire-wound resistors and are usable to 10 MHz. High resistance values can be obtained by forming the deposited layer in a spiral.

The method of obtaining a P-type resistor in an integrated circuit is shown in Fig. 9-2*a*. A *P-N* junction exists between the *P* material and the *N*

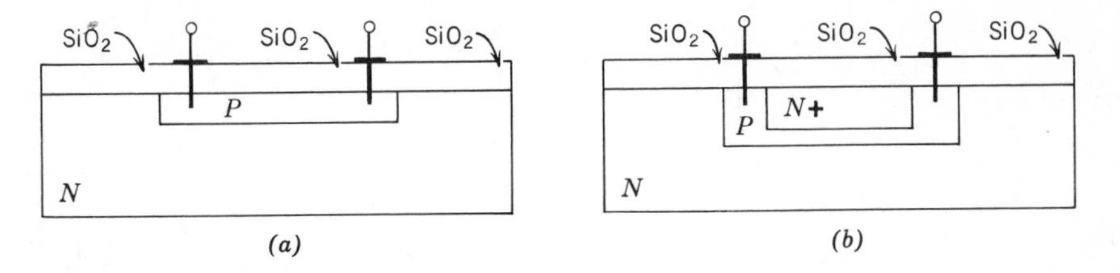

Figure 9-2 Resistors in integrated circuits. (*a*) Low resistance value. (*b*) High resistance value.

material that serves as the supporting substrate. If the resistor is maintained at a potential that is negative with respect to the substrate, the resistor is completely insulated from the substrate. The resistor could be *N* material and the substrate *P* material. Then the resistor must be positive with respect to the substrate.

The resistance value is determined from

$$R = \rho \frac{l}{Wd} \tag{9-1a}$$

in which ρ is the resistivity, l is the length, W is the width, and d is the thickness of the resistor material in compatible units. If high resistance values are required, an N^+ layer can be used to reduce the thickness of the *P*-type resistive material (Fig. 9-2*b*).

In thin-film technology and in integrated-circuit technology, the thickness of the resistive material is fixed and the value of the resistor is then a function of its length and width. Equation 9-1*a* is modified to

$$R = \left(\frac{\rho}{d}\right)\frac{l}{W} = R_S\left(\frac{l}{W}\right) \tag{9-1b}$$

R_S is the resistance of a square of this material (Fig. 9-3*a* and *b*). If the side of this square is tripled in size (Fig. 9-3*c*), its equivalent is shown in Fig. 9-3*d*. Obviously, the equivalent resistance of the network of Fig. 9-3*d* is

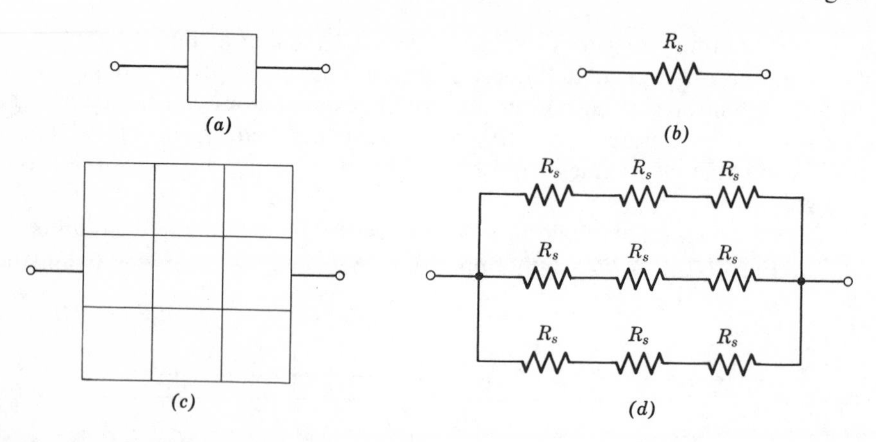

Figure 9-3 The resistance square. (*a*) A square section of resistance material. (*b*) The equivalent resistance of a square. (*c*) A three-by-three square. (*d*) The equivalent resistance of the three-by-three square.

R_S. Thus, the resistance value of a square of thin film or of a layer of *P* or *N* material in an integrated circuit is independent of the size of the square. A typical value for the resistance of a square of material used in an integrated circuit is 300 Ω. Therefore, a section of this material 10 μm wide and 40 μm long must be 4 × 300 or 1200 Ω.

Resistors in integrated circuits can range from 100 to about 50,000 Ω. They can be formed to a 10% tolerance between wafers, but within a wafer they are within 3% of each other in resistance value.

PROBLEMS

1. Without considering a derating factor, what are the voltage and current ratings of the following resistors:

1000 Ω, 1 W	68 kΩ, 2 W
1 MΩ, $\frac{1}{8}$ W	2.7 kΩ, $\frac{1}{2}$ W

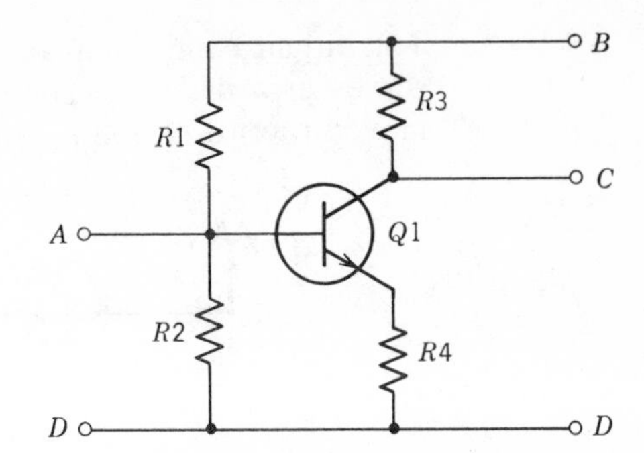

Circuit for Problem 9.

2. Without considering a derating factor, what are the voltage and current ratings of the following resistors:

   390 Ω, 2 W — 51 kΩ, $\frac{1}{4}$ W
   1000 Ω, $\frac{1}{8}$ W — 10 Ω, $\frac{1}{2}$ W

3. Wire-bound 10-kΩ resistors are available in 2-W, 5-W, 10-W, 50-W, and 100-W ratings. What are the maximum current ratings for each size without derating?
4. The voltage across a 39-Ω resistor is 15 V. Using a derating factor of 3, what is the power rating of the resistor?
5. A 15,000-Ω bleeder resistor is placed across 300 V. Using a derating factor of $2\frac{1}{2}$, what is the power rating of the unit? What is the nearest commercial size?
6. A 10,000-Ω 100-W rheostat is set at 3600 Ω in a particular application. What is the maximum voltage that can be placed across the terminals?
7. A 15,000-Ω 70-W potentiometer is used in a voltage-divider circuit. The supply voltage is 800 V. What is the maximum current that can be drawn from the slider when it is set at the midpoint?
8. The resistance material used for an IC has a value of 400 Ω/sq. The least dimension that can be used is 8 μm. What are the required sizes for 120-, 330-, 2000-, and 2400-Ω resistors?
9. The circuit for an integrated circuit contains four resistors and one *NPN* transistor. Leads are attached for external connections at *A*, *B*, *C*, and *D*. Show a top view of the IC and show a cross-sectional view of the IC. Label completely.

Section 9-2 Potentiometers

A *rheostat* is a continuously variable two-terminal resistance, and a *potentiometer* is a continuously variable three-terminal resistance. A potentiometer can be used as a rheostat, but a rheostat cannot be used as a potentiometer.

The circuit connections of these units are shown in Fig. 9-4. The rheostat serves as a device to control and limit the load current. As the slider is moved toward O, the resistance is *cut out*, and the load current increases.

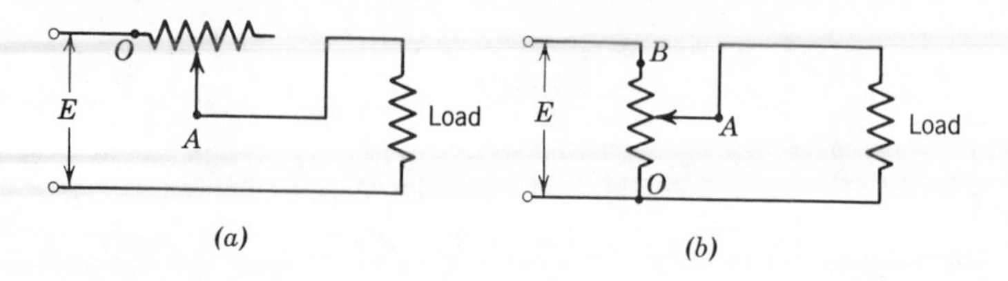

Figure 9-4 Circuits using continuously variable resistors. (*a*) Rheostat circuit. (*b*) Potentiometer circuit.

As an example, consider a rheostat that is 100 Ω total resistance and has a power rating of 100 W. When the entire resistance is in the circuit, 100 Ω, it has a current rating of one ampere. If the rheostat is made of uniform wire, the wire cannot handle more than 1 A even though the slider is set so that there is only 1 Ω between O and A. In the potentiometer circuit, the load current and load voltage increase as the slider is moved from O toward B. In this circuit, there is a current flow through the potentiometer at all times. This shunting current reduces the amount of current that may be available for the load. For example, if a 100-Ω 100-W potentiometer is used across a 100-V source, *any* current taken by the load overloads the potentiometer. In this case, the potentiometer may only be used to serve as a variable-voltage source which does not supply load current. In general, a rheostat serves to control current within a certain range, whereas a potentiometer is generally used as a voltage divider, giving a range of voltages from zero to maximum, the applied voltage.

A wire-wound potentiometer or rheostat is manufactured to handle a high-power dissipation, ranging from several watts to several hundred watts. For low-power ratings up to 2 or 4 W and for high values of resistance that cannot be obtained satisfactorily with a wire-wound unit, a carbon resistance element is used (Fig. 9-5*a*). The carbon control also has the advantage of being noninductive. These potentiometers are available with several *tapers*. The resistance between one end terminal and the wiper arm varies directly with rotation in a *linear-taper* potentiometer. As an example of a *nonlinear* taper, a potentiometer which has a total resistance of 1 mΩ at full rotation has a resistance of 500,000 Ω at three-fourths rotation, 250,000 Ω at one-half rotation, and 125,000 Ω at one-fourth rotation. Representative available tapers are illustrated in Fig. 9-5*b*. The carbon potentiometer is usually adapted so that a power line switch may be mounted on the back of the control. An

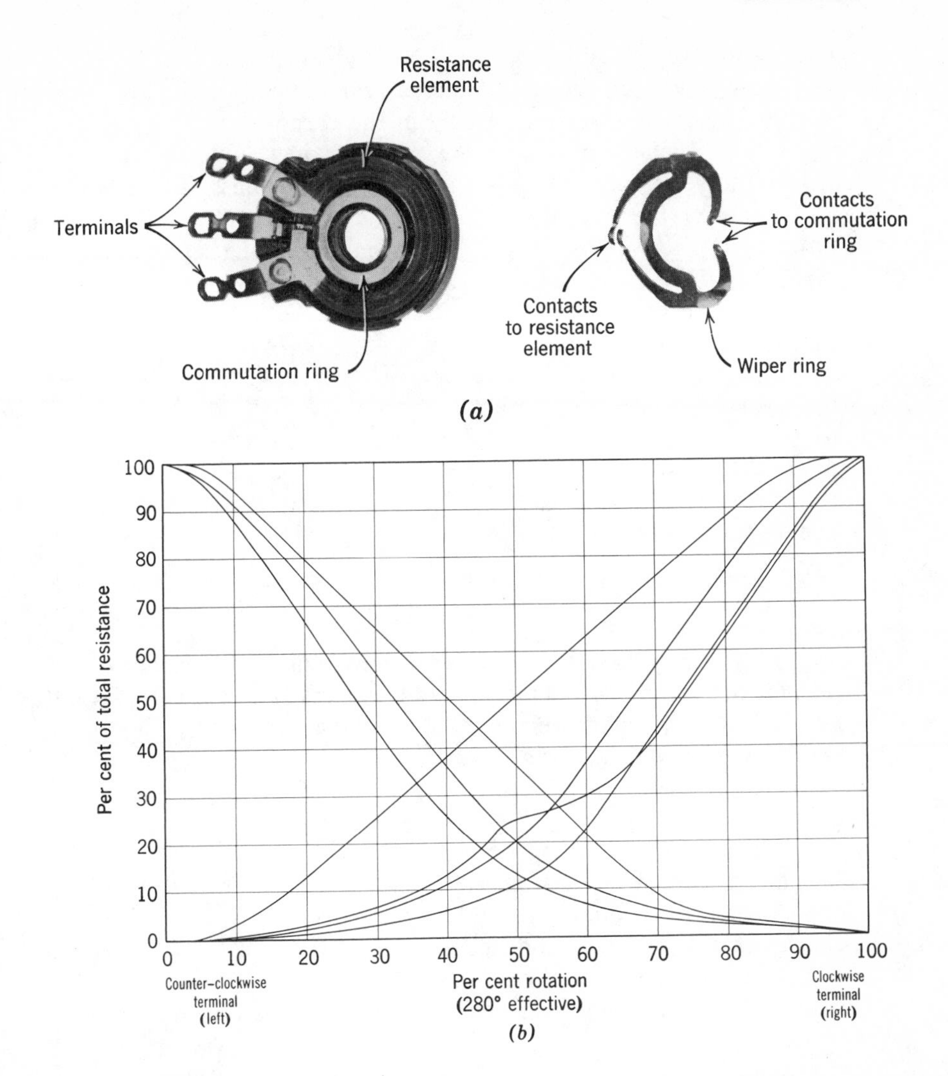

Figure 9-5 Carbon potentiometer and tapers. (*a*) Internal construction. (*b*) Standard tapers. (*Courtesy International Resistance Co.*)

example is the volume control on a radio receiver which is used also as an on-off switch.

Conventional practice requires that the rheostat or potentiometer be connected in a circuit so that a clockwise rotation produces an *increase* in whatever is being controlled and a counterclockwise rotation produces a *decrease*.

Wire-wound potentiometers can be made with a very high precision.

They also are available with high-power ratings and can be made with very low-resistance values. Tapered wire-wound units are expensive whereas tapers are obtained easily in the carbon type. Carbon potentiometers are available only with low-power ratings, but they can be manufactured to resistance values of the order of megohms. The carbon control has a shorter life than the wire-wound unit since the carbon resistance surface tends to wear off when subject to the scraping action of the contact wiper arm. When this happens, small arcs form in the carbon element and the control becomes noisy. This effect is quite noticeable in a worn volume control in a radio receiver.

Section 9-3 Inductors

The fundamental *definition* of inductance is given by

$$L \equiv \frac{N\phi}{I} 10^{-8} \tag{9-2}$$

where N is the number of turns, ϕ is the lines of flux, and I is the current.

This definition of inductance can be expressed in words as *flux linkages per ampere*. The phenomenon of a changing current in a coil producing an emf is given in equation form by Lenz's law:

$$e = -L\frac{di}{dt} = -L\frac{\Delta I}{\Delta T} \tag{9-3}$$

The negative sign associated with Eq. 9-3 merely indicates that e is an opposing or back voltage. If the inductance is of such value that a current change of 1 A in 1 sec produces an emf of 1 V, the unit of inductance is 1 H. From the magnetic-circuit analysis, the magnetic equivalent of Ohm's law states that the flux is the magnetomotive force divided by the reluctance:

$$\phi = \frac{\mathscr{F}}{\mathscr{R}} \tag{9-4a}$$

The magnetomotive force is

$$\mathscr{F} = 0.4\pi NI \tag{9-4b}$$

and the reluctance of the magnetic path is

$$\mathscr{R} = \frac{l}{\mu A} \tag{9-4c}$$

where l is the length of the magnetic path, A is the cross-sectional area of the magnetic circuit, and μ is the permeability. It should be pointed out that this value of reluctance $\mathscr{R}$ is an equivalent reluctance for the whole magnetic circuit and that the actual calculation of the reluctance for a particular coil can be a very complex procedure.

Substituting these values of $\mathscr{F}$ and $\mathscr{R}$ in Eq. 9-4a, we have

$$\phi = \frac{0.4\pi NI}{(l/\mu A)} = \frac{0.4\pi NI\mu A}{l}$$

and, using this expression for ϕ in Eq. 9-2, we find that

$$L = \frac{N\phi}{I} 10^{-8}$$

$$L = \frac{N}{I}\left(\frac{0.4\pi NI\mu A}{l}\right) 10^{-8}$$

$$L = \frac{0.4\pi N^2 A}{l} \mu 10^{-8} = k\mu N^2 \tag{9-4d}$$

Several general considerations can be formed from this last relation. For a coil in which the permeability is unity, such as an air-core coil, the inductance is strictly a geometric concept dependent on the number of turns and on the physical design. A handbook gives formulas for the value of inductance for many types of coils as a function of the turns and the dimensions. If the length of a solenoidal coil is increased without changing the number of turns, the inductance decreases. A larger cross section results in a larger value of inductance than a coil of small cross section, since the inductance is directly proportional to A. The inductance of a coil with a magnetic core is proportional to the permeability of the magnetic path.

When a magnetic material is introduced into the core of a coil, the flux is increased, and the value of inductance is increased. Since the resonant frequency of a tuned circuit is determined by the relation

$$f_0 = \frac{1}{2\pi\sqrt{LC}}$$

an increased inductance lowers the resonant frequency. Adjustable magnetic cores are manufactured which can be lowered into or removed from a coil to change the permeability of the coil, thus, changing its resonant frequency. These cores, called *tuning cores*, are manufactured of a powdered iron pressed in a binder to give the mixture a mechanical strength and stability. These cores or *slugs* can be cast or machined to provide threads and slots for adjustment by a plastic screwdriver called an *alignment tool.* The permeability of a powdered-iron core goes down as the frequency increases and, therefore, at some high frequency the core becomes ineffective.

The effect of a nonmagnetic metallic tuning core, such as brass or aluminium, is to act as a short-circuited turn on the secondary winding of a transformer (Section 14-1). The short-circuited turn reduces the inductance of the coil. Thus, the introduction of a brass core into a coil, by reducing the inductance, *increases* the resonant frequency of a tuned circuit in contrast to the powdered iron core which *decreases* the resonant frequency. The non-magnetic cores are used at high frequencies.

A very elemental application of the definition of inductance is its use in developing the explanation for *skin effect.* Let us consider a conductor (Fig. 9-6) in which the *current density* throughout the cross section is

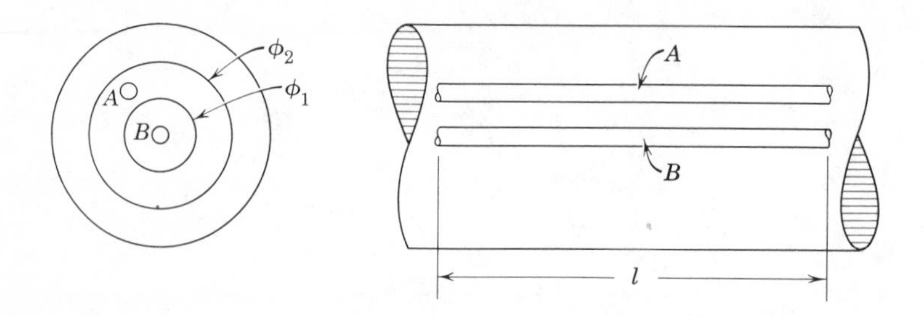

Figure 9-6 Conditions in wire showing skin effect.

uniform. Now consider two filaments or threads extending axially along the wire of equal cross section. One of these threads of the conductor B is located near the center of the wire, and the other A is located nearer the outside of the conductor. The flux ϕ_1 circles or links B but does not link A. The flux ϕ_2 circles around and links both A and B. Since this wire is in a complete electric circuit carrying current, we consider the circuit to be one turn. Then the flux linkages about B are greater than the flux linkages about A, or

$$N\phi_B > N\phi_A$$

The self-inductances of B and A from Eq. 9-2 are

$$L_B = \frac{N\phi_B}{I_B}10^{-8} \qquad \text{and} \qquad L_A = \frac{N\phi_A}{I_A}10^{-8}$$

In the original premise, we specified that the current density is uniform over the conductor and that the cross-sectional areas of A and B are equal. Thus, the current in A must equal the current in B. Then the inductance of B is greater than the inductance of A or

$$L_B > L_A$$

In an a-c circuit for a finite length l we have

$$Z_A = R + j\omega L_A$$

and

$$Z_B = R + j\omega L_B$$

As L_B is greater than L_A,

$$Z_B > Z_A$$

Since we have specified equal current through A and B, the voltage drop along B is greater than the voltage drop along A for the same length of conductor. This creates an impossible situation, and we must conclude that the voltage drops *must* be equal. Therefore, I_A cannot equal I_B. The current distribution cannot be uniform over the cross section of the conductor, but the current density must be greater on the surface than within the wire. This phenomenon is called *skin effect* since the current in an a-c circuit tends to travel toward the outside surface of the conductor.

As the frequency increases, this effect is more pronounced. At high frequencies it is not necessary to have a solid conductor. The conductor can be a hollow tube. At extremely high frequencies, the current is confined to the polished plating on the surface of the base material.

Let us assume that the resistances of A and B are each 2 Ω and that, since a direct current flows in the wire, the current is 3 A in A and 3 A in B. Since the effect of inductance on a steady-state current flow is zero, there is no skin effect. The power loss in A, by the relation I^2R, is 18 W and in B is 18 W, or a total of 36 W. The total current is 6 A, and the equivalent resistance of A and B in parallel is 1 Ω. Now, assume that by the uneven current distribution caused by skin effect, there is an alternating current in A of 4 A and in B of 2 A. The power loss in A is $4^2 \times 2$ or 32 W and the power loss in B is $2^2 \times 2$ or 8 W. The total loss for the a-c condition is 40 W whereas the same current, 6 A, caused a loss of only 36 W without regard to skin effect. This means that there is an *apparent* rise in the resistance of the conductor. This new equivalent resistance is called the *a-c effective resistance*. In the numerical example, the a-c resistance is $40/6^2$ or 1.11 Ω, whereas the d-c resistance is 1 Ω.

The remedy to minimize the skin-effect losses is to provide as great a surface area as possible. A special stranded wire, *litz wire*, is often used in which each strand is insulated from the other. The strands are properly *transposed* so that each wire is on the outside and in each strand position for the same length as each of the other strands. Stranded wire usually consists of either three strands or seven strands to maintain a geometric symmetry. Litz wire cannot be used at high frequencies were the capacitive reactance between strands tends to cancel the inductance of the coil.

The figure of merit or Q of a coil is defined as the ratio of the inductive reactance to the effective resistance:

$$Q \equiv \frac{2\pi fL}{R} \equiv \frac{X_L}{R} \tag{9-5}$$

The reactance of the coil is directly proportional to the frequency and the skin effect, as we have explained, depends on the frequency. If the value of the reactance increases faster than the effective resistance, the Q increases with an increase in frequency, curve a in Fig. 9-7. If the increase in reactance

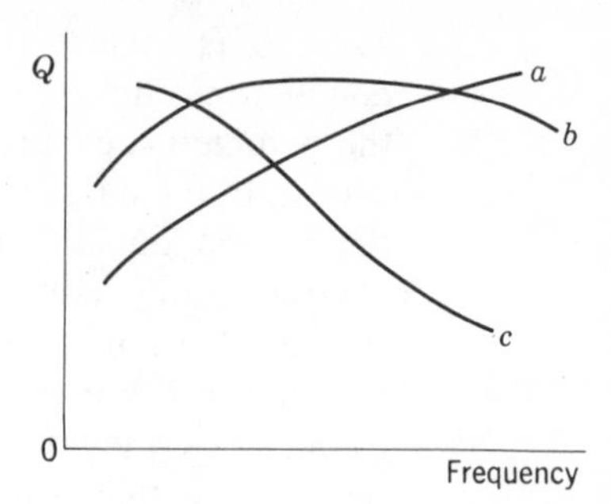

Figure 9-7 Variation of Q with frequency.

is in the same proportion as the increase in the effective resistance, the Q of the coil is fixed over a wide range in frequency, curve b. There is a capacitance between turns of a coil since the turns are near each other. This capacitive action causes the effective value of the inductive reactance to be smaller than ωL, and the effect increases with frequency. Thus, the Q of the coil decreases (curve c) with an increase in frequency when this capacitive shunting effect is appreciable. The exact curve of variation of the Q of a coil with frequency is a function of the size of the wire and the geometric design of the coil. Again, the function of the handbook is to provide the necessary equations for the calculation of the Q of a particular coil at different frequencies.

A coil has the equivalent circuit of a series resistance, R_{se}, and a series inductance, L_{se}, Fig. 9-8a. Let us confine this discussion to coils whose Q is ten or more which is usually the case in high-frequency applications. It is often advantageous to form an equivalent parallel circuit, Fig. 9-8b. By this we mean that, if the same voltage, V, is applied to both circuits, the resulting currents are the same. The current through L_p must be equal to the inductive component of current in the series circuit or must be equal to the line current times the sine of the phase angle:

$$\frac{V}{X_{pa}} = \frac{V}{\sqrt{R_{se}^2 + X_{se}^2}} \frac{X_{se}}{\sqrt{R_{se}^2 + X_{se}^2}} = \frac{VX_{se}}{R_{se}^2 + X_{se}}$$

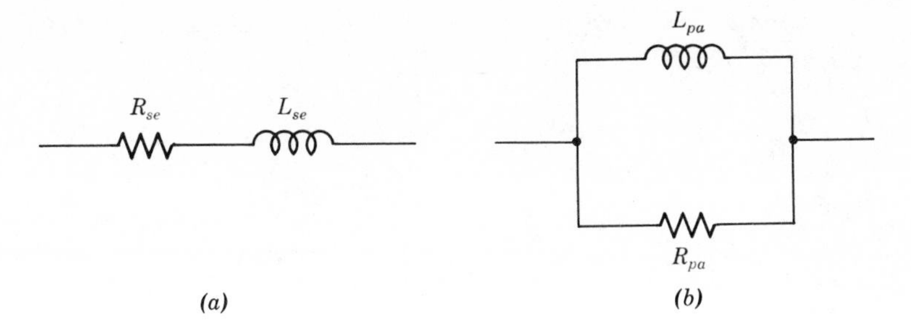

Figure 9-8 Equivalent circuit for a coil. (*a*) Series. (*b*) Parallel.

Canceling V and inverting

$$X_{\text{pa}} = \frac{R_{\text{se}}^2 + X_{\text{se}}^2}{X_{\text{se}}}$$

Since this is a high Q circuit

$$X_{\text{se}} \geqslant 10R_{\text{se}}$$

Then $$X_{\text{se}}^2 \gg R_{\text{se}}^2$$

and X_{pa} becomes

$$X_{\text{pa}} = \frac{X_{\text{se}}^2}{X_{\text{se}}} = X_{\text{se}} \qquad \text{or} \qquad L_{\text{pa}} = L_{\text{se}} \tag{9-6}$$

Therefore the value of inductance is the same, L, either in the series or in the parallel equivalent circuit.

Similarly, the current in R_p is equal to the in-phase current component in the series circuit. Letting X_{se} and X_{pa} be replaced by X_L,

$$\frac{V}{R_{\text{pa}}} = \frac{V}{\sqrt{R_{\text{se}}^2 + X_L{}^2}} \frac{R_{\text{se}}}{\sqrt{R_{\text{se}}^2 + X_L{}^2}} = \frac{VR_{\text{se}}}{R_{\text{se}}^2 + X_L{}^2}$$

Then $$R_{\text{pa}} = \frac{R_{\text{se}}^2 + X_L{}^2}{R_{\text{se}}}$$

As before $$X_L \geqslant 10R_{\text{se}}$$

$$R_{\text{pa}} = \frac{X_L{}^2}{R_{\text{se}}} = \frac{X_L}{R_{\text{se}}} X_L$$

Since $$Q = X_L/R_{\text{se}}$$

$$R_{\text{pa}} = QX_L = \frac{X_L{}^2}{R_{\text{se}}} \qquad \text{and} \qquad R_{\text{se}} = \frac{X_L{}^2}{R_{\text{pa}}} \tag{9-7}$$

It must be emphasized that the Q in a parallel circuit is R_{pa}/X whereas, in a series circuit, Q is X/R_{se}.

As a numerical example, if Q is 60 and X_L is 1200 Ω, then R_{se} is 20 Ω and R_{pa} is 72,000 Ω. Thus the equivalent series resistance R_{se} of a coil is a low numerical value whereas the equivalent parallel resistance of a coil is a high numerical value.

Coils with complete steel cores are used to provide high values of inductance of several or many henrys. The steel-core inductor is usually called *a choke.* Two forms of chokes are used in electronic applications. A choke that is used in low-frequency applications is called an *audio choke* and a choke that is used in power-supply circuits to aid in the conversion of a-c power into d-c power in order to operate other electronic circuits is called a *filter choke.*

The saturation curve for a choke (shown in Fig. 9-9) is the relation

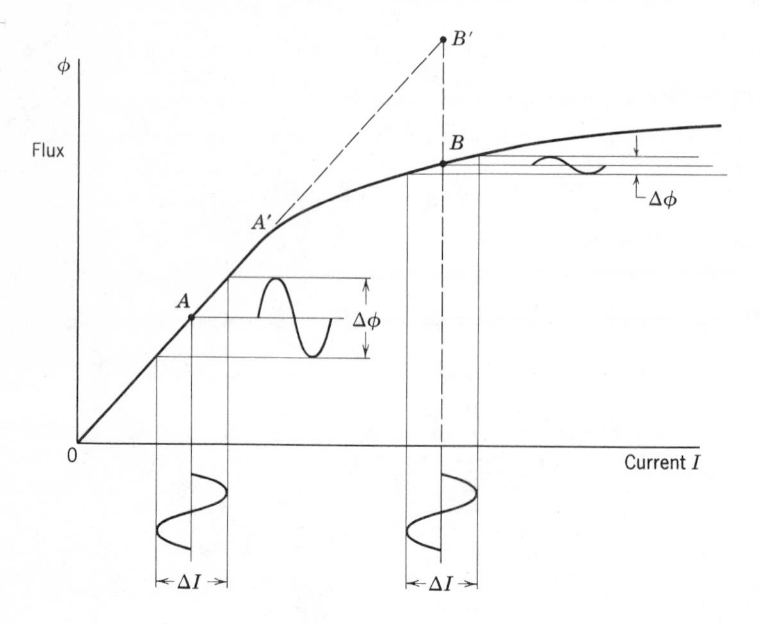

Figure 9-9 The saturation curve for a choke.

between the current I in a choke and the flux ϕ which this current produces. As the current increases, the core saturates, and the increase in ϕ is small compared to the increase in I. If the core did not saturate, the curve would rise from 0 to B'. Since the core saturates, the curve starts to bend at the *knee* of the curve A', and the maximum possible flux is only slightly greater than the flux at B.

The change of flux at A for a given current change ΔI is greater than the change of flux at B for the same incremental current change. By Eq. 9-2, the inductance at B is less than the inductance at A. The operating point, A or B, is established by the amount of fixed direct current through the choke. Thus, in order to specify the inductance of a choke, the value of the direct operating current at which the inductance is measured must be specified.

An inductance used as a component in an integrated circuit is formed by forming a metallic conductor in a spiral with lead connections at each end. Obviously, the resulting inductance values are quite small—of the order of several microhenries. Larger values of inductance require the use of external discrete coils, as in the circuit shown in Fig. 4-13.

PROBLEMS

In all problems, where applicable, if the number of turns is changed, assume that the length of the coil remains fixed.

1. A 1.8 mh coil has a resistance of 50 Ω. What are the values of reactance and power factor at 100, 1000, 10,000, 100,000, and 1,000,000 Hz?
2. For each frequency, determine the value of Q in Problem 1.
3. A wire-wound resistor is measured by an ohmmeter and the resistance is found to be 75 Ω. When this unit is placed across a 400-Hz 120-V a-c circuit, the unit takes 110 W at a power factor of 0.8. Determine the percentage increase in resistance caused by skin effect.
4. If a 3.4 mh coil has 800 turns and a μ of 15, what is the inductance if the μ is reduced to 11?
5. If the turns of the coil of Problem 4 are also reduced to 650, what is the new inductance?
6. If the turns of a coil are reduced by 20%, by what percentage is the inductance reduced?
7. A coil has 4800 turns and is wound on a core that has a μ of 60. The inductance is 1.4 H. The core is removed, and 800 turns are removed from the coil. What is the new inductance?
8. A 200 μh coil has 180 turns with a μ of 55. There are 60 turns added to the coil and the μ is reduced to 35. What is the new inductance of the coil?

Section 9-4 Transformers

The electrical analysis of a steel-core transformer is considered in detail in Section 14-1 through Section 14-3, and the air-core transformer is considered in Section 10-4 through Section 10-6. The ratings for power transformers used in electronic circuits are specified in terms of the voltage and current ratings for each secondary winding. The voltage rating of the primary winding is also given. Volt-ampere ratings are also specified for many electronic transformers. Power transformers are usually designed for a line frequency of 50 to 60 Hz. Aircraft equipment operates on 400 to 2400 Hz, and special transformers are available for these frequencies.

The core of a power transformer is formed by stacking insulated laminations (Fig. 9-10) in the same fashion as the core for a choke coil. Alternate

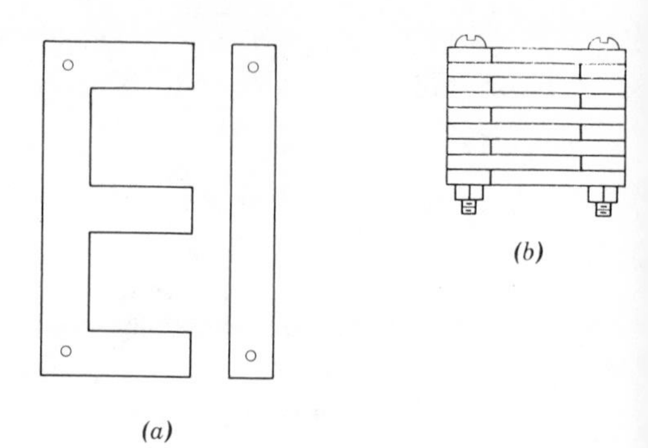

Figure 9-10 Typical transformer cores. (*a*) Typical *E-I* lamination. (*b*) Method of stacking laminations.

laminations are reversed in order to keep the effect of the air gap at a minimum. The laminations are stamped from special transformer steels.

A copper electrostatic shield serving as a Faraday screen is often placed between the primary and the secondary windings. This shield is a full turn of a copper sheet. The ends of the sheet overlap, but they are separated from each other by insulation to prevent the shield from acting as a short-circuited turn. A tab from this sheet extends and is clamped between the laminations to ground the shield. The use of the shield prevents energy transfer from the primary to the secondary winding by capacitive coupling. A cut-away view of a transformer, showing construction details, is shown in Fig. 9-11.

Section 9-5 Capacitors

A capacitor is formed when two parallel surfaces are separated by a dielectric. The basic equation for computing the value of the capacitance is

$$C = \frac{0.2448 \kappa S}{t} \text{ pF} \tag{9-8}$$

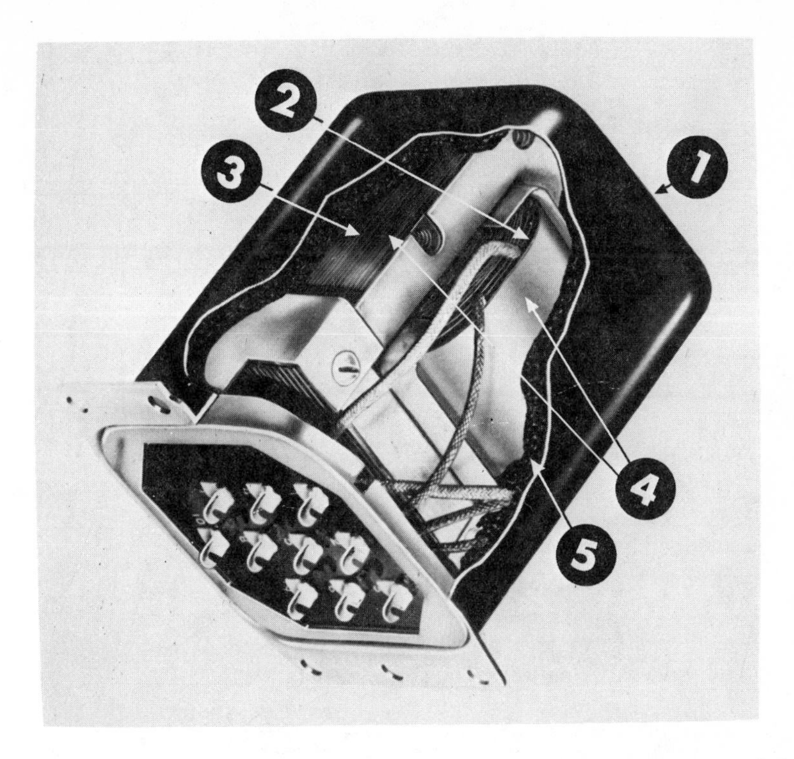

Figure 9-11 Details of transformer construction. (1) Drawn steel case for magnetic and electrostatic shielding. (2) Coil structure. (3) Laminations. (4) Core and coil are vacuum-impregnated with varnish. (5) Moisture-resistant compound. (*Courtesy Chicago Standard Transformer Co.*)

where S is the total parallel surface area in square inches, κ is the dielectric constant of the separating medium, and t is the distance between the active surfaces in inches.

The ideal capacitor has a perfect dielectric which has no losses. Actually, any dielectric indicates a measureable power loss if the applied a-c frequency is increased sufficiently. When a capacitor is measured on an a-c bridge, the result of the test gives the values for a series circuit (Fig. 9-12*b*). Some bridges measure susceptance and conductance, resulting in equivalent parallel arrangements. In many circuit calculations it is necessary to have an equivalent circuit which shows the resistance in parallel with the capacitance (Fig. 9-12*c*). When a capacitor is checked for leakage on an ohmmeter, the measurement of the ohms is a parallel resistance and not a series resistance. To determine the conversion between the series resistance and the parallel resistance, it is assumed that the power factor and the power losses of the capacitor are small.

In the series circuit, if the power factor is less than 10%, the current

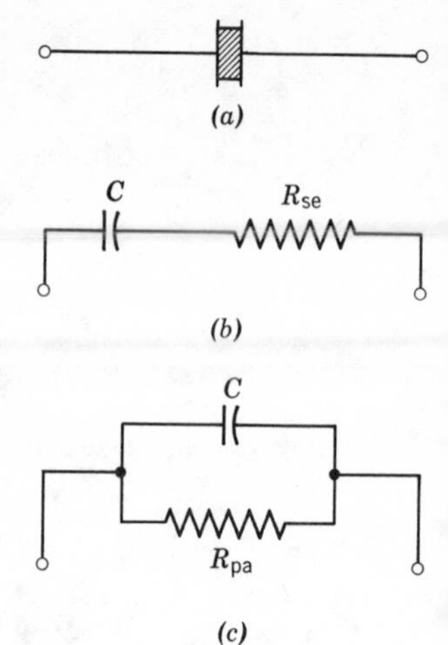

Figure 9-12. Equivalent circuit of a capacitor. (*a*) Capacitor with losses. (*b*) Series equivalent circuit. (*c*) Parallel equivalent circuit.

produced by an applied voltage E is

$$I = \frac{E}{X_C}$$

The power dissipation in the series resistance is

$$P_{se} = I^2 R_{se} = \frac{E^2}{X_C^2} R_{se}$$

The power dissipation in the parallel equivalent circuit is determined directly by the applied voltage E and the parallel resistance R_{pa}:

$$P_{pa} = \frac{E^2}{R_{pa}}$$

If these two circuits are to be equivalent, the powers are equal

$$P_{se} = P_{pa}$$

$$\frac{E^2}{X_C^2} R_{se} = \frac{E^2}{R_{pa}}$$

$$R_{pa} R_{se} = X_C^2 = \frac{1}{(2\pi f C)^2}$$

$$R_{pa} = \frac{X_C^2}{R_{se}} \tag{9-9a}$$

$$R_{se} = \frac{X_C^2}{R_{pa}} \tag{9-9b}$$

Notice that Eq. 9-9b has the same form as Eq. 9-7.

It is evident from Eq. 9-9*b* that a large parallel resistance is equivalent to a small series resistance. In an ideal capacitor, the parallel resistance is infinite and the series resistance is zero.

The dielectric used in a capacitor must have low losses and a high-voltage breakdown rating. The voltage breakdown rating is called the *dielectric strength* and is usually measured in volts per mil (0.001 in.) thickness. Dielectrics that are commonly used in commercial capacitors are mica, ceramic material such as titanium dioxide, paper, aluminum oxide, and air.

Sheets of metal foil separated by strips of mica form *mica capacitors*. The capacitor is enclosed in a plastic housing that keeps the unit both mechanically rigid and waterproof. Mica has very excellent characteristics at high frequencies and is used when the increased cost is justified by the need of a very low power factor. Usually micas are manufactured in sizes less than 0.01 μF.

The *ceramic capacitor* is manufactured by depositing directly on each side of the ceramic dielectric silver coatings which serve as the plates. The advantage of a ceramic dielectric is that the substance has a very high dielectric constant which can be in excess of 6000. The dielectric constant of mica is 6 or 7. By using ceramic dielectrics instead of mica, it is possible to reduce the required plate area by a factor of about 100 to 1000 for the same capacitance value. These ceramic capacitors are very popular because of their small size.

The paper capacitor is manufactured by rolling long narrow sheets of alternate aluminum foil and wax-impregnated paper into compact rolls. The completed unit is sealed to moisture-proof the capacitor. The aluminum sheet that is on the outside should be connected to the lower a-c potential, and the inside sheet to the higher a-c potential for good shielding qualities. The lead that is connected to the outside foil is marked "outside foil" or "ground." Paper capacitors are often enclosed in a metal can that is filled with oil. The oil improves the breakdown characteristic of the dielectric. Plastic films such as *Mylar* or *polystyrene* can improve the performance of a rolled-type capacitor.

An electrolytic capacitor is a capacitor in which an electrolytic process is used to form the dielectric layer (Fig. 9-13). Usually, proper polarity must

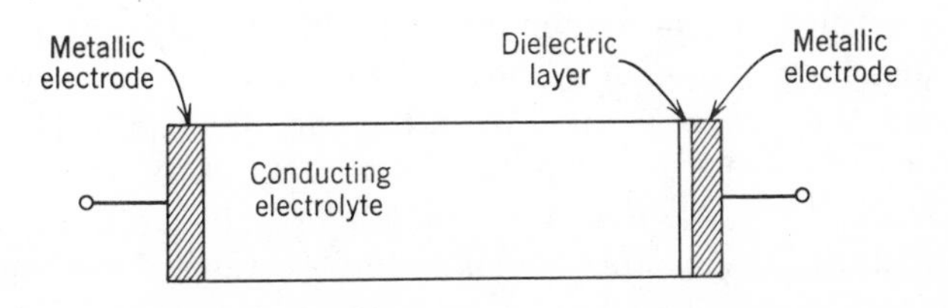

Figure 9-13 Cross-sectional view of an electrolytic capacitor.

be observed in the use of these capacitors or the capacitor will be destroyed. In an aluminum electrolytic capacitor, an aluminum electrode serves as the positive plate, and an alkaline electrolyte serves as the negative plate. The dielectric is an aluminum oxide film formed on the surface of the aluminum. This film is extremely thin and, from Eq. 9-8, the capacitance per unit volume is very high. Often the surface of the aluminum is *etched* to produce corrugations that further increase the working area of the capacitor. The relative dielectric constant of the aluminum oxide is 8 whereas the dielectric constant of tantalum oxide is 27.6. Consequently, an electrolytic capacitor made of *tantalum* has a smaller volume than an aluminum electrolytic capacitor. If an aluminum electrolytic is stored for more than six months, the dielectric layer deteriorates and must be reformed. On the other hand, the shelf life of a tantalum electrolytic is infinite. The electrolytic has an inherent high leakage current and cannot be used in an application in which a low power factor is required. When the applied voltage on an electrolytic is reversed, its resistance drops toward zero and short-circuits the circuit to which it is connected. The electrolytic capacitor does not function at high frequencies. As a result, the primary application of the electrolytic is to maintain a pure direct voltage, such as a bypass capacitor or as a power-supply filter capacitor.

A metalized capacitor (Fig. 9-14) can be a discrete component or a part of an integrated circuit. If SiO_2 is used as the dielectric, the relative dielectric

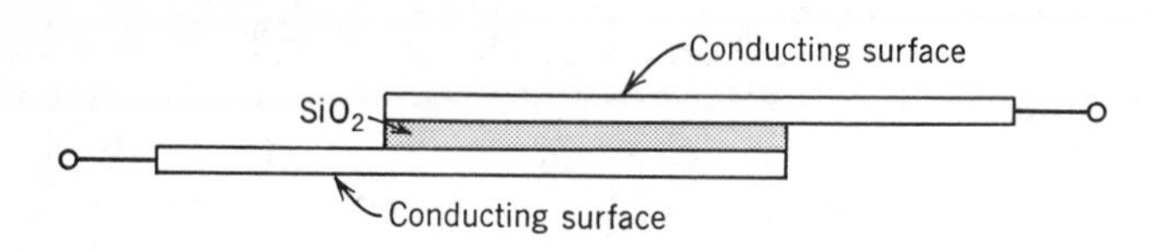

Figure 9-14 Cross-sectional view of a metalized capacitor.

constant is 3.8. This low value of dielectric constant can be compensated by the use of very thin films. The small thickness of the SiO_2 layer of the order of micrometers yields a relatively large value of capacitance up to the order of 50 pF in integrated or printed circuits.

A capacitor, used in integrated circuits, that makes use of the properties of a depletion region is shown in Fig. 9-15. The useful capacitance is obtained in the same manner as in the varactor diode. The whole capacitor is electrically isolated by means of the action of the substrate material. However, now there exists a capacitance between the *P-N* capacitor and the substrate which must be taken into account. This capacitive property exists between all elements of an integrated circuit and the substrate and, therefore, it is not a problem that is unique to microelectronic capacitors.

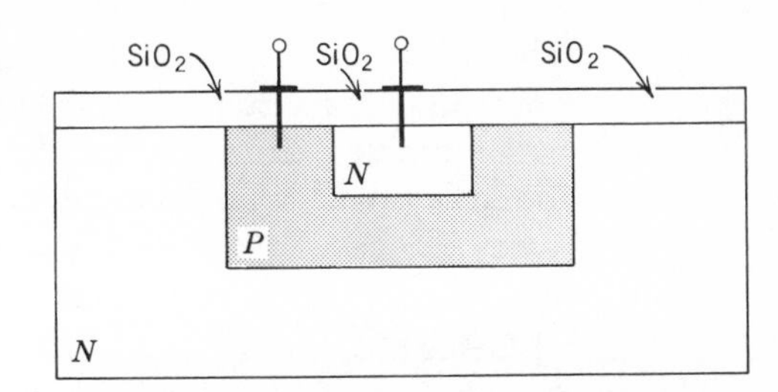

Figure 9-15 Cross-sectional view of an integrated-circuit capacitor.

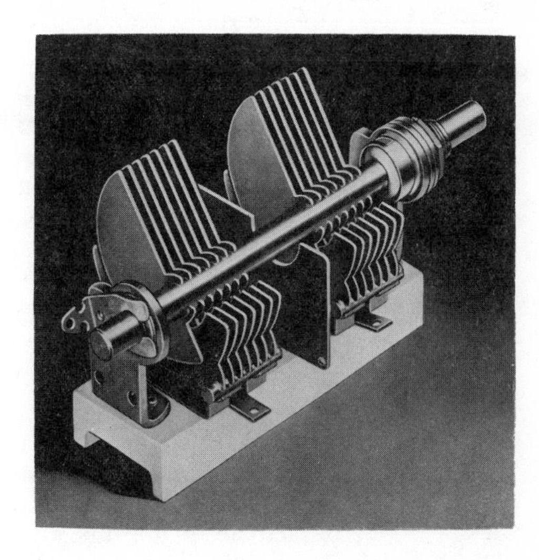

Figure 9-16 Typical air-dielectric variable capacitor. (*Courtesy Hammarlund Manufacturing Co.*)

Air-dielectric capacitors (Fig. 9-16) are used as tuning capacitances. The capacitance of the air-dielectric units can be varied either by changing the meshing areas of the plates or by changing the spacing between the plates. The *tuning capacitor* is varied by rotating the *rotor* section. The rotor is grounded to the frame through the shaft and the bearings. The *stator* plates are insulated from the frame.

Since air has a dielectric constant of unity and the spacing between the plates is greater than is found in a capacitor in which the dielectric is a thin sheet of mica, the capacitance per unit volume for the variable is less than for other forms. The rating of the variable capacitor is given in terms of both its maximum capacitance and its minimum capacitance. Small two- or three-plate capacitors are approximately 3 to 15 pF, whereas larger units of about 30 to 40 plates have correspondingly higher values of capacitance, for example 30 to 350 pF.

A *trimmer* or *padder* capacitor is usually made of metal plates separated by a mica dielectric. The capacitance is varied by changing the plate separation by adjusting a machine screw which compresses the plates together.

These capacitors are small, usually introducing not more than 150 pF and often only about 20 pF into the circuit. Trimmers are usually used in parallel with variable capacitors and padders are placed in series with coils to allow the circuit to be adjusted to resonance at a fixed frequency.

PROBLEMS

1. The plates of a fixed capacitor are 1.6″ by 2.9″ in area. The dielectric is 3.5 mils thick and has a dielectric constant of 12. Determine the capacitance.
2. The dielectric material used in Problem 1 is changed to a material having a thickness of 30 μm and a dielectric constant of 70. Determine the capacitance.
3. When a paper capacitor is unrolled, each sheet is found to be 3 yd long. The foil plates are 0.003 in thick and 1.75 in wide. The paper is 0.008 in thick and has a dielectric constant of 7.5. Determine the original capacitance of the unit.
4. A variable air-dielectric capacitor is constructed of 2.4-in diameter plates that are semicircular. There are 8 rotor plates and 7 stator plates each 0.022 in thick. The overall length of the rotor plates is $\frac{3}{4}$ in. Determine the maximum capacitance.
5. A tuning capacitor has 11 stator plates and 12 rotor plates. The plates are semicircular with a diameter of 1 in each. The plates are 0.016 in thick and the overall length of the rotor section is 0.500 in. Determine the maximum capacitance of this unit.
6. A bridge measurement at 1 MHz of a 0.002 μF capacitor indicates a shunting resistance of 0.68 MΩ. Determine the equivalent series resistance and the power factor of the capacitor.
7. A bridge measurement of a 150 pF capacitor at 500 kHz indicates a series resistance of 0.6 Ω. Determine the equivalent shunt resistance and the power factor of the capacitor.
8. A 1500 pF transmitter capacitor is rated at 450 V when it is used in a circuit where the frequency is 1 MHz. What are the voltage ratings at 5 MHz and at 10 MHz? If the direct voltage rating is 2000 V, at what frequency can this capacitor be used to obtain a maximum alternating voltage rating?
9. A particular capacitor when placed across a 350-V 18-MHz source draws 2 A, and its heating is equivalent to 15 W. Draw both the series and the parallel equivalent circuits.
10. At 1,000 Hz, an a-c bridge measurement give values for the capacitance and series resistance of a particular capacitor of 3600 pF and 2.8 Ω. What is the equivalent shunt resistance and what is the power factor at 1000 Hz?

Questions

1. Define or explain each of the following terms: (*a*) derating, (*b*) wire-wound, (*c*) fixed film, (*d*) ohms per square, (*e*) rheostat, (*f*) taper, (*g*) magnetomotive force, (*h*) reluctance, (*i*) permeability, (*j*) core, (*k*) Faraday screen, (*l*) lamination, (*m*) mylar, (*n*) outside foil, (*o*) trimmer, (*p*) padder.
2. Why is it necessary to determine the power rating of a resistor?
3. Why is a carbon resistor noninductive?
4. Why must different derating factors be taken into consideration?
5. Is a substrate (N^+) used with resistors in an IC?
6. How are resistors "floated" on a metallic chip?
7. If the value of the resistance per square is 200 Ω, how is a 600 Ω resistor formed?
8. Define inductance by both words and formula.
9. How does inductance vary with turns and permeability?
10. What is the permeability of copper? Of glass? Of brass? Of wood? Of air?
11. Describe the cause and result of skin effect.
12. Define Q.
13. Explain why, for a particular coil, the Q may decrease with an increase in operating frequency.
14. What is the equivalent parallel resistance of a coil in terms of Q and X_L?
15. What is the Q of a coil in terms of X_L and equivalent parallel resistance?
16. What is the Q of a coil in terms of X_L and equivalent series resistance?
17. Explain how a direct operating current can reduce the inductance of a choke.
18. How are multiple windings on a transformer distinguished from each other?
19. How does capacitance vary with the area of the plates? With the separation between plates? With the dielectric constant?
20. What are the restrictions on the use of an electrolytic capacitor?
21. Why does a metallized capacitor have a high capacitance?
22. How is a capacitor formed in an IC?
23. Why is the rotor rather than the stator connected to the frame of a variable capacitor?
24. What types of capacitors are tuned with a screwdriver or alignment tool?
25. How is the equivalent parallel resistance of a capacitor related to the equivalent series resistance?
26. What is the effect of a leakage resistance in a capacitor?

Chapter Ten

A-C CIRCUITS IN ELECTRONICS

The objective of this chapter is to review and to present certain basic circuit arrangements from a-c theory which form the foundation of electronic circuit analysis. Series (Section 10-1) and parallel (Section 10-2) tuned circuits are discussed from the viewpoint of their response to different frequencies. The nature of the term, bandwidth, is investigated (Section 10-3). The discussion of general coupled circuit theory (Section 10-4) leads to consideration of the single-tuned air-core transformer (Section 10-5) and the double-tuned air-core transformer (Section 10-6). Harmonics (Section 10-7) have major applications in electronics. A knowledge of four-terminal network theory (Section 10-8) is a prerequisite for transistor circuit analysis.

Section 10-1 Series Circuits

A series a-c circuit can be reduced to three elements; resistance R, inductance L, and capacitance C, in series with a source of emf E (Fig. 10-1). The equations for this circuit which do not have any restrictions or assumptions

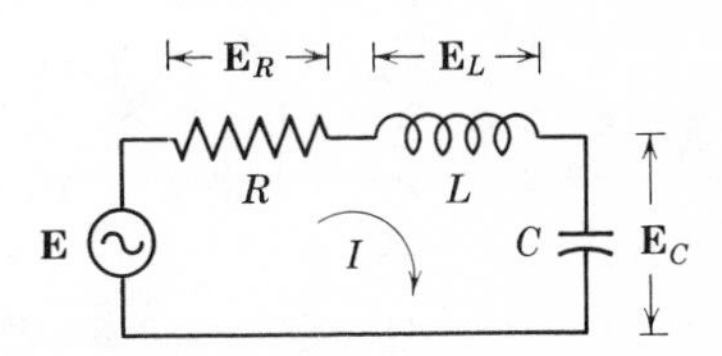

Figure 10-1 The series circuit.

for sinusoidal waveforms are

$$\mathbf{E} = \mathbf{E}_R + \mathbf{E}_L + \mathbf{E}_C$$

$$Z = R + j2\pi fL - j\frac{1}{2\pi fC}$$

$$= R + jX_L - jX_C$$

$$= R + j\left(2\pi fL - \frac{1}{2\pi fC}\right)$$

$$= R + j(X_L - X_C)$$

and

$$I = \frac{\mathbf{E}}{Z}$$

The Q of the circuit is defined as $2\pi fL/R$ or X_L/R if the Q of the capacitor is neglected.

In the first four sections of this chapter, there are certain equations in which the currents and voltages must be explicitly expressed as phasor quantities in order to prevent any confusion. At these places, the phasors are shown in boldface type. In a-c theory, it is a general policy to designate that all currents and voltages are phasors and should be in this boldface type unless specifically expressed as magnitudes. Electronic literature does not rigidly adhere to this rule. For instance, Eq. 10-14 should be written in phasor terminology in order to be technically correct. In this textbook we follow the loose interpretation with the understanding that the a-c equations have their main application in electronics to resistive circuits and that they are valid for complex impedances, should the need arise.

When a constant emf is applied to the circuit and when frequency is the only variable, at some one particular frequency, which is defined as resonance f_0, the inductive reactance equals the capacitive reactance. The circuit impedance is a minimum and is equal to the resistance R. The sum of $(X_L - X_C)$ is zero and is indicated as the condition of resonant frequency in Fig. 10-2.

As the frequency increases from zero (direct current), the inductive reactance X_L increases linearly, and the capacitive reactance X_C becomes smaller. The current is a maximum at resonance, and the voltage across the resistance equals the applied emf E, since E_C and E_L are equal but 180° out of phase at resonance. E_C reaches a peak value just before resonance, and E_L reaches its maximum value at a frequency just greater than resonance. When the Q of the circuit is high, these peaks are so very close to resonance that for all purposes they may be considered to be at resonance.

$$Z = R + j(X_L - X_C)$$

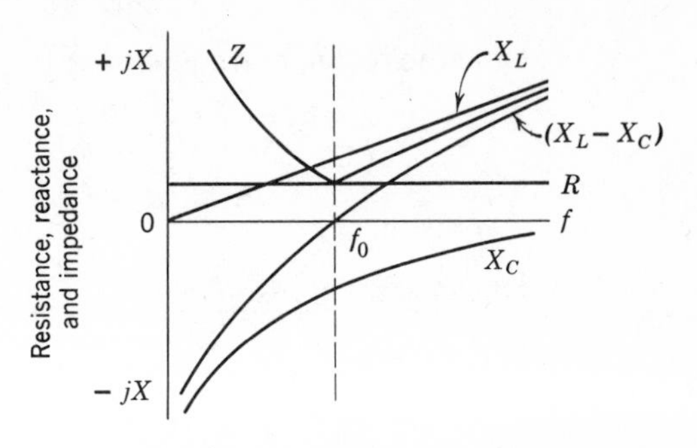

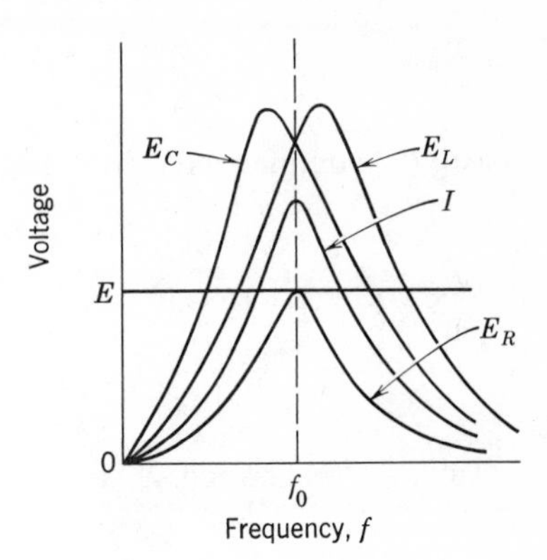

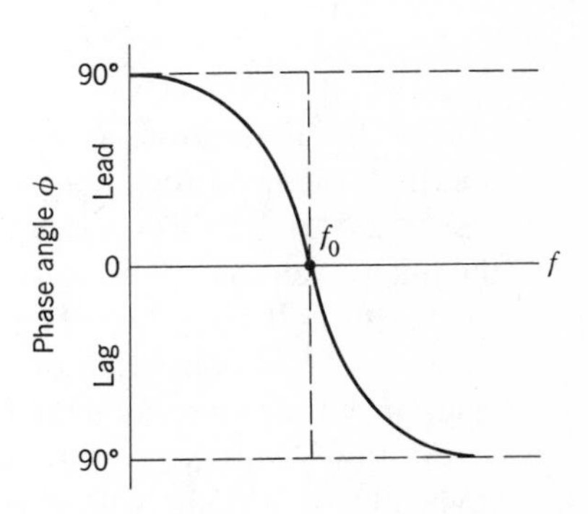

Figure 10-2 Response of the series circuit–variable frequency.

At resonance $$X_L = X_C$$

or $$2\pi f_0 L = \frac{1}{2\pi f_0 C}$$

Then $$f_0^2 = \frac{1}{4\pi^2 LC}$$

and the resonant frequency is

$$f_0 = \frac{1}{2\pi\sqrt{LC}} \tag{10-1}$$

At resonance

$$Z = Z_0 = R + j0 = R$$

the current at resonance is

$$I_0 = \frac{E}{Z_0} = \frac{E}{R}$$

The voltage across the inductance at resonance is

$$V_{L0} = I_0 X_L = \frac{EX_L}{R}$$

But Q is defined as X_L/R. Then

$$V_{L0} = EQ$$

The voltage across the capacitor at resonance is

$$V_{C0} = I_0 X_C = \frac{EX_C}{R}$$

Since, at resonance, $X_L = X_C$,

$$V_{C0} = \frac{EX_L}{R} = EQ$$

Then

$$V_{L0} = V_{C0} = EQ \tag{10-2}$$

By this derivation, in a series circuit, the voltage across the capacitor and the voltage across the inductance (if pure) are both equal in magnitude and equal to QE. For example, consider a series-resonant circuit in which the applied emf is 3 V, R is 10 Ω and X_L and X_C are each 220 Ω at resonance. The Q is 220/10 or 22. The voltage across the capacitor is, then, QE or 22×3 or 66 V. This rise of voltage across the inductance or capacitor at resonance is often termed the *Q gain* of a series-resonant circuit.

At low frequencies the circuit is primarily capacitive, and the phase angle of the circuit is leading at almost 90°. At high frequencies, above resonance, the circuit is inductive with a phase angle of almost 90° lagging. At resonance, the power factor is unity, and the phase angle is zero. The transition from frequencies just below resonance to frequencies just above resonance shows that the change of phase angle is linear with the change in frequency. A high-Q circuit has a greater linear range than a low-Q circuit.

In many applications of a tuned circuit such as a radio receiver, the source is a fixed frequency and the objective of using a variable L or C is to "tune in" the station. By this term we mean that the circuit is adjusted to resonance to produce a maximum output voltage at that frequency. Curves showing the response of a circuit similar to the ones of Fig. 10-2 can be developed for circuits having either a variable L or a variable C.

PROBLEMS

1. A 10 μh coil has an effective resistance of 25 Ω when it is resonated in a series circuit across a 35 μv source at 45 MHz. What size capacitor is required to resonate the circuit and what is the voltage across the capacitor?
2. A 100 μh coil has a Q of 40 when it is tuned (series resonance) to 3.5 MHz across a 4 mv source. What is the size of the required capacitor used to resonate the circuit and what is the voltage across the capacitor?
3. A 360 pF tuning capacitor is used to resonante a particular coil to 530 kHz. What is the minimum value required of the capacitor to tune the circuit to 1.6 MHz?
4. A 30-pF tuning capacitor tunes a particular coil to 87 MHz. What is the value of the tuning capacitor to tune the coil to 110 MHz?
5. A series circuit consisting of a 10-μh coil in series with a 25 Ω resistance is placed across a 30-V source. The frequency is varied. Plot current and coil voltage against frequency.
6. A series circuit consisting of a 100-μh coil in series with a resistance is placed across a 4-V 2.5-MHz source. The resistance is varied. Plot current and phase angle against resistance.

Section 10-2 Parallel Circuits

The basic parallel circuit is shown in Fig. 10-3. In electronic circuits, the capacitors used are usually of such quality that the series resistance of the capacitor can, for all practical purposes, be assumed to be zero. When the model for a capacitor is taken as a parallel circuit of R and C, the value of R

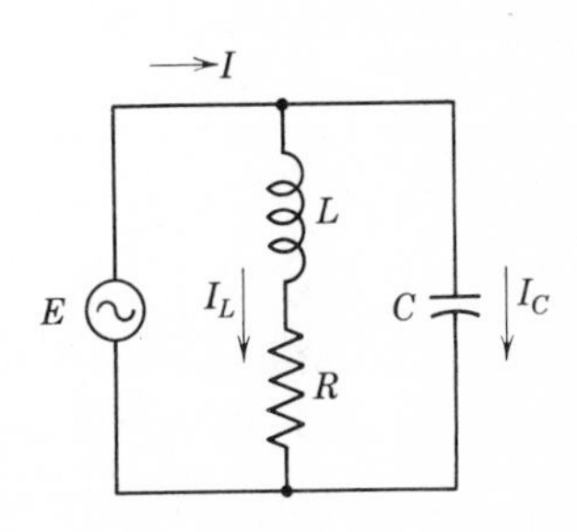

Figure 10-3 The parallel circuit.

is infinite for an ideal capacitor. The relations in the circuit are, from basic a-c circuit theory;

$$\mathbf{I} = \mathbf{I}_L + \mathbf{I}_C$$

$$Z_L = R_L + j2\pi f L$$

$$= R_L + jX_L$$

$$Z_C = -j\frac{1}{2\pi f C} = -jX_C$$

$$\mathbf{I}_L = \frac{\mathbf{E}}{Z_L}, \qquad \mathbf{I}_C = \frac{\mathbf{E}}{Z_C}$$

$$Z_T = \frac{Z_C Z_L}{Z_L + Z_C} = \frac{(-jX_C)(R_L + jX_L)}{R_L + jX_L - jX_C}$$

Let $$X_C = X_L = X \qquad \text{and} \qquad R = R_L$$

Then $$Z_T = \frac{-jX(R + jX)}{R} = \frac{X^2}{R} - j\frac{RX}{R}$$

$$= \frac{X^2}{R} - jX$$

but Q is X/R. Then

$$Z_T = QX - jX$$

$$= X(Q - j1)$$

If Q is equal to or greater than 10, the last relation may be written with negligible error as

$$Z_T = QX - j0 = QX \qquad \text{where} \qquad Q \geqslant 10 \tag{10-3}$$

This expression for Z_T is resistive and not reactive. The fundamental definition of resonance states that, at resonance, the power factor of the circuit is unity. Thus, when we specify a circuit in which the Q is not less than 10, we have the same relation for resonance in a parallel circuit that we had in the series circuit:

$$X_L = X_C$$

and $$f_0 = \frac{1}{2\pi\sqrt{LC}} \tag{10-4}$$

In a series circuit, the impedance at resonance is resistive and equal to R, whereas, in the high-Q parallel circuit, at resonance, the circuit is resistive and its impedance is QX.

If we have circuits where the Q is less than 10, we must use the rather laborious methods that have been studied in formal a-c circuit theory.

This relation of parallel impedance as QX at resonance is very useful in many applications of electronic-amplifier circuits. In the circuit of Fig. 10-4

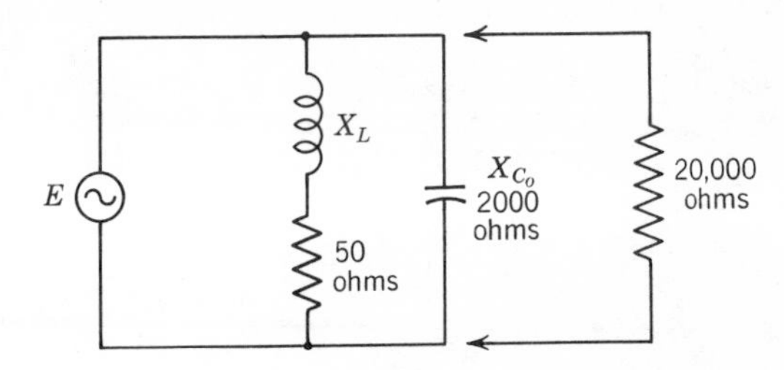

Figure 10-4 The parallel circuit with shunt loading.

at resonance, X_L equals X_C or 2000 Ω. Then the Q of the coil is 2000/50 or 40. The impedance at resonance is QX or 40×2000 or 80,000 Ω pure resistance. If a 20,000-Ω resistor is connected in parallel with the resonant circuit, we have at resonance two resistances in parallel:

$$R'_T = \frac{R_1 R_2}{R_1 + R_2} = \frac{80{,}000 \times 20{,}000}{80{,}000 + 20{,}000}$$

$$= \frac{80{,}000}{100{,}000} 20{,}000 = 16{,}000\ \Omega$$

The total impedance of the circuit is now 16,000 Ω. This total impedance is the effective Q, Q', times the reactance:

$$R'_T = Q'X$$

$$16{,}000 = Q' \times 2000$$

$$Q' = 8$$

Thus, placing a resistance in parallel with the tuned circuit effectively reduces the circuit Q. This *loading resistance* may be adjusted to obtain a particular overall circuit Q which is required for an application.

In Section 9-5, we treated an actual capacitor as being equivalent to a pure capacitor in parallel with its leakage resistance. If it is necessary to consider the resistance of the capacitor, the calculation should be first made for the ideal capacitor and then the leakage resistance considered as a shunt loading resistor which lowers the circuit Q and the circuit impedance.

The response of this circuit to a varying frequency is shown in Fig. 10-5. The current in the circuit at zero frequency (direct current) is determined by the values of E and R. As the frequency increases, the current through the capacitance increases linearly, and the current through the

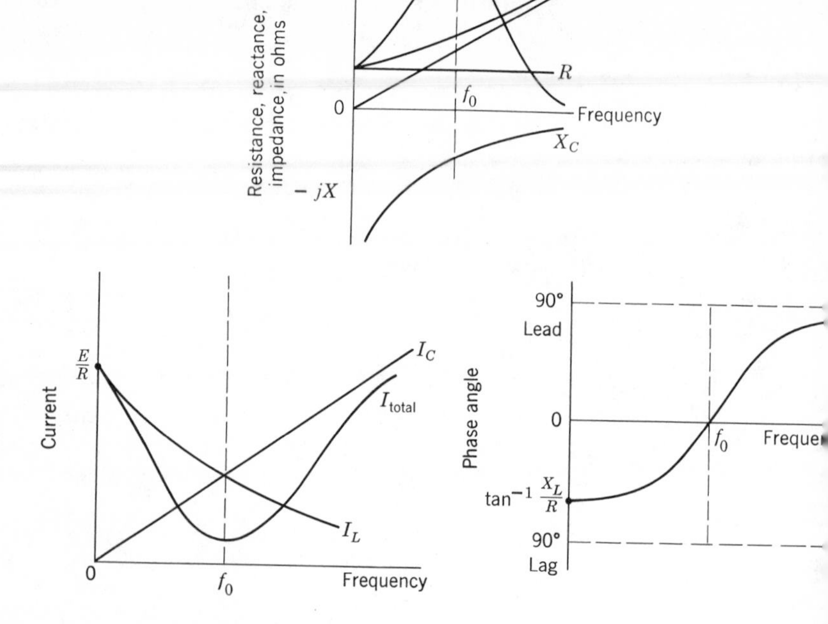

Figure 10-5 Response of the parallel circuit—variable frequency.

inductive branch decreases to zero. At resonance, the impedance is the maximum value QX, and the current is at a minimum value E/QX.

The series resistance R_{se} of the coil can be converted to an equivalent parallel resistance R_{pa} by Eq. 9-7:

$$R_{pa} = \frac{X_L^2}{R_{se}}$$

Now, if a voltage E is impressed across the parallel resonant circuit, the reactive power in the circuit is

$$P_{reactive} = \frac{E^2}{X_C} = \frac{E^2}{X_L}$$

and the resistive power (heating loss or dissipation) is

$$P_{resistive} = \frac{E^2}{R_{pa}} = \frac{E^2}{(X_L^2/R_{se})} = \frac{E^2 R_{se}}{X_L^2}$$

Taking the ratio of reactive power to resistive power

$$\frac{P_{reactive}}{P_{resistive}} = \frac{E^2/X_L}{E^2 R_{se}/X_L^2} = \frac{X_L}{R_{se}} = Q$$

This relationship can be phrased in an alternative form which is widely used to define Q:

$$Q \equiv \frac{\text{energy stored in a tuned circuit}}{\text{energy dissipated in a tuned circuit}} \tag{10-5}$$

Consider a coil (Fig. 10-6a) that has a sliding tap, B. The total inductance of the coil from A to C is independent of the position of the tap. If L_1 is the

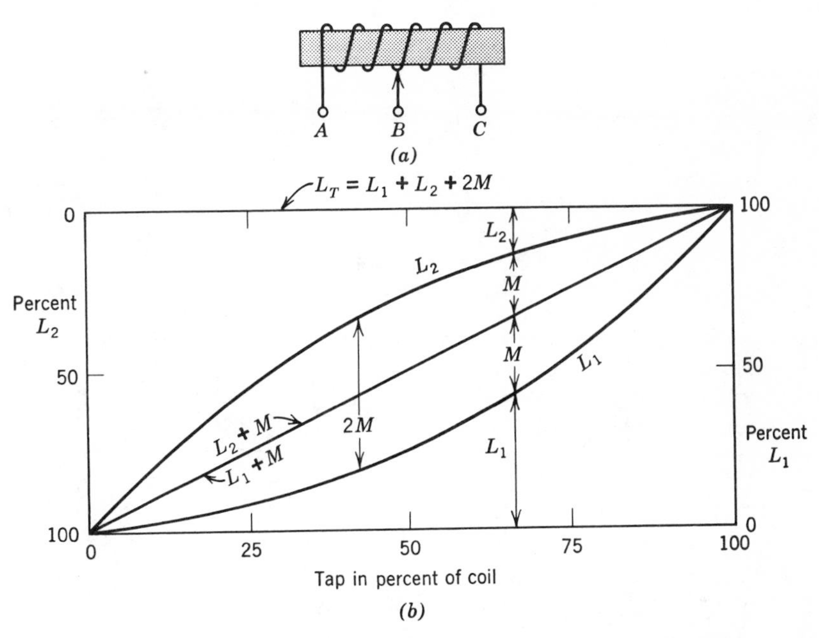

Figure 10-6 A tapped coil. (a) Construction. (b) Inductance properties.

inductance of the coil from A to B and if L_2 is the inductance of the coil from B to C, then the total inductance L of the coil between A and C is $L_1 + L_2 + 2M$ where M is the additive mutual inductance between L_1 and L_2. The total inductance, then, from A to B is $L_1 + M$, and the total inductance from B to C is $L_2 + M$. Assume that the tap B is located at a point where AB contains $\frac{1}{4}$ the number of turns of the coil. Since inductance is proportional to the square of the turns,

$$L_1 = (\tfrac{1}{4})^2 L = \tfrac{1}{16} L$$

and

$$L_2 = (\tfrac{3}{4})^2 L = \tfrac{9}{16} L$$

then

$$L_1 + L_2 = \tfrac{10}{16} L$$

and, by subtraction, $2M = \frac{6}{16} L$ or $M = \frac{3}{16} L$

Then $$L_1 + M = \tfrac{1}{16}L + \tfrac{3}{16}L = \tfrac{4}{16}L = \tfrac{1}{4}L$$

and $$L_2 + M = \tfrac{9}{16}L + \tfrac{3}{16}L = \tfrac{12}{16}L = \tfrac{3}{4}L$$

As a result, if $L_1 + M$ is considered as the inductance from A to B, the value of $L_1 + M$ varies linearly with the tap, as shown in Fig. 10-6*b*.

The input to a parallel resonant circuit of L and C is connected to a point of the tap on the coil (Fig. 10-7*a*). The impedance of the parallel circuit at

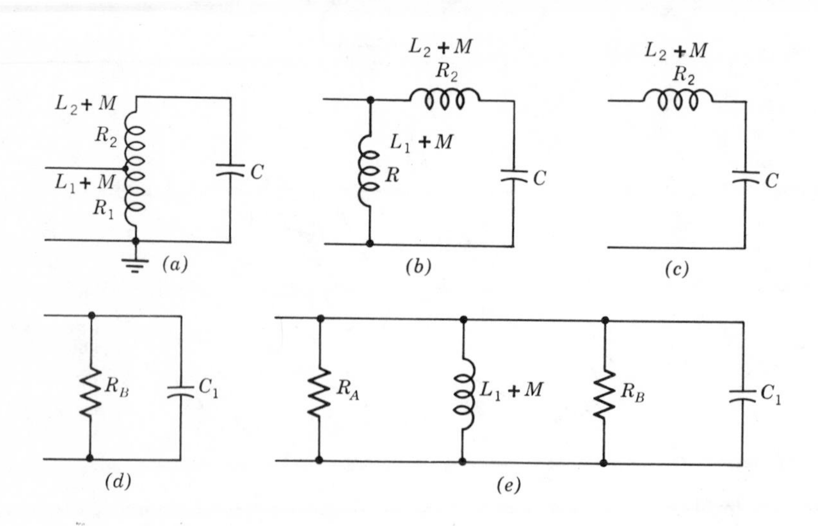

Figure 10-7 A tapped parallel tuned circuit. (*a*) Actual circuit. (*b*) Rearranged circuit. (*c*) Right half of circuit and (*d*) its equivalent circuit. (*e*) Total effective equivalent circuit.

resonance is resistive and is QX where X is either X_C or X_L, the total reactance of the coil. The resonance of the coil is determined by

$$\omega_0 = 1/\sqrt{LC} \qquad \text{or} \qquad \omega_0^2 = 1/LC.$$

The circuit can be rearranged as shown in Fig. 10-7*b*. The total resistance of the coil R is divided by the tap into two parts, R_1 and R_2. The right half of the circuit is redrawn in Fig. 10-7*c*. The reactance of $(L_2 + M)$ subtracts from the reactance of C, giving a new effective value of circuit reactance of X_{C_1}. The magnitude of the value of X_{C_1} is by subtraction, the magnitude of the reactance of $(L_1 + M)$. The series resistance R_2 is converted into a parallel resistance R_B by the use of Eq. 9-7:

$$R_B = \frac{[\omega_0(L_1 + M)]^2}{R_2}$$

Likewise R_1 is converted by Eq. 9-7 into a parallel resistance R_A:

$$R_A = \frac{[\omega_0(L_1+M)]^2}{R_1}$$

When R_A and R_B are combined into an equivalent parallel circuit R_{in} by using the product-over-sum rule, we simplify the result to

$$R_{\text{in}} = \frac{\omega_0{}^2(L_1+M)^2}{R_1+R_2} = \frac{\omega_0{}^2(L_1+M)^2}{R}$$

since $R = R_1 + R_2$. Thus R_{in} represents a pure resistance, R_{in}, looking into the tuned circuit between the tap and the ground. The impedance (resistive) of the whole tuned circuit is

$$R_0 = QX_L = \frac{X_L X_L}{R} = \frac{(\omega_0 L)^2}{R}$$

Forming a ratio

$$\frac{R_{\text{in}}}{R_0} = \frac{\omega_0{}^2(L_1+M)^2}{R} \Big/ \frac{(\omega_0 L)^2}{R} = \left(\frac{L_1+M}{L}\right)^2$$

But $(L_1+M)/L$ is the tap now represented as the fraction, η. Then

$$R_{\text{in}}/R_0 = \eta^2$$

or

$$R_{\text{in}} = \eta^2 R_0 \tag{10-6}$$

PROBLEMS

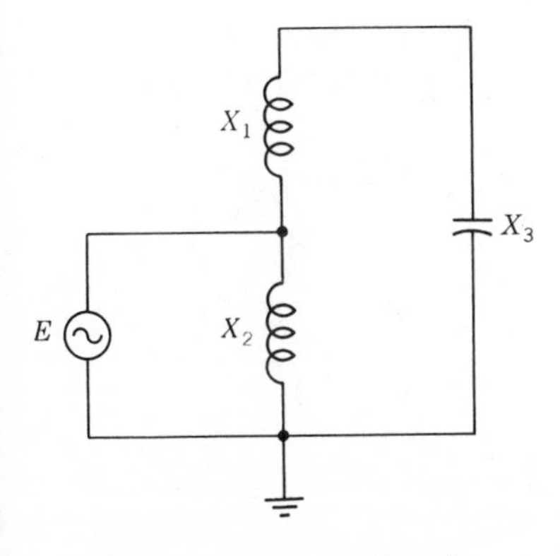

Circuit for Problems 6 and 7.

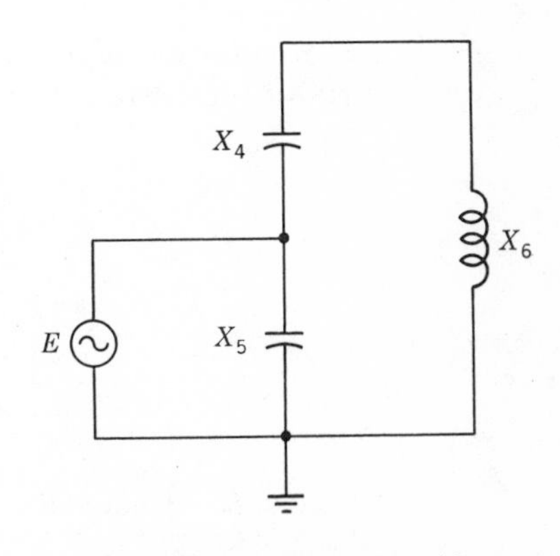

Circuit for Problems 8 and 9.

All problems in this group represent resonance in a parallel circuit.

1. Show how Eq. 9-9*a* can be used to arrive at Eq. 10-3 directly.
2. A coil has an inductance of 65 μh and a resistance of 35 Ω. The coil is resonated with a parallel capacitor at 5.0 MHz. What size capacitor is required, and what is the impedance of the tuned circuit at resonance?
3. A 230-pF capacitor resonates with a coil at 2.4 MHz. The Q of the coil is 45. What shunting resistance reduces the Q to an effective value of 30?
4. A coil has a Q of 60 and is resonant with an 80-pF capacitor at 455 kHz. The capacitor used has a power factor of 0.006. What is the resulting circuit impedance at resonance?
5. A parallel-tuned circuit is shunted with a 120-kΩ resistor. The circuit resonates at 1 MHz with a 150-pF capacitor. If the Q of the coil itself is 100, what is the impedance of the whole circuit at resonance?
6. $X_1 + X_2$ represents a 100-μh coil center-tapped. The whole coil has a Q of 50. The circuit is tuned to 3.0 MHz. What is the impedance looking into the tap? If E is 2 V, what is the current in X_3?
7. $X_1 + X_2$ represents a 30-turn, 50-μh coil with a Q of 40 tuned to resonance at 10 MHz. At what point is E connected so that the load on E is 15,000 Ω?
8. X_4 represents a 20-pF capacitor, X_5 represents a 60-pF capacitor, and X_6 represents a coil with a Q of 35 that tunes the circuit to 30 MHz. What is the inductance of the coil, and what is the load on the generator?
9. X_5 is 30 Ω, X_4 is 70 Ω, and X_6 is 100 Ω with a Q of 100. What is the impedance load on the generator at resonance. If E is 1 V, what is the tank current at resonance?

Section 10-3 Bandwidth

If we consider the current-response curve of a series-resonant circuit (Fig. 10-8), the current has a peak value of I_0 at the resonant frequency f_0. Let

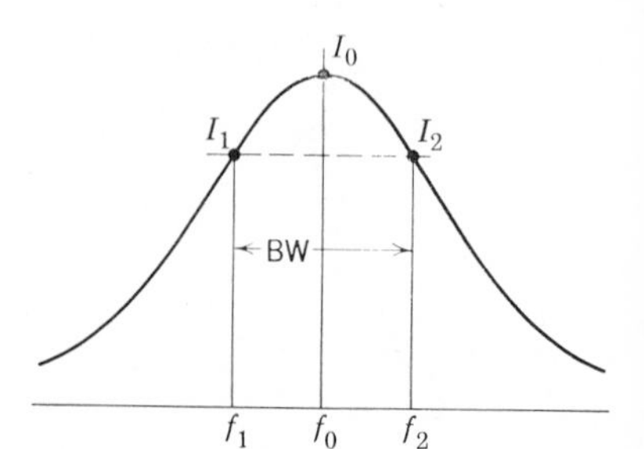

Figure 10-8 Bandwidth of a resonant circuit.

I_1 and I_2 be equal to $I_0/\sqrt{2}$ or 70.7% of I_0. The current value I_1 occurs at f_1, and the current value I_2 occurs at f_2. Now we define the *bandwidth* BW as

$(f_2 - f_1)$. The bandwidth, then, is defined as the range between the frequencies at which the response is down 3 db from peak response.

When considerations of bandwidth are involved in circuit applications, the Q of the coil is assumed constant over the frequency range (Section 9-3). Not only is this true of the frequencies used in communications, but it is also true of industrial electronic applications where we may have, for instance, an iron-core coil operating at a resonant frequency of 1300 cycles with a bandwidth of 50 cycles. If the Q is constant and, since Q equals $2\pi fL/R$, the resistance R of the coil must vary directly with frequency. This assumption is valid, since the bandwidth is small with respect to the resonant frequency. In the series circuit at resonance, the current is

$$\mathbf{I}_0 = \frac{\mathbf{E}}{R_0}$$

At any frequency, the current is

$$\mathbf{I} = \frac{\mathbf{E}}{R + j(X_L - X_C)}$$

Expressing this as a ratio:

$$\frac{\mathbf{I}_0}{\mathbf{I}} = \frac{R + j(X_L - X_C)}{R_0}$$

At any frequency,

$$R = R_0 \frac{f}{f_0}$$

Then

$$\frac{\mathbf{I}_0}{\mathbf{I}} = \frac{R_0(f/f_0) + j(X_L - X_C)}{R_0}$$

If we assume that $(X_L - X_C) = R_0 \dfrac{f}{f_0}$

Then

$$\frac{\mathbf{I}_0}{\mathbf{I}} = \frac{R_0(f/f_0) + jR_0(f/f_0)}{R_0}$$

$$= \frac{f}{f_0} \frac{1 + j1}{1}$$

The magnitude of the ratio is

$$\left|\frac{\mathbf{I}_0}{\mathbf{I}}\right| = \frac{f}{f_0}\sqrt{2}$$

and the phase angle is 45°.

If f_1 and f_2 are sufficiently close to f_0, this ratio then is $\sqrt{2}$. Now, to summarized what has been done up to this point, the bandwidth was defined as the interval between the 70.7% dropoff points, and these dropoff points occur when

$$R_0 \frac{f}{f_0} = X_L - X_C$$

The next step is to insert f_1 and f_2 into the relation and to solve the result for a final equation in terms of f_1, f_2, f_0, and Q. We have called f_2 the higher frequency and f_1 the lower frequency. Now, at f_2, X_L is greater than X_C and $(X_L - X_C)$ is positive. At f_1, X_C is greater than X_L and $(X_L - X_C)$ is negative. Making this substitution, we have, for f_1,

$$-R_0 \frac{f_1}{f_0} = 2\pi f_1 L - \frac{1}{2\pi f_1 C}$$

and, for f_2,

$$R_0 \frac{f_2}{f_0} = 2\pi f_2 L - \frac{1}{2\pi f_2 C}$$

If the first equation is multiplied through by f_0/f_1, and the second by f_0/f_2 and, then, if the resulting equations are divided through by R_0, Q may be substituted for $2\pi f_0 L/R_0$. When the resulting equations are simplified and one is subtracted from the other, we find that

$$2\pi R_0 C (f_2^2 - f_1^2) = f_0\left(\frac{1}{Q-1} - \frac{1}{Q+1}\right) = f_0 \frac{2}{Q^2 - 1}$$

Since Q is at least 10, $(Q^2 - 1)$ may be taken as Q^2. Then

$$2\pi R_0 C (f_2^2 - f_1^2) = \frac{2f_0}{Q^2}$$

Factoring,

$$2\pi R_0 C (f_2 + f_1)(f_2 - f_1) = \frac{2f_0}{Q^2}$$

but $(f_2 + f_1)/2$ is very close to f_0, and $(f_2 - f_1)$ is the bandwidth BW:

$$2\pi f_0 C R_0 \text{BW} = \frac{f_0}{Q^2}$$

Since $2\pi f_0 C = 1/2\pi f_0 L$ by the definition of resonance, we have an expression of the bandwidth in terms of the Q and the resonant frequency:

$$\text{BW} \frac{R_0}{2\pi f_0 L} = \frac{f_0}{Q^2}$$

$$\frac{\text{BW}}{Q} = \frac{f_0}{Q^2}$$

or

$$f_2 - f_1 = \text{BW} = \frac{f_0}{Q} \tag{10-7}$$

By a similar but considerably more lengthy algebraic process, we can show that Eq. 10-7 is also valid for high-Q parallel circuits.

Now let us refer back to the tapped tank of Fig. 10-7*a*. Since the voltage looking into the tap is the voltage across the whole tank multiplied by a constant η which is less than one, the response curve of the circuit looking into the tap has the same bandwidth as the whole tank. Therefore, by Eq. 10-7, the Q of the circuit looking into the tap is the Q of the whole circuit.

PROBLEMS

1. Determine the bandwidth of the tuned circuit given in Problem 2, Section 10-2.
2. Determine the bandwidth of the tuned circuit given in Problem 3, Section 10-2.
3. Determine the bandwidth of the tuned circuit given in Problem 6, Section 10-2.
4. Determine the bandwidth of the tuned circuit given in Problem 7, Section 10-2.

Section 10-4 Coupled Circuit Theory

When a signal is transferred through a network, the signal is said to be *coupled* from one point to another. The concepts of a coupled circuit are shown in Fig. 10-9*a*. A signal $\mathbf{E}_1$ develops a voltage $\mathbf{V}_2$ across the load

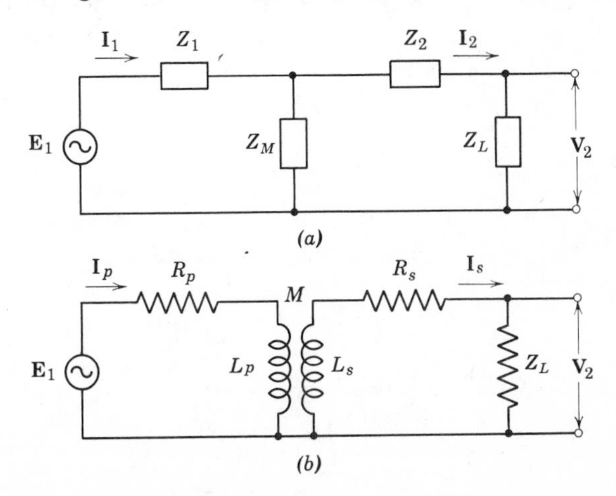

Figure 10-9 Basic coupled circuits. (*a*) General circuit. (*b*) The air-core transformer.

impedance Z_L. The impedance that is common to the input and the output circuits is called the *mutual* or *coupling impedance* Z_M. This coupling impedance can be made with any circuit element—a resistor, a capacitor, an inductance, or a transformer. The circuit analysis in this section and in the next two sections is devoted to the *air-core transformer* (Fig. 10-9*b*), used as the coupling device between the source and the load.

In the fundamentals of d-c and a-c circuit theory, certain definitions are made and basic relations established for transformer-coupled circuits. When two coils are linked by a common or mutual flux, the *coefficient of coupling* k is the ratio of the flux linking the second coil from the first coil to the total flux of the first coil. From this relation and from the definition of inductance, we can show that

$$M = k\sqrt{L_1 L_2}$$

or

$$\omega M = k\sqrt{(\omega L_1)(\omega L_2)}$$

where M is the mutual inductance between two coils which have self-inductance values, L_1 and L_2, and a common linking flux determined by k.

In an iron-core transformer, the objective of the design is to make k as close to unity as possible. In air-core transformers, the values of k are much less than unity and are usually of the order of 0.01 to 0.10 (1% to 10%).

In a transformer, a voltage is induced in the secondary winding by the mutual flux that is created from current flow in the primary winding. This induced voltage lags the current that produces it by 90° and may be expressed as

$$\mathbf{E}_s = -j\omega M \mathbf{I}_p \tag{10-8}$$

Notice from Fig. 10-9*b* that $\mathbf{V}_2$ equals $\mathbf{E}_s$ only when the transformer secondary circuit is open (Z_L is infinite). If Z_s is the total impedance of the secondary circuit including R_s, $j\omega L_s$, and Z_L, then

$$\mathbf{E}_s = \mathbf{I}_s Z_s \tag{10-9a}$$

and

$$\mathbf{V}_2 = \mathbf{I}_s Z_L \tag{10-9b}$$

When there is a current in the secondary circuit, a voltage $-j\omega M I_s$ is induced in the primary. The primary circuit voltage equation is

$$\mathbf{E}_1 = \mathbf{I}_p Z_p + j\omega M \mathbf{I}_s \tag{10-10}$$

Substituting Eq. 10-9*a* in Eq. 10-8, we have

$$\mathbf{I}_s Z_s = -j\omega M \mathbf{I}_p$$

$$\mathbf{I}_s = \frac{-j\omega M \mathbf{I}_p}{Z_s}$$

and, placing this in Eq. 10-10,

$$\mathbf{E}_1 = \mathbf{I}_p Z_p + \frac{(\omega M)^2}{Z_s}\mathbf{I}_p$$

$$\mathbf{E}_1 = \mathbf{I}_p\left[Z_p + \frac{(\omega M)^2}{Z_s}\right] \tag{10-11}$$

From these five fundamental equations of coupled circuits, we can describe the operation of an air-core transformer. When a voltage is applied across the primary terminals of the transformer, a voltage is induced in the secondary winding. The secondary voltage is a function of both the primary current and the mutual inductance. The current that flows in the secondary circuit is limited by the total series impedance Z_s of the secondary circuit.

The total primary impedance is $Z_p + (\omega M)^2/Z_s$. The term $(\omega M)^2/Z_s$ is the reflected impedance of the secondary into the primary. When the secondary circuit is open-circuited, its impedance is infinite and the reflected impedance is zero. When the secondary circuit is resistive, $R + j0$, the impedance reflected into the primary is resistive. This coupled or reflected resistance lowers the input Q of the transformer. When the secondary circuit impedance is inductive, $R_s + jX_s$, the reflected value of the impedance into the primary circuit is proportional to the reciprocal of Z_s. On rationalization, we find that

$$\frac{(\omega M)^2}{R_s + jX_s} = \frac{(\omega M)^2}{R_s^2 + X_s^2} R_s - j\frac{(\omega M)^2}{R_s^2 + X_s^2} X_s$$

The second term or reactive part of this expression is "$-j$". This means that an inductive reactance in the secondary circuit reflects into the primary circuit as a "negative inductive reactance" which acts as if it were a capacitive reactance in the circuit calculations. Correspondingly, a capacitive reactance in the secondary circuit reflects back into the primary circuit as a "positive capacitive reactance" acting as an inductive reactance in the calculations. In describing the operation of transformers, we shall say for simplicity that an inductive secondary impedance reflects into the primary circuit as a "capacitive reactance" and that a capacitive reactance reflects as an "inductive reactance." We shall retain the quotation marks around these terms to distinguish them from the concept of a literal inductance or capacitance.

The currents and voltages in the air-core transformers are phasors which have both direction and magnitude. Accordingly, they are placed in boldface type in this section. Since, in most practical applications, we are concerned only with the magnitudes of currents and voltages in the transformer circuit, we follow the usual electronic circuit notation and place them in ordinary type face in Sections 10-5 and 10-6. It should be remembered that these currents and voltages are phasors, although we do not keep the boldface notation.

PROBLEMS

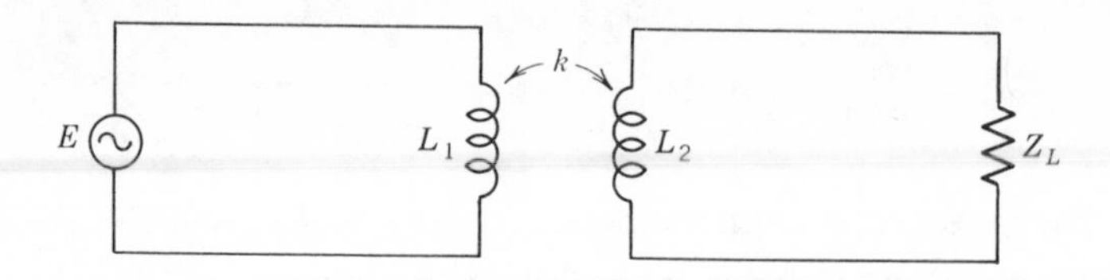

Circuit for Problems 1 to 5.

1. L_1 and L_2 each have an inductance of 60 mh and a Q of 50. The coefficient of coupling k is 25% and the load Z_L is 2000 Ω at 10 kHz. If the applied voltage E is 10 V, what is the input current to the circuit?
2. Repeat Problem 1 if the load Z_L is a short circuit.
3. L_1 has a Q of 50 and an inductance of 30 mh. L_2 has a Q of 40 and an inductance of 45 mh. The coefficient of coupling is 40%. The frequency is 5 kHz and the load is a 1200-Ω capacitor. If the applied voltage is 10 V, determine the input current.
4. L_1 and L_2 are each 60 mh with a Q of 35. The coefficient of coupling is 60% and the load on the secondary is 1500-Ω pure resistance. Determine the input current to the primary if the supply voltage is 20 V at 8 kHz.
5. Repeat Problem 4 if the load is a pure capacitance.

Section 10-5 Air-Core Transformers – Untuned Primary, Tuned Secondary

The untuned-primary, tuned-secondary air-core transformer (Fig. 10-10*a*) has many applications in that range of high frequencies which are referred to as RF (radio frequencies). At that frequency which resonates the secondary, the impedance of the secondary is resistive and is a minimum. The

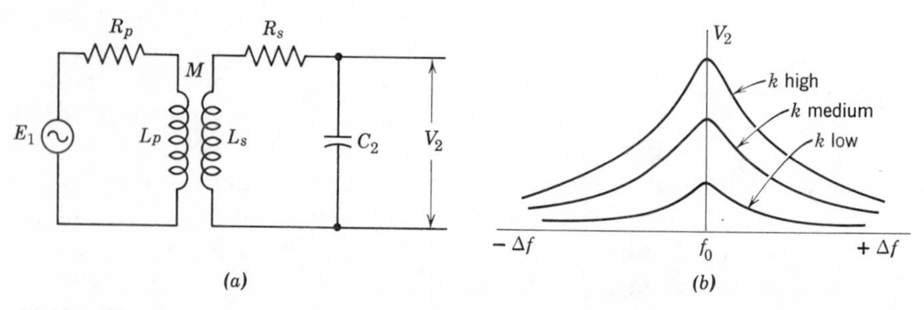

Figure 10-10 The air-core transformer – untuned primary, tuned secondary. (*a*) Circuit. (*b*) Response.

secondary current is then at maximum and the voltage across the capacitance, V_2, is a maximum.* When the coefficient of coupling k is very low, the mutual inductance M is low, and the secondary output voltage V_2 is low. As k is increased, this output voltage increases. The limitation on the output voltage is determined by the closest physical spacing that may be obtained with the particular coil structure.

The development of the resonant curves shown in Fig. 10-10*b* is quite complex but they have the shape of the bandpass characteristics of the simple resonant circuits. When data for a particular circuit are given, the curves can be computed by point-to-point calculations, using Eqs. 10-8 to 10-11.

When the circuit is used to feed power into or to *drive* an electronic device (such as a class-C amplifier), C_2 is adjusted to resonance. That is, it is tuned to produce a maximum output voltage V_2 at the resonant frequency. Then, the coefficient of coupling k is adjusted to produce the required output voltage for optimum operation of the circuit that is being driven. Typical curves showing the variation of the output voltage for different couplings at frequencies above and below resonance are given in Fig. 10-10*b*.

PROBLEMS

1. Solve Problem 1, Section 10-4, if the load is a capacitor that resonates the circuit.
2. Solve Problem 3, Section 10-4, if the load is a capacitor that resonates the circuit.
3. Solve Problem 4, Section 10-4, if the load is a capacitor that resonates the circuit.

Section 10-6 Air-Core Transformers—Tuned Primary, Tuned Secondary

In Fig. 10-11, we show both the primary and the secondary as series circuits. In most applications of this transformer, the primary is a parallel-resonant circuit and the secondary is a series-resonant circuit. We shall show later in the application of this circuit in an amplifier (Section 23-3) that an equi-

* An exact analysis of this circuit shows that the maximum value of V_2 occurs at a frequency that is very close to the resonant frequency of the secondary and that the error is negligible when it is assumed that they occur at the same point in a high-Q circuit.

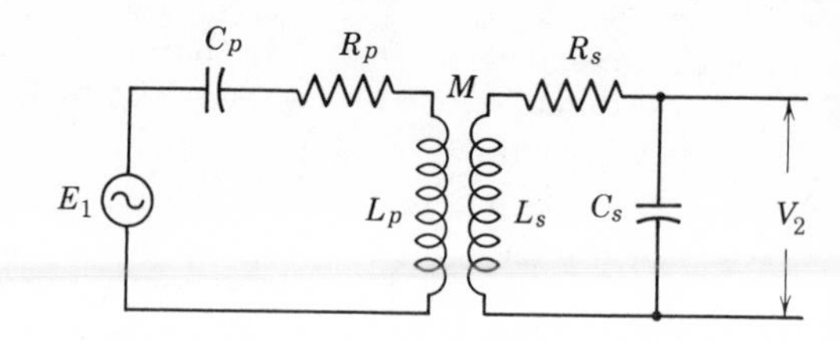

Figure 10-11 The air-core transformer–tuned primary, tuned secondary.

valent series circuit can be established by Thévenin's theorem from the actual parallel-resonance circuit. This transformer could be analyzed for a parallel-resonant circuit in the primary, but the algebra is much simpler when the primary and the secondary are both taken as series-resonant circuits.

Each coil L_p and L_s must be adjusted independently to resonate at the same frequency. At resonance the total secondary circuit is resistive, and the secondary impedance couples back into the primary as a reflected resistance, $(\omega M)^2/R_s$. As the coupling is increased from a very low value, the total resistance in the primary, $R_p + (\omega M)^2/R_s$, increases and the primary current decreases. As k increases, (ωM) increases from zero and the magnitude of the secondary voltage $\omega M I_p$ must increase from zero. However, since the secondary voltage is also proportional to I_p, it will reach a maximum value and then decrease back to zero. Since the output voltage V_2 is $I_s X_{Cs}$, it has the same response (Fig. 10-12).

In Fig. 10-12, we show k_x as that value of coupling which is the physical limit of closeness of coupling. That value of coupling which produces the maximum output is the *critical coupling* k_c. In terms of circuit components,

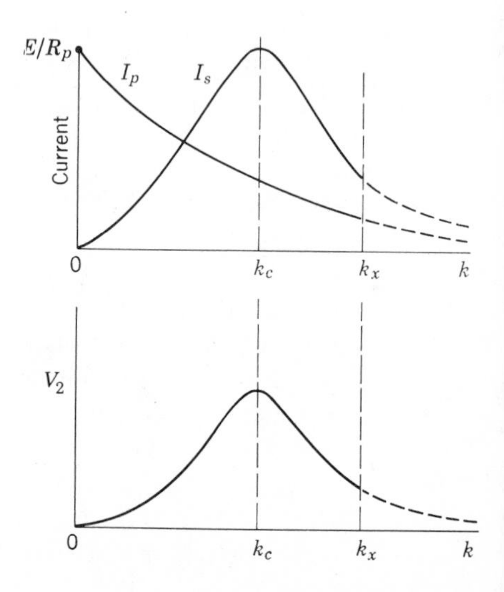

Figure 10-12 Response of the double-tuned transformer with variable coupling.

the critical coupling is

$$k_c = \frac{1}{\sqrt{Q_p Q_s}} \tag{10-12}$$

and, if $Q_p = Q_s = Q$, then $k_c = 1/Q$.

When the value of the coefficient of coupling is less than or equal to the critical coupling k_c, the frequency-response curves (Fig. 10-13) are

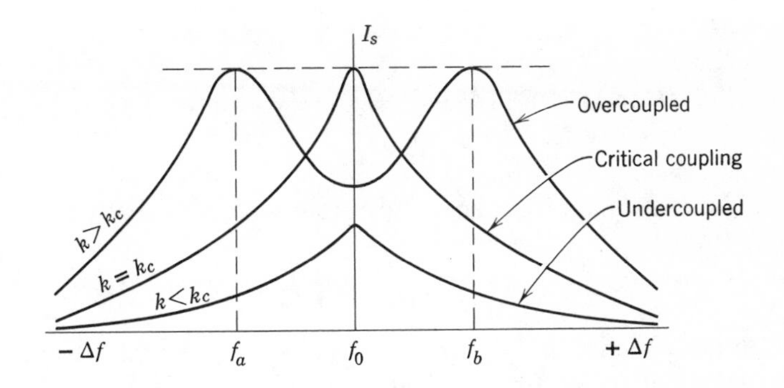

Figure 10-13 Secondary-current variation with frequency.

similar to the untuned-primary, tuned-secondary transformer curves. As the coefficient of coupling increases to the critical value, the bandwidth decreases. When the circuit is overcoupled, the output at the resonant frequency is reduced. In the overcoupled circuit, when the frequency is increased above resonance, the primary and secondary circuit impedances are both inductive because they are both series circuits. However, the inductive effect in the secondary reflects back into the primary as a "capacitance" and re-resonates the circuit at a new frequency f_b. Likewise, below resonance, the two circuits are capacitive. The capacitive secondary reflects back into the primary as an "inductance," causing a second resonance at f_a. Thus, in an overcoupled circuit, two points of apparent resonance are produced, f_a and f_b, and the bandwidth becomes large. Since X_{Cs} decreases with frequency, the value of V_2 which is $I_s X_{Cs}$ shows the effect of this variation of X_{Cs} with frequency (Fig. 10-14).

From a lengthy analysis of the circuit, we can show for identical overcoupled circuits in which Q_p equals Q_s that

$$\frac{f_a}{f_0} = 1 - \frac{k}{2}, \qquad \frac{f_b}{f_0} = 1 + \frac{k}{2} \tag{10-13a}$$

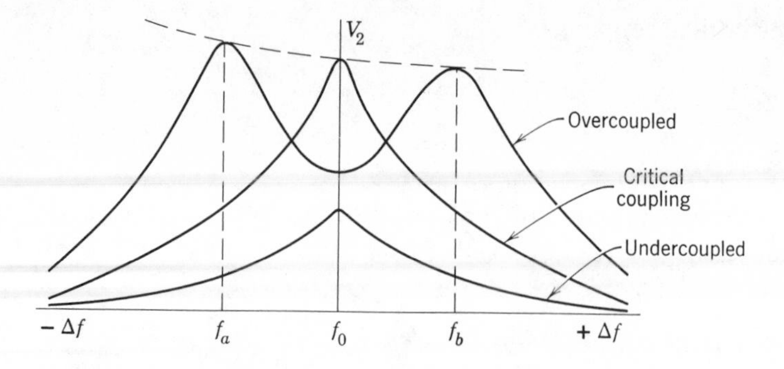

Figure 10-14 Secondary-voltage variation with frequency.

At the resonant frequency the output voltage is given by

$$V_2 = \sqrt{\frac{L_s}{L_p}} \frac{k}{k^2 + (1/Q_pQ_s)} E_1 \qquad (10\text{-}13b)$$

When the primary and secondary circuits are identical, L_s equals L_p and Q_p equals Q_s and, using Eq. 10-12, we have

$$V_2 = \frac{k}{k^2 + k_c^2} E_1 \qquad (10\text{-}13c)$$

and, for critical coupling (k is k_c), this equation simplifies to

$$V_2 = \frac{1}{2k_c} E_1 \qquad (10\text{-}13d)$$

which is the maximum possible output voltage at resonance.

Let us assume that a typical double-tuned transformer is manufactured with a coefficient of coupling of 5%. Also assume that the Q of both the primary and the secondary is 40. Then, the critical coefficient of coupling by Eq. 10-12 is 1/40 or 2.5%. It is evident that the transformer is overcoupled. In a normal resonant circuit, the usual procedure for tuning (varying either L or C) is to adjust the circuit for a maximum output-voltage reading. In this overcoupled transformer, we find that neither a peak nor a dip secures the proper tuning. If, by loading the primary and secondary circuits with shunt loading resistors which are noninductive, we can reduce the effective Q from 40 to 20 or less, the whole transformer is now operating at or below critical coupling. Now the primary and secondary circuits may be adjusted for a maximum reading of V_2. When the procedure of tuning is complete, the loading resistors are removed. If the shunt resistors are noninductive, the tuning is unchanged and the alignment is correct for the overcoupled condition.

PROBLEMS

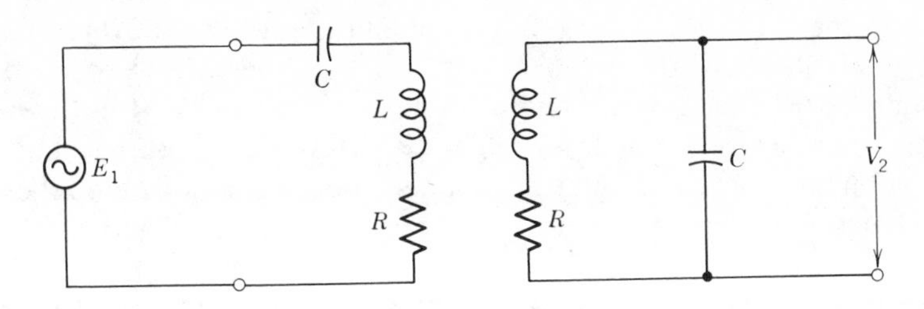

Circuit for Problems 1 to 5

1. C is 40 pF and the Q of the coils is 60 at the resonant frequency of 455 kHz. The coupling is adjusted to critical and the source voltage is 1 V. Find the source-current and the output voltage V_2 at resonance.
2. The coefficient of coupling in Problem 1 is increased to 150% of k_c. Calculate critical points to sketch the response curve.
3. The double-tuned circuit is critically coupled and resonates at 10.7 MHz. C is 20 pF and the Q of the coils is 50. What is the required value of inductance. What is the maximum output voltage of the circuit if E_1 is 100 μv?
4. It is desired to design the transformer of Problem 3 for a peak separation of 700 kHz in an overcoupled condition. What is the required value of coupling, and what is the response at 10.7 MHz when the transformer is overcoupled?
5. In the primary X_C is 70 Ω, X_L is 80 Ω, and R is 8 Ω. In the secondary X_C is 30 Ω, X_L is 60 Ω, and R is 46 Ω. If E_1 is 1 V and k is 60%, find the input current.

Section 10-7 Harmonics

A basic a-c circuits course is primarily concerned with the theory, the calculations, and the applications of sinusoidal waveforms. In practice, in electronic circuits, pure sinusoidal waveforms are not common. The waveforms as shown on an oscilloscope may be square waves, triangular waves, pulses, or completely nonrepetitive waveforms. Usually, we analyze the electronic circuit on the basis of sine-wave signal sources and tests and then apply this analysis to the nonsinusoidal waveforms. It is necessary to develop an understanding of harmonic content in order to consider these irregular shapes in terms of a sine-wave analysis. Later, the topic is extended to include a discussion of distortion and distortion analysis in amplifiers (Sections 11-7, 18-2, and 18-3).

The basic repetitive period of a wave establishes its form and its *fundamental* frequency. If we use a 20-cycle square wave, we mean that, over one second, 20 full cycles are completed. If two sine waves are added, one at a fundamental frequency f, and the other at double the fundamental frequency or at the *second harmonic* $2f$, we obtain the results shown in Fig. 10-15. If the amplitude of the second harmonic is correct with respect to the amplitude of the fundamental and if proper amounts of the 4th, the 6th, the 8th, etc., harmonics are added, we get the sawtooth waveforms shown. In order to obtain a pure sawtooth from mathematics, an infinite number of harmonics is required. Practically, we consider the waveform pure if all harmonics up to about the 10th or 12th are included.

It is obvious from Fig. 10-15 that not only is the relative amplitude of the harmonic to the fundamental important but also the phase angle of the

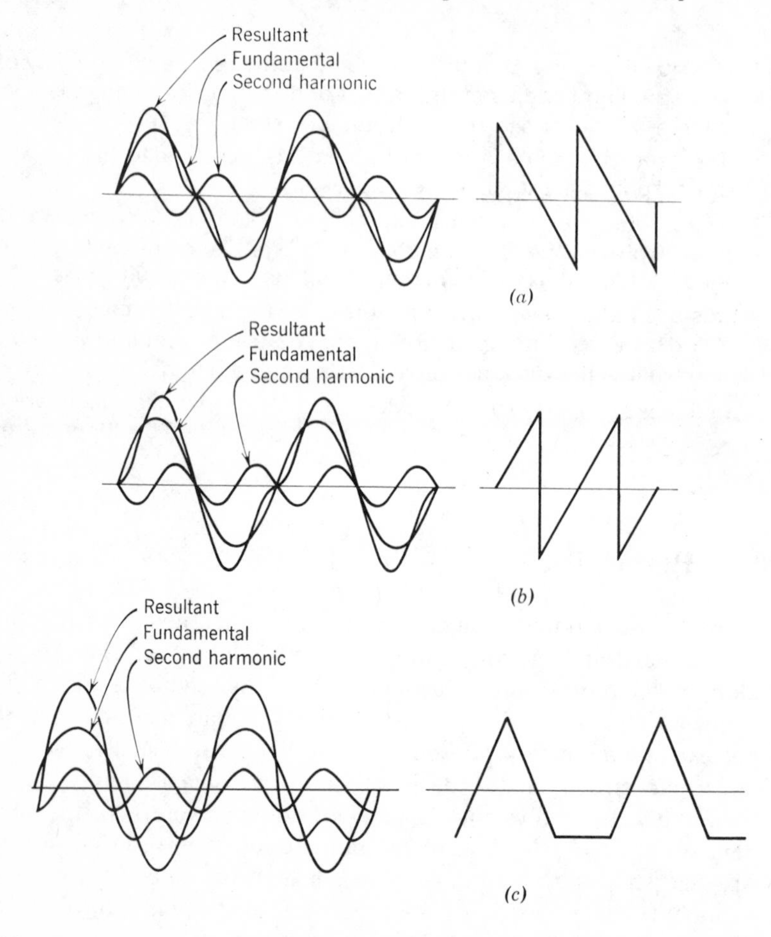

Figure 10-15 Fundamental with even harmonics. (*a*) Fundamental with even harmonics starting in phase. (*b*) Fundamental with even harmonics starting 180° out of phase. (*c*) Fundamental with even harmonics starting at a 90° lag.

phase relationship is very critical. The three different waveforms are produced, not by a change in amplitude, but by a change in phase-angle relationship only.

Figure 10-16 shows the results of adding odd harmonics to a fundamental. Again a phase difference radically affects the resultant waveform.

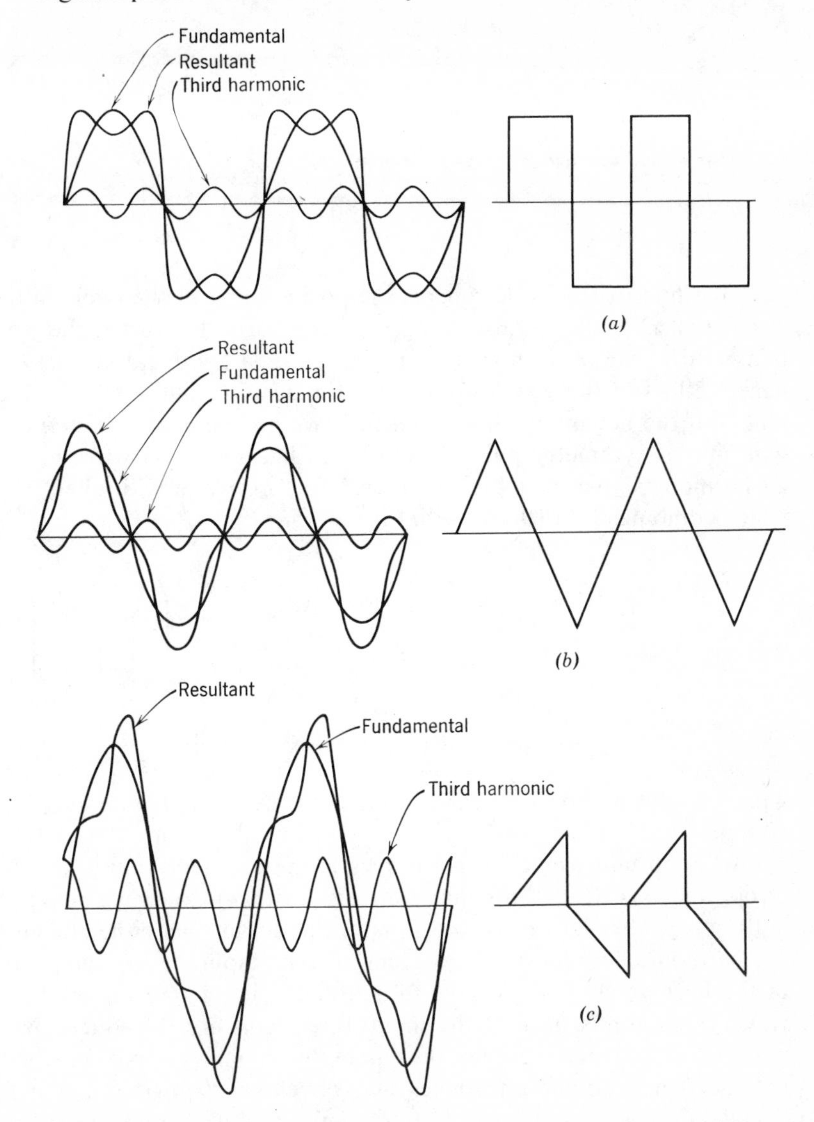

Figure 10-16 Fundamental with odd harmonics. (*a*) Fundamental with third harmonic starting in phase. (*b*) Fundamental with third harmonic starting 180° out of phase. (*c*) Fundamental with third harmonic starting at a 90° lead.

In Fig. 10-17, one cycle of a generalized irregular waveform is given. The positive half-cycle is *abcde*, and the negative half-cycle is *efghi*. If we use the results obtained from Fig. 10-15 and Fig. 10-16, we often can make a very useful analysis of the waveform under study (Fig. 10-17). If the

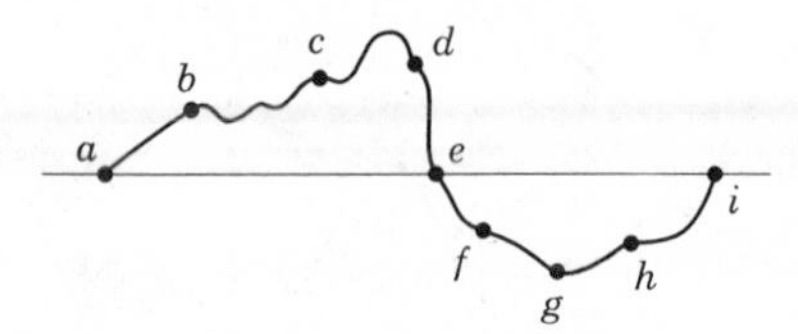

Figure 10-17 A general waveform used to show harmonic content.

negative half of the cycle repeats the positive half of the cycle in the same order, that is, if the negative loop is $-(abcde)$ in that order, the wave contains odd harmonics only. This is known as a *complete symmetry*. If the negative half of the waveform repeats the positive half in the reverse order, that is, if the negative half is $-(edcba)$, we have *mirror symmetry*. A wave with mirror symmetry must contain even harmonics. A wave may contain all harmonics. An example of this is a square wave with even harmonics which combines Fig. 10-16*a* with Fig. 10-15*a*:

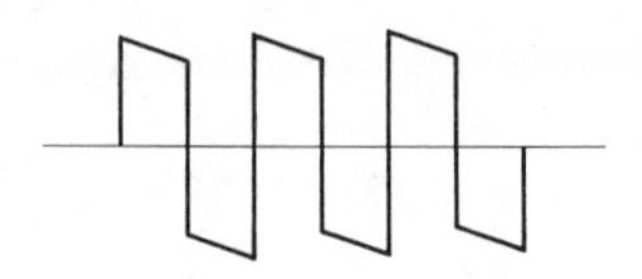

This waveform shows mirror symmetry, thereby having an even-harmonic content.

An important aspect of harmonics is the consideration of the converse to our analysis. In circuits that are tested with square-wave signal sources, if the output waveform is not square, the circuit has either failed to pass the harmonic content or it has changed the amplitude and phase relations of the harmonics. If the output of a circuit is not a sine wave when a sine-wave generator is used at the input, the circuit has introduced harmonics into the wave which were not present in the incoming signal. In other words, the electronic circuit introduces and develops *distortion*. Often from an examination of the symmetry characteristics of the output waveform, a quick analysis may be made of the kind of distortion that is being produced (Figs. 18-3 and 18-4).

Section 10-8 Four-Terminal Networks

In many electronic circuit applications, a generator E_s with an internal impedance R_s is the input to a circuit which produces a voltage E_2 across a load resistance R_L. The generator is connected to the input terminals A and B, and the load is connected across the terminals C and D (Fig. 10-18). The

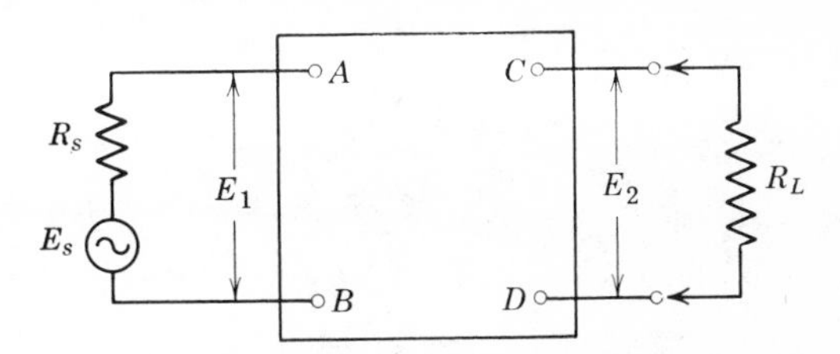

Figure 10-18 A four-terminal network.

internal circuit between A and B and between C and D may be a very complex network consisting of resistances, inductances, capacitances, transformers, vacuum tubes, and transistors. This internal circuit has no limit as to its complexity. It is often referred to as a *black box* since the nature of its components may be unknown. In order to analyze this circuit, we make two conditions on the circuit of the black box. First, there must be a direct connection between one input terminal and one output terminal which is the common ground. Second, all elements within the black box – resistances, inductances, capacitances, transformers, emf's and current sources – must be linear elements.

The black box itself (Fig. 10-19) may be reduced to a model by Thévenin's theorem. The input circuit between A and C is replaced by one model by

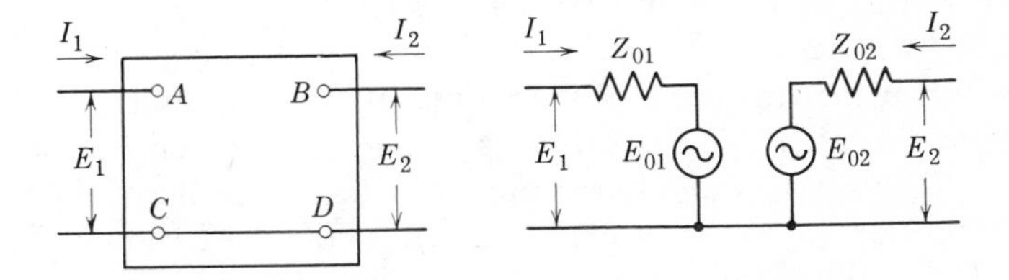

Figure 10-19 The black box and its model.

Thévenin's theorem, and the output circuit between B and D is replaced by a second model. This complete model is often called a *two-terminal-pair* network. Very obviously, E_{01} must be related to E_{02} since, in a linear network, a change in the input voltage produces a proportional change in the output voltage.

The output model is converted from the Thévenin form to the constant-current form of Norton's theorem (Fig. 10-20*a*). If we refer to the black box of Fig. 10-19 and short-circuit the output, the short-circuit current is

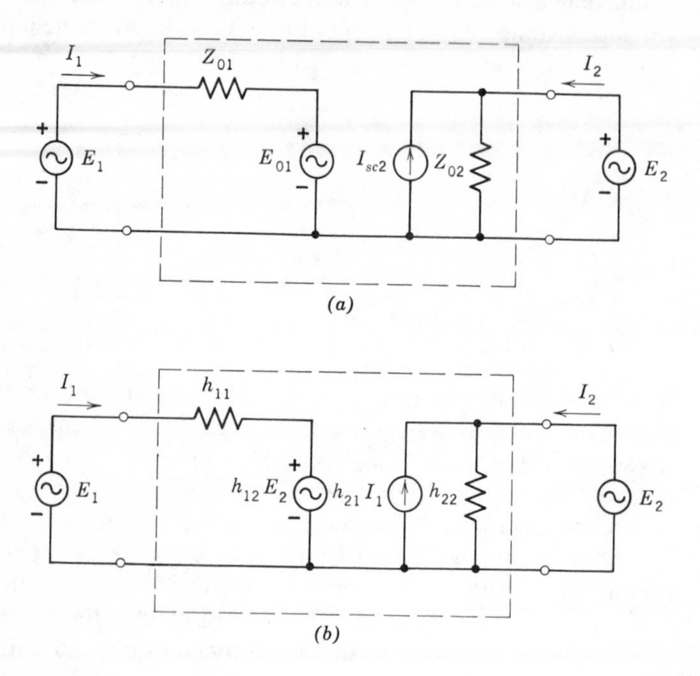

Figure 10-20 Models for four-terminal networks. (*a*) Using resistances. (*b*) Using hybrid parameters.

evidently determined by the magnitude of either E_1 or I_1. We can then replace $I_{sc,2}$ by $h_{21}I_1$ where h_{21} is a dimensionless constant of proportionality. When the input is open-circuited, there is a voltage in the input, E_{01}, which is directly determined by the value of E_2. Then E_{01} may be replaced by $h_{12}E_2$ where h_{12} is another dimensionless constant of proportionality. If Z_{01} is called h_{11} and h_{22} is defined as $1/Z_{02}$, the dimension of h_{11} is resistance in ohms, and the dimension of h_{22} is conductance in mhos. Making these substitutions, we obtain the circuit of Fig. 10-20*b*. The two equations from this new circuit are

$$E_1 = h_{11}I_1 + h_{12}E_2 \tag{10-14}$$

and

$$I_2 = h_{21}I_1 + h_{22}E_2 \tag{10-15}$$

This model has one voltage model and one current model. This mixed combination is a *hybrid* form, and the symbols, h_{11}, h_{12}, h_{21} and h_{22} used in this circuit and in the equations are termed *hybrid parameters*. Since h_{11} is in ohms, h_{22} is in mhos, and h_{12} and h_{22} are dimensionless, this combination of different units for the different h's also gives rise to the term hybrid parameters.

From Eq. 10-14 and Eq. 10-15, we may readily define these hybrid parameters in terms of E_1, I_1, E_2, and I_2.

$$h_{11} = \frac{E_1}{I_1} = Z_1 \quad \text{when} \quad E_2 = 0 \tag{10-16}$$

(output short-circuited)

$$h_{12} = \frac{E_1}{E_2} \quad \text{when} \quad I_1 = 0 \tag{10-17}$$

(input open-circuited)

$$h_{21} = \frac{I_2}{I_1} \quad \text{when} \quad E_2 = 0 \tag{10-18}$$

(output short-circuited)

$$h_{22} = \frac{I_2}{E_2} = \frac{1}{Z_2} \quad \text{when} \quad I_1 = 0 \tag{10-19}$$

(input open-circuited)

The significance of these definitions is that, by making a series of open- and short-circuit measurements on a four-terminal network, the model for the complex network may be established and evaluated.

PROBLEMS

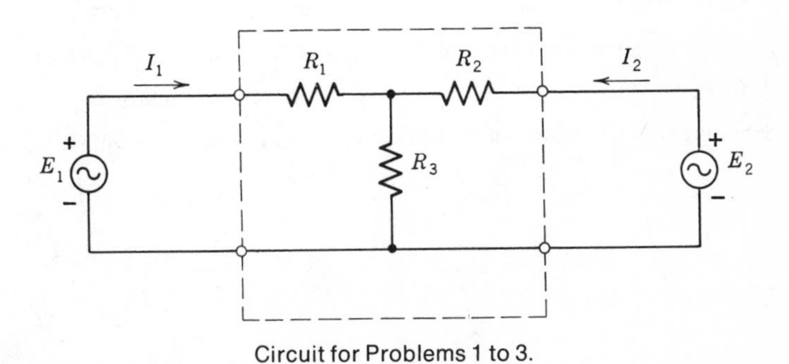

Circuit for Problems 1 to 3.

1. All the resistors are each 1000 Ω. Draw the hybrid-parameter model for the circuit and show all numerical values.
2. R_1 is 2000 Ω, R_2 is 5000 Ω, and R_3 is 1500 Ω. Draw the hybrid-parameter model for the circuit and show all the numerical values.

3. R_1 is 10 kΩ, R_2 is 20 kΩ, and R_3 is 680 Ω. Draw the hybrid-parameter model for the circuit and show all the numerical values.

Questions

1. Define or explain each of the following terms: (*a*) resonance, (*b*) Q gain, (*c*) shunt loading, (*d*) bandwidth, (*e*) Q in terms of energy, (*f*) Q in terms of bandwidth, (*g*) coefficient of coupling, (*h*) critical coupling, (*i*) over coupling, (*j*) h_{11}, (*k*) h_{12}, (*l*) h_{21}, (*m*) h_{22}.
2. Are there any assumptions in establishing the formula for series resonance as $f_0 = 1/2\pi\sqrt{LC}$?
3. Are there any assumptions in establishing the formula for parallel resonance as $f_0 = 1/2\pi\sqrt{LC}$?
4. What are the essential characteristics of series resonance?
5. What are the essential characteristics of parallel resonance?
6. What is the input impedance to a tapped resonant circuit?
7. Why is a tapped resonant parallel circuit used?
8. What is the effect of a secondary impedance on the primary winding of a coupled circuit?
9. On what factors does the secondary voltage depend in a coupled circuit?
10. Why does the output voltage of a double-tuned circuit decrease at resonance when the coupling exceed the critical value?
11. Explain how to tune an under-coupled double-tuned transformer.
12. Explain how to tune a critically-coupled double-tuned transformer.
13. Explain how to tune an over-coupled double-tuned transformer.
14. Explain how a fundamental and harmonics can produce a square waveform.

Chapter Eleven

THE LOAD LINE AND BIAS CIRCUITS

The load line concept is introduced by considering the graphical solution for two resistors in series (Section 11-1). One of the resistors is replaced by the common-base collector characteristic and the load line for the common-base amplifier is obtained (Section 11-2). Similarly, load lines are treated for the common emitter amplifier (Section 11-3), for the field effect transistor (Section 11-4), and for the vacuum tube (Section 11-5). The basis for the letter symbols for semiconductors and vacuum tubes is explained in Section 11-6. A graphical method for determining distortion is presented in Section 11-7. Fundamental bias circuit analysis for transistor amplifiers is covered in Section 11-8. The conversions given in the conversion table must be assimilated thoroughly. An analytic approach to load lines is taken in Section 11-9.

Section 11-1 The Load-Line Concept

Consider the series circuit of two resistors, R_1 and R_2, connected to a fixed supply voltage V (Fig. 11-1*a*). The graph on which the volt-ampere lines of the resistors are to be drawn is given in Fig. 11-1*b*. The total width of the graph is V volts, the supply voltage. The volt-ampere curve for R_2 is drawn from A to E. It is a straight line because R_2 is a fixed resistor, and it merely shows the Ohm's law relationship of the current and the voltage for the resistor. When the whole voltage V, is applied across R_2, the current value is V/R_2 and, thus, locates Point E. This line has a positive slope of value $+1/R_2$.

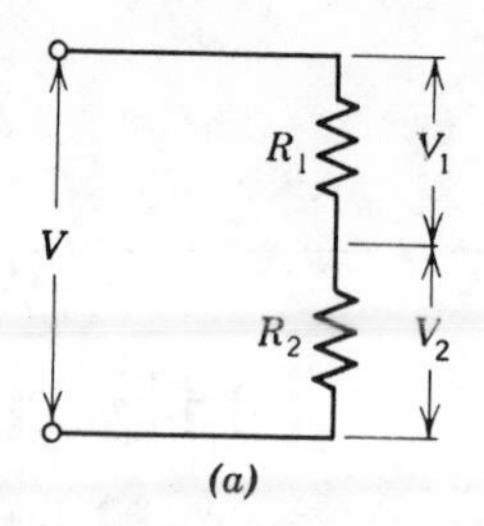

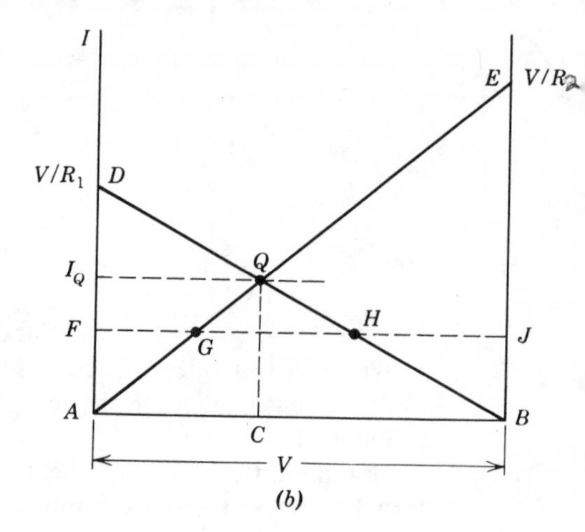

Figure 11-1 A graphical solution for a series circuit. (*a*) Circuit. (*b*) Graphical solution.

The volt-ampere curve for R_1 is drawn by considering point B to be zero volts and point A to be V volts; that is, the scale of V reads from B to A instead of from A to B. When the whole voltage V is applied across R_1, the current value is V/R_1 locating point D. In terms of the coordinate axes that have the zero at Point A, this line has a negative slope of value $-1/R_1$. This negative slope does not imply a negative resistance. It is only $-1/R_1$ because of the way in which the slope is being considered.

The line from A to E represents all values of current through R_2 when the voltage across R_2 is varied from 0 to V. Likewise the line from B to D represents all values of current through R_1 when the voltage across R_1 is varied from 0 to V. Consider the original circuit again. It is a series circuit and the requirement of a series circuit is that the current in all parts of the series circuit is the same. The horizontal line FJ represents a current F which is common to the two resistors. The voltage drop across R_2 for this current is FG and across R_1 is JH. Obviously, FG plus JH does not equal V so that this value of current F cannot be the graphical solution for the network.

The only value of current that can be the solution for the network is the value I_Q, given by the intersection of the two load lines, point Q. At Q, which is often called the *Q-point*, the voltage across R_2 is AC and the voltage across R_1 is BC. These two values properly add to the supply voltage V.

In using this graphical approach, notice that the slope and direction of the volt-ampere curve for R_2, AQE, does not change if the source voltage V changes. On the other hand, if the source voltage V changes, the location of B changes and the value of V/R_1, the intercept of the load line for R_1, changes. The slope of this load does not change; the slope remains at the value $-1/R_1$. Any line parallel to AE has the slope $1/R_2$, and any line parallel to BD has the slope $-1/R_1$.

PROBLEMS

1. A 30-Ω resistor and a 40-Ω resistor are connected in series across a 120-V source. By using a graphical approach, determine the current in the circuit and the voltage drop across each resistor.
2. A 10-Ω resistor and a 3-Ω resistor are connected in series across a 15-V source. By using a graphical approach, determine the current in the circuit and the voltage drop across each resistor.

Section 11-2 The Load Line for a Common-Base Amplifier

The concept developed in the last section can be extended easily to transistors. The resistor R_2 is replaced by the transistor, and the resistor R_1 is the load resistor R_c in the collector of the transistor (Fig. 11-2*a*). Instead of having a single line for R_2, we now have the composite family of transistor-curves (Fig. 11-2*b*). The load line for the collector load resistor R_c is a single straight line to be superimposed on the transistor characteristic. One end of the load line is the value of V_{CC}, 12 V, on the horizontal axis. The other end of the load line is the value of current on the vertical axis obtained from V_{CC}/R_c or 12/1.5 which is 8 ma. The load line is then drawn between these two points. It has a slope of $-1/R_c$. This particular load line is valid only for a collector supply voltage of 12 V and a load resistance of 1.5 kΩ.

The Kirchhoff's voltage equation around the right loop of the circuit of Fig. 11-2*a* is

$$V_{CC} = R_c I_C + V_{CB} \tag{11-1}$$

where V_{CB} is the voltage drop across the transistor from collector to base.

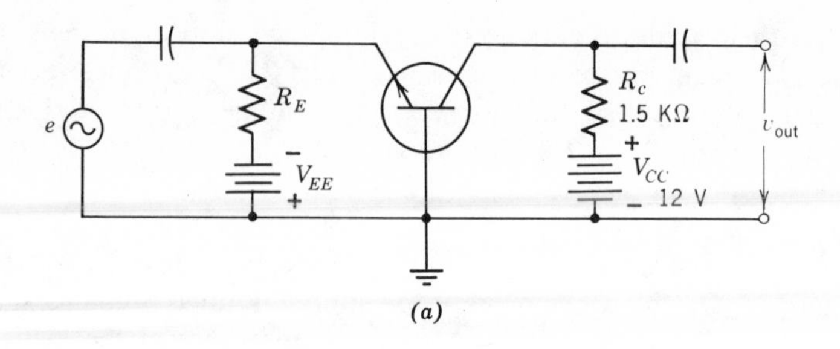

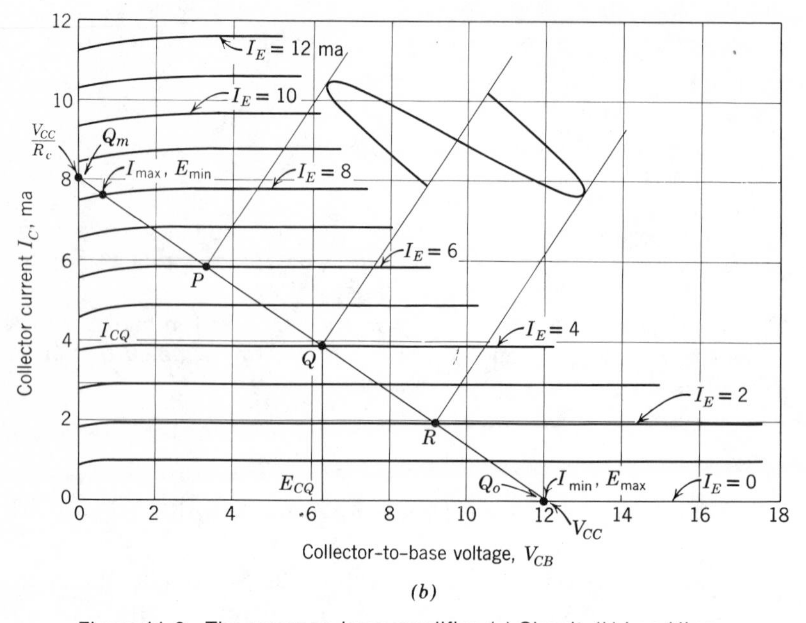

Figure 11-2 The common-base amplifier. (*a*) Circuit. (*b*) Load line.

The variables in the equation are I_C and V_{CB}. Equation 11-1 can be rearranged to

$$I_C = -\frac{1}{R_c} V_{CB} + \frac{V_{CC}}{R_c} \tag{11-2}$$

Equation 11-2 is in the standard form of the general equation of a straight line:

$$y = mx + b$$

where m is the slope of the line and b is the y-intercept. Consequently, the load line intersects the I_C axis at V_{CC}/R_c and has the slope $-1/R_c$. This approach is a mathematical proof that the load line must be a straight line on a collector characteristic.

It is evident from an inspection of Fig. 11-2b that there are an infinite number of points of intersection between the load line and the collector characteristic. The value of the d-c emitter current I_E that flows in the circuit establishes the particular intersection that is used for the circuit. This point is *the operating point* or *the Q-point* and determines the values of d-c currents and d-c voltages that exist in the circuit when the signal e_s is zero. The value of the emitter supply voltage V_{EE} and the value of the emitter resistor R_E determine the value of I_E. Consequently, in this case V_{EE} is the *bias supply voltage* and R_E is the *bias resistor.* Bias calculations are considered in detail in Section 11-8.

Inspection of the load line, Fig. 11-2b, shows that the collector current I_C can range from zero to 8 ma. The emitter current or *bias current* I_E can vary from zero to 8.6 ma. If the bias on the transistor is located at Q_0, the transistor is *cut off* and no current flows in the collector. If the bias on the transistor is located at Q_m, the transistor is *saturated* and maximum current flows in the collector. In computer circuits the bias is *switched* back and forth from Q_0 to Q_m so that the transistor is either in an "on" state or in an "off" state.

When transistors are used for linear signal amplification for communications purpose, the bias cannot be located at Q_o or Q_m but must be placed at an intermediate value. Usually, the design value locates the operating point at the center of the linear part of the load line. By this we mean that the line segments along the load line formed by equal steps of bias current are equal. An inspection of Fig. 11-2b shows that all the intercepts are approximately equal and, therefore, the center of the linear range along the load line is point Q shown on the characteristic at $I_E = 4$ ma. A small input a-c signal is shown on Fig. 11-2b. This signal variation between the points P and R on the load line causes changes in the collector circuit which can be tabulated as follows:

For the Emitter	For the Collector	
$I_{EP} = 6$ ma	$V_{CP} = 3.40$ V	$I_{CP} = 5.85$ ma
$I_{EQ} = 4$ ma	$V_{CQ} = 6.25$ V	$I_{CQ} = 3.90$ ma
$I_{ER} = 2$ ma	$V_{CR} = 9.10$ V	$I_{CR} = 1.95$ ma

The operating point is at Q and the maximum signal swing is from R to P. The distance from R to P represents the *peak-to-peak* distance of the a-c values. If we divide the peak-to-peak values by two, we obtain the *peak* values and, dividing again by $\sqrt{2}$ or 1.414, we obtain the *rms* or *effective* values.

The alternating emitter current is

$$I_e = \frac{I_{EP} - I_{ER}}{2\sqrt{2}} = \frac{6-2}{2\sqrt{2}} = 1.41 \text{ ma} \quad \text{(rms)} \tag{11-3a}$$

The alternating load current is

$$I_c = \frac{I_{CP} - I_{CR}}{2\sqrt{2}} = \frac{3.90}{2\sqrt{2}} = 1.38 \text{ ma} \quad \text{(rms)} \tag{11-3b}$$

The alternating load voltage is

$$V_c = \frac{V_{CR} - V_{CP}}{2\sqrt{2}} = \frac{9.10 - 3.40}{2\sqrt{2}} = 2.02 \text{ V} \quad \text{(rms)} \tag{11-3c}$$

The circuit current gain is

$$A_i = \frac{I_{CP} - I_{CR}}{I_{EP} - I_{ER}} = \frac{5.85 - 1.95}{6-2} = 0.975 \tag{11-3d}$$

When the signal is increased, the limiting values without serious overloading or clipping are represented by minimum and maximum values, as indicated on Fig. 11-2*b*. By tabulating these results for this circuit, we find the following:

For the Emitter	For the Collector	
$I_{E,\max} = 8$ ma	$I_{C,\max} = 7.60$ ma,	$V_{C,\max} = 12.00$ V
$I_{EQ} = 4$ ma	$I_{CQ} = 3.90$ ma,	$V_{CQ} = 6.25$ V
$I_{E,\min} = 0$	$I_{C,\min} = 0$,	$V_{C,\min} = 0.54$ V

Now the alternating output current is

$$I_2 = I_c = \frac{I_{C,\max} - I_{C,\min}}{2\sqrt{2}} = \frac{7.60 - 0}{2\sqrt{2}} = 2.69 \text{ ma} \quad \text{(rms)}$$

The alternating output voltage is

$$V_2 = V_c = \frac{V_{C,\max} - V_{C,\min}}{2\sqrt{2}} = \frac{12 - 0.54}{2\sqrt{2}} = 4.05 \text{ V} \quad \text{(rms)}$$

The alternating input current is

$$I_e = \frac{I_{E,\min} - I_{E,\min}}{2\sqrt{2}} = \frac{8-0}{2\sqrt{2}} = 2.83 \text{ ma (rms)}$$

The a-c power output is

$$P_o = \frac{(V_{C,\max} - V_{C,\min})\,(I_{C,\max} - I_{C,\min})}{8} \tag{11-4}$$

$$= \frac{(12.00 - 0.54)\,(7.60 - 0)}{8} = 10.9 \text{ mw}$$

The total circuit input power is

$$V_{CC} \times I_{CQ} = 12.00 \times 3.90 = 46.8 \text{ mw}$$

The total collector input is

$$V_{CQ} \times I_{CQ} = 6.25 \times 3.90 = 24.4 \text{ mw}$$

The efficiency of the circuit is

$$\eta_{ov} = \frac{10.9}{46.8} 100 = 23.3\%$$

The efficiency of the collector is

$$\eta_c = \frac{10.9}{24.4} 100 = 44.7\%$$

The load line and the Q point are redrawn on a set of axes without the transistor collector characteristic family (Fig. 11-3). The operating point is at 9 or Q, and the signal swing varies along the load line between points 1 and 2.

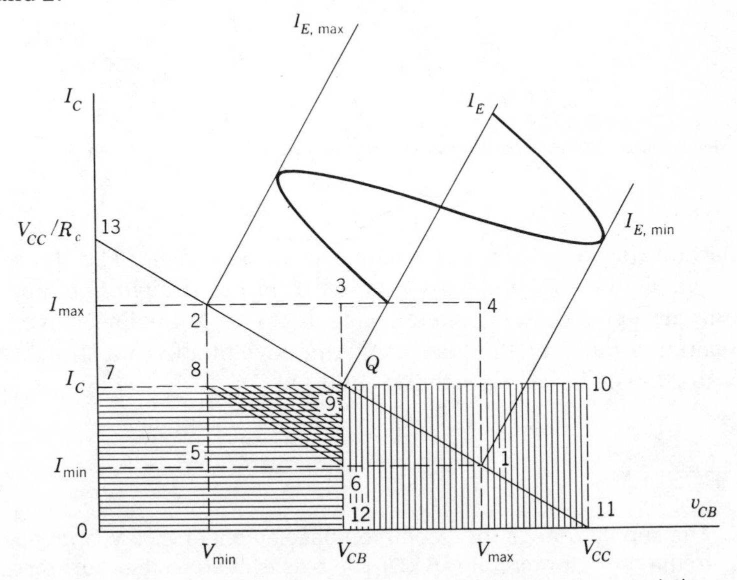

Figure 11-3 Representation of powers on a collector characteristic.

The total input power can be represented by the area of the rectangle 0-7-10-11 in Fig. 11-3. The area 0-7-9-12 represents the supply input power $V_{CC}I_C$. The rectangle 2-4-1-5 represents $(V_{max}-V_{min})(I_{max}-I_{min})$. The net output power is then one eighth of this area or triangle 6-8-9. If the amplifier were ideal, we would be able to make use of the entire load line for signal swing. Point 2 would approach point 13, point 1 would approach point 11, point 8 would approach point 7, and point 6 would approach point 12, with the operating point at the midpoint of the load line. Under these circumstances, the area of 0-7-9-12 would be one half the area of 0-7-10-11. The area of triangle 6-8-9 would be one half the area of 0-7-9-12. From this ideal set of conditions, the circuit has a maximum theoretical collector efficiency of 50% and a maximum theoretical overall efficiency of 25%.

It is possible to utilize almost the entire load line, using a common-base circuit. As a result, in order to achieve a maximum efficiency, this circuit is often used in power amplifiers where the load requires several watts of driving power.

If the set of values of I_C and I_E along the load line of Fig. 11-2*b* are plotted, we obtain the *dynamic transfer curve* of the circuit, Fig. 11-4. The

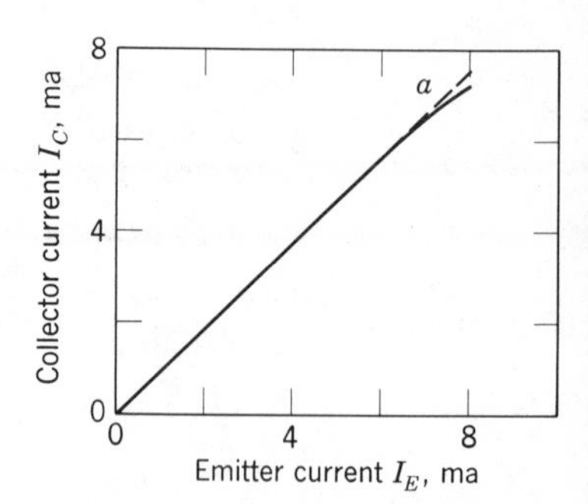

Figure 11-4 Dynamic line for the circuit of Figure 11-2.

dashed line to *a* is a straight line from the origin. The dynamic transfer curve shows only a slight deviation from the straight line, and this occurs only at high values of I_E and I_C. Thus the circuit is linear over most of the operating range of the load line. Any deviation from a straight line creates distortion in the output signal (Section 11-7).

PROBLEMS

1. The supply voltage for a common base amplifier is 30 V, and the load resistor in the collector circuit is 6 kΩ. The bias is at an emitter current of 2 ma. Draw the load line. The signal has a peak value of 1 ma. Locate the signal swing

on the load line. Determine the a-c emitter current, the a-c load current, and the a-c load voltage.

2. What is the input d-c power to the circuit of problem 1? Determine the collector efficiency and the overall circuit efficiency.
3. The collector can dissipate 40 mW. Draw the maximum collector dissipation curve on the collector characteristic.
4. Plot the dynamic load line, showing I_C plotted against I_E.

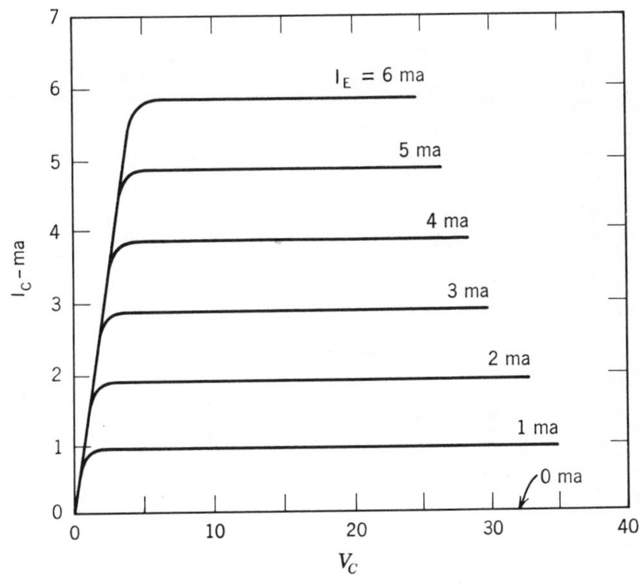

Common-base collector characteristics

Section 11-3 The Load Line for a Common-Emitter Amplifier

When the emitter is the common point between the input and the output circuits instead of the base, the resulting circuit is called the grounded-emitter or common-emitter amplifier (Fig. 11-5*a*). This circuit was discussed in Section 3-4 to the extent of showing a resulting gain along with a 180° phase shift between the input and output voltages. A material advantage of this circuit is that a single battery may be used as the supply for both V_{EE} and V_{CC}.

The graphical analysis of this circuit (Fig. 11-5*b*) is done in the same manner as in Section 11-2. The operating point Q is located at the center of the linear intercepts. It is evident from this load line that the changes in the collector voltage and collector current are about the same order as in

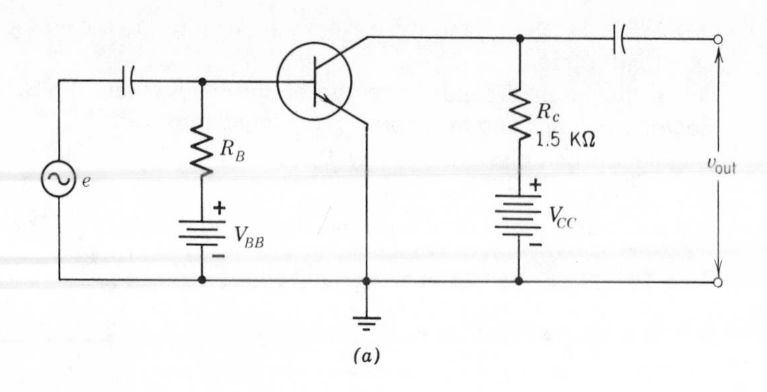

(a)

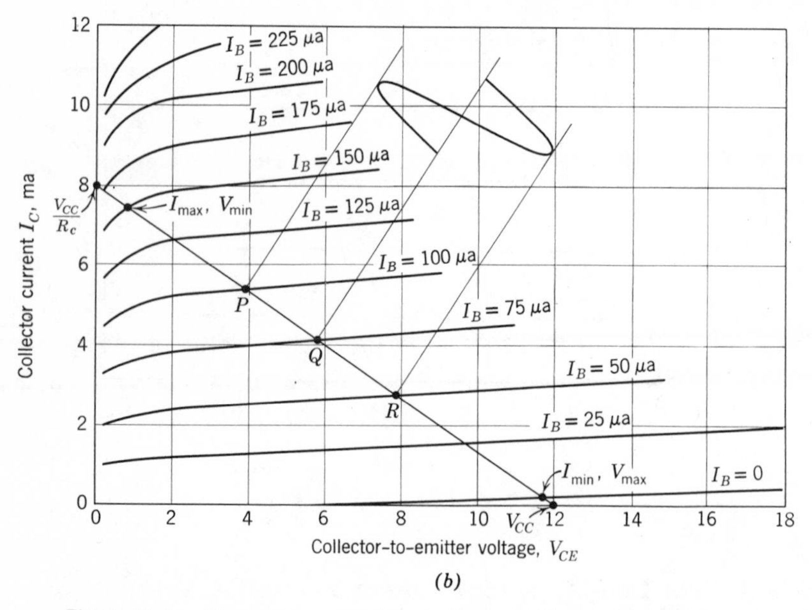

(b)

Figure 11-5 The common-emitter amplifier. (a) Circuit. (b) Load line.

the common-base circuit, but the signal changes are increments of microamperes instead of milliamperes. This means that we obtain the same outputs with very much less power in the signal. For a small signal which has a low distortion, the swing is between P and R. The current gain is

$$A_i = \frac{I_{CP} - I_{CR}}{I_{BP} - I_{BR}} = \frac{5.44 - 2.76}{100 - 50} = \frac{2.68 \text{ ma}}{50\ \mu\text{a}} = 53.6$$

The base bias current I_{BQ} for this operating point is 75 μa.

The values for the points on the load line for the condition of a maximum signal are as follows:

For the Base	For the Collector	
$I_{B,\max} = 150\ \mu a$	$V_{C,\max} = 11.7$ V,	$I_{C,\max} = 7.4$ ma
$I_{BQ} = 75\ \mu a$	$V_{CQ} = 5.85$ V,	$I_{CQ} = 4.10$ ma
$I_{B,\min} = 0$	$V_{C,\min} = 0.80$ V,	$I_{C,\min} = 0.2$ ma

The output signal is RMS

$$\frac{V_{C,\max} - V_{C,\min}}{2\sqrt{2}} = \frac{11.7 - 0.8}{2\sqrt{2}} = 3.85 \text{ V}$$

The output power is

$$P_o = \frac{(V_{\max} - V_{\min})(I_{\max} - I_{\min})}{8}$$

$$= \frac{(11.7 - 0.8)(7.4 - 0.2)}{8} = 9.81 \text{ mw}$$

The total collector circuit input power is

$$V_{CC} \times I_{CQ} = 12 \times 4.10 = 49.20 \text{ mw}$$

The total collector input power is

$$V_{CQ} \times I_{CQ} = 5.85 \times 4.10 = 23.99 \text{ mw}$$

The overall efficiency is

$$\frac{9.81}{49.20} 100 = 20.0\%$$

The collector efficiency is

$$\frac{9.81}{23.99} 100 = 40.9\%$$

These efficiencies are still close to the theoretical values, but not quite as high as in the common-base circuit. The dynamic-transfer characteristic (Fig. 11-6) shows that the linearity of the circuit falls off sooner than the curve for the common-base circuit. However, the increased power gain obtained, because the current gain is considerably greater than unity, more than offsets the lowered efficiency and increased nonlinearity. In stages where the signals are very small, the common-emitter amplifier is almost universally used to take advantage of this greater signal gain.

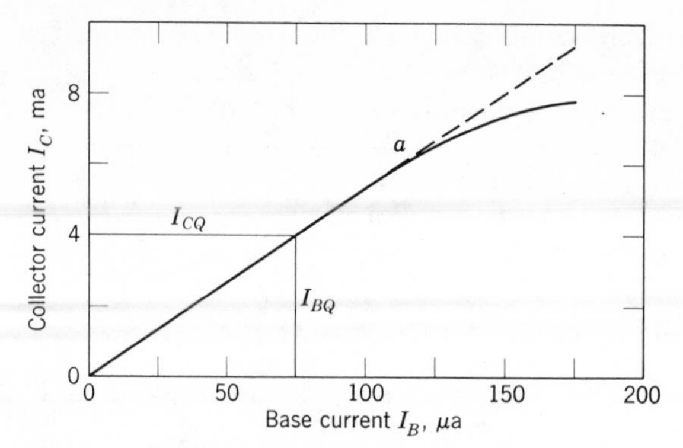

Figure 11-6 Dynamic load line for the circuit of Figure 11-5*a*.

PROBLEMS

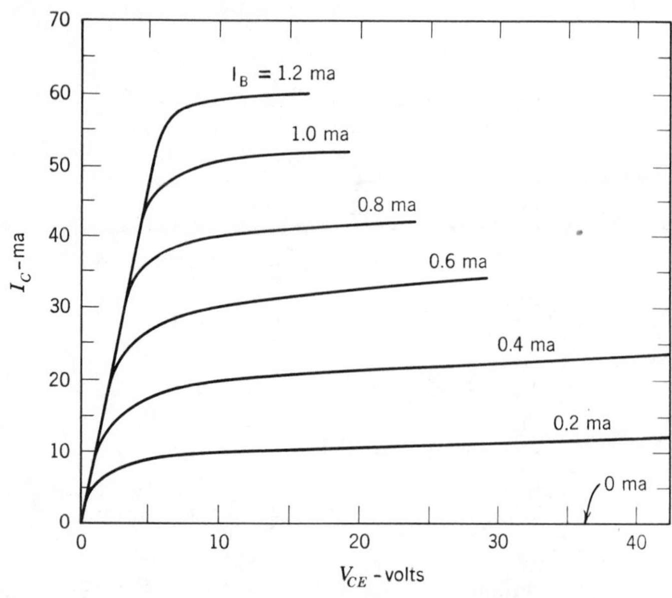

Common-emitter collector characteristics

1. The supply voltage for a common-emitter amplifier is 30 V, and the load resistor is 600 Ω. The circuit is biased at an emitter current of 0.4 ma. Draw the load line. The peak value of the signal is 0.2 ma. Determine the a-c emitter current, the a-c- load current, and the a-c output voltage.

2. What is the input d-c power to the circuit? Determine the collector efficiency and the overall efficiency.
3. The collector can dissipate 700 mw. Draw the maximum collector dissipation curve on the collector characteristic.
4. The supply voltage is now 30 V and the load resistance is 500 Ω. Draw the load line and then plot the dynamic load line for the circuit, showing I_C plotted against I_E.

Section 11-4 The Load Line for a Field Effect Transistor Amplifier

The circuit for a JFET is given in Fig. 11-7*a*. The loop equation on the drain side of the JFET is

$$V_{DD} = R_d I_D + V_{DS} \tag{11-5}$$

where V_{DS} and I_D are the variables. Solving for I_D,

$$I_D = -\frac{1}{R_d} V_{DS} + \frac{V_{DD}}{R_d} \tag{11-6}$$

Since Eq. 11-6 is in the form, $y = mx + b$, the load line is drawn from V_{DD} to V_{DD}/R_d as shown in Fig. 11-7*b* for the numerical values given on the circuit.

Since this FET is a depletion type, the gate-to-source voltage V_{GS} is always negative, and no current can flow in the gate circuit. Consequently, the component value of R_g is dictated by the use of the FET. Usually, R_g is of the order of megohms.

To locate the operating point at a particular value Q, the magnitude of V_{GG} is that value given by the intersection of the gate voltage with the load line. For the Q-point given on the load line, a value of -1.0 V for V_{GG} is required.

The circuit arrangement shown in Fig. 11-8 has the same Q point as Fig. 11-7 but the requirement for having a second power supply for the bias is eliminated. The load line is now drawn for the sum of R_s and R_d which, for these numerical values, is the same load line as drawn in Fig. 11-7*b*. The drain current I_D flows through R_s and produces a d-c voltage drop across R_s. The polarity of this voltage drop is shown on the circuit.

The voltage from the gate to the source is the voltage measured between the gate and the source terminals of the JFET. It is also the sum of the voltage drops across the resistors R_g and R_s. Therefore

$$V_{GS} = -I_G R_g - I_D R_s \tag{11-7a}$$

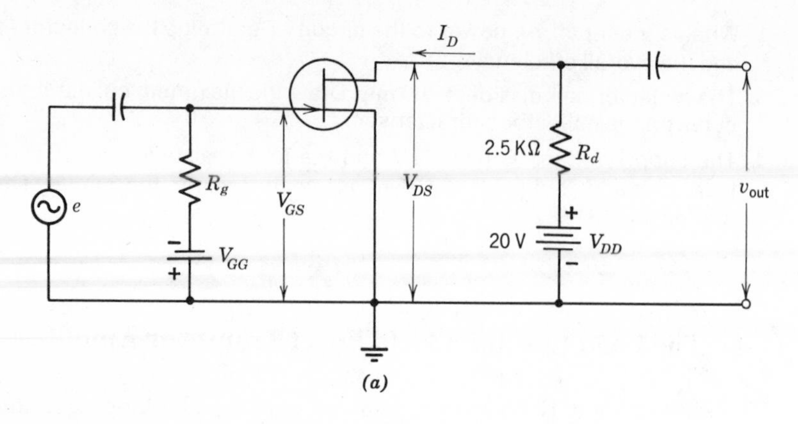

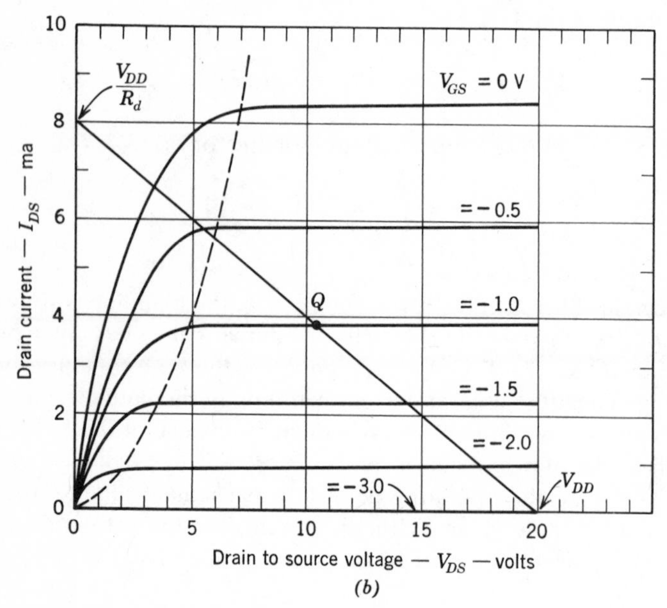

Figure 11-7 JFET load line. (*a*) Circuit. (*b*) Drain characteristic.

Since the gate current I_G is zero, the equation becomes simply

$$V_{GS} = -I_D R_s \tag{11-7b}$$

The Q point in Figure 11-7b has a value of −1.0 V for V_{GS} and a value of 3.8 ma for I_D. Then by Ohm's law R_s is 264 Ω and by subtraction R_d is 2336 Ω.

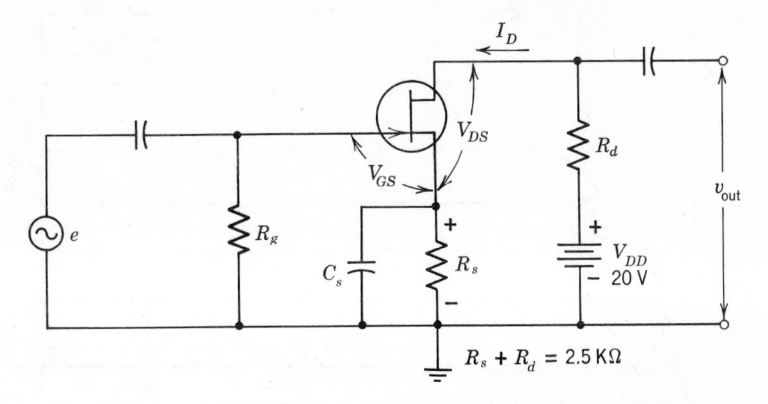

Figure 11-8 JFET using self bias.

Usually it is desired that the voltage drop across R_s be kept a pure d-c value. If there is a signal in the circuit, the drain current varies with the signal. This would produce a varying signal component of voltage across R_s. In order to eliminate it, a *bypass capacitor*, C_s, is placed in parallel with R_s. To bypass the signal components effectively, the reactance of the bypass capacitor should be no more than $\frac{1}{10}$ the value of R_s at the lowest frequency that the circuit is expected to handle.

In this example, since R_s is 264 Ω, the value of X_{C_S} must be no greater than 26 Ω. For 60 Hz, this value of reactance requires a capacitor of 100 μF. If it were desired to go as low as 20 Hz as in high-fidelity audio circuitry, the capacitor would have to be raised to a value of 300 μF.

The analysis of voltages, gains, and efficiencies is the same as the procedures of Sections 11-2 and 11-3 except that voltage gain is determined for the FET rather than the current gain of the transistor circuits.

PROBLEMS

1. Draw the load line for supply voltage of 20 V and a load resistance of 5,000 Ω. Determine the Q-point and the gain when the circuit is biased at −2.0 V.
2. Repeat Problem 1 for a supply voltage of 15 V and a load resistance of 3000 Ω.
3. Draw the load line for a supply voltage of 20 V and a load resistance of 20 kΩ. The bias is −3.5 V.
4. If self bias (R_s, Fig. 11-8) is used, determine the values of R_d and R_s for each of the Q-points in Problems 1 to 3.
5. Construct the dynamic transfer curve from the load line of Problem 1.

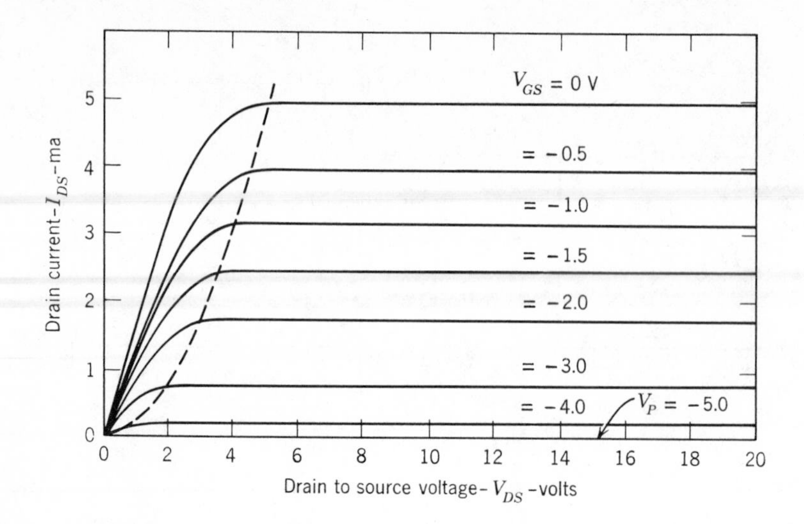

Characteristic for problems.

Section 11-5 The Load Line for a Vacuum Tube Amplifier

The load line for the vacuum-tube amplifier (Fig. 11-9) is drawn in the same manner as for the semiconductor circuits. The loop equation about the plate circuit is

$$E_{bb} = I_b R_c + E_b \tag{11-8}$$

or

$$I_b = -\frac{1}{R_c} E_b + \frac{E_{bb}}{R_c} \tag{11-9}$$

The intercepts of the load line are E_{bb} and E_{bb}/R_c and the load line has the slope $-1/R_c$, as shown in Fig. 11-10.

Let us assume that the desired Q-point is located at the intersection of the load line with the −6 V grid line. In Fig. 11-9a, then, a 6-V d-c bias source is required. The self-bias circuit (Fig. 11-9b) is very similar to that of the FET. In this case I_b is 4.3 ma and the desired voltage drop across R_K is 6 V. Therefore, R_K is, by Ohm's law, 1390 Ω. Thus for this load line to be exact, R_c in Fig. 11-9b would have to be reduced to 28.6 kΩ. The reactance of C_K should then be no greater than 139 Ω at the lowest frequency component of the signal.

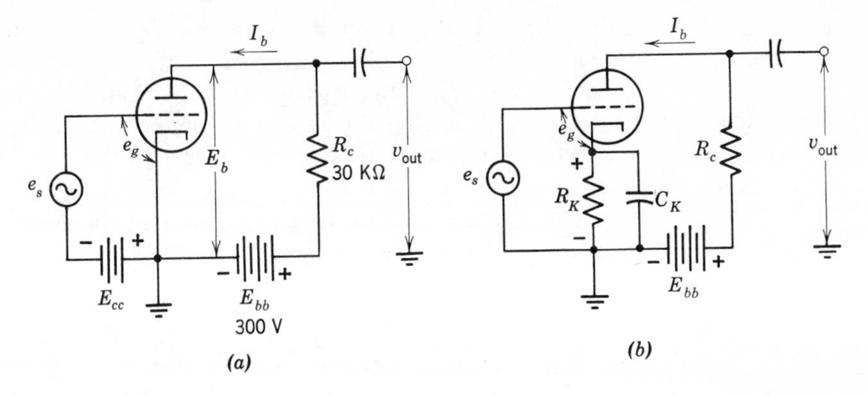

Figure 11-9 Vacuum-tube amplifier circuits. (a) With fixed bias. (b) With cathode bias.

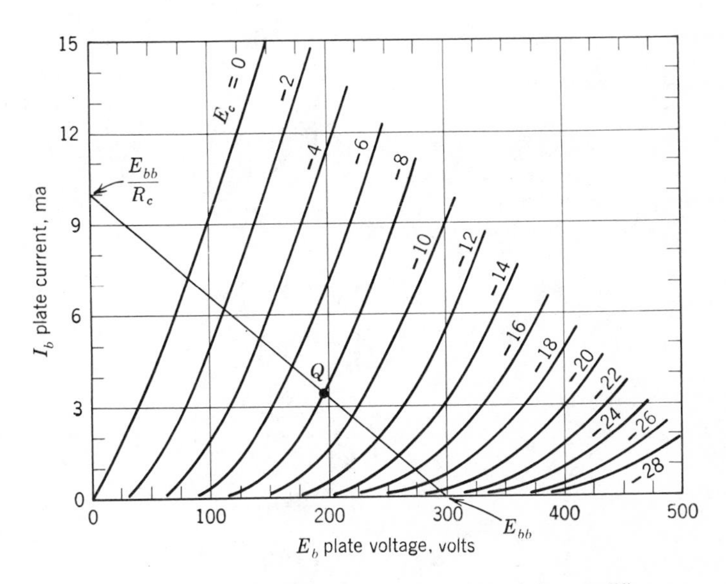

Figure 11-10 The load line for the vacuum-tube amplifier.

When a pentode is used, it is usual that the screen voltage is less than the plate voltage. The difference between the supply voltage E_{bb} and the desired screen voltage E_{c2} is obtained by placing a voltage dropping resistor R_2 in series with the screen (Fig. 11-11). The capacitor C_2 connected from screen to ground serves as a filter capacitor. The L filter configuration of R_2 and

C_2 gives rise to the nomenclature *screen decoupling filter*. C_2 is termed the *screen bypass capacitor*. Following the same procedure used for the emitter bypass capacitor and for the source bypass capacitor, the value of the bypass capacitor is determined by making the reactance of the capacitor at the lowest frequency to be amplified no greater than one tenth the value of the resistor it bypasses.

To show the method of calculation, let us assume that for a particular tube the operating values of current and voltage are as follows:

$$E_{bb} = 300 \text{ V}, \qquad I_b = 3 \text{ ma}$$

$$E_{c2} = 100 \text{ V}, \qquad I_{c2} = 2 \text{ ma}$$

$$E_{cc} = -4 \text{ V}, \qquad I_K = I_b + I_{c2} = 5 \text{ ma}$$

Then

$$R_K = \frac{E_{cc}}{I_b + I_{c2}} = \frac{4}{0.005} = 800\ \Omega$$

and

$$R_2 = \frac{E_{bb} - E_{c2}}{I_{c2}} = \frac{300 - 100}{0.002} = 100{,}000\ \Omega$$

For operation down to 30 cycles,

$$X_{CK} = 0.1 R_K$$

$$\frac{1}{2\pi f C_K} = 0.1 R_K$$

$$\frac{1}{2\pi 30 C_K} = 80$$

$$C_K = 66.5\ \mu\text{F}$$

$$X_{C2} = 0.1 R_2$$

$$\frac{1}{2\pi f C_2} = 0.1 R_2$$

$$\frac{1}{2\pi 30 C_2} = 10{,}000$$

$$C_2 = 0.053\ \mu\text{F}$$

The construction of the dynamic transfer curve is left as an exercise.

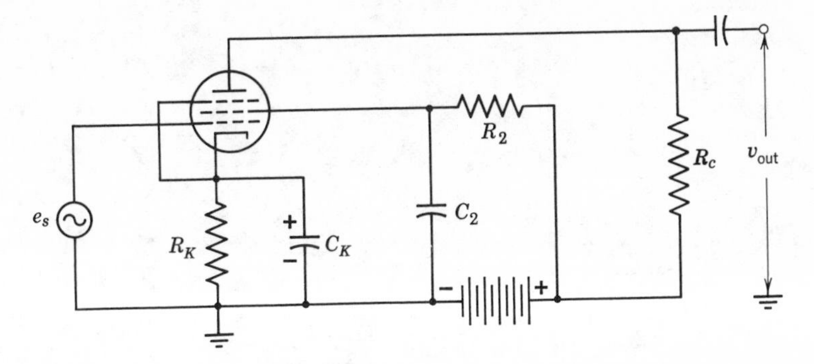

Figure 11-11 A pentode amplifier with a single direct voltage supply.

PROBLEMS

1. By using the plate characteristic shown in Fig. 11-10, draw a load line for a 400-V supply voltage and a plate load resistance of 80 kΩ. Determine the value of R_K required to establish the bias at −8.0 V. If the amplifier is to be used down to 20 Hz, what is the value of C_K? What is the voltage gain of the amplifier?
2. Repeat Problem 1 for E_{bb} 200 V, R_c 10 kΩ, and E_{cc} −4.0 V.
3. From data points taken from the load line given in Fig. 11-10, construct the dynamic transfer characteristic.
4. From the load line obtained in Problem 1, construct the dynamic transfer characteristic.

Section 11-6 Nomenclature

A JFET amplifier circuit is shown in Fig. 11-12. The load line on the drain characteristic is given in Fig. 11-13. A set of transfer curve axes V_{GS} and

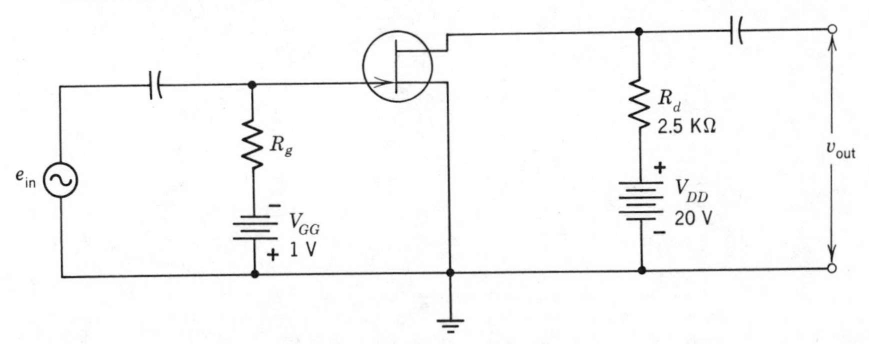

Figure 11-12 JFET with signal source.

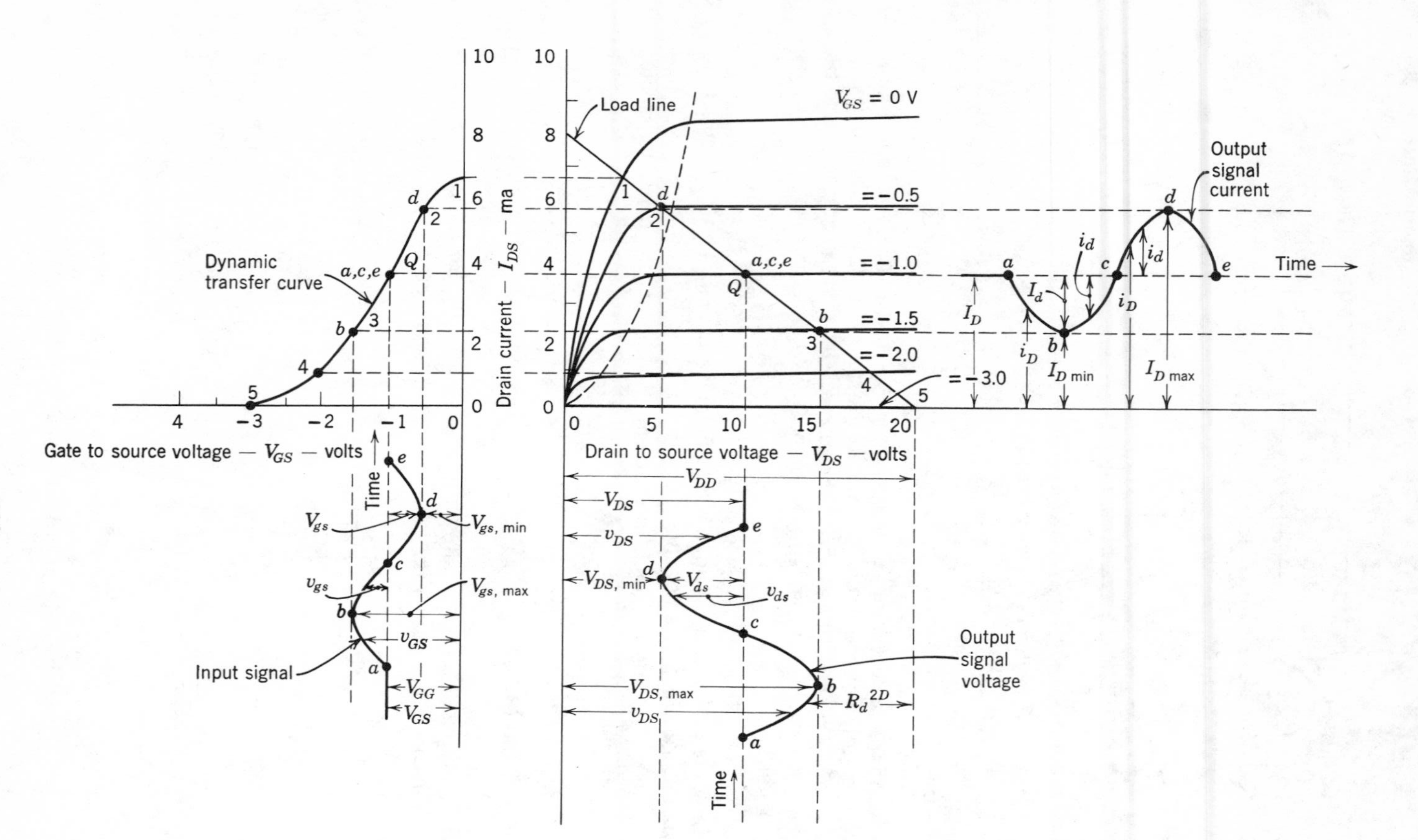

Figure 11-13 Signal superimposed on load lines.

I_{DS} is drawn to the left of this load line. The points of intersection of the load line with the drain characteristics, 1, 2, Q, 3, 4, and 5, are transferred over to the new set of axes. A smooth curve through them, the *dynamic transfer curve*, is drawn. A sinusoidal signal voltage e_{in} is applied to the circuit. This signal produces the sinusoidal waveform superimposed on the gate-to-source bias and is labeled *input signal.* It is drawn to a time base scale, and one cycle goes from a to b to c to d and to e.

These five points are projected up to the dynamic transfer curve and then over to the load line on the drain characteristic. From here they are extended both down and over to the right. The projection downward yields a point-to-point generation of the output voltage waveform, and the projection to the right produces the point-to-point time variation of the drain current. It is obvious from this set of curves that the output signal voltage is 180° out of phase with the input signal.

In semiconductor nomenclature, the voltages and currents are indicated by letters with subscripts. A capital letter represents a d-c value, a peak value, or an rms value. A lowercase letter represents an instantaneous quantity. A capital letter with a capital subscript represents a quantity that can be measured on a d-c indicating instrument. For example I_D is the d-c drain current and V_{DS} is the d-c drain-to-source voltage. A double capital subscript of the same letter repeated is a supply voltage. For example, V_{GG} is the gate supply voltage and V_{DD} is the drain supply voltage. Occasionally triple subscripts are used. V_{DDS} is a supply connected between drain and source. A lowercase subscript represents an a-c value. For example, I_d is the rms value of the signal component and i_d is the instantaneous signal value. A lowercase letter with a capital subscript is the instantaneous *total* value. For example, v_{GS} is the instantaneous sum of the signal v_{gs}, plus the bias, V_{GS}. The only time confusion can arise is in a symbol written as I_d. I_d can represent either the peak value of the a-c signal or the rms value of the a-c signal itself. In any event, in a situation where context is not clear and where confusion could arise, the symbols are written as $I_{d,\max}$ and $I_{d,\text{rms}}$.

There is zero gate current in the circuit and, as a result, V_{GG} equals V_{GS}. In the gate circuit the following equations may be written from an inspection of Fig. 11-13:

$$e_{\text{in}} = v_{gs} = E_{\text{in}} \sin \omega t = V_{gs} \sin \omega t \tag{11-10}$$

$$v_{GS} = V_{GS} + v_{gs} = V_{GS} + V_{gs} \sin \omega t \tag{11-11}$$

In the drain circuit the voltage relations are similar, but the IR drop across R_d must be taken into account:

$$V_{DD} = V_{DS} + R_d I_D \tag{11-12}$$

$$v_{ds} = -V_{ds} \sin \omega t = -R_d I_d \sin \omega t \tag{11-13}$$

$$v_{DS} = V_{DS} - V_{ds} \sin \omega t = V_{DS} - R_d i_d = V_{DS} - R_d I_d \sin \omega t \tag{11-14}$$

A transistor amplifier yields the same general form of sinusoidal waveforms as shown in Fig. 11-13. The subscripts used for the collector are C and c, for the base B and b, and for the emitter E and e. The circuit, the waveforms, and the standard symbols for transistors are illustrated in the frontispiece of this book.

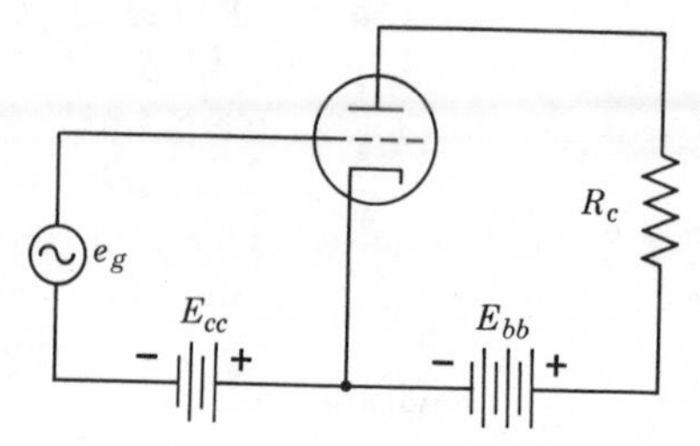

Figure 11-14 The basic vacuum-tube amplifier.

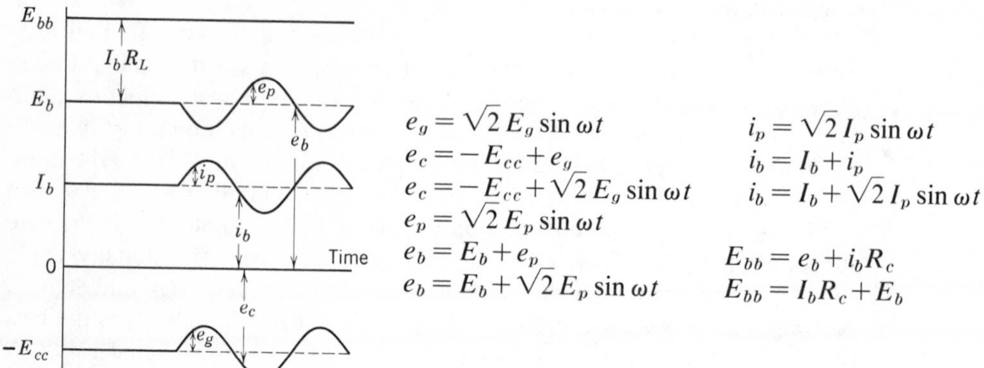

$$e_g = \sqrt{2}\, E_g \sin \omega t$$
$$e_c = -E_{cc} + e_g$$
$$e_c = -E_{cc} + \sqrt{2}\, E_g \sin \omega t$$
$$e_p = \sqrt{2}\, E_p \sin \omega t$$
$$e_b = E_b + e_p$$
$$e_b = E_b + \sqrt{2}\, E_p \sin \omega t$$

$$i_p = \sqrt{2}\, I_p \sin \omega t$$
$$i_b = I_b + i_p$$
$$i_b = I_b + \sqrt{2}\, I_p \sin \omega t$$

$$E_{bb} = e_b + i_b R_c$$
$$E_{bb} = I_b R_c + E_b$$

Figure 11-15 Waveforms and relations in a vacuum-tube amplifier for a sine-wave signal.

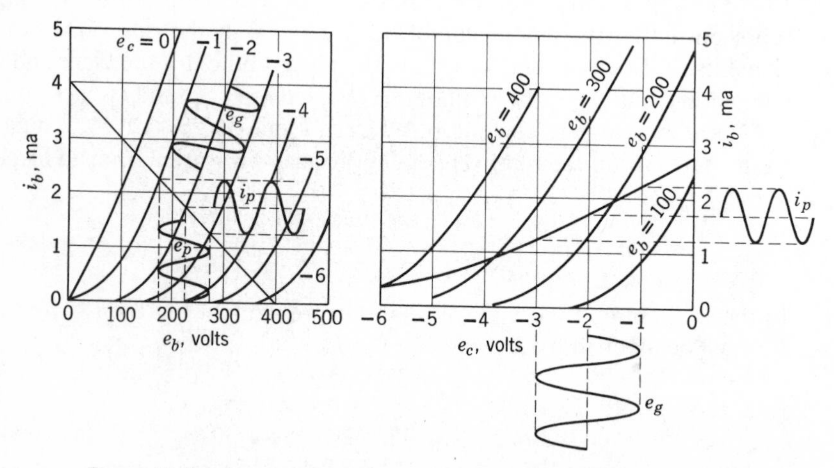

Figure 11-16 Dynamic load lines for a 6SL7 with 100-kΩ plate load.

The letter symbols used in a vacuum-tube amplifier do not follow these conventions. They predate the semiconductor letter symbols but, when transistors were developed, it was realized that a more compact and exact science was necessary. Also, as the vacuum tube had been in use for so many years, it was obvious that it would not be practical to conform the vacuum-tube nomenclature to the new arrangement. The circuit for a vacuum-tube amplifier is given in Fig. 11-14, and the waveforms are shown together with the nomenclature in Fig. 11-15. The generation of the waveforms showing the 180° phase inversion is given in Fig. 11-16.

PROBLEMS

1. Use the information given in Problem 4 of Section 11-3. Draw waveforms to scale for this circuit similar to the ones given in Fig. 11-13.
2. By using the results of Problems 1 and 5, Section 11-4, construct a dynamic transfer characteristic on a $V_{GS} - V_{DS}$ axis similar to that shown in Fig. 11-16.
3. Using the results of Problems 1 and 4, Section 11-5, construct a dynamic transfer characteristic on an $e_c - e_b$ axis similar to the one shown in Fig. 11-16.

Section 11-7 Distortion

Ideally an amplifier accepts an input signal and reproduces the signal in the output circuit without changing its shape or form, only its amplitude. An amplifier can be free of distortion only if all the output current and voltage intercepts are exactly equal along the load line. Also if these intercepts are equal, the dynamic transfer curve must be a straight line. Practically, a distortion-free amplifier is an impossibility. Thus, it is necessary to discuss distortion with a view to obtain both qualitative and quantitative evaluations of the problem. Qualitatively, the distortion occurs in three forms: amplitude, frequency, and phase.

Amplitude distortion is defined as the result of different amplifier gains at different signal levels. If a 2-V signal produces 10 V output, a 20-V signal without amplitude distortion is amplified to 100 V. If the power supply is only 30 V, an output of 100 V would not be possible. Nonlinearity of the dynamic load line, overloading, and cutoff, all contribute to amplitude distortion.

Frequency distortion is the distortion that results from different gains at different frequencies. To be free from frequency distortion, the amplifier must be *flat* over the necessary frequency range. As an example, lack of bass in an amplifier is a frequency distortion.

Phase distortion must be explained before it is defined. If a signal consists of a fundamental and a third harmonic (Fig. 11-17), in order for the plate wave to have the same shape, both the fundamental and the third harmonic must be zero at the same time. If the fundamental has a phase shift of 7° with respect to itself, the third harmonic which is three times the fundamental frequency must have a phase shift of 3×7 or 21°. Disregarding the natural 180° inversion of the signal in an amplifier stage, *no phase distortion occurs* if

1. The phase shift is zero, or
2. The phase shift is proportional to frequency.

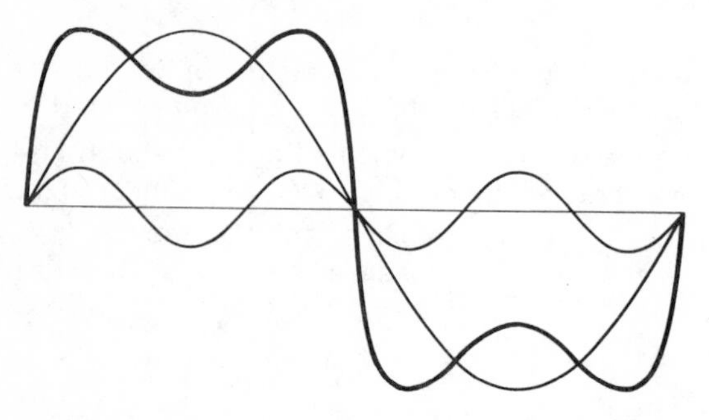

Figure 11-17 Signal waveform consisting of fundamental and third harmonic.

In audio work, little attention is paid to phase-shift distortion. In video amplifiers and in pulse amplifiers, a serious phase-shift distortion could change a square pulse into a triangular pulse. Fortunately, when the frequency response is satisfactory for passing all necessary harmonics without frequency distortion, the phase-shift distortion is almost negligible.

Notice that amplitude distortion is developed by the nonlinearity of the circuit components and by the nonlinearity of the amplifying device itself. Frequency distortion and phase distortion arise primarily from the variation of the impedances of the reactive elements in the circuit (inductance and capacitance). This distortion is present under circumstances where the amplitude distortion is negligible. As an example, an amplifier that is designed to deliver 35 W has a very small amplitude distortion when the output is only 1 W, but it still has limitations on the low- and on the high-frequency response.

A distortion that is often considered as independent of the amplitude, distortion discussed above is *intermodulation distortion*. Let us assume that an amplifier is distortionless when it is amplifying either a 200-Hz signal or a 4000-Hz signal. When both signals are simultaneously present in the input, any interaction between them is called intermodulation distortion.

This distortion is important when the complex waveforms of speech and music are considered.

The transfer curves for all the circuits we have considered thus far, show some degree of nonlinearity or deviation from a straight line as in Figs. 11-4, 11-6, and 11-13. In mathematics, any continuous curve can be represented by a power series that gives the current in terms of the input signal:

$$i = I_Q + a_1 e_{\text{in}} + a_2 e_{\text{in}}^2 + a_3 e_{\text{in}}^3 + a_4 e_{\text{in}}^4 + \cdots + a_n e_{\text{in}}^n + \cdots. \tag{11-15}$$

The exact values of the coefficients, a_1, a_2, a_3, etc., depend on the curve under consideration. If all coefficients a_n are zero for n equal or are greater than 2, the power series reduces to

$$i = I_Q + a_1 e_{\text{in}}$$

In this case, the amplifier is distortion-free and the output reproduces the input signal exactly. Alternatively, if it can be shown that a_2, a_3, a_4, etc., are negligible in comparison to the value of a_1, the amplifier is effectively free from distortion.

If we let $e_{\text{in}} = E \sin \omega t$, we have

$$i = I_Q + a_1 E \sin \omega t + a_2 E^2 \sin^2 \omega t + a_3 E^3 \sin^3 \omega t + a_4 E^4 \sin^4 \omega t + \cdots$$

From trigonometric tables of identities, we find that

$$\begin{aligned}
\sin^2 x &= \tfrac{1}{2} - \tfrac{1}{2}\cos 2x \\
\sin^3 x &= (\tfrac{1}{2} - \tfrac{1}{2}\cos 2x)\sin x \\
&= \tfrac{1}{2}\sin x - \tfrac{1}{2}\cos 2x \sin x \\
&= \tfrac{1}{2}\sin x - \tfrac{1}{4}\sin 3x + \tfrac{1}{4}\sin x \\
&= \tfrac{3}{4}\sin x - \tfrac{1}{4}\sin 3x \\
\sin^4 x &= (\tfrac{1}{2} - \tfrac{1}{2}\cos 2x)^2 \\
&= \tfrac{1}{4} - \tfrac{1}{2}\cos 2x + \tfrac{1}{4}\cos^2 2x \\
&= \tfrac{1}{4} - \tfrac{1}{2}\cos 2x + \tfrac{1}{8} + \tfrac{1}{8}\cos 4x \\
&= \tfrac{3}{8} - \tfrac{1}{2}\cos 2x + \tfrac{1}{8}\cos 4x
\end{aligned}$$

Then

$$\begin{aligned}
i = I_Q &+ \left(\frac{a_2 E^2}{2} + \frac{3a_4 E^4}{8} + \cdots\right) + \left(a_1 E + \frac{3a_3 E^3}{4} + \cdots\right)\sin \omega t \\
&- \left(\frac{a_2 E^2}{2} + \frac{a_4 E^4}{2} + \cdots\right)\cos 2\omega t - \left(\frac{a_3 E^3}{4} + \cdots\right)\sin 3\omega t \\
&+ \left(\frac{a_4 E^4}{8} + \cdots\right)\cos 4\omega t + \cdots
\end{aligned}$$

which may be simplified as

$$\begin{aligned}
i = I_Q + A_0 + A_1 \sin \omega t + A_2 \sin 2\omega t + A_3 \sin 3\omega t \\
+ A_4 \sin 4\omega t + A_5 \sin 5\omega t + \cdots
\end{aligned} \tag{11-16}$$

This derivation proves that an amplifier not only amplifies a signal but also introduces harmonics which are unwanted multiples of the signal frequency. A_1 is the amplitude of the fundamental in the output, A_2 is the amplitude of the second harmonic, A_3 is the amplitude of the third harmonic, A_4 is the amplitude of the fourth harmonic, and A_5 is the amplitude of the fifth harmonic. Fortunately, the magnitude of the harmonic amplitudes decreases rapidly to the point where we are ordinarily concerned only up to the fifth and usually only up to the second and third.

The percentage of the second harmonic is $(A_2/A_1) \times 100$; of the third, $(A_3/A_1) \times 100$; of the fourth, $(A_4/A_1) \times 100$; and of the fifth, $(A_5/A_1) \times 100$. Figure 11-18 shows the relation of the total harmonic content to the load

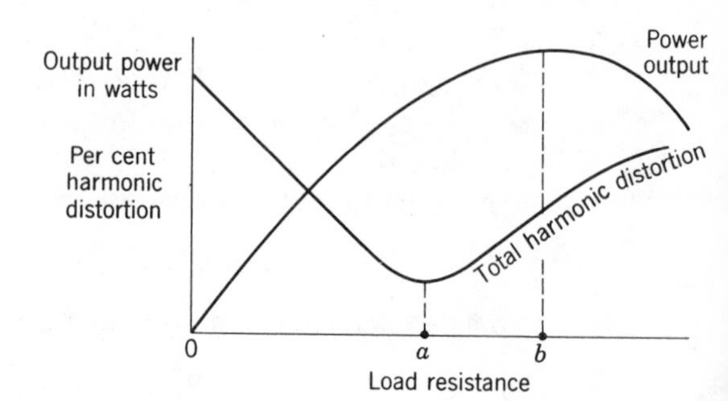

Figure 11-18 Output characteristics of power amplifier at full signal.

resistance for a power amplifier operated at maximum signal drive. It is found that the maximum power output and the minimum distortion do not occur at the same value of load resistance. Usually, an amplifier stage is operated at the point of minimum distortion and not at the point of maximum output.

One concept must be made very clear. In this discussion, the input signal is completely free of distortion. The amplifying device introduces distortion where none existed previously. The distortion is caused by the fact that the amplifier introduces distortion by virtue of its own nonlinearity. If the input signal already has distortion, the amplifier does not know it. It can only amplify and add its own distortion to the incoming signal. We are assuming the incoming signal is "clean" without distortion for this analysis.

Figure 11-19 shows the input and output signals for a JFET amplifier that has a marked curvature in the transfer characteristic. The final equation for i_D (Eq. 11-16) shows an A_0 term. Since the top loop of the output current waveform in Fig. 11-19 is larger than the bottom loop, the average value of the current is shifted upward by the amount of A_0. In deriving Eq. 11-16, the A_0 term was the result of d-c components produced by the presence of

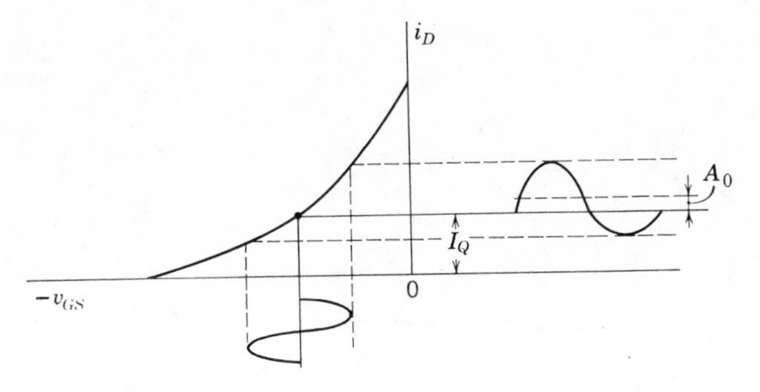

Figure 11-19 Distortion produced by the dynamic-transfer load line.

the even harmonic terms. The A_0 term accordingly goes to zero when the input signal is reduced to zero, and it increases when the signal is increased. The A_0 term is often referred to as "the rectification of the input signal in the output circuit."

A quantitative approach which has been developed for predetermining the harmonic distortion makes use of the following formulas:

Increase in I_Q (rectified component)
$$A_0 = \frac{\frac{1}{2}(I_{\max} + I_{\min}) + I_2 + I_3 - 3I_Q}{4} \tag{11-17a}$$

Rms value of fundamental
$$A_1 = \frac{\sqrt{2}(I_2 - I_3) + I_{\max} - I_{\min}}{4\sqrt{2}} \tag{11-17b}$$

Rms value of second harmonic
$$A_2 = \frac{I_{\max} + I_{\min} - 2I_Q}{4\sqrt{2}} \tag{11-17c}$$

Rms value of third harmonic
$$A_3 = \frac{I_{\max} - I_{\min} - 2\sqrt{2}A_1}{2\sqrt{2}} \tag{11-17d}$$

Rms value of fourth harmonic
$$A_4 = \frac{2A_0 - I_2 - I_3 + 2I_Q}{2\sqrt{2}} \tag{11-17e}$$

The location of the values of the symbols used is shown in Fig. 11-20.

To illustrate this method of harmonic-distortion analysis, a load line for 3000 Ω and a supply voltage of 600 V is drawn on the triode-connected plate characteristic of the 6L6 (Fig. 11-21). The operating point is at $E_{cc} = -45$ V. The grid swing is then ± 45 V. I_2 is located where the grid swing is 0.707 of the maximum value and is $(-45 + 31.8)$ or -13.2 V. I_3 is located at $(-45 - 31.8)$ or -76.8 V. A tabulation of values taken from the graph gives:

$$I_{\max} = 132 \text{ ma}, \qquad I_{\min} = 4 \text{ ma}$$
$$I_2 = 107 \text{ ma}, \qquad I_3 = 10 \text{ ma}$$
$$I_Q = I_b = 49 \text{ ma}$$

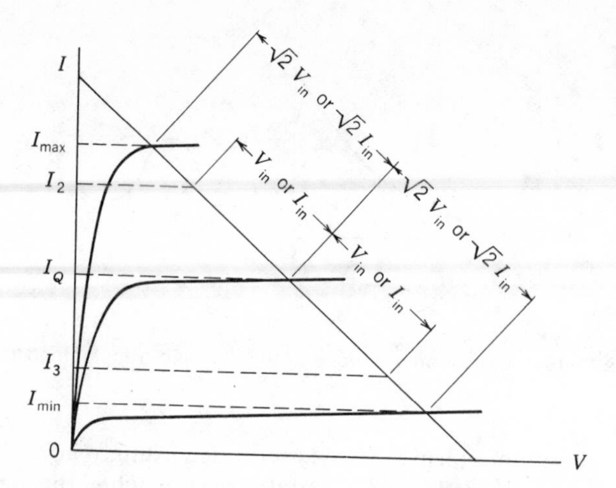

Figure 11-20 Graphical procedure for harmonic determination.

Putting these values in the equations and evaluating, we have the following:

$$A_0 = \frac{\frac{1}{2}(I_{max} + I_{min}) + I_2 + I_3 - 3I_Q}{4}$$

$$= \frac{\frac{1}{2}(132 + 4) + 107 + 10 - 3 \times 49}{4} = 9.5 \text{ ma}$$

$$A_1 = \frac{\sqrt{2}(I_2 - I_3) + I_{max} - I_{min}}{4\sqrt{2}}$$

$$= \frac{\sqrt{2}(107 - 10) + 132 - 4}{4\sqrt{2}} = \frac{65.4}{\sqrt{2}} = 46.9 \text{ ma}$$

$$A_2 = \frac{I_{max} + I_{min} - 2I_Q}{4\sqrt{2}} = \frac{132 + 4 - 2 \times 49}{4\sqrt{2}} = 6.7 \text{ ma}$$

$$A_3 = \frac{I_{max} - I_{min} - 2\sqrt{2}A_1}{2\sqrt{2}} = \frac{132 - 4 - 2\sqrt{2}\,46.9}{2\sqrt{2}} = 1.0 \text{ ma}$$

$$A_4 = \frac{2A_0 - I_2 - I_3 + 2I_Q}{2\sqrt{2}} = \frac{2 \times 9.5 - 107 - 10 + 2 \times 49}{2\sqrt{2}} = 0$$

$$P_{OUT} = A_1^2 R_L = 0.0469^2 \times 3000 = 6.56 \text{ W}.$$

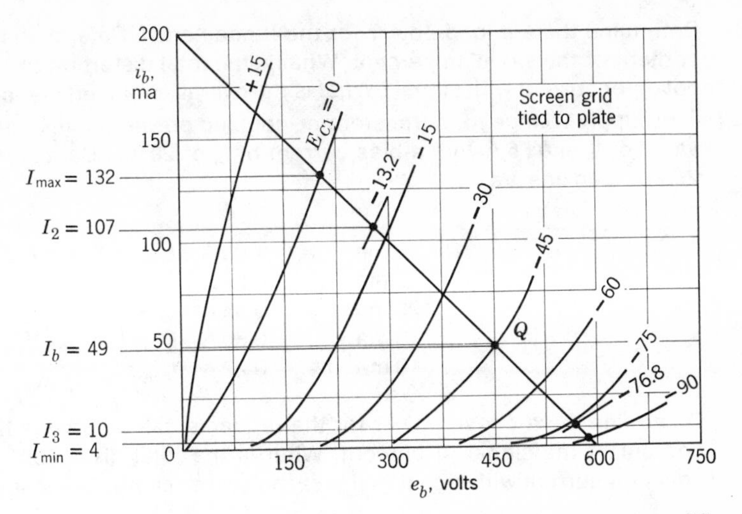

Figure 11-21 Calculation of distortion for a 6L6 triode-connected amplifier.

$$\% \text{ 2nd harmonic} = \frac{A_2}{A_1} \times 100 = \frac{6.7}{49.6} \times 100 = 14.3\%$$

$$\% \text{ 3rd harmonic} = \frac{A_3}{A_1} \times 100 = \frac{1.0}{46.9} \times 100 = 2.1\%$$

$$\% \text{ 4th harmonic} = \frac{A_4}{A_1} \times 100 = 0$$

$$\text{Total harmonic distortion} = \sqrt{(\% \text{ 2nd})^2 + (\% \text{ 3rd})^2 + (\% \text{ 4th})^2}$$

$$= \sqrt{(14.3)^2 + (2.1)^2}$$

$$= 14.4\%$$

PROBLEMS

1. The supply voltage on a power transistor is 20 V and the load is 2 Ω. Data points along the load line are

I_B	100 ma	I_C	15.8 A	I_{max}
	85 ma		14.1 A	I_2
	50 ma		9.3 A	I_Q
	15 ma		2.0 A	I_3
	0 ma		0 A	I_{min}

Determine the a-c load power for the fundamental. Determine the harmonic content of the signal in percent. What is the total distortion? What is the d-c collector current with signal? What is the overall circuit efficiency?

2. The supply voltage for a transformer-coupled power amplifier is 14 V and the value of I_Q is 0.55 A with a bias current of I_B of 20 ma. Data points along the 26.2-Ω load line are

I_B	30 ma	I_C	1.00 A	I_{max}
	27 ma		0.92 A	I_2
	20 ma		0.55 A	I_Q
	13 ma		0.30 A	I_3
	10 ma		0.22 A	I_{min}

Determine the a-c load power for the fundamental. Determine the harmonic content of the signal in percent. What is the total distortion? What is the collector current with signal? What is the overall circuit efficiency?

Section 11-8 Transistor Bias Circuits

The basic common-base amplifier is shown in Fig. 11-2*a*. The d-c Kirchhoff's voltage equation around the emitter loop is

$$V_{EE} = R_E I_E + V_{BE} \tag{11-18}$$

V_{BE} is the d-c voltage drop of a forward-biased diode that is approximately 0.7 V for silicon and 0.3 V for germanium. The operating point Q on the load line requires an emitter current I_E of 4.0 ma. Assuming a silicon transistor and 12 V for V_{EE}, the bias resistor R_E (see conversion Table) can be determined by substituting values in Eq. 11-18.

$$12 = 4R_E + 0.7$$

or

$$R_E = 11.3/4 = 2.825 \text{ k}\Omega$$

This approach to the determination of the value of the bias resistor is extended in a similar manner to the common-emitter amplifier shown in Fig. 11-5. The loop equation in the base circuit is

$$V_{BB} = R_B I_B + V_{BE} \tag{11-19a}$$

Conversion Table

Multiplying factors to convert — From \ To	I_B	I_C	I_E
I_B	1	β	$(1+\beta)$
I_C	$\frac{1}{\beta}$	1	$\frac{1+\beta}{\beta}$
I_E	$\frac{1}{1+\beta}$	$\frac{\beta}{1+\beta}$	1

By substituting values and assuming a silicon transistor and 12 V for V_{BB}

$$12 = 0.075\, R_B + 0.7$$

$$R_B = 11.3/0.075 = 162\ \text{k}\Omega$$

To establish a general procedure for bias analysis, the bias equations must be extended to include I_E and I_C. The transistor has a d-c current gain defined as

$$\beta_{\text{dc}} \equiv h_{FE} \equiv \frac{I_C}{I_B} \tag{3-2c}$$

If this value of β_{dc} is assumed constant over the whole collector characteristic, the theoretical maximum value of collector current is V_{CC}/R_c and, then, the corresponding theoretical maximum value of I_B is I_C/β_{dc} or $V_{CC}/(\beta_{\text{dc}} R_c)$. Substituting this into Eq. 11-19*a*, we have

$$V_{BB} = \frac{V_{CC}}{\beta_{\text{dc}} R_c} R_b + V_{BE}$$

Solving for R_B, the equation becomes

$$R_B = \beta_{\text{dc}} \frac{V_{BB} - V_{BE}}{V_{CC}} R_c \tag{11-19b}$$

Now, if V_{BB} is the same source as V_{CC} and if V_{BE} is small compared to V_{CC}, the result simplifies to

$$R_B \approx \beta_{\text{dc}} R_c \tag{11-19c}$$

The significance of this analysis is that there is a least value of base bias resistor that can be used. When R_B is this low or lower, the transistor amplifier is saturated, and a severe distortion of signal results. All of the circuits in this section have minimum values and the analysis of these other circuits will be left for the exercises at the end of the chapter.

Very often we desire to operate the circuit at the midpoint of the load line. The collector current is then $\frac{1}{2}V_{CC}/R_c$, and the base current is

$$I_B = \frac{V_{CC}}{2\beta_{dc} R_c}$$

Substituting this into Eq. 11-19*a*

$$V_{BB} = \frac{V_{CC}}{2\beta_{dc} R_c} R_B + V_{BE}$$

and solving for R_B,

$$R_B = 2\beta_{dc} \frac{(V_{BB} - V_{BE})}{V_{CC}} R_c \qquad (11\text{-}20a)$$

If V_{BE} is negligible

$$R_B = 2\beta_{dc} \frac{V_{BB}}{V_{CC}} R_c \qquad (11\text{-}20b)$$

and if $V_{BB} = V_{CC}$,

$$R_B = 2\beta_{dc} R_c \qquad (11\text{-}20c)$$

Again, this approach can be readily extended to the other circuits in this chapter.

A voltage divider arrangement providing the operating bias shown in Fig. 11-22*a* is quite commonly used. The circuit of R_1 and R_2 across V_{BB} is replaced by an equivalent circuit determined by the application of Thévenin's theorem. The open circuit voltage from base to ground is V'_{BB}

$$V'_{BB} = \frac{R_2}{R_1 + R_2} V_{BB} \qquad (11\text{-}21a)$$

and the back impedance of the network R_1 and R_2 in parallel is R_B:

$$R_B = \frac{R_1 R_2}{R_1 + R_2} \qquad (11\text{-}21b)$$

By substituting 11-21*a* in 11-21*b*, we find that

$$R_B = R_1 \frac{V'_{BB}}{V_{BB}} \qquad (11\text{-}21c)$$

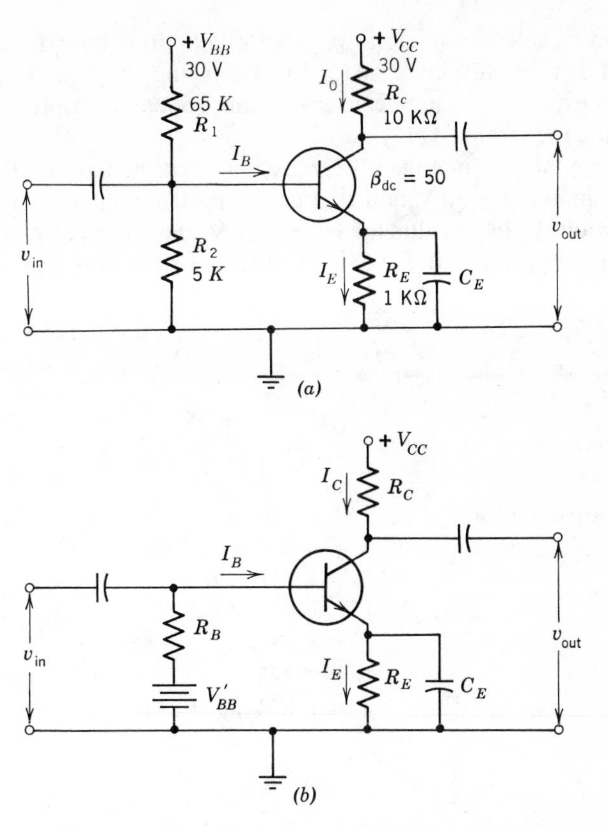

Figure 11-22 Voltage divider bias arrangement. (*a*) Circuit. (*b*) Converted circuit.

The loop voltage equation around the base-emitter circuit of Fig. 11-22*b* is

$$V'_{BB} = R_B I_B + V_{BE} + R_E I_E \tag{11-21d}$$

and the loop voltage equation around the collector-emitter circuit of Fig. 11-22*b* is

$$V_{CC} = I_C R_c + V_{CE} + R_E I_E \tag{11-21e}$$

These last two equations can be manipulated or solved by recalling that

$$I_C = \beta_{dc} I_B$$
$$I_E = (1+\beta_{dc}) I_B$$
$$I_C = \frac{\beta_{dc}}{1+\beta_{dc}} I_E$$

It is preferable to solve bias problems by approaching the circuit through the use of Kirchhoff's laws instead of by trying to generate precise formulas. This is especially true because there are so many variations in the method of securing the required bias.

For example, in Fig. 11-22*a* assume that the base and collector supplies are common at +30 V and that the numerical values are the ones shown on the circuit. Using a value for V_{BE} of 0.7 V, the values of I_C and V_{CE} are required. First, by Thévenin's theorem, the input circuit becomes

$$V'_{BB} = \frac{R_2}{R_1+R_2}V_{BB} = \frac{5}{5+65}30 = 2.145 \text{ V}$$

and

$$R_B = \frac{R_1R_2}{R_1+R_2} = \frac{5\times 65}{5+65} = 4.65 \text{ k}\Omega$$

The loop equation for the base circuit is*

$$\begin{aligned} 2.145 &= 4.65I_B + 0.7 + 1I_E \\ 2.145 &= 4.65I_B + 0.7 + (1+\beta_{dc})I_B \\ 2.145 &= 4.65I_B + 0.7 + 51I_B \\ 55.65I_B &= 1.445 \\ I_B &= 0.024 \text{ ma} = 24\ \mu\text{a} \end{aligned}$$

Then

$$I_C = \beta_{dc}I_B = 50\times 0.024 = 1.20 \text{ ma}$$

and

$$I_E = I_B + I_C = 1.22 \text{ ma}$$

The loop equation around the collector circuit is

$$\begin{aligned} V_{CC} &= R_cI_C + V_{CE} + R_EI_E \\ 30 &= 10\times 1.22 + V_{CE} + 1\times 1.22 \end{aligned} \qquad (11\text{-}21e)$$

or

$$V_{CE} = 16.6 \text{ V}$$

*If a calculation involved kilohms divided by kilohms, the "kilohms" cancel. For example 20,000 Ω/4,000 Ω is 20 kΩ/4 kΩ or 20/4 or 5. Similarly, kilohms times milliamperes is volts directly, and volts divided by kilohms yields milliamperes directly. For example, when the current in a 3.3 kΩ resistor is 2 ma, the voltage directly is 3.3 × 2 or 6.6 V.

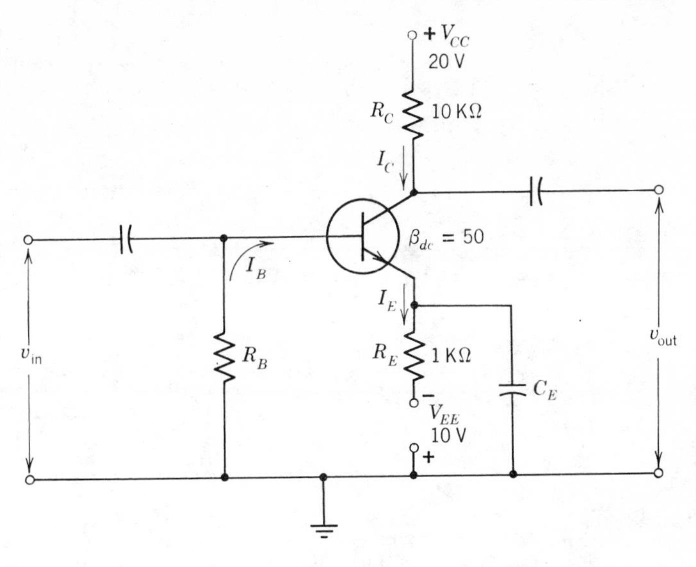

Figure 11-23 Biasing method using two supply voltages.

Another bias arrangement is shown in Fig. 11-23. Two separate supplies are used, V_{CC} and V_{EE} which are opposite in polarity to each other. The loop voltage equation in the base circuit is

$$R_B I_B + V_{BE} + R_E I_E = V_{EE} \tag{11-22a}$$

and, for the collector circuit,

$$R_c I_C + V_{CE} + R_E I_E = V_{CC} + V_{EE} \tag{11-22b}$$

Let us use the values for I_C, I_B, I_E, and V_{CE} used in the previous example and determine the value of R_B that is required for these circuit conditions. By substituting in Eq. 11-22*a*,

$$0.024\,R_B + 0.7 + 1 \times 1.22 = 10$$
$$0.024\,R_B = 8.08$$
$$R_B = 337\ \text{k}\Omega$$

In the circuit shown in Fig. 11-24, the bias is obtained by using a resistor R_B, connected from the collector to the base to obtain the desired bias. If the operating point of the collector changes, the value of the base current that is "fed back" into the base changes to compensate for any shift in the Q-point. This concept is covered in detail in Chapter 17. Two Kirchhoff's

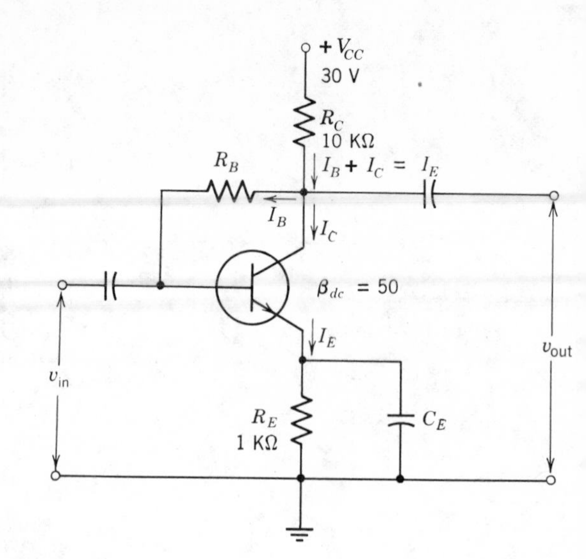

Figure 11-24 Biasing method using collector-to-base feedback.

voltage equations can be formed—one through the base resistor and the other through the transistor:

$$\begin{aligned} V_{CC} &= R_c(I_B + I_C) + R_B I_B + V_{BE} + R_E I_E \\ &= (R_c + R_E)I_E + R_B I_B + V_{BE} \end{aligned} \tag{11-23a}$$

and

$$\begin{aligned} V_{CC} &= R_c(I_B + I_C) + V_{CE} + R_E I_E \\ &= (R_c + R_E)I_E + V_{CE} \end{aligned} \tag{11-23b}$$

If these two equations are subtracted, a third equation results:

$$V_{CE} = R_B I_B + V_{BE} \tag{11-23c}$$

This last equation is the loop equation around the transistor and R_B.

As an illustrative example, we find the value of R_B in Fig. 11-24 that will set the value of V_{CE} at 15 V. Writing the equation for the voltage loop through the transistor:

$$\begin{aligned} V_{CC} &= 30 = (R_c + R_E)I_E + V_{CE} = (10 + 1)I_E + 15 \\ I_E &= 15/11 = 1.365 \text{ ma} \\ I_B &= I_E/(1 + \beta_{dc}) = 1.365/51 = 0.0268 \text{ ma} = 26.8\ \mu\text{a} \end{aligned}$$

Assuming a silicon transistor, these values are used in the voltage loop equation through R_B:

$$V_{CC} = (R_c + R_E)I_E + R_B I_B + V_{BE}$$
$$30 = (10+1)1.365 + R_B \times 0.0268 + 0.7$$
$$R_B = 14.3/0.0268 = 534 \text{ k}\Omega$$

A common-collector bias current is shown in Fig. 11-25*a*. The bias circuit is reduced by Thévenin's theorem to the circuit given in Fig. 11-25*b*. The loop voltage equation around the base-emitter circuit is

$$V'_{BB} = R_B I_B + V_{BE} + R_E I_E \tag{11-24}$$

Using the numerical values given in the circuit of Fig. 11-25*a*,

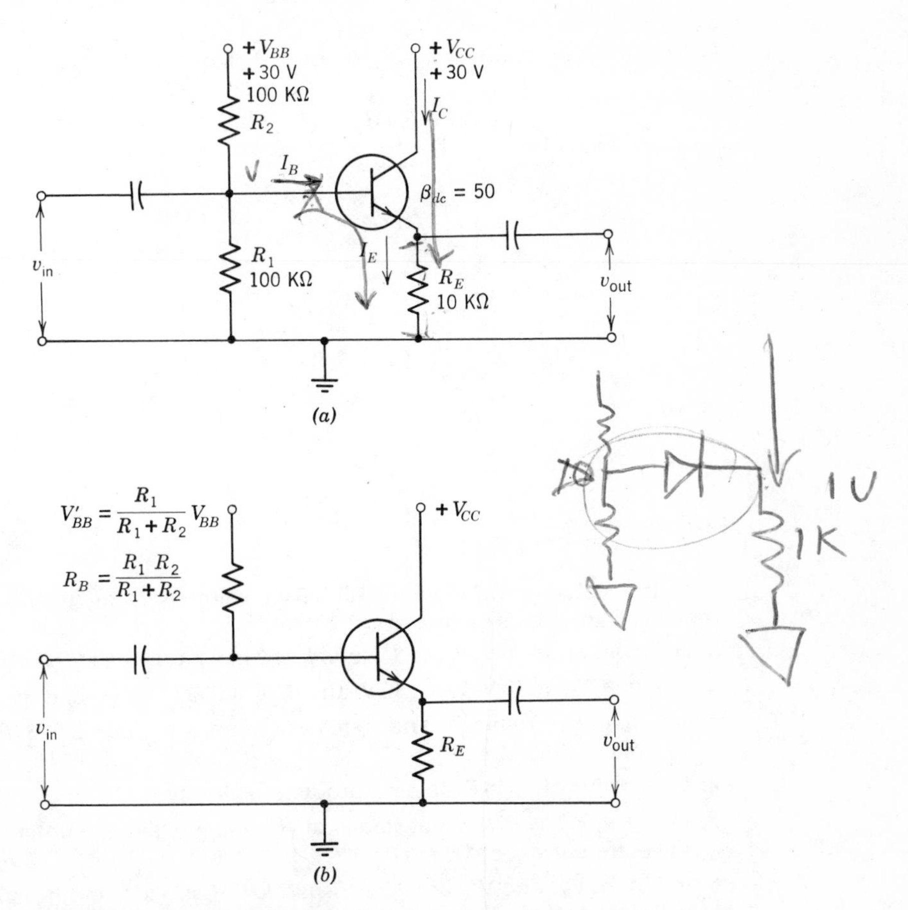

Figure 11-25 Common collector bias circuit. (*a*) Actual circuit. (*b*) Reduced circuit.

$$V'_{BB} = \frac{R_1}{R_1 + R_2} V_{BB} = \frac{100}{100 + 100} 30 = 15 \text{ V}$$

and

$$R_B = \frac{R_1 R_2}{R_1 + R_2} = \frac{100 \times 100}{100 + 100} = 50 \text{ k}\Omega$$

The voltage loop equation using numerical values is

$$\begin{aligned} 15 &= 50 I_B + 0.7 + 10 I_E \\ 15 &= 50 I_B + 0.7 + 10(1 + \beta_{dc}) I_B \\ 15 &= 50 I_B + 0.7 + 10(1 + 50) I_B \\ 560 I_B &= 14.3 \\ I_B &= 0.0256 \text{ ma} = 25.6 \ \mu\text{a} \end{aligned}$$

Using the conversion equation to obtain I_E from I_B,

$$I_E = (1 + \beta_{dc}) I_B = (1 + 50) 0.0256 = 1.3 \text{ ma}$$

and

$$R_E I_E = 10 \times 1.3 = 13 \text{ V}$$

Since

$$\begin{aligned} V_{CE} &= V_{CC} - R_E I_E, \\ V_{CE} &= 30 - 13 = 17 \text{ V} \end{aligned}$$

PROBLEMS

All transistors are silicon and all have a value of β_{dc} of 70. The value for V_{CE} when the transistor is saturated is 0.20 volt.

1. In Fig. *a*, V_{CC} is 10 V, V_{CE} is 3 V, and I_C is 2 ma. Find R_B and R_c.
2. In Fig. *a*, V_{CC} is 20 V, V_{CE} is 10 V, and R_c is 100 kΩ. Find I_C and R_B.
3. In Fig. *a*, determine R_B and R_c for a Q-point of V_{CE} equal to 3 V and I_C equal to 2 ma. V_{CC} is 8 V.
4. What value of R_B in Problem 2 produces saturation?
5. In Fig. *a*, V_{CC} is 6 V. What values of R_B and R_c give the optimum bias for a collector current of 1.5 ma?
6. In Fig. *b*, V_{BB} and V_{CC} are 20 V. R_E is 2 kΩ. R_c is 4 kΩ, and R_B is 750 kΩ. Find V_{CE} and I_C.

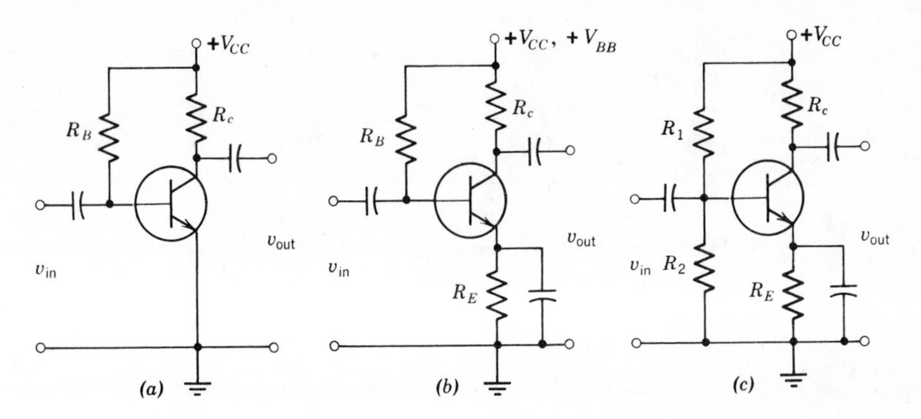
+V_{CC}
R_B
R_c
v_{in}
v_{out}
(a)
+V_{CC}, +V_{BB}
R_B
R_c
R_E
v_{in}
v_{out}
(b)
+V_{CC}
R_1
R_2
R_c
R_E
v_{in}
v_{out}
(c)

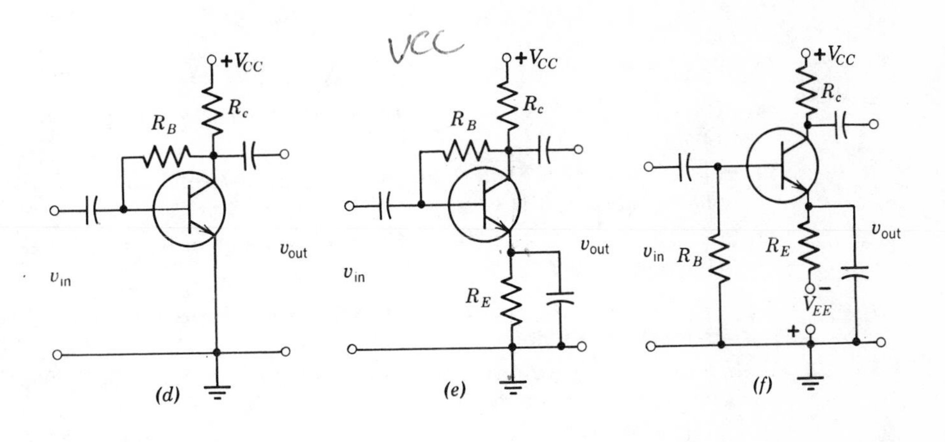
+V_{CC}
R_B
R_c
v_{in}
v_{out}
(d)
VCC
+V_{CC}
R_B
R_c
R_E
v_{in}
v_{out}
(e)
+V_{CC}
R_c
R_B
R_E
V_{EE}
v_{in}
v_{out}
(f)

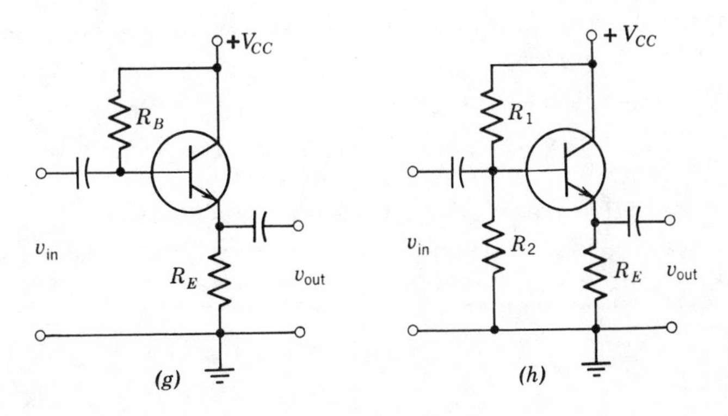
+V_{CC}
R_B
R_E
v_{in}
v_{out}
(g)
+ V_{CC}
R_1
R_2
R_E
v_{in}
v_{out}
(h)

7. In Fig. *b*, V_{BB} and V_{CC} are 10 V. R_E is 2 kΩ, R_c is 4 kΩ, and R_B is 750 kΩ. Find V_{CE} and I_C.
8. In Fig. *b*, V_{BB} is 24 V, and V_{CC} is 45 V. R_E is 5 kΩ, R_c is 8 kΩ, and R_B is 270 kΩ. Determine I_C and V_{CE}.
9. What value of R_B will give the optimum bias in Problem 7?
10. What value of R_B will give the optimum bias in Problem 8?
11. What value of R_B will saturate the transistor in Problem 7?
12. What value of R_B will saturate the transistor in Problem 8?
13. In Fig. *c*, V_{CC} is 10 V, I_C is 1 ma, and V_{CE} is 3 V. R_E is 1500 Ω and R_2 is 100 kΩ. Find R_c and R_1.
14. In Fig. *c*, V_{CC} is 20 V, R_c is 5 kΩ, R_E is 2 kΩ, R_1 is 70 kΩ, and R_2 is 30 kΩ. Find I_C and V_{CE}.
15. In Fig. *c*, V_{CC} is 4 V, R_E is 1 kΩ, R_c is 15 kΩ, R_1 is 150 kΩ, and R_2 is 50 kΩ. Find V_{CE}.
16. In Fig. *c*, V_{CC} is 10 V, R_E is 2 kΩ, R_c is 3 kΩ, and R_1 and R_2 are each 200 kΩ. Find I_C and V_{CE}.
17. In Fig. *d*, V_{CC} is 10 V and R_c is 10 kΩ. Find the value of R_B that makes V_{CE} 5 V.
18. In Fig. *d*, V_{CC} is 10 V and R_c is 10 kΩ. Find the value of R_B that saturates the transistor.
19. In Fig. *d*, V_{CC} is 6 V. Find R_c and R_B that establish the optimum *Q*-point for a collector current of 0.6 ma.
20. In Fig. *e*, V_{CC} is 10 V and R_E is 2 kΩ. What values of R_c and R_B yield a value of V_{CE} equal to 6 V.
21. In Fig. *e*, V_{CC} is 12 V and R_E is 1500 Ω. What values of R_c and R_B establish the optimum *Q*-point?
22. In Fig. *e*, V_{CC} is 12 V and R_E is 1800 Ω. What values of R_c and R_B make V_{CE} 3 V at an I_C equal to 2.0 ma?
23. In Fig. *f*, V_{CC} is +18 V and V_{EE} is −4 V. R_E is 2000 Ω and R_c is 4000 Ω. What value of R_B establishes an operating current of 1.5 ma for I_C? What is V_{CE}?
24. In Fig. *f*, V_{CC} is +12 V and V_{EE} is −5 V. R_E is 1 kΩ, R_B is 50 kΩ, and R_c is 2.4 kΩ. What are I_C and V_{CE}?
25. In Fig. *g*, V_{CC} is 10 V and R_E is 500 Ω. Determine R_B for optimum bias.
26. In Fig. *g*, V_{CC} is 20 V, R_E is 1 kΩ, and R_B is 80 kΩ. Determine V_{CE} and I_C.
27. What value of R_B will make V_{CE} 8 V for the transistor of Problem 26?
28. In Fig. *h*, V_{CC} is 12 V, R_E is 1500 Ω, and R_2 is 30 kΩ. Determine R_1 to make V_{CE} 6 V.
29. What value of R_1 makes V_{CE} 4V for the transistor in Problem 28?
30. In Fig. *h*, V_{CC} is 15 V, R_1 is 300 kΩ and R_2 is 120 kΩ. R_E is 1200 Ω. Determine I_C and V_{CE}.

Section 11-9 Analytic Approach to Load Lines

The common-emitter amplifier shown in Fig. 11-26*a* uses two power supplies, −6 V for V_{CC} and +4 V for V_{EE}. The bias resistor R_B is varied over a wide range, and the d-c voltages from collector to ground and from emitter to ground are measured. The values are plotted on the graph shown in Fig. 11-26*b*.

At point A, R_B is so low that the transistor is completely saturated. If

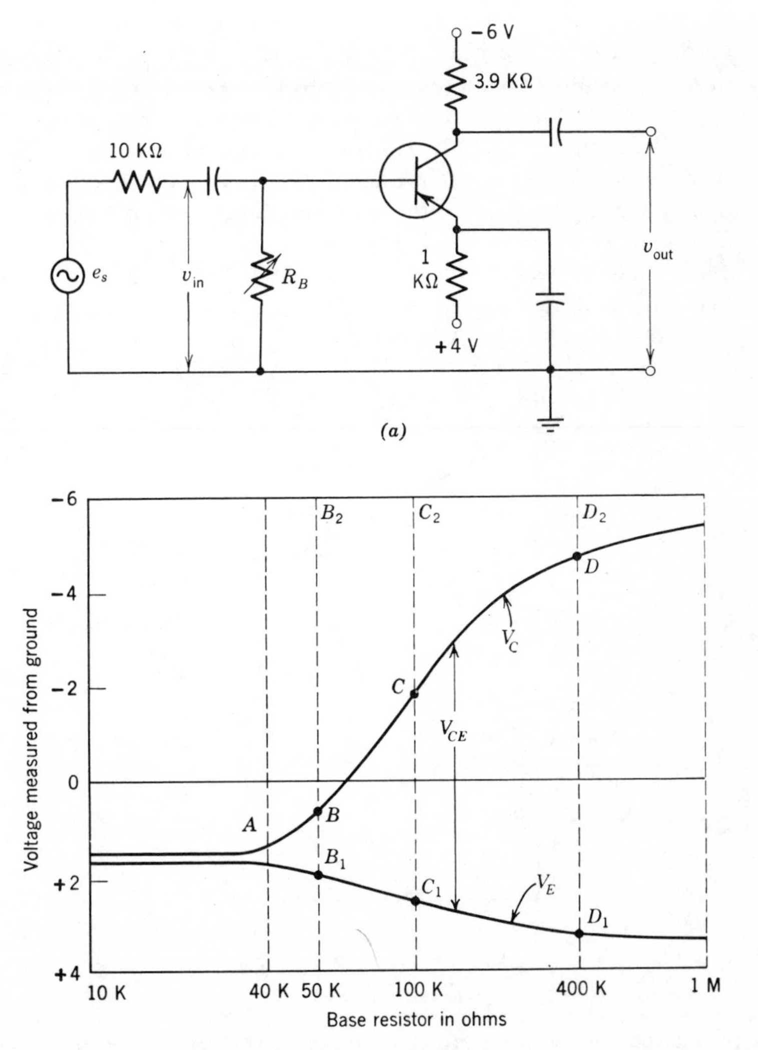

Figure 11-26 Effect of varying the bias of an amplifier stage. (*a*) Circuit. (*b*) Voltage relationships.

R_B is lowered still further, I_B increases toward a destructive value in a diode action without changing the collector current. Obviously, at point A no amplification of the signal occurs and v_{out} merely indicates a waveform that shows a half-wave rectification of the input signal. At operating point B, an output signal can be obtained. The maximum output signal without clipping is given by the distance measured in volts from B to B_1. When the operating point is at D, the maximum output signal is determined by the available swing from D to D_2.

At operating point C, the signal can be increased until simultaneous clipping occurs at points C_1 and C_2. This is the optimum bias for the stage if the largest available peak-to-peak output swing is desired. Optimum operation, then, is the Q point that establishes V_{CE} equal to $I_C R_c$.

In a series circuit where R_E is zero and the bias is obtained solely from R_B (for example, Fig. 11-5a) the optimum operating point is merely the center of the load line as shown in Fig. 11-27a. V_{CE} can be taken as $\frac{1}{2}V_{CC}$, and I_C is one half the y-intercept of the load line. All calculations for the circuit can be done easily by using a Kirchhoff voltage loop equation.

In a circuit where an emitter resistor R_E is included, as in Fig. 11-26a, a load line is drawn for $(R_c + R_E)$, using $(|V_{CC}| + |V_{EE}|)$ as the total supply voltage (Fig. 11-27b). Assuming that I_C and I_E are equal, a portion of the

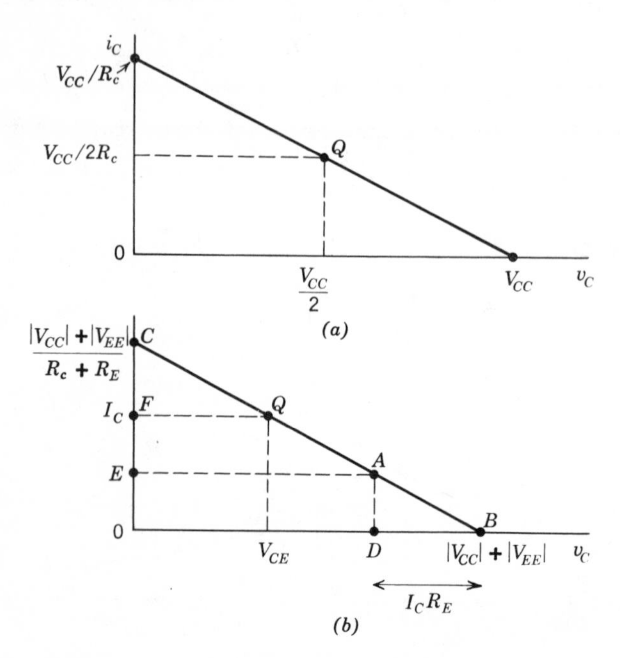

Figure 11-27 Analytic determination of operating point. (a) Circuit where R_E is zero. (b) Circuit with R_E.

load line from A to B must be used for the $I_C R_E (I_E R_E)$ voltage drop. Then the optimum bias point is located in the center of the remaining part of the load line between A and C. From the geometry of the figure,

$$OC = I_C + FC$$

The end point of the load line is

$$OC = \frac{|V_{CC}| + |V_{EE}|}{R_c + R_E}$$

The condition of optimum bias requires that

$$EF = FC$$

Then

$$OC = I_C + EF$$

From the geometry of the figure, triangles ABD and CBO are similar. By the ratio of the sides of similar triangles

$$\frac{DA}{OC} = \frac{DB}{OB}$$

Then

$$DA(= OE) = OC\frac{DB}{OB} = \frac{|V_{CC}| + |V_{EE}|}{R_c + R_E} \frac{I_c R_E}{|V_{CC}| + |V_{EE}|} = \frac{R_E}{R_c + R_E} I_C$$

From the figure

$$EF = I_C - OE = I_C - \frac{R_E}{R_c + R_E} I_C = \frac{R_c}{R_c + R_E} I_C$$

Now substituting back into

$$OC = I_C + EF$$

$$\frac{|V_{CC}| + |V_{EE}|}{R_c + R_E} = I_C + \frac{R_c}{R_c + R_E} I_C$$

Clearing of fractions and solving for I_C

$$I_C = \frac{|V_{CC}| + |V_{EE}|}{2R_c + R_E} \tag{11-25}$$

PROBLEMS

1. Solve Problem 9 of Section 11-8 by means of Eq. 11-25 and a graphical approach.
2. Solve Problem 10 of Section 11-8 by means of Eq. 11-25 and a graphical approach.

Questions

1. Define or explain each of the following terms: (*a*) load line, (*b*) *Q* point, (*c*) swing, (*d*) voltage gain, (*e*) current gain, (*f*) power gain, (*g*) collector dissipation, (*h*) dynamic transfer curve, (*i*) fixed bias, (*j*) self-bias, (*k*) amplitude distortion, (*l*) frequency distortion, (*m*) phase distortion, (*n*) total harmonic distortion, (*o*) cutoff, (*p*) saturation.
2. How is the slope of the load line related to the load resistance?
3. What are the end points of a load line?
4. How is the *Q* point varied along a load line?
5. Is a dynamic transfer characteristic linear? Explain.
6. What is the purpose of the bypass capacitor in a self-bias circuit? What determines the value of the capacitor used?
7. Explain what each of the following symbols indicates in a transistor amplifier circuit: i_b, I_b, i_B, i_c, i_C, I_c, I_C, v_{be}, V_{BE}, v_{CB}, V_{CE}, V_{EE}, V_{CC}.
8. Explain what each of the following symbols indicates in a field effect transistor amplifier circuit: I_G, V_{GG}, V_{GS}, I_S, i_s, i_d, i_D, I_D, V_{DD}, V_{DS}, v_{ds}, v_{DS}.
9. Explain what each of the following symbols indicates in a vacuum-tube amplifier circuit: e_g, E_g, I_b, e_b, E_{bb}, i_p, e_c, I_p, E_{cc}, E_p, i_b, E_b.
10. Explain intermodulation distortion.
11. What is the theoretical efficiency of a linear amplifier with a resistive load?
12. What is a power series?
13. How is total harmonic distortion obtained if the values for the individual harmonics are known?
14. Does the maximum power output occur at the point of minimum distortion?
15. Can amplitude distortion exist without frequency distortion?
16. When a signal is decreased from a maximum value toward zero in amplitude, what happens to the actual value of collector dissipation?
17. Show how an amplifying device converts d-c power into a-c power.
18. Show how the maximum undistorted output signal varies with bias in Fig. 11-26*b*.
19. What differentiates the optimum *Q* point from other *Q* points on a load line?

Chapter Twelve

MODELS AND SMALL SIGNAL ANALYSIS

In this chapter the use of α and β explicitly implies the a-c values α_{ac} and β_{ac}. When the d-c values of α and β are required, they are subscripted as follows: α_{dc} and β_{dc}. The concept of a-c emitter resistance, r'_e, is developed in Section 12-1. The common-base amplifier is analyzed in Section 12-2 from the viewpoint of ideal characteristics. The effects of actual deviations from the ideal characteristics are considered in Section 12-3. The fundamental common-emitter amplifier analysis (Section 12-4) is extended to cover those fundamental circuits where the bias is obtained from a resistor connected from collector to base (Section 12-5) and where the circuit has an additional unbypassed emitter resistor (Section 12-6). The model for the emitter-follower amplifier is analyzed in Section 12-7. The concept of the circuit model is extended to the FET (Section 12-8) and to the vacuum-tube amplifier (Section 12-9).

The conversions given in the conversion table must be assimilated thoroughly.

Section 12-1 Emitter Resistance in a Transistor

The circuit for a common-base amplifier is shown in Fig. 12-1. The d-c values of currents and voltages are established by using the d-c value of α following the procedure detailed in Section 11-2. The emitter current and the base currents are related by

$$I_E = (1+\beta_{dc})I_B \quad \text{and} \quad i_e = (1+\beta_{ac})i_b$$

Conversion Table

Multiplying factors to convert FROM \ TO	i_b	i_c	i_e
i_b	1	β	$(1+\beta)$
i_c	$\frac{1}{\beta}$	1	$\frac{1+\beta}{\beta}$
i_e	$\frac{1}{1+\beta}$	$\frac{\beta}{1+\beta}$	1

In the d-c analysis, the effect of the forward bias on the transistor is taken as a voltage drop V_{EB} and, as a rule of thumb based on experimental data, 0.3 V is used for the germanium and 0.7 V is used for silicon units. The a-c resistance of the emitter to base circuit is taken as the slope of the forward bias volt-ampere characteristic of a diode measured at the particular point of operation. Reference to a typical characteristic (Fig. 2-7) shows that this curve has a continuous change of slope. Accordingly, the value of this a-c resistance is a function of the d-c current flow, and the first procedure in analyzing a transistor circuit is to evaluate the d-c current I_E. Either from a mathematical analysis or from experimental determination, it can be shown that at room temperature, the a-c value of resistance r'_e lies between the

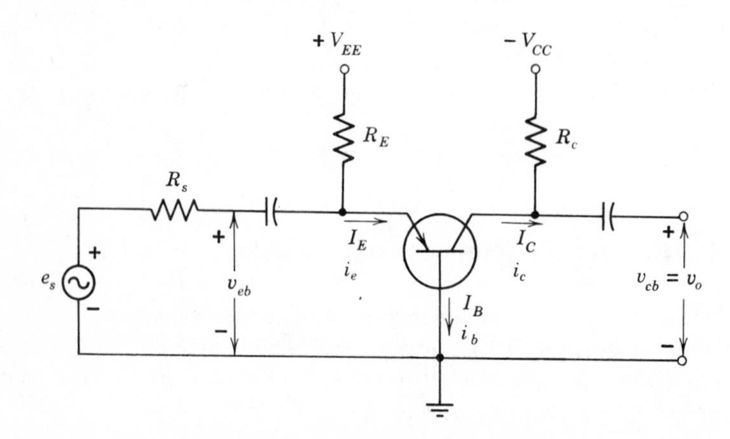

Figure 12-1 Common-base amplifier.

limits given by

$$\frac{25 \text{ mv}}{I_E} \leq r'_e \leq \frac{50 \text{ mv}}{I_E} \tag{12-1}$$

An alternative symbol for r'_e is h_{ib}.

When amplifiers are operated at low frequencies, this resistance can be assumed to be wholly within the emitter of the transistor, and the a-c resistance in the base itself can be considered to be zero. When amplifiers are treated at high frequencies, we must modify this concept slightly.

PROBLEM

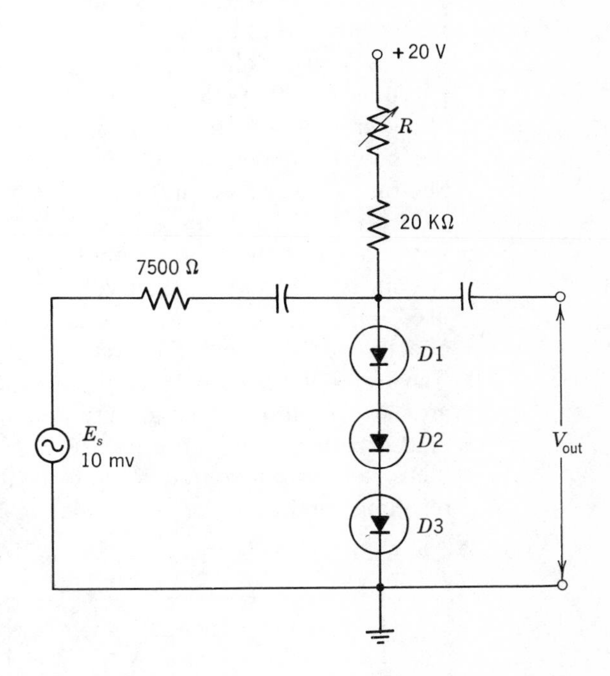

1. Three diodes are connected in series to serve as a variable attenuator. Assume that the d-c forward voltage drop across each diode is 0.5 V and that the a-c resistance of each diode is given by 25 mv/I. R is a potentiometer. Determine V_{out} when R is set to 0 Ω, to 80 kΩ, to 230 kΩ, to 500 kΩ, and to 2 MΩ. Neglecting the impedance-level correction terms, what is the loss of the attenuator in decibels for each given setting of the potentiometer?

Section 12-2 The Common-Base Amplifier Model

In this chapter we are concerned only with the operation of the circuit at small signal levels. Explicitly, distortion is to be avoided, and the operation of the circuit is within the linear part of the load line. The only need for a d-c analysis is to establish I_E from which r'_e is obtained by Eq. 12-1.

A very important concept in electronics is the function of the "ground" in Fig. 12-1. The "ground" is simply the zero-voltage reference point from which all other voltages are measured. The voltage measured between "ground" and the terminal marked $+V_{EE}$ is the voltage of the emitter supply, which is a pure d-c voltage. Likewise, the voltage measured between "ground" and the terminal marked $-V_{CC}$ is the voltage of the collector supply, which is again an ideal d-c voltage. No alternating current is involved. This means that *the a-c potential of the supply voltages is zero.* V_{EE} and V_{CC} are the voltages developed by one of the power supplies using a capacitor, an L, or a π filter, described in Chapter 6. In the power supply, relatively large values of capacitance are used to filter out or to bypass the a-c ripple present in the supply. This capacitance causes the output lead to be at a-c ground potential. In this and in succeeding chapters, we assume that the power supplies are ideal; that is, they have zero a-c impedance. In Section 20-11, we discuss the problem that exists when this is not the case.

An a-c model is desired for the circuit given in Fig. 12-1. The a-c model is to contain only those components that affect the signal operation of the circuit. All d-c currents and voltages are excluded. The capacitors that are used to block d-c currents and voltages are not included as long as they are effective a-c short circuits. Accordingly, in Fig. 12-1, the capacitors are not transferred to the model, and the points of application of the supply voltages are assumed to be at a-c ground potential. The a-c emitter resistance, r'_e, is placed between emitter and base. The physics of the transistor shows that there is a current source in the collector that is α (a-c alpha) times i_e. The resulting model is shown in Fig. 12-2. Based on our knowledge of the

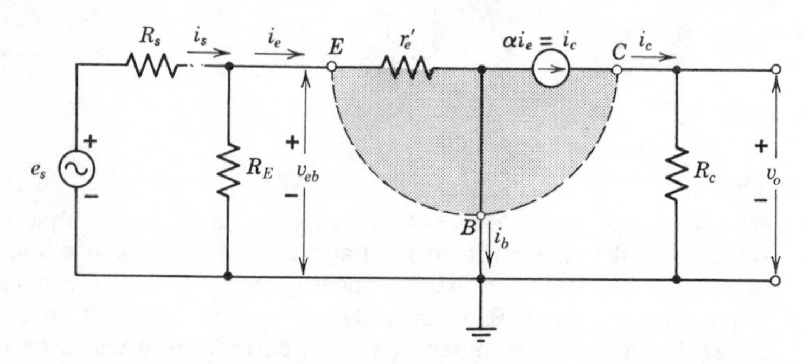

Figure 12-2 Model for common-base amplifier.

operation of the transistor, the current elements that are contained within the shaded half-circle *E-B-C* effectively describe the operation of the transistor as far as a signal is concerned. The polarity across e_s determines that i_e must be a current into the transistor. From the direction of i_e, the directions of i_b and i_c are automatically determined.

The gain of the transistor stage itself, disregarding the source resistance R_s and the bias resistor R_E, is

$$A_v = \frac{v_0}{v_{eb}} \tag{12-2}$$

For an inspection of the model, it is evident that

$$v_{eb} = i_e r'_e \qquad \text{and} \qquad v_0 = i_c R_c = \alpha i_e R_c$$

Substituting,

$$A_v = \frac{\alpha i_e R_c}{i_e r'_e} = \alpha \frac{R_c}{r'_e} \tag{12-3a}$$

Usually, α is very close to unity, with the result that Eq. 12-3a can be approximated as

$$A_v \approx \frac{R_c}{r'_e} \tag{12-3b}$$

The current gain for the circuit is

$$A_i = \alpha \approx 1 \tag{12-3c}$$

Multiplying, the a-c power gain is

$$A_p = \alpha^2 \frac{R_c}{r'_e} \approx \frac{R_c}{r'_e} \tag{12-3d}$$

To obtain the overall gain A_e, which is defined as the ratio of v_0/e_s, the effects of R_E and R_s must be considered. The bias resistor R_E in the a-c model is in parallel with r'_e. This parallel combination is r:

$$r = \frac{r'_e R_E}{r'_e + R_E} \tag{12-4a}$$

Now r and R_s form a voltage divider across the source e_s, and the portion of e_s that is across r is v_{eb}. From the voltage divider rule,

$$v_{eb} = \frac{r}{r + R_s} e_s$$

This expression is now written in the form:

$$\frac{v_{eb}}{e_s} = \frac{r}{r + R_s}$$

If both sides of this equation are multiplied by Eq. 12-2, we have

$$\frac{v_{eb}}{e_s} \times \frac{v_0}{v_{eb}} = \frac{r}{r + R_s} A_v$$

By canceling and substituting Eq. 12-3*a* for A_v, we have the overall gain equation for A_e:

$$A_e = \frac{v_0}{e_s} = \left(\frac{r}{r + R_s}\right)\left(\frac{\alpha R_c}{r'_e}\right) \qquad (12\text{-}4b)$$

The first bracketed expression is a *gain reduction factor* that must be less than unity. This gain reduction factor is due to the presence of an internal resistance R_s in the source. When the source is ideal, R_s is zero. This gain reduction factor is then unity, and Eq. 12-4*b* becomes Eq. 12-3*b*. Usually, R_E is very much larger than r'_e, and the parallel combination of R_E and r'_e is simply r'_e. Then, r'_e is substituted for r in Eq. 12-4*b* to yield

$$A_e = \left(\frac{r'_e}{r'_e + R_s}\right)\left(\frac{\alpha R_c}{r'_e}\right) \qquad (12\text{-}4c)$$

The first bracketed term is the reduction factor caused by R_s, and the second bracketed term is the gain of the stage without regard to R_s. If α is assumed to be unity, canceling the r'_e terms yields

$$A_e = \frac{R_c}{r'_e + R_s} \qquad (12\text{-}4d)$$

A comparison of Eq. 12-4*d* with Eq. 12-3*a* and Eq. 12-3*b* shows that the inclusion of the source resistance R_s reduces the gain. This gain reduction can be used to advantage in utilizing the process known as *swamping*. Let us assume that the value if I_E is 1 ma and that the value of R_c is 5000 Ω. By Eq. 12-1 the range of r'_e is

$$\frac{25}{1} \leqslant r'_e \leqslant \frac{50}{1} \qquad \text{or} \qquad 25\,\Omega \leqslant r'_e \leqslant 50\,\Omega$$

The exact value of r'_e can be expected to range from 25 to 50 Ω for a batch of transistors that are to be used in the production of quantities of this circuit.

Also, if a transistor is being replaced, this range variation of r'_e can be expected in the replacement. Accordingly, the gain of the circuit by Eq. 12-3*b* can range from 5000/50 to 5000/25 or from 100 to 200, which is a 100% variation in gain. If a source resistance of 450 Ω were deliberately introduced, the gain of the circuit will range from 5000/(450+50) to 5000/(450+25) or from 10 to 10.55, which is only a 5 to 6% variation. If two stages of the "swamped" design are connected in series, the overall gain is the product of the two, and the range of gain variation is from 10×10 to 10.55×10.55, which is an 11% range, instead of the 100% expected from a single stage without swamping.

By using swamping, two transistors are required to perform the gain accomplished by the use of a single transistor without swamping. On the other hand, the gain of the circuit is almost independent of the variations that are caused by different transistors. In most applications of semiconductors, it is extremely important to maintain a circuit performance that is independent of variations caused by the semiconductor element.

PROBLEMS

All transistors in the problems are germanium and have $\alpha = 0.98$. r'_e is determined by 25 mv/I_E. For each problem, draw the a-c model and label with numerical values.

1. V_{EE} and V_{CC} are each 20 V with the proper polarity. R_E is 2 kΩ and R_C is 810 Ω. Determine the voltage gain of the circuit. What is the maximum allowable value of E_s without clipping?
2. V_{EE} is +20 V and V_{CC} is −150 V. R_s and R_E are each 2 kΩ and R_c is 7.5 kΩ. R is infinite. Determine the operating point values and the voltage gain of the circuit. What is the maximum allowable value of E_s without clipping?
3. V_{EE} is +20 V and V_{CC} is −20 V. R_s is 20 Ω and R_E is 40 kΩ. R_c and R are each 20 kΩ. Determine the operating point and the voltage gain of the circuit.
4. V_{EE} is +20 V and V_{CC} is −20 V. R_E is 20 kΩ and R_c and R are each 10 kΩ. E_s is 5 mv rms and R_s varies from 0 to 1 kΩ. What are the minimum and maximum values of V_{out}?
5. E_s is 10 mv, V_{EE} is +4 V, R_E is 50 kΩ, V_{CC} is −4 V, R_c is 20 kΩ, and R is 30 kΩ. R_s is varied from 100 Ω to 1000 Ω. Determine the Q-point values of the circuit and determine the range of V_{out}.
6. Determine the value of V_{out}.

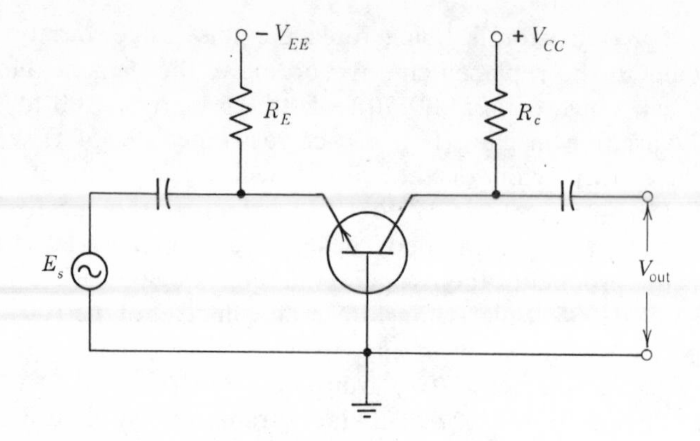

Circuit for Problem 1.

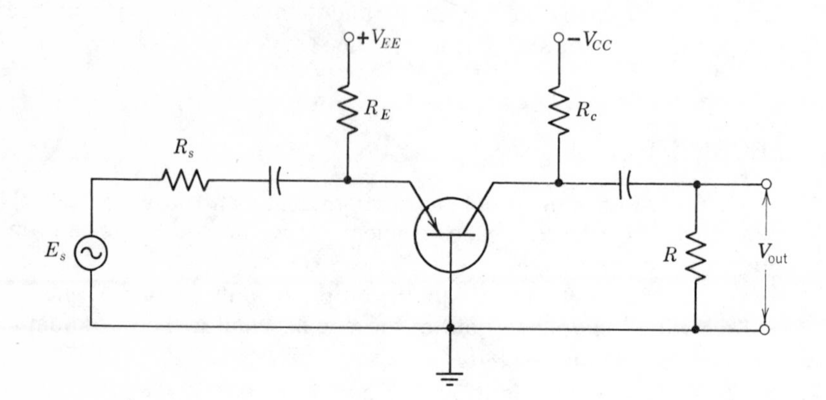

Circuit for Problems 2 to 5.

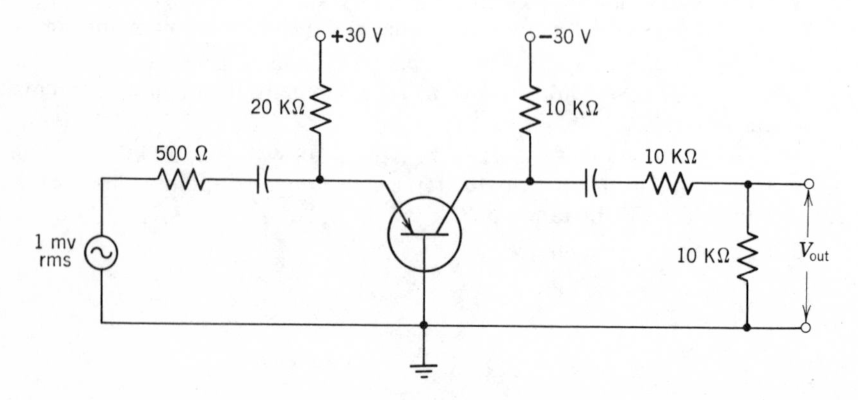

Circuit for Problem 6.

Section 12-3 Second-Order Considerations in the Common-Base Amplifier

Inspection of the model shown in Fig. 12-2 reveals that the input resistance to the amplifier itself is r'_e or h_{ib}. The subscripts on h_{ib} mean that "i" is the input and that the "b" refers to a common-base circuit connection. Similarly, $-h_{fe}$ is used to designate α in the hybrid parameter system of notation. The "f" designates the current amplification factor in the forward direction.

The collector characteristics of the common-base connection are not exactly flat but show a very slight rise as V_{CB} is increased (Fig. 3-7). The slope of the curves can be determined by

$$r_0 \equiv \frac{1}{h_{ob}} \equiv \frac{\Delta V_{CB}}{\Delta I_C} \text{ for a fixed } I_E \tag{12-5}$$

This resistance value is the reciprocal of the slope of the collector curves. A typical value of r_0 is 2 MΩ making h_{ob} 5×10^{-7} mho where h_{ob} is the output conductance in the common-base connection. The model is redrawn in Fig. 12-3 to include this output resistance. Evidently, r_0 is in parallel with R_c because the right side of r_0 goes to the collector and the left side of r_0 goes to the internal junction point which is ground through the base connection. If r_0 is of the order of a megohm or more, normally it is neglected. If, however, R_c is increased to a very high value, it would require using the value of the parallel combination of r_0 and R_c for R_c in Eq. 4-3 and Eq. 4-4.

When the value of R_c is increased, the gain of the stage increases directly as the gain is directly proportional to the value of R_c in any of the gain equations. On the other hand, when R_c is increased, the values of I_C and I_E decrease. Accordingly, by Eq. 4-1, r'_e will increase. If these factors were linear and proportional, the ratio R_c/r'_e would be constant yielding a constant gain for different values of R_c. All of these considerations and gain formulas assume that α is constant. However, reference to Fig. 3-8 shows that α is effectively constant only over a range of values of I_C and I_E. When the I_C and I_E fall off to very low values, α falls off toward zero. Also, at large values of I_C and I_E, α falls off.

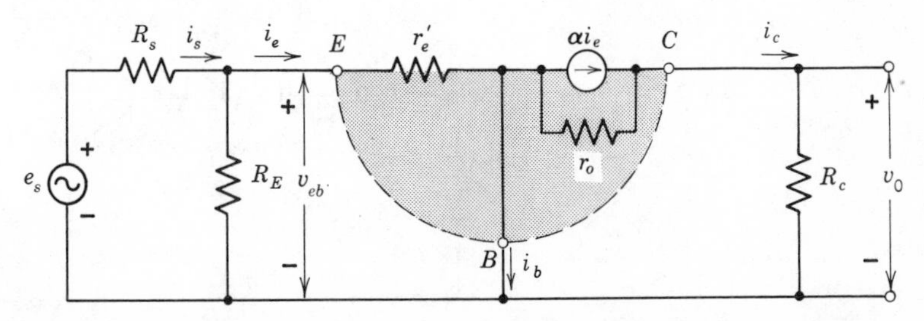

Figure 12-3 Model for common-base amplifier including output resistance.

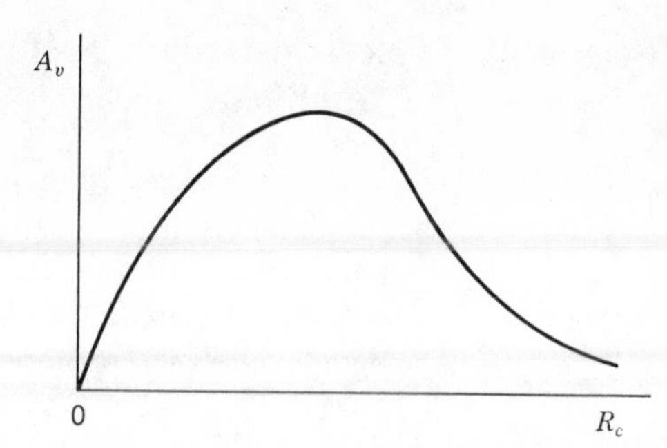

Figure 12-4 Variation of gain with R_c.

The result of the combination of these effects is to produce the gain curve shown in Fig. 12-4. The objective of good amplifier design is to locate the operation of the amplifier reasonably well up on the curve.

Section 12-4 The Common-Emitter Amplifier

A simple common-emitter circuit is shown in Fig. 12-5. The model (Fig. 12-6) can be drawn quite simply by rearranging the terminals of the transistor model as shown in Fig. 12-2. The current directions are again established by the relative polarity assumed for e_s. The value of r'_e is obtained from the d-c analysis explained in Section 12-1 (Eq. 12-1). The voltage v_{be} is directly the voltage drop across r'_e:

$$v_{be} = i_e r'_e$$

But i_e is related to i_b by

$$i_e = (1+\beta) i_b$$

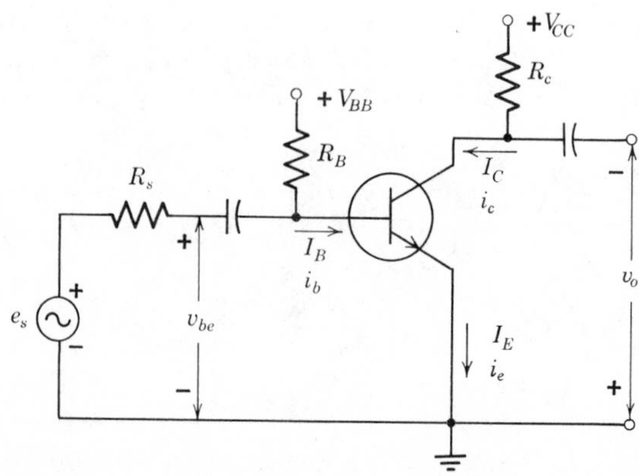

Figure 12-5 Common-emitter amplifier.

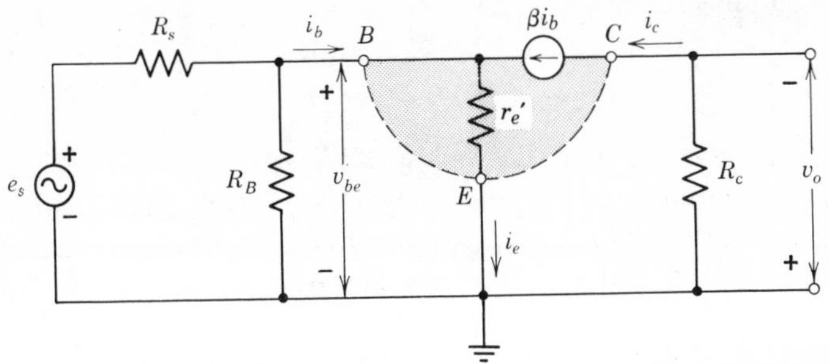

Figure 12-6 Model for common-emitter amplifier.

substituting

$$v_{be} = (1+\beta)i_b r_e'$$

If both sides of this equation are divided by i_b, v_{be}/i_b is the resistance, r_{in}, that is placed across v_{be} looking into the transistor

$$r_{\text{in}} = (1+\beta)r_e' \equiv h_{ie} \tag{12-6}$$

h_{ie} is the hybrid-parameter symbol used to designate the input resistance into a transistor connected in the common-emitter mode.

The input resistance to the common-base circuit was simply r_e'. If r_e' is 50 Ω and if the β of the transistor is 100, a numerical comparison of the input resistances of the two circuits can be made

$$h_{ib} = 50\ \Omega$$

and

$$h_{ie} = (1+100)50 = 5050\ \Omega$$

An immediate advantage of the common-emitter circuit over the common-base circuit is apparent. The input resistance values are much larger and the effect of R_s is much less on the circuit. Also, if h_{ie} is of the same order as R_c, it is not too great a problem to cascade common-emitter stages. Unless matching transformers are used, it is very difficult, if not impossible, to cascade common-base amplifiers. As a result, common-base circuits are usually reserved for power amplifiers where the longer load-line linearity of the common-base amplifier improves the operating efficiency.

We showed that the transistor input voltage v_{be} is

$$v_{be} = (1+\beta)i_b r_e'$$

The output voltage v_0 is

$$v_0 = -R_c i_c$$

To determine the gain, we ignore the negative sign but remember that this amplifier circuit configuration produces a 180° phase shift between the output and the input. This phase inversion can be observed from the relative polarities of e_s and v_0 in the model. The magnitude of the gain, A_v, taken from the input of the transistor to the output, is

$$A_v = \frac{v_0}{v_{be}} = \frac{R_c i_c}{(1+\beta) i_b r'_e}$$

Since $i_c = \beta i_b$,

$$A_v = \frac{\beta R_c}{(1+\beta) r'_e} = \frac{\beta R_c}{h_{ie}} \tag{12-7a}$$

Since β is normally much larger than one, $\beta/(1+\beta)$ is almost unity and the gain can be approximated as

$$A_v \approx \frac{R_c}{r'_e} \tag{12-7b}$$

For example, if β is 100, $(1+\beta)$ is 101 and $\beta/(1+\beta)$ is 100/101. When this is taken as unity, the error is 1%.

The voltage gain of this circuit is the same as the voltage gain obtained from the common-base circuit by Eq. 12-3*b*. However, the current gain now is not α which is less than unity but is β which is large:

$$A_i = \beta \tag{12-7c}$$

And the power gain is

$$A_p = \frac{\beta^2 R_c}{h_{ie}} = \frac{\beta^2 R_c}{(1+\beta) r'_e} \approx \beta \frac{R_c}{r'_e} \tag{12-7d}$$

The overall circuit gain A_e (v_0/e_s) is dependent on R_s and is determined in the same manner as for the common-base circuit. In this circuit, r_{in} is h_{ie} and r_{in} is in parallel with R_B in the model. This parallel combination is r given by

$$r = \frac{r_{\text{in}} R_B}{r_{\text{in}} + R_B} \tag{12-8a}$$

R_s and r form a voltage divider across e_s. The a-c voltage into the transistor is

$$v_{be} = \frac{r}{r+R_s} e_s$$

or

$$\frac{v_{be}}{e_s} = \frac{r}{r+R_s}$$

The gain A_e is obtained by multiplying this equation by Eq. 12-7a

$$\frac{v_{be}}{e_s} \times \frac{v_0}{v_{be}} = \frac{r}{r+R_s} \times A_v$$

to give

$$A_e \equiv \frac{v_0}{e_s} = \left(\frac{r}{r+R_s}\right)\left(\frac{\beta R_c}{h_{ie}}\right) \approx \left(\frac{r}{r+R_s}\right)\left(\frac{R_c}{r'_e}\right) \quad (12\text{-}8b)$$

Here, again, the first bracketed term is a gain reduction factor caused by R_s, and the second bracketed term is the gain of the amplifier stage itself without considering R_s. If R_B is much greater than h_{ie}, the parallel combination of R_B and h_{ie} is h_{ie}. Then, r reduces to r_{in} which is h_{ie}, and Eq. 12-8b becomes

$$A_e = \left(\frac{h_{ie}}{h_{ie}+R_s}\right)\left(\frac{\beta R_c}{h_{ie}}\right) \approx \left(\frac{h_{ie}}{h_{ie}+R_s}\right)\left(\frac{h_{ie}}{r'_e}\right) \quad (12\text{-}8c)$$

in which the first bracketed term is the gain reduction factor. If the h_{ie} terms are canceled, Eq. 12-8c reduces to

$$A_e = \frac{\beta R_c}{h_{ie}+R_s} \quad (12\text{-}8d)$$

Consider the circuit shown in Fig. 12-7a which is the same as the circuit given in 12-5 but with the addition of an external emitter resistor R_E. If this resistor were bypassed adequately with a capacitor C_E, the only effect of R_E would be in the determination of the value of I_E (and r'_e) by the methods of the last chapter. If this resistor is not bypassed, it appears in the a-c model, and the model given in Fig. 12-7b must include the resistor R_E between the terminal E and ground.

When this is done, the voltage v_b is measured across both r'_e and R_E in series:

$$v_b = (r'_e + R_E)i_e = (1+\beta)(r'_e + R_E)i_b$$

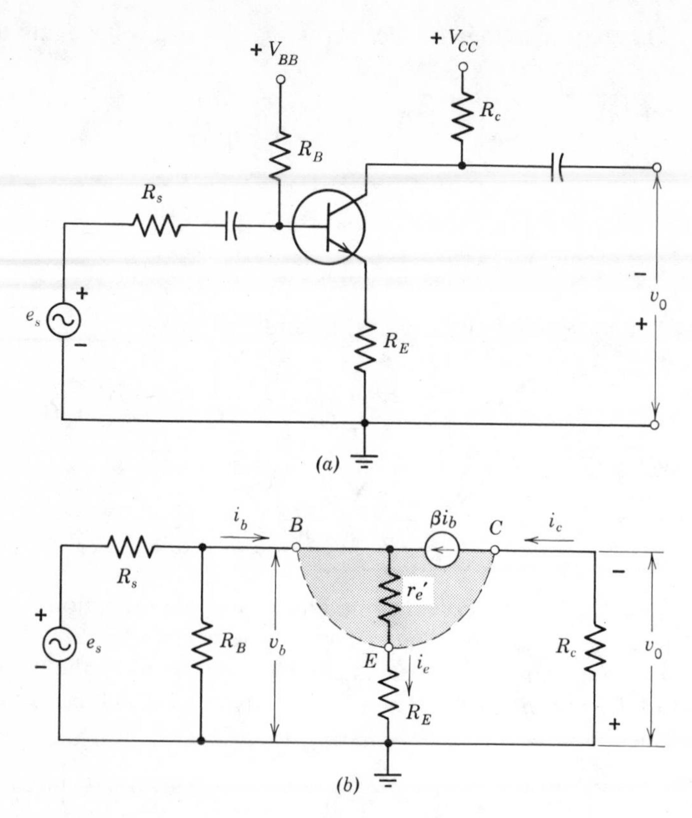

Figure 12-7 Common-emitter amplifier with an external emitter resistor. (*a*) Circuit. (*b*) Model.

The input resistance becomes

$$r_{\text{in}} = \frac{v_b}{i_b} = (1+\beta)(r_e'+R_E) = (1+\beta)r_e' + (1+\beta)R_E = h_{ie} + (1+\beta)R_E \tag{12-9}$$

Thus, an unbypassed emitter resistor can be used to raise materially the input resistance to a semiconductor amplifier. Using the values from the previous circuit of r_e' equal to 50 Ω and β equal to 100, let us assume a 200-Ω resistor is placed unbypassed in the emitter circuit. Without R_E the value of r_{in} was $(1+\beta)r_e'$ or 5050 Ω. Now with R_E, r_{in} is

$$r_{\text{in}} = (1+\beta)(r_e'+R_E) = (1+100)(50+200) = 25{,}250\ \Omega$$

As in the previous case, this external resistor R_E can be used to "swamp out" variations in r_e' between transistors to result in gain stabilization.

The gain equations given by Eq. 12-8b, Eq. 12-8c, and 12-8d can be converted by substituting r_{in} for h_{ie}. Then

$$A_v = \frac{\beta R_c}{r_{\text{in}}} = \frac{\beta R_c}{(1+\beta)(r_e'+R_E)} \approx \frac{R_c}{r_e'+R_E} \tag{12-10}$$

The value of R_B in parallel with r_{in} is r and the voltage divider action of R_s and r is accounted for by a gain reduction factor as before:

$$A_e = \left(\frac{r}{r+R_s}\right)\left(\frac{\beta R_c}{(1+\beta)(r_e'+R_E)}\right) \approx \left(\frac{r}{r+R_s}\right)\left(\frac{R_c}{r_e'+R_E}\right) \tag{12-11a}$$

Multiply this last gain equation by h_{ie}/h_{ie}. Then

$$A_e = \left(\frac{r}{r+R_s}\right)\left(\frac{h_{ie}}{h_{ie}}\right)\left(\frac{\beta R_c}{(1+\beta)\,(r_e'+R_E)}\right) \tag{12-11b}$$

Rearranging the order of the terms

$$A_e = \left(\frac{r}{r+R_s}\right)\left(\frac{h_{ie}}{(1+\beta)(r_e'+R_E)}\right)\left(\frac{\beta R_c}{h_{ie}}\right) = \left(\frac{r}{r+R_s}\right)\left(\frac{r_e'}{r_e'+R_E}\right)\left(\frac{\beta R_c}{h_{ie}}\right) \tag{12-11c}$$

The first bracketed term is the gain reduction factor caused by R_s. The second bracketed term is the gain reduction factor caused by the unbypassed emitter resistor R_E. The third bracketed term is the basic gain equation for the common-emitter amplifier.

When $(1+\beta)$ is taken as being effectively equal to β, Eq. 12-11c reduces to the approximation

$$A_e \approx \left(\frac{r}{r+R_s}\right)\left(\frac{r_e'}{r_e'+R_E}\right)\left(\frac{R_c}{r_e'}\right) \approx \left(\frac{r}{r+R_s}\right)\left(\frac{R_c}{r_e'+R_E}\right) \tag{12-11d}$$

The numerical value of r is determined by the circuit components, and whether r can be reduced to r_{in} can only be determined by comparing the numerical values of r_{in} and R_B. Therefore, the first bracketed term in Eq. 12-11c and in Eq. 12-11d cannot be simplified further if these equations are to be valid for all circuits as a general circuit gain equation.

PROBLEMS

All transistors are silicon, and r_e' is obtained from 25 mv/I_E unless specified. The a-c model is required for all problems.

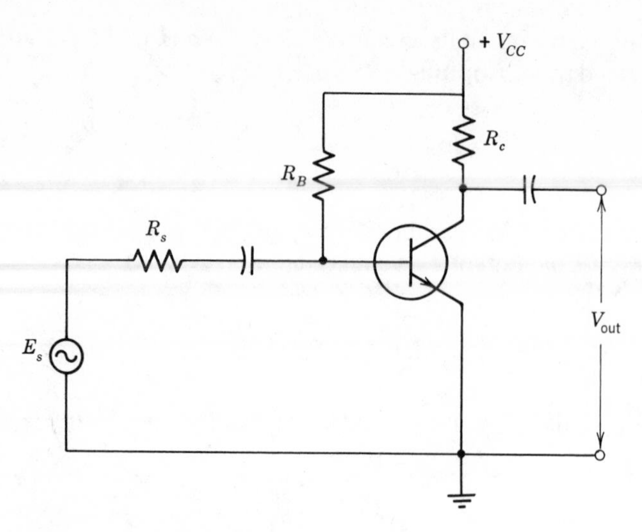

Circuit for Problems 1, 2, and 3.

1. R_s is 3600 Ω, R_B is 80 kΩ, and R_c is 3000 Ω. For the transistor, r_e' is 20 Ω and β is 100. Determine A_e, A_v, and the resistance that E_s sees.
2. Repeat Problem 1 if β is 50.
3. R_s is 600 Ω, R_B is 75 kΩ, R_c is 2.0 kΩ, V_{CC} is +7.5 V, and β is 20. Determine the voltage gain of the circuit.

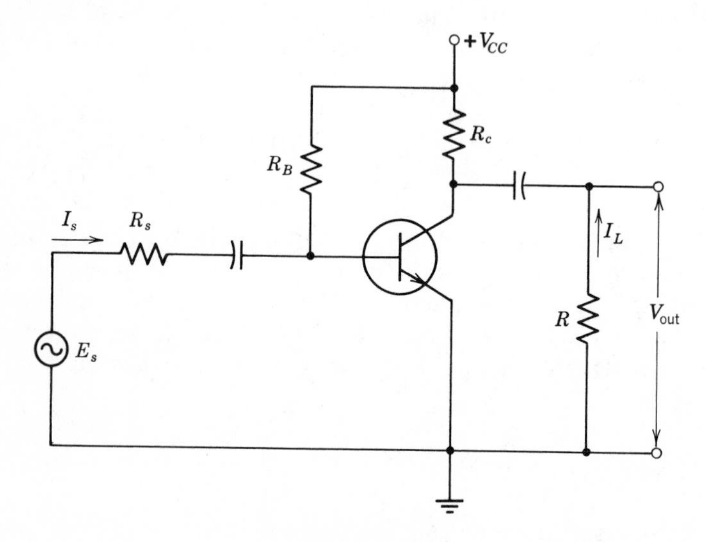

Circuit for Problem 4.

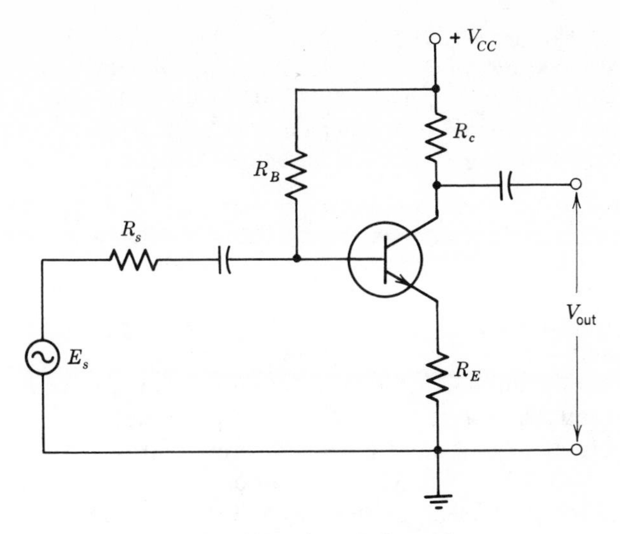

Circuit for Problems 5, 7, and 8.

4. R_s is 600 Ω, R_c is 4.7 kΩ, R is 4.3 kΩ, V_{CC} is +9.0 V, and β is 40. Determine the value of R_B for optimum bias that yields maximum peak-to-peak output voltage. Determine the maximum allowable value of E_s without clipping and determine the current gain I_L/I_s.
5. R_s is 10 kΩ, R_c is 2000 Ω, R_E is 75 Ω, V_{CC} is +4 V, and β is 40. R_B is adjusted to set I_C at 1 ma. Determine the load resistance on the source and determine the voltage gain of the circuit.
6. R_s is 4.3 kΩ, R_B is 68 kΩ, R_c and R are each 1000 Ω, R_E is 100 Ω, V_{CC} is +4.5 V, and β is 50. Determine the voltage gain of the circuit.

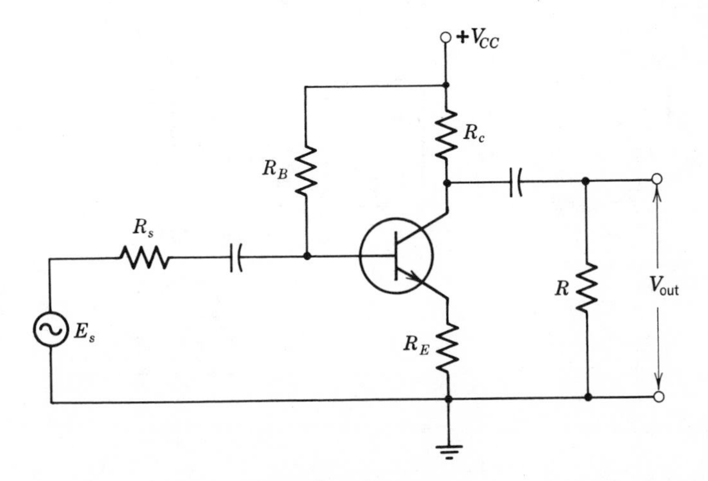

Circuit for Problem 6.

7. R_s is 10 kΩ, R_c is 12 kΩ, R_E is 3 kΩ, V_{CC} is +8 V. Determine R_B to provide maximum peak-to-peak output voltage swing and determine the value of E_s that provides this swing. The value of β for the transistor is 60.
8. Repeat Problem 7 assuming that the emitter resistor R_E is adequately bypassed with a capacitor C_E at the signal frequency.

Section 12-5 The Common-Emitter Amplifier Using Collector-to-Base Feedback

In the circuit shown in Fig. 12-8, the d-c operating bias is obtained through the feedback resistor, R_B, connected from collector to base. The value of r_e' is determined by the d-c analysis presented in the last chapter. The a-c model is given in Fig. 12-9. This model has an additional current, i_L, to account for the current division through R_B and the collector.

As in the previous circuit, the base current is determined from

$$v_{be} = r_e' i_e = (1+\beta)\, r_e' i_b = h_{ie} i_b$$

or

$$i_b = \frac{v_{be}}{(1+\beta)r_e'} = \frac{v_{be}}{h_{ie}} \tag{12-12}$$

The load current i_L, by Ohm's law, is simply

$$i_L = v_0/R_c$$

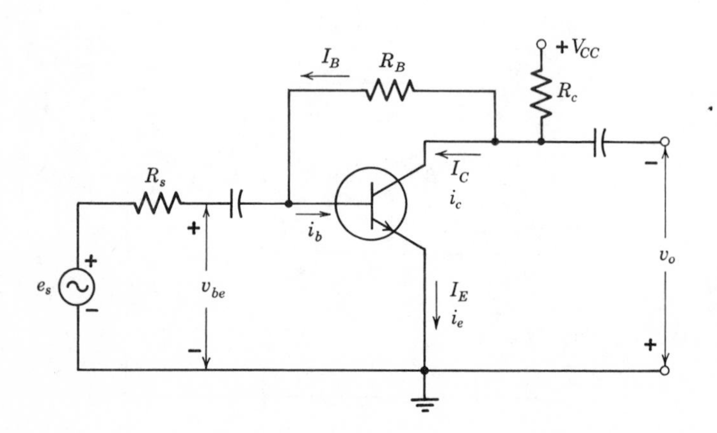

Figure 12-8 Common-emitter amplifier with bias derived from a feedback resistor.

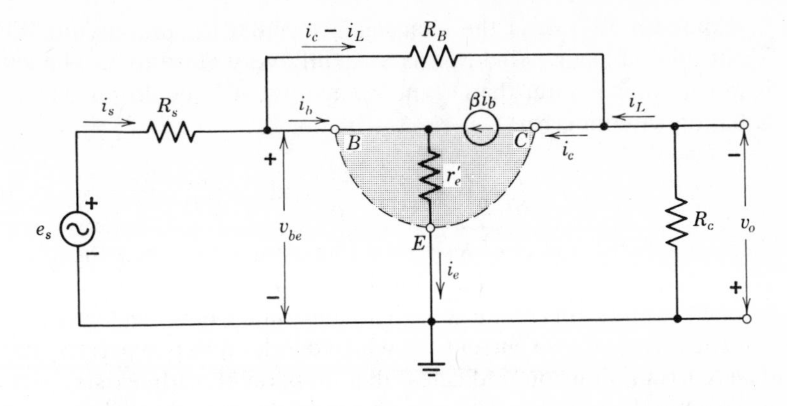

Figure 12-9 Model for the common-emitter amplifier of Figure 12-8.

The voltage across R_B is the *sum* of v_{be}, and v_0 because of the phase inversion of the amplifier and the current in R_B is

$$i_c - i_L = i_c - \frac{v_0}{R_B} = \frac{v_{be} + v_0}{R_B}$$

but $i_c = \beta i_b$ and, substituting Eq. 12-12 for i_b,

$$i_c = \beta \frac{v_{be}}{h_{ie}}$$

Then

$$\frac{\beta}{h_{ie}} v_{be} - \frac{1}{R_c} v_0 = \frac{1}{R_B} v_{be} + \frac{1}{R_B} v_0$$

Rearranging

$$\left(\frac{\beta}{h_{ie}} - \frac{1}{R_B}\right) v_{be} = \left(\frac{1}{R_B} + \frac{1}{R_c}\right) v_0$$

and the gain of the circuit becomes

$$A_v = \frac{v_0}{v_{be}} = \left(\frac{\beta}{h_{ie}} - \frac{1}{R_B}\right) \Big/ \left(\frac{1}{R_B} + \frac{1}{R_c}\right)$$

simplifying

$$A_v = \left(\frac{\beta R_B - h_{ie}}{h_{ie} R_B}\right) \frac{R_B R_c}{R_B + R_c} = \left(\frac{\beta R_B - h_{ie}}{R_B + R_c}\right) \frac{R_c}{h_{ie}}$$

$$A_v = \left(\frac{R_B - h_{ie}/\beta}{R_B + R_c}\right) \frac{\beta R_c}{h_{ie}} \tag{12-13a}$$

Equation 12-13a is the exact gain formula for the circuit. When h_{ie}/β is considered as $(1+\beta)r_e'/\beta$, it is evidently very close to r_e'. The value of R_B is much much greater than r_e' and r_e' can obviously be dropped from the gain equation without error:

$$A_v = \left(\frac{R_B}{R_B + R_c}\right)\frac{\beta R_c}{h_{ie}} \approx \left(\frac{R_B}{R_B + R_c}\right)\frac{R_c}{r_e'} \qquad \text{(12-13}b\text{)}$$

Accordingly, the effect of R_B on the transistor is to reduce the gain from that given by the basic gain equation by a gain reduction factor $R_B/(R_B+R_c)$. In most circuits, this factor is close to unity and can be neglected.

The signal source current is i_s which divides into two parts, i_b and $(i_c - i_L)$. The current division indicates that a parallel path exists, one through r_e' and one through R_B. The impedance of the path through r_e' is reflected to the source as h_{ie} or $(1+\beta)r_e'$.

Now the effect of the path through R_B must be determined. The current through R_B is

$$i_c - i_L = \frac{v_{be} + v_0}{R_B}$$

The fact that v_{be} and v_0 are additive can be observed from an inspection of the model. The phase inversion of the signal results in v_{be} and v_0 being additive rather than subtractive across R_B.

The output voltage in terms of A_v is

$$v_0 = A_v\, v_{be}$$

and substituting into the expression for $i_c - i_L$:

$$i_c - i_L = \frac{v_{be} + A_v\, v_{be}}{R_B} = \frac{(1+A_v)\, v_{be}}{R_B}$$

Then the impedance of this branch is

$$\frac{v_{be}}{i_c - i_L} = \frac{R_B}{1 + A_v}$$

It is important to stress that, as far as the input is concerned, R_B is reduced by the factor $(1+A_v)$. Thus, a seemly large value of R_B can be reduced to the point where it is of the same order as, or even less, than, h_{ie}. Then, as far as the signal is concerned, R_B can be regarded as a resistance of magnitude $R_B/(1+A_v)$ placed in parallel with h_{ie}.

In Fig. 12-10a, the action of the whole circuit between v_{be} and v_0 is

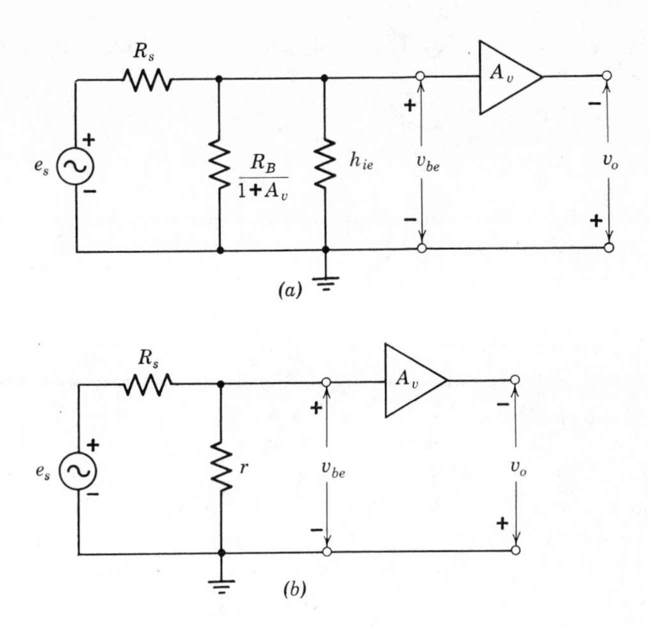

Figure 12-10 Model for the input to circuit of Figure 12-8.

represented by a symbol for an amplifier that has a gain A_v. The value of A_v is established by Eq. 12-13*a* or Eq. 12-13*b*. The resistance given by $R_b/(1+A_v)$ is shown as a shunt path in parallel with h_{ie}. These two parallel resistances can be combined to yield the single resistance r, shown in Fig. 12-10*b*:

$$r = \frac{h_{ie}R_B/(1+A_v)}{h_{ie}+R_B/(1+A_v)} = \frac{h_{ie}R_B}{(1+A_v)h_{ie}+R_B} \tag{12-14}$$

Now e_s and v_{be} can be related by a simple voltage divider

$$v_{be} = \frac{r}{R_s+r}e_s$$

and the output voltage is

$$v_0 = A_v v_{be} = \frac{r}{R_s+r}A_v e_s$$

Dividing, the gain of the whole circuit becomes

$$A_e = \frac{e_0}{e_s} = \left(\frac{r}{R_s+r}\right)A_v \tag{12-15a}$$

where A_v is given by Eq. 12-13a or Eq. 12-13b. Substituting in Eq. 12-13b,

$$A_e = \left(\frac{r}{R_s + r}\right)\left(\frac{R_B}{R_B + R_c}\right)\frac{\beta R_c}{h_{ie}} \approx \left(\frac{r}{R_s + r}\right)\left(\frac{R_B}{R_B + R_c}\right)\frac{R_c}{r'_e} \tag{12-15b}$$

Here we have two reduction factors – one created by the presence of a source resistance R_s, and the other caused by the feedback effect of R_B on the amplifier. As noted before, this last factor is usually unity and as such is neglected.

It must be understood that in order to obtain the reduced value of R_B that is reflected into the input $R_B/(1+A_v)$, it is necessary to obtain the gain first by means of Eq. 12-13, which does not account for R_s.

The analysis and models of transistor circuits use r'_e which is a parameter of the common-base circuit arrangement. Likewise, when it is necessary to consider a-c collector resistance, the parameter that is used is r_{ob} or h_{ob}. If it is necessary to consider h_{ob} in the circuit arrangement of Fig. 12-5, the method of analysis of this section must be used wherein $1/h_{ob}$ is substituted for R_B. If h_{ob} must be considered in the circuit of Fig. 12-8, a new value R'_B must be used in the equations which is the parallel combination of R_B and $1/h_{ob}$.

A quick approach to this circuit is to consider that as far as the load on the collector is concerned, from an examination of Fig. 12-8, R_B is effectively returned to a-c ground. When this is assumed, R_B and R_c are in parallel, and the load on the collector is not R_c but the parallel combination:

$$\frac{R_B R_c}{R_B + R_c}$$

When this load resistance is used in the basic gain equation for R_c, the gain of the stage is

$$A_v = \frac{\beta\left(\dfrac{R_B R_c}{R_B + R_c}\right)}{h_{ie}} = \frac{\beta R_B R_c}{h_{ie}(R_B + R_c)}$$

This equation is rearranged to

$$A_v = \left(\frac{R_B}{R_B + R_c}\right)\left(\frac{\beta R_c}{h_{ie}}\right)$$

This is Eq. 12-13b. Equation 12-13b was obtained by neglecting h_{ie}/β. Returning R_B to ground does the same thing.

PROBLEMS

The a-c model is required for all problems.

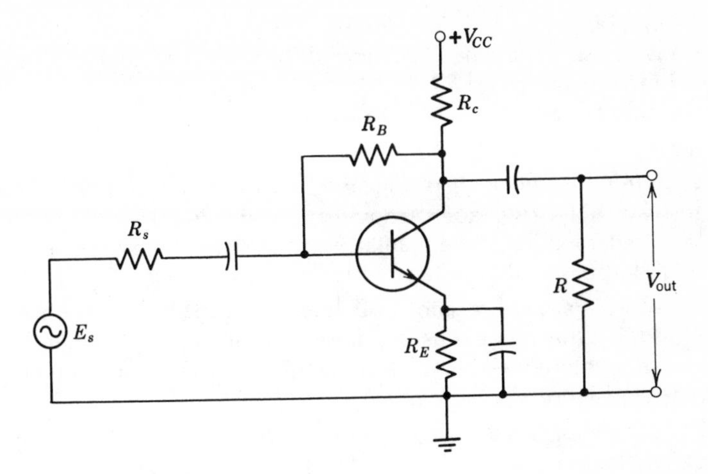

The transistor is silicon and r_e' is determined from 50 mv/I_E. Circuit for Problems 1 to 6.

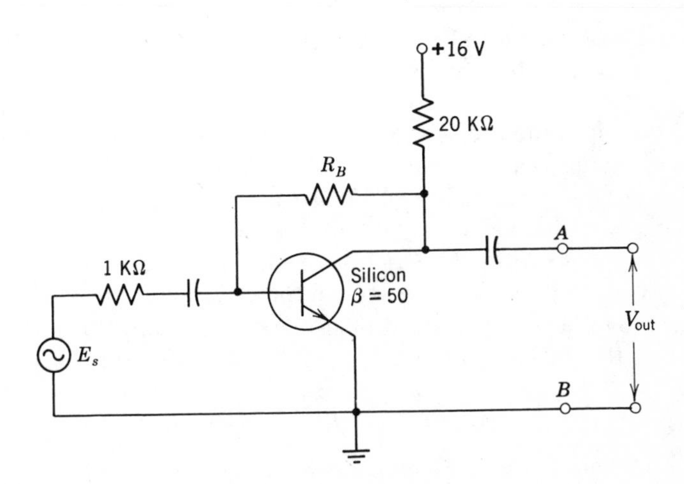

Circuit for Problems 7 to 9.

1. R_s is 1000 Ω, R_B is 1 MΩ, R_c is 10 kΩ, V_{CC} is 12 V, and R_E is 1000 Ω. R is infinite and the value of β is 100. Determine the load on the source and find the voltage gain of the circuit.

2. Repeat Problem 1 for R equal to 6.2 kΩ.
3. E_s is 1 mv and R_s is 4.7 kΩ. R_B is 750 kΩ and R_c and R are each 47 kΩ. Assume that r'_e is 100 Ω. β for the transistor varies between 40 and 100. What is the variation in V_{out}? R_E is adequately bypassed.
4. Repeat Problem 3 if R_s is reduced to 470 Ω.
5. Recalculate Problem 1, Section 12-4, assuming that the transistor has a value of 10 μmhos for h_{ob}.
6. Recalculate Problem 3, assuming that the transistor has a value of 1 μmho for h_{ob}.
7. Determine R_B when R_B is adjusted to provide a maximum undistorted peak-to-peak output voltage. Assume r'_e is 25 mv/I_E. Determine the value of E_S required to drive the amplifier when it delivers the maximum undistorted peak-to-peak output voltage. What is the overall circuit gain?
8. An additional load resistor of a value of 15 kΩ is placed between terminals A and B. What value must R_B have to establish an optimum Q-point for the circuit? Determine the value of E_s that develops a maximum peak-to-peak output voltage.
9. The transistor has a value of 2 μmhos for h_{ob}. How does this affect the results of Problem 1?

Section 12-6 The Common-Emitter Amplifier with Collector Feedback and Emitter Feedback

The circuit in the previous section (Fig. 12-8) is now modified by the addition of an unbypassed resistor in the emitter circuit. Two different versions of the circuit are given in Fig. 12-11*a* and *b*. The d-c analysis that establishes the bias and the value of r'_e differs for the two circuits, but the model (Fig. 12-11*c*) applies without restriction to either circuit.

The same procedure in deriving the equations is used as in the last section. Now, however, instead of using r'_e, the sum of $(r'_e + R_E)$ is used. One result is that the input impedance to the transistor base is now

$$r'_{in} = h_{ie} + (1+\beta)R_E = (1+\beta)(r'_e + R_E)$$

Also the gain reduction factor caused by the action of R_E must be included. The development of the final equations is the same as in the previous section, to yield

$$A_v = \left(\frac{R_B - r'_{in}/\beta}{R_B + R_c}\right)\left(\frac{r'_e}{r'_e + R_E}\right)\left(\frac{\beta R_c}{h_{ie}}\right) \tag{12-16a}$$

r'_{in}/β is approximately $(r'_e + R_E)$ which is negligible compared to R_B, making

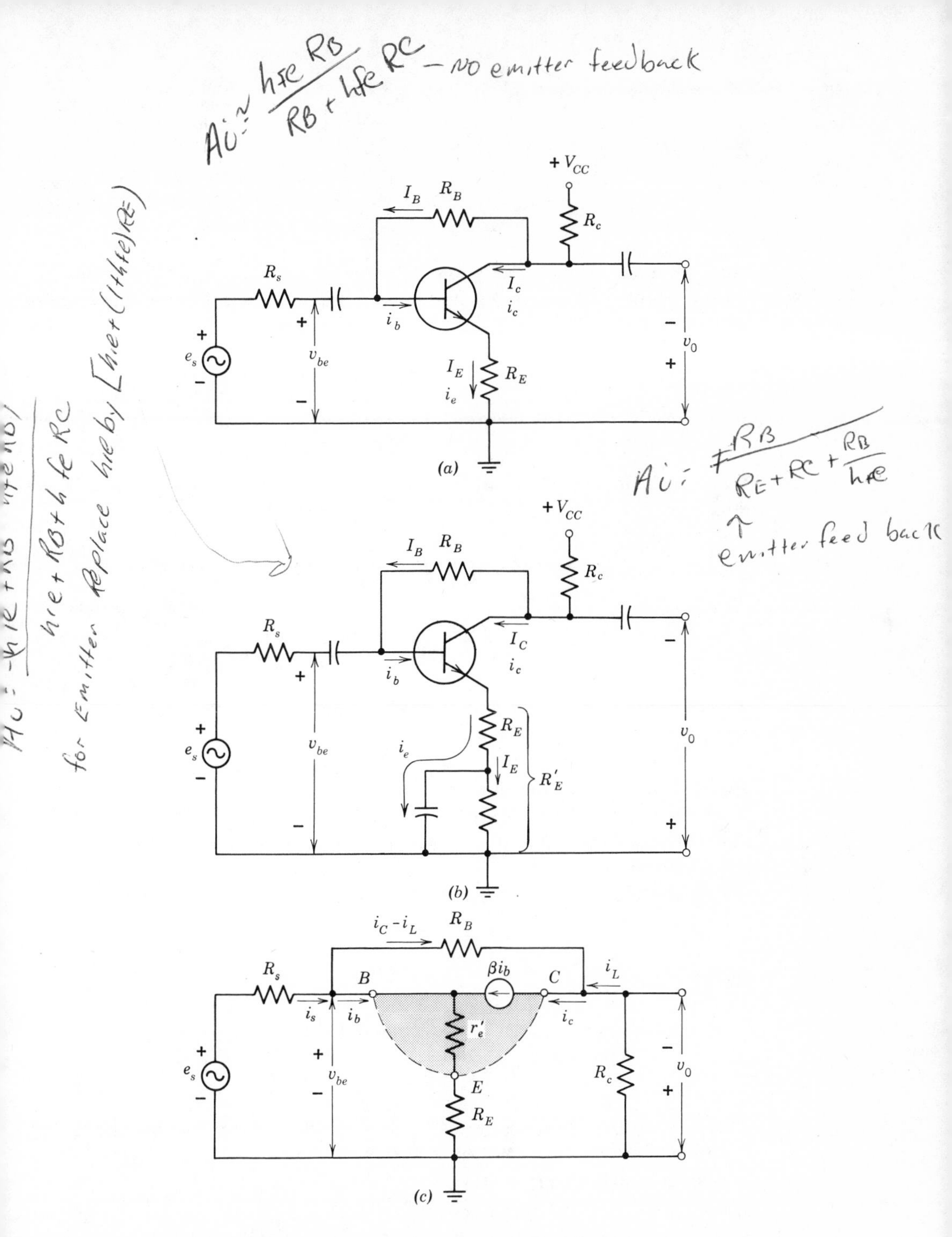

Figure 12-11 Common-emitter circuit using both collector feedback and emitter feedback. (*a*) and (*b*) Circuits. (*c*) Model.

Eq. 12.16*a*

$$A_v = \left(\frac{R_B}{R_B+R_c}\right)\left(\frac{r'_e}{r'_e+R_E}\right)\left(\frac{\beta R_c}{h_{ie}}\right) \tag{12-16b}$$

Now we have two gain reduction factors. The first is caused by the collector-to-base feedback resistor R_B, and the second is the effect of the unbypassed emitter resistor R_E. When the last bracketed term is reduced by assuming that β equals $(1+\beta)$, the r'_e terms cancel, reducing Eq. 12-16*b* to

$$A_v = \left(\frac{R_B}{R_B+R_c}\right)\left(\frac{R_c}{r'_e+R_E}\right) \tag{12-16c}$$

When the effect of R_s is considered, a third gain reduction factor is introduced:

$$A_e = \left(\frac{r}{r+R_s}\right)\left(\frac{R_B}{R_B+R_c}\right)\left(\frac{r'_e}{r'_e+R_E}\right)\left(\frac{\beta R_c}{h_{ie}}\right)$$

$$\approx \left(\frac{r}{r+R_s}\right)\left(\frac{R_B}{R_B+R_c}\right)\left(\frac{R_c}{r'_e+R_E}\right) \tag{12-16d}$$

The value of r is obtained from finding the parallel combination of $(1+\beta)(r'_e+R_E)$ and $R_B/(1+A_v)$.

In most circuits, R_B is very much greater than R_c and the factor $R_B/(R_B+R_c)$ is considered unity.

If the actual circuit is that of Fig. 12-7 and if h_{ob} must be considered, the analysis presented in this section must be used. If h_{ob} is to be considered in the circuits of this section, given in Fig. 12-11, R_B must be replaced with the parallel combination of R_B and $1/h_{ob}$ before division by $(1+A_v)$.

PROBLEMS

All transistors are silicon, and r'_e is determined from 50 mv/I_E. An a-c model is required for all problems.

1. Determine the input resistance to the circuit and the voltage gain of the circuit. The β for the transistor is 50. R is infinite.
2. Repeat Problem 1 if R is 10 kΩ.
3. Repeat Problem 2 if the source resistance R_s is 2000 Ω.
4. If V_{out} is 2 V, determine E_s and I_s. β is 100.
5. Repeat Problem 4 if the resistance of the source R_s is 10 kΩ and h_{ob} is 4 μmhos.

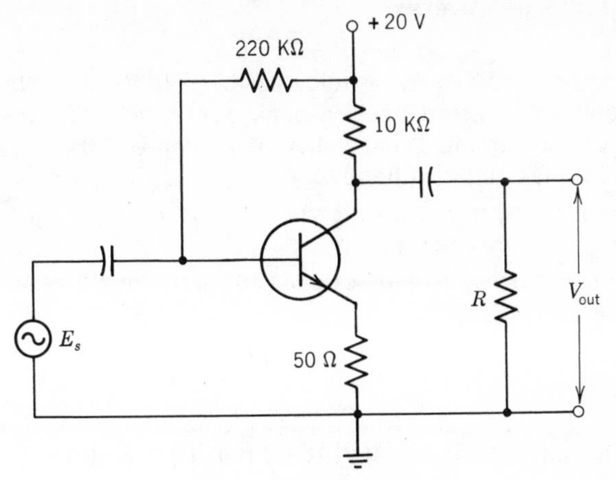

Circuit for Problems 1 to 3.

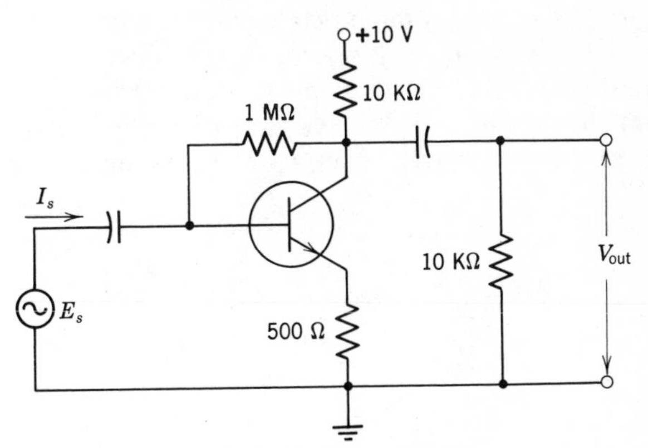

Circuit for Problems 4 and 5.

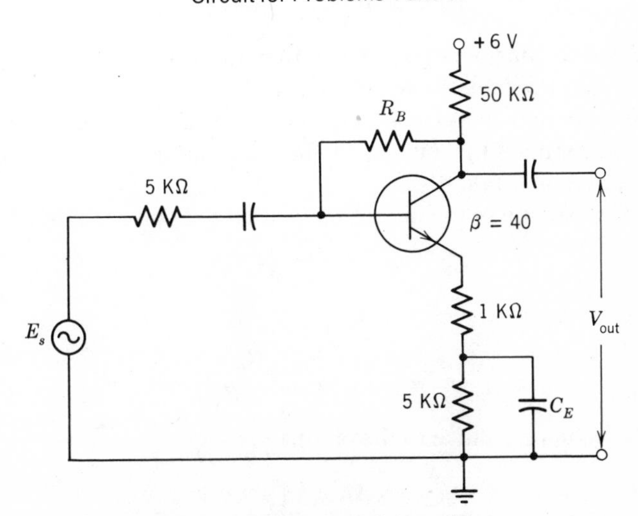

Circuit for Problems 6, 7, and 8.

6. Determine R_B when R_B is adjusted to make V_{CE} 2 V. What is the maximum available undistorted peak-to-peak output voltage? Determine the overall gain of the circuit. What value of E_s delivers the maximum undistorted peak-to-peak output voltage?
7. The emitter bypass capacitor C_E is removed from the circuit. Recalculate problem 6 for this new circuit.
8. The transistor has a value of 0.5×10^{-6} mho for h_{ob}. How does this affect the results of Problem 6?

Section 12-7 The Common-Collector or Emitter-Follower Amplifier

Two versions of the common-collector amplifier stage are given in Fig. 12-12*a* and *b*. The different arrangements affect only the d-c calculations. The model (Fig. 12-12*c*) is the same for both circuits.

From the model, we observe that v_b is the voltage from the base to ground and that v_0 is derived from r_e' and R_E, acting as a voltage divider. Then

$$v_0 = \frac{R_E}{r_e' + R_E} v_b$$

or

$$A_v = \frac{v_0}{v_b} = \frac{R_E}{r_e' + R_E} \qquad (12\text{-}17a)$$

if, and only if,

$$R_E \gg r_e' \qquad A_v \approx 1 \qquad (12\text{-}17b)$$

Usually one considers that the emitter-follower circuit has a voltage gain of 1. However, circuits in which R_E is of the same magnitude as r_e' are used. In this situation the gain is less than 1. Therefore, the size of R_E given by an inspection of the circuit is the determining factor as to whether the approximation is taken.

An inspection of the model shows that the current gain is

$$A_i = (1+\beta) \qquad (12\text{-}17c)$$

The power gain is

$$A_p = (1+\beta)\frac{R_E}{r_e' + R_E} \approx (1+\beta) \qquad (12\text{-}17d)$$

and the input resistance to the circuit r_{in} is

$$r_{in} = \frac{v_b}{i_b} = (1+\beta)(r_e' + R_E) \qquad (12\text{-}17e)$$

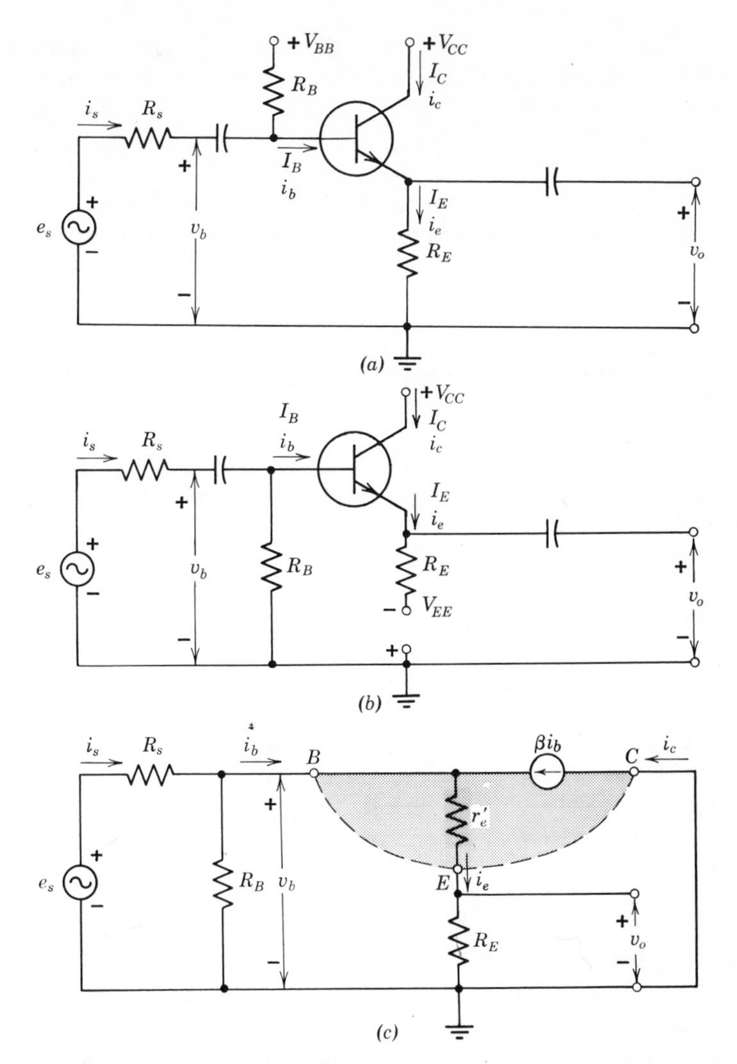

Figure 12-12 The emitter follower. (*a*) and (*b*) Actual circuits. (*c*) Model.

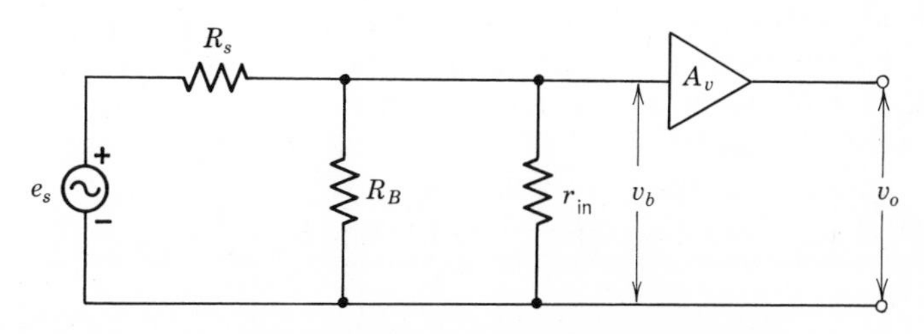

Figure 12-13 Input model for the emitter follower.

The effect of the source resistance R_s on the circuit can be observed from Fig. 12-13. A voltage divider action attenuates the input from e_s to v_b.

$$v_b = \frac{\left(\dfrac{R_B r_{\text{in}}}{R_B + r_{\text{in}}}\right)}{R_s + \left(\dfrac{R_B r_{\text{in}}}{R_B + r_{\text{in}}}\right)} e_s = \frac{R_B r_{\text{in}}}{R_B R_s + R_s r_{\text{in}} + R_B r_{\text{in}}} e_s$$

$$= \frac{1}{1 + (R_s/R_B) + (R_s/r_{\text{in}})} e_s$$

Since $v_0 = A_v v_b$ and $A_e = v_0/e_s$,

$$A_e = \frac{1}{1 + (R_s/R_B) + (R_s/r_{\text{in}})} A_v \tag{12-18a}$$

Usually in a circuit $R_B \gg R_s$ and we can make a first approximation of Eq. 12-18a as

$$A_e = \frac{1}{1 + (R_s/r_{\text{in}})} A_v$$

$$A_e = \left(\frac{r_{\text{in}}}{r_{\text{in}} + R_s}\right) A_v = \frac{r_{\text{in}}}{r_{\text{in}} + R_s} \frac{R_E}{r'_e + R_E} \tag{12-18b}$$

A particular circuit determines the values for r_{in} and R_s and these values determine if the factor $r_{\text{in}}/(r_{\text{in}} + R_s)$ can be approximated as 1. Also the particular circuit values for r'_e and R_E determine if the factor $R_E/(r'_e + R_E)$ can be approximated as 1. When both these approximations can be made, then, Eq. 12-18b becomes

$$A_e \approx 1 \tag{12-18c}$$

To determine the output impedance of the emitter-follower, a driving voltage e_d is applied across the load resistor R_E. The source voltage e_s is shorted, but the source resistance R_s is retained. The model for this analysis is given in Fig. 12-14. Inspection of the model indicates that the current i_d, produced by this driving voltage, divides into two branches i_L and i_e. Accordingly, the load on the driving source is two resistances in parallel. One of these resistances is R_E. The other resistance value is that seen looking back up into the transistor and its value must be ascertained from an analysis of the circuit.

The voltage from base to collector v_b is given by Ohm's law as the voltage drop of i_b in the parallel circuit of R_s and R_B:

$$v_b = \left(\frac{R_s R_B}{R_s + R_B}\right) i_b$$

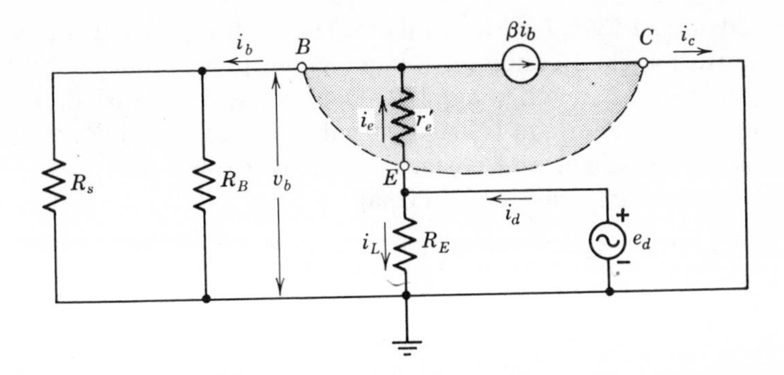

Figure 12-14 Model used to determine output impedance of the emitter follower.

Then

$$e_d = r'_e i_e + v_b = r'_e i_e + \left(\frac{R_s R_B}{R_s + R_B}\right) i_b$$

replacing i_b by $i_e/(1+\beta)$

$$e_d = r'_e i_e + \left(\frac{R_s R_B}{R_s + R_B}\right)\frac{i_e}{1+\beta} = \left[r'_e + \left(\frac{R_s R_B}{R_s + R_B}\right)/(1+\beta)\right] i_e$$

Thus the resistance r_0 looking back into the transistor is

$$r_0 = r'_e + \left(\frac{R_s R_B}{R_s + R_B}\right)\frac{1}{1+\beta} \qquad (12\text{-}19a)$$

Therefore, the resistance, r'_0, seen by e_d is the parallel combination of r_0 and R_E:

$$r'_0 = \frac{r_0 R_E}{r_0 + R_E} \qquad (12\text{-}19b)$$

Now, having derived the set of equations for the emitter follower, the operation and features of this circuit can be considered. The voltage gain of the amplifier is always less than one, but in most applications the value one can be used with negligible error. There is, however, a current gain $(1+\beta)$ that develops a substantial power gain. Consequently, the circuit is often used as an output power amplifier.

The input resistance to the circuit is high and can be made very high by the use of a large value of R_E. In this type of application, the stage can be used to approach the high values of input resistance of FET circuits.

The output resistance of the circuit is inherently low. Component values can be selected so that r_0 (Eq. 12-19*a*) can be made equal to R_E. Thus the circuit is *matched* to provide a maximum power transfer into R_E. If the impedance of a load is low, of the order of a few ohms, it is only possible to secure a matched condition for maximum power transfer by the use of a transformer when other amplifier configurations are used. Therefore, the emitter-follower circuit is usually used in a power amplifier when it is urgent to save the weight and cost of a matching transformer. Amplifiers using transformers are considered in Chapter 14.

PROBLEMS

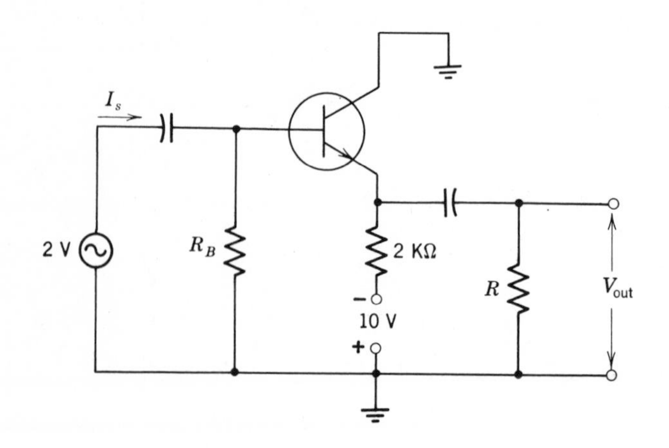

Circuit for Problems 1 and 2.

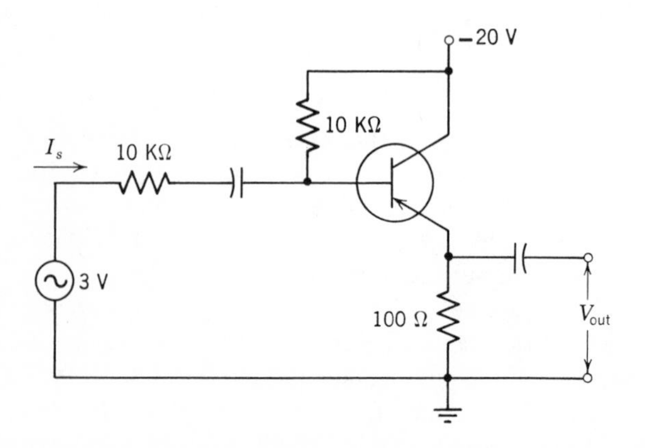

Circuit for Problem 3.

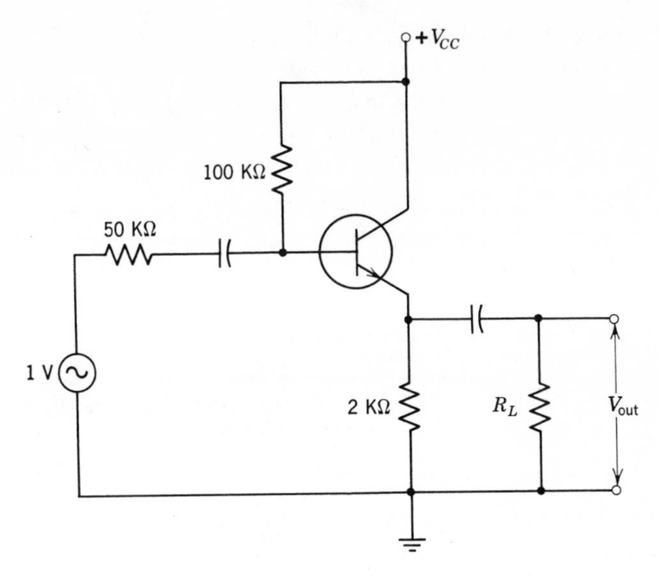

Circuit for Problem 4.

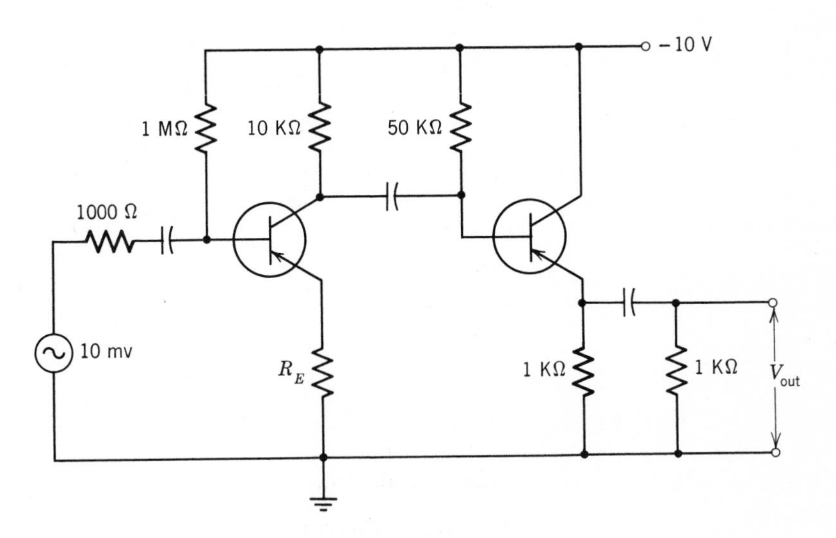

Circuit for Problems 5 and 6.

The a-c model is required for all problems.

1. R_B is selected to establish V_{CE} at 6 V. R is infinite and the value of β for the silicon transistor is 100. Find I_s and V_{out}. r_e' is 50 mv/I_E.

2. Repeat Problem 1 if R is 3000 Ω.
3. Find I_s and V_{out}. β is 50 and r'_e is 10 Ω.
4. The value of β for the transistor is 100, and r'_e is 10 Ω. R_L is adjusted for maximum power transfer into R_L. Find R_L and the voltage across R_L.
5. Determine V_{out} when R_E is zero. For each transistor, β is 50 and r'_e is 30 Ω.
6. Repeat Problem 5 using a value of 1000 Ω for R_E.

Section 12-8 The Field Effect Transistor

From Chapter 4 (Eq. 4-1) the equation for the drain current in a junction field effect transistor for the pinched-off condition is given by

$$i_D = I_{DSS}\left(1 - \frac{v_{GS}}{V_P}\right)^2 \tag{4-1}$$

The total instantaneous voltage from drain to ground is

$$v_{DS} = V_{DD} - i_D R_d = V_{DD} - R_d I_{DSS}\left(1 - \frac{v_{GS}}{V_P}\right)^2$$

But

$$v_{GS} = V_{gs} + v_{gs}$$

and

$$v_{DS} = V_{DS} + v_{ds}$$

Substituting

$$V_{DS} + v_{ds} = V_{DD} - R_d I_{DSS}\left[\left(1 - \frac{V_{GS}}{V_P}\right) - \frac{v_{gs}}{V_P}\right]^2$$

Expanding

$$V_{DS} + v_{ds} = V_{DD} - R_d I_{DSS}\left(1 - \frac{V_{GS}}{V_P}\right)^2 + \frac{2R_d I_{DSS}}{V_P}\left(1 - \frac{V_{GS}}{V_P}\right)v_{gs} - \frac{R_d I_{DSS}}{V_P^2}v_{gs}^2$$

Let us assume that v_{gs} is a cosinusoidal function

$$v_{gs} = E\cos\omega t$$

Substituting

$$V_{DS} + v_{ds} = V_{DD} - R_d I_{DSS}\left(1 - \frac{V_{GS}}{V_P}\right)^2 + \frac{2R_d I_{DSS}}{V_P}\left(1 - \frac{V_{GS}}{V_P}\right)E\cos\omega t$$
$$- \frac{R_d I_{DSS}}{V_P^2}E^2\cos^2\omega t$$

the $\cos^2 \omega t$ term must be reduced by using the trigometry identity

$$\cos^2 x = \tfrac{1}{2} + \tfrac{1}{2}\cos 2x$$

$$V_{DS} + v_{ds} = V_{DD} - R_d I_{DSS}\left(1 - \frac{V_{GS}}{V_P}\right)^2 + \frac{2R_d I_{DSS}}{V_P}\left(1 - \frac{V_{GS}}{V_P}\right)E\cos\omega t$$
$$-\frac{R_d I_{DSS}}{2V_P^2}E^2 - \frac{R_d I_{DSS}}{2V_P^2}E^2\cos 2\omega t$$

This last equation can be separated into two equations, one for the d-c circuit and one for the a-c circuit by the use of the concept of the superposition theorem:

$$V_{DS} = V_{DD} - R_d I_{DSS}\left(1 - \frac{V_{GS}}{V_P}\right)^2 - R_d\frac{I_{DSS}E^2}{2V_P^2} \tag{12-20}$$

$$v_{ds} = \frac{2R_d I_{DSS}}{V_P}\left(1 - \frac{V_{GS}}{V_P}\right)E\cos\omega t - \frac{R_d I_{DSS}}{2V_P^2}E^2\cos 2\omega t \tag{12-21}$$

Equation 12-21 shows that the presence of a second harmonic is an inherent characteristic of the FET. The magnitude of the harmonic is the distortion in the output and can be determined by taking the ratio of the coefficients as a percentage:

$$D = 100\frac{R_d I_{DSS}E^2}{2V_P^2}\bigg/\left[\frac{2R_d I_{DSS}}{V_P}\left(1 - \frac{V_{GS}}{V_P}\right)E\right]$$

$$D = \frac{25E}{V_P(1 - V_{GS}/V_P)} \text{ percent second harmonic distortion} \tag{12-22}$$

The third term in Eq. 12-20 represents the effect in the FET that is the same as the increase in current due to "signal rectification" that was discussed in Section 11-7. The term $I_{DSS}E^2/2V_P^2$ represents the increase in d-c drain current with the presence of a signal. By making the ratio of E/V_P much less than one and, since the effect of the harmonic is proportional to $(E/V_P)^2$, it is not difficult to make the effect of this third term in Eq. 12-20 quite negligible.

As a result, the FET is primarily a small signal amplifier. Its main advantage arises from the fact that the gate current is zero and, therefore, the input impedance to the circuit is effectively the value of the external resistor connected from gate to source.

Now, assume that the signal is small enough relative to V_P that the distortion terms can be neglected. Then, Eq. 12-20 and Eq. 12-21 become

$$V_{DS} = V_{DD} - R_d I_D = V_{DD} - R_d I_{DSS}\left(1 - \frac{V_{GS}}{V_P}\right)^2 \tag{12-23a}$$

and

$$v_{ds} = A_v v_{gs} = A_v E \cos \omega t = \frac{2R_d I_{DSS}}{V_P}\left(1 - \frac{V_{GS}}{V_P}\right) E \cos \omega t \qquad (12\text{-}23b)$$

Immediately, an expression for circuit voltage gain is obtained

$$A_v = \frac{2R_d I_{DSS}}{V_P}\left(1 - \frac{V_{GS}}{V_P}\right) \qquad (12\text{-}24)$$

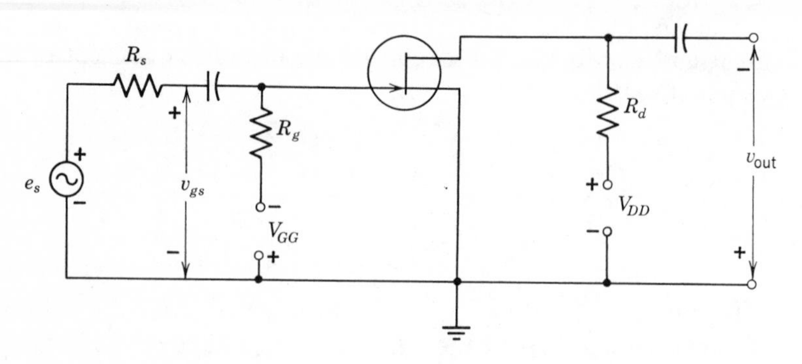

Figure 12-15 The basic JFET amplifier.

As an example, consider a JFET circuit (Fig. 12-15) that has the following values:

$$I_{DSS} = 10 \text{ ma} \qquad R_d = 8 \text{ k}\Omega$$
$$V_P = -6 \text{ V} \qquad V_{GG} = -3 \text{ V}$$
$$V_{DD} = 40 \text{ V} \qquad e_s = 0.1 \cos \omega t \text{ volt}$$

Assuming that R_g is much greater than R_s,

$$e_s = v_{gs}$$

Substituting in Eq. 12-24

$$A_v = \frac{2R_d I_{DSS}}{V_P}\left(1 - \frac{V_{GS}}{V_P}\right) = \frac{2 \times 8 \times 10}{-6}\left(1 - \frac{-3}{-6}\right) = -\frac{80}{6} = -13.3$$

and

$$v_{out} = A_v e_{gs} = A_v e_s = -1.33 \cos \omega t \text{ volts}$$

The negative sign indicates the 180° phase inversion of the amplifier.

From Eq. 12-23b, the output voltage is

$$v_{out} = v_{ds} = \frac{2R_d I_{DSS}}{V_P}\left(1 - \frac{V_{GS}}{V_P}\right) v_{gs}$$

This equation can be arranged to

$$v_{out} = \left[\frac{2I_{DSS}}{V_P}\left(1 - \frac{V_{GS}}{V_P}\right)\right] R_d v_{gs}$$

The term within the square brackets must have the dimensions of mhos in order for the product mhos × ohms (R_d) × volts (v_{gs}) to be equal to v_0 (volts). Now we form a definition for the *transconductance* g_m of the JFET:

$$g_m \equiv \frac{2I_{DSS}}{V_P}\left(1 - \frac{V_{gs}}{V_P}\right) \qquad (12\text{-}25)$$

The output voltage of the amplifier is

$$v_{out} = g_m R_d v_{gs} = g_m R_d e_s \qquad (12\text{-}26a)$$

The gain is found by dividing both sides of the equation by e_s as

$$A_v = g_m R_d \qquad (12\text{-}26b)$$

If the dividing network of R_s and R_g is considered, the overall gain is

$$A_e = \frac{R_g}{R_s + R_g} g_m R_d \qquad (12\text{-}26c)$$

Usually R_s is negligible as compared to R_g. Then the gain equations become

$$A_e = A_v = g_m R_d \qquad (12\text{-}26d)$$

Now an a-c model can be drawn for the FET (Fig. 12-16). The model is quite simple as compared to the junction transistor and it is valid also for both the depletion and the enhancement MOSFET provided a proper definition is made for g_m. This equation for g_m can be obtained mathematically by defining g_m as

$$g_m \equiv \frac{\Delta i_D}{\Delta v_{gs}} = \frac{di_D}{dv_{gs}} \qquad \text{for constant } V_{DS} \qquad (4\text{-}3)$$

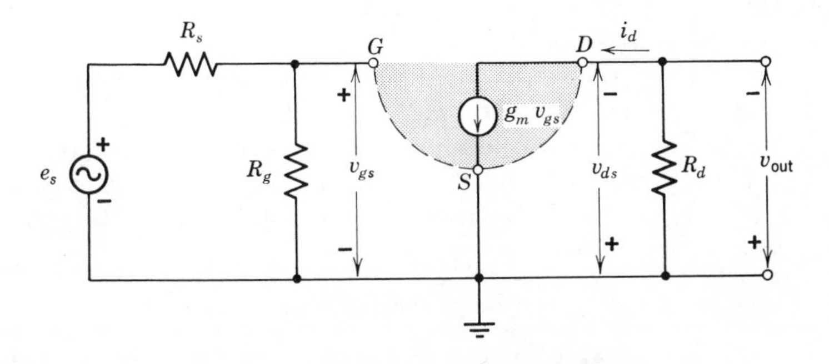

Figure 12-16 The model for the FET amplifier.

and differentiating Eq. 4-1 to obtain

$$g_m = -\frac{2I_{DSS}}{V_P}\left(1-\frac{V_{GS}}{V_P}\right) \tag{4-4}$$

and then evaluating at $v_{GS} = V_{GS}$. Graphically, g_m is the slope of the transfer curve at the operating point Q.

If V_{GS} is zero in Eq. 12-25, the value of the transconductance is evaluated at $I_D = I_{DSS}$. This particular transconductance has the symbol g_{mo}. Very often, instead of g_m or g_{mo}, FET data sheets use the symbol y_{fs}.
Using

$$y_{fs} \equiv g_{mo} \equiv -\frac{2I_{DSS}}{V_P} \tag{12-27a}$$

The transconductance at other values of I_D becomes

$$y_{fs} \equiv g_m = g_{mo}\left(1-\frac{V_{GS}}{V_P}\right) \tag{12-27b}$$

Since $I_D = I_{DSS}\left(1-\frac{V_{GS}}{V_P}\right)^2$, Eq. 12-27$b$ can be transformed to

$$g_m = g_{mo}\sqrt{I_D/I_{DSS}} \tag{12-27c}$$

Using the numerical values in the last example,

$$g_m = -\frac{2I_{DSS}}{V_P}\left(1-\frac{V_{GS}}{V_P}\right) = -\frac{2\times 10}{-6}\left(1-\frac{-3}{-6}\right)$$

$$= 1.667 \text{ millimhos} = 1667\ \mu\text{mhos}$$

The negative sign indicates the condition of the 180° phase reversal in the circuit. The gain then, directly from the model is

$$A_e = g_m R_d = 1.667 \times 8 = 13.36$$

The analysis presented thus far in this section is valid for either the JFET or the depletion type MOSFET. The enhancement type MOSFET requires a somewhat different analytic procedure, although the model given in Fig. 12-16 is valid for all three of the FET types.

As can be recalled, the enhancement type MOSFET does not have a value for I_{DSS}, since the MOSFET is cut off when the gate is less than the positive voltage V_T that is called the threshold voltage. The equation for the pinched-off drain current is

$$i_D = K(v_{GS} - V_T)^2 \tag{4-9}$$

Since this equation shows that i_D is a function of a squared term, a second harmonic content must be present in the output. Following the procedure used for the JFET,

$$v_{GS} = V_{GS} + v_{gs}$$

$$v_{DS} = V_{DS} + v_{ds}$$

and

$$V_{DD} = v_{DS} + i_D R_d$$

therefore

$$\begin{aligned} v_{DR} = V_{DS} + v_{ds} = V_{DD} - i_D R_d &= V_{DD} - KR_d(v_{GS} - V_T)^2 \\ &= V_{DD} - KR_d[(V_{GS} - V_T) + v_{gs}]^2 \\ &= V_{DD} - KR_d(V_{GS} - V_T)^2 - 2KR_d(V_{GS} - V_T)v_{gs} - KR_d v_{gs}^2 \end{aligned}$$

(12-28)

The last term, $KR_d v_{gs}^2$, determines the second harmonic content. If $v_{gs} = E \cos \omega t$, this term expands to

$$-\frac{KR_d E^2}{2} - \frac{KR_d E^2}{2} \cos 2\omega t$$

The term $KR_d E^2/2$ is the change in the d-c current in the drain due to the "rectification of signal." The amount of second harmonic can be calculated from

$$D = 100 \frac{KR_d E^2}{2KR_d(V_{GS} - V_T)E} = \frac{25E}{V_{GS} - V_T} \qquad (12\text{-}29)$$

which is the second harmonic distortion in percent. Let us assume that the distortion term is negligible in Eq. 12-28 and, from Eq. 12-28, form two equations – one for the d-c condition and the other for the a-c signal:

$$V_{DS} = V_{DD} - KR_d(V_{GS} - V_T)^2 \qquad (12\text{-}30a)$$

$$v_{ds} = -2KR_d(V_{GS} - V_T)v_{gs} \qquad (12\text{-}30b)$$

The circuit gain is

$$A_e = A_v = \frac{v_{ds}}{v_{gs}} = -2KR_d(V_{GS} - V_T) \qquad (12\text{-}31)$$

and

$$g_m = 2K(V_{GS} - V_T) \qquad (12\text{-}32a)$$

Since $I_D = K(V_{GS} - V_T)^2$, Eq. 12-32a can be converted to

$$g_m = 2\sqrt{KI_D} \qquad (12\text{-}32b)$$

If the value of g_m is given as g_{mr} for a particular reference value of current, I_{Dr}, the g_m at any value of I_D can be found from

$$g_m = g_{mr}\sqrt{I_D/I_{Dr}} \tag{12-32c}$$

When an actual family of drain curves is considered, there is a finite value for the slope of the curves other than zero. The slope at the operating point defines the drain transconductance as

$$g_d \equiv \frac{1}{r_d} \equiv \frac{\Delta I_D}{\Delta V_{DS}} \quad \text{(for constant } V_{GS}\text{)} \tag{12-33a}$$

The resistance r_d is placed in the model (Fig. 12-16) in parallel to the current source $g_m v_{gs}$ *within* the dashed boundary of the FET. Now r_d is effectively in parallel with R_d, and it tends to reduce the gain. The equation for drain current now is

$$i_d = g_m v_{gs} + g_d v_{ds} \tag{12-33b}$$

PROBLEMS

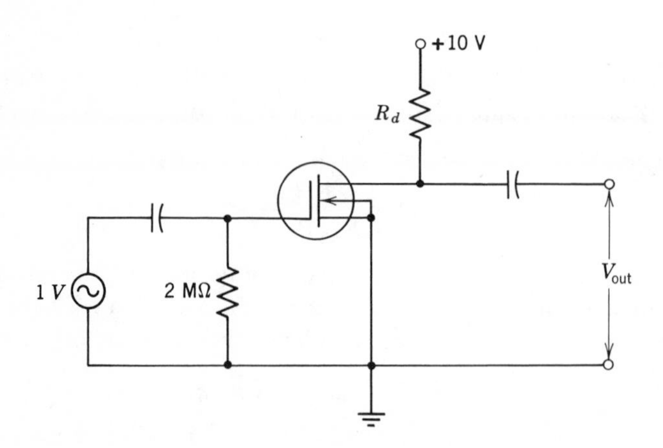

Circuit for Problems 1 and 2.

The a-c model is required for all problems.

1. The equation for the current in the *N*-channel depletion-type MOSFET is

$$I_D = 2.0\left(1 - \frac{V_{GS}}{V_P}\right)^2 \text{ma}$$

R_d is 2.5 kΩ. V_P is −4 V, and r_d is 10 kΩ. Find V_{DS}, g_m, and V_{out}.

2. What is the harmonic content of the output signal in Problem 1?

3. The MOSFET of Problem 1 is operated in a circuit with a self-bias arrangement of R_s suitably bypassed with a capacitor C_s to operate the circuit at V_{GS} equal to -2.0 V. What is the value of R_s? What is the value of V_{DS}? Determine the circuit gain. Draw the circuit. R_d and r_d are each 10 kΩ.
4. What is the harmonic content of the output in Problem 3?
5. An N-channel enhancement-type MOSFET is used in an amplifier with a 10 kΩ load. The supply voltage is +20 V and the signal input is 200 mV. The equation for the drain current for the MOSFET is

$$I_D = 1.2(V_{GS} - V_T)^2 \text{ma}$$

where V_T is 1.0 V.
The amplifier is operated at V_{GS} equal to +2.0 V.
Determine V_{DS}, g_m, and the gain of the circuit. Draw the circuit.
6. Determine the harmonic content in the output of the signal in Problem 5.
7. Repeat Problem 5 if r_d is 15 kΩ.
8. Repeat Problem 5 if the MOSFET is operated at a quiescent point of I_D equal to 0.8 ma.

Section 12-9 The Vacuum-Tube

By Thévenin's theorem, the model for the vacuum-tube amplifier (Fig. 12-17) can be represented by an open-circuited emf in series with the internal impedance. Since R_g is very much larger than R_s in vacuum-tube circuits, we neglect R_s in the analysis. For a vacuum tube, this emf is the amplifying action of the tube itself. The tube accepts a signal e_g, multiplies the signal by the inherent amplification factor μ, and presents this resultant voltage μe_g to the plate circuit. The internal impedance of the plate circuit is the a-c plate resistance r_p. The inclusion of the load resistance R_c in series completes the model (Fig. 12-17b).

A vacuum tube is theoretically a four-terminal network (Section 10-10). However, we have specified that the grid is negative at all times and does not have any direct grid current. If we also specify that there is no alternating grid current which might be produced from the reactances of the interelectrode capacitances, it is valid to use this simple model. The assumption that there is no grid current is valid at low frequencies. The modifications necessitated by the consideration of the interelectrode capacitances will be discussed later in the development of the theory of the cascaded amplifier (Section 15-5).

From this simple circuit we notice that

$$-\mu e_g = i_p(r_p + R_c)$$

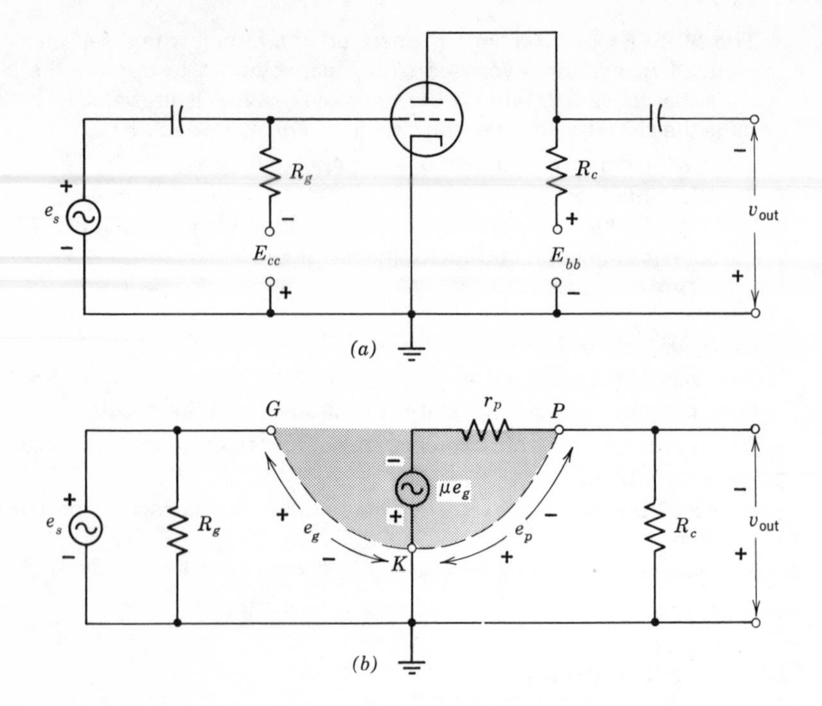

Figure 12-17 The vacuum-tube amplifier. (*a*) Circuit. (*b*) Model.

or

$$i_p = \frac{-\mu e_g}{r_p + R_c}$$

Since

$$e_p = i_p R_c$$

By substitution

$$e_p = \frac{-\mu e_g R_c}{r_p + R_c}$$

and the voltage gain is defined by

$$A_e = \frac{e_p}{e_g} = \frac{v_{out}}{e_s}$$

Then

$$A_e = \frac{e_p}{e_g} = \frac{v_{out}}{e_s} \tag{12-34a}$$

This equation may be made more general by replacing R_c with Z_L,

$$A_e = \frac{-\mu Z_L}{r_p + Z_L} \tag{12-34b}$$

When both the numerator and denominator of Eq. 12-34*a* are divided by R_c, the equation becomes

$$A_e = -\frac{\mu}{1 + r_p/R_c} \tag{12-34c}$$

When μ and r_p are fixed, as R_c becomes very large, the term $(1 + r_p/R_c)$ approaches unity. This shows that *the voltage gain of an amplifier cannot exceed* μ. When R_c equals r_p, the gain is $\frac{1}{2}\mu$. The gain of an amplifier in which the only variable is R_c is plotted in Fig. 12-18.

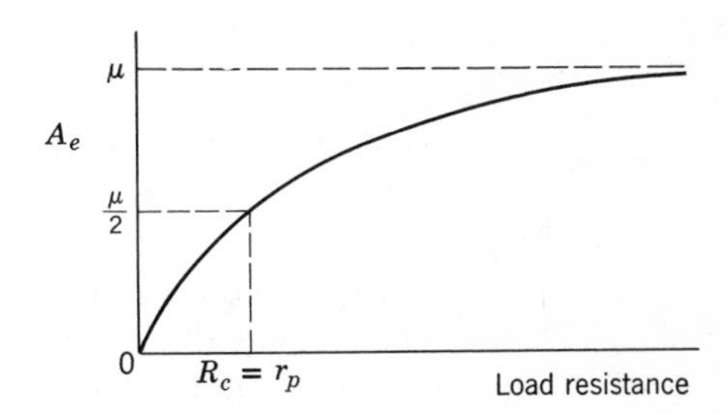

Figure 12-18 Gain variation with load resistance.

The results shown in Fig. 12-18 are predicated on a fixed μ and a fixed r_p. We established in Section 3-10 (Fig. 3-29) that r_p varies inversely with I_b. To keep I_b constant with an increasing R_c requires an ever-increasing value of E_{bb}. Obviously, there is a practical limitation to this increase in E_{bb}. The actual gain obtained in an amplifier circuit is a compromise between these two considerations.

A second form of the model may be obtained from the first model. If we substitute

$$\mu = r_p g_m \tag{3-4d}$$

in

$$e_p = \frac{-\mu e_g R_c}{r_p + R_c}$$

we obtain

$$e_p = \frac{-r_p g_m e_g R_c}{r_p + R_c}$$

which may be arranged in the form

$$e_p = -(g_m e_g)\left(\frac{r_p R_c}{r_p + R_c}\right)$$

The units of $(g_m e_g)$ are mhos × volts, or volts/ohms, or amperes. Thus $(g_m e_g)$ represents a current. The expression $r_p R_c/(r_p + R_c)$ is the parallel combination of the two resistances. e_p results from the IR drop of this current, $g_m e_g$, flowing through the parallel resistance combination. Using these relations, we can represent an alternate form of the model in Fig. 12-19.

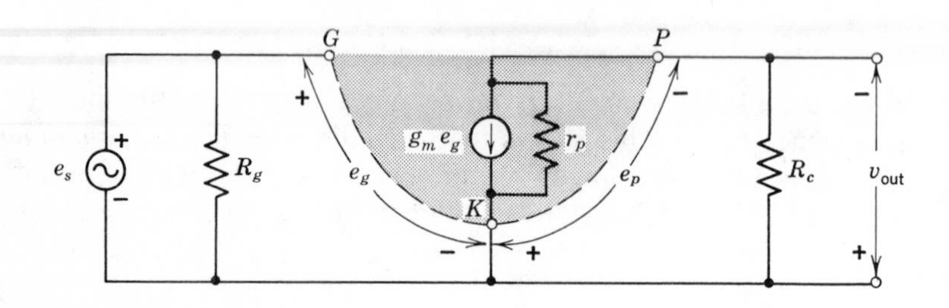

Figure 12-19 Constant current model or vacuum-tube amplifier.

This new model can be obtained from Fig. 12-17 directly by application of Norton's theorem. By Norton's theorem the short-circuit current is $\mu e_g/r_p$ or $g_m e_g$, and the shunt impedance is r_p. The load resistance R_c is then placed in parallel with r_p.

The gain may be expressed as

$$A_e = -g_m \frac{r_p R_c}{r_p + R_c} \qquad (12\text{-}35a)$$

If the value of plate resistance r_p is at least ten times the load resistance, r_p may be neglected:

$$A_e = -g_m R_c \qquad \text{when} \qquad r_p \geqslant 10R_c \qquad (12\text{-}35b)$$

If R_c is replaced by Z_L, we have

$$A_e = -g_m Z_L \qquad \text{when} \qquad r_p \geqslant 10Z_L \qquad (12\text{-}35c)$$

Either model may be used for the same amplifier, since they are equivalent to each other. We find that the first model, the so-called constant-voltage model, is best suited for analyzing triodes where low values of r_p are found as compared to the magnitude of R_c. The second model, the constant-current model, is more useful where the value of r_p is very high compared to R_c or Z_L as in most tetrode and pentode circuits.

PROBLEMS

The a-c model is required for all problems.

1. An amplifier has a 25,000 Ω load. The amplifier tube has a g_m of 1800 μmho and a plate resistance of 50,000 Ω. Determine the voltage gain of the circuit.
2. Solve problem 1 for a plate load-resistance value of 7500 Ω.
3. Solve problem 1 for a plate load-resistance value of 30,000 Ω.
4. Solve problem 1 for a plate load-resistance value of 100,000 Ω.
5. Solve problem 1 for a plate load that consists of a 140-mh coil in series with a 2000-Ω resistance. Calculate the gain at 1000 Hz and at 10 kHz.
6. Solve problem 5 when the plate load is a 0.01-μF capacitor in parallel with a 16,000-Ω resistance.

Questions

1. Define or explain each of the following terms: (*a*) model, (*b*) r'_e, (*c*) A_v, (*d*) A_e, (*e*) A_i, (*f*) h_{ob}, (*g*) h_{oe}, (*h*) h_{ie}, (*i*) swamping, (*j*) gain reduction factor, (*k*) I_{DSS}, (*l*) g_{mo}.
2. How is h_{ie} related to r'_e?
3. What is the effect of a source resistance on voltage gain?
4. What is the effect of considering h_{ob} on voltage gain?
5. Is the voltage gain of a common-base amplifier the same as the voltage gain of a common-collector amplifier.
6. Compare the input resistance of a common-base amplifier to the input resistance of a common-emitter amplifier.
7. What is the effect of an unbypassed emitter resistor R_E on the gain of a common-emitter amplifier?
8. Can the collector efficiency be obtained from an analysis of the model? Explain.
9. Show how h_{ie}/β is negligible as compared to R_B.
10. How is a resistor connected between collector and base reflected into the input circuit?
11. What gain reduction factors are considered when a common-emitter amplifier with collector feedback and emitter feedback is driven from a source that has an appreciable source resistance?
12. Under what conditions is the voltage gain of an emitter follower considerably less than unity?
13. What is the input resistance to an emitter-follower amplifier?

14. What is the current gain of an emitter-follower amplifier?
15. What is the output resistance of an emitter-follower amplifier if the source resistance is zero?
16. What is the output resistance of an emitter-follower amplifier if the source resistance is appreciable?
17. Is the representation of voltage gain for an FET amplifier as $g_m R_d$ without assumptions?
18. Why is an FET amplifier unsuited for large signal operation?
19. What is the significance of a minus sign in a gain equation?
20. What is the voltage gain of a vacuum-tube amplifier, using the voltage form? The current form?
21. What is the condition that must be met if the gain of a vacuum-tube amplifier is taken as $g_m Z_L$?

Chapter Thirteen

HYBRID PARAMETER ANALYSIS

This chapter is concerned solely with the hybrid parameter model developed in Section 10-10 and with how it is applied to the small signal analysis of the common-base amplifier (Section 13-1), the common-emitter amplifier (Section 13-2), and the common-collector amplifier (Section 13-3). The interrelation of the h parameters, the h^b parameters, the h^e parameters, and the h^c parameters, is carefully developed. Numerical examples are carried through the development. All results, both analytic and numeric, are summarized at the end of the chapter as a quick and convenient cross-reference.

Section 13-1 The Common-Base Amplifier

In Section 10-8, we developed the model for a four-terminal network. It is repeated here together with the associated equations for convenience, using standard notation in Table A. These h factors may be measured directly by test circuits and bridges and are listed for the transistor by the manufacturer.

By agreement with established standards for transistor nomenclature, the direction of I_2 is taken toward the transistor and not from the transistor. When α was defined as I_2/I_1, the two currents were in the same direction. Thus, h_{21} is $-\alpha$. Also V_2 is $(-I_2R_L)$, and P_o, the output power, which is the product of V_2 and I_2 is a negative number. As long as it is understood that by a negative power we mean that the power is delivered *to* the load and not taken *from* the load in the transistor circuit and that a negative V_2 indicates that the load is a sink (a resistance) and not a source (an emf), no confusion should result. All the equations for the transistor models are completely consistent with each other.

TABLE A Hybrid Parameter Summary

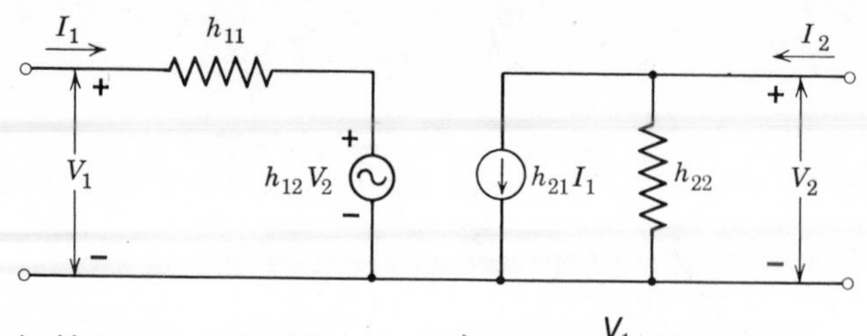

$$V_1 = h_{11}I_1 + h_{12}V_2 \tag{10-14}$$

$$I_2 = h_{21}I_1 + h_{22}V_2 \tag{10-15}$$

$$h_{11} = \frac{V_1}{I_1} = Z_1 \quad (V_2 = 0) \tag{10-16}$$

$$h_{12} = \frac{V_1}{V_2} \quad (I_1 = 0) \tag{10-17}$$

$$h_{21} = \frac{I_2}{I_1} \quad (V_2 = 0) \tag{10-18}$$

$$h_{22} = \frac{I_2}{V_2} = \frac{1}{Z_2} \quad (I_1 = 0) \tag{10-19}$$

When $I_1 = 0$, the input is open-circuited and, when $V_2 = 0$, the output is short-circuited.

h_{11} is called the input resistance with the output short-circuited.

h_{12} is called the reverse-voltage ratio (or feedback-voltage ratio) with the input open-circuited.

h_{21} is called the forward-current ratio with the output short-circuited.

h_{22} is called the output conductance (or admittance) with the input open-circuited.

The actual circuit for a common-base amplifier is shown in Fig. 13-1*a*. The transistor is replaced by the model for a four-terminal network as shown in Table A. In order to simplify the algebra of this section considerably, the network of E_s', R_s', and R_E is replaced by the use of Thévenin's theorem to give the two components, E_s and R_s, shown in Fig. 13-1*c*. Accordingly, the circuit that is analyzed in this section is the converted model (Fig. 13-1*c*).

A load resistance R_c, placed across the output terminal, does not change the hybrid parameter network equations. The objective of the analysis is to solve the two equations in such a manner as to develop equations for the various gains and impedances from

$$V_1 = h_{11}I_1 + h_{12}V_2 \tag{10-14}$$

and

$$I_2 = h_{21}I_1 + h_{22}V_2 \tag{10-15}$$

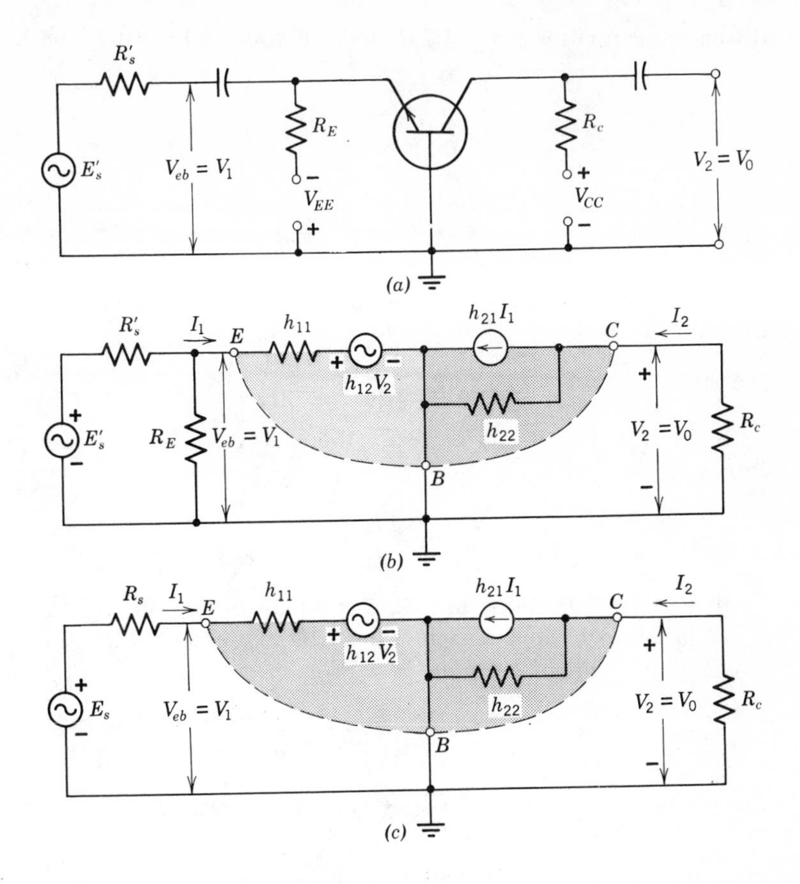

Figure 13-1 Common-base amplifier. (a) Circuit. (b) Model using hybrid parameters. (c) Model with input circuit simplification.

By Ohm's law, the relation across the load is

$$V_2 = -I_2 R_c$$

Then, Eq. 10-14 and Eq. 10-15 become

$$V_1 = h_{11}I_1 - h_{12}R_cI_2 \tag{A}$$

$$I_2 = h_{21}I_1 - h_{22}R_cI_2 \tag{B}$$

or

$$h_{21}I_1 - (1 + h_{22}R_c)I_2 = 0 \tag{B}$$

If we use the relation $I_2 = -V_2/R_c$ and substitute in Eq. 10-15, we have

$$h_{21}I_1 + h_{22}V_2 = -\frac{V_2}{R_c} \qquad (C)$$

or

$$h_{21}I_1 + \left(h_{22} + \frac{1}{R_c}\right)V_2 = 0 \qquad (C)$$

To simplify the algebra and to reduce the length of the resulting equations, we define Δ^h as

$$\Delta^h = h_{11}h_{22} - h_{12}h_{21} \qquad (13\text{-}1)$$

If we take Eq. B and solve for the ratio of I_2/I_1, we obtain the *circuit current gain* which is

$$A_i = \frac{I_2}{I_1} = \frac{h_{21}}{1 + h_{22}R_c} \qquad (13\text{-}2)$$

If we take Eq. 10-14 and Eq. C and solve the pair to eliminate I_1, the result involves V_1 and V_2 which may be expressed as a ratio V_2/V_1. This ratio is the *circuit voltage gain*

$$A_v = \frac{V_2}{V_1} = -\frac{h_{21}R_c}{\Delta^h R_c + h_{11}} \qquad (13\text{-}3)$$

The product of A_i and A_v gives the *power gain of the circuit*

$$A_p = -\frac{h_{21}^2 R_c}{(1 + h_{22}R_c)(\Delta^h R_c + h_{11})} \qquad (13\text{-}4)$$

As we pointed out, the minus sign merely shows that power is delivered to the load and not by the load.

If we take Eq. A and divide by I_1, we have

$$h_{11} - h_{12}R_c\frac{I_2}{I_1} = \frac{V_1}{I_1}$$

I_2/I_1 can be replaced by the expression for current gain (Eq. 13-2), and V_1/I_1 is the *circuit input resistance* r_{in}, which simplifies to

$$r_{in} = \frac{R_c\Delta^h + h_{11}}{1 + h_{22}R_c} \qquad (13\text{-}5a)$$

From this equation, it is apparent that the input resistance to a transistor-amplifier circuit is very definitely a function of the output resistance and cannot be neglected.

If the generator is turned off in Fig. 13-1*c*, the input voltage V_1 must be zero. The input-voltage loop equation is then

$$(R_s + h_{11})I_1 + h_{12}V_2 = 0$$

or

$$I_1 = -\frac{h_{12}}{R_g + h_{11}} V_2$$

If this is substituted in Eq. 10-15, the ratio of V_2/I_2 is the *output circuit resistance*, r_{out}:

$$r_{out} = \frac{R_s + h_{11}}{h_{22}R_s + \Delta^h} \tag{13-5b}$$

From this equation, we see that r_{out} depends on the signal source impedance R_s. Thus the action of a transistor is complex, since the dual consideration exists where the input impedance depends on the load impedance and the output impedance depends on the source impedance. It also means that an input matching is exact only for one value of load and that exact output matching will not be maintained if the generator impedance changes. Also, exact matching requires that both the input and the output circuits must be *dually matched.*

In Fig. 13-1*c*, the circuit input voltage is the generator voltage less its internal impedance drop

$$V_1 = E_s - I_1 R_s$$

If we substitute this relation for V_1 in Eq. 10-14 and solve this equation with Eq. *C* for V_2 by eliminating I_1, we have

$$V_2 = -\frac{R_c h_{21} E_s}{h_{22}R_s R_c + R_s + \Delta^h R_c + h_{11}}$$

Solving this equation for V_2/E_s, we have an expression for the overall circuit gain A_e

$$A_e = -\frac{R_c h_{21}}{h_{22}R_s R_c + R_s + \Delta^h R_c + h_{11}} \tag{13-6}$$

The load power is $V_2{}^2/R_c$ or

$$P_L = \frac{V_2{}^2}{R_c} = \frac{R_c h_{21}^2 E_s{}^2}{(h_{22}R_s R_c + R_s + \Delta^h R_c + h_{11})^2} \tag{13-7}$$

When the generator is matched to its load ($R_s = r_{in}$), we substitute Eq. 13-5a into Eq. 13-7 for R_s to obtain the load power as

$$P_L = \frac{R_c h_{21}^2 E_s^2}{4(\Delta^h R_c + h_{11})^2} \qquad (13\text{-}8a)$$

Note that, since R_s equals r_{in}, there is a 3-db loss in the generator itself.

When the circuit is matched to the load ($R_c = r_{out}$), Eq. 13-5b is substituted into Eq. 13-7 for R_c to obtain the load power as

$$P_L = \frac{h_{21}^2 E_s^2}{4(h_{22}R_s + \Delta^h)(R_s + h_{11})} \qquad (13\text{-}8b)$$

When the circuit is matched,

$$r_{in} = R_s \qquad \text{and} \qquad r_{out} = R_c$$

Substituting in Eq. 13-5 and in Eq. 13-6,

$$R_s = \frac{R_c \Delta^h + h_{11}}{1 + h_{22} R_c}$$

$$R_c = \frac{R_s + h_{11}}{h_{22} R_s + \Delta^h}$$

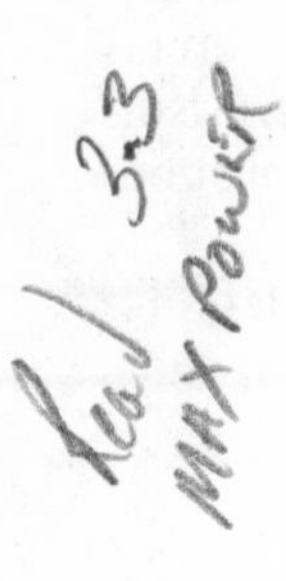

If each of these equations is cleared of fractions and then the two results are subtracted, we have

$$R_s = \Delta^h R_c$$

Putting this back in the equations for r_{in} and r_{out}, we find that

$$R_s = \left(\frac{\Delta^h h_{11}}{h_{22}}\right)^{1/2} \qquad (13\text{-}9a)$$

and

$$R_c = \left(\frac{h_{11}}{\Delta^h h_{22}}\right)^{1/2} \qquad (13\text{-}9b)$$

If Eq. 13-9a is substituted into Eq. 13-8a or if Eq. 13-9b is substituted into Eq. 13-8a, the load power for a *dually-matched* condition is found as

$$P_L = \frac{h_{21}^2 E_s{}^2}{4(\sqrt{\Delta^h} + \sqrt{h_{11} h_{22}})^2} \sqrt{\frac{h_{22}}{\Delta^h h_{11}}} \qquad (13\text{-}10)$$

Again, there is a 3-db loss in the generator itself. In a generator that is connected to a matched load ($R_s = r_{in}$), the current is $E_s/2R_s$. The power delivered to the circuit by I^2R is $E_s{}^2/4R_s$. Forming a ratio of the load power (Eq. 13-10) and the power delivered to the circuit and simplifying by use of

Eq. 13-9*b*, we have the *maximum available gain*:

$$\text{MAG} = \frac{h_{21}^2}{(\sqrt{h_{11}h_{22}} + \sqrt{\Delta^h})^2} \tag{13-11}$$

The transistor has, for the circuit operating conditions of Fig. 13-1, the following h-parameter values given in the technical data sheet:

$$h_{11} = 48.6\ \Omega$$

$$h_{21} = -0.982$$

$$h_{12} = 3.58 \times 10^{-4}$$

$$h_{22} = 2.5 \times 10^{-7}\ \text{mho}$$

In order to continue the analysis which was started in the examination of the properties of the load line in Section 11-2, the following calculations based on Eq. 13-1 through Eq. 13-11 are made with a discussion of the significance of the results.

For Eq. 13-1,

$$\begin{aligned}\Delta^h &= h_{11}h_{22} - h_{12}h_{21} \\ &= 48.6(2.5 \times 10^{-7}) + 0.982(3.58 \times 10^{-4}) \\ &= 36.42 \times 10^{-5}\end{aligned}$$

The value used for the load resistance R_c was 1500 Ω. Assume a source resistance R_s of 200 Ω. This value of 200 Ω for the Fig. 13-1*c* includes both the generator resistance and the shunting effect of R_E.

For Eq. 13-2,

$$\begin{aligned}A_i &= \frac{h_{21}}{1 + h_{22}R_c} \\ &= \frac{-0.982}{1 + (2.5 \times 10^{-7})1500} \\ &= -0.982\end{aligned}$$

For Eq. 13-3,

$$\begin{aligned}A_v &= \frac{-h_{21}R_c}{\Delta^h R_c + h_{11}} \\ &= \frac{0.982 \times 1500}{(36.42 \times 10^{-5})1500 + 48.6} \\ &= 30\end{aligned}$$

The positive sign of A_v indicates that this circuit arrangement has no phase shift between the input and the output voltages. Since the product of A_i and A_v is the power gain

$$A_p = -0.982 \times 30$$
$$= -29.46 = 14.65 \text{ db}$$

From Eq. 13-5a, $r_{in} = \dfrac{R_c \Delta^h + h_{11}}{1 + h_{22} R_c}$

$$= \frac{(36.42 \times 10^{-5})(1.5 \times 10^3) + 48.6}{1 + (2.5 \times 10^{-7})(1.5 \times 10^3)}$$

$$= 49.15\ \Omega$$

From Eq. 13-5b, $r_{out} = \dfrac{R_s + h_{11}}{h_{22} R_s + \Delta^h}$

$$= \frac{200 + 48.6}{(2.5 \times 10^{-7})200 + 36.42 \times 10^{-5}}$$

$$= 600\ \Omega$$

In order to have a perfect impedance match, the source resistance and the load resistance would have to be changed to

$$R_s = \left(\frac{\Delta^h h_{11}}{h_{22}}\right)^{1/2}$$

$$= \left[\frac{(36.42 \times 10^{-5})48.6}{2.5 \times 10^{-7}}\right]^{1/2}$$

$$= 266\ \Omega$$

$$R_c = \left(\frac{h_{11}}{\Delta^h h_{22}}\right)^{1/2}$$

$$= \left[\frac{48.6}{(36.42 \times 10^{-5})(2.5 \times 10^{-7})}\right]^{1/2}$$

$$= 730\ \text{k}\Omega$$

Using these values of R_s and R_c, we find by substituting in Eq. 13-2 and Eq. 13-3 that

$$A_v = 2280$$

and

$$A_i = -0.83$$

Instead of substituting in Eq. 13-11, we may find the maximum available gain by multiplying together these values of A_v and A_i, since they have been evaluated by using the dually matched values of R_s and R_c

$$\text{MAG} = 2280(-0.83)$$
$$= -1892 = 32.77 \text{ db}$$

It appears that there is a very large discrepancy between the voltage gain of 30 and the power gain of 29.46, from the results of the graphical analysis done in Section 11-2 for this transistor, and a voltage gain of 2280 and a maximum available gain of 1892 under the optimum matched conditions. The conditions of matching provide a *maximum power gain.* For a power amplifier, we are seeking a *large power output* and not a large power gain. If V_{CC} is fixed, and the load resistance is increased from 1500 Ω to higher values, the numerical values of I_{max} and I_C must decrease. This means that the a-c power output decreases but the power gain for this transistor increases. Accordingly, as for vacuum tubes, two types of transistor units are available. One class is intended for use as a small-signal amplifier (a voltage amplifier or a current amplifier), and the other class is intended for use as a large-signal power amplifier. Since we have found that a high value of load resistance is required for matching this transistor, it is evident that it is intended primarily as a small-signal amplifier rather than as a power amplifier.

From the results of the graphical analysis, the alternating output voltage is 4.05 V, and the output power is 10.9 mw. Since the voltage gain is 30, the input signal V_1 required is 4.05/30 or 0.135 V. Likewise, the a-c power input to the emitter is 10.9/29.4 or 0.370 mw or 370 μw. The current-gain equation (Eq. 13-4) and the maximum available gain equation (Eq. 13-11) do *not* include either the a-c power loss in the generator or the a-c power loss in the bias resistor R_E. The gain figures consider as the input power $V_1 \times I_1$ only. Therefore, it is important not to forget to include the losses in the input coupling circuit as a supplementary calculation if necessary.

In the preceding discussion, we used the symbols h_{11}, h_{22}, h_{12} and h_{21} as the values for the hybrid parameters for the common-base circuit. According to *IEEE* Standards, these symbols are not strictly correct. These h values using double-subscript notation apply to an equivalent circuit for a four-terminal network and not for a common-base transistor amplifier. A special set of h-parameter symbols is used for the common-base circuit

$$h_{ib} \quad \text{for} \quad h_{11}$$
$$h_{rb} \quad \text{for} \quad h_{12}$$
$$h_{fb} \quad \text{for} \quad h_{21}$$
$$h_{ob} \quad \text{for} \quad h_{22}$$

The i refers to the *i*nput (resistance); the r refers to *r*everse (voltage ratio);

the f refers to *f*orward (current ratio); the o refers to *o*utput (conductance), and the b refers to the common-*b*ase circuit. It is recognized that, because of the nature of the development of the equations from the four-terminal network, there will be a tendency for the carry-over of h_{11}, h_{22}, h_{12}, and h_{21} into common-base nomenclature. Thus, it is understood that, if a numerical value is listed for h_{22}, for example, it means and refers to h_{ob} only. This set may be called either h parameters or h^b parameters.

Tables B and C at the end of this chapter contain all the equations and the numerical results obtained for the example as a convenient reference for comparing this and the next two transistor circuits.

At this point it is obvious that the approach to a circuit by means of hybrid parameter analysis is considerably more complex than the approach used in Chapter 12. This complexity is a result of the fact that we are now considering a factor h_{12} that was *not* considered in Chapter 12. Also, these hybrid parameter equations include a value for the output resistance ($1/h_{22}$) of the transistor. When it is assumed that the feedback within the transistor is negligible ($h_{12} = 0$) and when it is assumed that the output resistance of the transistor is much higher than R_c ($1/h_{22} \gg R_c$), these equations can be reduced quite easily to the ones developed in Chapter 12. These reductions are included in the problems.

PROBLEMS

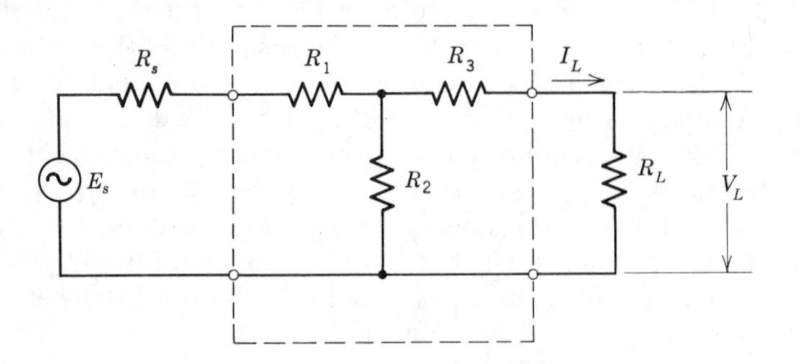

Circuit for Problems 1 to 6.

1. Determine the hybrid parameter values for the network that contains the values of 1000 Ω for each of R_1, R_2, and R_3. E_s is 10 V and R_s is 500 Ω. R_L is 500 Ω. Using the hybrid-parameter analysis, determine the current and the voltage of the load R_L.
2. For Problem 1, determine I_L and V_L when R_L is set to obtain maximum power transfer from the network.

3. For Problem 1, R_s and R_L are adjusted to obtain a maximum power transfer. Determine the power in the load and the values of V_L and I_L.
4. Repeat Problem 1 if R_2 is 100 Ω, R_1 is 50 Ω, R_2 is 500 Ω, R_3 is 100 Ω, and R_L is 400 Ω.
5. What value of R_L extracts the maximum power from the circuit? For this condition what are V_L and I_L?
6. R_s and R_L are adjusted in the circuit of Problem 4 to produce an optimumly matched condition. What are the values of R_s and R_L and what are the values of I_L and V_L?
7. Show how Eq. 12-4*b* can be developed from Eq. 13-6.
8. The hybrid parameter values for a particular transistor are:

$$h_{ie} = 4800\ \Omega, \qquad h_{re} = 9.1 \times 10^{-4}$$

$$h_{fe} = 45, \qquad h_{oe} = 12.4\ \mu\text{mhos}$$

The source resistance R_s is 300 Ω, and the load resistance is 60,000 Ω. Calculate A_v, A_i, A_p, r_{in}, and r_{out} for the common-base connection.

9. Determine the values of R_s and R_L required for perfect matching in problem 8, and determine A_v, A_i, and A_p for this condition.
10. The hybrid parameter values for a particular transistor are:

$$h_{ie} = 1667\ \Omega, \qquad h_{re} = 4.95 \times 10^{-4}$$

$$h_{fe} = 44, \qquad h_{oe} = 22.8\ \mu\text{mhos}$$

The source resistance R_s is 80 Ω and the load resistance is 130,000 Ω. Calculate A_v, A_i, A_p, r_{in}, and r_{out} for the common-base connection.

11. Determine the values of R_s and R_L required for perfect matching in problem 10, and determine A_v, A_i, and A_p for this condition.
12. A transistor connected as a common-base amplifier is connected to a 1200-Ω source and to a 50,000-Ω load. Determine the voltage gain, the current gain, and the power gain if the transistor parameters are:

$$h_{ib} = 32\ \Omega \qquad h_{\text{rb}} = 3 \times 10^{-4}$$

$$h_{fb} = -0.95 \qquad h_{ob} = 1.0\ \mu\text{mho}$$

13. If the amplifier of Problem 12 is properly matched to source and load, determine the source resistance required, the load resistance required, and the power gain.
14. A transistor connected as a common-base amplifier is connected to a 2500-Ω source and to a 100,000-Ω load. Determine the voltage gain, the current gain, and the power gain if the transistor parameters are:

$$h_{ib} = 29\ \Omega \qquad h_{rb} = 4 \times 10^{-4}$$

$$h_{fb} = -0.989 \qquad h_{ob} = 0.5\ \mu\text{mho}$$

15. If the amplifier of Problem 14 is properly matched to source and load, determine the source resistance required, the load resistance required, and the power gain.

Section 13-2 The Common-Emitter Amplifier

It is possible to develop a set of equations similar to Eq. 13-1 to 13-11 for the common-emitter circuit. If this were done, there would be eleven more equations for this circuit and then eleven more equations for the circuit in the next section, the common collector circuit. On the other hand, if it is possible to convert the model for the common-emitter amplifier to a new model which is in the same form as the model for the common-base amplifier, then, all the equations that have been developed for the common-base amplifier will automatically be valid for the common-emitter amplifier. Now the method of attack is to determine a set of h^e parameters for the common-emitter circuit which are the h_{11}, the h_{22}, the h_{12}, and h_{21} values used in Eq. 13-1 to Eq. 13-11. Manufacturer's data for a transistor, by *IEEE* recommendations, is given for h^b parameters. Therefore, what is sought in the development of the common-emitter model is a set of equations that will convert h^b parameters to h^e parameters for use in Eq. 13-1 to Eq. 13-11.

The model for the circuit of Fig. 13-2*a* is obtained by changing the ground connection and the input signal connection of Fig. 13-1*c*. But the equations developed for impedance and gain are only valid for the model in the form of Fig. 13-3. From the discussion on the concept of the "black box" (Section 10-10), as long as I_1, V_1, I_2, and V_2 do not change, it does not matter whether the internal circuit of the transistor is that of Fig. 13-2*b* or that of Fig. 13-3. Since they are the same as far as the external input and output circuits are concerned, the objective is to express the components of one in terms of the other. We define the h^e parameters for Fig. 13-3 in the same manner as we did for the h^b parameters:

$$h_{ie} = \frac{V_1}{I_1} \qquad \text{where} \qquad V_2 \text{ is zero}$$

$$h_{fe} = \frac{I_2}{I_1} \qquad \text{where} \qquad V_2 \text{ is zero}$$

$$h_{re} = \frac{V_1}{V_2} \qquad \text{where} \qquad I_1 \text{ is zero}$$

$$h_{oe} = \frac{I_2}{V_2} \qquad \text{where} \qquad I_1 \text{ is zero}$$

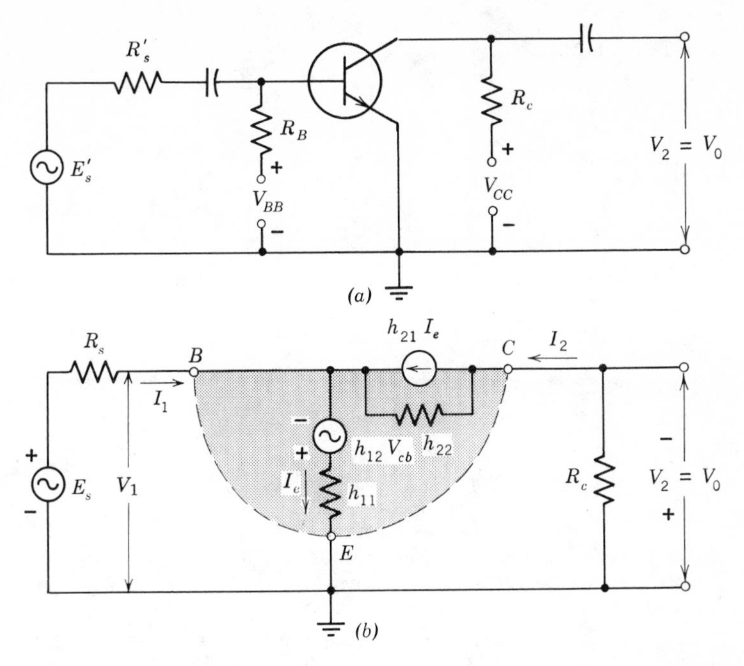

Figure 13-2 Common-emitter amplifier. (*a*) Actual circuit. (*b*) Rearrranged model.

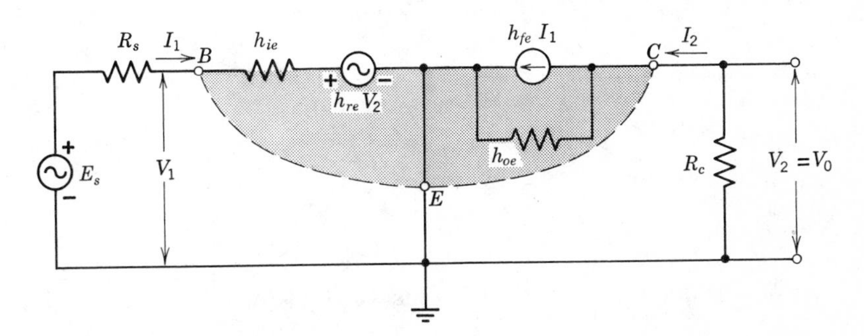

Figure 13-3 Model for the common-emitter amplifier using h^e parameters.

When the load is short-circuited, V_2 is zero and h_{ie} and h_{fe} can be obtained from Fig. 13-2*b*. The loop equation for Fig. 13-2*b* with V_2 zero is

$$V_1 = -I_e h_{11} - h_{12} V_{cb}$$

and, if we call I_3 the current through h_{22},

$$V_1 = \frac{I_3}{h_{22}} = \frac{h_{21} I_e - I_2}{h_{22}}$$

From Fig. 13-2*b*,

$$I_1 + I_2 = -I_e$$

and, when R_c is short-circuited,

$$V_1 = -V_{cb}$$

When these two substitutions are made, the equations can be solved for V_1/I_1 and for I_2/I_1:

$$\frac{V_1}{I_1} = h_{ie} = \frac{h_{11}}{\Delta^h + h_{21} - h_{12} + 1}$$

and

$$\frac{I_2}{I_1} = h_{fe} = -\frac{\Delta^h + h_{21}}{\Delta^h + h_{21} - h_{12} + 1}$$

We find that, when actual circuit measurements are made,

$$(1 + h_{21}) \gg (\Delta^h - h_{12})$$

and

$$h_{21} \gg \Delta^h$$

Making these simplifications, we have

$$h_{ie} = \frac{h_{11}}{1 + h_{21}} \qquad (13\text{-}12)$$

and

$$h_{fe} = \frac{-h_{21}}{1 + h_{21}} \qquad (13\text{-}13)$$

From the definition of β made in Section 3-3 (Eq. 3-2*b*),

$$h_{fe} = \beta$$

When the generator E_s is removed from the circuit of Fig. 13-2*b*, the input current I_1 is zero. The voltage equations are

$$V_2 = \frac{I_2 - h_{21} I_e}{h_{22}} - h_{12} V_{cb} - h_{11} I_e$$

and

$$V_1 = -h_{12} V_{cb} - h_{11} I_e$$

In these, we may substitute

$$I_e = -I_2$$

and

$$V_{cb} = \frac{I_2 - h_{21} I_e}{h_{22}}$$

When the resulting equations are solved for V_1/V_2 and I_2/V_2, we find that

$$\frac{V_1}{V_2} = h_{re} = \frac{\Delta^h - h_{12}}{\Delta^h + 1 + h_{21} - h_{12}}$$

and

$$\frac{I_2}{V_2} = h_{oe} = \frac{h_{22}}{\Delta^h + 1 + h_{21} - h_{12}}$$

Again, using the numerical simplifications which show that

$$(1 + h_{21}) \gg (\Delta^h - h_{12})$$

we have

$$h_{re} = \frac{\Delta^h - h_{12}}{1 + h_{21}} \tag{13-14}$$

and

$$h_{oe} = \frac{h_{22}}{1 + h_{21}} \tag{13-15}$$

At this point, we will evaluate the same set of calculations for the circuit of Fig. 13-2a that we made for the common-base circuit. In Section 13-1, we used for the transistor these h^b parameters:

$$h_{11} = h_{ib} = 48.6\ \Omega$$

$$h_{12} = h_{rb} = 3.58 \times 10^{-4}$$

$$h_{21} = h_{fb} = -0.982$$

$$h_{22} = h_{ob} = 2.5 \times 10^{-7}\ \text{mho}$$

To determine the h^e parameters, we substitute these values in Eq. 13-12 to Eq. 13-15:

$$\begin{aligned} \Delta^h &= h_{11} h_{22} - h_{12} h_{21} \\ &= 48.6(2.5 \times 10^{-7}) + 0.982(3.58 \times 10^{-4}) \\ &= 3.64 \times 10^{-4} \end{aligned}$$

From Eq. 13-12,

$$h_{ie} = \frac{h_{11}}{1+h_{21}}$$

$$= \frac{48.6}{1-0.982}$$

$$= 2720\ \Omega$$

From Eq. 13-13,

$$h_{fe} = \frac{-h_{21}}{1+h_{21}}$$

$$= \frac{0.982}{1-0.982}$$

$$= 55$$

From Eq. 13-14,

$$h_{re} = \frac{\Delta^h - h_{12}}{1+h_{21}}$$

$$= \frac{3.64\times10^{-4} - 3.58\times10^{-4}}{1-0.982}$$

$$= 3.23\times10^{-4}$$

From Eq. 13-15,

$$h_{oe} = \frac{h_{22}}{1+h_{21}}$$

$$= \frac{2.5\times10^{-7}}{1-0.982}$$

$$= 14\times10^{-6}\ \text{mho}$$

In the actual circuit in Section 11-3, we used the value of 1500 Ω for the load resistance R_c, and assumed in the last section that the value of the source resistance in parallel with the bias resistance R_B to be 200 Ω. When these values are put in Eq. 13-1 to Eq. 13-6, using h_{ie} for h_{11}, h_{re} for h_{12}, h_{fe} for h_{21}, and h_{oe} for h_{22}, we find that

$$\Delta^h = 758.5\times10^{-4} = 0.07585$$

$$A_v = -29.15$$

$$A_i = 53$$

$$A_p = 1541 = 31.88\ \text{db}$$

$$r_{\text{in}} = 2730\ \Omega$$

$$r_{\text{out}} = 37.1\ \text{k}\Omega$$

The negative sign for A_v indicates there is a 180° phase shift between the input and output voltages. The graphical result of 53.6 for A_i obtained in Section 11-3 is very close to the value of 53 from the model. We notice a severe mismatch between the load and the transistor output, but again we found from the graphical method a power output that has a low distortion at high level rather than a maximum power-gain condition. The results of the graphical analysis gave an output power of 10.38 mw at 3.85 V. Using the voltage-gain value of 29.15 and the power-gain value of 1541, an input signal level of 132 mv and 6.75 μw is required. This compares to the 135 mv and the 370 μw required for the input to the common-base circuit.

When the circuit is properly matched, we find from Eqs. 13-9*a*, 13-9*b*, and 13-11 that

$$R_s = 3830\ \Omega$$

$$R_c = 50{,}600\ \Omega$$

$$\text{MAG} = 13{,}700 = 41.87\ \text{db}$$

When these values of R_c and R_s are substituted in Eq. 13-2 and in Eq. 13-3, we have

$$A_i = 32.2$$

and

$$A_v = -425$$

Under ideal conditions, the gain figures for the common-base circuit were

$$A_i = -0.83$$

$$A_v = 2280$$

and

$$\text{MAG} = 1893$$

The equations and results developed in this section are summarized in Tables B and C for convenient reference at the end of this chapter.

PROBLEMS

1. Show how Eq. 12-7*a* and Eq. 12-8*b* can be developed from Eq. 13-6.
2. A transistor connected as a common-emitter amplifier is connected to a 250-Ω source and to a 500-Ω load. Determine the voltage gain, the current gain, and the power gain if the transistor parameters are:

$$h_{ib} = 30\ \Omega \qquad h_{rb} = 6.5 \times 10^{-4}$$

$$h_{fe} = 64 \qquad h_{ob} = 0.42\ \mu\text{mho}$$

3. If the amplifier of Problem 2 is properly matched to source and load, determine the source resistance required, the load resistance required, and the power gain.
4. A transistor connected as a common-emitter amplifier is connected to a 1500-Ω source and to a 2200-Ω load. Determine the voltage gain, the current gain, and the power gain if the transistor parameters are:

$$h_{ib} = 29\ \Omega \qquad h_{rb} = 4 \times 10^{-4}$$

$$h_{fe} = 115 \qquad h_{oe} = 0.5\ \mu\text{mho}$$

5. If the amplifier of Problem 4 is properly matched to source and load, determine the source resistance required, the load resistance required, and the power gain.
6. The hybrid parameter values for a particular transistor are:

$$h_{ie} = 2880\ \Omega \qquad h_{re} = 5.5 \times 10^{-4}$$

$$h_{fe} = 55 \qquad h_{oe} = 16.3\ \mu\text{mho}$$

   The source resistance R_s is 1200 Ω, and the load resistance is 10,000 Ω. Calculate A_v, A_i, A_p, r_{in}, and r_{out} for the common-emitter connection.
7. Determine the values of R_s and R_L required for perfect matching in problem 6, and determine A_v, A_i, and A_p for this condition.
8. The hybrid parameter values for a particular transistor are:

$$h_{ie} = 6040\ \Omega \qquad h_{re} = 17.2 \times 10^{-4}$$

$$h_{fe} = 32 \qquad h_{oe} = 11.1\ \mu\text{mho}$$

   The source resistance R_s is 800 Ω, and the load resistance is 12,000 Ω. Calculate A_v, A_i, A_p, r_{in}, and r_{out} for the common-emitter connection.
9. Determine the values of R_s and R_L required for perfect matching in problem 8, and determine A_v, A_i, and A_p for this condition.

Section 13-3 The Common-Collector Amplifier

The common-collector circuit, one of the three basic transistor arrangements, is analyzed by means of the equations developed in Section 13-1 when the h^c parameters are known. The h^c parameters are evaluated in terms of the $h(h^b)$ parameters in the same manner as was done for the common-emitter circuit. The model of Fig. 13-1*c* is rearranged as a common-collector circuit (Fig. 13-4*b*). The model which must be used, if the equations of Section 13-1 are to be valid, is shown in Fig. 13-4*c*. The h^c parameters are defined as follows:

$$h_{ic} = \frac{V_1}{I_1} \qquad \text{and} \qquad h_{fc} = \frac{I_2}{I_1} \qquad \text{when } V_2 \text{ is zero}$$

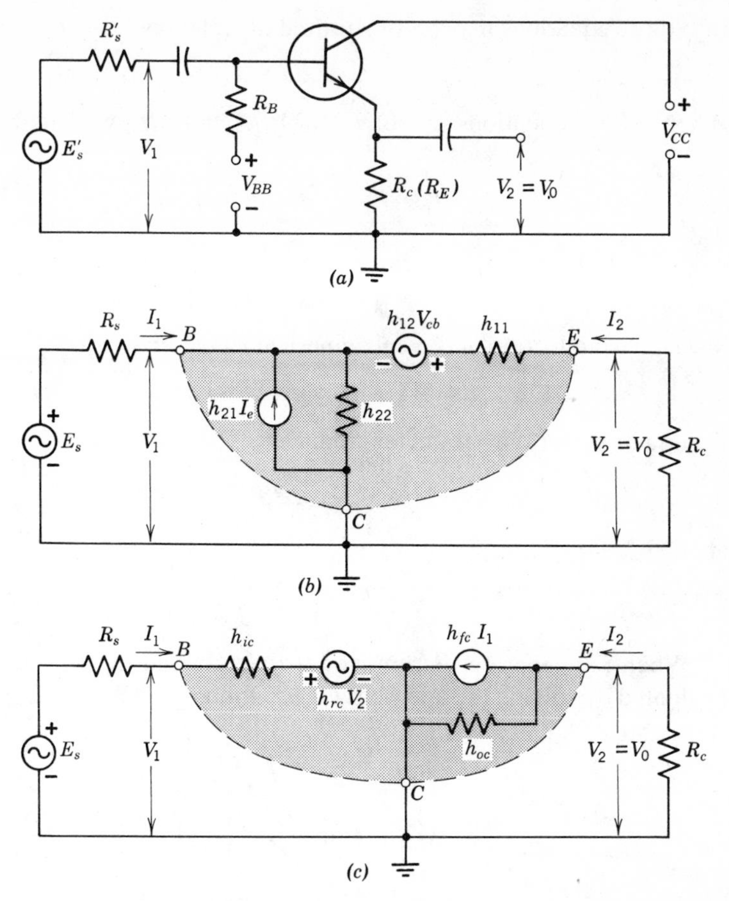

Figure 13-4 Common-collector amplifier. (*a*) Actual circuit. (*b*) Rearranged model. (*c*) Model using h^c parameters.

and

$$h_{rc} = \frac{V_1}{V_2} \qquad \text{and} \qquad h_{oc} = \frac{I_2}{V_2} \qquad \text{when } I_1 \text{ is zero}$$

When the load resistance R_c is short-circuited, V_2 is zero. We are using R_c as the symbol for the load resistance in the emitter lead in place of the usual R_E in order to keep all the equations developed here for the common-collector amplifier, compatible to the equations developed for the common-base and common-emitter circuits. The resulting equations from Fig. 13-4*b* are as follows:

$$V_1 = \frac{I_1 + h_{21}I_e + I_2}{h_{22}}$$

and

$$V_1 = -h_{12}V_{cb} - h_{11}I_1$$

Into these equations may be substituted the relations:

$$I_e = I_2 \qquad \text{and} \qquad V_{cb} = -V_1$$

When the two equations are solved for V_1/I_1 and I_2/I_1, we find that

$$\frac{V_1}{I_1} = h_{ic} = \frac{h_{11}}{\Delta^h - h_{12} + 1 + h_{21}}$$

and

$$\frac{I_2}{I_1} = h_{fc} = -\frac{1 - h_{12}}{\Delta^h - h_{12} + 1 + h_{21}}$$

Since an examination of typical numerical values shows that

$$(1 + h_{21}) \gg (\Delta^h - h_{12})$$

the equations for the parameters reduce to

$$h_{ic} = \frac{h_{11}}{1 + h_{21}} \tag{13-16}$$

and

$$h_{fc} = -\frac{1}{1 + h_{21}} \tag{13-17}$$

When the generator is removed, I_1 is zero, and then h_{rc} and h_{oc} may be evaluated by solving the open-circuit equations:

$$V_1 = \frac{h_{21} I_e + I_2}{h_{22}}$$

and

$$V_2 = I_2 h_{11} + h_{12} V_{cb} + V_1$$

Since

$$I_e = I_2 \qquad \text{and} \qquad V_{cb} = -V_1$$

The solutions for V_1/V_2 and for I_2/V_2 are

$$\frac{V_1}{V_2} = h_{rc} = \frac{h_{21} + 1}{\Delta^h - h_{12} + h_{21} + 1}$$

$$\frac{I_2}{V_2} = h_{oc} = \frac{h_{22}}{\Delta^h - h_{12} + h_{21} + 1}$$

Using the simplification

$$(1 + h_{21}) \gg (\Delta^h - h_{12})$$

we find that

$$h_{rc} = \frac{h_{21} + 1}{h_{21} + 1} = 1 \tag{13-18}$$

and

$$h_{oc} = \frac{h_{22}}{1 + h_{21}} \tag{13-19}$$

The h^c parameters for the transistor in the example are evaluated, as were the h^e parameters, from the given h (h^b) parameters:

$$h_{11} = h_{ib} = 48.6\ \Omega$$

$$h_{12} = h_{rb} = 3.58 \times 10^{-4}$$

$$h_{21} = h_{fb} = -0.982$$

$$h_{22} = h_{ob} = 2.5 \times 10^{-7}\ \text{mho}$$

Substituting these values in Eq. 13-16 to Eq. 13-19, we find that

$$h_{ic} = \frac{h_{11}}{1+h_{21}}$$

$$= \frac{48.6}{1-0.982}$$

$$= 2720\ \Omega$$

$$h_{rc} = 1$$

$$h_{fc} = -\frac{1}{1+h_{21}}$$

$$= -\frac{1}{1-0.982}$$

$$= -56$$

$$h_{oc} = \frac{h_{22}}{1+h_{21}}$$

$$= \frac{2.5 \times 10^{-7}}{1-0.982}$$

$$= 14 \times 10^{-6}\ \text{mho}$$

To be consistent with the previous two cases, we use 200 Ω as the source resistance and 1500 Ω as the load resistance. When these values are put in Eq. 13-1 to Eq. 13-6 using h_{ic} for h_{11}, h_{rc} for h_{12}, h_{fc} for h_{21}, and h_{oc} for h_{22}, we find that

$$\Delta^h = 56.04$$
$$A_v = 0.968$$
$$A_i = -54.8$$
$$A_p = -53.1 = +17.3\ \text{db}$$
$$r_{\text{in}} = 85\ \text{k}\Omega$$
$$r_{\text{out}} = 52.1\ \Omega$$

The voltage gain which is less than one is a positive number which indicate the inphase relation between the input and the output voltages. Even thoug the voltage gain is less than one, the power gain is large, since there is current gain of the same order as in the common-emitter amplifier. Th input resistance of the circuit is very high, and the output resistance i very low.

When optimum conditions are evaluated by Eqs. 13-9a, 13-9b, and 13-11 we find that

$$R_s = 104 \text{ k}\Omega$$

$$R_c = 1865\ \Omega$$

$$A_i = -54.54$$

$$A_v = 0.964$$

$$\text{MAG} = -52.45 = 17.21 \text{ db}$$

In the common-base and common-emitter amplifiers, the results for duall matched conditions were greatly different from the results obtained fron the operation of the amplifier with a source resistance of 200 Ω and a loa resistance of 1500 Ω. In the common-collector circuit, there is a very sligh difference in the results. Thus, the operation of the common-collecto

TABLE B Tabulation of Results for the Transistor Using $R_s = 200\ \Omega$ and $R_c = 1500\ \Omega$

	Common Base	Common Emitter	Common Collector
h_{11} ohms	48.6	2720	2720
h_{12}	3.58×10^{-4}	3.23×10^{-4}	1
h_{21}	−0.982	55	−56
h_{22} mhos	2.5×10^{-7}	14×10^{-6}	14×10^{-6}
Δ^h	36.42×10^{-5}	0.07585	56.04
r_{in}	49.15	2730	85 K
r_{out}	600K	37.1K	52.1
A_v	30	−29.15	0.968
A_i	−0.982	53	−54.8
A_p	−29.46 = 14.65 db	−1541 = 31.88 db	−53.1 = 17.3 db
	Results when the circuits are dually matched		
R_c	730K	50.6K	1865
R_s	266	3830	104K
A_v	2280	−425	0.964
A_i	−0.83	32.2	−54.54
MAG	−1893 = 32.77 db	13700 = 41.87 db	−52.45 = 17.21 db

amplifier becomes substantially independent of the external circuit components and is one of its most useful properties.

TABLE C Summary of Equations

	Common Base	Common Emitter	Common Collector
Input resistance	$h_{11} = h_{ib}$	$h_{ie} = \dfrac{h_{11}}{1+h_{21}}$	$h_{ic} = \dfrac{h_{11}}{1+h_{21}}$
Reverse voltage ratio	$h_{12} = h_{rb}$	$h_{re} = \dfrac{\Delta^h - h_{12}}{1+h_{21}}$	$h_{rc} = 1$
Forward current ratio	$h_{21} = h_{fb}$	$h_{fe} = \dfrac{-h_{21}}{1+h_{21}}$	$h_{fc} = -\dfrac{1}{1+h_{21}}$
Output conductance	$h_{22} = h_{ob}$	$h_{oe} = \dfrac{h_{22}}{1+h_{21}}$	$h_{oc} = \dfrac{h_{22}}{1+h_{21}}$

$\Delta^h = h_{11}h_{22} - h_{12}h_{21}$

Input resistance:

$$r_{\text{in}} = \frac{R_c\Delta^h + h_{11}}{1 + h_{22}R_c}$$

Output resistance:

$$r_{\text{out}} = \frac{R_s + h_{11}}{h_{22}R_s + \Delta^h}$$

Circuit voltage gain:

$$A_v = \frac{-h_{21}R_c}{\Delta^h R_c + h_{11}}$$

Circuit load power:

$$P_L = \frac{V_2^2}{R_c} = \frac{R_c h_{21}^2 E_s^2}{(h_{22}R_sR_c + R_s + \Delta^h R_c + h_{11})^2}$$

Circuit current gain:

$$A_i = \frac{h_{21}}{1 + h_{22}R_c}$$

Circuit power gain:

$$A_p = \frac{-h_{21}^2 R_c}{(\Delta^h R_c + h_{11})(h_{22}R_c + 1)}$$

For perfect matching:

$$R_c = \sqrt{\frac{h_{11}}{\Delta^h h_{22}}}$$

$$R_s = \sqrt{\frac{\Delta^h h_{11}}{h_{22}}}$$

Circuit load power when R_s is matched to r_{in}:

$$P_L = \frac{R_c h_{21}^2 E_s^2}{4(\Delta^h R_c + h_{11})^2}$$

TABLE C (cont.)

Circuit load power when R_c is matched to r_{out}:

$$P_L = \frac{h_{21}^2 E_s^2}{4(h_{22}R_s + \Delta^h)(R_s + h_{11})}$$

Circuit load power when $R_s = r_{\text{in}}$ and $R_c = r_{\text{out}}$:

$$P_L = \frac{h_{21}^2 E_s^2}{4(\sqrt{\Delta^h} + \sqrt{h_{11}h_{22}})^2}\sqrt{\frac{h_{22}}{\Delta^h h_{11}}}$$

Maximum available gain when $R_s = r_{\text{in}}$ and $R_c = r_{\text{out}}$:

$$\text{MAG} = \frac{h_{21}^2}{(\sqrt{h_{11}h_{22}} + \sqrt{\Delta^h})^2}$$

From algebraic manipulation of the conversion equations, we find that

$$h_{ib} = \frac{h_{ie}}{1 + h_{fe}} \qquad h_{fb} = -\frac{h_{fe}}{1 + h_{fe}}$$

$$h_{rb} = \frac{\Delta^{he} - h_{re}}{1 + h_{fe}} \qquad h_{ob} = \frac{h_{oe}}{1 + h_{fe}}$$

and $\quad h_{ic} = h_{ie} \qquad h_{rc} = 1 \qquad h_{fc} = -(1 + h_{fe}) \qquad h_{oc} = h_{oe}$

PROBLEMS

1. Show how Eq. 12-18*b* can be developed from Eq. 13-6.
2. In a common-collector amplifier, the source resistance R_s is 1800 Ω, and the load resistance is 30,000 Ω. The hybrid parameter values for this transistor are:

   $$h_{ie} = 3200\ \Omega \qquad h_{re} = 6.2 \times 10^{-4}$$
   $$h_{fe} = 48 \qquad h_{oe} = 18\ \mu\text{mhos}$$

   Determine the values for A_v, A_i, A_p, r_{in} and r_{out}.
3. Determine the dually matched resistances and gains for the common-collector amplifier of Problem 2.
4. If the amplifier of Problem 2, Section 13-2, is arranged in a common-collector connection, determine the power gain in decibels.
5. If the amplifier of Problem 4, Section 13-2, is arranged in a common-collector connection, determine the power gain in decibels.

Questions

1. Explain each of the following symbols:

V_{EE} v_c I_C V_C V_B v_E I_c V_{BB}
V_E P_o v_C I_E I_e Z_o V_e v_B
V_b Z_i p_C v_e r_{in} I_B V_{CC} v_C
I_b V_{EEB} z_{out} z_{in} V_{CCB} r_{out} p_o v_b

2. Explain each of the following symbols:

h_{fc} h_{22} h_{rc} h_{rb}
h_{re} h_{ic} h_{oc} h_{ib}
h_{ob} h_{fe} h_{11} h_{12}
h_{ie} h_{21} h_{fb} h_{oe}

3. What is a sink, and what is a source?

4. Define maximum available gain.

5. What is the principle behind a dually matched circuit?

6. Can collector dissipation be obtained from hybrid parameter analysis? Explain.

7. Can the actual output power be obtained from hybrid parameter analysis? Explain.

8. Compare the voltage and the current gains for the three basic circuit arrangements.

APPROXIMATIONS

① $A_i = \dfrac{h_F}{1 + h_o R_c}$ if $h_o R_C \leq .05$

$A_i \cong h_F$

② $A_v = \dfrac{-h_f R_c}{h_i + R_c(h_o h_i - h_r h_f)}$ if $R_c(h_o h_i - h_r h_f) < .05 h_i$

$A_v \cong -\dfrac{h_f}{h_i} R_c$

③ $Z_i = r_{in} = h_i - \dfrac{h_f h_r R_c}{1 + h_o R_c}$ for $\dfrac{h_f h_r R_c}{1 + h_o R_c} < .05 h_i$

$r_{in} \cong h_i$

④ $A_P = A_v A_i \cong \dfrac{-h_f^2 R_C}{-h_i}$ Approx for A_v + A_I

Chapter Fourteen

TRANSFORMERS AND TRANSFORMER-COUPLED AMPLIFIERS

The transformer is one of the basic components of an electronic circuit. An understanding of the theory of operation (Section 14-1) leads to the development of the model (Section 14-3). Transformers are used as impedance-changing devices (Section 14-2) to match one value of impedance to another. An analysis of the operation of the transformer together with the equivalent circuit involves the factors that determine the frequency response (Section 14-4). A discussion of the load-line construction and load-line calculations for the transistor amplifier (Section 14-5) is extended to the vacuum-tube circuit (Section 14-6).

Section 14-1 Theory of the Transformer

Figure 14-1 shows a simple transformer consisting of a primary winding of N_1 turns and a secondary winding of N_2 turns on a magnetic core. When an alternating line voltage V_1 is applied to the transformer winding, an exciting current I_0 flows. I_0 creates an exciting flux ϕ_{11}. Part of this flux, ϕ_{12}, passes completely through the core linking the secondary winding of N_2 turns. This linking a-c flux induces a voltage E_2 in the secondary winding. We may show (refer to any standard textbook on transformers and a-c machines) that, for a given core and winding arrangement, the voltage per turn is a constant providing that there is no magnetic saturation. For

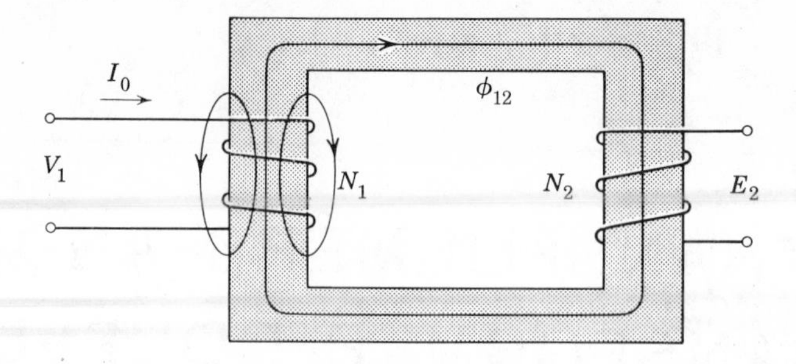

Figure 14-1 The elementary transformer.

example, if the applied voltage is 120 V and N_1 is 240 turns, this figure is 0.5 V per turn. If N_2 is 30 turns, the secondary voltage E_2 is 15 V. If N_2 is 800 turns, E_2 is 400 V.

Ideally, all the primary flux should link the secondary or ϕ_{12} should be equal to ϕ_{11}. Practically, ϕ_{12} must be less than ϕ_{11} because there are air gaps and fringing effects in the magnetic path. The ratio on ϕ_{12}/ϕ_{11} is defined as the coefficient of coupling k. The flux of the primary which does not link the secondary is $(\phi_{11}-\phi_{12})$ or $\phi_{11}(1-k)$. The inductance this flux produces in the primary is

$$\frac{N_1(\phi_{11}-\phi_{12})}{I_1 10^8} = \frac{N_1\phi_{11}(1-k)}{I_1 10^8} = L_1(1-k)$$

This inductance is called the *primary leakage inductance*. By the same process of reasoning, the *secondary leakage inductance* is

$$\frac{N_2(\phi_{22}-\phi_{21})}{I_2 10^8} = \frac{N_2\phi_{22}(1-k)}{I_2 10^8} = L_2(1-k)$$

When a load is placed across the secondary terminals, the voltage E_2 produces a current I_2 in the load. This secondary current develops a flux in the secondary winding which opposes the primary flux. The net flux in the core is the primary flux less the secondary flux. However, the transformer is an inherent constant potential device, and, for a fixed supply voltage, the core flux is constant. If a constant net value of flux is required in the core, when there is an opposing flux caused by the secondary current, there must be an increase in the primary flux to oppose the secondary flux. This increase in primary flux is produced by a rise in the primary current I_2'. In this manner, the application of a load on a transformer secondary winding causes a rise in input current while maintaining constant-voltage values throughout the transformer.

Section 14-2 Impedance Ratios

Impedance ratios are determined by considering the ideal transformer of Fig. 14-2. If a transformer is ideal, the energy transfer is complete and the efficiency is 100%. We may analyze the unit by considering that the primary

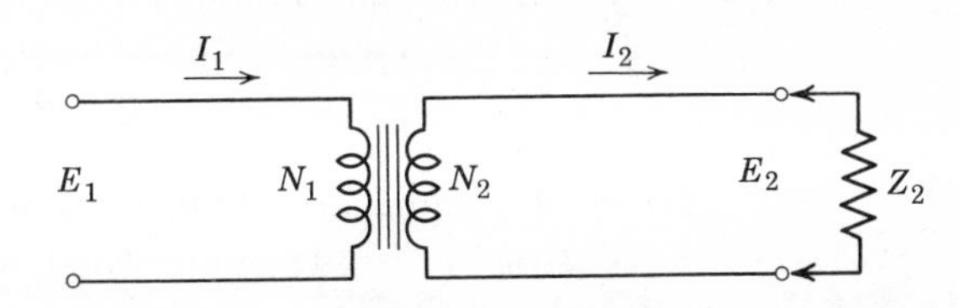

Figure 14-2 Currents and voltages in a transformer under load.

volt-amperes (or kva) equals the secondary volt-amperes (or kva) or that the value of the primary power in watts (or kw) numerically equals the secondary power in watts (or kw). The *turns ratio* α of a transformer is defined as the ratio of the secondary turns to the primary turns:

$$\alpha \equiv \frac{N_2}{N_1} \tag{14-1}$$

If α exceeds unity, the transformer is step-up in voltage; if α is less than unity, the transformer is step-down. Since the voltage per turn of the ideal transformer is a constant, the voltage ratio must be the same as the turns ratio:

$$\alpha = \frac{E_2}{E_1} \tag{14-2a}$$

By applying the principle of conservation of energy that the primary volt-amperes equal the secondary volt-amperes, the current ratio is inverse to the voltage ratio:

$$\alpha = \frac{I_1}{I_2} \tag{14-2b}$$

Then

$$\alpha E_1 = E_2 \qquad \text{and} \qquad I_1 = \alpha I_2$$

Dividing one by the other,

$$\frac{\alpha E_1}{I_1} = \frac{E_2}{\alpha I_2}$$

But E_1/I_1 is Z_1, and E_2/I_2 is Z_2. Then

$$\alpha Z_1 = \frac{Z_2}{\alpha}$$

$$\alpha^2 Z_1 = Z_2$$

Finally

$$\frac{Z_2}{Z_1} = \alpha^2 \quad \text{or} \quad \alpha = \left(\frac{Z_2}{Z_1}\right)^{1/2} \tag{14-3}$$

We can put this relation into words:

1. The impedance ratio is the square of the turns ratio or,
2. The turns ratio is the square root of the impedance ratio.

Very often, a transformer is referred to as an impedance-matching device or an impedance-changing device. As an example, if power is obtained from a 100-Ω source and must be matched to a 16-Ω load, the required transformer would be step-down with a turns ratio of

$$\alpha = \left(\frac{16}{100}\right)^{1/2} = \left(\frac{1}{6.25}\right)^{1/2} = \frac{1}{2.5}$$

Transformers are often constructed with multiple windings on the secondary. The calculation of impedances is based on the relation that the total power of the secondary loads is equal to the primary input power. For example, assume that a transformer (Fig. 14-3) operates from a 40-Ω source and delivers 12 W into a 16-Ω load and 10 W into a 500-Ω load. The turns ratios are required.

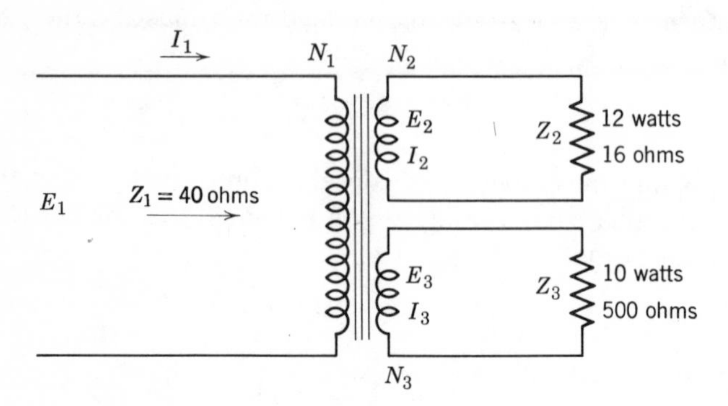

Figure 14-3 Example of a transformer with multiple loads.

Since power is E^2/R we have

$$P_2 = E_2{}^2/R_2, \qquad P_3 = E_3{}^2/R_3$$

$$12 = E_2{}^2/16, \qquad 10 = E_3{}^2/500$$

$$E_2 = 13.88 \text{ V}, \qquad E_3 = 70.7 \text{ V}$$

For the primary, by reflection, we have

$$\begin{aligned} P_1 &= P_2 + P_3 \\ &= 12 + 10 = 22 \text{ W} \end{aligned}$$

Then, since

$$\begin{aligned} P_1 &= E_1{}^2/R_1 \\ 22 &= E_1{}^2/40, \\ E_1 &= 29.6 \end{aligned}$$

The turns ratio is directly determined by the ratios of the voltages:

$$\begin{aligned} N_1 : N_2 : N_3 &= 29.6 : 13.88 : 70.7 \qquad \text{or} \\ &= 1 : 0.458 : 2.39 \qquad \text{or} \\ &= 2.14 : 1 : 5.10 \qquad \text{or} \\ &= 0.418 : 0.196 : 1 \end{aligned}$$

PROBLEMS

1. The primary voltage of a transformer is 2300 and the secondary is 220 V. The load on the secondary is 15 kW resistive. If there are 165 turns on the secondary winding, determine the primary current, the primary turns, and the impedance ratio.
2. A transformer has 1200 turns on the primary and 100 turns on the secondary winding. A 50-Ω load is connected across the secondary. How much current does the transformer take when the primary is connected to a 220-V line? What is the impedance ratio?
3. A transformer is designed to transfer 10 W from a 3000-Ω circuit to a 16-Ω circuit. What turns ratio is required? What is the primary current, and what is the secondary voltage?
4. A transformer is designed to supply power from an 2500-Ω source to multiple loads that must be fed by individual windings. The load requirements are: (*a*) 4 W at 6 Ω, (*b*) 2 W at 8 Ω, (*c*) 10 W at 16 Ω, (*d*) 3 W at 500 Ω. Determine the turns ratio required for the transformer.
5. If the transformer secondary in Problem 4 has a tapped secondary winding with a common return for the loads, specify the turns and the taps in the secondary winding.
6. The primary of a transformer is connected to a 500-Ω source and the secondary load impedance is 600 Ω. If the load power is 8 W, determine the turns ratio, the primary voltage, and the secondary voltage.
7. A transformer primary winding is 1500 Ω. It has three simultaneous secondary loads on three windings: 3 W at 15 Ω, 8 W at 32 Ω, and 4 W at 500 Ω. If there is 0.5 V per turn in the transformer, determine the number of turns in each winding.

Section 14-3 The Model for the Transformer

The model for a transformer (Fig. 14-4) like the models for a transistor and a vacuum tube is a fictitious approach, but it is very useful in terms of circuit performance and analysis. For purposes of simplification, we consider a

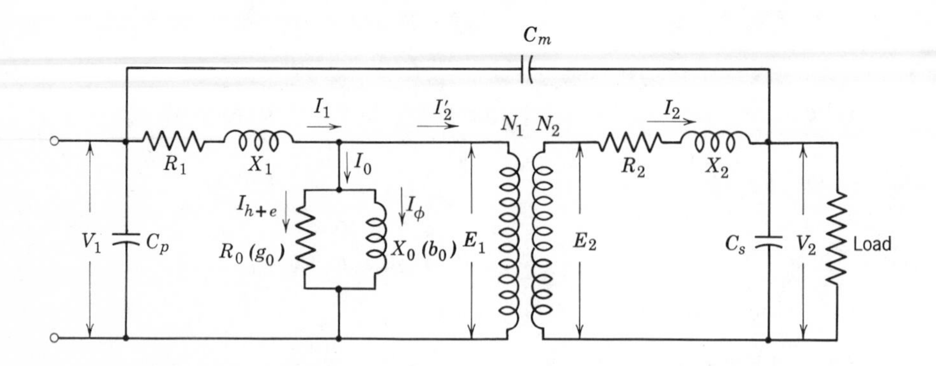

Figure 14-4 The model for a transformer.

transformer that has a single primary winding and a single secondary winding. We showed in Section 14-1 that the primary current consisted of two parts: the flux-producing part I_0, plus the component I_2', which is the reflected value of the load current. This suggests a parallel-circuit arrangement wherein I_1 divides into two branches, one for I_0 and the other for I_2'.

The exciting current I_0 may be broken into two components. There is an inphase or resistive component which produces the heat energy of the hysteresis and eddy-current loss of the core material. The inductive or wattless component produces the operating flux. The exciting current is represented by I_0, the resistive component by I_{h+e}, and the flux-producing component by I_ϕ. The a-c resistance of the windings is denoted by R_1 in the primary and R_2 in the secondary. As the coefficient of coupling of a transformer is, in practice, less than unity, X_1 represents the primary leakage reactance and X_2 is the secondary leakage reactance. The step-up or step-down action of the transformer is represented by an ideal transformer having N_1 turns in the primary and N_2 turns in the secondary winding.

Capacitance effects within the transformer are separated into three parts. The winding capacitance, the lead capacitance, and the capacitance between turns in the primary are together denoted by C_p. C_s represents the sum of these three capacitances in the secondary. The mutual capacitance between the primary and the secondary windings is C_m.

From Eq. 14-2a, the ratio of E_2/E_1 is the turns ratio α. A direct measurement of voltage to obtain the turns ratio gives α as V_2/V_1, whereas E_2/E_1 is the true value of the turns ratio. If I_2 is reduced toward zero, the error in the measurement diminishes considerably. For practical purposes, a turns-ratio measurement is made by two voltmeter readings, V_1 and V_2, taken under no-load conditions with a reduced primary voltage which prevents any saturation in the transformer core. Certain specialized measurements require that V_1 and V_2 be corrected to give the true turns ratio, E_2/E_1.

Section 14-4 Frequency Response

Reference to the model of the transformer (Fig. 14-4) assists in analyzing the distortion created by the use of a transformer. In an ideal transformer, C_p, C_m, and C_s shrink to zero as do R_1, R_2, X_1, and X_2. R_0 and X_0 become infinite in order to reduce I_0 to zero.

C_m permits an energy transfer from the primary to the secondary winding. In a transformer, energy should be transferred only by means of the changing magnetic field and not by capacitance effects. In Section 5-1, we discussed the function of a capacitance shield. In a transformer, the capacitance shield consists of a single turn of copper sheeting placed between the primary and the secondary windings. The ends of the sheet overlap, but they must be insulated to prevent the shield from acting like a short-circuited turn. This copper sheet is mechanically bonded to the frame of the transformer which also serves as the ground point, since the transformer is both electrically and mechanically fastened to the chassis.

R_1 and R_2 may be reduced by using a larger wire size. For a given number of turns, an increase in wire size necessitates a larger available window area. This increases the amount of steel used in the transformer.

X_1 and X_2 depend directly on the value of the coefficient of coupling between the primary and the secondary windings. An increase in k may be obtained by any one or all of several ways. The physical layout of the primary- and secondary-coil windings may be improved to secure a greater mutual flux. A lower operating flux density tends to force fewer lines of the flux out of the magnetic path. A lower flux density is obtained by using a larger core cross section. Increasing core area increases the cost and the weight and lengthens the turns of the copper wire. The use of special steels may increase k by operating at lower values of magnetomotive force. Careful core construction can reduce air gaps and flux fringing. Toroidal cores without air gaps are used in transformers in which the leakage reactance must be kept to an absolute minimum.

The exciting current I_0 offers a shunt path to the primary current. An increase in the number of turns in the primary reduces I_ϕ. Improved steels permit the required value of operating flux to be obtained with a lower value of magnetomotive force, thus decreasing I_ϕ. The method used in stacking and interleaving the laminations plays an important part in reducing air gaps in the core. R_{h+e} represents the equivalent of the iron losses: eddy currents and hysteresis. Eddy currents are reduced by laminating the core and insulating the laminations for each other. If the laminations become very thin, the total thickness of the varnish or scale insulation takes up too great a percentage of the core cross-section area. Hysteresis losses are reduced by using a better-grade steel. Both the eddy-current and the hysteresis losses may be reduced by operating at a lower flux density.

The input and output capacitances, C_p and C_s, are produced by capacitance between turns and leads. A reduction of these capacitances calls for special winding methods along with larger transformer window areas.

Good transformer design is predicated on a careful balance among the various factors discussed. Some transformers are designed for the least cost, whereas in others the designer seeks the ideal without regard to cost as, for example, transformers used for pulse amplifiers and for extreme high-fidelity amplifiers.

The shunting action of X_0 at low frequencies causes the decrease in output at A in Fig. 14-5. The shunt capacitances C_p and C_s cause the

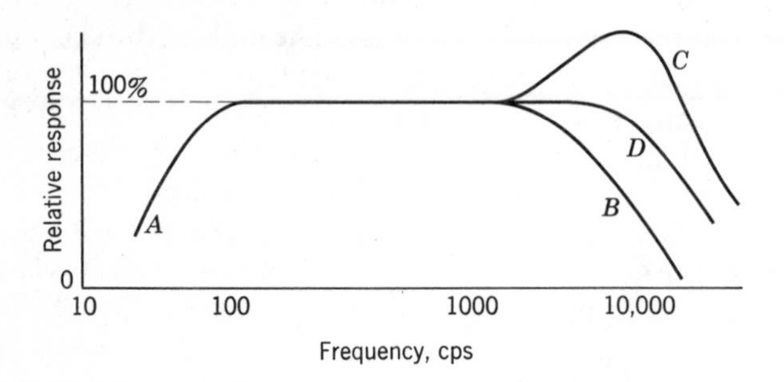

Figure 14-5 Frequency response of a transformer.

frequency response to fall off at the high frequencies, at B in Fig. 14-5. The transformer secondary voltage is developed *within* the secondary winding and not *across* the secondary winding. Thus, if the Q of the secondary winding is sufficiently high, there may be a pronounced effect of a series resonance in the secondary circuit which produces a *rise* in the response characteristics at high frequencies, at C in Fig. 14-5. Usually, transformers are designed to reduce this effect to give the extended flat response of D in Fig. 14-5.

Section 14-5 The Transformer-Coupled Amplifier and the Load Line

A transformer-coupled power amplifier using an input transformer $T1$ and an output transformer $T2$ is shown in Fig. 14-6. $T1$ is an input transformer whose main function is to match the relatively high output impedance of a

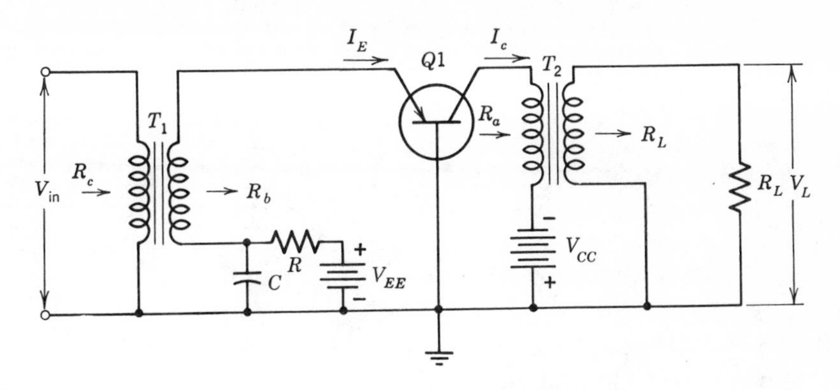

Figure 14-6 A transformer-coupled power amplifier.

source V_{in}, which usually is the output of a common-emitter amplifier or an FET amplifier to the low impedance input of the common-base stage, $Q1$. The output transformer $T2$ matches the output of $Q1$ to the load impedance R_L, which is often the low impedance of a loudspeaker.

The bias arrangement for $Q1$ is somewhat different from the circuits previously considered. The d-c resistance of the secondary of transformer $T1$, although it is not zero, is not high enough to establish a proper operating point for the transistor. An additional resistor R is required in the emitter circuit. To prevent R from affecting the signals, a bypass capacitor C is used. The combination of R and C thus forms a decoupling filter.

The gain falls off at high and low frequencies for the reasons outlined in the previous section. In addition to these considerations, there is a direct current I_E in the secondary of $T1$ and there is a direct current I_C in the primary of $T2$. These direct currents complicate the analysis, as they could produce saturation within the core even for small signals. A saturation clips the flux during that part of the a-c cycle when the a-c flux is additive to the d-c flux, thereby causing a severe distortion in the developed transformer secondary voltage.

Usually, the low value of the secondary load impedances on $T1$ and $T2$ reduce the Q's to such low values that any resonant rise effect can be ignored in these loaded transformers.

When the d-c resistance of the primary $T2$ is taken into consideration, a d-c load line AB is drawn on the characteristic curves for the transistor (Fig. 14-7) in the same manner as the load lines in Chapter 11. Usually, this

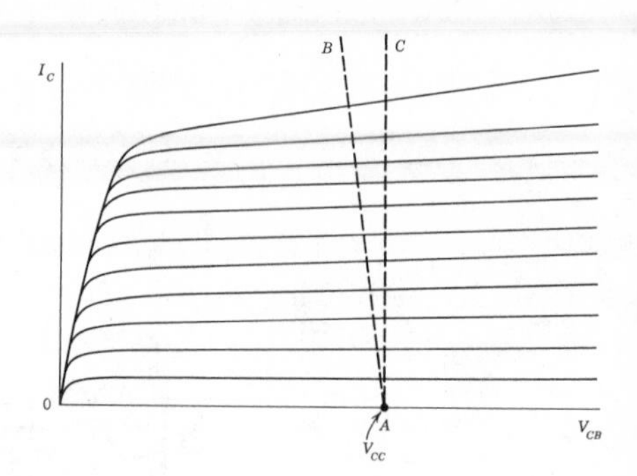

Figure 14-7 Determination of the operating point.

d-c resistance is neglected, and the load line for this value of zero d-c resistance is the vertical load line AC. The operating point must lie on the d-c load line. For the d-c load line AC, we notice that the operating collector voltage V_C is identical with the supply voltage V_{CC}. The operating point is then determined by the value of the bias current—in this example I_E (Fig. 14-6). Through the operating point, an a-c load line equal to the value of the reflected resistance R_a is drawn.

Figure 14-8 shows the method of drawing the load line on a set of collector characteristics. For example, assume that the operating point for a

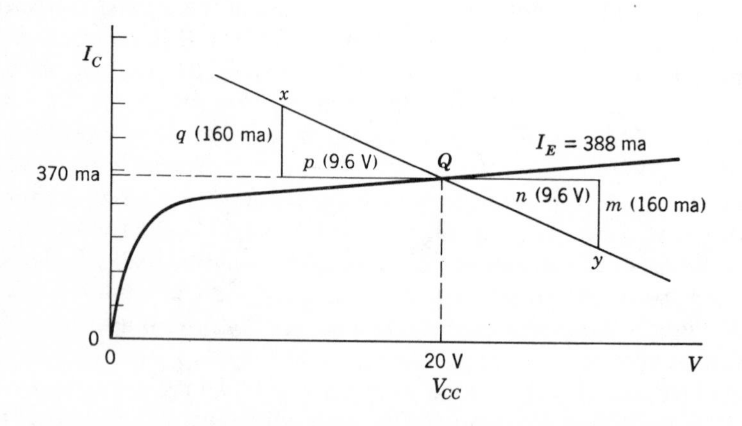

Figure 14-8 Load line for a transformer-coupled amplifier.

PNP germanium transistor is located at

$$V_{CC} = V_C = -20 \text{ V}$$

$$V_{EE} = +10 \text{ V}$$

$$R = 25\ \Omega$$

Also assume that the transformer windings have zero d-c resistance. The Kirchhoff's loop equation for the emitter circuit is

$$V_{EE} = RI_E + V_{BE}$$

$$10 = 0.025 I_E + 0.3$$

$$I_E = 9.7/0.025 = 388 \text{ ma}$$

At the collector supply voltage, 20 V, a vertical line is drawn (Fig. 14-8). The operating point Q must lie on this vertical line. From the bias calculation, it must be the intersection of this vertical line with the $I_E =$ 388 ma curve. Assume that the reflected impedance of R_L into the collector (R_a) is 60 Ω. A simple method used to draw the a-c load line is to assume a small convenient current change and, by Ohm's law, to determine the corresponding voltage change. For 60 Ω, an assumed current change of 160 ma gives a voltage change of 160×0.060 or 9.6 V. On Fig. 14-8, if we shift 9.6 V to the left along path p and up 160 ma on the i_C scale along path q, we locate point x. Likewise, a shift of 9.6 V to the right along path n and down 160 ma along path m locates point y. The line that is drawn through points Q, x, and y and extended beyond x and y is the a-c load line. The load line is extended beyond x and beyond y to give the length of the load line that is required to handle the signal swing.

The calculations for output current, output voltage, output power, and the distortion are identical with the methods outlined in Chapter 11. From Fig. 14-8, we notice that V_C is the same as V_{CC}. This means that no d-c power is lost in the transformer primary. The d-c power supplied to the circuit and the d-c power supplied to the transistor are the same. We conclude then that, for transformer coupling, the collector efficiency and the overall efficiency are identical and have maximum theoretical values of 50%. The universal use of transformers in power amplifiers is customary because of this increased efficiency. In certain special load applications, such as a d-c motor winding, an output transformer cannot be used and an inherent lower overall efficiency is tolerated.

A typical two-stage amplifier using transformers is shown in Fig. 14-9. The operating bias is obtained from V_{CC} by means of the base bias resistors, R_1 and R_3, which are bypassed by C_1 and C_3.

The load resistance, R_L, reflects back through $T3$ as R_a ohms. The input resistance R_b of transistor Q_z is determined by the h^e parameters and R_a.

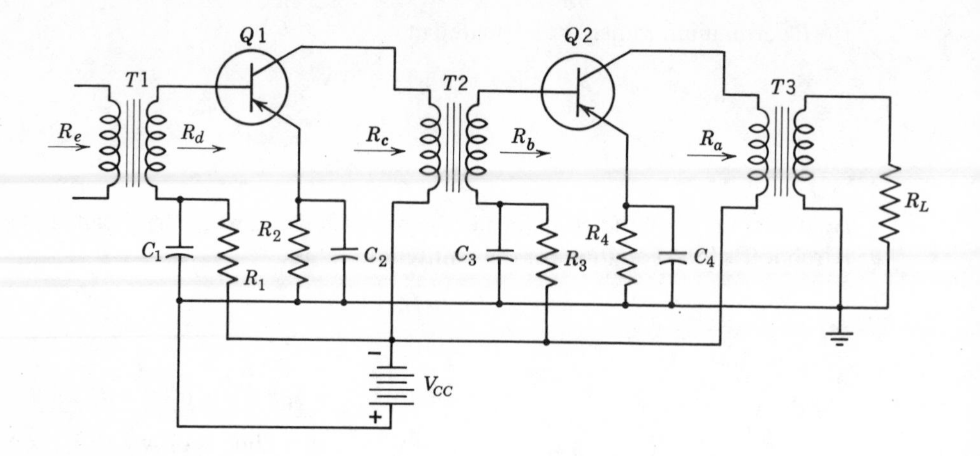

Figure 14-9 Transformer-coupled amplifier.

Now, with R_a and R_b known, the gain of the stage $Q2$ may be determined. R_b reflects through transformers $T2$ as R_c which is the load on transistor $Q1$. Again, by using the h^e parameters and the value for R_c, R_d may be evaluated. Then the gain of the stage using $Q1$ may be found. It is not possible in this circuit to start from the input and work toward the output. The sequence must be to begin at the load and, by calculating each stage in turn, to proceed toward the signal source.

When ideal transformers are considered, the power gain of the transformers themselves is unity or zero decibels. Then the total gain is the product of the power gains of the individual stages. When decibels are used, the total gain is the sum of the decibel gains of the individual stages. It is very difficult to try to interpret the overall voltage or current gains in terms of the voltage or current gains of the individual stages because of the different transformer turns ratios. Let us assume that the circuit has a power gain of 1000 and that the power in the load resistance is 1 W. Then the power input to transformer $T1$ must be 1 mw. If R_e is 100,000 Ω, by using $P = E^2/R$, the input voltage is 0.1 V. When R_e is 100,000 Ω, the input voltage is 10 V. Thus, the overall voltage and current gains depend on the turns ratios of the transformers, assuming that the source is matched for all values. However, the power gain is fixed. Thus, it should be apparent that it is much easier to talk of power gains in transistor circuits instead of in terms of current and voltage gains.

It is important to observe that it is a simple problem to obtain a gain equal to the maximum available gain from each of the stages. The only requisite is that the turns ratios of the transformers be correct.

PROBLEMS

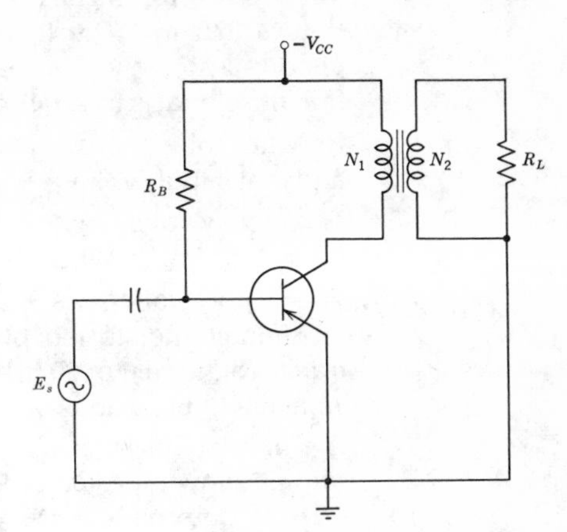

Circuit for Problems 1 to 6.

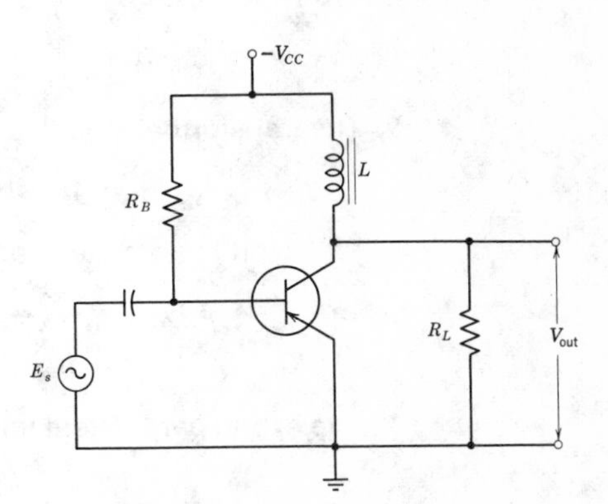

Circuit for Problem 7.

In all problems assume that the semiconductor characteristic is ideal.
For Problems 1 to 6, draw the d-c and the a-c load lines on ideal characteristics. Draw and dimension the collector current and voltage waveforms.

1. The supply voltage is −20 V, I_{CQ} is 150 ma, and R_L is 5 Ω. The germanium transistor has a β of 75. Determine the maximum undistorted power that can be supplied to the load. Determine the turns ratio and R_B.
2. The supply voltage is −12 V and the germanium transistor has a β of 60. The required maximum undistorted power in the 8-Ω load resistor is 4 W. Determine I_{CQ}, the required turns ratio, and R_B.
3. The supply voltage is −40 V, the load resistance is 40 Ω, and the germanium transistor has a β of 100. The turns ratio N_1/N_2 is 1.4. What is the maximum undistorted power that can be transmitted to the load? What are I_{CQ} and R_B?

4. The maximum voltage rating BV_{CE} is 35 V, the maximum allowable current in the collector is 250 ma, and the maximum permissible collector dissipation is 875 mw. The load R_L is 20 Ω. What is the maximum possible undistorted power in the load? For this condition what is V_{CC} and I_{CQ}. Assume that the signal is intermittently ON and OFF.
5. Repeat Problem 4, assuming that the signal is always at the maximum value and that supply voltage will be turned off the instant the signal voltage fails.
6. The supply voltage is −30 V, R_L is 4 Ω, and the maximum undistorted power in the load is $\frac{1}{2}$ W. What are N_1/N_2 and I_{CQ}?
7. The supply voltage is −40 V, R_L is 100 Ω, and L is very large. What is the maximum undistorted power than can be delivered to R_L. If the silicon transistor has a β of 60, what are I_{CQ} and R_B? In order to perform this function, what minimum values of $I_{C,\max}$ and BV_{CE} must the transistor have?
8. A single stage transistor amplifier is composed of $T2$, $Q2$, and $T3$ of the circuit shown in Fig. 14-9. R_c is 20,000 Ω and R_L is 4 Ω. The load power is 50 mW. Determine the impedance ratios of $T2$ and $T3$ and the input voltage required at the primary of $T2$ if the transistor has the following parameters:

$$h_{ib} = 30\ \Omega \qquad h_{rb} = 6.5 \times 10^{-4}$$
$$h_{fe} = 64 \qquad h_{ob} = 0.42\ \mu\text{mho}$$

9. The transformer-coupled circuit shown in Fig. 14-9 has the following values:

$$R_L = 16\ \Omega \quad \text{and} \quad \text{Load power} = 20\ \text{mw}$$

The hybrid parameters for $Q2$ are:

$$h_{ib} = 29\ \Omega \qquad h_{rb} = 4 \times 10^{-4}$$
$$h_{fb} = -0.991 \qquad h_{ob} = 0.5\ \mu\text{mho}$$

The hybrid parameters for $Q1$ are:

$$h_{ib} = 32\ \Omega \qquad h_{rb} = 3 \times 10^{-4}$$
$$h_{fb} = -0.950 \qquad h_{ob} = 1.0\ \mu\text{mho}$$

The internal resistance of the device that drives $T1$ is 55 Ω. Determine the turns ratio required for $T1$, $T2$, and $T3$. What is the overall gain, and how much power is required from the source to obtain full output power? What are the input and the output voltages?

10. In Fig. 14-9, R_L is 6 Ω and R_e is 10,000 Ω. The load power is 0.25 W. Determine the impedance ratios of the transformers and the input voltage required at the primary of $T1$ if the transistors have the following parameters:

$$\text{for } Q1 \quad h_{ib} = 29\ \Omega \qquad h_{rb} = 4 \times 10^{-4}$$
$$h_{fe} = 115 \qquad h_{ob} = 0.5\ \mu\text{mho}$$
$$\text{for } Q2 \quad h_{ib} = 30\ \Omega \qquad h_{rb} = 6.5 \times 10^{-4}$$
$$h_{fe} = 64 \qquad h_{ob} = 0.42\ \mu\text{mho}$$

Section 14-6 Vacuum-Tube Circuits

The circuit of Fig. 14-10 uses two transformers, each showing slightly different considerations. $T1$ is an input transformer in which the main function is to provide a stepped-up signal voltage to the grid of V_2. If the

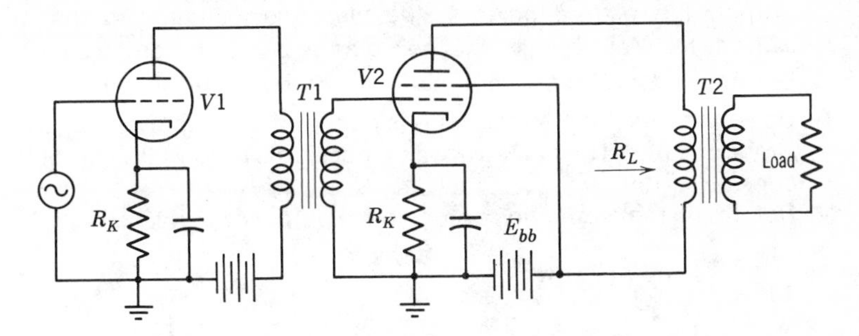

Figure 14-10 Amplifier using vacuum tubes.

operation of V_2 is class A without grid current, V_2 will not draw power from $T1$. $T2$ is a power-transferring device. The principle purpose of this transformer is to reflect the low load impedance back into the primary as a high impedance. Thus, a high-impedance source V_2 works into a low-impedance load, such as a loudspeaker, by using the transformer as an impedance-changing device.

The input transformer, since it is not loaded, may tend to show the high-Q characteristic of C in Fig. 14-5. The impedance that this transformer presents to $V1$ at middle-range audio frequencies (400 to 1000 Hz) is very high. Ideally, this impedance is infinite. As we recall from Fig. 12-18, if the external load impedance in the plate circuit of an amplifier stage is infinite, the stage voltage gain approaches the value of the amplification factor μ. In addition, the transformer itself develops a voltage gain equal to its turns ratio α. If we consider both factors, the voltage gain of a transformer-coupled stage at middle frequencies is given by

$$A_e = \mu\alpha \tag{14-4}$$

The load line for the output stage, $V2$ and $T2$ of Fig. 14-10, is developed in the same manner as for the transistor in the last section. The output and input calculations, the powers, and the efficiency calculations are also the same. The vacuum-tube amplifier is simpler to analyze because, when the grid is negative, there is no input power to the stage. The input resistance of the vacuum tube is infinite and we do not have the complexity of analysis that is required in a transistor circuit where the output load impedance is reflected all the way through a cascaded circuit to the input.

PROBLEMS

1. Using the tube manual, draw a load line for a 6L6 which is transformer-coupled to a load. The load reflects 3000 Ω into the transformer primary. The tube is operated at a 15-V bias, and the plate supply voltage is 300 V. Assume that the signal has a peak-to-peak value of 30 V. Determine the power output, the plate efficiency, the plate dissipation, and the value of the cathode resistor necessary for self-bias. Assume the screen current is 3 ma.
2. Determine the harmonic content in the output of problem 1.
3. Solve problem 1 when the tube is operated at a supply voltage of 250 V with a bias of −10 V. The peak-to-peak input signal is now 20 V. Assume the screen current is 5 ma.
4. Determine the harmonic content in the output of problem 3.

Questions

1. Define or explain each of the following terms: (*a*) exciting flux, (*b*) leakage flux, (*c*) hysteresis, (*d*) eddy currents, (*e*) volt-ampere, (*f*) mutual capacitance, (*g*) turns ratio, (*h*) voltage ratio, (*i*) impedance ratio, (*j*) reflected value.
2. Explain why placing a load on the secondary of a transformer produces an increase in primary current.
3. Refer to Fig. 14-4 and explain what factors in the design and construction of a transformer influence: $R_1, R_2, X_1, X_2, R_0, X_0, C_p, C_m$, and C_s.
4. Explain the influence of each term in Question 3 on the frequency response of a transformer.
5. Why are transformers laminated?
6. Why is the turns ratio not exactly the voltage ratio, V_2/V_1?
7. How does the turns ratio compare to the impedance ratio?
8. Explain the method of establishing the a-c load line for a transformer-coupled amplifier.
9. Compare the overall efficiency of a transformer-coupled amplifier with a resistive load to collector efficiency. Make this comparison for a circuit that uses a resistance load only.
10. Why does the response of a transformer-coupled amplifier fall at low frequencies?
11. Why does the response of a transformer-coupled amplifier fall at high frequencies?
12. Why, and under what conditions, does the transformer response peak at high frequencies?
13. How is the gain calculated for a three-stage transistor amplifier using transformers?

Chapter Fifteen

RESISTANCE-CAPACITANCE-COUPLED AMPLIFIERS

In order to complete the theory of earlier chapters, the method of calculating gain graphically for an R-C coupled amplifier is presented (Section 15-1). The operation and the model for the resistance-capacitance coupled amplifier is presented from a generalized viewpoint (Section 15-2). The method for predeterminating the gain at low frequencies (Section 15-3) and at high frequencies (Section 15-4) makes certain simplifications in the general model and yields gain reduction factors that are used to modify the midband gain. Input loading (Section 15-5) has a definite bearing on the high-frequency response. Amplifier gains at midband and at the low and high frequencies can be handled easily by the use of decibels. Bode plots simplify the presentation of gain information (Section 15-6).

Volume, tone, and equalizing circuits are generally associated with these amplifier circuits, (Section 15-7).

Section 15-1 Graphical Determination of the Load Line

Before an investigation is made into the characteristics of resistance-capacitance-coupled amplifiers, it is necessary to have available all the essential tools by which the circuit given in Fig. 15-1 can be analyzed. If β_{dc} and R_B are known, r_e' can be determined, and the circuit gain can be evaluated by the procedure presented in Section 12-4. If the hybrid parameters of the transistor are given, the calculations presented in Chapter 13 can be performed.

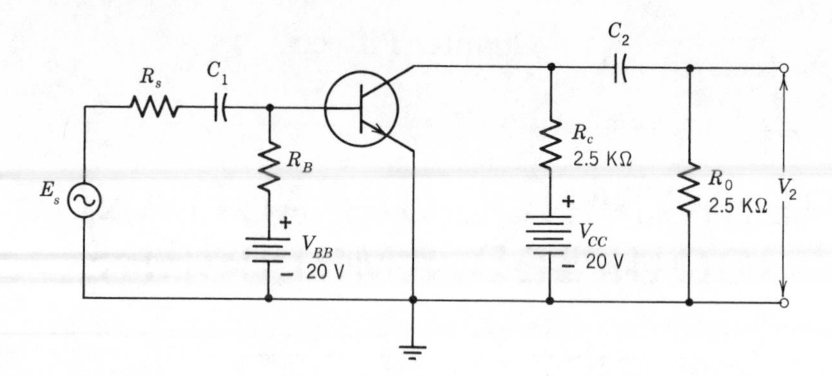

Figure 15-1 Amplifier having a resistance–capacitance–coupled load.

On the other hand, if it is essential to establish the Q point as a function of peak-to-peak output, a graphical approach must be used. The load line for R_c is drawn on the collector characteristic (Fig. 15-2) as the line A-B. One end of the load line, A, is V_{CC}, 20 V, and the other end of the load line

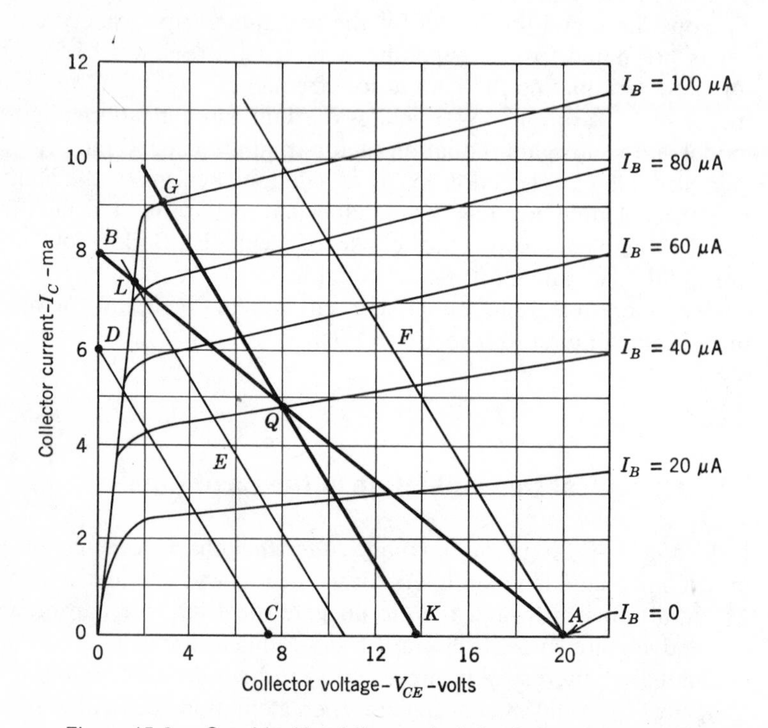

Figure 15-2 Graphical load-line analysis for R-C coupled circuit.

is B, which is V_{CC}/R_c or 20/2.5 or 8 ma. The Q point *must* lie on this line, since R_c is the resistance in the collector circuit equation

$$V_{CC} = R_c \times I_C + V_{CE}$$

This load line from A to B is called the *static load line.*

Assuming that C_1 and C_2 are coupling capacitors and that the reactances of C_1 and C_2 are negligible at the frequencies considered, the a-c load r_L on the transistor is not R_c, but is the parallel combination of R_c and R_0

$$r_L = \frac{R_c R_0}{R_c + R_0}$$

The a-c signal must vary across a load line given by r_L. The load line for r_L is called the *dynamic load line* or the *dynamic operating path.*

In this circuit r_L, the parallel combination of R_c (2.5 kΩ) and R_0 (2.5 Ω) is numerically, 1.25 kΩ. Assume a current that is convenient to the current scale of the characteristic, say, 6 ma. By Ohm's law, the voltage drop of 6 ma in a 1.25 kΩ resistor is 7.5 V. Now locate on the axes of the characteristic 6 ma, point D, and 7.5 V, point C, and connect the two points with a straight line. This line has a slope of $-1/r_L$. All of the lines parallel to this line have the same slope and, therefore, represent an a-c resistance of 1.25 kΩ. CD, E, KG, and F all represent 1.25-kΩ load lines.

The intersection of the a-c load line with the static load line is the operating point of the circuit. Dynamic load line E intersects the static load line at L. At L, the transistor is saturated and a signal cannot be placed on the transistor without clipping. Dynamic load line F intersects the static load line at A. At A, the transistor is cut off and, again, clipping would occur with a signal. Therefore, the optimum load line lies between E and F. By sliding a triangle on the characteristic, an optimum load line KQG can be located to yield the largest available signal swing. The values from this graph are

$$V_{CEQ} = 8 \text{ V} \qquad I_{CQ} = 4.8 \text{ ma}$$

$$V_{CE,\max} = 13.5 \text{ V} \qquad V_{CE,\min} = 2.6 \text{ V}$$

$$I_{C,\max} = 9.1 \text{ ma} \qquad I_{C,\min} = 0$$

$$I_{B,\max} = 100\ \mu\text{a} \qquad I_{B,\min} = 0$$

$$I_{BQ} = 40\ \mu\text{a}$$

If two additional values I_B are used, a complete analysis of the circuit can be made using distortion calculations (Section 11-7).

An analytic approach must also be used for the a-c load line to determine optimum peak-to-peak operation. The required load lines for a transistor amplifier are drawn in Fig. 15-3*a*. From the geometry of the diagram, equal

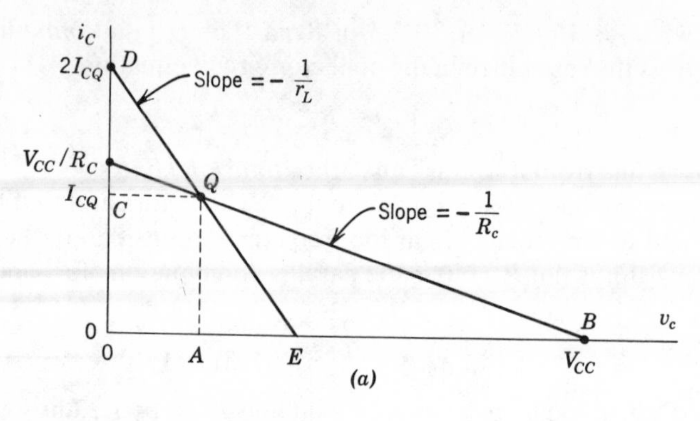

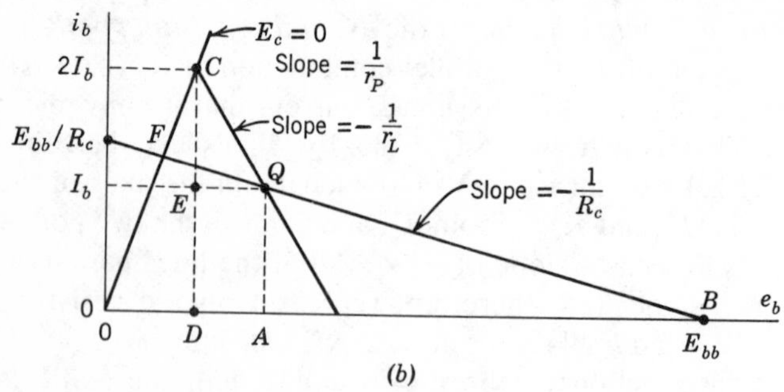

Figure 15-3 Ideal Q-point location. (a) Transistor load line. (b) Vacuum-tube triode load line.

peak-to-peak swings for this optimum operation requires that

$$OA = AE \quad \text{and} \quad OC = CD \quad \text{where} \quad OC = I_{CQ}$$

From an inspection of the diagram

$$V_{CC} = OA + AB$$

Since $OA = CQ$ and $CD = I_{CQ}$, we have from triangle QCD and Ohm's law

$$CQ = I_C r_L$$

Likewise, triangle AQB involves I_{CQ} and R_c to give by Ohm's law

$$AB = I_{CQ} R_c$$

Substituting

$$V_{CC} = I_{CQ} r_L + I_{CQ} R_c$$

Solving for I_{CQ}

$$I_{CQ} = \frac{V_{CC}}{R_c + r_L} \tag{15-1}$$

In words, the optimum Q-point current is found by dividing the supply voltage by the sum of the d-c and the a-c load resistance values.

A similar technique can be applied to a vacuum-tube triode (Fig. 15-3b). Now, there is the restriction on the load line that we cannot operate to left of point F on the load line or unwanted grid current will flow. From the diagram

$$E_{bb} = OD + DA + AB$$

From triangle ODC, $$OD = (2I_b)r_p$$

From triangle EQC, $$EQ = DA = I_b r_L$$

From triangle AQB, $$AB = I_b R_c$$

Substituting $$E_{bb} = 2r_p I_b + r_L I_b + R_c I_b$$

Solving for I_b, $$I_b = \frac{E_{bb}}{R_c + r_L + 2r_p} \qquad (15\text{-}2)$$

PROBLEMS

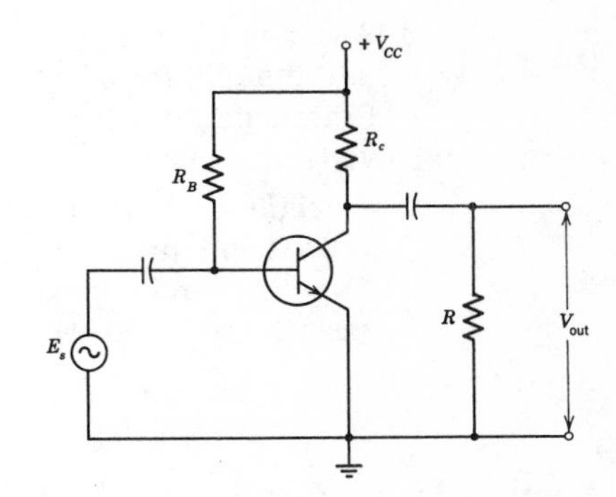

Circuit for Problems 2 and 3.

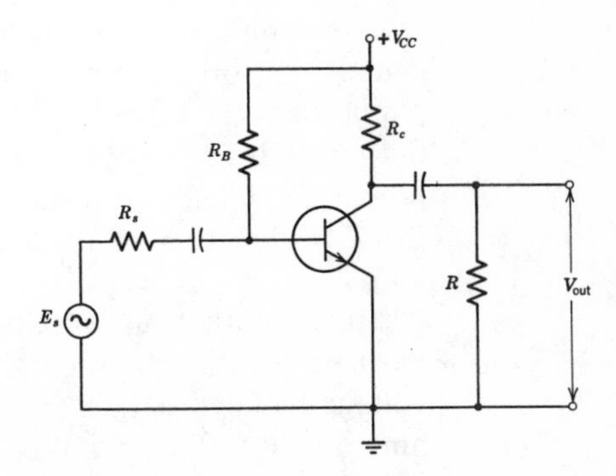

Circuit for Problems 4 and 5.

All transistors are silicon and r_e' is determined from 25 mv/I_E.

1. What is the effect of considering a properly bypassed emitter resistor in the development of Fig. 15-3*a* and Eq. 15-1?
2. V_{CC} is 30 V, R_c is 10 kΩ, and R is 15 kΩ. β is 80. Determine the value of R_B that provides optimum bias to give the maximum available peak-to-peak output voltage. What is this value of V_{out} and what value of E_s is required to produce it?
3. Repeat Problem 1 for V_{CC} equal to 8 V, R_c equal to 8 kΩ, R equal to 12 kΩ and β equal to 60.
4. R_s is 1000 Ω, V_{CC} is 40 V, R_c is 10 kΩ, R is 10 kΩ, and β is 100. Determine R_B for optimum bias conditions. What are the maximum values of E_s and V_{out}?
5. Repeat Problem 4 if R is changed to 5 kΩ.
6. A properly bypassed 2000-Ω emitter resistor R_E is added to the circuit of Problem 2. What is the optimum value for R_B and what is the maximum value of V_{out}?
7. A properly bypassed 2000-Ω emitter resistor R_E is added to the circuit of Problem 4. What is the optimum value of R_B and what is the maximum value of V_{out}?
8. A vacuum-tube amplifier stage is operated from a supply of 70 V. R_c is 10 kΩ and the grid resistor to the next stage is 12,000 Ω. The value of r_p is 5000 Ω. Determine the optimum Q-point and the maximum peak-to-peak plate voltage.
9. A vacuum-tube amplifier stage is operated from a supply of 350 V with R_c 100 kΩ and the grid resistor to the next stage 150 kΩ. Determine the optimum Q-point and the maximum available peak-to-peak output voltage. The plate resistance r_p is 50 kΩ.

Section 15-2 General Considerations of the RC-Coupled Amplifier

In most applications a single stage amplifier does not have sufficient voltage gain. Therefore, two or more stages are necessary. For example, if a microphone delivers 30 mv into the input of a stage which has a gain of 40, the output voltage is 30 × 40 or 1200 mv. This voltage is insufficient to drive a power amplifier, and a second stage with a gain of 10 raises the signal level to 1.2 × 10 or 12 V. The combined voltage gain of the two stages is 40 × 10 or 400.

In order that an amplifier stage can drive a following stage, the signal must be fed into the second stage as a pure a-c signal with the direct supply voltage blocked out. A transformer-coupled stage (Fig. 14-9) performs this function, but a cheaper alternative method is in general use (Fig. 15-4*a*).

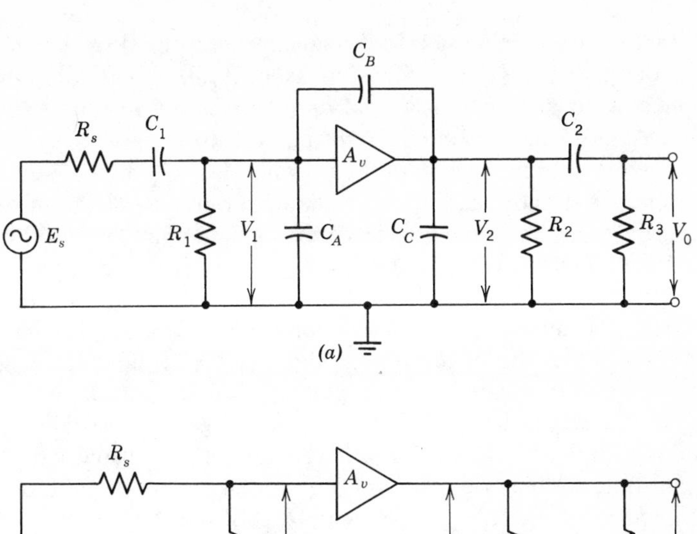

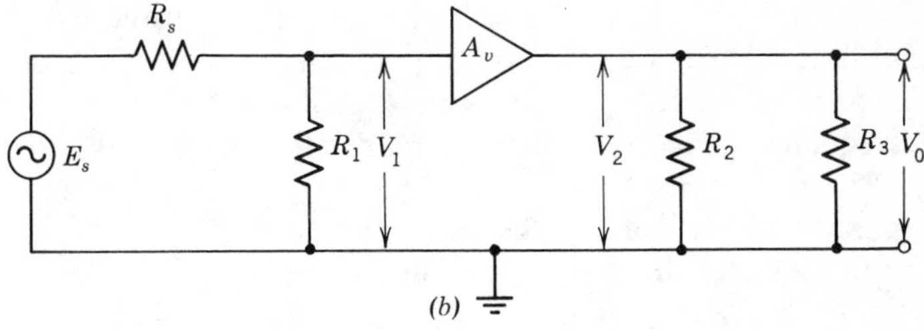

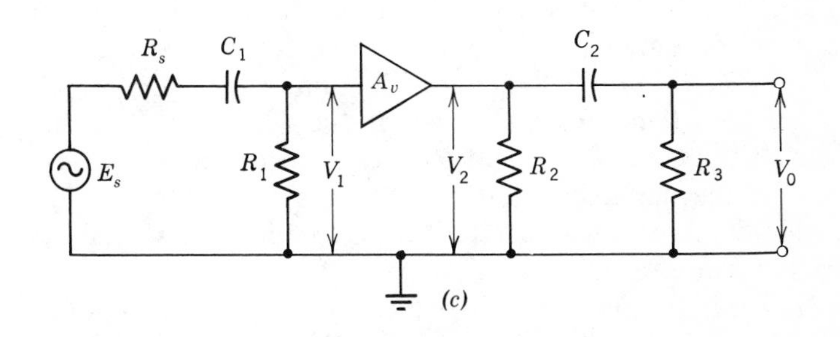

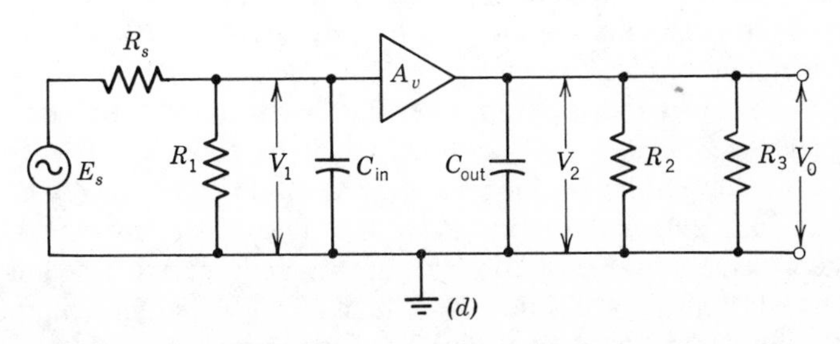

Figure 15-4 Capacitance effects in amplifiers. (*a*) Complete circuit. (*b*) Middle-frequency equivalent. (*c*) Low-frequency equivalent. (*d*) High-frequency equivalent.

In Fig. 15-4*a*, the amplifier itself is represented by a triangular symbol in order to make this discussion as general as possible and to make it applicable to all types of amplifying devices both solid state and electron tube. A blocking capacitor C_1 is required to keep any d-c components in the driving source E_s from affecting the bias, and conversely, in transistor circuits, we do not want a low resistance source to short out the d-c bias on the transistor. The capacitance C_A is not a component, but it represents the combined value of the wiring capacitance of the input circuit to ground plus the input capacitance of the amplifying device. A capacitance C_B exists between the input and the output terminals of the amplifying device. Likewise, C_C is the sum of the output capacitance of the amplifying device plus the wiring capacitance of the output circuit to ground. R_2 is the load resistor (R_c) of the amplifying device. C_2 is a blocking capacitor that isolates any d-c potential on R_2 from any d-c potential on R_3. R_3 is usually the parallel combination of the bias resistor and the input resistance to the following stage.

Middle frequencies, approximately 300 to 3000 Hz, cover that range of the frequency response where all capacitors can be ignored in a model (Fig. 15-4*b*). The reactances of the coupling capacitors, C_1 and C_2, are small compared with R_1 and R_3. The reactances of C_A, C_B, and C_C are sufficiently high that any currents flowing through them are negligible compared to the normal signal currents in the circuit. The voltage gain of the circuit over this middle-frequency range is $A_{e,\text{MF}}$.

The middle frequency gain of the circuit can be determined by any of the methods thus far discussed. The model we are using is that given in Fig. 15-4*b*. R_1 is the parallel combination of the bias resistor and the input resistance of the amplifier shown by the symbol labeled A_v. R_2 is the load resistor for the amplifier, and R_3 is the effective input resistance to the next stage. If the amplifier itself has an output resistance value that must be considered, its output resistance forms a parallel combination with the actual load resistance to yield R_2.

Low frequencies, below approximately 300 Hz, cover that frequency range below the middle frequencies where the reactances of the coupling capacitors, C_1 and C_2, cannot be considered negligible. Naturally, if C_A, C_B, and C_C do not affect middle frequency gain, they certainly cannot affect the low-frequency gain. The model that includes C_1 and C_2 is given in Fig. 15-4*c*. The voltage gain of the circuit over this low-frequency range is $A_{e,\text{LF}}$.

High frequencies, above approximately 3000 Hz, cover the frequency range above middle frequencies where the currents in C_A, C_B, and C_C are appreciable and cannot be neglected. The model for the high-frequency range is given in Fig. 15-4*d*. At first appearance, C_B seems to have disappeared. This is not true; the effect of C_B is rather complex, and a full section in this chapter (Section 15-5) is devoted to the analysis. In any event, we shall show that the three capacitors can be treated as the two capacitors,

C_{in} and C_{out}, shown in Fig. 15-4*d*. The voltage gain of the circuit over the high-frequency range is $A_{e,HF}$.

A typical frequency response curve of an amplifier is shown in Fig. 15-5.

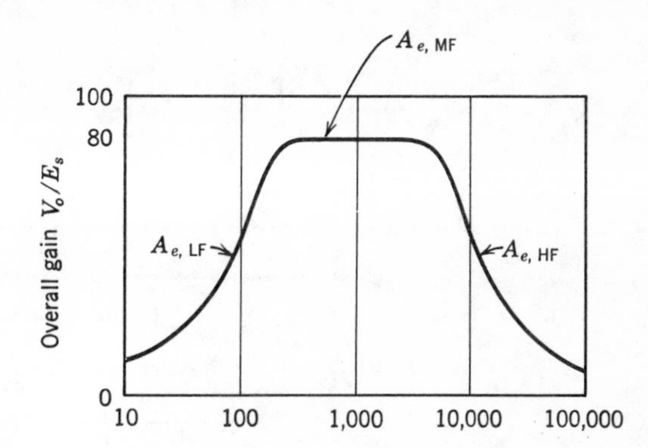

Figure 15-5 Amplifier frequency response.

At low frequencies, the reactances of the coupling capacitors become large, and a voltage divider action across the coupling capacitors occurs that reduces the output voltage. At high frequencies the low shunting reactance of C_{in}, acting in parallel with R_1, reduces V_1. Also, the low reactance of C_{out} reduces the load impedance on the amplifier which, in turn, causes the gain to drop.

When the size of the coupling capacitors is increased, there is less falloff or *attenuation* at the low frequencies. However, as these capacitance values are increased, the cost and the physical size of the capacitors themselves also increase. Larger size circuit components increase the wiring capacitance that can adversely affect the high-frequency response. The loss at the higher frequencies is caused by the shunt capacitors C_{in} and C_{out}. Careful wiring techniques and short leads reduces the wiring capacitances.

PROBLEMS

The a-c model is required for all problems.

1. All the resistors R are 10 kΩ each. r_e' is 100 Ω and β is 100 for each transistor. Determine the gain V_{out}/E_s.
2. The transistors in the two-stage amplifier are germanium and they have a value for β of 30. r_e' is obtained from 30 mv/I_E. Determine the values of the base resistors to establish the operation of the stages in the center of the load lines. Determine the value of E_s required to deliver maximum peak-to-peak undistorted output voltage, V_{out}.
3. The transistors are silicon and have a value for β of 100. r_e' is determined from 50 mv/I_E. Determine all the d-c voltages and currents in the circuit. Determine

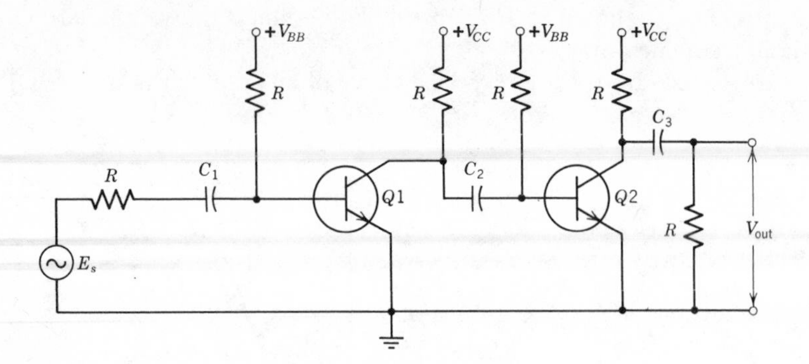

Circuit for Problem 1.

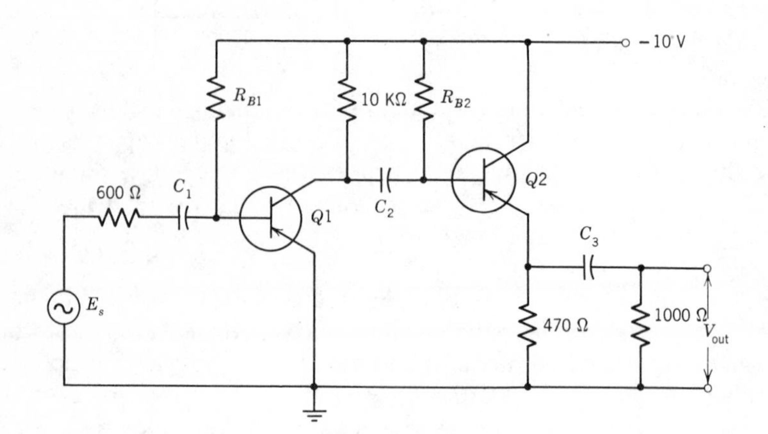

Circuit for Problem 2.

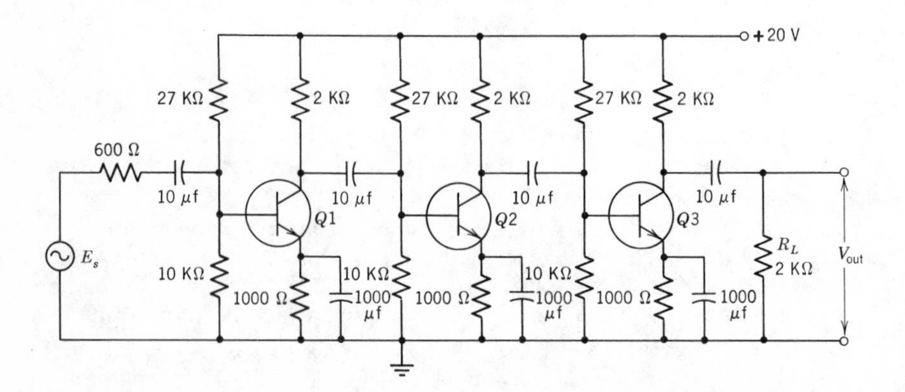

Circuit for Problems 3 and 4.

the value of E_s that provides an output signal across R_L of 2 V peak-to-peak. What is the maximum available peak-to-peak voltage across R_L without clipping? What is the overall voltage gain?

4. Repeat Problem 3 if all the emitter bypass capacitors are removed.

Section 15-3 Low-Frequency Response

The analysis of the effect of the coupling capacitors can be made by considering either C_1 or C_2. Let us analyze the effect of C_1 on the input section of the circuit of Fig. 15-4. The mid-frequency model for the input is shown in Fig. 15-6*a*. V_1 is the output of a simple voltage divider and V_1 is multiplied by A_v to yield V_2:

$$V_1 = \frac{R_1}{R_1 + R_s} E_s \qquad \text{and} \qquad V_2 = V_1 A_v$$

Then

$$A_{e,\text{MF}} = \frac{V_2}{E_s} = \frac{R_1}{R_1 + R_s} A_v \tag{15-3}$$

The phasor diagram is shown in Fig. 15-6*b*. It is relatively simple because only pure resistance values are involved. V_2 is drawn to show a 180° phase angle for θ because the amplifier A_v introduces the conventional 180° phase inversion.

At a low frequency, the coupling capacitor C_1 is introduced into the model (Fig. 15-6*c*). The voltage drop across C_1 reduces the voltage to V_1' from V_1 and consequently, V_2' is less than V_2. Now, by a voltage divider action,

$$V_1' = \frac{R_1}{R_1 + R_s - jX_{c1}} E_s$$

and

$$V_2' = V_1' A_v$$

then

$$A_{e,\text{LF}} = \frac{V_2'}{E_s} = \frac{R_1}{R_1 + R_s - jX_{c'}} A_v \tag{15-4}$$

The phasor diagram is shown in 15-6*d*. The length of E_s is not changed, but V_1' and V_2' are reduced from V_1 and V_2, showing a falloff in the magnitude of the circuit output at the low frequency. Also, a phase shift, ϕ, is introduced. The angle θ between E_s and V_2' is no longer 180°. The output voltage is supposed to be 180° out of phase with E_s along the dashed line *OM*. However, V_2' *leads* this dashed line by ϕ degrees. Therefore, a leading angle

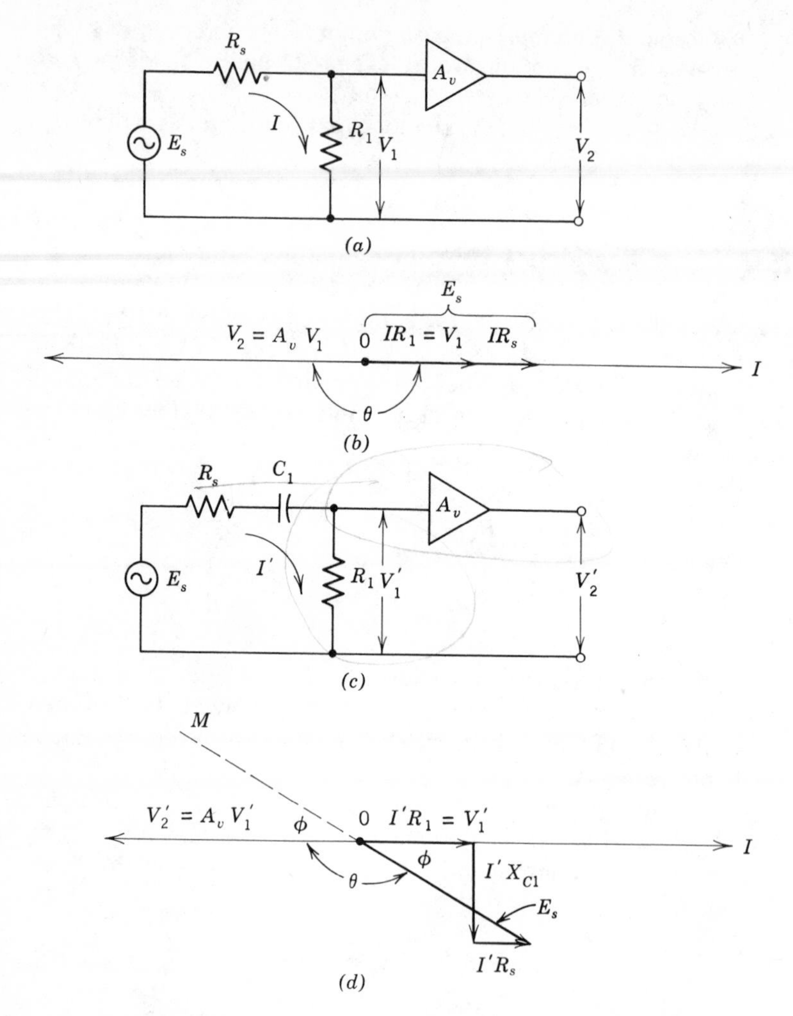

Figure 15-6 Low-frequency analysis. (*a*) Model and (*b*) phasor diagram at mid-frequencies (*c*) Model and (*d*) phasor diagram at low frequencies.

ϕ has been introduced in the circuit. At mid-frequencies ϕ is zero and, as the frequency of the signal is decreased, ϕ ranges from zero up to a full 90° lead. To develop a simpler analytical approach toward low-frequency gain calculation, divide Eq. 15-4 by Eq. 15-3:

$$\frac{A_{e,\mathrm{LF}}}{A_{e,\mathrm{MF}}} \equiv K_{\mathrm{LF}} = \frac{R_1 + R_s}{R_1 + R_s - jX_{C_1}} \tag{15-5}$$

K_{LF} is defined as a *low-frequency gain reduction factor*. Now divide the numerator and denominator by (R_1+R_s)

$$K_{\text{LF}}=\frac{1}{1-jX_{C_1}/(R_1+R_s)} \tag{15-6a}$$

where $0 \leq |K_{\text{LF}}| \leq 1$

or, as a magnitude,

$$K_{\text{LF}}=\frac{1}{\sqrt{1+[X_{C_1}/(R_1+R_s)]^2}} \tag{15-6b}$$

There is a frequency that we shall define as f_1 where, at f_1,

$$X_{C_1}=(R_1+R_s)$$

therefore

$$R_1+R_s=\frac{1}{2\pi f_1 C_1} \tag{15-7a}$$

But X_{C_1} in Eq. 15-6*a* and *b* is the value of the reactance at *any* frequency f:

$$X_{C_1}=\frac{1}{2\pi f C_1} \tag{15-7b}$$

Now substitute Eq. 15-7*a* and Eq. 15-7*b* back into Eq. 15-6 to obtain

$$K_{\text{LF}}=\frac{1}{1-j(f_1/f)} \tag{15-8a}$$

where f_1 from Eq. 15-7*a* is

$$f_1=\frac{1}{2\pi C_1(R_1+R_s)} \tag{15-8b}$$

Eq. 15-8*a* can be placed in a rectangular form

$$K_{\text{LF}}=\frac{1}{\sqrt{1+(f_1/f)^2}} \tag{15-9a}$$

at a leading phase angle ϕ where

$$\tan\phi=\frac{f_1}{f} \tag{15-9b}$$

The same analysis can be applied to the output circuit of Fig. 15-4*c*. We obtain a new K_{LF} which is determined by (R_2+R_3) and C_2. As a result, the overall gain of the whole circuit at low frequencies is the product of the two K's and the mid-frequency gain

$$A_{e,\text{LF}}=K_{\text{LF},1}\times K_{\text{LF},2}\times A_{e,\text{MF}} \tag{15-10}$$

It is not necessary to perform these calculations in detail for an amplifier.

The results of Equations 15-8a, 15-8b, 15-9a, and 15-9b can be plotted as a universal curve (Fig. 15-7) which gives the value of K_{LF} and ϕ, provided that f_1 is obtained from Eq. 15-7a or 15-7b. It is important to observe on the universal curve that the horizontal axis for low frequencies is f/f_1 instead of the f_1/f, which is the factor in the developed equations.

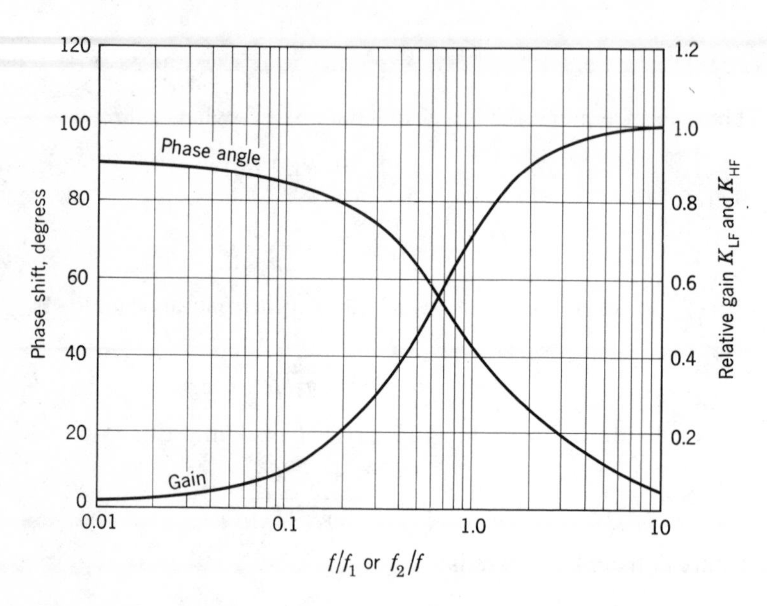

Figure 15-7 Universal amplification curve. Angles are lead for low frequencies and lag for high frequencies in addition to normal 180° shift.

PROBLEMS

1. R_s is 9 kΩ, R_B is 600 kΩ, R_c is 3000 Ω, and R is 2400 Ω. r_e' is 10 Ω and β is 100. Determine the midband gain. Find C_1 and C_2 that establish f_1 at 60 Hz for both the input and the output circuits. What is the magnitude of the gain and the phase angle at f_1?
2. R_s is 10 kΩ, R_c is 4.7 kΩ, R_E is 1500 Ω, and R is 3600 Ω. V_{CC} is +10 V, and β for the silicon transistor is 80. R_B is adjusted to establish the optimum operating point. Use r_e' as 50 mv/I_E. C_1 and C_2 are selected to give a value for f_1 at 100 Hz for each circuit. What is the midband gain? What is the gain and phase shift at 100 Hz and at 10 Hz?
3. What values of C_1, C_2, and C_3 will set f_1 at 20 Hz for each stage in the circuit for the amplifier given in Problem 1, Section 15-2?

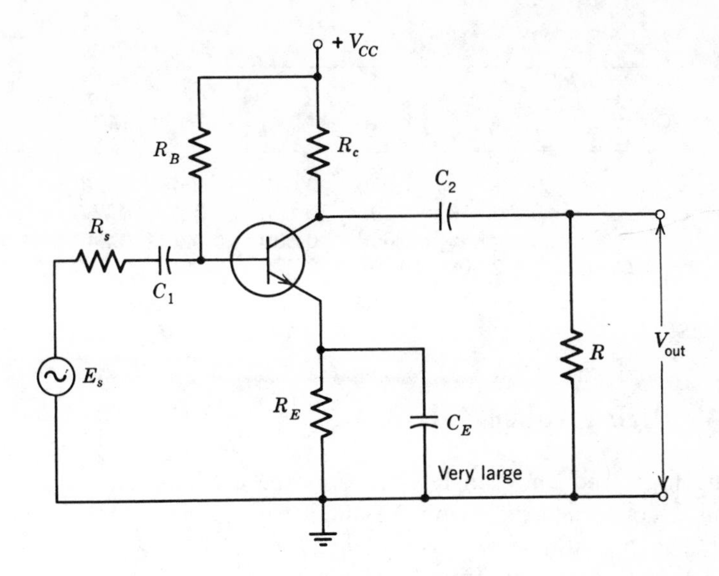

Circuit for Problems 1 and 2.

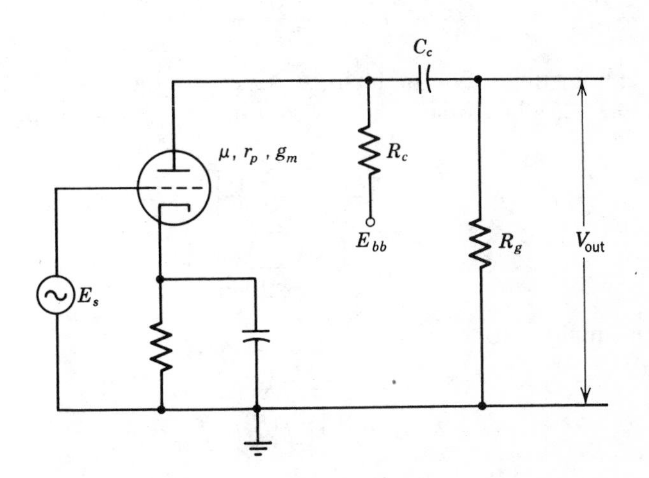

Circuit for Problems 5 and 6.

4. What values of C_1, C_2, and C_3 will set f_1 at 20 Hz for each stage in the circuit for the amplifier given in Problem 2, Section 15-2?

5. Determine the midband gain and f_1 for amplifiers A and B in Table B.

6. Determine the midband gain and f_1 for amplifiers C and D in Table B.

TABLE B

Circuit	μ	r_p, MΩ	g_m, μmhos	R_c, MΩ	C_c, μF	R_g, MΩ
A	65		1200	0.047	0.050	0.130
B		0.070	4000	0.100	0.001	0.750
C	30		2600	0.330	0.020	0.240
D		2.000	1900	0.270	0.005	0.240

Section 15-4 High-Frequency Response

The high frequency analysis ignores the coupling capacitors, C_1 and C_2, and is concerned only with the shunt capacitance values, Fig. 15-4*d*. The input section of the circuit is shown in Fig. 15-8.

As before, the mid-frequency gain is

$$A_{e,\,\mathrm{MF}} = \frac{V_2}{E_s} = \frac{R_1}{R_s + R_1} A_v \qquad (15\text{-}3)$$

The voltage divider in the high-frequency circuit (Fig. 15-8*c*) is R_s in series with the parallel combination of R_1 and $X_{C\,\mathrm{in}}$. Using the voltage divider rule,

$$V_1' = \frac{\left[\dfrac{R_1(-jX_{C\,\mathrm{in}})}{R_1 - jX_{C\,\mathrm{in}}}\right]}{R_s + \left[\dfrac{R_1(-jX_{C\,\mathrm{in}})}{R_s - jX_{C\,\mathrm{in}}}\right]}$$

which simplifies to

$$V_1' = \frac{-j\,R_1\,X_{C\,\mathrm{in}}}{R_1 R_s - j\,(R_s + R_1) X_{C\,\mathrm{in}}} E_s$$

Then

$$A_{e,\,\mathrm{HF}} \equiv \frac{V_2'}{E_s} = \frac{-j\,R_1\,X_{C\,\mathrm{in}}}{R_1 R_s - j(R_s + R_1) X_{C\,\mathrm{in}}} A_v \qquad (15\text{-}11)$$

The phasor diagram is given in Fig. 15-8*d*. The mid-frequency normal 180° phase inversion is indicated by the dashed line *OM*. *V_2' lags OM* by ϕ degrees and, consequently, at high frequencies a lagging angle is introduced. The angle ϕ can range from zero to 90°.

Dividing Eq. 15-11 by Eq. 15-3 and defining K_{HF}, the *high-frequency gain reduction factor*, we have

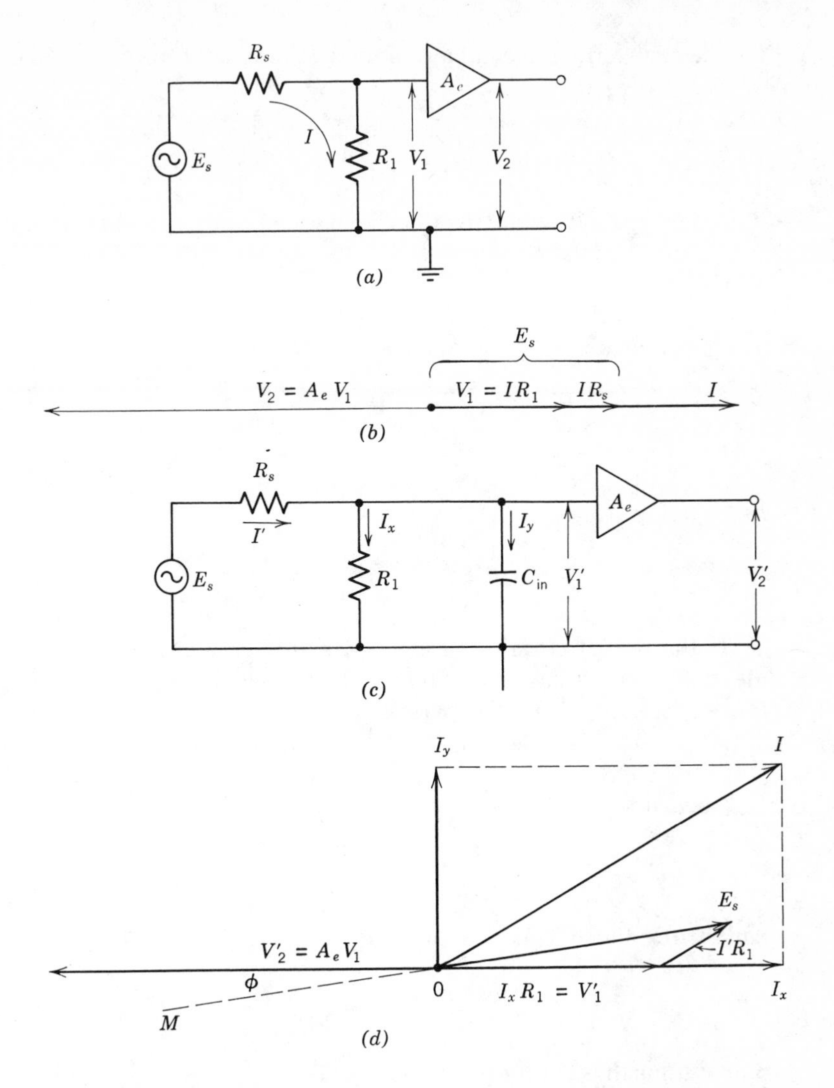

Figure 15-8 High-frequency analysis. (*a*) Model and (*b*) phasor diagram at mid-frequencies. (*c*) Model and (*d*) phasor diagram at high frequencies.

$$K_{HF} \equiv \frac{A_{e,HF}}{A_{e,MF}} = \frac{-jX_{C\,in}(R_s + R_1)}{R_1 R_s - j(R_s + R_1)X_{C\,in}}$$

Multiply each term in the numerator and denominator by j

$$K_{HF} \equiv \frac{A_{e,HF}}{A_{e,MF}} = \frac{X_{C\,in}(R_s + R_1)}{(R_s + R_1)X_{C\,in} + jR_1 R_s}$$

Divide each term in the numerator and denominator by $X_{C\,\text{in}}(R_s+R_1)$

$$K_{\text{HF}} \equiv \frac{A_{e,\text{HF}}}{A_{e,\text{MF}}} = \frac{1}{1+j\left(\dfrac{R_1R_s}{R_1+R_s}\right)\dfrac{1}{X_{C\,\text{in}}}}$$

Now define R_{eq} (R-equivalent) as the parallel combination of R_1 and R_s

$$R_{\text{eq}} \equiv \frac{R_1R_s}{R_1+R_s} \tag{15-12}$$

Then K_{HF} becomes

$$K_{\text{HF}} \equiv \frac{A_{e,\text{HF}}}{A_{e,\text{MF}}} = \frac{1}{1+j\,R_{\text{eq}}/X_{C\,\text{in}}} \tag{15-13a}$$

where

$$0 \leqslant |K_{\text{HF}}| \leqslant 1$$

In magnitude form

$$K_{\text{HF}} = \frac{1}{\sqrt{1+(R_{\text{eq}}/X_{C\,\text{in}})^2}} \tag{15-13b}$$

If the output circuit of Fig. 15-4 had been considered in this development, R_{eq} would have been the equivalent of R_3, R_2, and the output resistance of the amplifier, all in parallel.

At a particular frequency, f_2, the magnitude of $X_{C\,\text{in}}$ is the same as the magnitude of R_{eq}.

Then at f_2

$$R_{\text{eq}} = X_{C\,\text{in}} = \frac{1}{2\pi f_2 C_{\text{in}}} \tag{15-14a}$$

and the reactance of $X_{C\,\text{in}}$ at any frequency, f, is

$$X_{C\,\text{in}} = \frac{1}{2\pi f C_{\text{in}}} \tag{15-14b}$$

Substituting these into Eq. 15-13*a*

$$K_{\text{HF}} = \frac{1}{1+jf/f_2} \tag{15-15a}$$

where

$$f_2 = \frac{1}{2\pi C_{\text{in}} R_{\text{eq}}} \tag{15-15b}$$

Eq. 15-15*a* can be transformed to the rectangular form

$$K_{\text{HF}} = \frac{1}{\sqrt{1+(f/f_2)^2}} \tag{15-16a}$$

at a lagging phase angle, ϕ where

$$\tan \phi = \frac{f}{f_2} \tag{15-16b}$$

These factors are already available on the universal amplification curve provided the horizontal axis is now taken as the ratio f_2/f.

In a transistor circuit, the current gain, α or β, is defined as the short-circuit current gain. The circuit shown in Fig. 15-9*a* has a current gain α

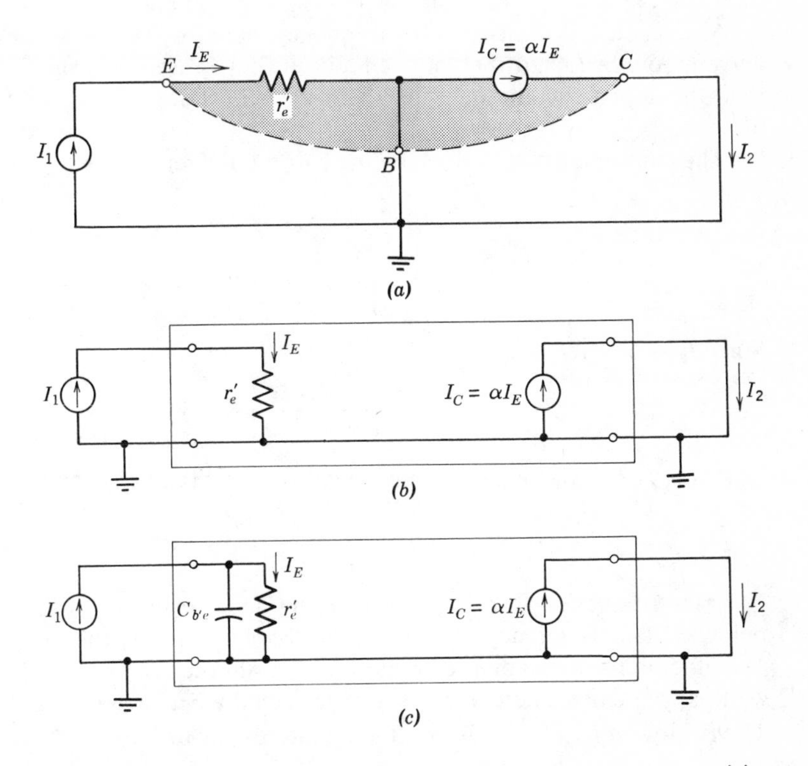

Figure 15-9 Transistor models. (*a*) Low-frequency model. (*b*) Low-frequency model arranged as a four-terminal network. (*c*) Effect of input capacitance.

given by I_2/I_1, or h_{21} is $-I_2/I_1$. This model can be arranged into the general configuration of a four-terminal network as shown in Fig. 15-9*b*.

The input capacitance of a transistor, $C_{b'e}$ is the result of complex physical factors within the transistor that involve the dielectric constant of the depletion region, charges within the transistor caused by current carriers, and the barrier layer. A typical value of $C_{b'e}$ is of the order of 100 pF for small high-frequency transistors up to several hundred or thousand pF for large power transistors. This input capacitance is shown in Fig. 15-9c as a shunting reactance across the input terminal. The effects of transistor capacitances are detailed in Section 23-1, but the consideration of only $C_{b'e}$ here is sufficient for the immediate requirement.

In Fig. 15-9c, the ratio of I_2/I_1 is now affected by the presence of the shunting capacitance $C_{b'e}$. In a transistor, it is the emitter current in r_e' that is multiplied by α to produce the collector current. Now, at high frequencies, I_1 divides between $C_{b'e}$ and r_e'. Consequently, as the frequency of the source increases, more and more current goes through $C_{b'e}$ and less and less through r_e'. As a result I_2 decreases as the frequency increases. As far as the constant-current source I_1 and the short-circuit current I_2 are concerned, it appears that the alpha-current gain of the transistor falls off as the frequency increases.

The emitter current is given by the current divider rule as

$$I_E = I_1 \frac{-j\dfrac{1}{\omega C_{b'e}}}{r_e' - j\dfrac{1}{\omega C_{b'e}}} = \frac{1}{1 + j\omega C_{b'e} r_e'} I_1$$

Since $I_2 = I_C = \alpha I_E$

$$I_2 = \frac{\alpha}{1 + j\omega C_{b'e} r_e'} I_1$$

The ratio I_2/I_1 is the apparent current gain of the circuit α'. Then

$$\alpha' = \frac{\alpha}{1 + j\omega C_{b'e} r_e'} = \frac{\alpha}{1 + j2\pi f C_{b'e} r_e'} \tag{15-17a}$$

At some frequency f_α, $\omega C_{b'e} r_e'$ is one and, as a result, α' is three decibels below α. This frequency f_α is the *alpha-frequency cutoff* that is given in the manufacturer's data for the transistor. An alternative symbol for f_α is $f_{\alpha 0}$. At this frequency, the current I_1 divides equally between $C_{b'e}$ and r_e' in Fig. 15-9c. Now at f_α, $\omega C_{b'e} r_e'$ is $2\pi f_\alpha C_{b'e} r_e'$ and solving for $2\pi C_{b'e} r_e'$, we have

$$2\pi C_{b'e} r_e' = \frac{1}{f_\alpha}$$

Substituting this back into Eq. 15-17a, we have

$$\alpha' = \frac{\alpha}{1 + jf/f_\alpha} \tag{15-17b}$$

and in rectangular form

$$\alpha' = \frac{\alpha}{[1 + (f/f_\alpha)^2]^{1/2}} \tag{15-17c}$$

We can use the four-terminal network approach and write

$$h'_{21} = \frac{h_{21}}{1 + jf/f_\alpha} \tag{15-17d}$$

We evaluated the problem on the basis of α. We could have started this problem on the basis of the common-emitter circuit which would have yielded the *beta-frequency cutoff*, f_β. An alternative symbol for f_β is $f_{\beta 0}$. In either case, the cutoff frequency is defined as that frequency at which the current gain falls 3 db or decreases to 70.7% of its value at mid-frequencies.

A conversion between f_α and f_β can be made simply by

$$f_\alpha = \frac{f_\beta}{1 - \alpha} \tag{15-18a}$$

or by

$$f_\beta = f_\alpha(1 - \alpha) \tag{15-18b}$$

For instance, the transistor used in the example in Chapter 13 has an f_α equal to 700 kHz. α equals 0.982. Then

$$f_\beta = f_\alpha(1 - \alpha) = 700{,}000(1 - 0.982) = 12{,}600 \text{ Hz or } 12.6 \text{ kHz.}$$

The effect of the drop in α and in β at high frequencies can be taken as a ratio K to be used the same as K_{LF} and K_{HF}.

$$K_{\mathrm{HF},\alpha} = \frac{1}{\sqrt{1 + (f/f_\alpha)^2}} \tag{15-19a}$$

and

$$K_{\mathrm{HF},\beta} = \frac{1}{\sqrt{1 + (f/f_\beta)^2}} \tag{15-19b}$$

The "gain" curve of the universal amplification curve can be used directly for the evaluation of Eq. 15-19*a* and Eq. 15-19*b*.

We see that the transistor connection used radically affects the frequency response. This lowered frequency response of the common-emitter circuit creates a serious problem in its use as a high-frequency amplifier. The methods that enable transistors to be used at high frequencies are studied in Chapter 23.

In conclusion, it is important to remember that, whereas at low frequencies one correction is made to account for the effect of the coupling capacitor, at high frequencies in transistor circuits there are two correction factors, one for the network because of the shunt capacitance and the other for the transistor itself because of the decrease in α and β.

PROBLEMS

1. f_α is 4 MHz for a transistor with a β of 120 and a value of r'_e of 50 Ω. What are the values of $C_{b'e}$ and f_β?
2. A high-gain transistor has an alpha-frequency cutoff of 7 MHz and a value for h_{ib} of 80 Ω. h_{fb} is −0.987. What are the values of $C_{b'e}$ and f_β?
3. $C_{b'e}$ is 1000 pF, and r'_e is 50 Ω in a transistor that has a value of 80 for β. What are the f_α and f_β cutoff frequencies?
4. $C_{b'e}$ is 200 pF and r'_e is 100 Ω. The β for this transistor ranges from 40 to 100. What are the ranges of variation for f_α and f_β?

Section 15-5 Input Loading

Now let us go back and examine the interelectrode capacitance, C_A, C_B, and C_C, in Fig. 15-4*a*, which we lumped into C_{in} and C_{out} in the high frequency model (Fig. 15-4*d*). The essential part of the model circuit is redrawn in Fig. 15-10*a*. The amplifying device, (A_v), has a gain A_v at the frequency at

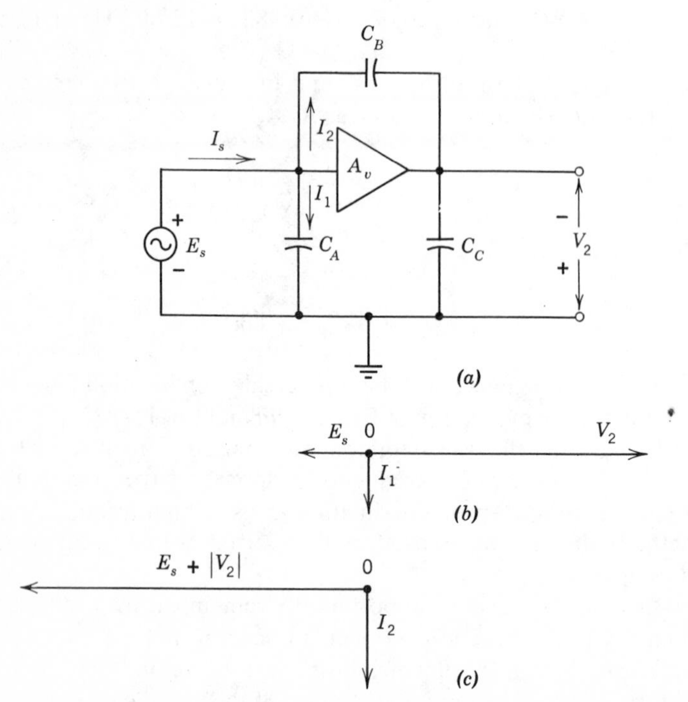

Figure 15-10 Input loading. (*a*) Circuit model. (*b*) Phasor diagram for I_1. (*c*) Phasor diagram for I_2.

which the calculation is being made. Where this is done at a high frequency, the output voltage of the amplifier is K_{HF} times the midband value of output voltage.

The source generator E_s has a current I_s. Therefore, there is a load on E_s, and the load is a parallel circuit because I_s divides into two branch currents, I_1 and I_2. The load presented by I_1 is obviously C_A. The phasor diagram for I_1 is given in Fig. 15-10*b*. The voltage from the left terminal of C_B to ground is E_s. The voltage from the right terminal of C_B to ground has the magnitude of V_2, but it is 180° out of phase with E_s. Therefore, the magnitude of the voltage across C_B is $E_s + |V_2|$. This voltage across C_B can be expressed as

$$E_s + A_v E_s = (1 + A_v)E_s$$

The current I_2 is the current through the capacitor C_B and is shown in the phasor diagram (Fig. 15-10*c*). Since the reactance of C_B is $1/2\pi f C_B$, the magnitude of I_2 is then

$$I_2 = \frac{(1 + A_v)E_s}{1/2\pi f C_B} = (1 + A_v)E_s 2\pi f C_B$$

This equation can be rearranged to

$$I_2 = [2\pi f(1 + A_v)C_B]E_s = \frac{E_s}{1/2\pi f(1 + A_v)C_B}$$

Therefore, the effective reactance of C_B is

$$\frac{1}{2\pi f(1 + A_v)C_B}$$

From this expression it is evident that, as far as the source is concerned, I_2 is flowing through a capacitor that has a value not of C_B but a value of $(1 + A_v)C_B$.

From this analysis, it is evident that C_B affects the input circuit and not the output circuit. Therefore, the input capacitance C_{in} that E_s sees is given by

$$C_{in} = C_A + (1 + A_v)C_B \qquad (15\text{-}20a)$$

The output capacitance is simply

$$C_{out} = C_C \qquad (15\text{-}20b)$$

The concept that C_B is multiplied by $(1 + A_v)$ given in Eq. 15-20*a* is known as the *Miller effect*. We assume in this development that E_s and V_2 are exactly 180° out of phase. If there is a phase angle other than 180°, the analysis becomes more complex and a resistive component is additionally reflected back into the input. This resistive component can be either positive or negative.

A tube manual gives the following data on two typical low-frequency amplifier tubes:

6SL7 Triode	6SJ7 Pentode
$C_{gp} = 2.8$ pF	$C_{gp} = 0.005$ pF
$C_{gk} = 3.0$ pF	$C_{gk} = 6.0$ pF
$C_{pk} = 3.8$ pF	$C_{pk} = 7.0$ pF

If each of these tubes is used in an amplifier stage with a gain of 30, the input capacitances of the two stages are as follows:

$$C_{\text{in}} = 3.0 + 2.8(30+1) = 3.0 + 86.8 = 89.8 \text{ pF}$$

For the pentode:

$$C_{\text{in}} = 6.0 + 0.005(30+1) = 6.0 + 0.155 = 6.2 \text{ pF}$$

Assume that the output capacitance of the previous stage plus the wiring capacitance amounts to 20 pF. In a circuit using the triode, C_s is 20 + 89.8 or 109.8 pF. If a pentode is used in the same circuit, C_s is reduced to 20 + 6.2 or 26.2 pF. The high effective shunting capacitance of the triode materially reduces the high-frequency response of the circuit. Therefore, when several stages are cascaded, it is imperative that at least one pentode stage be used in order to maintain a suitable frequency response. In high-frequency amplifiers, pentodes are normally used to the exclusion of triodes.

The interelectrode capacitances of the transistor determine its usefulness as a high-frequency amplifier. The usual circuit in a high frequency amplifier is the common emitter connection. Accordingly C_{ob} is the capacitance value of interest since it is to be multiplied by the $(1 + A_v)$ factor of the Miller effect. The capacitance values of some typical transistors are

$C_{ob} = 200$ pF	Germanium high-power audio amplifier
$C_{ob} = 40$ pF	Germanium audio amplifier
$C_{ob} = 7$ pF	Silicon for broadcast band use to 2 MHz
$C_{ob} = 2.2$ pF	Silicon for FM broadcast to 100 MHz
$C_{ob} = 0.32$ pF	Silicon high frequency to 250 MHz
$C_{ob} = 0.55$ pF	Silicon high frequency to 500 MHz

The interelectrode capacitances of FET's, especially MOSFET's, are very small, therefore, making them especially adapted for high frequency work. For example, in a MOSFET intended for general purpose applications, audio, video, and high frequencies, the input capacitance is 7 pF and the capacitance between gate and drain is 0.30 pF. This MOSFET is usable to 200 MHz.

The input capacitance in a common-emitter amplifier circuit is

$$C_{\text{in}} = C_{b'e} + C_{ob}(1 + A_v) \tag{15-20c}$$

and in a field effect transistor amplifier the input capacitance is

$$C_{\text{in}} = C_{gs} + C_{gd}(1 + A_v) \tag{15-20d}$$

PROBLEMS

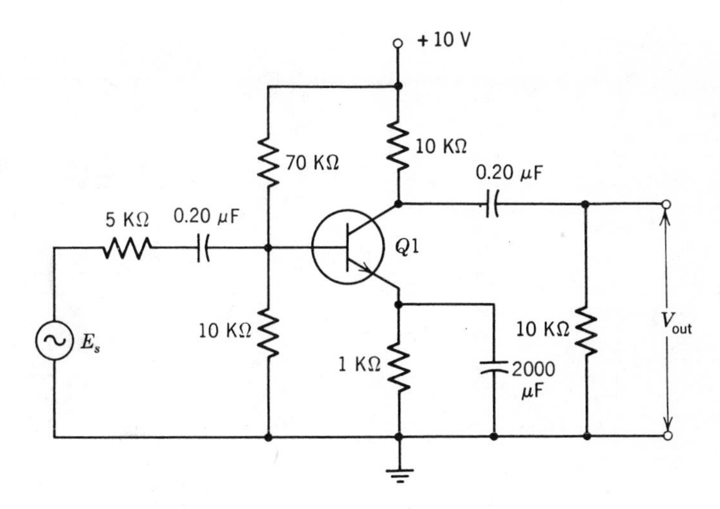

Circuit for Problem 1.

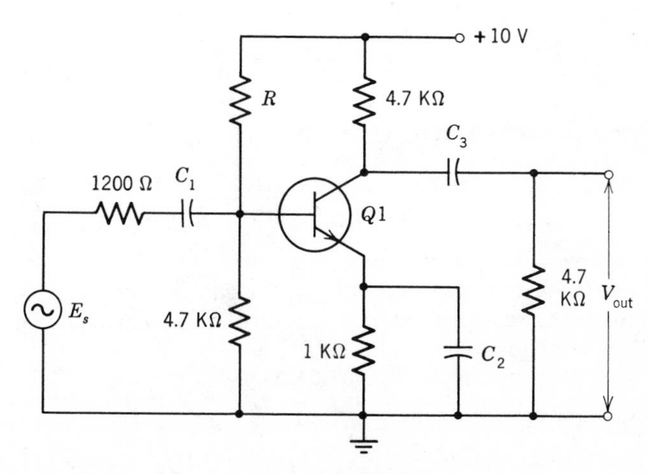

Circuit for Problem 2.

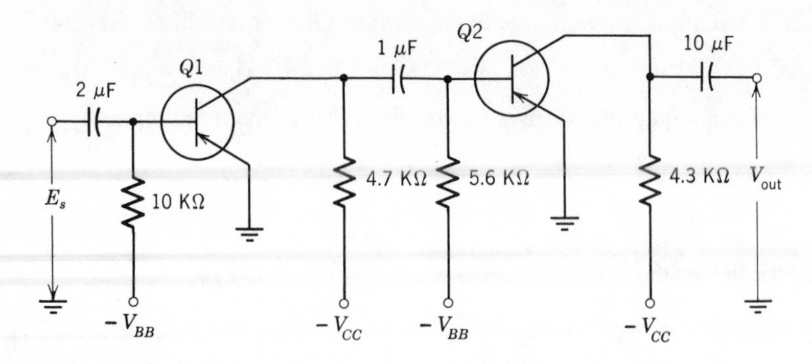

Circuit for Problems 3 and 4.

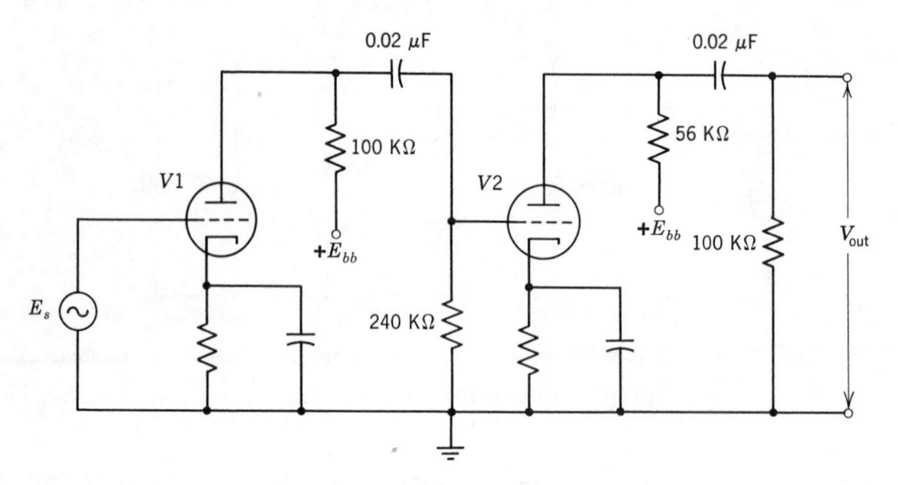

Circuit for Problems 5 and 6.

1. The data for the silicon transistor $Q1$ are

$$\beta = 100, \quad r'_e = 50\ \Omega, \quad C_{ob} = 20\text{ pF},$$
$$C_{b'e} = 200\text{ pF}, \quad \text{and} \quad f_\alpha = 5\text{ MHz}$$

a. Determine the mid-frequency gain.
b. Determine the low-frequency, f_1 for each coupling capacitor and determinte the overall circuit gain at each of these two frequencies.
c. Determine the high-frequency f_2.
d. Determine the circuit gain at 50 kHz.

2. The data for the germanium transistor $Q1$ are

$$\beta = 75, \quad r'_e = 35\text{ mv}/I_E, \quad C_{ob} = 30\text{ pF}$$
$$C_2 \text{ is very large}, \quad f_\alpha = 4\text{ MHz}$$

a. Determine the value of R to establish I_C at 1 ma.
b. Determine the mid-frequency gain of the circuit.
c. Determine the value of C_1 that sets f_1 at 100 Hz.
d. Determine the value of C_3 that sets f_1 at 150 Hz.
e. Determine the circuit gain at 60 Hz.
f. Calculate $C_{b'e}$ and evaluate C_{in}, neglecting wiring capacitances. What f_2 results from this value of C_{in}?
g. What f_2 results from a consideration of the collector load circuit only. Assume that the output capacitance is C_{ob}.
h. What is the beta-frequency cutoff?
i. What is the circuit gain at 80 kHz?

3. For the circuit diagram shown, the transistors have the following hybrid parameter values:

$h_{ie} = 1600\ \Omega$, $h_{re} = 4 \times 10^{-4}$
$h_{fe} = 40$ $h_{oe} = 25\ \mu\text{mhos}$
$C_{ob} = 40$ pF Wiring capacitance = 10 pF
Alpha-cutoff frequency = 1 MHz

Determine the stage gain at midband, at 20 Hz, and at 20 kHz.

4. Solve Problem 3 if the transistors each have the following values:

$h_{ie} = 2700\ \Omega$ $h_{re} = 5 \times 10^{-4}$
$h_{fe} = 50$ $h_{oe} = 15\ \mu\text{mhos}$
$C_{ob} = 20$ pF Wiring capacitance = 10 pF
Alpha-cutoff frequency = 700 kHz

5. Data for the two vacuum tubes are
$V1$: $\mu = 40$, $r_p = 65$ kΩ, $C_{gk} = 6$ pF, $C_{pk} = 8$ pF, $C_{gp} = 2$ pF, and $C_{\text{wiring}} = 10$ pF
$V2$: $\mu = 20$, $r_p = 10$ kΩ, $C_{gk} = 5$ pF, $C_{pk} = 4$ pF, and $C_{gp} = 1.5$ pF.
Find the mid-frequency gain and f_1 and f_2.

6. Repeat Problem 5 when the data for the two vacuum tubes are
$V1$: $\mu = 50$, $r_p = 100$ kΩ, $C_{gk} = 8$ pF, $C_{pk} = 6$ pF, and $C_{gp} = 2$ pF
$V2$: $\mu = 25$, $r_p = 30$ kΩ, $C_{gk} = 5$ pF, $C_{pk} = 3$ pF, and $C_{gp} = 1.8$ pF

Section 15-6 Decibels and Bode Plots

The fundamental relationship for determining a decibel value is

$$\text{db} = 10 \log P_2/P_1 = 20 \log E_2/E_1 + 10 \log R_1/R_2 \qquad \text{(8-1 and 8-2}b\text{)}$$

The mid-frequency gain can be calculated and converted to the decibel value. The gain at a frequency other than midband is determined by multiplying the midband gain by the proper K's. Since the gains are measured on the same circuit at different frequencies, the value of $10 \log R_1/R_2$ is zero. As forming decibels is a logarithmic operation, the multiplication by K's results in a process of adding the decibel values of the K's to the decibel value of midband gain. Also, as the numerical values of the K's lie between 0 and 1, the resulting decibel values for the K's are negative which is a loss in gain. The various K's we determined in previous sections in this chapter become

$$K_{\text{LF}} = -20 \log \sqrt{1+[X_{C_1}/(R_1+R_s)]^2} = -10 \log \{1+[X_{C_1}/(R_1+R_s)]^2\}$$
$$= -10 \log [1+(f_1/f)^2] \quad (15\text{-}21a)$$

$$K_{\text{HF}} = -10 \log [1+(R_{\text{eq}}/X_{C,\text{in}})^2] = -10 \log [1+(f/f_2)^2] \quad (15\text{-}21b)$$

$$K_{\text{HF},\alpha} = -10 \log [1+(f/f_\alpha)^2] \quad (15\text{-}21c)$$

and

$$K_{\text{HF},\beta} = -10 \log [1+(f/f_\beta)^2] \quad (15\text{-}21d)$$

In the factors of Eq. 15-21a and Eq. 15-21b a special case arises when $f=f_1$ or $f=f_2$. The ratios are one and each equation reduces to

$$K \text{ in db} = -10 \log 2 = -3 \text{ db}$$

When the gain is down 3 db the voltage ratio is $1/\sqrt{2}$ or 0.707 and the power ratio is $\frac{1}{2}$. Therefore, f_1 and f_2 are called the *3-db frequencies* or the *half-power frequencies*. The *bandwidth* is defined as (f_2-f_1).

A Bode plot is intended as a quick and simple method to draw gain curves and phase-shift curves. The major difference between the frequency response curves we have discussed and the Bode plot is that the horizontal axis can be either frequency or $\omega(2\pi f)$.

In Section 15-4, f_2 was the high frequency at which R_{eq} equals $X_{C\text{in}}$. Thus, the high-frequency 3-db point occurs at

$$f_2 = \frac{1}{2\pi C_{\text{in}} R_{\text{eq}}} \quad (15\text{-}15b)$$

Eq. 15-15b can be written in terms of ω as

$$\omega_2 = 2\pi f_2 = \frac{1}{C_{\text{in}} R_{\text{eq}}} = \frac{1}{\tau_2} \text{ when } \tau_2 = C_{\text{in}} R_{\text{eq}} \quad (15\text{-}22a)$$

In this manner the 3-db point in terms of ω is merely the reciprocal of the time constant of the circuit involved. Similarly the low frequency, f_1, can be expressed in terms of ω_1 as

$$\omega_1 = \frac{1}{\tau_1} \quad \text{when} \quad \tau_1 = C_1(R_s+R_1) \quad (15\text{-}22b)$$

The coefficients for K_{LF} and K_{HF} can be written in terms of ω as

$$K_{LF} \text{ in db} = -10 \log [1 + (\omega_1/\omega)^2] \tag{15-23a}$$

and

$$K_{HF} \text{ in db} = -10 \log[1 + (\omega/\omega_2)^2] \tag{15-23b}$$

If we consider values of K_{LF} and K_{HF} only when the factors (ω_1/ω) and (ω/ω_2) are large compared to one, these equations reduce to

$$K_{LF} \text{ in db} = -10 \log (\omega_1/\omega)^2 = -20 \log (\omega_1/\omega) \tag{15-24a}$$

and

$$K_{HF} \text{ in db} = -10 \log(\omega/\omega_2)^2 = -20 \log (\omega/\omega_2) \tag{15-24b}$$

Now in the region where Eq. 15-24*a* and 15-24*b* are valid, a ratio of the ω's of two yields a value for K of -6 db. Also, when the ratio of the ω's is ten, the value of K is -20 db. From these results, we can formulate two very useful conclusions. A ratio of two for ω or for f is an *octave*. Therefore, over an octave change in either f or ω, there is a *6 db per octave* change in K. Likewise, a ratio of ten for ω is a *decade*. Therefore, over a decade change in either f or ω, there is a change in K of *20 db per decade*.

Consider the circuit shown in Fig. 15-11*a*. The time constant for the circuit is

$$\tau = (1 \times 10^{-6})(0.1 \times 10^6) = 0.1 \text{ sec}$$

therefore, the 3-db point is

$$\omega_1 = \frac{1}{\tau} = 10 \text{ rad/sec}$$

At frequencies much higher than $f_1(\omega_1)$, the output voltage V is 10 V. At these mid-frequencies, the gain is zero db and the phase shift is zero degrees.

A horizontal line AB is drawn on the ω scale (Fig. 15-11*b*). The line CD has a slope of 20 db per decade (or 6 db per octave) and is drawn to intersect line AB at $\omega_1 = 10$, which intersection is called the *breakpoint*, or the *break-frequency*. The actual response curve must be asymptotic to these two lines, and at ω_1 the response is down 3*db*. The solid curve is the actual response of the network.

On the Bode phase plot (Fig. 15-11*c*) line AB represents the zero phase shift of the circuit at very high values of ω and line CD represents the theoretical 90° lag of the network at very low values of ω. At the breakpoint ω_1 the phase shift is exactly 45°. This is point F. Through point F a dotted line EG is drawn to give the full 90° phase change over *two* decades. The actual phase variation is shown as the solid curve that goes through point F and is asymptotic to AB and to CD.

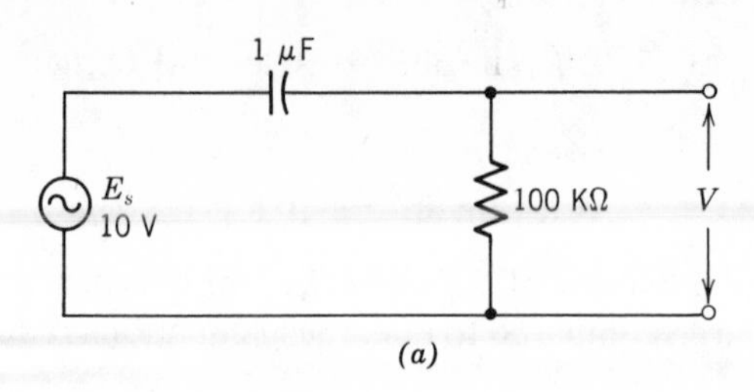

(a)

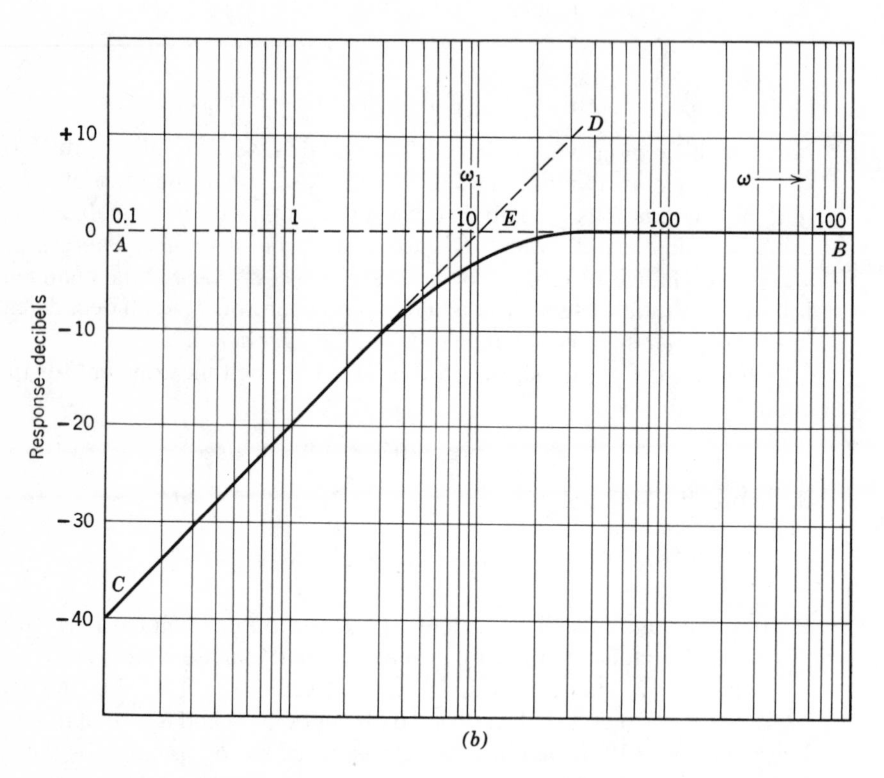

(b)

Figure 15-11 Bode response of a network. (a) Circuit. (b) Bode plot of response.

The curves given in Fig. 15-12 and Table A detail some of the corrections between the actual curve and the straight-line approximations. In practice, the Bode plots are straight lines only. Those who use them realize the fact that the actual curves deviate slightly from the straight lines.

Let us consider a transistor amplifier that has the following specifications.

$$A_{e,\text{MF}} = +60 \text{ db} \qquad \omega_1 = 10 \text{ rad/sec}$$
$$\omega_2 = 10{,}000 \text{ rad/sec} \qquad f_\beta = 100{,}000/2\pi \text{ Hz}$$

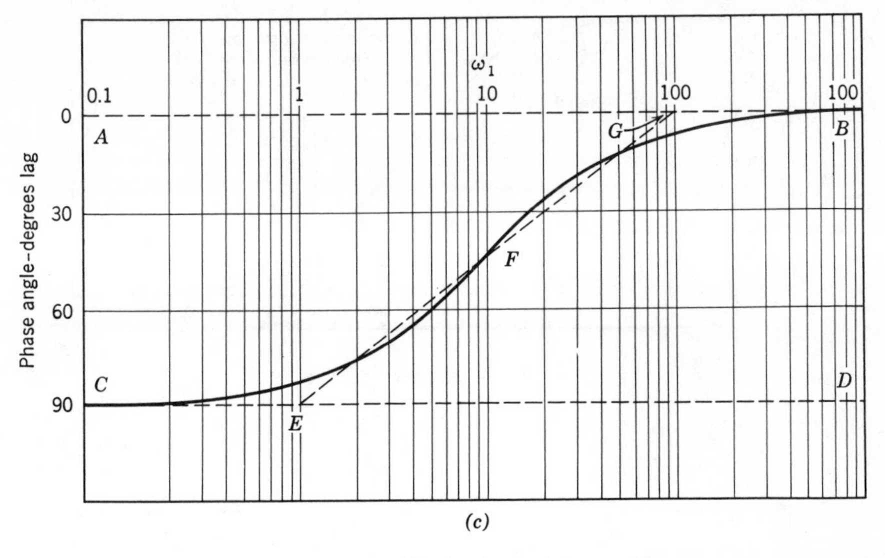

(c) Bode phase plot.

TABLE A Corrections to Approximate Straight-line Phase-shift or Phase-Angle Curve

ω	Corrections
$0.05\omega_c$	$-3°$
$0.1\omega_c$	$-6°$
$0.3\omega_c$	$+5°$
$0.5\omega_c$	$+5°$
$1.0\omega_c$	$0°$
$2.0\omega_c$	$-5°$
$3.0\omega_c$	$-5°$
$10.0\omega_c$	$+6°$
$20.0\omega_c$	$+3°$

The mid-frequency gain of 60 db is a horizontal straight line on the Bode plot, Fig. 15-13a, because there is no variation in the response with frequency. The breakpoints are ω_1 and ω_2 and, accordingly, the low-frequency gain and the high-frequency gain each fall off at the rate of 20 db per decade. On the phase plot, the phase angles are each 45° at these breakpoints. At ω_3 the effect of the fall off in the beta is treated. At an ω of 100,000 rad per

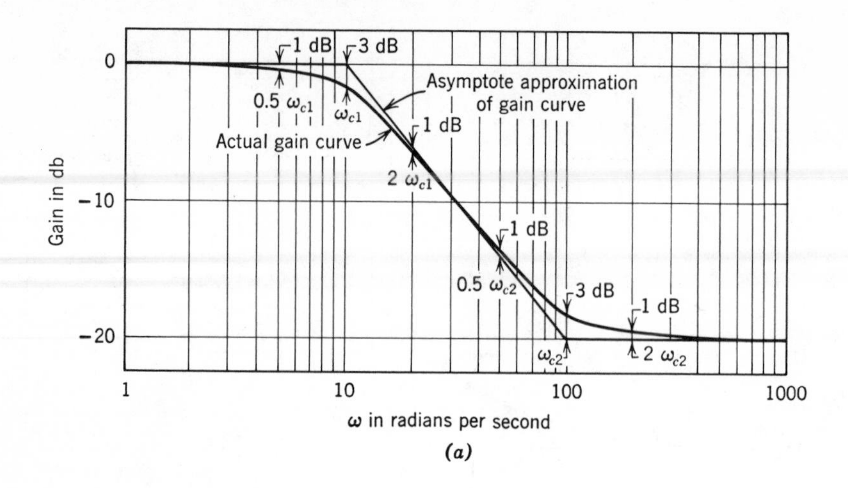

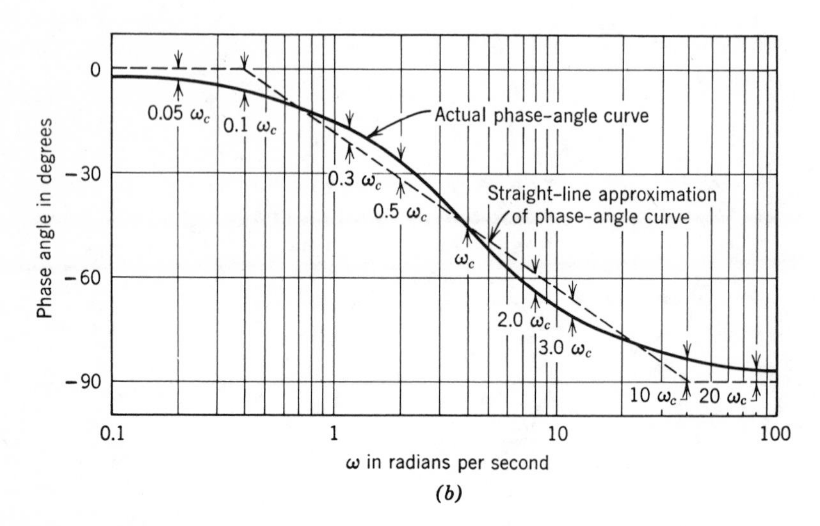

Figure 15-12 Corrections to approximate straight-line Bode plots (*Courtesy B. I. DeRoy, Automatic Control Theory, Wiley, 1966.*)

sec, the beta is off by 3 db. By Eq. 15-21*d*, there is a 20 db per decade *roll-off* by this effect in beta. Since there already is a roll-off of 20 db per decade starting at ω_2, we now have the combined roll-off caused by ω_2 and now ω_3. Consequently, the roll-off beyond ω_3 is at the rate of $20+20$ or 40 db per decade. The drop in beta does not affect the phase diagram since a drop in beta does not introduce phase effects.

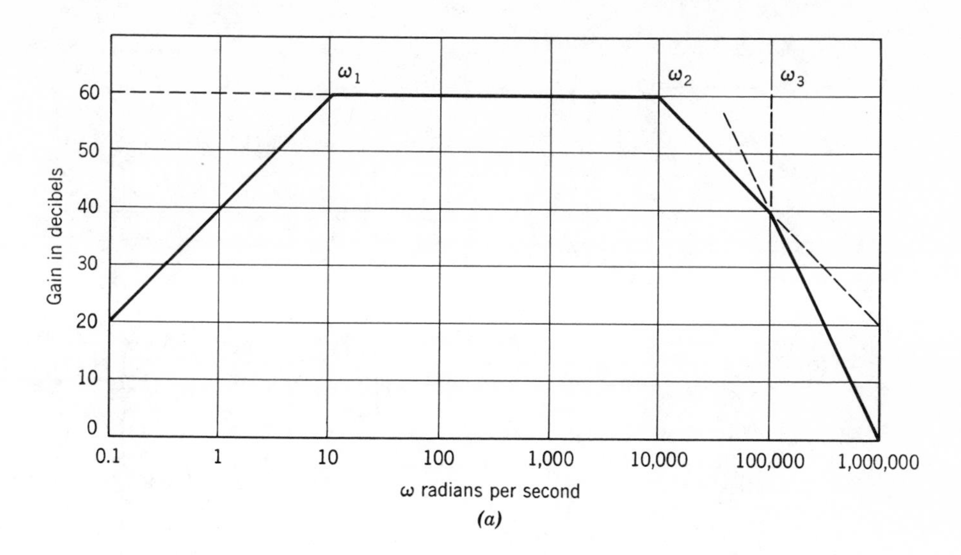

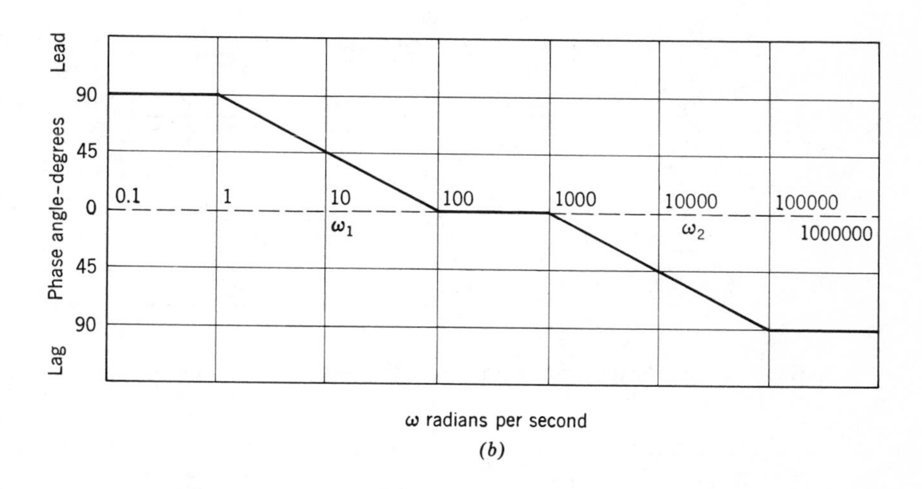

Figure 15-13 Amplifier response bode plots. (*a*) Gain plot. (*b*) Phase plot.

PROBLEMS

1. Sketch a Bode plot showing gain and phase variations for V_{out}/V_{in} for each network.
2. Sketch a Bode plot showing gain and phase variations for I_2/I_1.

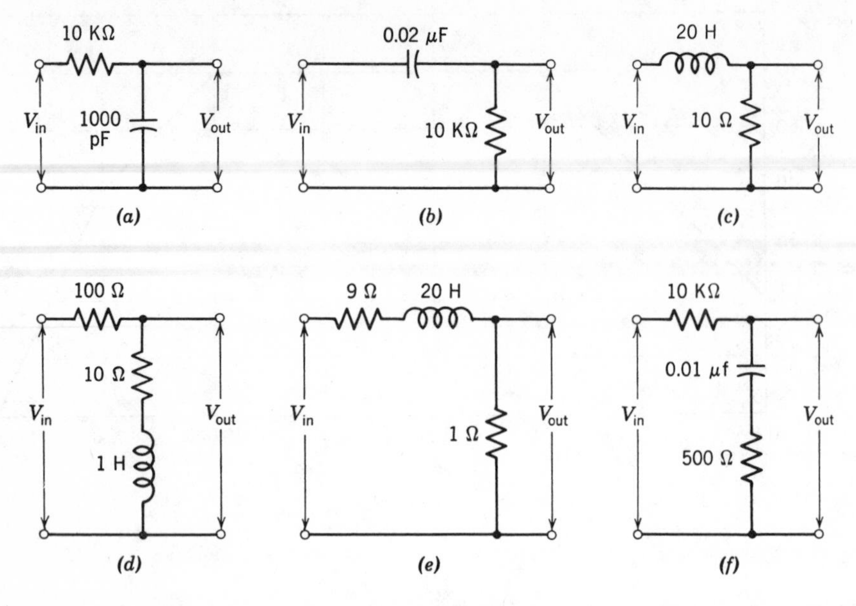

W-682 Circuits for Problem 1.

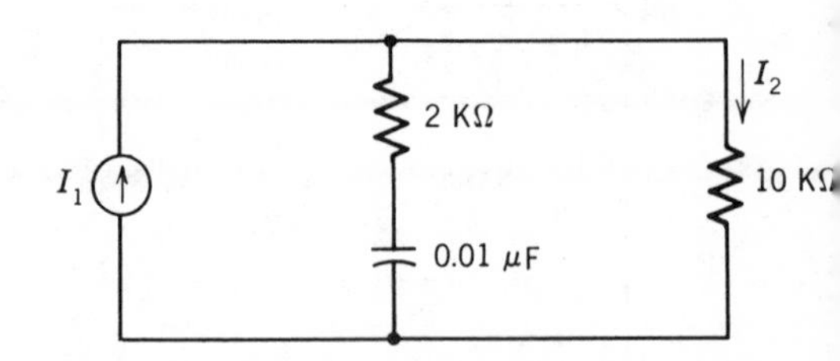

Circuit for Problem 2.

3. Sketch a Bode plot showing gain and phase variations for A_e for each of the following cases:

(a) $\tau_a > \tau_b > \tau_c > \tau_d$
(b) $\tau_d > \tau_c > \tau_b > \tau_a$
(c) $\tau_a > \tau_c > \tau_b > \tau_d$
(d) $\tau_c > \tau_a > \tau_d > \tau_b$

$$\frac{V_{out}}{V_{in}} = \frac{(1+\omega\tau_a)(1+\omega\tau_b)}{(1+\omega\tau_c)(1+\omega\tau_d)}$$

Section 15-7 Volume and Tone Controls

An amplifier is usually equipped with a manually operated potentiometer which serves as a voltage divider to control the signal level at the grid of one of the stages (Fig. 15-14).

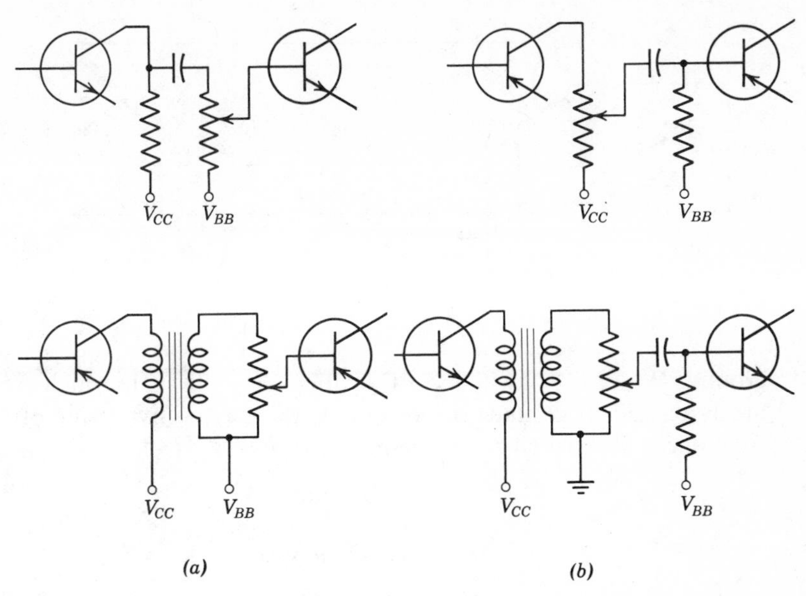

Figure 15-14 Volume-control placement in transistor circuits. (*a*) Incorrect level control circuits. (*b*) Correct level control circuits.

Conventional wiring practice requires that a clockwise rotation produces an increase in level output for volume controls. Since the level of hearing is logarithmic (Chapter 8), the resistance taper of the audio control cannot be linear but must be logarithmic, in order to have the desired perceptible and smooth increase in listening level with a uniform clockwise rotation of the control shaft. Many applications of amplifier circuits do require linear tapers on the level controls.

Volume controls are available in which the level is not continuously variable but is switched in equal decibel steps. These are called *attenuators*. Fixed-resistance networks which reduce the level are called *pads* and are usually inserted in or removed from the circuit by special switches called *keys*. Both attenuators and pads are designed in various network configurations: *L pads*, *T pads*, *H pads*, *lattices*, etc.

In transistor circuits using level controls, the control must be placed in the circuit in such a position that it does not change the operating bias values when the setting is changed. Figure 15-14*a* shows incorrect methods of control connections. In both examples, a change in the variable arm of the potentiometer changes the bias. In Fig. 15-14*b*, the bias does not depend on the setting of the level control.

In the circuit of the three-position tone control (Fig. 15-15), when C is switched to position 1, most of the high frequencies are bypassed to

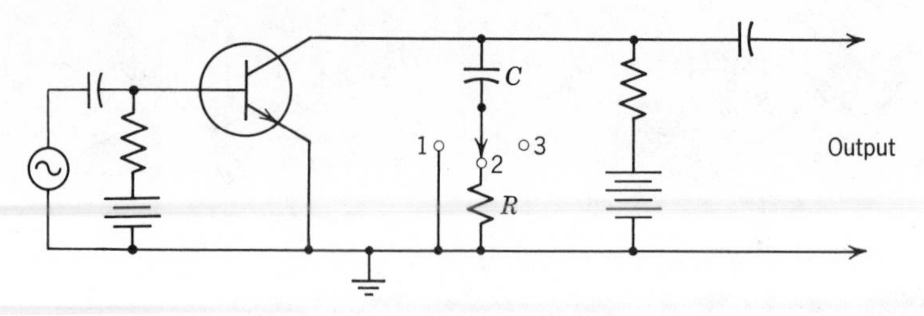

Figure 15-15 Three-position tone control.

ground. When C is switched to position 2, some of the high frequencies are bypassed to ground. In position 3, there is no bypassing of the signal frequency. The switch positions may be labeled:

No. 1 Bass
No. 2 Normal
No. 3 Treble

The circuit of Fig. 15-16 uses a volume control which has a tap T, from which point a capacitor is placed to ground. When the volume control is set at a high level, the effect of the capacitor is negligible on the high-frequency response. As the level is reduced, the bypassing action of the

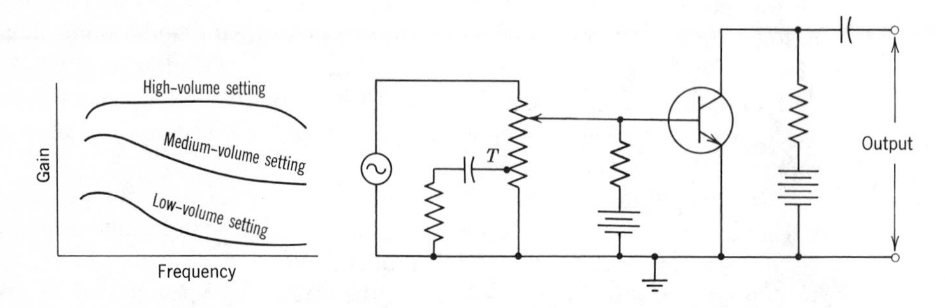

Figure 15-16 Method of tapping a volume control for tone compensation.

capacitor at high frequencies becomes greater. The resulting curves are an attempt to approximate the equal-loudness contours shown in Fig. 8-3.

If equal-loudness contours are duplicated in an amplifier, theoretically there is no need for a tone control. There are available a number of circuits and commercial units which appear under the designation of *loudness controls* to serve this purpose. Some arrangements, such as the circuit of Fig. 15-16, compensate only for the low- and middle-frequency ranges whereas other more complex circuits approach the loudness contours at all frequencies and at all levels.

Very often audio circuits become quite complex in order to allow for separate bass and treble controls. Other circuits provide special response characteristics to *equalize* or compensate for the special frequency-response characteristics of transducers, records, speakers, etc. A few of the infinite variety of circuit arrangements that are used to obtain special frequency responses are shown in Fig. 15-17.

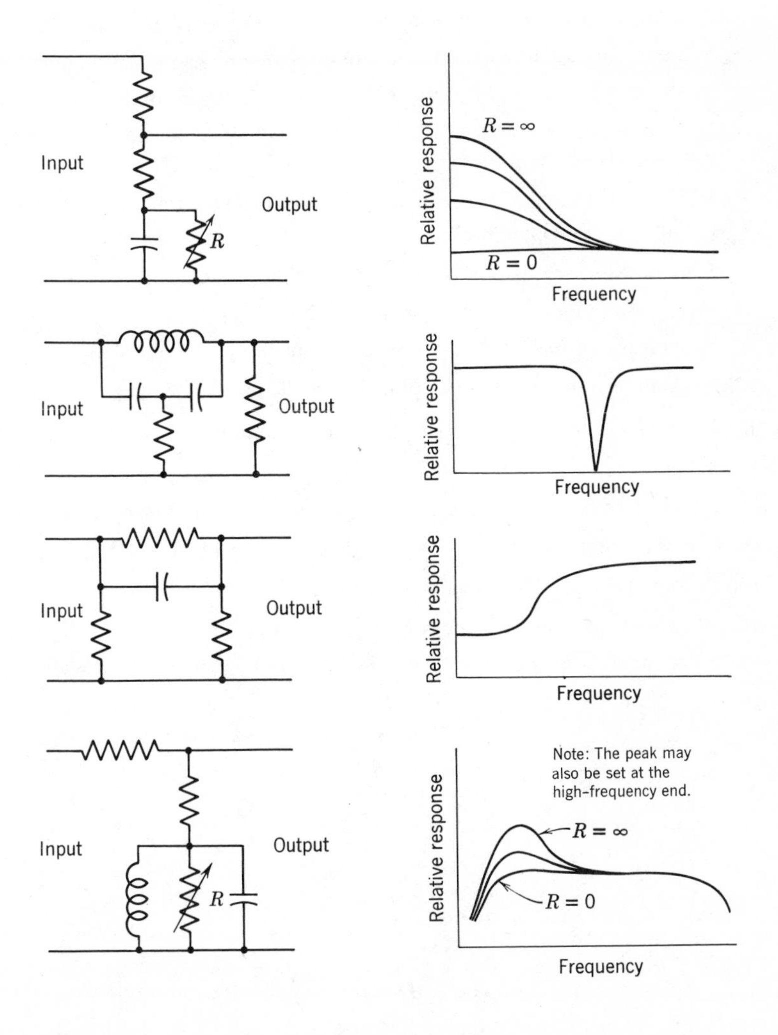

Figure 15-17 Typical passive networks designed to provide special frequency responses.

Questions

1. Define or explain each of the following terms: (*a*) static load line, (*b*) dynamic load line, (*c*) optimum Q point, (*d*) middle frequencies, (*e*) low frequencies, (*f*) high frequencies, (*g*) attenuation, (*h*) K_{LF}, (*i*) K_{HF}, (*j*) f_α, (*k*) f_β, (*l*) Miller effect, (*m*) universal curve, (*n*) Bode plot, (*o*) rolloff.
2. What is the optimum bias point for a transistor amplifier stage?
3. What is the optimum bias point for a vacuum-tube amplifier stage?
4. What factors influence the low-frequency response of an amplifier?
5. What factors influence the high-frequency response of an amplifier?
6. Why are the individual voltage gains of two stages multiplied and not added?
7. What information does a load-line analysis give that cannot be obtained from the analysis of a model?
8. What information does the analysis of a model give that cannot be obtained from the analysis of a load line?
9. Why does α decrease for a transistor at high frequencies?
10. Why is input loading a problem and what are its effects?
11. What errors are made in assuming a Bode plot is made of straight-line segments?
12. What determines the break-points in a Bode plot?
13. What is the slope of the phase plot at a break point?
14. What is the change of slope of the gain plot at a break point?
15. Does 20 db per decade and 6 db per octave represent the same slope?
16. Can the slope in a gain plot be as high as 60 db per decade? Explain.
17. Point out the need for equalization circuits.
18. Why are tone controls necessary?
19. Explain why the circuits of Fig. 15-17 yield the response curves shown.

Chapter Sixteen

SPECIAL AMPLIFIERS

Complementary symmetry (Section 16-1) has no counterpart in electron-tube circuitry and is a method that permits the successive cascading of transistor stages while using a low-voltage power source. The Darlington pair (Section 16-2) is a standard semiconductor circuit wherein two transistors are connected in tandem. Darlington pairs are often packaged into a single case. Differential or difference amplifiers are used in both discrete circuitry and in microelectronics. They can be used to compare two signals or to convert a single-ended signal into a balanced signal (Section 16-3), to amplify a balanced signal (Section 16-4), and to convert a balanced signal back into a single-ended output signal (Section 16-5). The limitations of differential amplifiers are considered (Section 16-6). A third transistor is used to provide current stabilization in integrated circuit design.

The function of a clipping circuit (Section 16-8) and the operation of the vacuum-tube grounded-grid amplifier (Section 16-9) complete the chapter.

Section 16-1 Complementary Symmetry

The concept of *complementary symmetry* is unique in its application to transistors. The circuit (Fig. 16-1) requires two transistors. In this example, $Q1$ is a *PNP* transistor and $Q2$ is an *NPN* transistor. In order to understand the circuit operation, the d-c operating potentials are given on the diagram. Since the base voltage on $Q1$ is -3.2 V and the emitter voltage is -3 V, the emitter is 0.2 V positive with respect to the base which is the correct polarity for the *PNP* transistor. The collector, at -10 V, is directly coupled to the base of $Q2$. The emitter voltage of $Q2$ is -10.2 V, making the emitter of $Q2$, the *NPN* unit, negative with respect to the base. The collector-

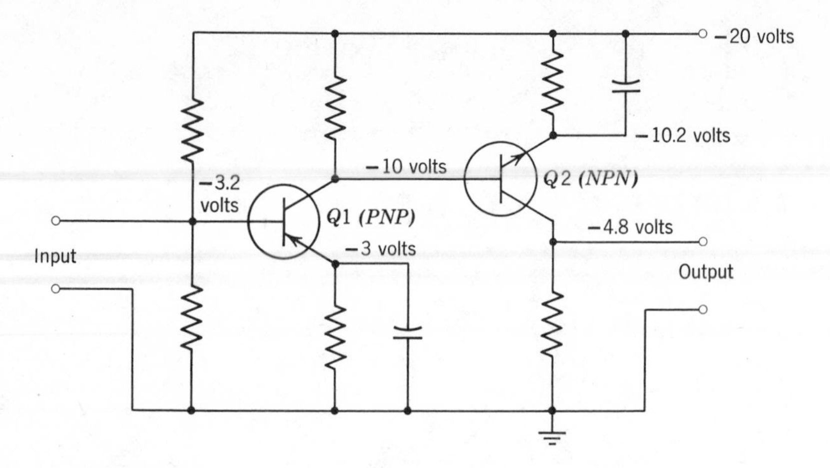

Figure 16-1 Circuit showing complementary symmetry.

voltage of $Q2$ is -4.8 V, making the collector less negative or positive wit
respect to the emitter. The voltage distribution of this circuit is sketched i
Fig. 16-2, showing the relative voltages of the electrodes of the transistors

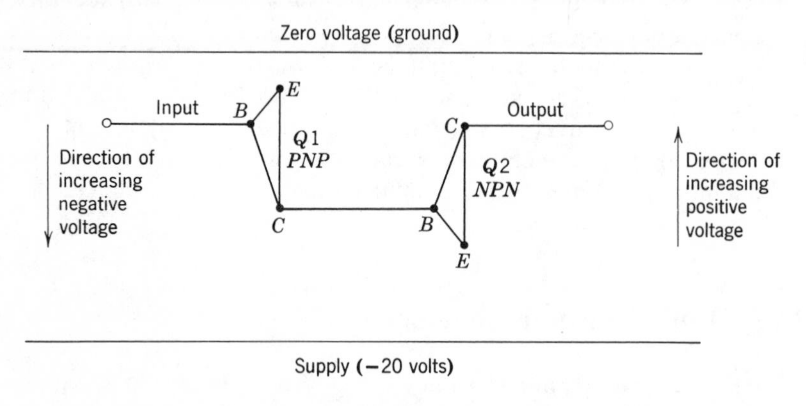

Figure 16-2 Operating voltage levels in complementary symmetry.

By adjusting the values of the resistors, the output d-c potential can be mad
the same as the input d-c potential.

A positive input signal reduces the emitter to base voltage on $Q1$. Thi
reduction occurs because this positive signal is applied to the negative bas

instead of the positive emitter. This reduction of emitter-to-base voltage causes the collector current in $Q1$ to decrease. A decreased collector current lowers the voltage drop across the collector load resistance. A decreased load-voltage drop causes the collector voltage of $Q1$ to become more negative. Since the base of $Q2$ is directly coupled to the collector of $Q1$, the base of $Q2$ becomes more negative. Now the signal on $Q2$ is smaller, and this reduced signal decreases the collector current in $Q2$. The load-voltage drop across the output resistor becomes smaller, making the output (the collector voltage of $Q2$) less negative. A voltage that becomes less negative shows a change in the positive direction. In this manner, a positive signal on the input of $Q1$ produces a positive output signal from $Q2$. This information is evident, since the circuit consists of two common-emitter amplifier stages, each of which produce a phase shift of 180°, and which, when taken together, shift the signal 360° or back in phase.

Semiconductor manufacturers have the complementary-symmetry pairs of transistors available—one a *PNP* and the other an *NPN*. The pairs are available in a series of different current and power ratings. A further discussion of complementary-symmetry is in Section 18-7. The use of the complementary-symmetry principle in push-pull amplifiers is most important.

PROBLEMS

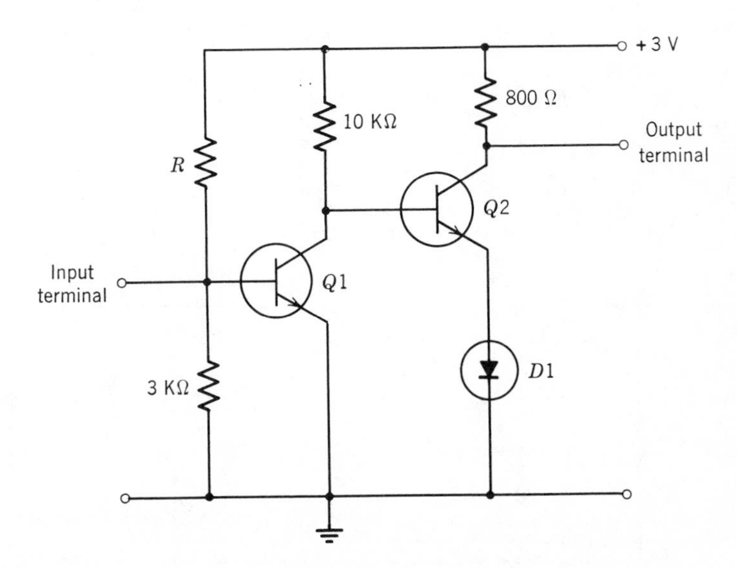

Circuit for Problems 1 and 2.

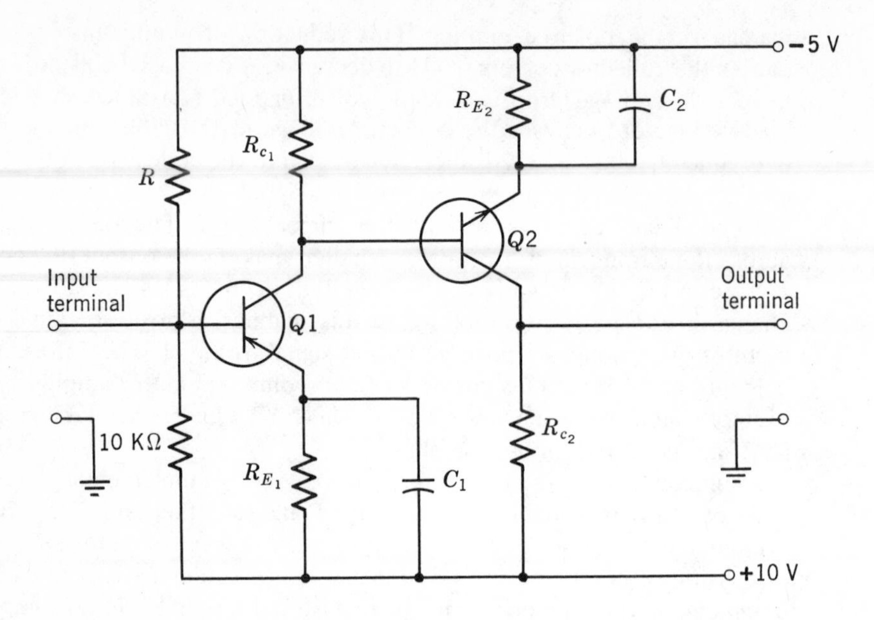

Circuit for Problems 3 and 4.

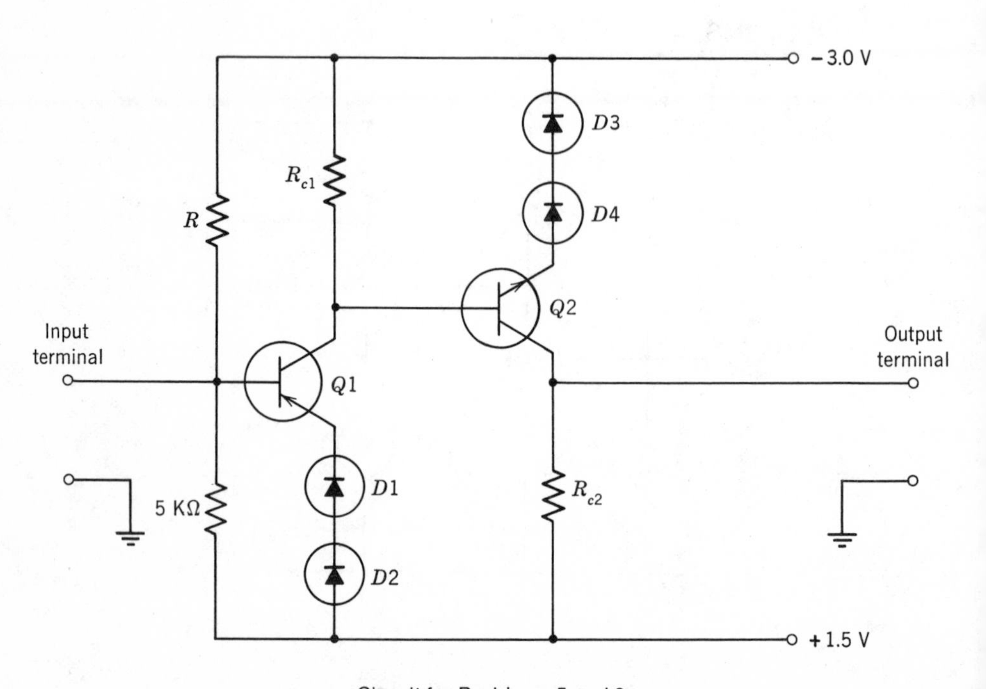

Circuit for Problems 5 and 6.

1. The active elements of the two-stage direct-coupled *IC* amplifier have the following values

$Q1$:	$V_{BE} = 150$ mv,	$r'_e = 50$ mv$/I_E$,	$\beta = 40$
$Q2$:	$V_{BE} = 180$ mv,	$r'_e = 50$ mv$/I_E$,	$I_C = 1.4$ ma, $\beta = 35$
$D1$:	$r'_e = 50$ mv$/I$,	forward direct voltage drop $= 700$ mv	

   Determine the value of R and the d-c levels at the input and the output terminals. What is the circuit voltage gain and what is the maximum peak-to-peak output voltage without clipping?
2. An *IC* made to the specifications of Problem 1 now has a value of 50 for β for each transistor. What is the effect of the increase in β on the performance of the *IC*?
3. The d-c levels at the input terminal and at the output terminal must be zero without signal in the complementary-symmetry amplifier. Both transistors have the characteristics:

$$I_C = 5 \text{ ma}, \quad r'_e = 50 \text{ mv}/I_E, \quad V_{BE} = 0.25 \text{ V}, \quad V_{CE} = 3 \text{ V, and } \beta = 50$$

   Determine the values of the various resistors and the circuit voltage gain.
4. If C_2 is omitted from the amplifier of Problem 3, what is the circuit voltage gain?
5. The complementary-symmetry amplifier uses diodes in the emitter circuit. The diodes have a forward voltage drop of 600 mv and an a-c resistance given by $r'_e = 50$ mv$/I$. For the transistors, β is 50, V_{BE} is 300 mv, and $r'_e = 50$ mv$/I_E$. $I_{C,1}$ is 1 ma and $I_{C,2}$ is 5 ma. The d-c input and output levels are at ground potential. Determine the required resistance values and the circuit voltage gain.
6. Repeat Problem 5 if three diodes are used in series in each emitter circuit and if the positive voltage supply is raised to 2.1 V.

Section 16-2 The Darlington Pair

The Darlington pair is the name given to a circuit (Fig. 16-3*a*) in which the emitter of one transistor is connected directly to the base of a second transistor. The emitter current of the first transistor is the base current of the second transistor. Darlington pairs are available commercially in one case enclosure that has only three leads, a collector connection, the base input of the first transistor, and the emitter output lead from the second transistor. The Darlington pair connection is readily fabricated from two adjacent transistors in microcircuits.

In considering the Darlington pair, first a d-c analysis must be made.

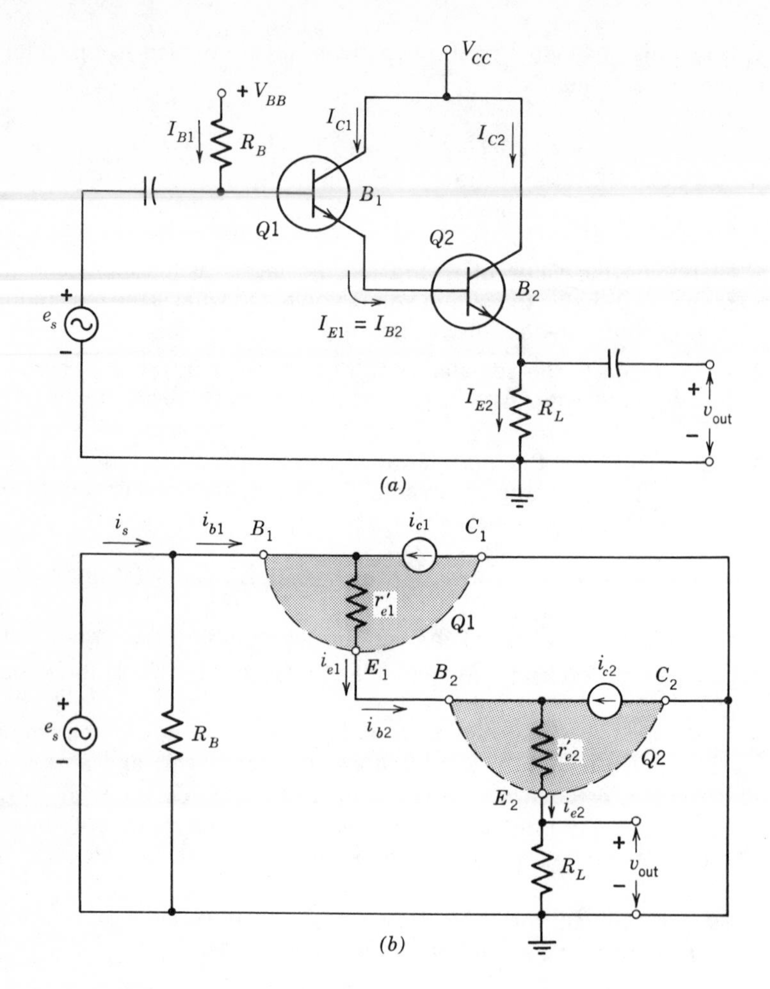

Figure 16-3 The Darlington pair. (*a*) Circuit. (*b*) Model.

A Kirchhoff's voltage loop equation through the base resistor, R_B, is

$$I_{B_1}R_B + V_{BE_1} + V_{BE_2} + I_{E_2}R_L = V_{BB}$$

but

$$I_{B_2} = \frac{I_{E_2}}{1+\beta_{dc_1}} = I_{E_1}$$

and

$$I_{B_1} = \frac{I_{E_1}}{1+\beta_{dc_1}} = \frac{I_{E_2}}{(1+\beta_{dc_1})(1+\beta_{dc_2})}$$

or

$$I_{E_2} = (1+\beta_{dc_1})(1+\beta_{dc_2})I_{B_1}$$

Substituting

$$I_{B_1}R_B + (1+\beta_{dc_1})(1+\beta_{dc_2})\,R_L I_{B_1} = V_{BB} - V_{BE_1} - V_{BE_2}$$

$$I_{B_1} = \frac{V_{BB} - V_{BE_1} - V_{BE_2}}{R_B + (1+\beta_{dc_1})(1+\beta_{dc_2})R_L} \approx \frac{V_{BB}}{R_B + \beta_{dc_1}\beta_{dc_2}R_L} \tag{16-1a}$$

Assume that β_{dc_1} and β_{dc_2} are equal. This is especially true in microcircuitry where adjacent transistors have very close parameters.
Then

$$I_{B_1} \approx \frac{V_{BB}}{R_B + \beta_{dc}^2 R_L} \tag{16-1b}$$

This same loop equation can be solved easily for I_{E_2} instead of I_{B_1}

$$I_{E_2} = \frac{V_{BB} - V_{BE_1} - V_{BE_2}}{R_L + R_B/(1+\beta_{dc_1})(1+\beta_{dc_2})} \approx \frac{V_{BB}}{R_L + R_B/\beta_{dc_1}\beta_{dc_2}} \tag{16-1c}$$

and when

$$\beta_{dc_1} = \beta_{dc_2}$$

$$I_{E_2} = \frac{V_{BB}}{R_L + R_B/\beta_{dc}^2} \tag{16-1d}$$

For an a-c signal analysis of the Darlington pair, the model (Fig. 16-3*b*) must be used. An inspection of the circuit and the model shows that the circuit is effectively two cascaded emitter-follower amplifiers. Since the emitter follower does not introduce a phase inversion of signal, the output and the input of the Darlington pair are in phase.

The emitter-follower stage was analyzed in Section 12-7. If the same methods of analysis are applied to the Darlington pair, the algebra becomes rather complex. If the results of Section 12-7 are applied directly to the Darlington configuration, the final equations can be obtained in a much simpler fashion. All β's are the a-c values.

The input resistance r_2 to $Q2$ is $(1+\beta_2)(r'_{e_2}+R_L)$. This value is the load resistance in the emitter circuit of $Q1$. Since $Q1$ is also an emitter follower, its input resistance r_1 is

$$\begin{aligned} r_1 &= (1+\beta_1)(r'_{e_1}+r_2) = (1+\beta_1)[r'_{e_1} + (1+\beta_2)(r'_{e_2}+R_L)] \\ &= h_{ie_1} + (1+\beta_1)h_{ie_2} + (1+\beta_1)(1+\beta_2)R_L \end{aligned} \tag{16-2a}$$

If the values of the betas are both high and equal, the equation for r_1 simplifies to

$$r_1 = h_{ie_1} + \beta h_{ie_2} + \beta^2 R_L \approx \beta^2(r'_{e_2}+R_L) \tag{16-2b}$$

Then, the input resistance to the circuit is the parallel combination of R_B and r_1:

$$r = \frac{r_1 R_B}{r_1 + R_B} \tag{16-2c}$$

The voltage gain of the second stage is simply the voltage divider

$$A_{v2} = \frac{R_L}{r'_{e2} + R_L} \tag{16-3a}$$

If, and only if, R_L is much larger than r'_{e2}, Eq. 16-3*a* simplifies to

$$A_{v2} \approx 1 \tag{16-3b}$$

If the value of R_L were as low as zero ohms, r'_{e2} would be reflected up to $Q1$ as $(1+\beta_2)r'_{e2}$. Then the gain of the $Q1$ stage is

$$A_{v1} = \frac{(1+\beta_2)r'_{e2}}{r'_{e1} + (1+\beta_2)r'_{e2}}$$

Remembering that we are considering a small value for R_L and that the value of β_2 is large, this gain must be extremely close to one. Consequently, the total voltage gain of the Darlington pair is the same as that given by Eq. 16-3*a*:

$$A_v = \frac{R_L}{r'_{e2} + R_L} \tag{16-3c}$$

and when R_L is much larger than r'_{e2}

$$A_v \approx 1 \tag{16-3d}$$

If the input current to the base of $Q1$ is i, the input current to the base of $Q2$ is the emitter current of $Q1$, which is $(1+\beta_1)i$. Then the load current is the emitter current of $Q2$, which is $(1+\beta_2)(1+\beta_1)i$. The current gain of the circuit is

$$A_i = (1+\beta_1)(1+\beta_2) \approx \beta_1\beta_2 \tag{16-4a}$$

When β_1 equals β_2, Eq. 16-4*a* reduces to

$$A_i \approx \beta^2 \tag{16-4b}$$

The power gain of the circuit is approximately

$$A_p \approx \beta^2 \tag{16-5}$$

If the effect of R_B is to be considered in taking the current gain, then a reduction factor must be used:

$$A_i = \frac{R_B}{R_B + r_1}\beta_1\beta_2 \tag{16-4c}$$

wherein r_1 is given by Eq. 16-2*a* or by Eq. 16-2*b*.

To determine the output impedance of the circuit, we first look up into the emitter of $Q1$. Any resistance in the source is placed in parallel with R_B to give the equivalent resistance r_s. r_s reflects through $Q1$ into the emitter as $r_s/(1+\beta_1)$. The total resistance seen looking into the emitter of $Q1$ is then

$$r'_{e_1}+r_s/(1+\beta_1)$$

This resistance reflects through $Q2$, so that the resistance seen looking up into the emitter of $Q2$ is

$$r'_{e_2}+[r'_{e_1}+r_s/(1+\beta_1)]/(1+\beta_2)$$

From a practical viewpoint, r_s cannot be greater than a few thousand ohms at the most, and if β_1 and β_2 are 100 or more, the last term of the expression must be much less than 1 Ω. Consequently, the resistance looking up into $Q2$ is r'_{e_2}. Then the output impedance of the Darlington pair is the parallel combination of R_L and r'_{e_2}:

$$r_{\text{out}}=\frac{r'_{e_2}R_L}{r'_{e_2}+R_L} \tag{16-6}$$

In summary, the essential characteristics of a Darlington pair are

1. Very high input resistance,
2. Very low output resistance,
3. Unity voltage gain, and
4. A high current gain (β^2_{ac}).

PROBLEMS

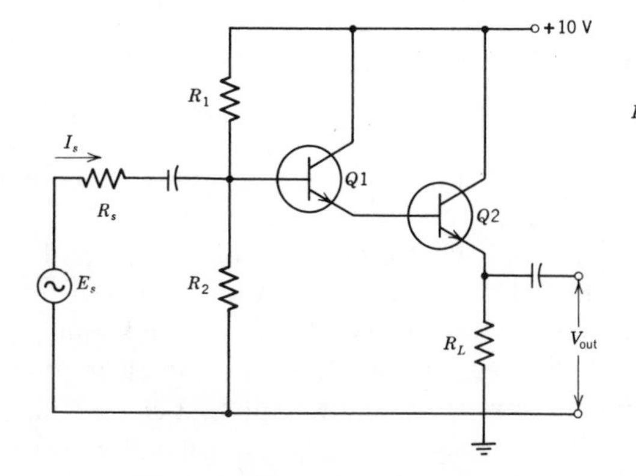

Circuit for Problems 1 to 3.

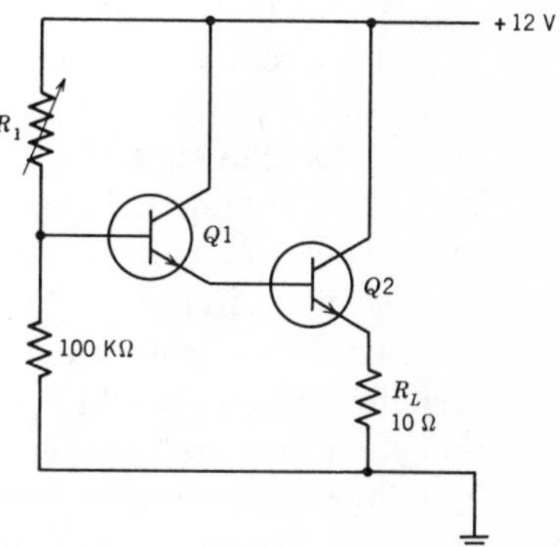

Circuit for Problems 4 and 5.

For all problems, r'_e is 50 mv/I_E.

1. R_L is 1000 Ω, R_2 is infinite, and R_s is 10 kΩ, V_{CE} for $Q2$ is 5 V. V_{BE} is 700 mv and β is 50 for $Q1$ and $Q2$. Find R_1. Find E_s and I_s that produces the largest output signal without clipping. Draw the a-c model.
2. R_L is 10 Ω, R_s is 50 kΩ, R_2 is 1 MΩ, V_{BE} is 600 mv, and β is 60. V_{CE} is 5 V for $Q2$. Find the current gain, the voltage gain, and the output impedance. Draw the a-c model.
3. Repeat Problem 2 if transistors that have a β of 120 are used. Compare the results.
4. The load R_L is a control winding of a d-c generator. The transistors have a β of 100 and a value of 500 mv for V_{BE}. R_1 is a control resistor that is used to vary the current in the generator control winding from 0.3 A to 0.9 A. What is the required range of variation in R_1?
5. Repeat Problem 4 if the transistors with a β of 140 are used.

Section 16-3 Differential Amplifier—Single Input, Balanced Output

Before proceding with the circuit analysis in this section, some general comments on the *differential* or *difference* amplifier are in order. These comments apply to the circuits considered in Sections 16-4 and 16-5 as well. Two semiconductors are required for the circuit. We are assuming they are matched in all respects. If this is not done, the algebra of the analysis becomes very complex. Fortunately, in microcircuitry the two adjacent transistors used for a difference amplifier are very closely matched. Likewise resistance components formed next to each other in microcircuitry are closely matched.

The circuits used in this section and in Sections 16-4 and 16-5 show coupling capacitors used to block d-c voltages. It is not necessary to use these blocking capacitors if the proper values of circuit biasing are used. If this were done here, the circuit layout would become complex, and nothing would be gained in developing the concepts and the a-c equations.

In microcircuitry it becomes somewhat difficult to mix *NPN* and *PNP* transistors on the same wafer. Thus, the concept of complementary-symmetry is not used to produce a multistage amplifier. A direct connection between successive stages of differential amplifiers to form a multistage amplifier is common practice. First, a direct coupling between stages removes the low-frequency roll-off caused by a coupling capacitor. Second, large physical coupling capacitors negates the advantage of the small size capability of microcircuitry.

Differential amplifiers are widely used in instrumentation to generate the difference between two inputs. A direct application is the differential input connection to the more advanced oscilloscope.

The d-c bias calculations of differential amplifiers do not present any new concepts and the analysis presented here will be concerned only with the signal circuits. *Therefore, the symbol β will refer to β_{ac} and not to β_{dc} in this section and in Sections 16-4 and 16-5.*

The circuit for a differential amplifier with a balanced output and a single input is shown in Fig. 16-4*a* and the model is given in Fig. 16-4*b*. The

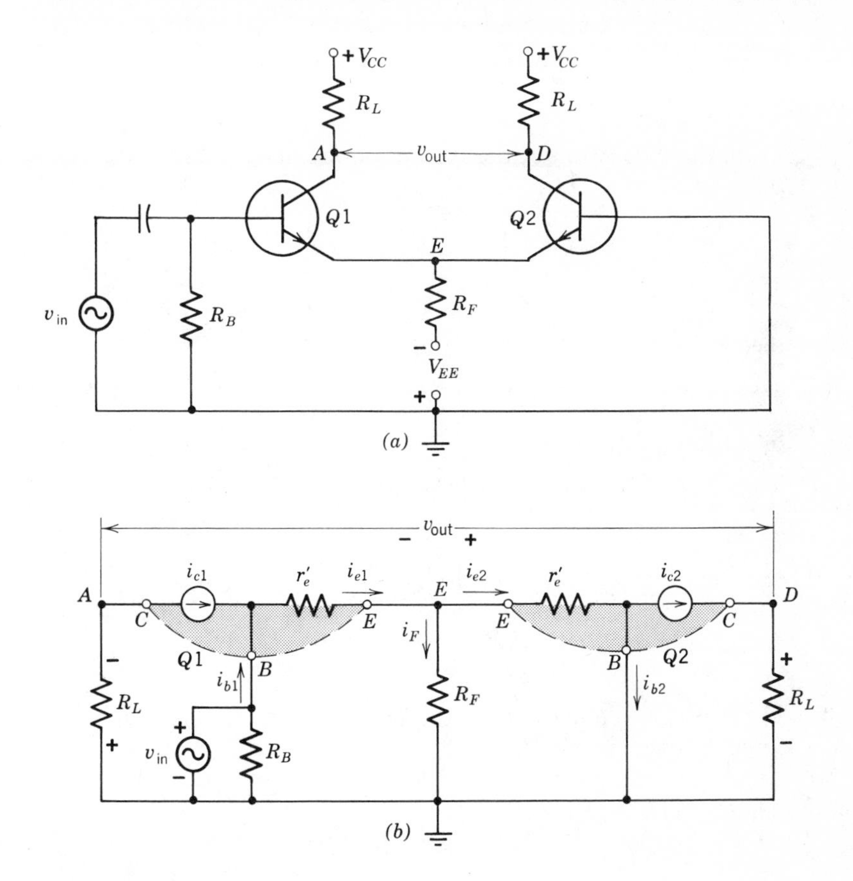

Figure 16-4 Differential amplifier – single input, balanced output. (*a*) Circuit. (*b*) Model.

concept of the term *balanced* in electric circuits means that *each* of the two signal terminals (A and D) are each above the reference ground by the same potential. The input signal, v_{in}, is *unbalanced* because one side of v_{in} is grounded and the other side is above ground. In addition to making the reasonable assumption that the two transistors have equal parameters, let us also assume for the discussion that h_{oe} is large enough to be neglected.

The signal v_{in} looks into a parallel circuit formed of R_B and the input resistance r_{in} to the transistor circuit

$$r = \frac{R_B r_{\text{in}}}{R_B + r_{\text{in}}} \tag{16-7a}$$

and by Ohm's law

$$r_{\text{in}} = \frac{v_{\text{in}}}{i_{b_1}}$$

The Kirchhoff's voltage loop around v_{in} and R_F is

$$v_{\text{in}} = r'_e i_{e_1} + R_F i_F$$

From the inspection of the circuit and by making use of the current divider rule

$$i_F = \frac{r'_e}{r'_e + R_F} i_{e_1}$$

as

$$i_{e_1} = (1+\beta)\, i_{b_1}$$

then

$$i_F = \frac{r'_e}{r'_e + R_F}(1+\beta)\, i_{b_1}$$

Substituting,

$$v_{\text{in}} = \left[(1+\beta) r'_e + \frac{(1+\beta) r'_e R_F}{r'_e + R_F}\right] i_{b_1}$$

Then

$$r_{\text{in}} = (1+\beta) r'_e \left[1 + \frac{R_F}{r'_e + R_F}\right] = (1+\beta) r'_e \frac{r'_e + 2R_F}{r'_e + R_F}$$

$$= \frac{r'_e + 2R_F}{r'_e + R_F} h_{ie} \tag{16-7b}$$

Now, as an explicit part of the design of the circuit, make R_F sufficiently large so that

$$R_F \gg r'_e$$

Then Eq. 16-7*b* becomes

$$r_{\text{in}} = 2(1+\beta) r'_e = 2h_{ie} \tag{16-7c}$$

At point E, by the current divider rule,

$$i_{e2} = \frac{R_F}{r'_e + R_F} i_{e_1}$$

Then

$$i_{c_2} = \frac{\beta}{1+\beta} i_{e2} = \frac{\beta}{1+\beta} \frac{R_F}{r'_e + R_F} i_{e_1} = \frac{\beta}{1+\beta} \frac{R_F}{r'_e + R_F} (1+\beta) i_{b_1}$$

$$= \frac{R_F}{r'_e + R_F} \beta i_{b_1}$$

and $i_{c_1} = \beta i_{b_1}$

The output voltage is

$$v_{\text{out}} = v_A + v_D = R_L i_{c_1} + R_L i_{c2} = R_L\ (i_{c_1} + i_{c_2})$$

Substituting the current relations

$$v_{\text{out}} = R_L \left[\beta i_{b_1} + \frac{R_F}{r'_e + R_F}\beta i_{b_1}\right] = R_L \left[1 + \frac{R_F}{r'_e + R_F}\right]\beta i_{b_1}$$

$$= \frac{r'_e + 2R_F}{r'_e + R_F}\beta R_L i_{b_1}$$

But

$$i_{b_1} = \frac{v_{\text{in}}}{r_{\text{in}}}$$

Substituting Eq. 16-7*b*,

$$i_{b_1} = \frac{r'_e + R_F}{(r'_e + 2R_F)h_{ie}} v_{\text{in}}$$

and by putting this into the expression for v_{out} and by dividing through by v_{in}

$$A_e = \frac{v_{\text{out}}}{v_{\text{in}}} = \frac{\beta R_L}{h_{ie}} = \frac{\beta R_L}{(1+\beta) r'_e} \approx \frac{R_L}{r'_e} \tag{16-8a}$$

If external resistors R_E are added to each emitter lead, the gain becomes

$$A_e = \frac{R_L}{r'_e + R_E} \tag{16-8b}$$

and the new input resistance becomes

$$r_{\text{in}} = 2(1+\beta)(r'_e + R_E) = 2h_{ie} + 2(1+\beta)R_E \tag{16-8c}$$

One of the advantages of this circuit is that the input resistance (Eq. 16-7*c*) is double the input resistance of the conventional one transistor circuit. This circuit provides the same voltage gain (Eq. 16-8*a*) that is obtained from the conventional circuit and also provides a balanced output. Equation 16-8*a* shows that the circuit gain is independent of R_F. Added emitter resistors (Eq. 16-8*b*) can be effectively used to increase the input resistance and to reduce the variations in gain that result from different values of r'_e as long as the relation for R_F is kept to

$$R_F \gg (r'_e + R_E)$$

PROBLEMS

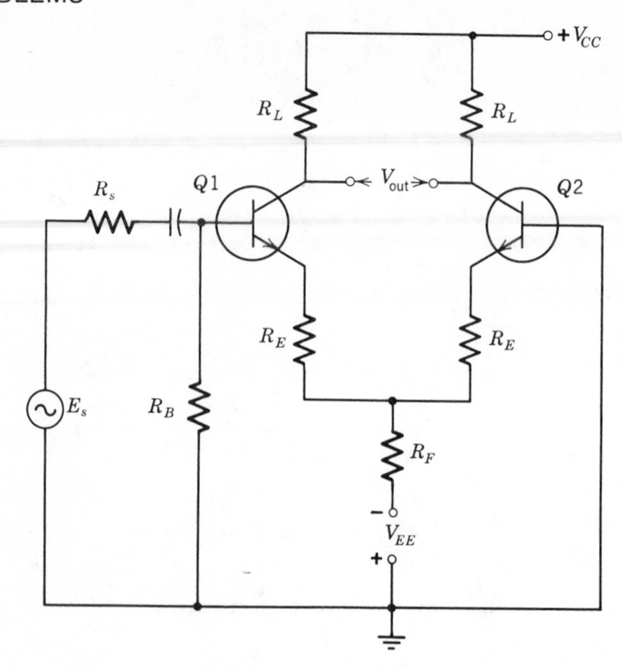

Circuit for Problems 1 to 4.

The a-c model is required for each problem.

1. R_L and R_F are each 10 kΩ. R_E is zero. R_B is 47 kΩ. V_{CC} and V_{EE} are each 30-V supplies. The transistors have a β of 80 and are silicon. $r'_e = 25$ mv/I_E. Determine the d-c currents and voltages in the circuit. Find the circuit gain and the maximum value of E_s before clipping occurs in the output. R_s is 4.3 kΩ.
2. Repeat Problem 1 if R_E is 250 Ω.
3. V_{CC} is +10 V and V_{EE} is −4 V. R_L is 5.6 kΩ, R_E is zero, R_F is 2 kΩ, R_B is 10 kΩ, and R_s is 4.3 kΩ. The transistors are silicon and have a β of 100. Determine the voltage gain of the circuit and the maximum available peak-to-peak output voltage of the circuit. r'_e is 50 mv/I_E.
4. Repeat Problem 3 if R_E is 100 Ω.

Section 16-4 Differential Amplifier – Dual Input, Balanced Output

Two signals, $v_{1_{in}}$ and $v_{2_{in}}$, are fed into the two bases of the differential circuit (Fig. 16-5*a*), and a single balanced output v_{out} is derived from the amplifier. The model is given in Fig. 16-5*b*.

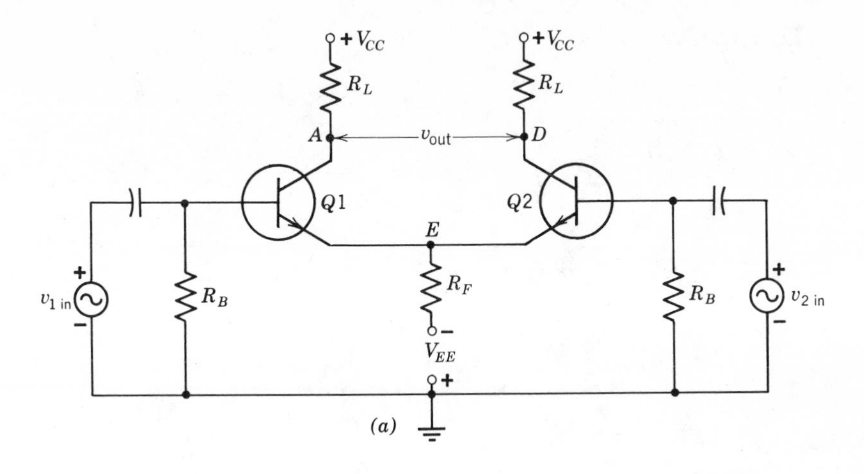

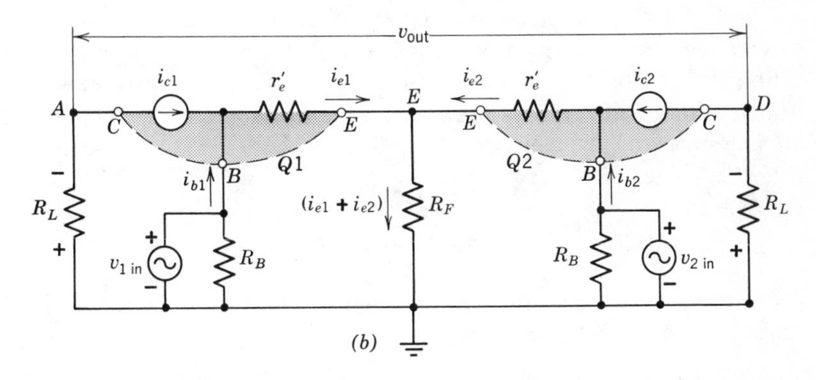

Figure 16-5 Differential amplifier – dual input, balanced output. (*a*) Circuit. (*b*) Model.

Kirchhoff's voltage equations can be written for the signal voltages:

$$v_{1_{in}} = r'_e i_{e_1} + R_F(i_{e_1} + i_{e_2}) = (r'_e + R_F)i_{e_1} + R_F i_{e_2}$$

and

$$v_{2_{in}} = r'_e i_{e_2} + R_F(i_{e_1} + i_{e_2}) = R_F i_{e_1} + (r'_e + R_F)i_{e_2}$$

These two equations are simultaneous and can be solved for i_{e_1} and i_{e_2} (the use of determinants is a direct approach) to give

$$i_{e_1} = \frac{(r'_e + R_F)v_{1_{in}} - R_F v_{2_{in}}}{r'_e(r'_e + 2R_F)}$$

and

$$i_{e_2} = \frac{-R_F v_{1_{in}} + (r'_e + R_F)v_{2_{in}}}{r'_e(r'_e + 2R_F)}$$

The output voltage is

$$v_{\text{out}} = v_D - v_A = R_L i_{c_2} - R_L i_{c_1} = R_L \frac{\beta}{1+\beta} i_{e_2} - R_L \frac{\beta}{1+\beta} i_e$$

$$= \frac{\beta R_L}{1+\beta}(i_{e_2} - i_{e_1})$$

Substituting the current relations

$$v_{\text{out}} = \frac{\beta R_L}{1+\beta}\left[\frac{-R_F v_{1_{\text{in}}} + (r'_e + R_F)v_{2_{\text{in}}} - (r'_e + R_F)v_{1_{\text{in}}} + R_F v_{2_{\text{in}}}}{r'_e(r'_e + 2R_F)}\right]$$

$$v_{\text{out}} = \frac{\beta R_L}{(1+\beta)r'_e}(v_{2_{\text{in}}} - v_{1_{\text{in}}})$$

Now, at this point, we must very clearly establish the purpose of the circuit and must explicitly define the gain. A differential or difference amplifier is intended to amplify only the difference between two inputs. It is *not* intended to amplify the signals themselves. This concept is extended in Section 16-6. The gain of this amplifier is

$$A_e \equiv \frac{v_{\text{out}}}{v_{2_{\text{in}}} - v_{1_{\text{in}}}} = \frac{\beta R_L}{(1+\beta)r'_e} = \frac{\beta R_L}{h_{ie}} \approx \frac{R_L}{r'_e} \tag{16-9a}$$

By using external resistors, R_E, in each emitter the dependence of gain on variations in r'_e can be reduced

$$A_e = \frac{\beta R_L}{(1+\beta)(r'_e + R_E)} \approx \frac{R_L}{r'_e + R_E} \approx \frac{R_L}{R_E} \tag{16-9b}$$

The last approximation is valid when $R_E \gg r'_e$. Notice that the gain of this circuit is independent of R_F, since R_F did not appear in any final gain equation.

At the start of this section two equations were solved to yield i_{e_1} and i_{e_2}. If these equations are divided by $(1+\beta)$, the base currents i_{b_1} and i_{b_2} are obtained:

$$i_{b_1} = \frac{(r'_e + R_F)v_{1_{\text{in}}} - R_F v_{2_{\text{in}}}}{(1+\beta)r'_e(r'_e + 2R_F)}$$

and

$$i_{b_2} = \frac{-R_F v_{1_{\text{in}}} + (r'_e + R_F)v_{2_{\text{in}}}}{(1+\beta)r'_e(r'_e + 2R_F)}$$

We define

$$r_{\text{in}_1} = \frac{v_{1_{\text{in}}}}{i_{b_1}} \qquad \text{and} \qquad r_{\text{in}_2} = \frac{v_{2_{\text{in}}}}{i_{b_2}}$$

When the signals are exactly equal

$$v_{1_{\text{in}}} = -v_{2_{\text{in}}}$$

and

$$v_{2_{\text{in}}} = -v_{1_{\text{in}}}$$

Substituting these values,

$$r_{\text{in}_1} = (1+\beta)r'_e = h_{ie}$$

and (16-10a)

$$r_{\text{in}_2} = (1+\beta)r'_e = h_{ie}$$

Now these values of r_{in} are only equal for exactly equal signals. When one signal becomes larger than the other, the input resistance increases to the side of the circuit where the signal becomes larger. The input resistance of the other side decreases. The signal source sees r_{in} in parallel with R_B. Obviously, the use of external emitter resistors will change the input resistance to

$$r_{\text{in}_1} = (1+\beta)(r'_e + R_E)$$

and (16-10b)

$$r_{\text{in}_2} = (1+\beta)(r'_e + R_E)$$

PROBLEMS

For these problems use the circuit diagram of Fig. 16-5. All transistors have a β of 80 and are silicon. Use $r'_e = 25\text{ mv}/I_E$.

1. V_{CC} and V_{EE} are each 15 V. R_L is 20 kΩ, R_F is 20 kΩ, and R_B is 47 kΩ. Determine the d-c operating levels in the circuit. Determine the circuit voltage gain, the input resistance, and the maximum available peak-to-peak output voltage without clipping.
2. Repeat Problem 1 if 200-Ω resistors, R_E, are placed externally in series with each emitter.
3. V_{CC} is +10 V, V_{EE} is −4 V, R_L is 10 kΩ, R_F is 3.9 kΩ, and R_B is 10 kΩ. Determine the d-c operating levels of the transistors. Determine the circuit voltage gain, the input resistance, and the maximum available peak-to-peak output voltage without clipping.
4. Repeat Problem 3 if 120-Ω resistors, R_E, are placed externally in series with each emitter.

Section 16-5 Differential Amplifier – Dual Input, Unbalanced Output

A differential amplifier having a balanced input and an unbalanced output is given in Fig. 16-6. The output voltage is

$$v_{\text{out}} = -R_L i_{c_2} = -\frac{\beta R_L}{1+\beta} i_{e_2}$$

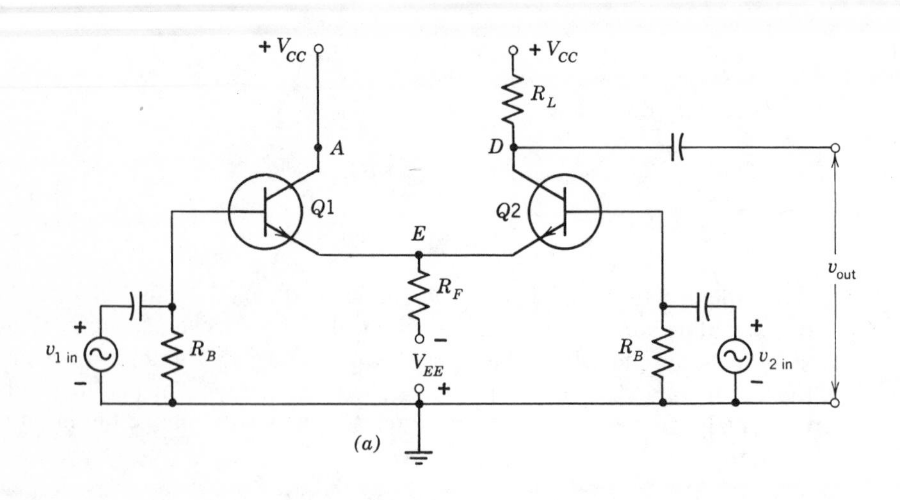

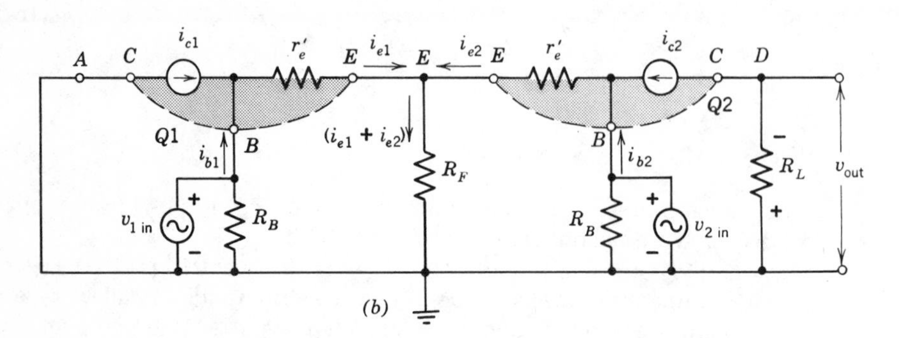

Figure 16-6 Differential amplifier – dual input, unbalanced output. (*a*) Circuit. (*b*) Model.

The loop equations for the input signals are

$$v_{1_{\text{in}}} = r'_e i_{e_1} + R_F(i_{e_1} + i_{e_2}) = (r'_e + R_F) i_{e_1} + R_F i_{e_2}$$

and

$$v_{2_{\text{in}}} = r'_e i_{e_2} + R_F(i_{e_1} + i_{e_2}) = R_F i_{e_1} + (r'_e + R_F) i_{e_2}$$

Solving for i_{e2},

$$i_{e2} = \frac{(r'_e + R_F)v_{2\text{in}} - R_F v_{1\text{in}}}{r'_e(r'_e + 2R_F)}$$

Substituting in the first equation

$$v_{\text{out}} = -\frac{\beta R_L}{(1+\beta)r'_e}\left[\frac{(r'_e + R_F)v_{2\text{in}} - R_F v_{1\text{in}}}{r'_e + 2R_F}\right]$$

Now let us place a restriction on the design of the circuit

$$R_F \gg r'_e$$

Then

$$v_{\text{out}} = -\frac{\beta R_L}{(1+\beta)r'_e}\left[\frac{R_F v_{2\text{in}} - R_F v_{1\text{in}}}{2R_F}\right]$$

$$v_{\text{out}} = -\frac{\beta R_L}{2(1+\beta)r'_e}(v_{1\text{in}} - v_{2\text{in}})$$

Then the magnitude of the circuit gain is

$$A_e \equiv \frac{v_{\text{out}}}{v_{1\text{in}} - v_{2\text{in}}} = \frac{\beta R_L}{2(1+\beta)r'_e} = \frac{\beta R_L}{2h_{ie}} \approx \frac{R_L}{2r'_e} \qquad (16\text{-}11a)$$

If an external R_E is added that meets the restriction that $R_F \gg R_E$, Eq. 16-11*a* becomes

$$A_e = \frac{\beta R_L}{2(1+\beta)(r'_e + R_E)} \approx \frac{R_L}{2(r'_e + R_E)} \qquad (16\text{-}11b)$$

and if R_E can be selected to meet two restrictions,

$$R_F \gg R_E \qquad \text{and} \qquad R_E \gg r'_e$$

the gain becomes

$$A_e \approx \frac{R_L}{2R_E} \qquad (16\text{-}11c)$$

The gain of this circuit is half the gain of the previous two circuits, but it does accomplish the objective of accepting a balanced input and of converting it to a single-ended output.

Since the current equations are in the same form as the ones for the circuit in the last section, the same discussion for input resistance is applicable here also. When the inputs are identical, the input resistance to the transistor circuit to each source is

$$r_{\text{in}} = (1+\beta)r'_e = h_{ie} \qquad (16\text{-}11d)$$

and, when external emitter resistors are used for stabilization, the input load resistance on each source is

$$r_{\text{in}} = (1+\beta)(r'_e + R_E) \qquad (16\text{-}11e)$$

PROBLEMS

For these problems use the circuit diagram of Fig. 16-6.
All transistors have a β of 100 and are silicon.
Use $r'_e = 25$ mv/I_E.

1. V_{CC} and V_{EE} are each 20 V. R_L is 6.8 kΩ, R_F is 6.8 kΩ, and R_B is 24 kΩ. Determine the d-c operating levels of the transistors. Determine the circuit voltage gain, the input resistance, and the maximum available peak-to-peak output voltage without clipping.
2. Repeat Problem 1 if 130-Ω external resistors, R_E, are placed in each emitter lead.
3. V_{CC} is + 10 V, V_{EE} is − 4 V, R_L is 20 kΩ, R_F is 7.5 kΩ, and R_B is 20 kΩ. Determine the d-c operating levels of the transistors. Determine the circuit voltage gain, the input resistance, and the maximum available peak-to-peak output voltage without clipping.
4. Repeat Problem 4 if 270-Ω resistors, R_E, are placed in each emitter lead.

Section 16-6 Common-Mode Rejection

The two signals into a differential amplifier are v_{in_1} and v_{in_2}. The gain of the amplifier is

$$A_e \equiv \frac{v_{\text{out}}}{v_{\text{in}_1} - v_{\text{in}_2}}$$

Let us examine the circuit by means of some numerical values. Assume the gain, A_e, of the amplifier is 1000 and that the supply voltages are such that the maximum peak-to-peak output voltage obtainable from the amplifier is 10 volts. If these numbers are substituted into the gain equation, the maximum value, peak-to-peak for the difference of the input signals is

$$(v_{\text{in}_1} - v_{\text{in}_2}) = 10 \text{ mv, max}$$

If v_{in_1} is 5 mv and v_{in_2} is zero, the output is 5 V in the same phase as v_{in_1}. If v_{in_2} is 5 mv and v_{in_1} is zero, the output is also 5 V but it is negative indicating a phase reversal. This concept is carried to the circuit where the $v_{1_{\text{in}}}$ terminal is the positive or + input and the $v_{2_{\text{in}}}$ terminal is the negative or − input.

According to the derivations we have made, if $v_{1_{\text{in}}}$ were 1000.005 V and if $v_{2_{\text{in}}}$ were 1000.000 V, the difference is 5 mv and the output would be a 5-V signal. Without question a 1000-V signal would destroy the whole circuit. Therefore, a d-c analysis of the circuit is required to determine the limits of A in

$$v_{1_{in}} = A + e_1$$

and

$$v_{2_{in}} = A + e_2$$

The input signal to the circuit for the gain equation is

$$v_{1_{in}} - v_{2_{in}} = (A + e_1) - (A + e_2) = e_1 - e_2$$

We stated at the start of this section that $(e_1 - e_2)$ was limited to 10 mv peak-to-peak maximum. The terms, A, which are eliminated by subtraction are called the *common mode*. Theoretically, the common mode does not appear in the output at all. Actually, because of inherent unbalances in differential amplifier circuitry, there is some effect of A in the output.

Now let us assume that the common-mode limit on this circuit is two volts. If $v_{1_{in}}$ is a 15-mv peak-to-peak signal and if $v_{2_{in}}$ is a 10-mv peak-to-peak signal, $(v_{1_{in}} - v_{2_{in}})$ is a 5-mv peak-to-peak signal and the output v_{out} is a 5 V peak-to-peak signal. The common-mode signal is 10 mv peak-to-peak. This value of the common mode is small compared to the case where $v_{1_{in}}$ is 2.005 V peak-to-peak and $v_{2_{in}}$ is 2.000 V peak-to-peak. In this case, the common mode is a 2-V signal which is presented to the inputs of an amplifier that has a gain of 1000. This is the worst case. To determine how much of this signal gets into the output, two identical sinusoidal signals equal to the common mode limit (in this case 2 V) are connected to the two inputs. The amount of unwanted signal is any v_{out} that can be measured. The maximum output signal is 10 V according to specification. If the *common-mode rejection ratio* for this amplifier is 60 db, the maximum tolerable signal in the output in this test is found from

$$60 = 20 \log \frac{10}{\chi}$$

or

$$\chi = \frac{10}{1000} \text{ V} = 10 \text{ mv}$$

Thus, when the common-mode signal is 2 V, a 10-mv output is permissible in the amplifier output. Very obviously, if a common-mode value of 2 V produces an unwanted 10 mv in the output, when $v_{1_{in}}$ is 15 mv and $v_{2_{in}}$ is 10 mv, the common-mode level of 10 mv cannot produce any noticeable signal in the output.

The common-mode rejection ratio (CMRR) can be expressed in the form of an equation as

$$\text{CMRR} = \frac{(\text{Common-mode signal}) \times (\text{Amplifier gain})}{(\text{Common-mode output error signal})} \tag{16-12}$$

Usually the common-mode rejection ratio is specified in decibels.

PROBLEMS

1. A 20-V peak-to-peak signal is connected simultaneously to the two inputs of a differential plug-in to an oscilloscope. In the differential mode the signal on the oscilloscope is 100 μv peak-to-peak. What is the common-mode rejection ratio specification in decibels?
2. Another plug-in unit produces a 20 mv peak-to-peak signal on the oscilloscope when the common-mode signal is 5 V peak-to-peak. What is the common-mode rejection ratio of this plug-in unit in decibels?
3. A difference amplifier has a difference gain of 80 db and a common-mode rejection ratio of 86 db. One signal is $3.000+0.001$ V and the other signal is $3.000-0.001$ V. What is the desired output voltage and what is the actual output voltage? Assume no phase shift.
4. A difference amplifier has a difference gain of 66 db and a common-mode rejection ratio of 80 db. One signal is $1.000+0.001$ V and the other signal is $1.000-0.001$ V. What is the desired output voltage, and what is the actual output voltage?

Section 16-7 Constant-Current Stabilization

The differential amplifier is one of the fundamental building blocks of the linear amplifiers used in integrated circuits. The cost of high-gain microamplifiers is currently so low that they have effectively displaced the conventional wired amplifiers with discrete components. The integrated circuits do not use the common resistor R_F which we have used in the previous circuits. We required that R_F be much larger than r'_e and much larger than (r'_e+R_E) if external emitter resistors are used. The resistor R_F is replaced by the transistor $Q3$, as shown in Fig. 16-7*a*. The d-c bias circuit of $R1$ and $R2$ determines the d-c collector current in $Q3$. This collector current must be the sum of the d-c collector currents of $Q1$ and $Q2$ or $(I_{C_1}+I_{C_2})$.

This d-c collector current $(I_{C_1}+I_{C_2})$ is fixed and must be invariant, since there is no signal injected into either the base or the emitter of $Q3$. Consequently, the collector current in $Q3$ is an ideal current generator. This ideal current generator which can have no a-c components is represented in the a-c model as an infinite resistance or an open circuit. Accordingly, in the model (Fig. 16-7*b*) there is no resistor connected between E and ground. Actually, there should be a resistance shown between E and ground which would be the value of $1/h_{oe}$ for $Q3$. This value of resistance is so high in comparison with the other resistance values in the circuit that it can be ignored without any serious consequence, except in a calculation that evaluates the common-mode rejection ratio.

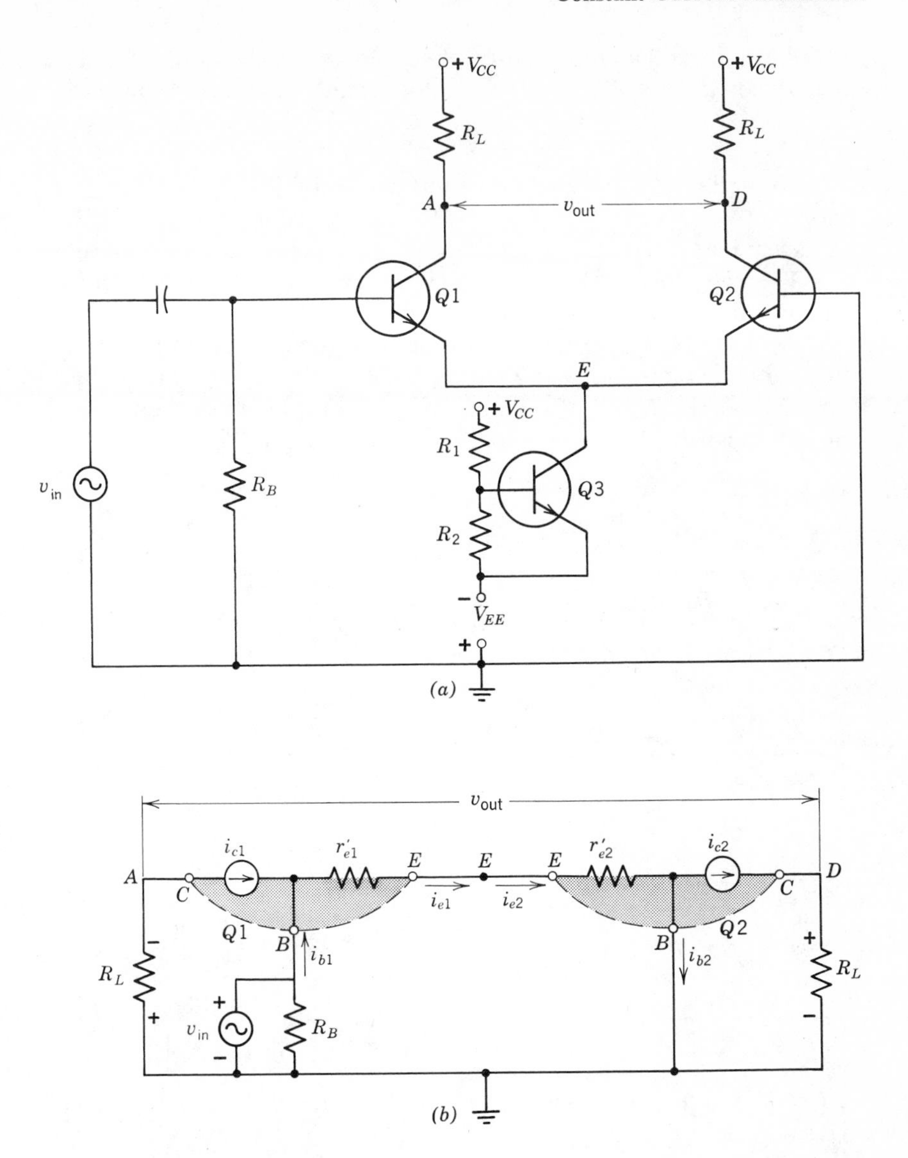

Figure 16-7 Differential amplifier—single input, balanced output using a constant-current stabilization. (*a*) Circuit. (*b*) Model.

As a result of this introduction of $Q3$, the various gain and impedance relations can be written directly from an inspection of the model. In Fig. 16-7*b*, v_{in} generates the base current i_{b_1}. The direction of i_{b_1} arises from the assigned instantaneous polarity of v_{in}. Then the direction of i_{b_1} is known, and the directions of i_{c_1} and i_{e_1} are automatically determined, since

$$i_e = i_b + i_c$$

The direction of i_{e_1} is leaving the terminal E of $Q1$. Since no part of i_{e_1} can go into $Q3$, i_{e_2} is i_{e_1}, but as far as $Q2$ is concerned, i_{e_2} is *entering* $Q2$ whereas i_{e_1} is leaving $Q1$. Now the directions of i_{c_2} and i_{b_2} are fixed and determined.

The directions of the $i_c R_L$ drops show that they are additive for v_{out}. Since the magnitude of i_{e_1} equals the magnitude of i_{e_2}, the value of i_{c_1} equals the value of i_{c_2}. Now the individual transistor subscripts can be dropped, and the output voltage is $2i_c R_L$ and v_{in} is $2i_e r'_e$. The circuit gain is $\beta R_L/h_{ie}$ or, approximately, R_L/r'_e. The input circuit to the signal source consists of the inner loop of $(r'_e + r'_e)$ which reflects into the base as $2h_{ie}$.

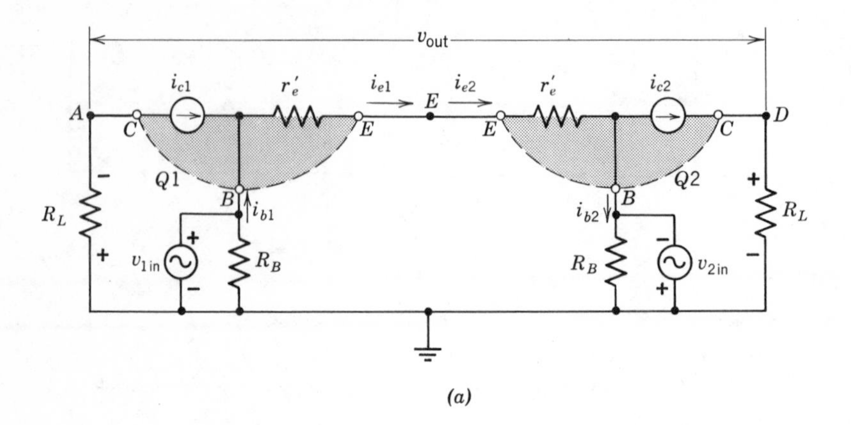

(a)

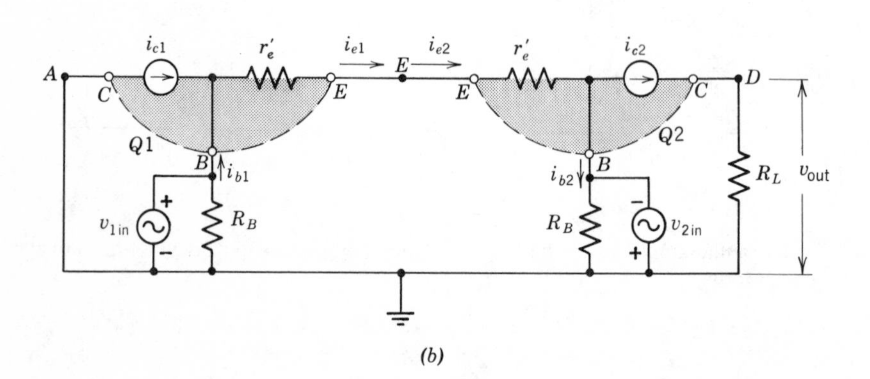

(b)

igure 16-8 Models for differential amplifiers using constant-current stabilization. (*a*) Dual input, alanced output. (*b*) Dual input, unbalanced output.

The model for the differential amplifier having a dual input with a balanced output, using $Q3$ in place of R_F, is shown in Fig. 16-8*a*. Now $v_{1_{in}}$ and $v_{2_{in}}$ are shown with reversed polarities, following the central concept of using the circuit in a difference sense. Accordingly, an examination of the circuit shows that $(v_{1_{in}} + v_{2_{in}})$ is applied across $(h_{ie} + h_{ie})$. Thus, as far as either input is concerned, the input resistance to the transistor circuit for either source is h_{ie} in parallel with R_B.

The total output voltage is $i_c(R_L + R_L)$, and the total input voltage is $i_e(r'_e + r'_e)$. Consequently, the circuit gain is $\beta R_L/h_{ie}$ or, approximately, R_L/r'_e. The load on either source is R_B in parallel with h_{ie}.

The model for the differential amplifier having a dual input with an unbalanced output, using $Q3$ in place of R_F, is shown in Fig. 16-8*b*. Here, again, an inspection of the model shows that the total input is $(v_{1_{in}} + v_{2_{in}})$, and the total input resistance is $(h_{ie} + h_{ie})$. Each source, then, sees the resistance h_{ie} in parallel with R_B. The output voltage is across one resistor only, i_cR_L, but the input voltage is $i_e(r'_e + r'_e)$, making the gain $\beta R_L/2h_{ie}$ or, approximately, $R_L/2r'_e$.

Now, having gone through an analysis of the three fundamental circuits for the differential amplifier, let us see how they are used in an integrated circuit application. Consider the three-stage amplifier shown in Fig. 16-9.

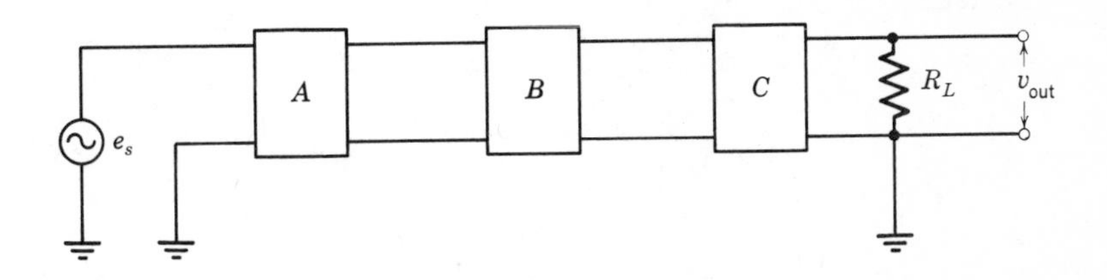

Figure 16-9 Three-stage integrated-circuit amplifier.

The external signal source is single-ended. Therefore, the first stage, *A*, must be the circuit configuration that accepts an unbalanced input while producing a balanced output. The intermediate stage (or stages), *B*, uses the balanced-input, balanced-output circuit configuration. The final stage, *C*, must convert its balanced-input signal to a single-ended or unbalanced output. The use of the balanced intermediate stage permits the use of direct coupling between stages. If large coupling capacitors between stages were required, the whole objective of microelectronics would be defeated. A summary of the properties of the three basic circuit configurations is given in Table A.

TABLE A Properties of the Basic Differential Amplifier Circuit Configurations

	Unbalanced Input, Balanced Output	Balanced Input, Balanced Output	Balanced Input, Unbalanced Output
Circuit	Fig. 16-5, Fig. 16-8	Fig. 16-6, Fig. 16-9*a*	Fig. 16-7, Fig. 16-9*b*
Voltage gain without external R_E	$\frac{\beta R_L}{(1+\beta)r'_e} \approx \frac{R_L}{r'_e}$	$\frac{\beta R_L}{(1+\beta)r'_e} \approx \frac{R_L}{r'_e}$	$\frac{R_L}{2(1+\beta)r'_e} \approx \frac{R_L}{2r'_e}$
Input resistance without external R_E[a]	$2(1+\beta)r'_e = 2h_{ie}$	$(1+\beta)r'_e = h_{ie}$	$(1+\beta)r'_e = h_{ie}$
Voltage gain with external R_E	$\frac{\beta R_L}{(1+\beta)(r'_e+R_E)} \approx \frac{R_L}{r'_e+R_E}$	$\frac{\beta R_L}{(1+\beta)(r'_e+R_E)} \approx \frac{R_L}{r'_e+R_E}$	$\frac{\beta R_L}{2(1+\beta)(r'_e+R_E)} \approx \frac{R_L}{2(r'_e+R_E)}$
Input resistance with external R_E[a]	$2(1+\beta)(r'_e+R_E)$	$(1+\beta)(r'_e+R_E)$	$(1+\beta)(r'_e+R_E)$

[a]This value is in parallel with any R_B that is used.

PROBLEMS

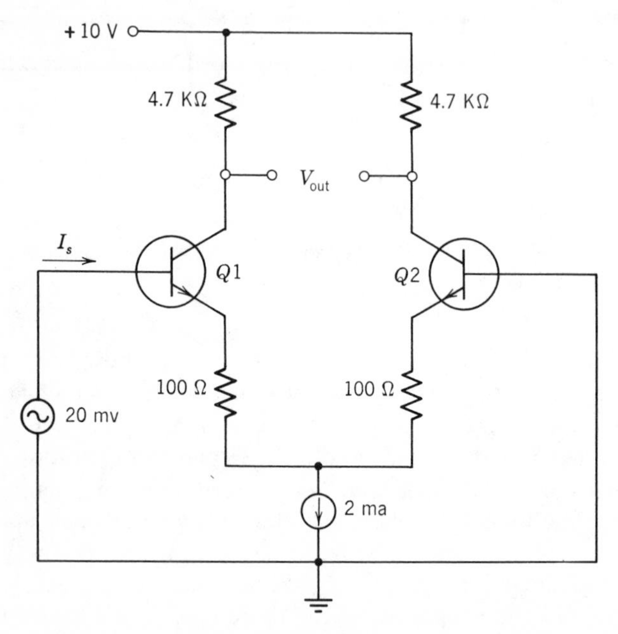

Circuit for Problem 1.

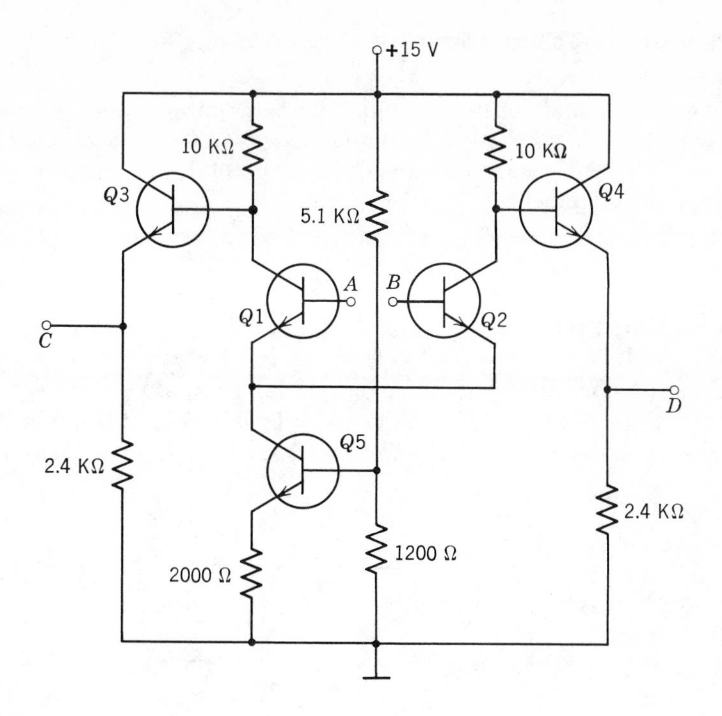

Circuit for Problem 2.

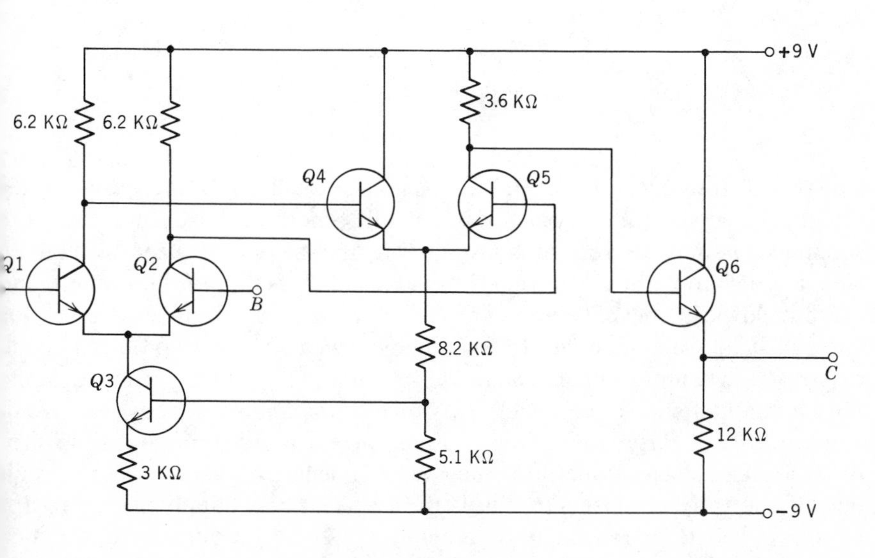

Circuit for Problem 3.

All transistors are silicon with a β of 80. $r'_e = 25\ \text{mv}/I_E$.

1. Determine V_{out} and I_s.
2. For the given circuit of the IC chip, determine (a) quiescent currents and voltages, (b) input resistance, (c) differential gain, (d) output voltage swing. *Hint.* Neglect the loading effects of the base currents.
3. Repeat the requirements of Problem 2 for this IC circuit.

Section 16-8 Clipping Circuits

Two transistors are arranged in the general configuration of complementary symmetry, Fig. 16-10. In this circuit, when the input signal is positive, there is a reverse voltage on the *PNP* transistor. There can be no collector current

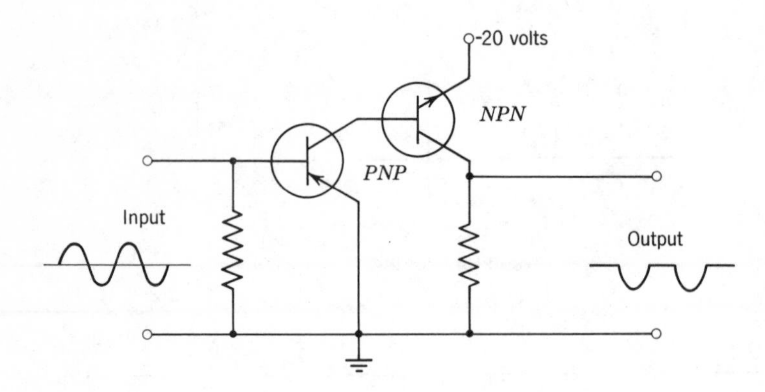

Figure 16-10 Transistor clipper circuit.

in a transistor unless the transistor is forward-biased in the base-emitter circuit. Thus, the first transistor is cut off. Since the base current of the second stage must be zero, it is also cut off, and the output is zero. When the signal is negative, the *PNP* transistor is biased in the forward direction, and current flows in the collector. This current, in turn, produces a collector current in the second stage. In this manner, we show that an output signal exists only when the signal is negative. If the input signal is a sine wave, the output is a half sine wave. The term *class-B amplifier* describes this circuit condition. In a class-B amplifier, output current flows through just 180° of the full cycle. Class-B operation may be established in a vacuum-tube circuit only if the tube is biased to cutoff by a special bias supply. In transistor circuits, class-B operation is obtained very simply: a zero bias (no bias) automatically produces class-B operation.

The circuit of Fig. 16-10 is very useful as a clipper in microelectronic applications. It allows only the positive or negative side of a wave to pass, depending on whether the first transistor is a *PNP* or an *NPN* type. As an example, the complex input signal of Fig. 16-11 can be changed to an output signal which consists only of short pulses.

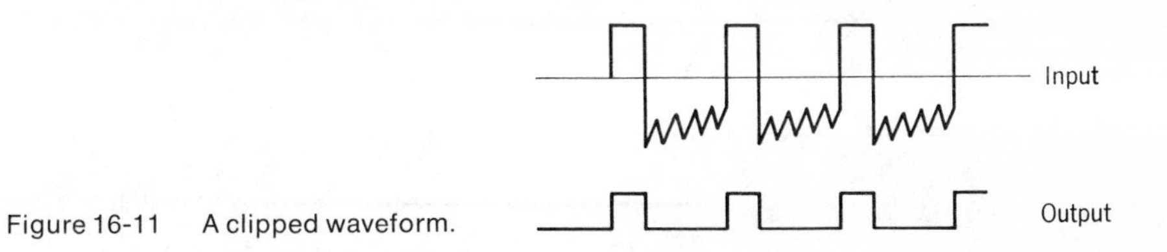

Figure 16-11 A clipped waveform.

The simple single-phase rectifier (Fig. 16-12) can be considered a clipping circuit. One half of the a-c waveform appears across the load resistance, and the other half is the inverse voltage on the diode. The connection of the diode determines which half of the input is the output voltage across the load.

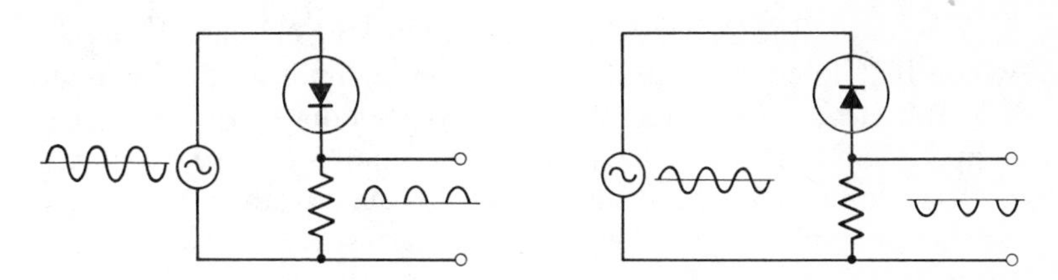

Figure 16-12 The diode clipper.

The biased diodes (Fig. 16-13) allow that part of the waveform which is less than the bias to appear as the voltage across the load. When the incoming waveform exceeds the bias voltages, the diodes effectively short-circuit the signal. The circuit can be used to "square off" a waveform. If the output

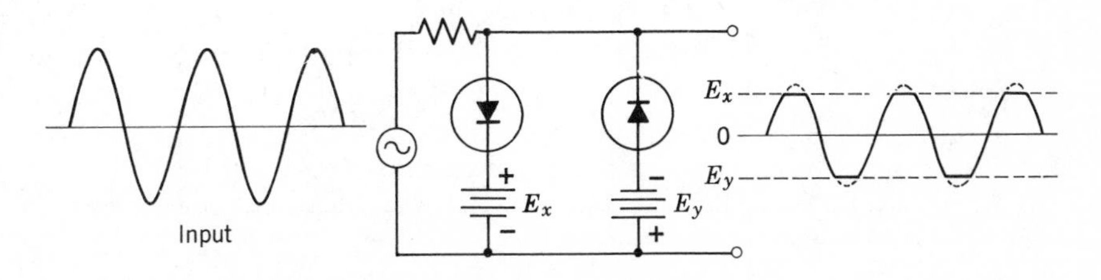

Figure 16-13 The diode limiter.

waveform is amplified and fed into a second diode-limiting circuit, the resulting waveform has very square fronts.

Zener diodes (Section 2-4) are often used in the clipping circuit (Fig 16-14). The back impedance of a Zener diode is very high before the break

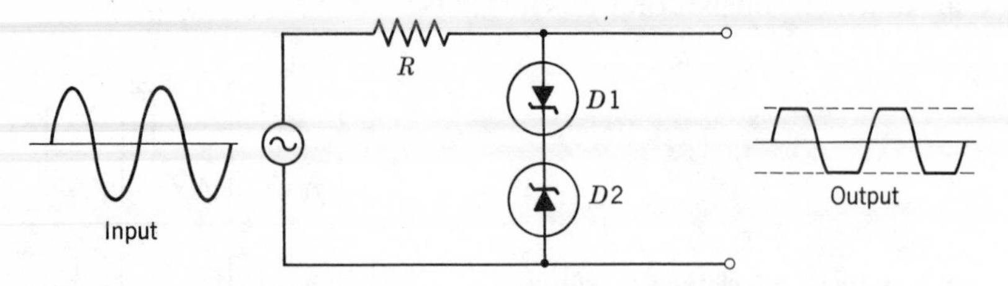

Figure 16-14 Clipping circuit using Zener diodes.

down potential is reached, and the forward resistance is very low at all voltages. Connecting two Zener diodes back-to-back places a very high impedance across the output, but, when the Zener potential is reached, the resistance of the diode combination is very low, and the diode current flowing through R produces an IR drop which absorbs any source voltage that is in excess of the Zener potential. The resistor R limits the diode current to values that do not exceed the safe dissipation limits of the Zener diodes.

Multielectrode semiconductors and vacuum tubes can also be used as clippers and limiters by overdriving the input signal so that the device cuts off for one polarity of the signal and saturates for the reverse polarity of the signal. An application is the action of a limiter used in frequency-modulation receivers (Section 25-4).

PROBLEMS

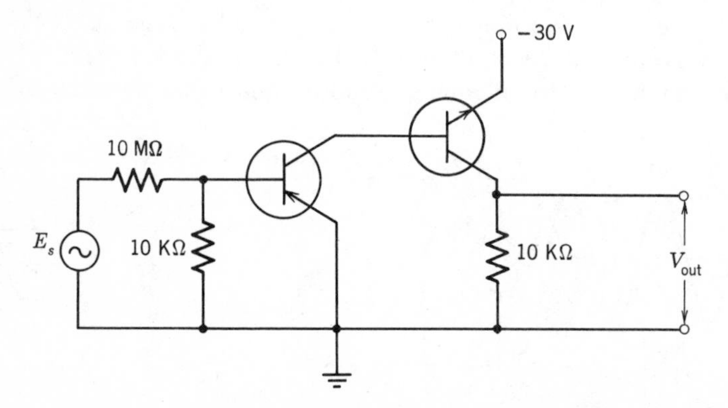

Circuit for Problems 1 and 2.

1. What is the maximum peak value of E_s, a sinusoidal signal, which will produce clipping on one side of the output waveform only? What is the peak value of the output waveform? β is 80 and $r'_e = 50$ mv/I_E.
2. Draw the circuit that will produce clipping on the negative half of E_s.

Section 16-9 The Grounded-Grid Amplifier

In the study of transistors, three circuit connections for a transistor were considered: the common-base amplifier, the common-emitter amplifier, and the common-collector amplifier. There are three vacuum-tube circuits that are analogous to these three transistor forms. The conventional vacuum-tube amplifier may be thought of as equivalent to the common-emitter amplifier. The cathode follower (Section 20-10) is equivalent to the common-collector amplifier. The grounded-grid amplifier is very similar to the common-base amplifier. We studied the three transistor circuits collectively as a unit topic since all three forms are commonly used. This is not true of the vacuum-tube amplifiers. The grounded-grid amplifier, because of its special characteristics, is encountered only in very high-frequency circuits. A grounded grid serves to act as a shield between the input and the output circuits, and this shielding action reduces the interelectrode capacitance between the input and the output circuits. In the study of the vacuum tube, we found that the approach used to reduce this input to output mutual capacitance was the development of the multigrid tubes, and, accordingly, our study at that time went in that direction. The pentode amplifier cannot be used at high frequencies where the *transit time* of travel of an electron from the cathode to the plate becomes an appreciable part of the a-c cycle. A special triode designed for use as a grounded-grid amplifier has a very close spacing between the cathode and the plate and can be used at frequencies at which the pentode is unsatisfactory.

The circuit and the model for the grounded-grid amplifier are shown in Fig. 16-15. When the incoming signal is positive on the cathode, the total bias on the tube increases negatively. The plate current decreases and, since the IZ_L drop decreases, the plate voltage increases. Thus, a positive input signal produces a positive output voltage, and the amplifier does not develop a phase inversion. Then, in the equivalent circuit, the signal voltage e_g and μe_g are additive. The plate current is

$$i_p = \frac{\mu e_g + e_g}{r_p + Z_L} = \frac{e_g(\mu + 1)}{r_p + Z_L}$$

The output voltage is

$$e_o = i_p Z_L = \frac{e_g Z_L(\mu + 1)}{r_p + Z_L}$$

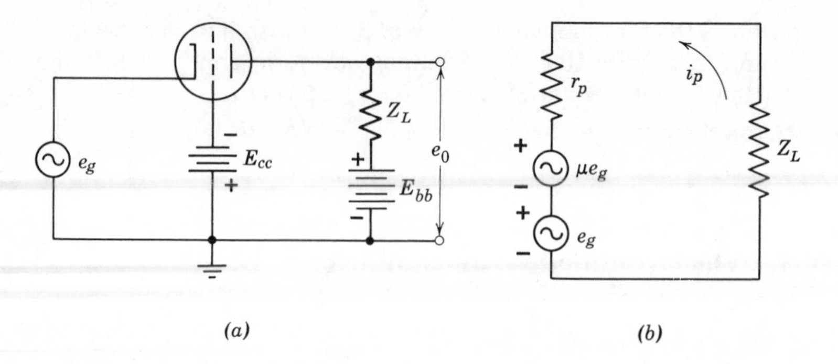

Figure 16-15 The grounded-grid amplifier. (a) Actual circuit. (b) Model.

The voltage gain is

$$A_e = \frac{e_o}{e_g} = \frac{(\mu+1)Z_L}{r_p + Z_L} \tag{16-13}$$

The input impedance is

$$Z_{\text{in}} = \frac{e_g}{i_p} = \frac{r_p + Z_L}{\mu + 1} \tag{16-14}$$

The grounded-grid amplifier has a higher gain than the conventional amplifier. On the other hand, even though the operation of the tube is in class A, the load on the driving voltage is finite and is relatively a low impedance. This means that there is a power-transfer consideration in the input circuit as in the case of the transistor, and this matching problem cannot be ignored in determining the overall gain.

Questions

1. Define or explain each of the following terms: (*a*) complementary symmetry, (*b*) Darlington pair, (*c*) single input, (*d*) dual input, (*e*) balanced output, (*f*) unbalanced output, (*g*) common mode signal, (*h*) common-mode rejection ratio, (*i*) constant current stabilization, (*j*) clipping circuit, (*k*) class B, (*l*) grounded-grid amplifier.
2. Can a complementary symmetry circuit respond to d-c signals?
3. Explain how complementary symmetry can be obtained by transistors and not by vacuum tubes.
4. What is the input resistance to a Darlington amplifier?
5. What is the output resistance of a Darlington amplifier?

6. What is the current gain of a circuit using a Darlington amplifier?
7. What is the voltage gain of a circuit using a Darlington amplifier?
8. What are the characteristics of an amplifier using a Darlington pair?
9. What are some applications of the differential amplifier?
10. Why are differential amplifiers particularly adapted for an IC?
11. Why is the common-mode rejection ratio important in a differential amplifier?
12. If the common-mode rejection ratio in decibels equals the gain of a differential amplifier in decibels, what limitations in the use of the amplifier are encountered?
13. What is the advantage of using constant current stabilization in a differential amplifier?
14. Compare the input resistances of the three basic differential amplifier circuits? What is the effect of adding an R_E to each emitter?
15. Compare the voltage gains of the three basic differential amplifier circuits. What is the effect of adding an R_E to each emitter?
16. How is class-B operation obtained in a transistor circuit? In an FET circuit? In a vacuum-tube circuit?
17. What is the effect of reversing the diode in a diode clipper?
18. Why is a resistor placed in series with the input of a diode clipper?
19. Why is a resistor placed in series with the input of a diode limiter?
20. What are the advantages in using Zener diodes in a limiter?
21. What is the advantage of a grounded-grid amplifier?
22. Is the voltage gain of a grounded-grid amplifier the same as that of a conventional vacuum-tube amplifier?
23. What is the input resistance to a grounded-grid amplifier? To a conventional vacuum-tube amplifier?

Chapter Seventeen

STABILITY, COMPENSATION, AND TEMPERATURE

When transistors are mass produced the value of β_{dc} can vary between units by as much as four to one. This variation can radically change the operating point of a circuit. The problem is examined (Section 17-1), and then specific basic circuits are analyzed (Section 17-2). A somewhat similar problem is encountered with the leakage current of a semiconductor (Section 17-3). This leakage current can be amplified into the collector circuit. This phenomenum is considered in the various basic circuits configurations (Section 17-4).

A graphical approach is taken with the field effect transistor (Section 17-5), and the general problem of collector dissipation is also treated graphically. Circuits that use diode biasing and diode compensation are examined (Section 17-6). The thermal properties of semiconductors and heat sinks are explained with the objective of the selection of a suitable heat sink (Section 17-7).

Section 17-1 General Concepts of Beta Stability

The mass production methods of the manufacture of electronics equipment for the home consumer market, in particular, can easily result in the construction of many thousands of copies of the same circuit. An electronic item used in the automotive industry could approach a production run of a million units. Let us assume that two transistor types are available for the same circuit application. One transistor has its variation of β_{dc} or h_{FE} held to $100 \pm 10\%$ and costs 30¢ each in lots of 10,000. The other has a spread in

β_{dc} from 50 to 150 and costs 6¢ each in lots of 10,000. A method that ca
utilize the cheaper transistor results in the savings of $24,000 in the cost c
one transistor in a production run of 100,000 units. If the circuit has man
transistors, a considerable sum of money is involved.

This same concept extends to servicing. If a transistor is replaced, th
overall operation of the equipment should not change radically when th
replacement transistor has a different value of β_{dc}.

When the application becomes very critical, for instance, in space c
military equipment, the design often requires semiconductors that ar
controlled within very close tolerances. Also, the production runs ar
usually not very large. Consequently, the cost factor takes a place that i
secondary to performance and reliability.

Consider the simple amplifier that is shown in Fig. 17-1*a*. The bas
current is found from

$$V_{BB} = R_B I_B + V_{BE}$$

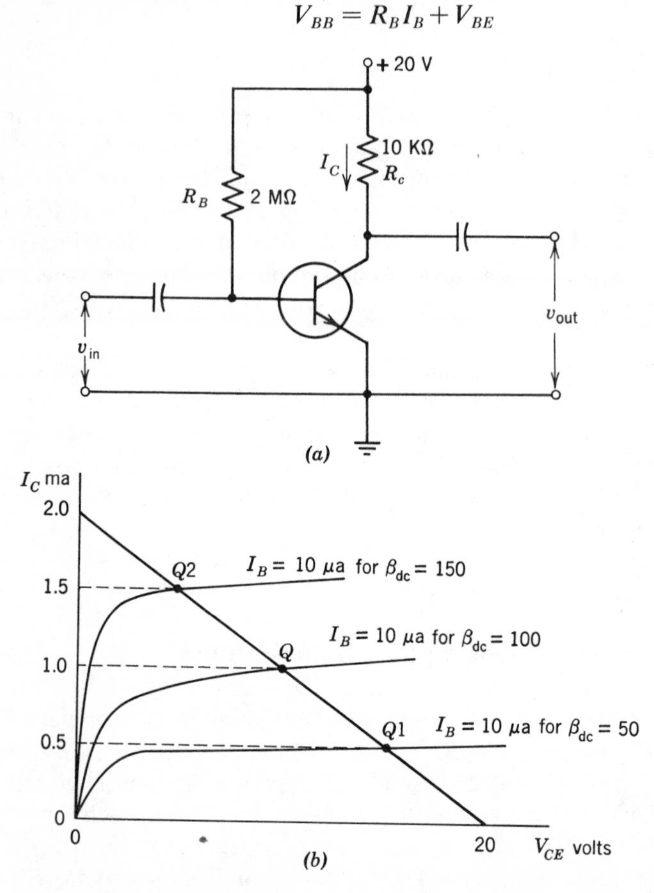

Figure 17-1 Effect of change in β_{dc}. (*a*) Circuit. (*b*) Load line.

If we neglect V_{BE},

$$V_{BB} = R_B I_B$$

and the base current is given by Ohm's law as simply

$$I_B = V_{BB}/R_B = 20/2 = 10\ \mu a$$

If V_{BE} were not neglected and were as large as 1 V, the base current would be

$$I_B = 19/2 = 9.5\ \mu a$$

This calculation shows that the maximum conceivable change of I_B in this circuit caused by a variation of V_{BE} is 0.5 μa or about $\pm 2.5\%$ from a center value. We are concerned with a transistor that has a variation in β_{dc} from 50 to 150. Therefore, we are not concerned in this circuit with the contribution of any small variation in V_{BE} that may occur from transistor to transistor.

Now let us assume that the nominal value of β_{dc} for the transistor is 100. The nominal collector current is

$$I_c = \beta_{dc} I_B = 100 \times 10 = 1000\ \mu a = 1\ ma$$

This value of I_C locates the operating point of the circuit at Q (Fig. 17-1*b*).

The least value for β_{dc} expected in a sample of this particular transistor type is 50. The resulting collector current is

$$I_C = \beta_{dc} I_B = 50 \times 10 = 500\ \mu a = 0.5\ ma$$

which is $Q1$ on the load line.

The maximum expected value of β_{dc} is 150, and the resulting collector current is

$$I_C = \beta_{dc} I_B = 150 \times 10 = 1500\ \mu a = 1.5\ ma$$

which is $Q2$ on the load line.

The load line is fixed on the collector characteristic. The end points are determined by V_{CC} and R_c. When the β_{dc} of a transistor decreases, the family of curves shrinks toward the horizontal axis. Where β_{dc} is high, the family of curves *spreads* upward. The operating point of the circuit shifts materially along the load line. What is an acceptable signal swing at the mid-value Q, obviously, could produce cutoff at $Q1$ and saturation at $Q2$. The objective of this analysis is to investigate this shift of operating point. From our numerical values, we take β_{dc} as the nominal mid-value. $+\Delta\beta_{dc}$ is the upward increase in β_{dc} that produces an upward increase $+\Delta I_C$ in collector current. $-\Delta\beta_{dc}$ is the downward change in β_{dc} that results in a decrease in I_C of $-\Delta I_C$.

In this analysis presented β will be used without subscript. This β then represents β_{dc}.

The definition of beta stability, K, becomes

$$K \equiv \left(\frac{\Delta I_C}{I_C}\right) \bigg/ \left(\frac{\Delta\beta}{\beta}\right) \tag{17-1}$$

where

$$0 \leqslant K \leqslant 1$$

If K is zero, a change in β produces no change in I_C. This is ideal. The worst case is a value for K of unity. Then a particular percentage change in β produces the same percentage change in I_C.

At this point, let us digress to mathematics in order to have a better understanding of the significance of the formulas to be developed in the next section. Consider the equation plotted in Fig. 17-2:

$$y = x^2 \tag{17-2a}$$

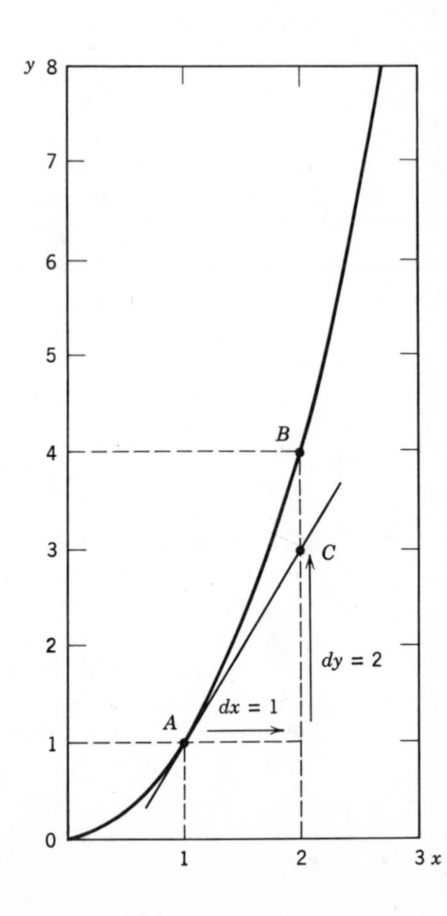

Figure 17-2 Plot of $y = x^2$.

If the value of x is changed by the amount Δx, the new value of x is $(x+\Delta x)$. If x is changed, a change Δy must occur in y, so that the new value of y is $(y+\Delta y)$. Substituting into $y=x^2$,

$$y+\Delta y=(x+\Delta x)^2=x^2+2x\Delta x+(\Delta x)^2$$

Now subtract $y=x^2$ to obtain

$$\Delta y=2x\Delta x+(\Delta x)^2 \tag{17-2b}$$

In differential calculus, we divide through by Δx,

$$\frac{\Delta y}{\Delta x}=2x+\Delta x \tag{17-2c}$$

and take the limit of the ratio $\Delta y/\Delta x$ as Δx and Δy each get smaller and smaller. Obviously, the value of $2X$ does not change, but the value of the Δx on the right becomes very small to the point that it is neglected. Then we say that the ratio of *small changes* $\Delta y/\Delta x$ is written in the new symbol dy/dx with the result that

$$\frac{dy}{dx}=2x \tag{17-2d}$$

Now, if we rewrite Equation 17-1, we have

$$K=\frac{\beta}{I_C}\frac{\Delta I_C}{\Delta \beta}$$

then for *small changes*

$$K=\frac{\beta}{I_C}\frac{dI_C}{d\beta} \tag{17-2e}$$

Most textbooks set up an equation for I_C in terms of β and, using the methods of differential calculus, perform the operation of differentiation and multiply the result by β/I_C to determine K. This is fine for *small changes* in β and in I_C. A laboratory test will not duplicate this value of K. The actual transistor has *large changes* in β and in I_C. Therefore, the term Δx in Equation 17-2c cannot be neglected.

Let us return to Fig. 17-2. At point A, $x=1$ and $y=1$. Now let the change in x, which is Δx, be one. Therefore $(x+\Delta x)$ is 2, and now from the curve (or Equation 17-2a) the true value of $(y+\Delta y)$ is 4. Thus $\Delta x=1$ and $\Delta y=3$.

The calculus approach is Equation 17-2d which gives

$$\frac{dy}{dx}=2x=2$$

or

$$dy=2\,dx$$

Substituting 1 for dx, $dy=2$.

If we add $dx = 1$ and $dy = 2$ to point A on Fig. 17-2, we obtain point C Point C is not the true value at $B(y + \Delta y)$ that is desired. Mathematically, i we draw a tangent to the curve at A, the tangent goes through C. Th methods of differential calculus require that the curve between A and C is straight line, and this straight-line approximation is given by the tangent t the curve at A. This is completely valid when the changes are *small* but nc when the changes are large.

The distances from point C to point B is the Δx term in Equation 17-2c which cannot be neglected for accurate calculations.

PROBLEMS

1. In Fig. 17-1, R_B is 300 kΩ, R_c is 2 kΩ, and β is 50. The transistor is germanium and the supply voltage is −20 V. What is the Q-point and what is the nev Q-point if β is doubled? Sketch the shift on the load line.
2. In Fig. 17-1, R_B is 10 kΩ, R_c is 75 Ω, and β is nominally 100. The transistor is an *NPN* unit and the supply voltage is +3 V. If β varies from 50 to 150, what is the shift of the Q-point from the nominal value? Show this shift on a sketch on the load line for the germanium transistor.

Section 17-2 Beta Stability Circuit Analysis

Before analyzing any circuits, a general formula for beta stability will be given, and then the procedure of this section will be to show that this genera formula is valid for the various circuit configurations. The objective is to reduce the number of necessary equations from a theoretical infinite numbe to a single equation.

Consider a network (Fig. 17-3) in which I_B and I_C flow from terminal to terminal 2. When a parallel branch is met, the two currents divide a shown. Let us define R_j as the sum of those resistors in circuit in whicl

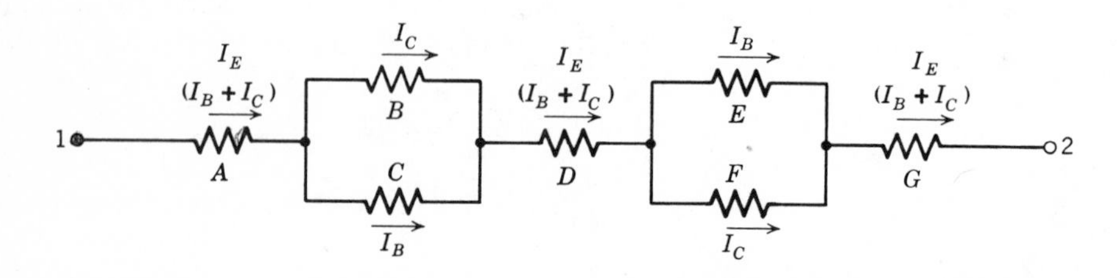

Figure 17-3 Network carrying I_B and I_E.

I_E, $(I_B + I_C)$ flows

$$R_j = A + D + G$$

and let us define R_a as the path in which I_B flows:

$$R_a = A + C + D + E + G$$

Eq. 17-1 is rearranged to a more convenient form

$$\frac{\Delta I_C}{I_C} \equiv K\frac{\Delta\beta}{\beta} \qquad 0 \leqslant K \leqslant 1 \tag{17-1}$$

and K can be found from a circuit as

$$K = \frac{1}{1 + (\beta + \Delta\beta)(R_j/R_a)} \tag{17-3}$$

Now let us examine the circuit of Fig. 17-1 and not neglect V_{BE}:

$$I_B = \frac{V_{BB} - V_{BE}}{R_B}$$

$$I_C = \beta I_B = \beta\frac{V_{BB} - V_{BE}}{R_B} \tag{17-4a}$$

Now replace β with $\beta + \Delta\beta$ and replace I_C with $I_C + \Delta I_C$.

Then

$$I_C + \Delta I_C = \beta\frac{V_{BB} - V_{BE}}{R_B} + \Delta\beta\frac{V_{BB} - V_{BE}}{R_B} \tag{17-4b}$$

Subtract Eq. 17-4a from Eq. 17-4b:

$$\Delta I_C = \Delta\beta\frac{V_{BB} - V_{BE}}{R_B} \tag{17-4c}$$

Divide Eq. 17-4c by Eq. 17-4a to obtain

$$\frac{\Delta I_C}{I_C} = \frac{\Delta\beta}{\beta} \tag{17-4d}$$

Therefore,

$$K = 1. \tag{17-4e}$$

As far as the general formula (Eq. 17-3) is concerned, R_j is zero, since there is no resistance path for I_E outside the transistor and R_B is the value of R_a. Thus, directly,

$$K = 1$$

This circuit yields the worst case value for K. A given percentage change in β causes the same percentage change in the operating point value of I_C in a circuit. The numerical values we used in Section 17-1 bear this out.

Now consider a circuit (Fig. 17-4) using an external emitter resistor

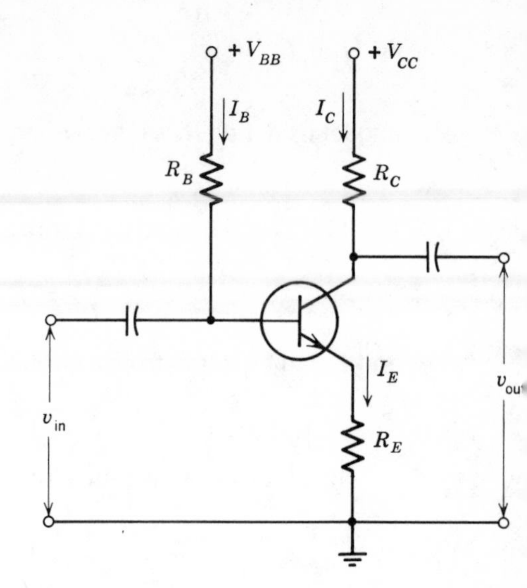

Figure 17-4 Circuit using emitter resistor.

R_E. R_E may or may not be bypassed with a capacitor. We are concerned only with a d-c analysis. A bypass capacitor affects only the a-c characteristic of the amplifier. The Kirchhoff voltage loop equation through the base resistor is

$$V_{BB} = I_B R_B + V_{BE} + I_E R_E$$

or

$$V_{BB} - V_{BE} = I_B R_B + (I_B + I_C) R_E = (R_B + R_E) I_B + R_E I_C$$

Then

$$V_{BB} - V_{BE} = (R_B + R_E) \frac{I_C}{\beta} + R_E I_C \tag{17-5a}$$

Replace I_C with $I_C + \Delta I_C$ and replace β with $\beta + \Delta\beta$.

$$V_{BB} - V_{BE} = (R_B + R_E) \frac{I_C + \Delta I_C}{\beta + \Delta\beta} + R_E (I_C + \Delta I_C) \tag{17-5b}$$

Subtract Eq. 17-5a from Eq. 17-5b:

$$(R_B + R_E) \frac{I_C + \Delta I_C}{\beta + \Delta\beta} - (R_B + R_E) \frac{I_C}{\beta} + R_E \Delta I_C = 0$$

Clearing fractions and collecting terms yields

$$[\beta (R_B + R_E) + \beta R_E (\beta + \Delta\beta)] \Delta I_C = (R_B + R_E) I_C \Delta\beta$$

Solving for $\Delta I_C / I_C$,

$$\frac{\Delta I_C}{I_C} = \frac{R_B + R_E}{R_B + R_E + \beta R_E + \Delta\beta R_E} \frac{\Delta\beta}{\beta}$$

dividing through by $(R_B + R_E)$

$$\frac{\Delta I_C}{I_C} = \frac{1}{1 + (\beta + \Delta\beta)\dfrac{R_E}{R_E + R_B}} \frac{\Delta\beta}{\beta} \qquad (17\text{-}5c)$$

and the beta stability factor K is

$$K = \frac{1}{1 + (\beta + \Delta\beta)\dfrac{R_E}{R_E + R_B}}$$

By use of Eq. 17-3 the general formula, R_j, in which I_E flows, is R_E, and the path of I_B, R_a, is $(R_E + R_B)$, giving Eq. 17-5*d* directly.

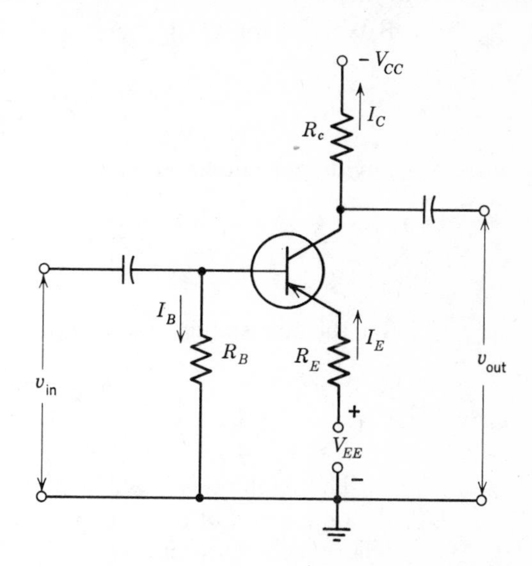

Figure 17-5 Circuit using collector and emitter-bias supplies.

To make K as small as possible, R_E must be made relatively large and R_B must be minimized in a design.

Consider the circuit shown in Fig. 17-5 wherein R_B is returned to ground. The Kirchhoff voltage loop equation around the emitter-base circuit is

$$V_{EE} = I_E R_E + V_{BE} + I_B R_B = (I_B + I_C) R_E + V_{BE} + I_B R_B$$

$$= R_E \frac{I_C}{\beta} + R_E I_C + V_{BE} + R_B \frac{I_C}{\beta}$$

Collecting terms

$$V_{EE} - V_{BE} = (R_E + R_B) \frac{I_C}{\beta} + R_E I_C \qquad (17\text{-}6a)$$

Substituting $I_C + \Delta I_C$ for I_C and $\beta + \Delta\beta$ for β

$$V_{EE} - V_{BE} = (R_E + R_B)\frac{I_C + \Delta I_C}{\beta + \Delta\beta} + R_E(I_C + \Delta I_C) \qquad (17\text{-}6b)$$

Subtracting Eq. 17-6a from 17-6b,

$$(R_E + R_B)\frac{I_C + \Delta I_C}{\beta + \Delta\beta} - (R_E + R_B)\frac{I_C}{\beta} + R_E \Delta I_C = 0$$

Cross-multiplying and collecting terms yields

$$[R_E + R_B + (\beta + \Delta\beta)R_E]\beta \Delta I_C = (R_E + R_B) I_C \Delta\beta$$

By solving for $\Delta I_C / I_C$,

$$\frac{\Delta I_C}{I_C} = \frac{R_E + R_B}{R_E + R_B + (\beta + \Delta\beta)R_E}\frac{\Delta\beta}{\beta}$$

Dividing the numerator and denominator by $(R_E + R_B)$,

$$\frac{\Delta I_C}{I_C} = \frac{1}{1 + (\beta + \Delta\beta)\dfrac{R_E}{R_E + R_B}}\frac{\Delta\beta}{\beta} \qquad (17\text{-}6c)$$

and the beta stability factor K is

$$K = \frac{1}{1 + (\beta + \Delta\beta)\dfrac{R_E}{R_E + R_B}} \qquad (17\text{-}6d)$$

This is the result which is obtained from the general formula, Eq. 17-3. A comparison of the circuits of Fig. 17-4 and Fig. 17-5 does not show any difference in current paths; the only difference is the method of applying the bias voltages.

The bias in the circuit shown in Fig. 17-6 is derived from a resistor R_B placed from collector to base. The Kirchhoff's voltage equation around the base is

$$V_{CC} = (I_B + I_C)R_c + R_B I_B + V_{BE}$$

$$V_{CC} = (R_c + R_B) I_B + R_c I_C + V_{BE} \qquad (17\text{-}7a)$$

$$V_{CC} = (R_c + R_B)\frac{I_C}{\beta} + R_c I_C + V_{BE}$$

Substitute $(I_C + \Delta I_C)$ for I_C and $(\beta + \Delta\beta)$ for β:

$$V_{CC} = (R_c + R_B)\frac{I_C + \Delta I_C}{\beta + \Delta\beta} + R_c(I_C + \Delta I_C) + V_{BE} \qquad (17\text{-}7b)$$

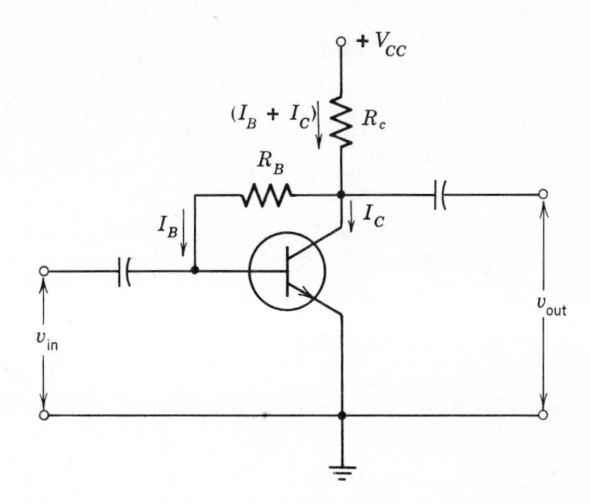

Figure 17-6 Circuit using collector-to-base feedback.

Subtract Eq. 17-7*a* from Eq. 17-7*b*:

$$(R_c + R_B)\frac{I_C + \Delta I_C}{\beta + \Delta\beta} - (R_c + R_B)\frac{I_C}{\beta} + R_c \Delta I_C = 0$$

By cross-multiplying and collecting terms,

$$[R_L + R_B + (\beta + \Delta\beta) R_c]\beta \Delta I_C = (R_c + R_B) I_C \Delta\beta$$

Solving for $\Delta I_C / I_C$,

$$\frac{\Delta I_C}{I_C} = \frac{R_C + R_B}{R_C + R_B + (\beta + \Delta\beta) R_C} \frac{\Delta\beta}{\beta}$$

Dividing numerator and denominator by $(R_c + R_B)$,

$$\frac{\Delta I_C}{I_C} = \frac{1}{1 + (\beta + \Delta\beta)\dfrac{R_c}{R_c + R_B}} \frac{\Delta\beta}{\beta} \tag{17-7c}$$

And the beta stability factor K is

$$K = \frac{1}{1 + (\beta + \Delta\beta)\dfrac{R_c}{R_c + R_B}} \tag{17-7d}$$

This equation can be obtained directly from the general formula (Eq. 17-3), since I_E flows in R_c and I_B flows in $(R_c + R_B)$.

The circuit given in Fig. 17-7 uses both collector-to-base feedback and an emitter resistor. By the use of the general formula (Eq. 17-3), the beta stability factor K is

$$K = \frac{1}{1 + (\beta + \Delta\beta)\dfrac{R_c + R_E}{R_c + R_E + R_B}} \tag{17-8}$$

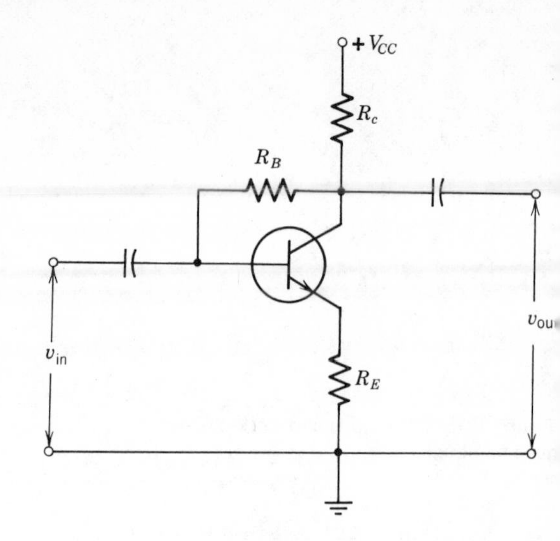

Figure 17-7 Circuit using collector-to-base feedback with additional emitter resistor.

The formal derivation of this by circuit analysis is left as an exercise.

The circuit shown in Fig. 17-8 uses a voltage divider in the base circuit

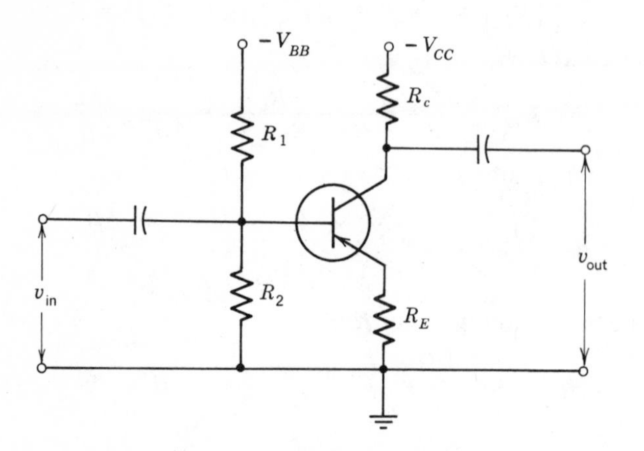

Figure 17-8 Circuit deriving base bias from a voltage divider.

V_{BB}, R_1, and R_2 are converted by the use of Thévenin's theorem into a circuit similar to that of Fig. 17-4, and this new circuit is used to evaluate K.

Now let us carefully consider the consequences of a circuit design that is concerned with a shift in β. The circuit shown in Fig. 17-9 has the following center values:

$$\beta = 100 \qquad R_c = 10\ \text{k}\Omega$$
$$R_E = 2\ \text{k}\Omega \qquad V_{CC} = 20\ \text{V}$$
$$I_B = 10\ \mu\text{a} \qquad I_C = 1\ \text{ma}$$
$$V_{BE} = 0.7\ \text{V} \qquad (\beta + \Delta\beta)\ \text{max} = 150$$

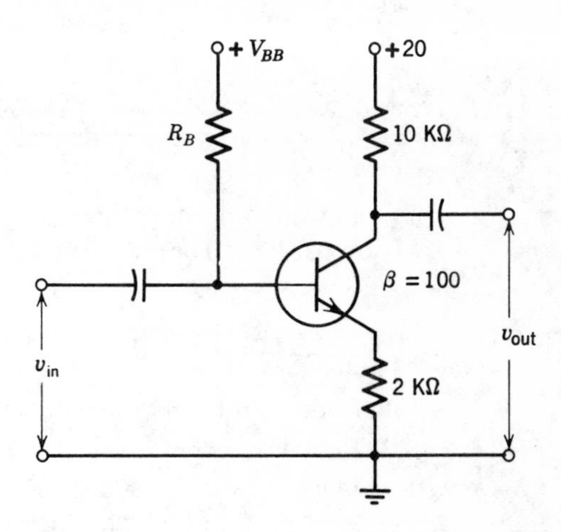

Figure 17-9 Circuit used to illustrate β stability, K.

Assume that the bias supply voltage V_{BB} is the same 20 V as V_{CC}. The value of R_B is evaluated from

$$V_{BB} - V_{BE} = R_B I_B + (I_B + I_C) R_E$$

as

$$20 - 0.7 = 0.01 R_B + (0.01 + 1.00)2$$

or

$$R_B = 1.77\ \text{M}\Omega$$

The beta stability factor K for the worst case of an increase in β to 150 from the nominal value of 100 is

$$K = \frac{1}{1 + (\beta + \Delta\beta)\dfrac{R_E}{R_E + R_B}} \tag{17-8}$$

$$K = \frac{1}{1 + 150\dfrac{2}{2 + 1730}} = \frac{1}{1 + 0.17} = \frac{1}{1.17} = 0.85$$

Consequently, when the beta of the transistor changes from 100 to 150 (a 50% increase), we can expect the change in I_C to be $0.85 \times 50\%$ or to increase 43% to 1.43 ma.

Let us rework the design assuming that V_{BB} is derived from a voltage divider circuit that gives a value for V_{BB} of 3.5 V. Then the required value of R_B to maintain the listed current values, when beta is 100, is determined from

$$V_{BB} - V_{BE} = R_B I_B + (I_B + I_C) R_E$$

as

$$3.5 - 0.7 = 0.01 R_B + (0.01 + 1.00)2$$

or

$$R_B = 78 \text{ k}\Omega$$

Then K evaluates as

$$K = \frac{1}{1 + 150 \dfrac{2}{2+78}} = \frac{1}{1 + 3.75} = \frac{1}{4.75} = 0.21$$

Now a transistor whose β is 50% greater than the nominal value increases I_C by 0.21×50 or only 10.5%.

PROBLEMS

1. Repeat Problem 1, Section 17-1, making use of Eq. 17-3.
2. Repeat Problem 2, Section 17-1, making use of Eq. 17-3.
3. In Fig. 17-4, R_B is 300 kΩ, R_c is 2000 Ω, and R_E is 1000 Ω. The silicon transistor has a β of 50, and the supply voltage is 20 V. Determine the operating point and K. If the range of variation in β is from 30 to 75, what is the change in I_C and what is the change in V_{CE}?
4. In Fig. 17-4, R_B is 750 kΩ, R_c is 3.6 kΩ, and R_E is 2000 Ω. The silicon transistor has a β of 100 and the supply voltage is 10 V. Calculate K and the change in I_C for a variation in β of $\pm 20\%$.
5. Use the data given in Problem 4. If the maximum allowable shift in operating current for a particular application is $\pm 20\%$, what range of β is acceptable for the transistor?
6. In Fig. 17-5, a *PNP* germanium transistor with a β of 30 is used in the circuit in which R_B is 10 kΩ, R_c is 10 kΩ, and R_E is 3.9 kΩ. V_{CC} is −30 V and V_{EE} is +10 V. Find K and, using this value of K, determine the change in I_C when a transistor having a β of 50 is substituted. Verify this result by solving each condition for the Q-point, using Kirchhoff's voltage-loop equations.
7. In Fig. 17-5, V_{CC} is −20 V, V_{EE} is +5 V, R_c is 20 kΩ, and R_E is 2000 Ω. The β of the germanium *PNP* transistor is 60. R_B is 20 kΩ. What range of β will keep ΔI_C to $\pm 10\%$?
8. In Fig. 17-6, V_{CC} is 20 V, R_c is 3.9 kΩ, R_B is 390 kΩ, and β is 100 for the silicon transistor. Determine the Q-point and K. If a transistor with a β of 150 is substituted, what is the new value of I_C?
9. If a transistor with a β of 50 is used in the circuit of Problem 8, what is the new value of I_C?

10. Derive Eq. 17-8.

11. In Fig. 17-7, V_{CC} is 10 V, V_{CE} is 4 V, R_E is 1500 Ω, and I_C is 1 ma. V_{BE} is 0.7 V, and β is 100. Find R_c and R_B. What is K for the circuit? If a transistor having a β of 80 is substituted, what is the new value of I_C and what is the new value of V_{CE}?

12. In Fig. 17-7, V_{CC} is 10 V, R_c is 4 kΩ, R_B is 750 kΩ, R_E is 2000 Ω, β is 100, and V_{BE} is 0.7 V. Find I_C and K. If the β of the transistor varies from 50 to 150, what is the range of I_C?

13. In Fig. 17-8, R_2 is 200 kΩ, R_c is 3000 Ω, R_E is 2000 Ω, and β is 100 for the silicon transistor. The supply voltage is 10 V. R_1 is adjusted to set V_{CE} to 5 V. Determine R_1 and I_C. What is the variation of I_C if β varies from 60 to 140? What is K?

14. In Fig. 17-8, R_1 is 75 kΩ, R_2 is 33 kΩ, R_c is 4.7 kΩ, and R_E is 1800 Ω. The supply voltage is 20 V, and the β of the silicon transistor is 30. Find I_C. What is I_C when β is 20? What is K?

Section 17-3 Leakage Currents

Consider an *NPN* transistor that is connected in a common-base circuit. A positive voltage is applied to the collector. A forward bias on the emitter injects electrons into the base. These extra electrons in the base reduce the number of holes in the base. The number of holes in the base is established by doping in manufacture. The base rejects electrons out of the base into the collector. The collector tries to maintain the number of free electrons to the value determined by its doping. Consequently, electrons flow out of the collector to the collector supply. These electrons make up the collector current that has the symbol I_C.

Minority current carriers in the base are electrons, and minority current carriers in the collector are holes. These minority current carriers also constitute a current flow in the collector in a similar manner. Now this current flow is a function of the number of minority current carriers, and it is not a function of signal or bias values. This current is called leakage current and has the symbol, I_{CBO}.

The physics of a transistor explains that minority current carriers are generated in several ways. A component is produced by impurities present during manufacture. A component is produced because of the discontinuities at the surface of the crystals where the uniform lattice structure stops. Other components are created by the breaking of covalent bonds in different parts of the transistor.

Minority current carriers must be present in any transistor. The more highly refined types, which consequently are the most expensive, have much lower values of the I_{CBO} than the cheaper units. Also, the value of I_{CBO} in a silicon transistor is much less than the value of I_{CBO} in a germanium transistor.

If a transistor is ideally made with no unwanted impurities whatsoever, at absolute zero, all the electron-hole pairs (covalent bonds) are complete. There exists the number of free electrons in the *N* and the number of free holes in the *P* that were specifically put there in the doping process. Where the transistor is in an ambient that is above absolute zero some covalent bonds are broken. The higher the ambient, the greater the number of broken covalent bonds and, consequently, the higher the value of I_{CBO}. If the temperature is high enough, I_{CBO} can swamp out I_C and can eliminate any possible signal action in a transistor.

A reverse voltage is applied to the collector of a transistor in which the base is grounded. The emitter is open or floating, so that there can be no current in the collector produced by the α action of the transistor. Then the current that is measured must be leakage current only and is given the symbol I_{CBO} which is very often shortened to I_{CO}. This is the value of leakage current that is given in a transistor specification sheet. The temperature at which this value of I_{CBO} is valid is usually specified. If not, 25°C is assumed.

To give some concept of the size of I_{CBO}, Table A gives the maximum values at 25°C as taken from transistor specification sheets.

TABLE A

I_{CBO}	$I_{C,\max}$	Transistor
10 μa	50 ma	Germanium *PNP* for audio service
3 ma	3 A	Germanium *PNP* audio power amplifier
12 μa	10 ma	Germanium *PNP* for broadcast receivers
at 25°C 0.01 μa at 150°C 1 μa	1.5 A	Silicon *NPN* power amplifier to 150 MHz
20 na	8 ma	Silicon *NPN* small signal amplifiers to 500 MHz
50 na	200 ma	Silicon *NPN* for critical industrial amplifiers to 10 MHz

An inspection of the values of I_{CBO} and I_C in the table might indicate that I_{CBO} is so much smaller than I_C that it can be neglected without further consideration. It usually can be ignored for transistors used in a common base circuit and for very high grade silicon units used at room temperatures. However, there are two factors that create a serious problem. One consideration is that, since I_{CBO} is effectively a current generator and since I_{CBO} is a current in the base, there is the usual beta amplification amplifying I_{CBO} when the transistor is used in the common-emitter configuration. Now there is a current in the collector that is $\beta_{dc} I_{CBO}$ in addition to the expected Q-point value of I_C plus the original value of I_{CBO} itself. The second consideration is that I_{CBO} is temperature sensitive in accordance to rules derived from the physics of semiconductors:

1. *I_{CBO} doubles for each 10°C rise in germanium units.*
2. *I_{CBO} doubles for each 6°C rise in silicon units.*

These two rules should be memorized.

A reverse voltage is applied between the collector and the emitter of a transistor. This is the normal polarity of V_{CC}. The base of the transistor is left open or floating. Any current that exists in the collector I_{CEO} is the current that is produced by the self-amplification of I_{CBO} within the transistor, and the two leakage currents are related by

$$I_{CEO} = (1 + \beta_{dc}) I_{CBO} \tag{17-9}$$

As an example, consider the germanium transistor listed in Table A where I_{CBO} is 12 μa and I_C is 10 ma maximum. Assume that the Q-point of the circuit is 6 ma and that the d-c beta of the transistor is 50. The circuit, a common-emitter connection, is used in an environment where the ambient is 65°C. The rise is 40°C. This means that I_{CBO} doubles four times to $12 \times 2 \times 2 \times 2 \times 2$ or 96 μa. The effect in the collector is β_{dc} times this value or 50×96 or 4800 μa. Then I_{CEO} is $(4800 + 96)$ μa or approximately 4.9 ma. Clearly, the effect of I_{CBO} in this circuit is to generate an I_{CEO} which is of the order of the Q-point itself.

For the same numerical example, let us use the silicon unit rated at 8 ma I_C and 20 na for I_{CBO}. Also, assume that the temperature rise is now 60°C to an 85°C ambient. The rise of 60° means that I_{CBO} doubles 10 times, which is 2^{10}. Then the new I_{CBO} is

$$I_{CBO} = 20 \times 2^{10} = 20 \times 1024 = 20480 \text{ na} = 20.5\ \mu\text{a}$$

Now the value of I_{CEO} in the collector is

$$I_{CEO} = (1 + \beta_{dc}) I_{CBO} = (1 + 50)20.5 = 1045.5\ \mu\text{a} \approx 1 \text{ ma}$$

These two calculations show that there is a large difference between germanium and silicon transistors. This difference is the effect of I_{CBO} in the collector. A germanium transistor has a large I_{CBO} as compared to the silicon unit. A temperature rise increases I_{CBO} more in a silicon transistor, but the advantage of a silicon transistor in the first place is that at 25°C the value of I_{CBO} is extremely small as compared to I_C max. Consequently, silicon transistors must be used in applications where the equipment is subjected to high temperatures.

The effect of I_{CBO} is usually negligible on a common-base characteristic since there is no multiplication by β_{dc}, but it is definitely not negligible on a common-emitter characteristic where I_{CBO} is multiplied by β_{dc}. Consequently, when I_B is zero, there is a finite curve, not zero, for I_C, as shown in Fig. 3-10. The presence of $\beta_{dc} I_{CBO}$ *shifts* all the characteristic curves upward. In Section 17-2 an increased value of β_{dc} *spreads* the curves upward.

Temperature changes also affect other transistor parameters. r'_e is determined at room temperature (25°C) from

$$\frac{25}{I_E} \leqslant r'_e \leqslant \frac{50}{I_E} \qquad (12-1$$

r'_e increases with an increase in ambient temperature at the rate of 1% for a change of 3°C. The effect of this change in r'_e can be minimized by using an external "swamping" resistor. The value of V_{BE} is taken as 0.3 V for a germanium junction and 0.7 V for a silicon junction. These values are valid for room temperature (25°C) and must be corrected for other temperatures. In germanium units, V_{BE} decreases 1.6 mv/°C, and in silicon units, V_{BE} decreases 2.0 mv/°C. Corrections need only be made when V_{BE} and V_{EE} are very low values. The values of β_{dc} and β_{ac} can change radically with temperature showing a variation that can be as much as four to one. There is no general approach that can be used for β. The data given in the transistor specification sheet for the particular transistor must be used.

PROBLEMS

For each transistor listed in Table A, assume that the leakage current is established at 25°C. For each listed transistor determine the temperature at which I_{CEO} equals the listed value of I_C. Assume β is 49 for each unit.

Section 17-4 Temperature Sensitivity

In this section all values of β are β_{dc}, and for the simplification of equation the subscript on β_{dc} will be omitted.

Temperature sensitivity, S, is defined as

$$S \equiv \frac{\Delta I_C}{\Delta I_{CBO}} \qquad (17-10$$

where

$$1 \leqslant S \leqslant (1+\beta)$$

As in the Section 17-2, a general formula will be given. Then the analysis of the basic circuits will show that the results could have been obtained directly from the general formula

$$S = \frac{\text{sum of resistors in which } I_B \text{ flows}}{\text{same sum except } R_B \text{ is divided by } (1+\beta)} \qquad (17\text{-}11)$$

The leakage current, I_{CBO} is shown in all of the circuits to be analyzed as a current generator external to the transistor for convenience. Actually, it is a current generator within the transistor. In a model for the transistor it would be placed within the limits of the transistor model. We have not used models for transistors in analyzing the d-c circuit. We have confined our approach to the use of $I_C = \beta I_B$ and $I_E = (1+\beta)I_B$, and we have taken the approach using Kirchhoff's voltage loop equations. There is no need in this analysis to change to a d-c model approach, since the problem can be treated similarly to the approach taken in Section 17-2 as long as the concept of an external current generator for I_{CBO} is taken to serve merely as an aid or as a reminder in establishing voltage equations directly from an examination of the circuit itself.

In all of the circuits we shall examine, a resistor placed externally to the transistor in an emitter lead will be shown without a bypass capacitor. The capacitor does not affect this d-c analysis. If it is needed in an actual circuit application for a-c gain considerations, it may be present.

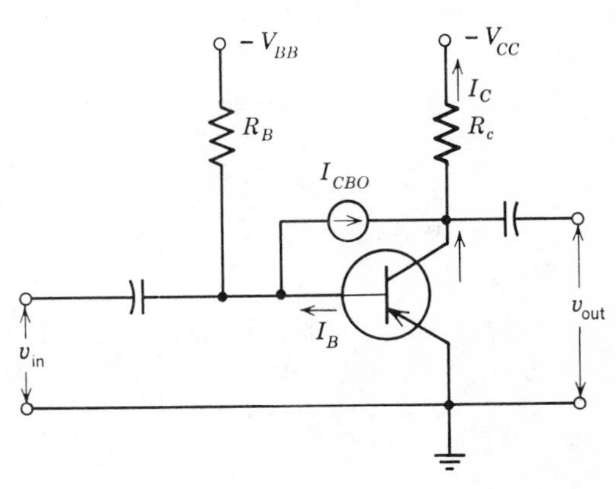

Figure 17-10 Simple common-emitter amplifier.

The simplest basic amplifier circuit is shown in Fig. 17-10. The effective base current inside the transistor is

$$I_B = \frac{V_{BB} - V_{BE}}{R_B} + I_{CBO}$$

then

$$I_C = \beta I_B + I_{CEO}$$

This last equation should be examined very carefully. I_C has two components: the value of I_{CBO} plus βI_B. Since I_B contains I_{CBO} and βI_B amplifies this I_{CBO} by β, the total effect of I_{CBO} in the collector is

$$I_{CEO} = (1+\beta)\, I_{CBO} \tag{17-9}$$

Usually, for a large β, the effect can be taken approximately as

$$I_{CEO} = \beta I_{CBO}$$

Now, making the substitution of the equation for I_B into the equation for I_C,

$$I_C = \beta\left(\frac{V_{BB} - V_{BE}}{R_B} + I_{CBO}\right) + I_{CBO}$$

$$I_C = \beta\left(\frac{V_{BB} - V_{BE}}{R_B}\right) + (1+\beta)\, I_{CBO}$$

When I_{CBO} increases by ΔI_{CBO}, I_C increases by ΔI_C:

$$I_C + \Delta I_C = \beta\left(\frac{V_{BB} - V_{BE}}{R_B}\right) + (1+\beta)\,(I_{CBO} + \Delta I_{CBO})$$

Subtracting

$$\Delta I_C = (1+\beta)\,\Delta I_{CBO} \tag{17-12a}$$

Then by the definition of S (Eq. 17-10)

$$S = (1+\beta) \tag{17-12b}$$

As in the problem of the beta stability factor, this circuit represents a worst-case condition.

The general formula (Eq. 17-11) evaluates S as

$$S = \frac{R_B}{R_B/(1+\beta)} = (1+\beta)$$

which checks the result of Eq. 17-12b.

The circuit shown in Fig. 17-11 has an emitter resistor R_E added. The voltage loop equation through the base is

$$V_{BB} = R_B I_B + V_{BE} + R_E I_E$$

substituting

$$I_E = (1+\beta) I_B + (1+\beta)\, I_{CBO}$$

$$V_{BB} = R_B I_B + V_{BE} + (1+\beta)\, R_E I_B + (1+\beta)\, R_E I_{CBO}$$

Solving for I_B,

$$I_B = \frac{V_{BB} - V_{BE}}{R_B + (1+\beta) R_E} - \frac{(1+\beta) R_E}{R_B + (1+\beta) R_E} I_{CBO}$$

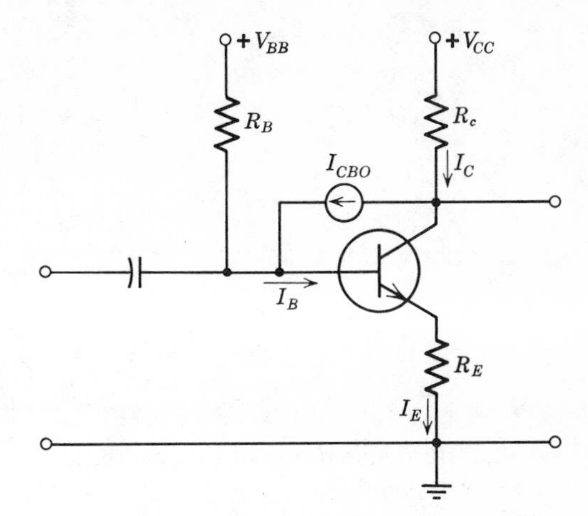

Figure 17-11 Common-emitter amplifier with emitter resistor.

Substitute this into

$$I_C = \beta I_B + (1+\beta) I_{CBO}$$

$$I_C = \beta \frac{V_{BB} - V_{BE}}{R_B + (1+\beta) R_E} - \frac{\beta(1+\beta) R_E}{R_B + (1+\beta) R_E} I_{CBO} + (1+\beta) I_{CBO}$$

Now let I_{CBO} increase to $I_{CBO} + \Delta I_{CBO}$, causing I_C to rise to $I_C + \Delta I_C$:

$$I_C + \Delta I_C = \beta \frac{V_{BB} - V_{BE}}{R_B + (1+\beta) R_E} - \frac{\beta(1+\beta) R_E}{R_B + (1+\beta) R_E} (I_{CBO} + \Delta I_{CBO})$$

$$+ (1+\beta)(I_{CBO} + \Delta I_{CBO})$$

From this subtract the expression for I_C to obtain ΔI_C,

$$\Delta I_C = \left[(1+\beta) - \frac{\beta(1+\beta) R_E}{R_B + (1+\beta) R_E}\right] \Delta I_{CBO} = \left[\frac{(1+\beta) R_B + (1+\beta) R_E}{R_B + (1+\beta) R_E}\right] \Delta I_{CBO}$$

dividing numerator and denominator by $(1+\beta)$

$$\Delta I_C = \frac{R_E + R_B}{R_E + R_B/(1+\beta)} \Delta I_{CBO} \tag{17-13a}$$

and

$$S = \frac{R_E + R_B}{R_E + R_B/(1+\beta)} \tag{17-13b}$$

This result is the same as the one obtained by an inspection from the general formula.

The circuit shown in Fig. 17-12 uses a voltage divider to develop the bias. When V_{BB}, R_1, and R_2 are converted into an equivalent network by Thévenin's theorem, the network reduces to the circuit of Fig. 17-11, and Eq. 17-13*b* becomes valid for the new equivalent circuit.

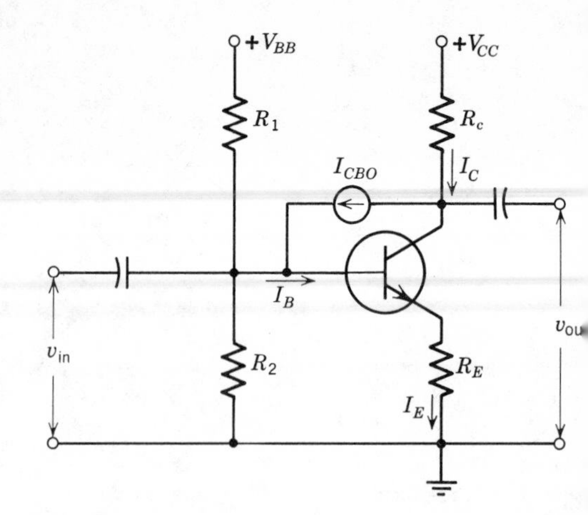

Figure 17-12 Amplifier with bias derived from a divider network.

The circuit given in Fig. 17-13 makes use of a feedback resistor, R_B connected from collector to base to develop the bias. The voltage loop equation through R_B is

$$V_{CC} = R_C I_E + R_B I_B + V_{BE}$$

$$I_C = \beta I_B + (1+\beta) I_{CBO}$$

or

$$I_B = \frac{I_C - (1+\beta) I_{CBO}}{\beta}$$

and

$$I_E = I_C + I_B$$

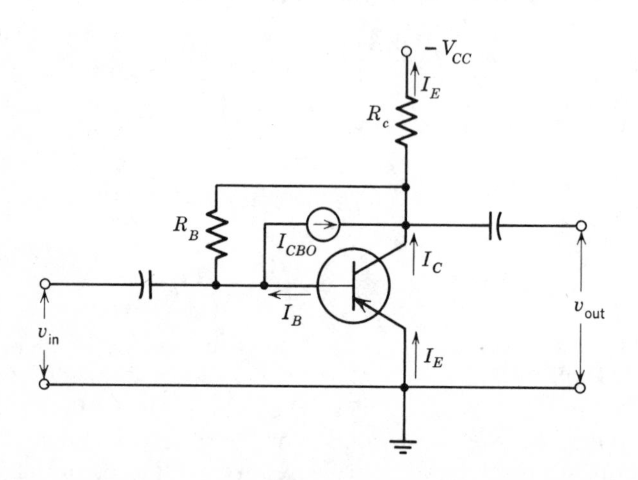

Figure 17-13 Amplifier having collector-to-base feedback bias.

Substituting these into the loop equation

$$V_{CC} = R_c\left[\beta\frac{I_C - (1+\beta)I_{CBO}}{\beta} + \frac{I_C - (1+\beta)I_{CBO}}{\beta}\right]$$

$$+ R_B\left[\frac{I_C - (1+\beta)I_{CBO}}{\beta}\right] + V_{BE}$$

This equation reduces to

$$[R_B + (1+\beta)R_c]I_C = \beta(V_{CC} - V_{BE}) + (1+\beta)(R_B + R_c)I_{CBO}$$

I_C is replaced by $I_C + \Delta I_C$ when I_{CBO} is increased to $I_{CBO} + \Delta I_{CBO}$.

$$[R_B + (1+\beta)R_c](I_C + \Delta I_C) = \beta(V_{CC} - V_{BE}) + (1+\beta)(R_B + R_c)(I_{CBO} + \Delta I_{CBO})$$

Subtracting the last two equations yields

$$[R_B + (1+\beta)R_c]\Delta I_C = (1+\beta)(R_B + R_c)\Delta I_{CBO}$$

rearranging

$$\Delta I_C = \frac{(1+\beta)(R_c + R_B)}{(1+\beta)R_c + R_B}\Delta I_{CBO}$$

dividing both numerator and denominator by $(1+\beta)$

$$\Delta I_C = \frac{R_c + R_B}{R_c + R_B/(1+\beta)}\Delta I_{CBO} \qquad (17\text{-}14a)$$

making the temperature sensitivity, S,

$$S = \frac{R_c + R_B}{R_c + R_B/(1+\beta)} \qquad (17\text{-}14b)$$

The application of the general formula (Eq. 17-11) gives this result of Eq. 17-14*b* directly.

The circuit of Fig. 17-14 is the circuit of Fig. 17-13 with a resistor R_E added external to the emitter. Since I_E flows in R_E as well as in R_c, the derivation given for the last circuit is valid merely by replacing $(R_E + R_c)$ for R_c throughout. Then the temperature sensitivity S is

$$S = \frac{R_E + R_c + R_B}{R_E + R_c + R_B/(1+\beta)} \qquad (17\text{-}15)$$

The circuit analysis of the amplifier shown in Fig. 17-15 is left as an exercise at the end of the chapter. The value of temperature sensitive sensitivity given directly by the general formula is

$$S = \frac{R_E + R_B}{R_E + R_B/(1+\beta)} \qquad (17\text{-}16)$$

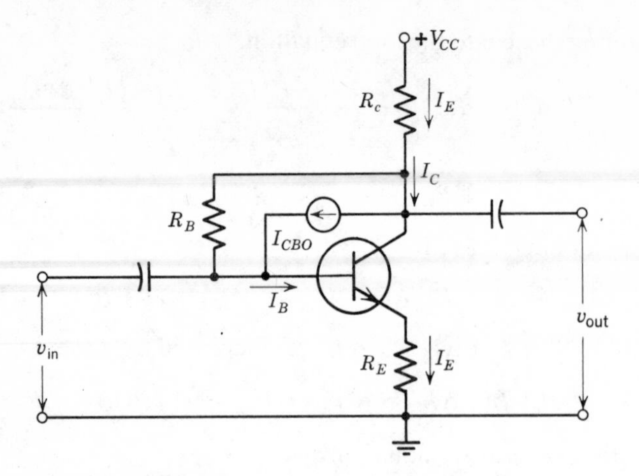

Figure 17-14 Amplifier with collector-to-base bias and an emitter resistor.

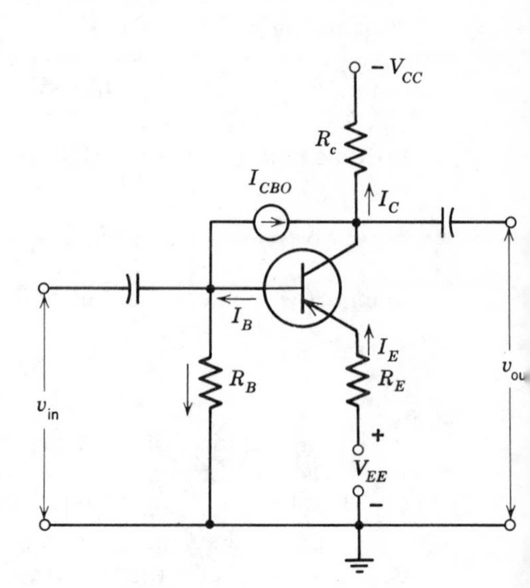

Figure 17-15 Amplifier with grounded-base resistor.

In this last circuit, if the values can be arranged so that R_B and R_E are equal, the sensitivity becomes for a circuit with a beta of 100

$$S = \frac{R_E + R_E}{R_E + R_E/101} \approx 2$$

This means that an increase of ΔI_{CBO} is only multiplied by $(1+2)$ into the collector circuit. Consequently, the effect of a temperature change on ΔI_C can be readily made negligible. This contrasts sharply with the first case

(Fig. 17-10) in which a ΔI_{CBO} would be multiplied by a factor of 101 into the collector circuit.

The circuit shown in Fig. 17-16 uses a transformer to drive the transistor. Ideally, the resistance of the secondary winding is zero. Consequently,

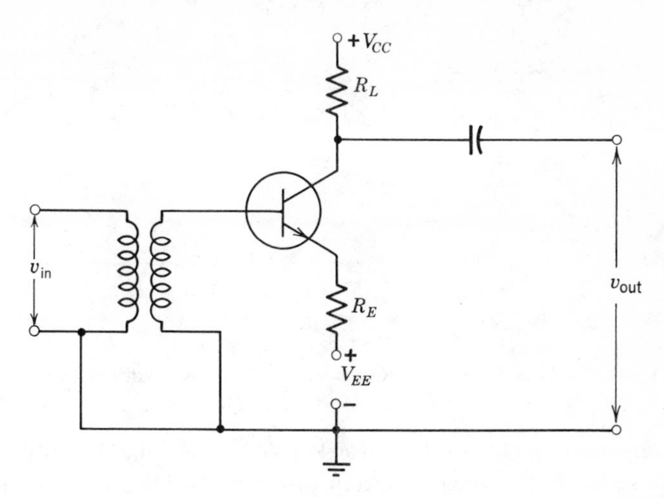

Figure 17-16 Amplifier circuit having $S = 1$.

S has the value one. Now there is no multiplication of a ΔI_{CBO} into the collector circuit. This is the only simple circuit configuration for which $S = 1$. Consequently, ΔI_C is merely ΔI_{CBO}.

To gain a better understanding of the significance of S, let us examine the circuits given in Fig. 17-17. The sensitivity of the circuit of Fig. 17-17a is the worst case $(1+\beta)$. The base current I_B is determined from the voltage loop equation

$$I_B R_B + V_{BE} = V_{CC}$$

The base current is determined solely from this equation and, therefore, it is a function of the numerical values of V_{CC}, R_B, and V_{BE}. The collector current and the emitter current are not involved in this calculation. Consequently, the leakage current I_{CBO} and any increase in leakage current ΔI_{CBO} that results from an increase in temperature is multiplied by $(1+\beta)$ and adds to the collector current. The change in collector current ΔI_C changes V_{CE}.

Now let us consider the circuit shown in Fig. 17-17b. The value of S for the circuit is, assuming a β of 100,

$$S = \frac{5 + 1000 + 1}{5 + 1000/(1+100) + 1} = 63$$

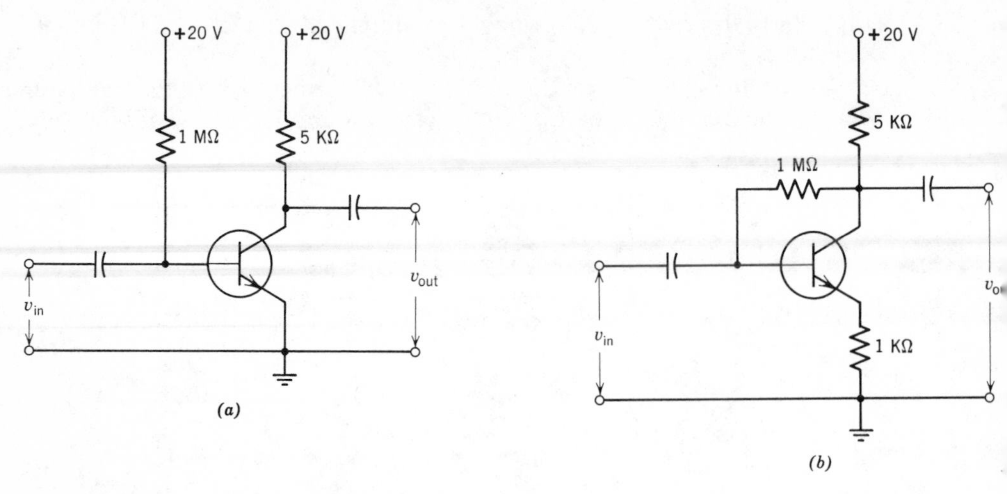

Figure 17-17 Circuits used to show effect of I_{CBO}. (*a*) Without feedback. (*b*) With feedback.

Let us use the value of 0.7 V for V_{EB} and determine the values for I_B, I_E, an
V_{CE}, assuming that the leakage current is zero. The loop equation is

$$R_c I_E + R_B I_B + V_{CE} + R_E I_E = V_{CC}$$

and substituting

$$5(101 I_B) + 1000 I_B + 0.7 + 1(101 I_B) = 20$$

Solving $$I_B = 12\ \mu a$$

Then $$I_E = 1212\ \mu a = 1.212\ ma \qquad \text{and} \qquad V_{CE} = 12.728\ V$$

As a first case, assume that I_{CBO} is 1 μa. The effect in the collector is t
multiply this 1 μa by S to produce 63 μa in the collector. As a result, I_E i
$1212 + 63 \times 1$ or 1275 μa and V_{CE} is 12.350 V. At an elevated temperature
I_{CBO} rises to 4 μa. The change in I_{CBO} is 3 μa, and the effect is a ΔI_C o
$S \Delta I_{CBO}$ or 63×3 or 189 μa. Now I_E is 1474 μa or 1.474 ma. V_{CE} is 11.216 V
The relation between I_B and V_{CE} is

$$V_{CE} = I_B R_B + V_{BE}$$

By using the values when I_{CBO} was 1 μa and solving for I_B in this last equa
tion

$$12.350 = 1000 I_B + 0.7 \qquad \text{or} \qquad I_B = 11.65\ \mu a$$

Now by using the values when I_{CBO} was 4 μa and solving for I_B, we find that

$$11.216 = 1000 I_B + 0.7 \qquad \text{or} \qquad I_B = 10.52\ \mu a$$

If S is the worst-case value $(1 + \beta)$ as in the circuit of Fig. 17-17*a*, th
base current I_B is unaffected by I_{CBO} or by ΔI_C. In a circuit in which S is les

than the worst case (for example, Fig. 17-17*b*) there is a compensation effect against the increase in I_{CBO}. This compensation is shown by the decrease in I_B. I_B is 12 μa when I_{CBO} is zero. I_B is 11.65 when I_{CBO} is 1 μa, and I_B is 10.52 μa when I_{CBO} rises to 4 μa. If the value of S had been lower, this compensation would have been more pronounced.

PROBLEMS

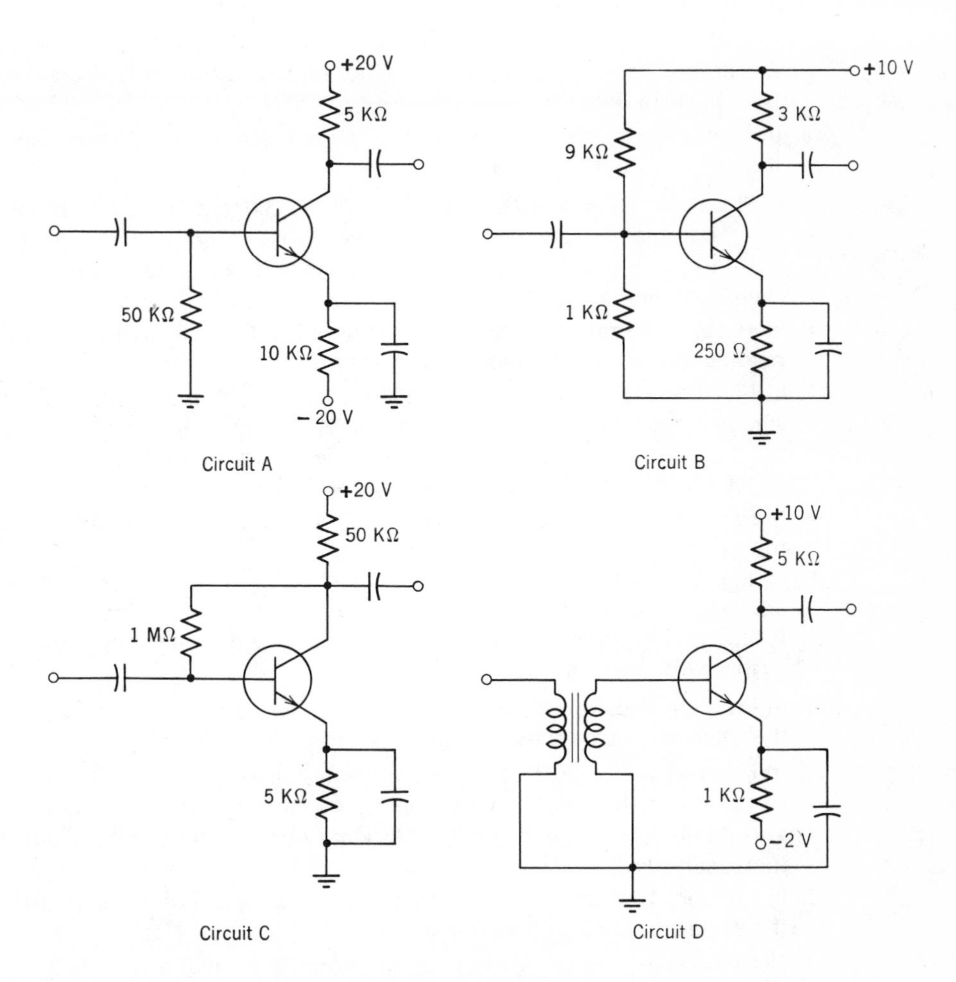

Circuits for Problem 16.

Use 25°C for room temperature.

1. In Fig. 17-10, R_B is 200 kΩ,R_c is 2 kΩ, and β is 50. The *PNP* transistor is germanium and the supply voltage is −20 V. At room temperature I_{CBO} is 0.1 μa, and its effect on I_C is negligible. At what elevated temperature will it cause I_C to increase by 50%?
2. In Fig. 17-10, R_B is 10 kΩ, R_c is 75 Ω, and β is 60. The *NPN* transistor is silicon, and the supply voltage is +3 V. At room temperature I_{CBO} is 50 na, and its effect on I_C is negligible. At what elevated temperature will it cause I_C to rise by 40%?
3. In Fig. 17-11, R_B is 300 kΩ, R_c is 2000 Ω, R_E is 1000 Ω, and β is 75. The transistor is silicon and the supply voltage is 20 V. At room temperature I_{CBO} is 20 na, and its effect on I_C is negligible. At what elevated temperature will I_C increase by 50%.
4. In Problem 3, determine the external circuit value of I_B and compare it to the value of I_B at room temperature.
5. In Fig. 17-11, R_B is 750 kΩ, R_c is 3.9 kΩ, R_E is 2000 Ω, and β is 100. The transistor is germanium and the supply voltage is 12 V. At room temperature I_{CBO} is 0.1 μa, and its effect on I_C at room temperature is negligible. At what elevated temperature will it cause I_C to increase 30%?
6. In Problem 5, determine the external circuit value of I_B and compare this result to the value of I_B at room temperature.
7. In Fig. 17-12, R_2 is 100 kΩ, R_c is 3000 Ω, R_E is 1000 Ω, and β is 150 for the silicon transistor. R_1 is adjusted to set V_{CE} to 4 V with a 10-V supply. At room temperature I_{CBO} is 5 na, and its effect on I_C at room temperature is negligible. At what elevated temperature will it cause I_C to rise 20%?
8. In Problem 7, determine the external circuit value of I_B and compare it to the value of I_B at room temperature.
9. In Fig. 17-13, R_c is 5 kΩ and β if 60 for the germanium transistor. R_B is adjusted to set V_{CE} to 2 V with a −4-V supply. At room temperature I_{CBO} is 0.1 μa, and its effect is negligible on I_C. At what elevated temperature will I_C be increased by 15%?
10. In Problem 9, determine the external circuit value of I_B and compare it with the value of I_B at room temperature.
11. In Fig. 17-13, R_C is 39 kΩ and β is 200 for the silicon transistor. R_B is selected to set V_{CE} to 10 V with a 20-V supply. At room temperature I_{CBO} is 30 na, and its effect on I_C is negligible. At what elevated temperature will I_C be increased 20%?
12. In Problem 11, determine the external circuit value of I_B and compare it with the value of I_B at room temperature.
13. Derive Eq. 17-15 by using Kirchhoff's voltage loop equations.
14. Repeat Problems 9 and 10 if a 300-Ω external emitter resistor, R_E is added to the circuit.
15. Repeat Problems 11 and 12 if a 300-Ω external emitter resistor, R_E, is added to the circuit.

16. The transistor is silicon and has a value of 0.7 for V_{BE} and a value of 100 for β. At 25°C, I_{CBO} is 1 μa, and its effect on I_C is negligible. The maximum allowable ambient temperature for the transistor is 67°C. The maximum permissible shift in I_C is 50%. What is the maximum allowable operating ambient temperature for each of the four circuits?

Section 17-5 Graphical Analysis

The circuit for a simple JFET amplifier using self bias is shown in Fig. 17-18*a*. The load line is drawn on the drain characteristic (Fig. 17-18*b*) in the usual fashion by locating V_{DD} as one end point and $V_{DD}/(R_d+R_s)$ as the other end point. The sum of R_d and R_s is used for this, the static load line. The transfer curve for a middle-value JFET is drawn as curve *A* on the $i_D - v_{GS}$ axes.

The volt-ampere curve for the resistor Rs is drawn on the transfer curve. This is a straight line obeying Ohm's law and labeled on the transfer curve (Fig. 17-18*b*) as the *bias line*. The intersection of the bias line with the transfer curve *A* must be operating point *Q*. *Q* is projected over to the static load line. The dynamic load line is drawn through *Q*. The resistance value for this dynamic operating path is R_d, the resistor that is actually the drain load resistor.

Now let us say that a batch of these JFET's yields transfer characteristics that can be found to vary between the limits *B* and *C*. For example, a typical FET has a rating stating that I_{DSS} can range from 6 to 14 ma with 10 ma the nominal value. The bias line intersects curve *B* at *Q*1, and curve *C* at *Q*2. These limit values are projected over the static load line. As a result, the dynamic operating path can shift anywhere from *Q*1 to *Q*2. Thus a saturation or cutoff can occur on these new dynamic load lines which would not occur on the nominal value load line. This problem is almost identical to the expected variation in the β_{dc} of a transistor.

The circuit shown in Fig. 17-19 is a method used to minimize this shift of operating point. Since the gate current in an FET is zero, R_2 and R_1 form a simple voltage divider making the voltage from gate to ground

$$V_{GG} = \frac{R_1}{R_1+R_2} V_{DD}$$

The voltage from source to ground is

$$V_s = R_s I_D$$

The gate-to-source voltage is

$$V_{GS} = V_{GG} - V_S = V_{GG} - R_s I_D$$

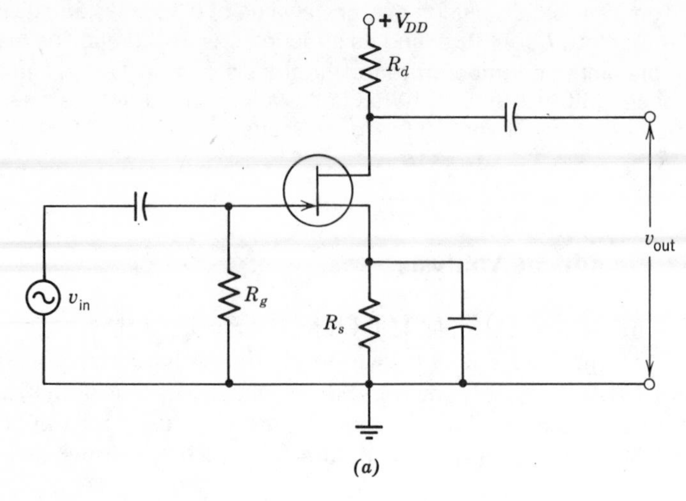

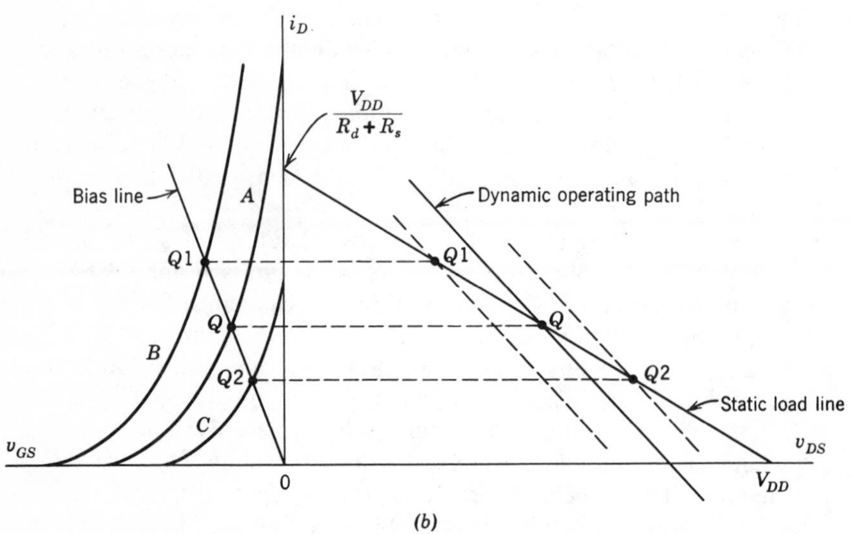

Figure 17-18 JFET using self-bias. (*a*) Circuit. (*b*) Load line.

Solving this equation for I_D,

$$I_D = -\frac{1}{R_s} V_{GS} + \frac{V_{GG}}{R_s}$$

This equation is in the form $y = mx + b$ and is, therefore, a straight line o[n] the $v_{GS} - i_D$ axes. The slope of the line is $-1/R_s$ and the i_D intercept i[s] V_{GG}/R_s. Also, when i_D is zero, $v_{GS} = V_{GG}$. This bias line is now shown o[n]

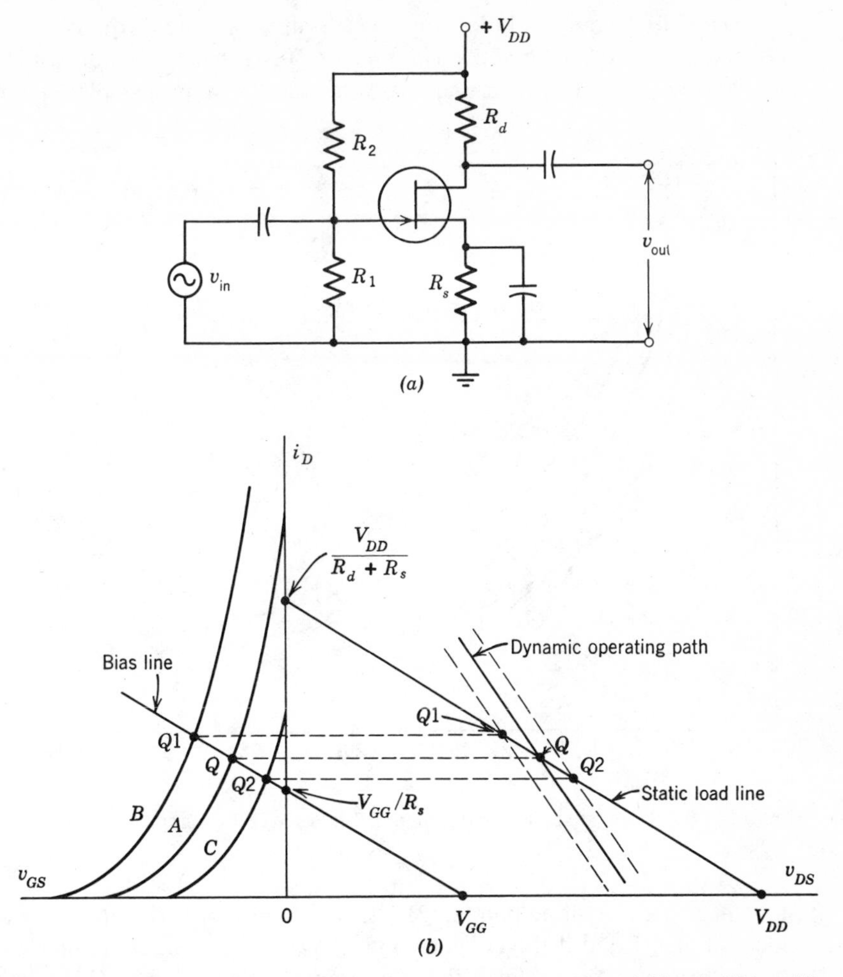

Figure 17-19 JFET using a voltage-divider to establish the bias. (*a*) Circuit. (*b*) Load line.

Fig. 17-19*b*. The static load line is drawn as before, and the operating point Q is the intersection of this new bias line with transfer curve A.

The bias line intersects transfer curves B and C at $Q1$ and $Q2$, respectively. $Q1$ and $Q2$ are projected over to the static load line. Now they are much closer to Q than the results shown in Fig. 17-18*b*, and the effect of the differences between FET's of the same type is minimized.

The dissipation at the collector of a transistor without signal is the product $V_C I_C$. A constant collector dissipation line is determined from

$$P_C = V_C I_C$$

and is a hyperbola. Three constant dissipation lines are drawn on th collector characteristic shown in Fig. 17-20. A load line is drawn betwee V_{CC} and V_{CC}/Rc. The load line intersects the constant dissipation curves a

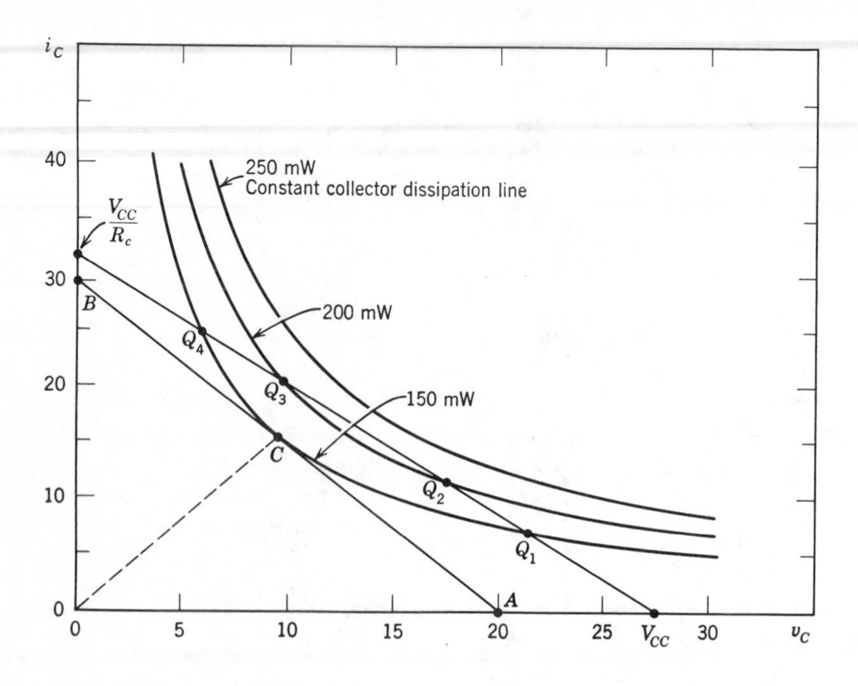

Figure 17-20 Constant collector dissipation lines on a collector characteristic.

four possible operating points, $Q1$, $Q2$, $Q3$, and $Q4$. If the operating point i located at $Q1$ and if the collector current rises because of an increased β_0 or because of a ΔI_{CBO} effect, the operating point will shift to $Q2$. This mean that the transistor heats up, and self-destruction could result. If the operatin point were located at $Q3$ and if I_C increases, the operating point shifts t $Q4$. Now the heating within the transistor decreases, and no overheatin occurs.

Now let us draw a load line from A to B that is just tangent to the 150-mW curve at C. Any shift in operating point in either direction reduces the collec tor dissipation. From this viewpoint, point C is the ideal operating point The equation of the curve is

$$V_C I_C = 150$$

or

$$I_C = \frac{150}{V_C}$$

differentiating

$$\frac{dI_C}{dV_C} = -\frac{150}{V_C} = -\frac{1}{V_C}\left(\frac{150}{V_C}\right)$$

since

$$I_C = \frac{150}{V_C},$$

we substitute

$$\frac{dI_C}{dV_C} = -\frac{I_C}{V_C}$$

These values of I_C and V_C are the values of the tangent point C of the curve. Now draw a line from 0 to C. The slope, m, of this line is

$$m = \frac{I_C}{V_C}$$

The slope from 0 to C is $+I_C/V_C$, and the slope from A to C is $-I_C/V_C$. Therefore, angle COA and angle OAC are equal and OC equals AC. The tangent point at C has the voltage coordinate $A/2$. As a result, when the operating point of a transistor is set to be $\frac{1}{2}V_{CC}$, a shift in operating point by a change in β_{dc} or in ΔI_{CBO} cannot in itself cause overheating in the transistor.

PROBLEMS

1. A FET has a nominal value for I_{DSS} of 10 ma with V_P rated at −4 V. The minimum expected value for I_{DSS} is 5 ma with a V_P of −2 V. The maximum expected value for I_{DSS} is 15 ma with a V_P of −6 V. The FET is used in an amplifier circuit (Fig. 17-18*a*) that has the values $V_{DD} = 24$ V, $R_g = 100$ kΩ, $R_s = 500$ Ω, and $R_d = 1500$ Ω. Construct a graph similar to Fig. 17-18*b*, giving the location of the operating points $Q1$, Q, and $Q2$ that are the limits for this FET. What are the voltage gains of the circuit at the operating points?
2. For Problem 1, determine the peak-to-peak output voltage available at each operating point $Q1$, Q, and $Q2$ for a sinusoidal input signal.
3. The FET used in Problem 1 is now used in the circuit of Fig. 17-19*a*. Now, $R_1 = 100$ kΩ, $R_2 = 300$ kΩ, $R_s = 1500$ Ω, and $R_d = 1000$ Ω for a supply voltage of 24 V. Draw the curves similar to the ones of Fig. 17-19*b* and determine the operating points $Q1$, Q, and $Q2$. What is the voltage gain of the circuit at each operating point?
4. For Problem 3, determine the peak-to-peak voltage available across the output terminals at each operating point, $Q1$, Q, and $Q2$, for a sinusoidal input signal.
5. A FET has a nominal value for I_{DSS} of 4 ma with V_P rated at −4 V. The minimum expected value for I_{DSS} is 2 ma with a V_P of −2 V. The maximum expected value for I_{DSS} is 7 ma with a V_P of −7 V. The FET is used in an amplifier circuit (Fig. 17-18*a*) that has the values $V_{DD} = 30$ V, $R_g = 100$ kΩ, $R_s = 1000$ Ω, and

$R_d = 6500\ \Omega$. Construct a graph similar to Fig. 17-18*b*, giving the location o the operating points *Q*1, *Q*, and *Q*2 that are the limits for this FET. What are the voltage gains of the circuit at the operating points?

6. In the amplifier of Problem 5, determine the peak-to-peak voltage available a each operating point *Q*1, *Q*, and *Q*2 for a sinusoidal input signal.
7. The FET of Problem 5 is now used in the circuit of Fig. 17-19*a*. Now, $R_1 =$ 100 kΩ, $R_2 = 500$ kΩ, $R_s = 5000\ \Omega$, and $R_d = 5000\ \Omega$ for a supply voltage o 30 V. Draw curves similar to the ones of Fig. 17-19*b* and determine the operating points *Q*1, *Q*, and *Q*2. What is the voltage gain of the circuit at each operating point?
8. In the amplifier of Problem 7, determine the peak-to-peak voltage available across the output terminals at each operating point, *Q*1, *Q*, and *Q*2, for a sinusoidal input signal.

Section 17-6 Diode Biasing and Compensation

The amplifier shown in Fig. 17-21*a* has a diode *D*1 placed in the emitter circuit that compensates against changes in V_{BE} that occur as the ambient temperature changes. In order to facilitate analysis, the bias circuit comprising V_{CC}, R_2, and R_1 is converted by Thévenin's theorem to an equivalent

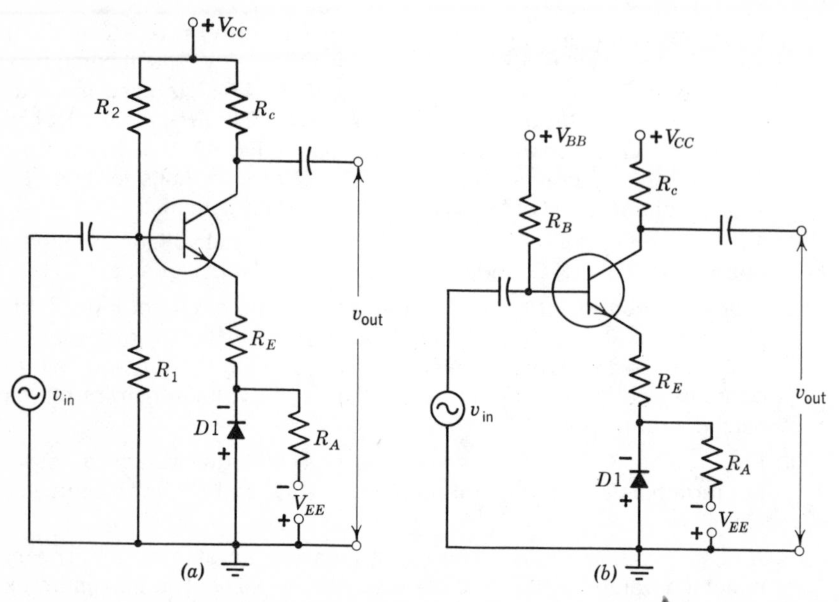

Figure 17-21 Diode compensation for variations in V_{BE}. (*a*) Actual circuit. (*b*) Modified circuit for analysis.

V_{BB} and R_B as shown in Fig. 17-21*b*. The Kirchhoff's voltage loop equation through the base is

$$V_{BB} = R_B I_B + V_{BE} + R_E I_E - V_{D1}$$

where V_{D1} is the voltage drop across the diode $D1$, developed by the value of V_{EE} and R_A. Then

$$V_{BB} = R_B I_B + R_E I_E + (V_{BE} - V_{D1})$$

when

$$V_{BE} = V_{D1}$$

$$V_{BB} = R_B I_B + R_E I_E$$

Thus, when a diode is selected that will have the identical characteristics of variation of V_{D1} that V_{BE} has, the equation is independent of V_{BE} and perfect compensation is achieved. The practical difficulty that arises is in the selection of a diode that has the exact variation required for compensation. Therefore, a compromise is made to get the compensation as close as possible. It is possible to use two diodes in series and to obtain a degree of compensation for changes in β_{dc} also.

The circuit shown in Fig. 17-22*a* uses a diode to compensate for a shift in the operating point caused by I_{CBO}. When there is a change in I_{CBO}, ΔI_{CBO}, the change is reflected into the collector as $(1+\beta_{dc})\Delta I_{CBO}$, and it is the effect of $\beta_{dc}\Delta I_{CBO}$ that requires compensation, since $I_{CBO} + \Delta I_{CBO}$ in itself is not large compared to the quiescent value of I_C. The bias circuit, V_{CC}, R_2, and R_1, is converted by Thévenin's theorem to V_{BB} and R_B, as shown in Fig. 17-22*b*. The compensation is based on the fact that there is a reverse current I_R in the diode.

The Kirchhoff's voltage loop equation through the base is

$$V_{BB} = R_B(I_B + I_R) + V_{BE} + R_E I_E$$

Since

$$I_E = (1+\beta_{dc})(I_B + I_{CBO})$$

$$V_{BB} = R_B(I_B + I_R) + V_{BE} + (1+\beta_{dc})R_E(I_B + I_{CBO})$$

Rearranging

$$[R_B + (1+\beta_{dc})R_E]I_B = V_{BB} - V_{BE} - R_B I_R - (1+\beta_{dc})R_E I_{CBO}$$

and solving for I_B,

$$I_B = \frac{V_{BB} - V_{BE} - R_B I_R - (1+\beta_{dc})R_E I_{CBO}}{R_B + (1+\beta_{dc})R_E}$$

$$I_B = \frac{V_{BB} - V_{BE}}{R_B + (1+\beta_{dc})R_E} - \frac{R_B I_R + (1+\beta_{dc})R_E I_{CBO}}{R_B + (1+\beta_{dc})R_E}$$

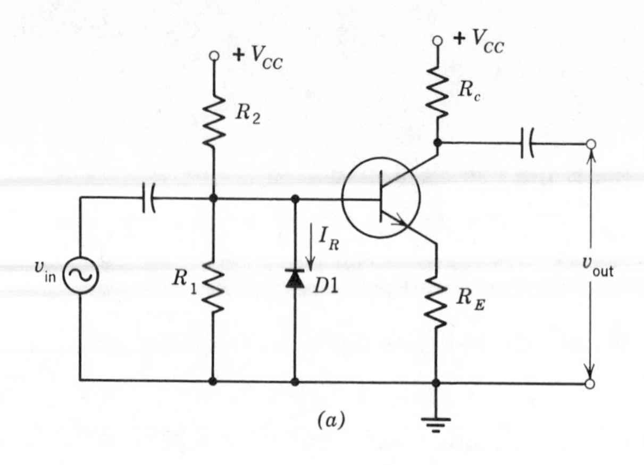

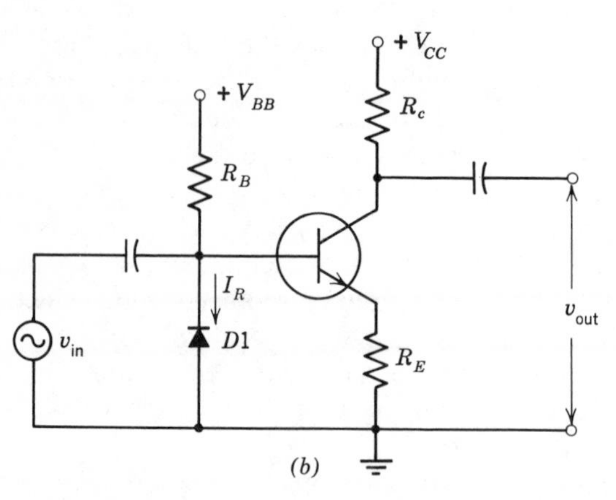

Figure 17-22 Diode compensation for I_{CBO}. (a) Actual circuit. (b) Modified circuit for analysis.

But

$$I_C = \beta_{dc} I_B + (1+\beta_{dc}) I_{CBO}$$

$$I_C = \beta_{dc} I_B + I_{CBO} + \beta_{dc} I_{CBO} \tag{17-17a}$$

substituting

$$I_C = \beta_{dc} \frac{V_{BB} - V_{BE}}{R_B + (1+\beta_{dc}) R_E} - \beta_{dc} \frac{R_B I_R + (1+\beta_{dc}) R_E I_{CBO}}{R_B + (1+\beta_{dc}) R_E} + I_{CBO} + \beta_{dc} I_{CBO} \tag{17-17b}$$

The term that we wish to compensate out in Eq. 17-17a is $\beta_{dc} I_{CBO}$. Thus we accept the presence of the first two terms in Eq. 17-17a as the ideal, allowing the presence of a small I_{CBO}. In Eq. 17-17b the first term corresponds to the first term of Eq. 17-17a. The second term in Eq. 17-17a is the third term in Eq. 17-17*b*. Consequently, if we wish to eliminate the third term in Eq. 17-17*a*, this elimination is accomplished by making the sum of the second term and the fourth term in Eq. 17-17*b* zero:

$$-\beta_{dc}\frac{R_B I_R + (1+\beta_{dc})R_E I_{CBO}}{R_B + (1+\beta_{dc})R_E} + \beta_{dc} I_{CBO} = 0$$

Simplifying

$$R_B I_R + (1+\beta_{dc})R_E I_{CBO} = [R_B + (1+\beta_{dc})R_E] I_{CBO}$$

$$R_B I_R + (1+\beta_{dc})R_E I_{CBO} = R_B I_{CBO} + (1+\beta_{dc})R_E I_{CBO}$$

$$R_B I_R = R_B I_{CBO}$$

$$I_R = I_{CBO} \tag{17-17c}$$

The conclusion then is quite simple: the compensation for I_{CBO} is achieved when a diode $D1$ is selected which has a reverse current that is identical to the I_{CBO} of the transistor over the expected range of the ambient temperature variation. Again, in a practical situation, a compromise must be accepted.

A circuit that shows both methods of diode compensation is given in Fig. 17-23. An adjustable resistor (a potentiometer) used for R_{A_2} permits individual circuit compensations to be made.

Very often, *thermistors* are used in place of diodes. A thermistor is a

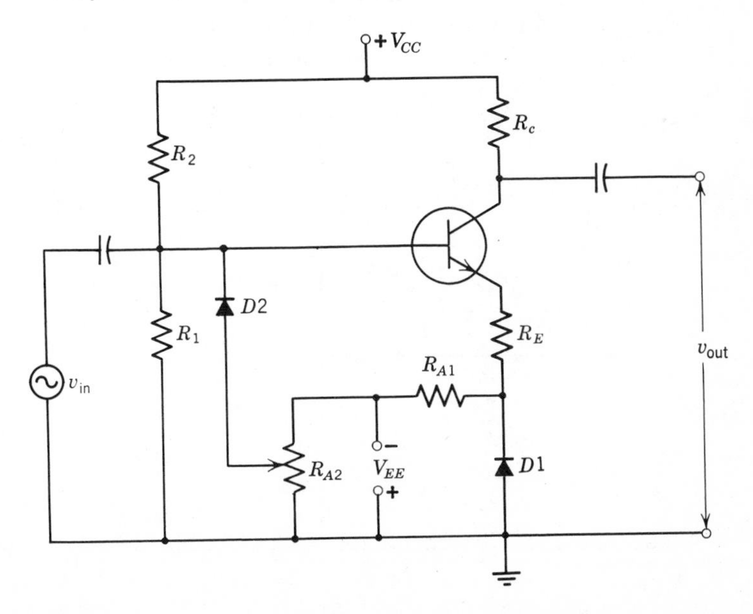

Figure 17-23 Circuit using diodes to compensate for I_{CBO} and for variations in V_{BE}.

resistor that has a negative temperature coefficient. Thermistors are available over a wide range of resistance and current ranges.

The process of microcircuits inherently permits many units to be produced at one operation. Assume that a 1 sq in. wafer yields 400 transistors or FET's and that twenty of these 1 sq in. wafers are simultaneously produced. If the whole production cost is $500, the cost of each unit is less than 3 cents.

Let us assume that two semiconductors are required for a basic amplifier. Also, for purposes of economy, diodes can be mass produced together on wafers, transistors can be mass produced together on wafers, and FET's can be mass produced together on wafers. When FET's and transistors or transistors and diodes are mass produced together on wafers, the costs increase considerably. The net result is that it is much cheaper to use two transistors instead of a transistor and a diode in microcircuitry. The same holds true for the FET.

The circuits considered thus far in this section require technically a transistor and one or more diodes. However, since this application is to microcircuits, adjacent transistors are used and needed diodes are formed by using transistors connected for diode operation.

A FET amplifier circuit widely used in microelectronics is shown in Fig. 17-24. The algebra and circuit analysis presented is for enhancement MOSFET's. The depletion FET can be used toward the same end result.

The coupling capacitors and the gate resistor R_g are used in the circuit to simplify the circuit for analysis. In actual practice the various supply voltages would be such as to eliminate the requirement of R_g and the

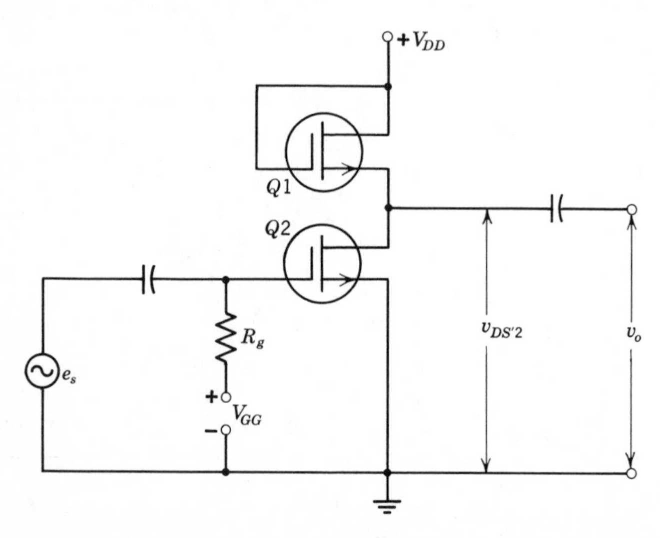

Figure 17-24 Microcircuit amplifier using enhancement-type MOSFET semiconductors.

coupling capacitors. The objective of the circuit is the attainment of overall micro-miniaturization so that many of these circuits can be mounted complete within one small package.
The total signal of the input gate is

$$v_{gs_2} = V_{GG} + e_s.$$

The drain current, i_{D_2}, in $Q2$ is

$$i_{D_2} = K_2\,(v_{GS_2} - V_{T_2})^2$$
$$= K_2(V_{GG} - V_{T_2} + e_s)^2$$

The signal voltage on $Q1$ is the voltage from gate to source. The gate is connected to V_{DD}, and the source is connected to the drain of $Q2$. Therefore, the total gate-to-source voltage on $Q1$ is

$$v_{GS_1} = V_{DD} - v_{DS_2}$$

Then
$$i_{D_1} = K_1(V_{DD} - v_{DS_2} - V_{T_1})^2$$

Since the two MOSFET's are in series

$$i_{D_1} = i_{D_2}$$

Then
$$K_2(V_{GG} - V_{T_2} + e_s)^2 = K_1(V_{DD} - v_{DS_2} - V_{T_1})^2$$

$$\sqrt{\frac{K_2}{K_1}}(V_{GG} - V_{T_2} + e_s) = V_{DD} - v_{DS_2} - V_{T_1}$$

Rearranging

$$v_{DS_2} = \left\{V_{DD} - \sqrt{\frac{K_2}{K_1}}V_{GG} - V_{T_1} + \sqrt{\frac{K_2}{K_1}}V_{T_2}\right\} - \sqrt{\frac{K_2}{K_1}}e_s$$

Now this last equation shows that v_{DS_2} consists of two parts. The first part, which is bracketed, contains d-c operating point values only. The second is the signal component, and the negative sign indicates the characteristic 180° phase inversion of the circuit. This result is expected as v_{DS_2} is an instantaneous total voltage, which means that it has a d-c component and an a-c component. The a-c output voltage v_0 is

$$v_0 = -\sqrt{\frac{K_2}{K_1}}e_s$$

and the gain of the circuit is

$$A_e = \frac{v_0}{e_s} = -\sqrt{\frac{K_2}{K_1}} \qquad (17\text{-}18a)$$

The MOSFET's used are naturally adjacent with the result that the properties of the semiconductor materials in $Q1$ and $Q2$ are very close. The result is that the value of $\sqrt{K^2/K_1}$ becomes a function of the geometry of the channels of $Q1$ and $Q2$. An analysis of the physics of the two adjacent units results in the relation

$$\sqrt{\frac{K_2}{K_1}} = \sqrt{\frac{W_2 L_1}{W_1 L_2}} \tag{17-18b}$$

in which W and L are the physical dimensions of the widths and the lengths of the respective channels. It is possible to secure easily a voltage gain of ten in the production of these amplifiers.

Section 17-7 Heat Sinks

All semiconductors have a collector dissipation rating which is stated as a function of temperature. These dissipation ratings are established by the manufacturer as the result of extensive destructive testing. As may be recalled, at a certain temperature, the crystalline structure is destroyed, and there is no recovery or second chance once this has happened. The critical point in the transistor is the junction between the collector and the base. The maximum allowable junction temperature is given as T_J in degrees centigrade. The lower limit of a semiconductor is taken as $-65°C$. Therefore the fundamental restriction on the operating temperature range of a semiconductor is

$$-65°\text{C} \leqslant T_J \leqslant T_{J,\text{MAX}}$$

Before transistors are used they are stored, and consideration must be given to storage temperature, T_{STG}. The storage concept also applies to a completed piece of equipment that is turned off or that is serving as a spare. Military equipment, for instance, can be placed in a warehouse in Alaska or can be stored in a metal container in the Sahara Desert. The limit of storage temperature is usually the limit of T_J, but for some semiconductors the limit can be somewhat higher. The specification may read, for example,

$$-65°\text{C} \leqslant T_J = T_{STG} \leqslant 150°\text{C}$$

A heat sink (Fig. 17-25) is a mechanical device that is connected to the case of the semiconductor and that provides a path for the developed heat. The heat flows through the heat sink and is carried off to the surrounding air. If a heat sink is not used, all the heat must transfer from the case to the surrounding air. The heat sink causes the temperature of the case to be lowered.

If all the heat that is generated at a collector junction could be transferred out of the transistor instantaneously, the allowable collector dissipation would be infinite. There is, however, a finite *thermal lag*, and heat can only flow in a path where there is a temperature difference. The concept is the same as Ohm's law. The actual temperature at a point corresponds to an

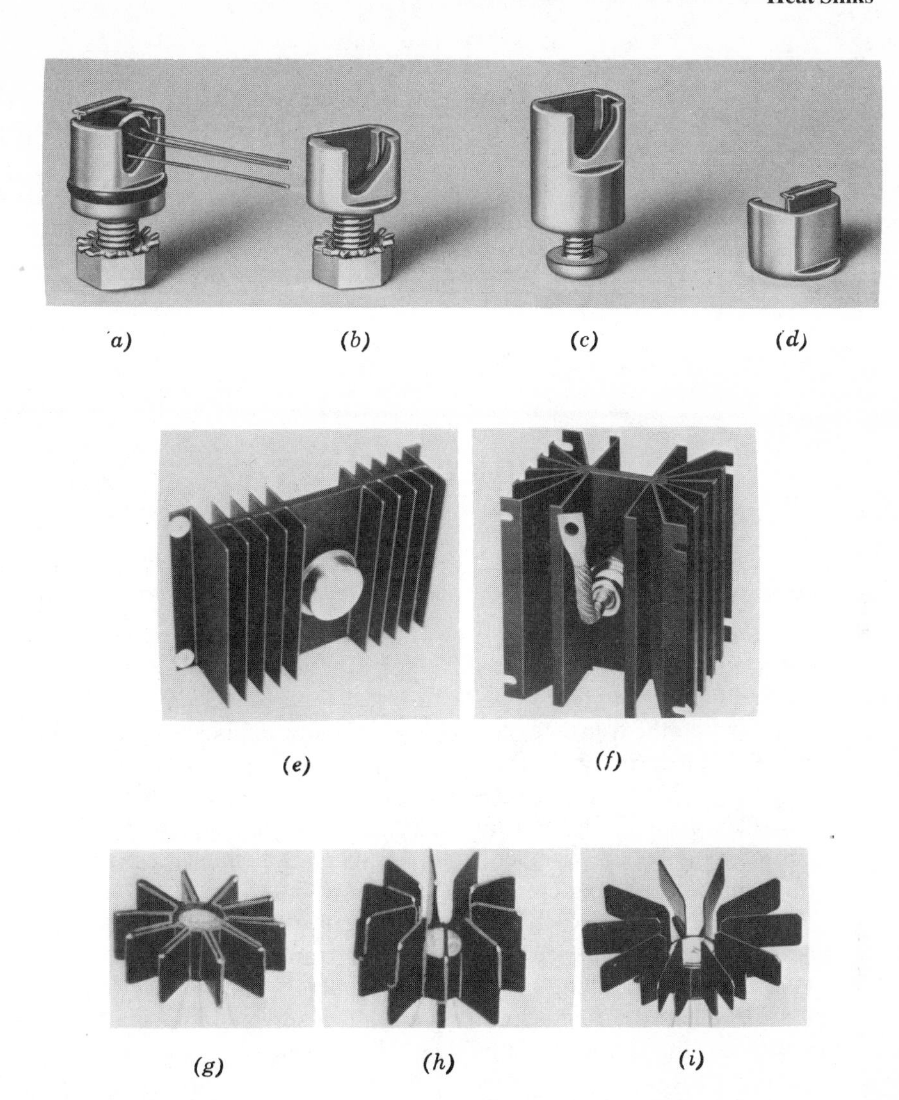

Figure 17-25 Typical heat sinks (*a*), (*b*), (*c*), and (*d*) Thermal retainers for TO-5 case. (*e*) and (*f*) Natural convection coolers. (*g*), (*h*), and (*i*) Typical snap-on heat dissipators for various case sizes. (*Courtesy Wakefield Engineering, Inc.*)

electric potential, and the *thermal resistance* θ is equivalent to electrical resistance. The units of θ are °C/W and where the temperature difference $T_2 - T_1$ is divided by θ, the heat flow in watts is determined by

$$P = \frac{T_2 - T_1}{\theta} \tag{17-19}$$

In a complex circuit, Eq. 17-19 responds to all of the rules and techniques that were developed for Ohm's law. However, in working with heat-sink problems, the usual circuit is a series circuit (Fig. 17-26). The junction

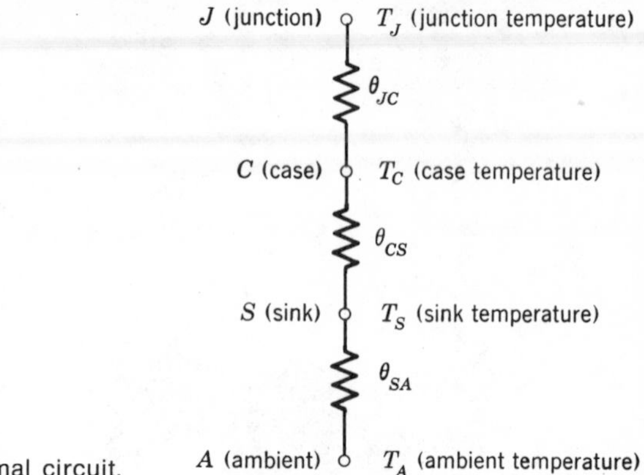

Figure 17-26 Series thermal circuit.

temperature (T_J) at the base-collector junction (J) of a transistor is the specified maximum value. The heat of the collector dissipation, P, in watts flows through the transistor to the case (C) and establishes a case temperature (T_C). There is some separation between the transistor case and the heat sink that creates a thermal resistance. In many units, a washer often is used for insulation, since the case of the transistor itself frequently serves as the electrical connection to the collector. A special silicone grease often is used to establish a good heat conducting path between the case and the heat sink. Consequently, the sink temperature (T_S) differs from the case temperature. The sink is provided with fins designed to transfer the heat to the ambient (A) or to the surrounding air that is at the ambient temperature (T_A).

As in electric circuits, the property of a series circuit is that the total resistance is the sum of the individual resistances

$$\theta_{JA} = \theta_{JC} + \theta_{CS} + \theta_{SA} \tag{17-20a}$$

and an overall equation can be written from Eq. 17-19 as

$$T_J = T_A + \theta_{JA} P_D \tag{17-20b}$$

where P_D is the power dissipated in the collector.

Let us consider the concept of derating as it applies to variations in the ambient temperature T_A. Assume that a transistor is rated at 1 W in free air at 25°. The allowable junction temperature is 100°C, and the transistor may be used from −65°C to 100°C. The maximum allowable collector dissipation P_D is plotted in Fig. 17-27a. The rating given at 25°C ambient is the rating

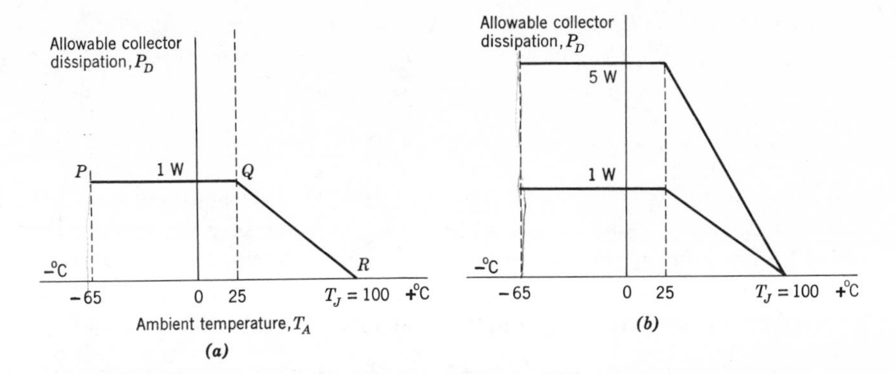

Figure 17-27 De-rating curves. (*a*) Without heat sink. (*b*) With heat sink.

at *all* temperatures lower than 25°C. Hence, the rating curve is horizontal from P to Q. When the transistor is operated in an ambient of 25°, a collector dissipation of 1 W causes the temperature of the collector-base junction (T_J) to rise to 100°C. When the transistor is in an ambient temperature of 100°C, the whole transistor is at 100°C, including the junction. Consequently, the transistor cannot tolerate any collector dissipation. Hence, point R is located at T_J (100°C) on the T_A axis. Since the heat flow is linear, a straight line is drawn between Q and R. The midpoint between 25°C and 100°C is $25+\frac{1}{2}(100-25)$ or 62.5°C. Therefore, when the ambient is 62.5°C, the maximum allowable power dissipation in the transistor is 0.5 W. This is the concept of *derating* a semiconductor. The downward slope of the straight line is the derating of the transistor usually specified in mw/°C.

Let us assume in the previous example that, when the transistor is operated in free air (25°C) at Q in Fig. 17-27*a*, the actual case temperature of the transistor is 90°C. Then the value of the thermal resistance from junction to case (θ_{JC}) is

$$\theta_{JC} = \frac{100-90}{1} = 10°\text{C/W}$$

The effective thermal resistance from case to ambient must be

$$\theta_{CA} = \frac{90-25}{1} = 65°\text{C/W}$$

A heat sink is now provided. The heat sink is relatively large compared to the transistor, and has a large surface to the ambient air. Consequently, the value for its thermal resistance θ_{SA} is low, say, 5°C/W. Now the total thermal resistance is

$$\theta_{JA} = \theta_{JC} + \theta_{SA} = 10+5 = 15°\text{C/W}$$

From Eq. 17-20*b*,

$$T_J = T_A + \theta_{JA} P_D$$

$$100 = 25 + 15 P_D$$

$$P_D = 5 \text{ W}$$

Now, with this heat sink, the maximum power dissipation of the transistor is raised from a value of 1 W in free air to a value of 5 W with the heat sink. The derating of the combination for ambient temperatures above 25°C is shown in Fig. 17-27*b*. Again, the same principles apply. We omitted any washer between the case and the heat sink to simplify the calculations.

TABLE B Specific Thermal Resistance ρ of Interface Materials, °C inches per watt

Material	ρ
Still air	1200
Silicone grease	204
Mylar film	236
Mica	66
Wakefield Type 120 Compound	56
Wakefield Delta Bond 152	47
Anodize	5.6
Aluminum	0.19
Copper	0.10

$\theta = \rho t/A$ where t is thickness in inches and A is in square inches.

Courtesy of Wakefield Engineering, Inc.

In practice the thermal resistance θ_{CS} cannot be neglected unless the value of θ for the heat sink is for a device that is not used with a washer but that fits on or clips on the transistor case (Fig. 17-25*g*, *h*, and *i*.). A list of typical materials used to separate a case from a heat sink is given in Table B with their thermal resistivities. By using this table, the value of θ for a mica washer 0.75 sq in. and 0.003 in. thick is

$$\theta = \frac{\rho t}{A} = \frac{66 \times 0.003}{0.75} = 0.25\text{°C/W}$$

If this washer is used on a semiconductor that dissipates 100 W, the temperature drop across the washer alone is 25°C, which certainly is not negligible

A silicone grease, by filling in scratches and airgaps, can reduce the thermal temperature drop from the stud mounting of a 200-W semiconductor to its heat sink by about 15°C.

A fan can be used to improve heat transfer from the heat sinks of large semiconductors. A fan with a guaranteed life of 5 years costs approximately 5 dollars and, when the total power dissipation is of the order of 300 W or more, the use of the cooling fan becomes practical.

A small transistor has the following ratings from its specification sheet:

$$T_A \text{ to } 25°\text{C} \qquad P_D = 0.8 \text{ W}$$

$$T_C \text{ to } 25°\text{C} \qquad P_D = 3.0 \text{ W}$$

$$-65°\text{C} < T_J = T_{STG} < 200°\text{C}$$

$$\theta_{JC} = 58.3°\text{C/W} \qquad \theta_{JA} = 219°\text{C/W}$$

In free air the transistor can be used up to a collector dissipation of 0.8 W, but above ambient temperatures above 25°C it must be derated linearly to 200°C. Let us assume that heat sink *B* from Table C is used. The table shows this heat sink has a value for θ of 4.1°C/W. Also, assume that the chassis is large and thick to the extent that it can be considered to be at a temperature of 30°C. The total temperature difference from junction to

TABLE C Typical Heat-Sink Data

Figure[a]	degrees Centigrade per watt Case to Chassis	Approximate Cost
a	6.0	25 ¢
b	4.1	18 ¢
c	5.3	20 ¢
d	4.0	13 ¢

Figure[b]	Type	Size HWD Inches	Approximate Cost
e	NC 401	1.50 × 4.81 × 1.25	$1.50
e	NC 403	3.00 × 4.81 × 1.25	$1.75
e	NC 413	3.00 × 4.81 × 1.87	$2.40
e	NC 421	3.00 × 4.81 × 2.63	$2.50
e	NC 423	5.50 × 4.81 × 2.63	$3.60
f	NC 441	5.50 × 4.75 × 4.50	$6.40

[a]Illustrated in Fig. 17-25.
[b]Illustrated in Fig. 17-25.
Courtesy of Wakefield Engineering, Inc.

chassis is (200 − 30) or 170°C. The total thermal resistance is (58.3 + 4.1) or 62.4°C/W. The allowable power dissipation then is 170/62.4, or 2.73 W. Although this capability of a 2.73-W dissipation might meet the actual circuit requirement, this heat sink is not large enough to enable this transistor to be used at its maximum capability of a 3-W dissipation.

Now consider a large power transistor that has the ratings

$$-65°\text{C} \leqslant T_J = T_{STG} \leqslant 100°\text{C}$$

$$P_D = 30 \text{ W for } T_{MF} \leqslant 55°\text{C}$$

T_{MF} is the temperature of the mounting frame (the heat sink). No rating is given for free air dissipation because at this power level a heat sink is specifically required. The large units shown in Fig. 17-25*e* and *f* must be considered. The characteristics of these heat sinks are given in Table C and in Fig. 17-28.

If the temperature of the heat-sink mounting surface T_{MF} is 55°C, it is 30°C above an ambient temperature of 25°C. The intersection of 30°C and 30 W is located on the graph of Fig. 17-28. This point shows that the NC 401, NC 403 and NC 421 heat sinks are too small. The NC 423 heat sink provides some margin. At 30 W dissipation the ambient could rise another 5°C without semiconductor damage.

When heat is generated at a collector junction, there is a finite time lag for it to flow away from the junction. Transient pulses can destroy a semiconductor because of this thermal lag. If the heat is not drawn off, the temperature increases and more current is produced. Therefore, the dissipation increases. This upward spiraling of heat at a junction is called *thermal runaway*. If an ammeter is in the collector circuit, the meter reading steadily increases to the destruction of the transistor.

According to the analysis developed from Fig. 17-20, when the operating point is in the center of the load line at $V_{CE} = \frac{1}{2}V_{CC}$, any change in the operating point from this tangent point to the constant collector-dissipation line results in a reduced collector dissipation. If R is the resistance value of the load line, and if the operating point is at $\frac{1}{2}V_{CC}$, the collector dissipation P_D is

$$P_D = (\tfrac{1}{2}V_{CC})^2/R$$

Solving for R,

$$R = \frac{V_{CC}^2}{4P_D}$$

If P_D is the maximum allowable collector dissipation under a given set of operating conditions for a particular ambient temperature with a particular heat sink, R becomes the least value of d-c resistance required in the collector circuit to protect the semiconductor from a thermal runaway:

$$R_{\min} = \frac{V_{CC}^2}{4P_D} \qquad (17\text{-}2$$

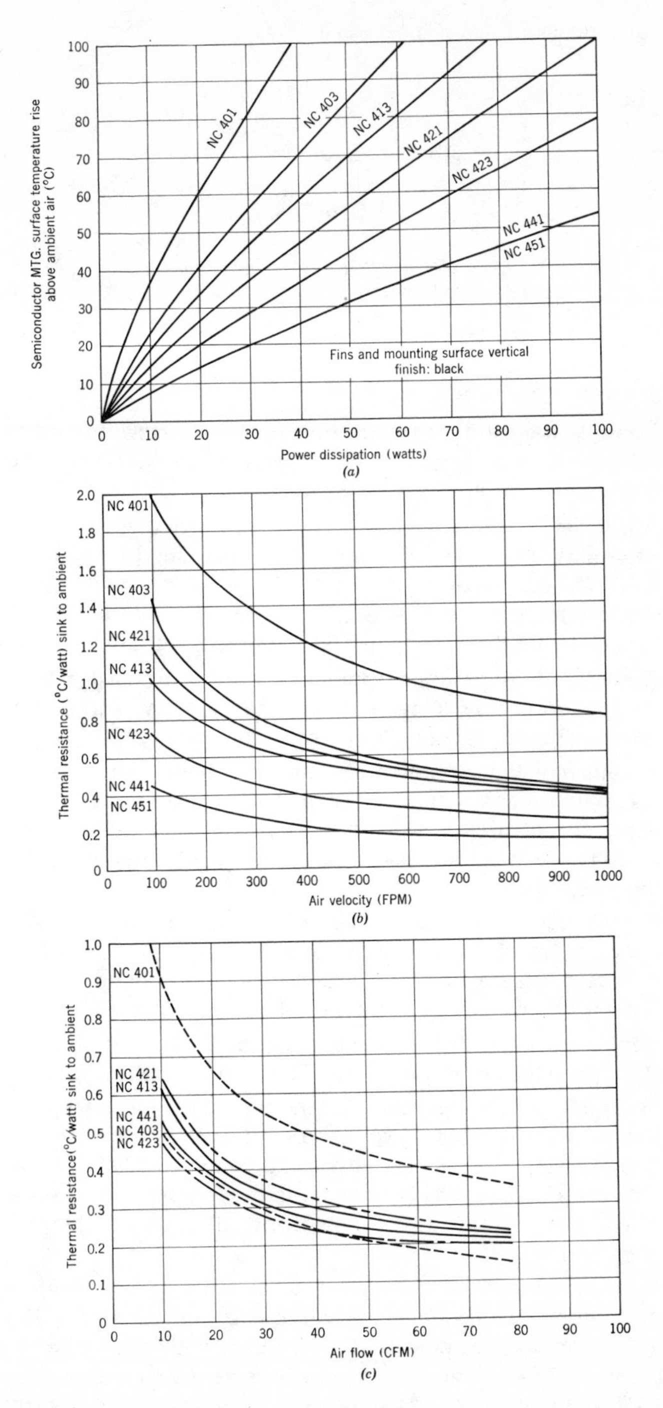

Figure 17-28 Convection characteristics. Data for typical large heat sinks listed in Table C and illustrated in Figure 17-24 (*e*) and (*f*). (*a*) Natural convection. (*b*) Forced convection for air velocity. (*c*) Forced convection for air flow. (*Courtesy Wakefield Engineering, Inc.*)

PROBLEMS

1. A transistor has the following ratings

$$-65°C \leqslant T_J = T_{STG} \leqslant 100°C$$
$$\text{to } 25°C\ T_A, \quad P_D = 1.0\ W$$
$$\text{to } 25°C\ T_C, \quad P_D = 7.5\ W$$

Find θ_{JA} and θ_{JC}.

2. Using the data given for the transistor in Problem 1, find the power ratings for an ambient temperature of 60°C and also for a case temperature of 60°C.
3. A 2N404 transistor is rated in free air at 150 mw and, at this power, has a maximum value for T_J of 85°C. What is the derating value in mw/°C for elevated temperatures?
4. A transistor in free air has a value for θ_{JA} of 0.25°C/mw and for an infinite heat sink θ_{JC} is 0.11°C/mw. The maximum junction temperature T_J is 85°C. The heat sink *C* listed in Table C is used in an ambient of 35°C. What is the maximum allowable collector dissipation? What is the maximum allowable collector dissipation in a 35°C ambient without a heat sink?
5. The operating junction temperature of a transistor is 125°C. The total dissipation at a 25°C case temperature is 0.5 W and at a 25°C ambient the total dissipation is 0.2 W. What is the value of θ_{CA}?
6. A transistor that operates at a junction temperature of 170°C has a power dissipation capability of 12 W when the case is at 25°C. The transistor is used with heat sink *B* listed in Table C. The ambient temperature is 60°C. What is the maximum allowable collector dissipation for the transistor with the heat sink?
7. The allowable power dissipation of a transistor at or below a mounting frame temperature of 70°C is 60 W. Select a heat sink from Fig. 17-27 that will dissipate this power by natural convection cooling in an ambient of 25°C. What is the reserve for this heat sink in ambient rise?
8. Repeat problem 7 if the ambient is 50°C. Forced air cooling is required. Determine the air flow required. The air temperature is also 50°C.
9. The surface area of a transistor in contact with a heat sink is 1.20 sq. in. The transistor dissipates 40 W. Irregularities in the surfaces of the transistor and its heat sink create an effective air gap of 0.0002 in. What is the temperature drop across the gap? What is the temperature drop across the gap if silicone grease is used? If Delta Bond 152 is used? Thermal resistances are given in Table B.
10. A transistor can be operated with a junction temperature of 200°C. The transistor can dissipate 1 W without a heat sink in a 25°C ambient. The transistor can dissipate 10 W when used with an infinite heat sink in a 25°C ambient. The transistor is used with a heat sink that is rated at 10°C/W to a 25°C ambient. What dissipation can be tolerated for this transistor with this heat sink?

Questions

1. Define or explain each of the following terms: (*a*) beta stability, (*b*) leakage current, (*c*) I_{CBO}, (*d*) I_{CEO}, (*e*) temperature sensitivity, (*f*) thermistor, (*g*) heat sink, (*h*) junction temperature, (*i*) storage temperature, (*j*) thermal lag, (*k*) thermal resistance, (*l*) specific thermal resistance, (*m*) thermal runaway.
2. Define K as used in beta stability.
3. What are the limits of K and what is the ideal value of K?
4. What is the worst-case value of K?
5. What is the effect of an increase in beta on collector characteristics?
6. What is the effect of an increase in I_{CEO} on collector characteristics?
7. Define S as used in temperature sensitivity.
8. What are the limits of S and what is the ideal value of S?
9. What is the worst-case value of S?
10. Explain how a circuit can compensate for an increase in I_{CEO} by changing its biasing point.
11. What is the effect of temperature on I_{CEO} in a germanium unit? In a silicon unit?
12. How does a diode or a thermistor compensate for a temperature change?
13. What is Ohm's law for a thermal circuit?
14. How are transistors derated at high temperatures?
15. What is the effect of using a silicone grease in a heat sink?
16. How can thermal runaway be prevented?
17. Why is the use of a blower effective with a heat sink? Explain.

Chapter Eighteen

PUSH-PULL AND PHASE INVERSION

Push-pull amplifiers are widely used for applications possessing requirements that a single semiconductor or tube cannot meet or cannot conveniently meet. The discussion in this chapter gives the basic class-A circuit (Section 18-1), analyzes the circuit (Sections 18-2 and 18-3), and compares this push-pull circuit with the previously studied single-ended amplifiers (Section 18-4). Section 18-5 treats the necessary auxiliary circuitry which is used in conjunction with push-pull amplifiers. Class-AB and class-B amplifiers (Section 18-6) are presented both for transistors and for vacuum tubes. The use of complementary symmetry in push-pull (Section 18-7) is restricted to transistor circuits. The chapter concludes with a discussion of representative circuits that are used commercially in audio amplifiers.

Section 18-1 The Basic Circuit

The signal voltage is the primary voltage of the input transformer $T1$ in Fig. 18-1. The secondary winding of the input transformer is grounded at the center tap. When the center tap on the winding is made the reference point (in this case, ground), the voltage from the center tap to the top of the winding is 180° out of phase with the voltage from the center tap to the bottom of the winding. By use of a center tap, the number of turns in the top half of the winding equals the number of turns in the bottom half of the winding, and e_1 is exactly equal in magnitude to e_2. Thus, if we consider the input voltage to $Q1$ to be instantaneously $+1$ V, the input voltage to $Q2$ must be -1 V at that instant. Accordingly, when e_1 is positive, the forward bias on

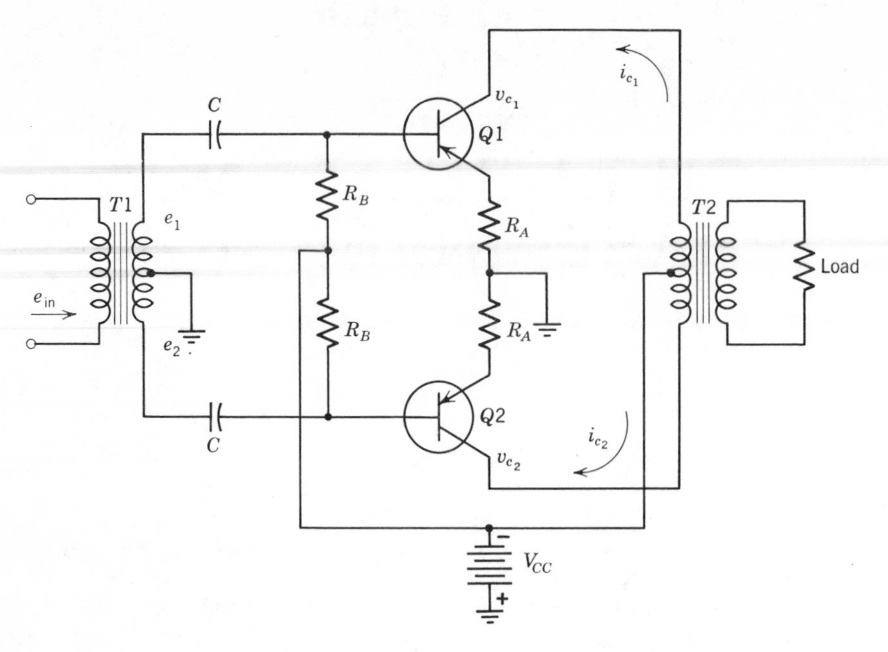

Figure 18-1 The basic push-pull circuit.

transistor $Q1$ decreases and its collector current i_{C_1} decreases. Simultaneously, e_2 is negative increasing the forward bias on $Q2$ which causes the collector current in $Q2$ to increase in magnitude. If we assume the circuit is ideally linear, the decrease in i_{C_1} equals in magnitude the increase in i_{C_2}. Correspondingly, v_{C_1} and v_{C_2} are out of phase with each other. Since the action of a transistor as a common-emitter amplifier introduces a 180° phase inversion, v_{C_1} is in phase with e_2 and v_{C_2} is in phase with e_1. Since i_{C_1} decreases as i_{C_2} increases, the sum of i_{C_1} and i_{C_2} is a constant and does not vary with signal.

Let us assume on the diagram that the flux in the primary of $T2$ caused by i_{C_1} acts upward and that the flux caused by i_{C_2} acts downward. Without a signal, i_{C_1} and i_{C_2} are equal, and the two fluxes produced are equal and cancel, with the result that the net flux in the transformer is zero. With a signal, i_{C_1} and i_{C_2} differ. Then $(i_{C_1} - i_{C_2})$ produces the net primary flux which develops the load voltage and the load power in the secondary winding of $T2$.

In checking the operation of a push-pull circuit with a test signal, the magnitude of e_1 should equal the magnitude of e_2, using either an oscilloscope or an a-c meter. The observed alternating collector voltages on $Q1$ and $Q2$ should also be equal in magnitude.

The very small resistors R_A are placed in the emitter of $Q1$ and $Q2$ to provide stability and to prevent thermal runaway at low operating temperatures. The proper operating bias on the transistors is obtained by means of a base resistor R_B for each transistor. To prevent the bias currents from being short-circuited, blocking capacitors C are required.

The circuits shown in Fig. 18-2 illustrate other basic push-pull circuits. The bias can be obtained from a voltage-dividing network (Fig. 18-2*a*) which is simultaneously applied to both bases. In the circuit of Fig. 18-2*b*, the bias circuit is modified to provide a temperature compensation. If the ambient temperature rises, the collector characteristics shift in the direction of an increased collector current. In order to keep the operating point at the center of the load line, the bias current must decrease when the curves rise. The resistor R_3 is a *thermistor* which is used for the compensation. As the ambient increases, the resistance of the thermistor decreases, causing more current to be shunted to ground and less bias current to enter into the transistor. Proper design of this circuit keeps the operating point at the center of the load line at different temperatures. The circuit given in Fig. 18-2*c* employs vacuum tubes as the power amplifiers. The IR drop of $(i_{b_1}+i_{b_2})$ flowing through R_K develops the cathode bias voltage. A bypass capacitor across the cathode resistor is not necessary because this voltage drop is a pure direct voltage.

Some push-pull amplifiers have a potentiometer connected in series with the emitters or cathodes. The variable arm goes to ground or in the case of the vacuum tubes to R_K. This circuit arrangement is called a *balance control*. The potentiometer is adjusted until the d-c currents in the two amplifiers are equal. This procedure compensates for the actual differences that are found in the characteristics of the two transistors or tubes used. In addition, we often find that $Q1$ and $Q2$ or $V1$ and $V2$ are special units which have been closely matched to give a better balance.

Section 18-2 Quantitative Analysis of Harmonics

In Section 11-7, the expression for the output current in an amplifier was shown to be represented by the power series

$$i = a_0 + a_1 e_{\text{in}} + a_2 e_{\text{in}}^2 + a_3 e_{\text{in}}^3 + a_4 e_{\text{in}}^4 + \cdots$$

The current of each transistor in the push-pull amplifier may be expressed by this power series with the same coefficients, since the circuit and the transistors are matched. As the input signal voltages are 180° out of phase, we may replace e_{in} in the power series by

$$e_1 = E \sin \omega t$$

and

$$e_2 = -E \sin \omega t$$

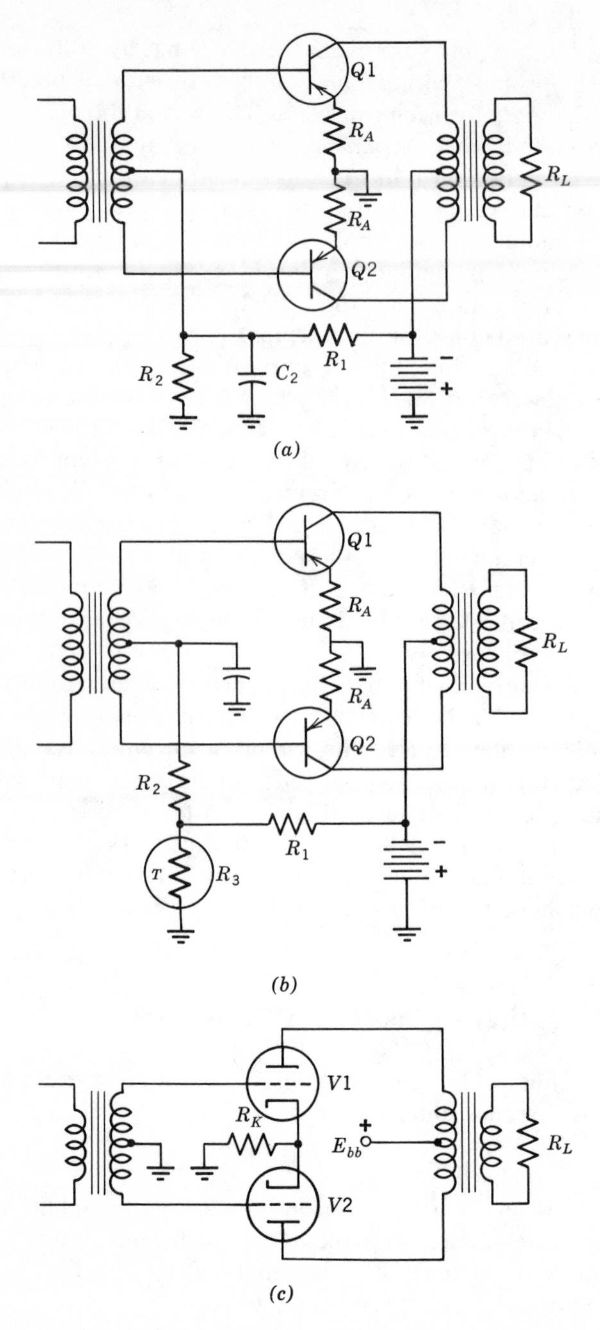

Figure 18-2 Typical push-pull circuits. (*a*) Bias obtained from voltage divider. (*b*) Temperature-compensated bias. (*c*) Electron-tube circuit.

Then

$$i_{C_1} = a_0 + a_1E \sin \omega t + a_2E^2 \sin^2 \omega t + a_3E^3 \sin^3 \omega t + a_4E^4 \sin^4 \omega t + \cdots$$

and

$$i_{C_2} = a_0 - a_1E \sin \omega t + a_2E^2 \sin^2 \omega t - a_3E^3 \sin^3 \omega t + a_4E^4 \sin \omega t - \cdots$$

By subtracting these two power series, we have the expression for the push-pull output:

$$i_{C_1} - i_{C_2} = 2a_1E \sin \omega t + 2a_3E^3 \sin^3 \omega t + \cdots$$

The discussion of the power series in Section 11-7 showed that the direct current is the a_0 term; the fundamental is the a_1 term; the second harmonic is the a_2 term; the third harmonic is the a_3 term; and the fourth harmonic is the a_4 term. The expression for the output of a push-pull amplifier $(i_{C_1} - i_{C_2})$ does not contain an a_0 term, an a_2 term, or an a_4 term. It contains only the a_1 and a_3 terms. Several conclusions can be drawn from this. The absence of the a_0 term confirms mathematically that there is no d-c flux in the transformer primary winding which could cause saturation. Also, there is no second harmonic or fourth harmonic (or any even harmonic) in the output of the push-pull amplifier. This feature of reduced harmonic distortion is the main advantage of the push-pull amplifier.

The coefficient of the fundamental term in the power series for each transistor or tube is a_1E. In the combined push-pull output, the coefficient is $2a_1E$. Thus, the power from a push-pull circuit is twice that of a single transistor or tube. In a single transistor or tube, the ratio of the third harmonic to the fundamental, which is the third-harmonic distortion, is a_3E^3/a_1E. In push-pull, this ratio has the same value, $2a_3E^3/2a_1E$. This shows that the percentage of odd harmonics is neither increased nor decreased by using push-pull.

Section 18-3 Qualitative Analysis of Harmonics

In Section 10-7, we showed that, if a wave has true symmetry, it contains the fundamental and odd harmonics only. Now we show that this symmetry must exist in the push-pull circuit. Figure 18-3 illustrates the typical dynamic-transfer characteristic which is obtained from a single-JFET amplifier stage. The input-signal and the output-current waveforms are shown for the operating point at a. The asymmetry in the output current wave is evident. Thus, the output wave consists of the fundamental with both odd and even harmonics.

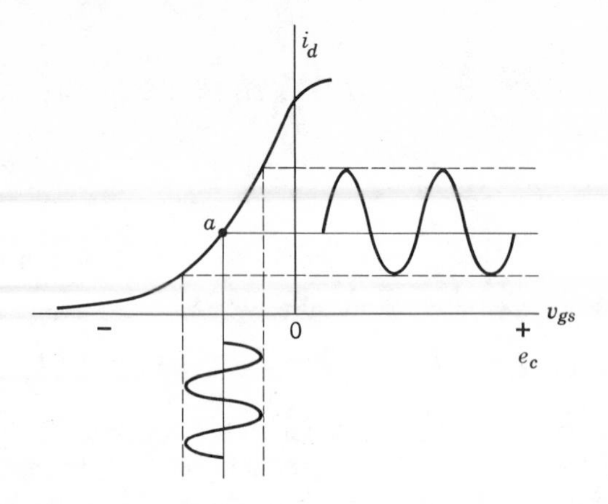

Figure 18-3 Single FET dynamic-transfer characteristic.

A push-pull characteristic is the *composite* or resultant of the difference $(i_{d_1} - i_{d_2})$ of the two individual semiconductors. One JFET characteristic is reversed to care for the 180° push-pull phase relationship and is so placed that the gate-voltage value of the operating point of one JFET coincides with the gate-voltage value of the operating point of the second JFET. This is shown in Fig. 18-4 wherein point *e* is the operating point of one JFET and

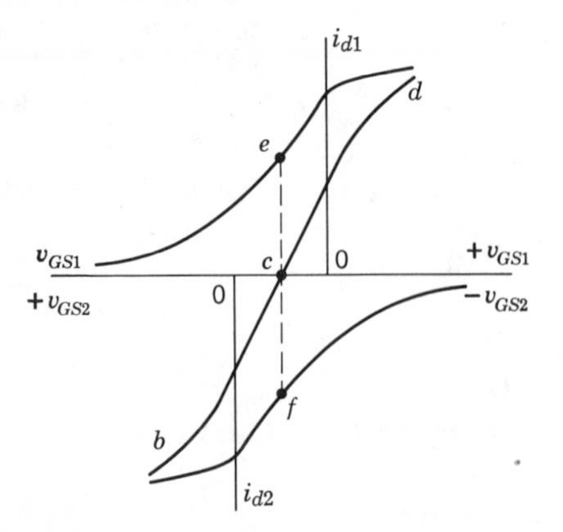

Figure 18-4 Push-pull dynamic-transfer characteristic.

point *f* is the operating point of the other. If one curve is added point by point from the other, we obtain the resultant composite curve *bcd*. This resultant is redrawn as the composite transfer curve in Fig. 18-5 in which $(i_{d_1} - i_{d_2})$ is the ordinate and the gate signal is the abscissa. If the individual

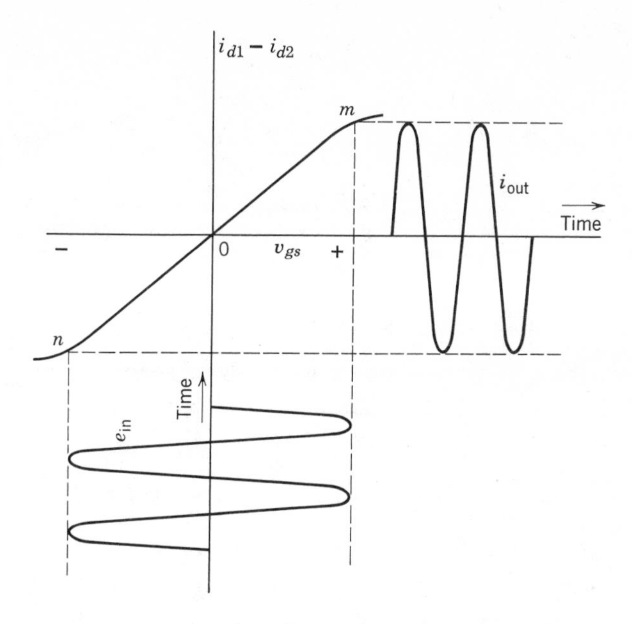

Figure 18-5 Resultant push-pull dynamic-transfer characteristic.

JFET's are identical, the curvature at *m must* be identical with the curvature at *n* but reversed in sign. This curvature produces a symmetrical output waveform which cannot have any even harmonics.

ection 18-4 Comparison between Push-Pull and Single-Ended Operation

As we recall from Section 11-7, for a fixed signal input voltage and for a fixed supply voltage, the output power and the distortion vary with the effective load resistance, as shown in the curves for a power vacuum-tube amplifier (Fig. 18-6). The single-ended amplifier is normally operated at point *a* in order to obtain the benefits of minimum distortion. However, push-pull operation permits a shift in the location of the operating point from the point of minimum distortion *a* to the point of maximum power output *b*. This can be illustrated by using the numerical values on the graph. When two single-ended tubes are used in parallel and operated at point *a*, the output power is 2×5 or 10 W at 9% distortion. By using the load resistance which produces maximum power output, a push-pull connection, by operating at point *b* instead of at point *a*, eliminates the second and fourth harmonics and delivers 2×6 or 12 W at only 4% harmonic distortion.

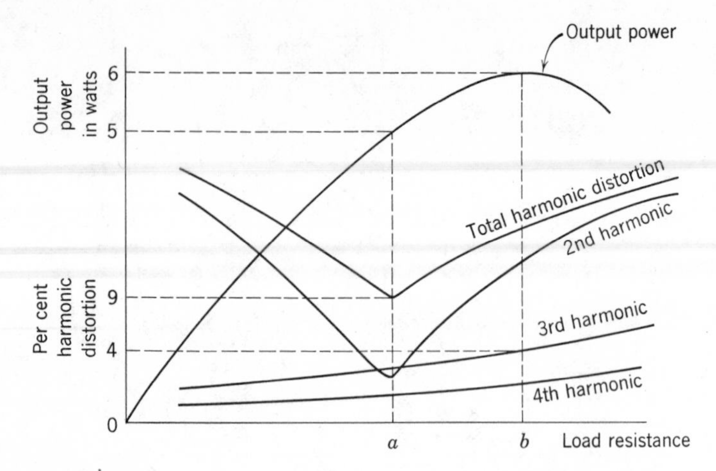

Figure 18-6 Output characteristic of a power pentode or tetrode.

Let us assume that the turns ratio of an output transformer (Fig. 18-7) is n to one from each half of the primary to the secondary winding. Then, each transistor has n^2R_L as a load resistance. The full primary-to-secondary turns ratio is $2n$ to one. Since the impedance varies as the square of the turns

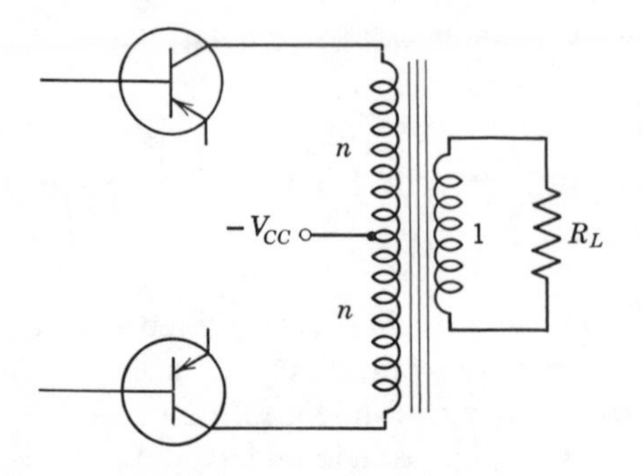

Figure 18-7 Output transformer loading.

ratio, the collector-to-collector impedance is $4n^2R_L$. The alternating current in the top transistor is i_{c_1}, and in the bottom transistor $i_{c_2} \cdot i_{c_2}$ is equal in magnitude to i_{c_1}, but it is 180° out of phase. Actually i_{c_2} flows through the primary winding in a direction opposite to i_{c_1}. The effective current in the primary winding may be considered to be i_{c_1}. Then, the total collector-to-collector voltage drop is i_{c_1} $(4n^2R_L)$. The alternating voltage across one tran sistor is one half this or $2n^2i_{c_1}R_L$. In a single transistor this load-voltage drop is $i_{c_1}n^2R_L$. Thus, it appears in the push-pull circuit that the value of alternating collector current is doubled. Actually, this means that there is an effective reduction by one half in the internal resistance of the transistors when used in push-pull as compared to the single-ended operation. By the maximum power transfer theorem, a reduction in internal resistance means that a greater maximum output power can be obtained from the circuit.

The factors influencing response of output transformers detailed in Section 14-4 apply also to the output transformers used in push-pull circuits. The problem of d-c saturation in the primary winding, however, does not exist for the push-pull connection.

Advantages of the use of push-pull operation include the following:

1. A marked reduction of developed harmonic distortion is obtained. The rigid specification for low distortion in high-fidelity amplifiers requires the use of a push-pull power amplifier.
2. Since the no-signal flux in the core is zero, the transformer cannot become saturated because of too high a value of operating current. A smaller transformer could be used.
3. Hum in the power supply does not affect the operation of the push-pull circuit. The hum is simultaneously present in each output current, and the difference in the output currents $(i_1 - i_2)$, cancels out the effect of the hum voltage. A less well-filtered power supply may be used for push-pull circuits.
4. An increased output power is obtained by having a reduced internal impedance in the output of the stage.
5. At full signal drive, the overall operating efficiency is increased, since the operating point has been shifted from the point of minimum distortion to the point of maximum power output.
6. A large bypass capacitor is not needed when both d-c operating currents flow through the same resistor in a bias circuit.

Disadvantages of push-pull operation include:

1. A complex driving circuit is needed to produce the balanced voltage 180° out of phase.
2. Two semiconductors or vacuum tubes with sockets and the associated wiring may be more expensive than a single large transistor or vacuum tube.

Section 18-5 Phase Inverters

A circuit arrangement to produce balanced voltages which are 180° out of phase for the inputs of the push-pull stage is termed a *phase inverter* or *driver*. Many circuit variations have been developed for this purpose. We will consider some of the fundamental designs that are in common use.

The circuit of Fig. 18-8 provides a simple and effective means of obtaining the balanced driving voltages. The balance in this circuit is determined by

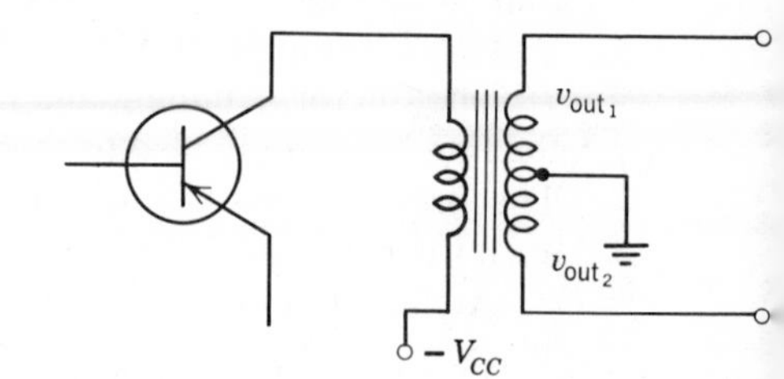

Figure 18-8 Transformer phase inverter.

the exactness of the location of the center tap. The simplicity of the circuit is often outweighed by the expense of the driver transformer. When there exists a large current requirement for the input of push-pull stage, the use of the driver transformer cannot be avoided without producing a very large distortion which is caused by the *IZ* drop in the driving circuit and a shift in the operating point.

In the cascade phase inverter (Fig. 18-9) two identical amplifier stages are used. In order to understand the operation of this circuit, assume that the input signal at *a* is instantaneously positive. $Q1$ amplifies this signal with a 180° phase reversal. Thus the signal at *b* is negative and $v_{out,1}$ is negative. The capacitor *C* is large and functions as a blocking capacitor feeding the signal from *b* to *c* without phase change. The signal at *c* is negative, and the amplifier $Q2$ amplifies this signal with a phase inversion to *d*. Now the signal at *d* is positive, and the proper phase relations for a driving signal for a push-pull amplifier are obtained.

The input signal, v_{in}, produces an a-c signal at the base, i_{b_1}. The value of *R* is selected, so that the base current, i_{b2}, in $Q2$ is identical to i_{b_1}. Therefore if $Q1$ and $Q2$ are identical in characteristics, the outputs, v_{out} and v_{out_2}, are balanced. An experimental adjustment of *R* can provide an exact balance.

The vacuum-tube version of this circuit is shown in Fig. 18-10. Here, the sum of R_1 and R_2 equals R_3, and the ratio $R_2/(R_1+R_2)$ is the reciprocal of the voltage gain of the stage.

The split-load phase inverter (Fig. 18-11*a*) uses the concept of simultaneous collector and emitter outputs. The signal from the collector, v_{out_1}, is 180° out of phase with v_{in}, and the signal from the emitter, v_{out_2}, is in phase with v_{in}. Since the emitter follower amplifier cannot have a voltage gain exceeding unity, the value of R_3 must be decreased (or R_4 increased), to match v_{out_1} to v_{out_2} in amplitude to provide the requisite balance of the two outputs

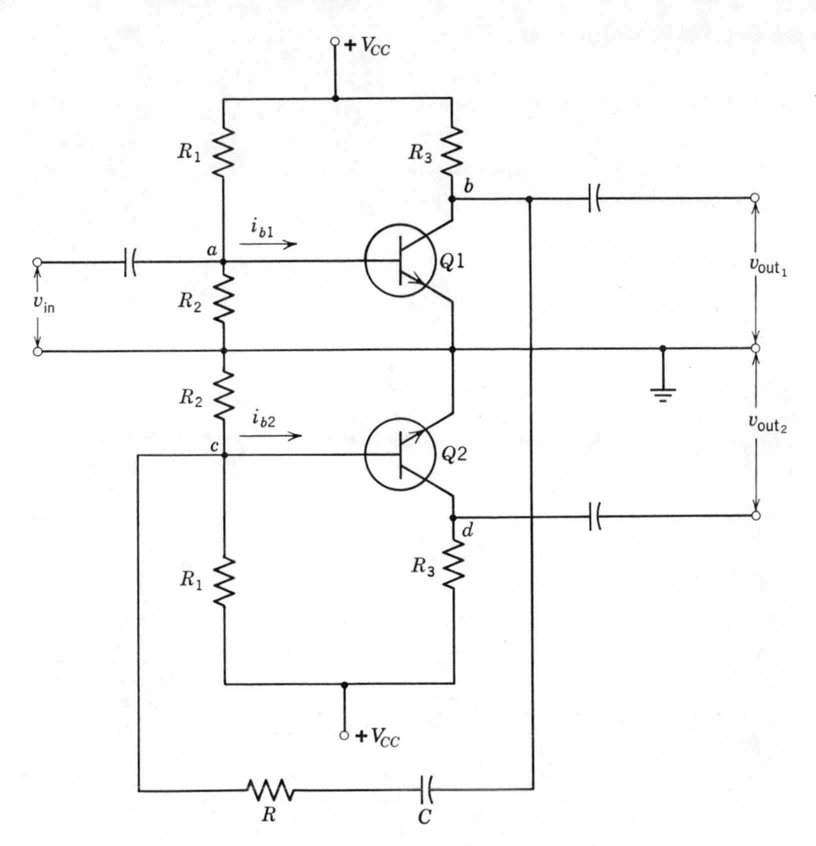

Figure 18-9 Cascade phase inverter.

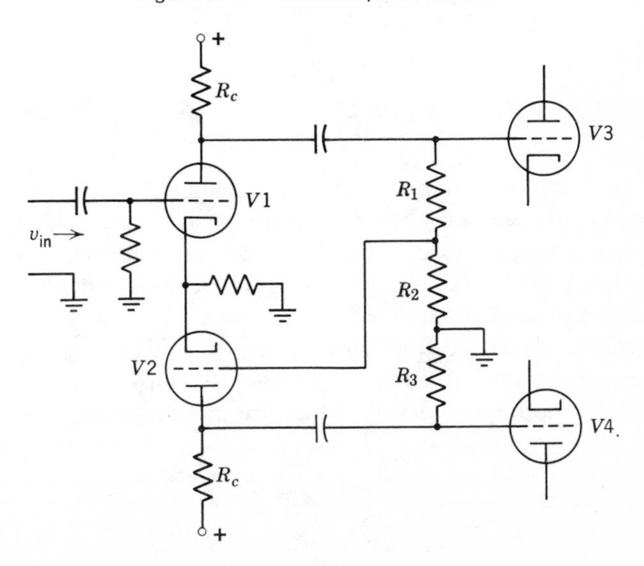

Figure 18-10 The Vacuum-tube phase inverter.

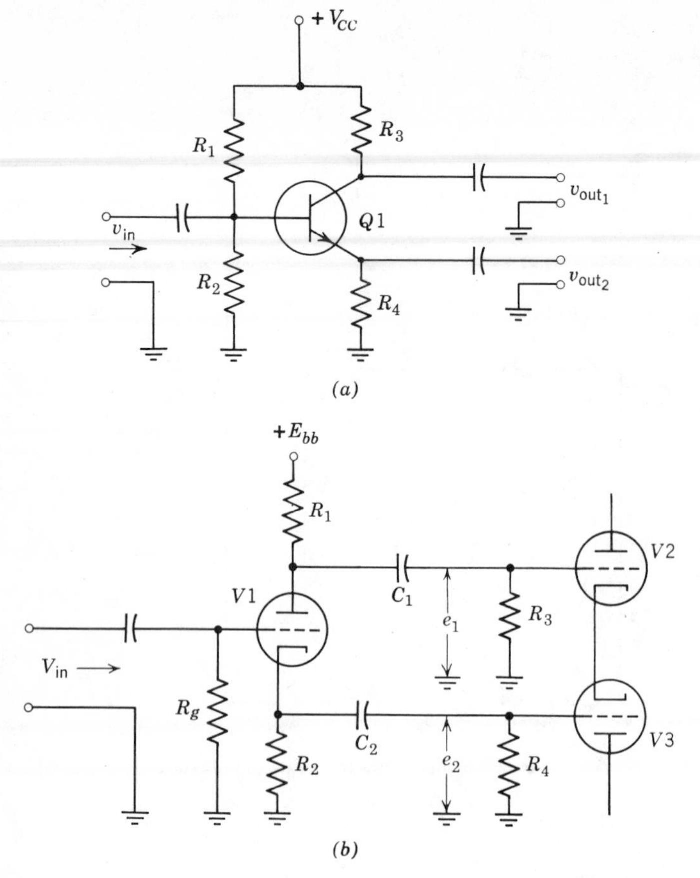

Figure 18-11 Split-load phase inverters. (*a*) Transistor version. (*b*) Vacuum-tube version.

If R_3 and R_4 are equal, the outputs are almost exactly balanced. An exact balance can be obtained by adjusting or by selecting either R_3 or R_4.

Since the base currents of the push-pull amplifiers vary instantaneously with signal, the loads on v_{out_1} and on v_{out_2} are not constant. When this effect is sufficient to create distortion in the output of the power amplifier stage, resistors are often placed in series with the leads to v_{out_1} and v_{out_2}. Although these resistors lower the drive, they can "swamp out" this undesired effect.

The vacuum-tube version of this circuit is shown in Fig. 18-11*b*.

PROBLEMS

Use 50 mv/I_E for r'_e for the transistors.

1. In the circuit of Fig. 18-9, the silicon transistors have a β of 100. The supply voltages are 10 V each. R_3 is 10 kΩ, R_2 is 10 kΩ, and V_{CE} is 5 V. Find R_1 and R. What is the gain of the circuit?

2. In the circuit of Fig. 18-10, the vacuum tubes have a μ of 40 and a g_m of 3500 μmhos. R_c is 20 kΩ. R_3 and (R_1+R_2) are 100 kΩ. Find R_1, R_2, and the gain of the circuit.
3. In the circuit of Fig. 18-11, the silicon transistor has a β of 60. The supply voltage is 15 V, and V_{CE} is 5 V. If R_2, R_3, and R_4 are 10-kΩ precision resistors, what are the exact values of the output voltages when the input signal is 2 V?

Section 18-6 Class-AB and Class-B Amplifiers

The term *class* describes the operation of an amplifier by specifying the conditions of collector current flow for an a-c cycle of the signal (18-12). In a class-A amplifier, the collector current flows for the full a-c cycle (360°).

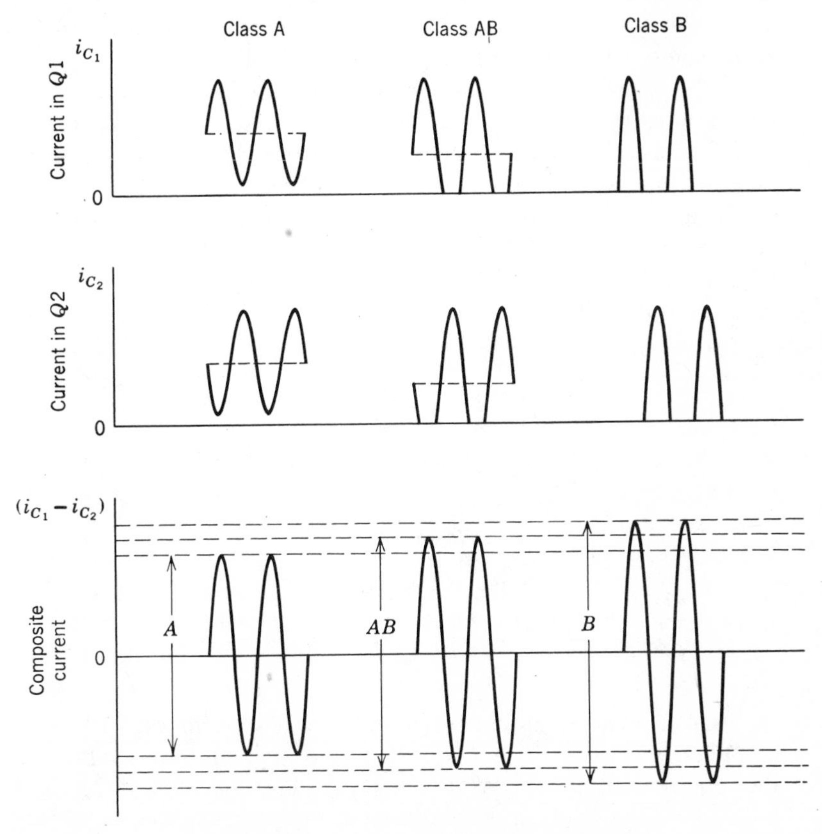

Figure 18-12 Currents in a push-pull circuit.

In a *class-AB* amplifier, collector current flows for more than half the a-c cycle but less than a full a-c cycle. In a *class-B* amplifier, the collector current flows for exactly 180° of the full a-c cycle. In *class-C* operation, the collector current flows for less than 180° of the full a-c cycle.

All of the amplifiers that we have considered to this point have been class-A amplifiers. Particular efforts were taken to insure that the amplifiers were linear over the full a-c signal cycle. In a vacuum-tube circuit, a class-A amplifier draws no grid current because the signal is less than the amount required to drive the grid positive. In a class-A circuit in which there is grid current, it is necessary to specify the operation as *class-*A_2. In vacuum-tube amplifiers the designation *class-AB* must be clarified. *Class-*AB_1 is a circuit without grid current and *class-*AB_2 is a circuit in which grid is driven positive to produce grid current flow. Class-B and class-C operations of a vacuum-tube automatically assume a grid current. Chapter 19 is devoted to the study of the class-C amplifier.

Class-AB and class-B amplifiers are used almost exclusively in push-pull amplifier circuits. The waveforms in Fig. 18-12 show the collector currents in the two output transistors of a push-pull circuit. The peak value of current is the same for all three classes, but when a difference, $i_{c_1} - i_{c_2}$, is taken that gives the composite output a-c waveform to the load, the class-A waveform has the smallest peak-to-peak value. Since peak output power is $(i_{c_1} - i_{c_2})^2 R_L$, a 20% increase in peak-to-peak amplitude yields a 44% (1.20^2) increase in peak output power. Consequently, the class-AB and class-B amplifiers are used exclusively in high-power amplifiers such as the popular stereo amplifiers. A comparison of these classes of operation can also be made from a tabulation (Table A) of the tube manual ratings for two 6L6 beam-power tubes in push-pull.

TABLE A

Class	Bias	Output Power
A_1	Fixed	18.5
	Cathode resistor	17.5
AB_1	Fixed	26.5
	Cathode resistor	24.5
AB_2	Fixed	47.0

Typical class-B amplifier circuits are shown in Fig. 18-13. A class-B amplifier is biased at cutoff. For transistors, this is very simple. Zero bias is the required cutoff bias. When the bias is zero, one half of the signal cycle is a forward bias, causing collector current, while the other half of the cycle is

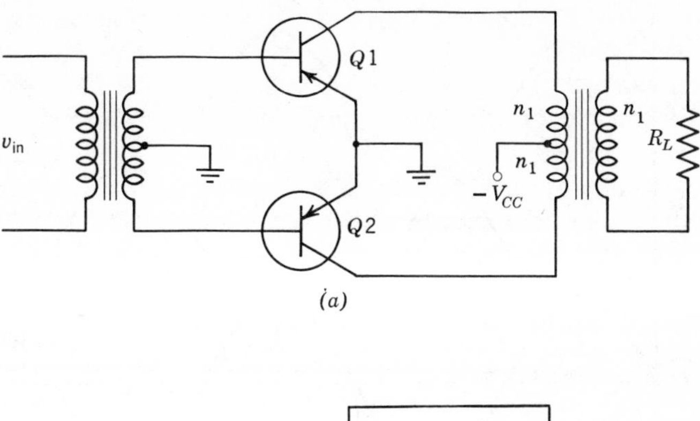

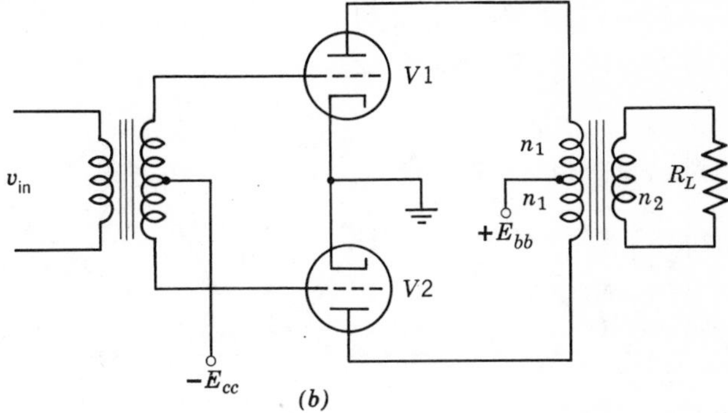

Figure 18-13 Class-B amplifier. (a) Transistor circuit. (b) Vacuum-tube circuit.

a reverse bias preventing collector current. On the other hand, a separate fixed bias is required for class-B operation for the vacuum-tube and for the FET. The fixed bias E_{cc} of the vacuum-tube circuit must be set exactly at cutoff to insure a 180° current flow in each tube.

The waveforms for two cycles of a class-B amplifier are shown in Fig. 18-14. The direct current in each transistor is by Eq. 6-1:

$$I'_{dc} = \frac{I_m}{\pi}$$

Then, for two transistors, the direct current is

$$I_{dc} = \frac{2I_m}{\pi}$$

and the power-supply demand is

$$P_{in} = V_{CC} \times I_{dc} = \frac{2I_m V_{CC}}{\pi} \tag{18-1}$$

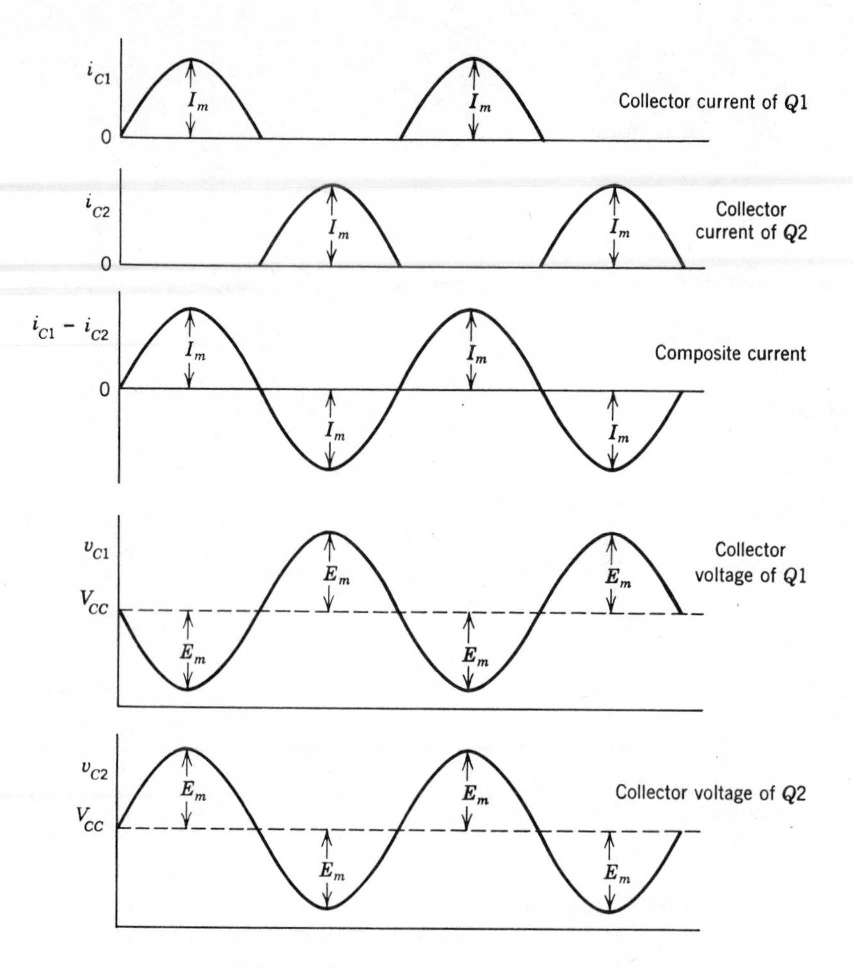

Figure 18-14 Waveforms of a class-B amplifier.

The a-c power in the load is

$$P_{\text{out}} = \frac{V_m}{\sqrt{2}} \times \frac{I_m}{\sqrt{2}} = \frac{V_m I_m}{2} \tag{18-2}$$

The collector dissipation is

$$P_{\text{diss}} = P_{\text{in}} - P_{\text{out}} = \frac{2 I_m V_{CC}}{\pi} - \frac{V_m I_m}{2} = 2 I_m \left(\frac{V_{CC}}{\pi} - \frac{V_m}{4} \right) \tag{18-3}$$

If we consider a circuit in which the power is supplied to a load by an output transformer, there is no d-c power lost in the load, and the collector efficiency is

$$\eta = \frac{V_m I_m/2}{2I_m V_{CC}/\pi} = \frac{\pi}{4}\frac{V_m}{V_{CC}} \tag{18-4}$$

When the operation of the circuit is ideal, the collector voltage is zero at the instant of the peak of the alternating collector voltage V_m equals V_{CC}. Then, the *maximum theoretical efficiency of a class-B amplifier is $\pi/4$ or 78.5%*. This figure of 78.5% compares to the 50% value obtained for the class-A amplifier. The efficiencies of the class-AB_1 and the class-AB_2 amplifiers are intermediate values dependent on the bias conditions.

Equation 18-3 can be written in the form

$$P_{\text{diss}} = P_{\text{in}} - P_{\text{out}} = \frac{2I_m V_{CC}}{\pi} - \frac{I_m{}^2 R_{\text{ac}}}{2} \tag{18-5}$$

where R_{ac} is the reflected value of the load resistance. Now Eq. 18-5 is differentiated with respect to I_m to determine the maximum power dissipation:

$$\frac{dP_{\text{diss}}}{dI_m} = \frac{2V_{CC}}{\pi} - R_{\text{ac}} I_m = 0$$

Solving for I_m,

$$I_m = \frac{2V_{CC}}{\pi R_{\text{ac}}}$$

This value of I_m is substituted back into Eq. 18-3:

$$P_{\text{diss, max}} = \frac{2V_{CC}}{\pi}\left(\frac{2V_{CC}}{\pi R_{\text{ac}}}\right) - \left(\frac{2V_{CC}}{\pi R_{\text{ac}}}\right)^2 \frac{R_{\text{ac}}}{2}$$

$$= \frac{4V_{CC}^2}{\pi^2 R_{\text{ac}}} - \frac{2V_{CC}^2}{\pi^2 R_{\text{ac}}} = \frac{2V_{CC}^2}{\pi^2 R_{\text{ac}}} \approx 0.20\frac{V_{CC}^2}{R_{\text{ac}}} \tag{18-6a}$$

This value of dissipation, given by Eq. 18-6a, is the value for two transistors. The maximum dissipation for one transistor is

$$P_{\text{diss, max}} = \frac{V_{CC}^2}{\pi^2 R_{\text{ac}}} \approx 0.10\frac{V_{CC}^2}{R_{\text{ac}}} \tag{18-6b}$$

which occurs when $I_m = 2V_{CC}/\pi R_{\text{ac}}$
The input power for this value of current is the supply voltage times the d-c value of I_m

$$P_{\text{in}} = V_{CC}\left[\frac{1}{\pi}\left(\frac{2V_{CC}}{\pi R_{\text{ac}}}\right)\right] = \frac{2V_{CC}^2}{\pi^2 R_{\text{ac}}} \tag{18-6c}$$

The output power for two transistors is

$$P_{\text{out}} = \frac{1}{2}\left(\frac{2V_{CC}}{\pi R_{\text{ac}}}\right)^2 = \frac{2V_{CC}^2}{\pi^2 R_{\text{ac}}} \tag{18-6d}$$

The efficiency is

$$\eta = P_{\text{out}}/P_{\text{in}} = \tfrac{1}{2} \text{ or } 50\% \tag{18-6e}$$

By Ohm's law

$$E_m = I_m R_{\text{ac}} = \frac{2}{\pi} V_{CC} \tag{18-6f}$$

Therefore, maximum collector dissipation occurs when the peak output voltage is $2/\pi$ times V_{CC}.

According to Eq. 18-1, the input power in the class-B amplifier increases linearly as I_m becomes larger. From Eq. 18-5, the output power increases as the square of I_m. A general expression for efficiency is, then,

$$\eta = \frac{I_m{}^2 R_{\text{ac}}/2}{2I_m V_{CC}/\pi} = \frac{\pi}{4}\frac{I_m R_{\text{ac}}}{V_{CC}} \tag{18-7}$$

The *maximum* value of efficiency (78.5%) occurs when I_m reaches its greatest value. Thus, this high efficiency is *only* obtained with the maximum driving signal. When the signal decreases from maximum, the efficiency drops and the greatest value of collector dissipation occurs at an intermediate signal level. Consequently, the transistor power rating is established from Eq. 18-6*b* and not at the maximum signal condition. This is especially important in audio amplifier work, since the signal varies continuously with the program information.

Class-AB and especially class-B amplifiers create problems of power-supply regulation. In class B, with a maximum signal, the drain on the power supply is a maximum. When the signal is zero, the power supply is at no-load conditions. To keep the harmonic content of the output low, the voltage regulation of the power supply must be kept as low as possible by providing large line transformers, large chokes, and large filter capacitors.

The transistor circuit of Fig. 18-1 and the vacuum-tube circuit of Fig. 18-13*b* can be operated in class AB by proper choice of the biasing components and the signal level. When the bases of the transistors in Fig. 18-1 are cut off, there may be a tendency for the charge on the coupling capacitors to discharge into the transistors, upsetting the circuit balance. Consequently, the base resistors are often paralleled with diodes to short out the capacitors from the transistors on the reverse cycle of the signal.

The presence of collector cutoff current, which is a collector current due to minority current carriers at zero bias, creates a problem in transistor class-B circuits that does not occur in vacuum-tube circuits. The individual collector currents of the two transistors (Fig. 18-15) are never quite zero during the full cycles of the signal. The difference of the two collector currents which is the output shows a slight irregularity or deviation from a sinusoidal waveform as the wave crosses the zero or time axis. This irregul-

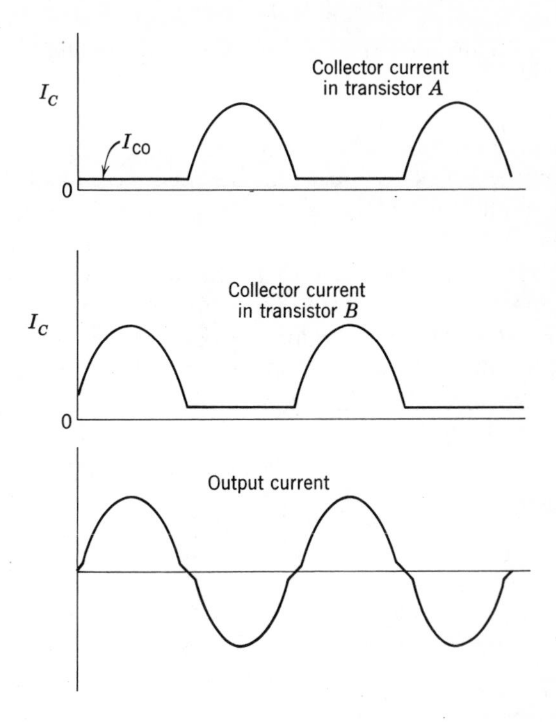

Figure 18-15 Cross-over distortion.

arity is called *cross-over distortion.* This distortion is an inherent characteristic of the class-B transistor circuit, and it can be avoided only by keeping the operating point slightly off class B and into class AB.

Technically then, it is a class-AB amplifier but, since it is only a method of removing the effect of leakage current, this circuit with an offset bias is accepted as a class-B amplifier. The class-B power amplifier is almost universally used in portable transistor radios. When the signal is zero, there is no d-c collector current, and the d-c collector current is a function of signal level. This conservation of battery power is most important in the portable unit. A class-A amplifier would require a constant high d-c drain on the battery.

Let us consider a typical class-B calculation. A 30-V supply is to be used with a class-B amplifier that delivers 4 W to an 8-Ω load. The circu[i]t used is shown in Fig. 18-13a, and typical waveforms are given in Fig. 18-1[4]. An average power of 4 W in an 8-Ω load gives the voltage and current value[s]

$$P = E^2/R \qquad \text{or} \qquad E = \sqrt{P \times R} = \sqrt{4 \times 8} = 5.66 \text{ V}$$

and

$$I = P/E = 4/5.66 = 0.707 \text{ A.}$$

These values of current and voltage are rms values that convert to pea[k] values as $E_m = 5.66\sqrt{2} = 8.00$ V and $I_m = 0.707\sqrt{2} = 1.00$ A.

When a maximum signal is applied to the input of the class-B amplifie[r] the a-c signal on the collector becomes larger with the result that E_m in th[e] waveforms of Fig. 18-14 just touches the zero axis of the voltage, provide[d] that the small value of forward voltage drop from collector to emitter i[s] neglected. Therefore, in this example, at the instant the value of E_m acros[s] the load is 8.00 V, the value of E_m at the collector must be 30 V. Conse[-]quently the transformer turns ratio n_1/n_2 must be 30/8.00, and then I_m i[n] the collector is the peak value of load current divided by this factor o[f] 1.00/(30/8.00) or 267 ma. Having these values, any further circuit calcula[-]tions that we make are quite simple.

PROBLEMS

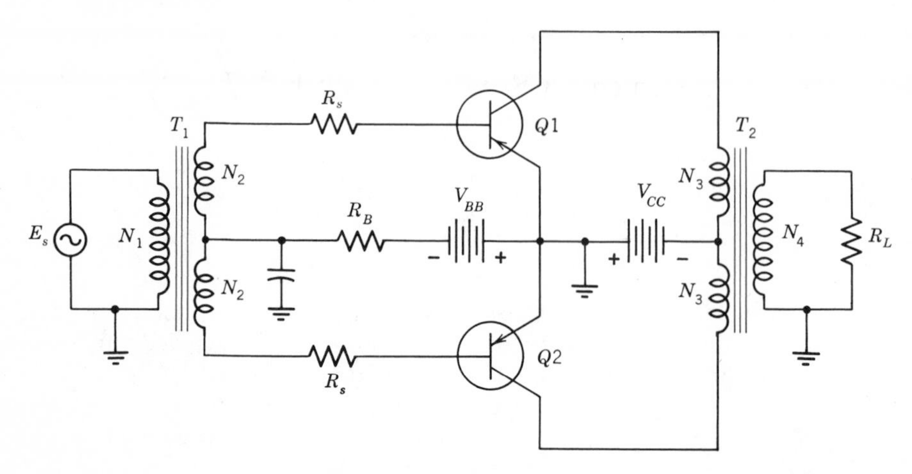

Circuit for Problems 1 to 5.

1. The load on a class-A amplifier is 200 Ω. In the transformers, $N_1 = N_2 = N_3 =$ N_4. The supply voltages are 30 V, and the germanium transistors have a β o[f] 120. What is the maximum available undistorted load power? What is th[e] value of I_C? Calculate R_B. If R_s is 50 Ω, what is the required input voltage E_s t[o]

develop maximum undistorted output power? What is the input power requirement? What is the power gain in decibels? Use $r'_e = 50$ mv/I_E. Draw to scale the current and voltage waveforms for the collector circuit.

2. In Problem 1, if the input resistance R_s is reduced to 10 Ω, what drive power is required and what is the power gain of the circuit in decibels?
3. The supply voltage for a class-B amplifier ($V_{BB} = 0$ and $R_B = 0$), using germanium transistors having a β of 74, is 80 V. $N_3 = N_4$ and $N_1 = 4N_2$. R_s is 10 Ω. The load resistance, a loudspeaker, is 16 Ω. Determine the maximum available undistorted power in the loudspeaker, the d-c supply current, and the maximum collector dissipation in the power transistors. What is the rms value of E_s that is required to obtain the maximum power in the loudspeaker? Determine the power required from the source and the gain of the circuit in decibels. Use $r'_e = 50$ mv/I_E. Draw to scale the current and voltage waveforms for the collectors of the power transistors.
4. Solve Problem 3 when the supply voltage is 50 V and the load resistance is 5 Ω. $R_s + h_{ie}$ is 2 Ω. $N_1 = 5N_2$ and $N_3 = 2N_4$.
5. In a class-B amplifier ($V_{BB} = 0$ and $R_B = 0$), $N_1 = 2.5N_2$ and $N_3/N_4 = 4.6$. The supply voltage is 5 V and the germanium transistors have a β of 100. R_L is 24 Ω and R_s is 10 kΩ. What value of E_s produces maximum undistorted power in the load? Use $r'_e = 50$ mv/I_E. What is the output power and what is the gain of the circuit in decibels? What is the maximum average power dissipated in each transistor?

Section 18-7 Complementary Symmetry in Push-Pull

The principles of complementary symmetry described in Section 16-1 may be applied to the push-pull amplifier (Fig. 18-16). This circuit uses neither an

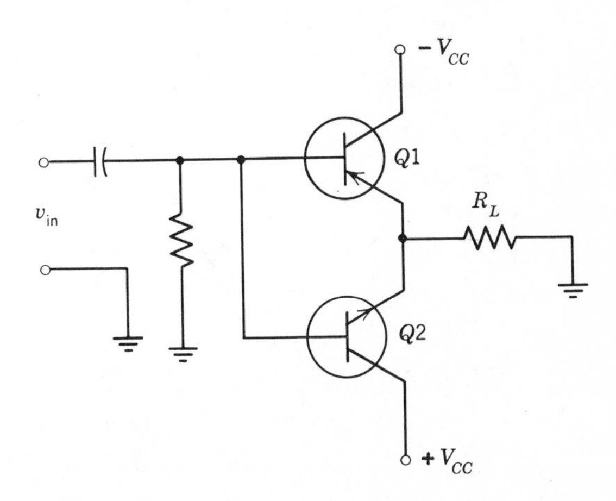

Figure 18-16 Push-pull using complementary symmetry.

input nor an output transformer. The current through the load is the sum c the two collector currents, i_{C_1} and i_{C_2}. Since one of the transistors is a *PN* unit and the other an *NPN* unit, one current is positive and the other nega tive, so that the load current is $(i_{C_1} - i_{C_2})$ which is a true push-pull concep When the signal is positive, the emitter-to-base voltage on the *PNP* transis tor is reduced while the emitter-to-base voltage on the *NPN* transisto increases. This positive signal increases the collector current in the *NP* unit and decreases the collector current in the *PNP* unit. Correspondingly, negative signal increases the output current in the *PNP* transistor an decreases the output current in the *NPN* transistor. The *PNP* and the *NP* transistors must be identical but with opposite polarity in all current an voltage ratings to secure a balance.

The waveforms for the circuit are shown in Fig. 18-17. The vertica

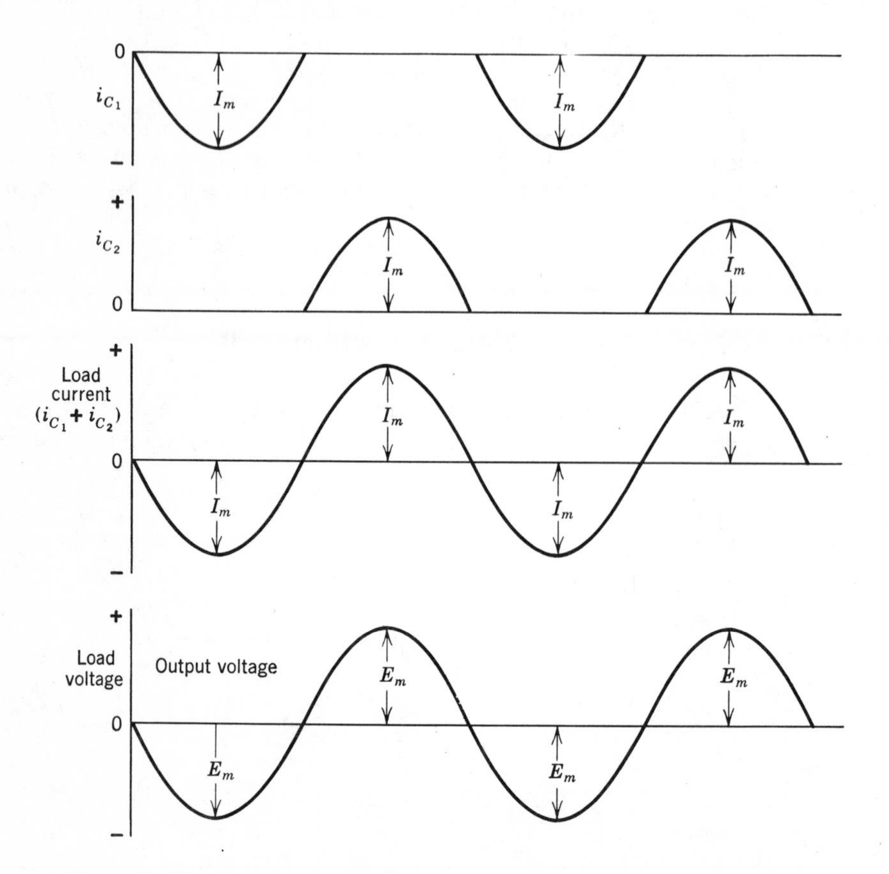

Figure 18-17 Waveforms for the complementary-symmetry class-B amplifier.

scales of the waveforms have opposite polarities. Consequently, when they are added, the d-c terms cancel out leaving only a pure a-c signal in the load resistor. There is no d-c heating loss in the load resistor even though an output transformer is not being used.

The two transistors are effectively connected as emitter followers. The input drive signal is slightly in excess of the load voltage. For approximate calculations they can be considered as equal. Thus, the peak value of the load is the required peak value of the drive for a calculation. The emitter follower circuit, then, has a voltage gain of one, and a current gain of $(1+\beta)$. The circuit used in 18-13*a* has both a voltage and a current gain but requires the use of an output transformer.

Another circuit arrangement using complementary symmetry in conjunction with class-B amplifiers is shown in Fig. 18-18. When there is no

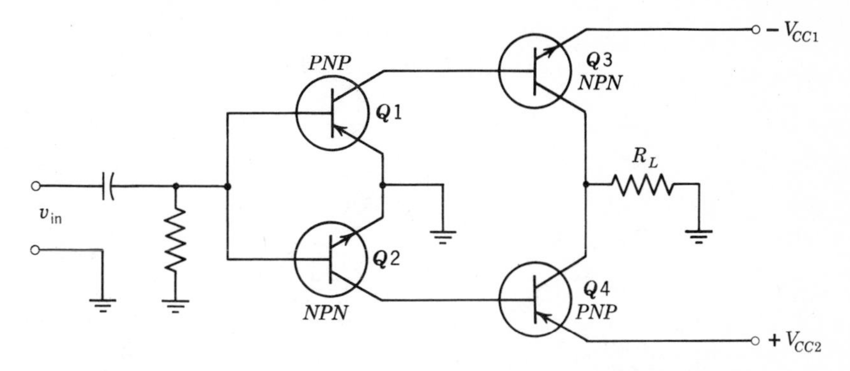

Figure 18-18 Class-B amplifier using complementary symmetry.

signal at the input, both driver transistors, $Q1$ and $Q2$, are cut off since the bias voltages on them are zero. Also, there is no bias current in the output transistors, $Q3$ and $Q4$, and their collector currents are zero. In this manner, without a signal, there is no drain on the supply batteries, V_{CC1} and V_{CC2}. When the signal is positive, there is a bias current and a collector current in $Q2$ but not in $Q1$. The collector current in $Q2$ is the base current for $Q4$, allowing $Q4$ to supply energy into the load resistance R_L. When the signal is negative, there are currents in $Q1$ and $Q3$, but not in $Q2$ and $Q4$.

Again, in this circuit, matched characteristics are necessary for both the driver transistors, $Q1$ and $Q2$, and the output transistors, $Q3$ and $Q4$. The relation of ratings between the driver transistors and the output transistors must be such that the magnitude of the collector currents of $Q1$ and $Q2$ is of the order of the base currents of $Q3$ and $Q4$.

PROBLEMS

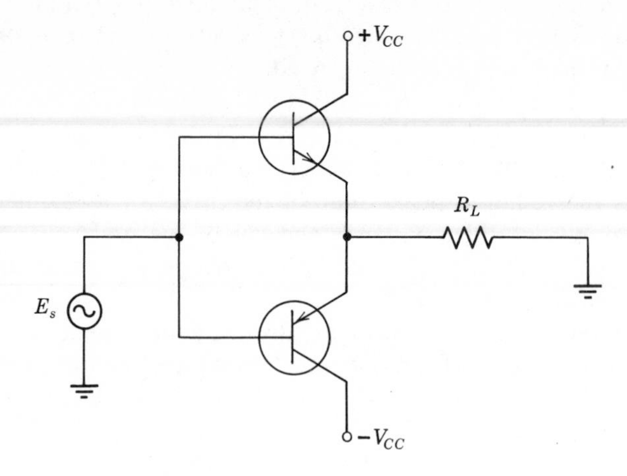

Circuit for Problems 1 and 3.

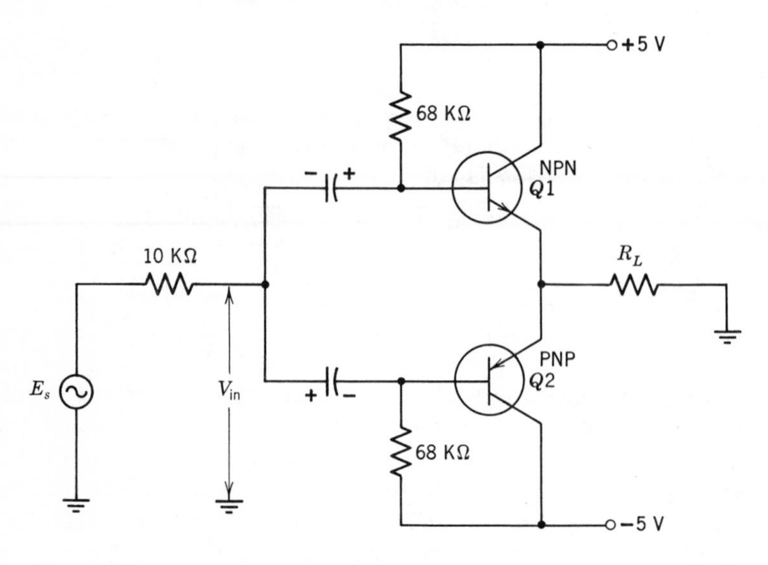

Circuit for Problem 4.

1. A complementary-symmetry pair of power transistors is used to drive a 16-Ω speaker. The rating of the channel is 60 W. What supply voltages are required? If the average program material produces 5 W in the speaker, what is the current drain on the power supplies, what is the collector dissipation in each transistor, and what is the overall efficiency? Assume class-B operation.

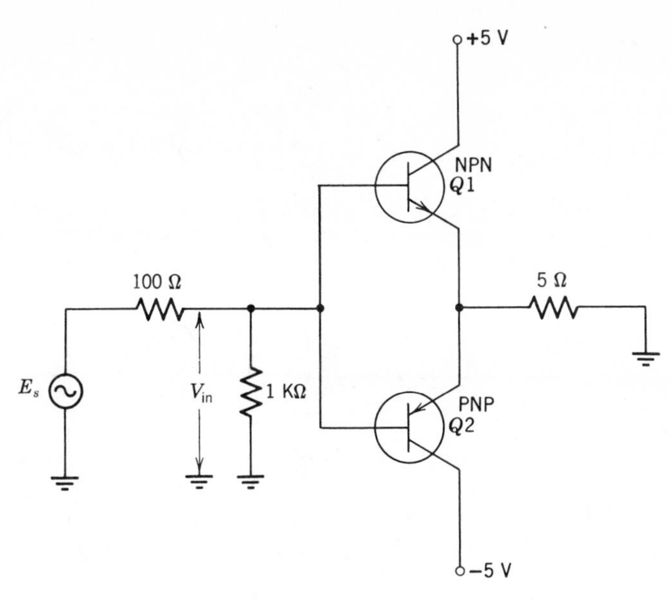

Circuit for Problem 5.

What drive (current, voltage, and power) is required for the peak load power, and what drive is required for the average load power? Assume that the value of β for the transistors is 120.

2. In the circuit used for Problem 1, the Darlington pair arrangement of Fig. 18-18 is used. Assuming all transistors have a β of 120, what are the requirements for the input signal drive?
3. R_L is a 16-Ω loudspeaker that handles 20 W average power. Determine the required supply voltages and the maximum d-c supply current. Specify E_s.
4. Find the value of R_L for maximum undistorted power. The operation is Class-A. The transistors have a β of 80 and are silicon. What is the a-c power in the load? For this condition, what are the values of V_{in} and E_s? What is the overall power gain of the circuit in decibels? Use $r'_e = 50$ mv/I_E.
5. Find the a-c load power for maximum undistorted output power conditions. What are the values of V_{in} and E_s? What is the overall power gain of the circuit in decibels? β for the silicon transistor is 49.

Section 18-8 Commercial Audio Amplifiers

Three commercially used circuits are given to conclude this chapter. The amplifier shown in Fig. 18-19*a* is the complete electronic circuit used in a portable record player. A driver transformer and an output transformer are

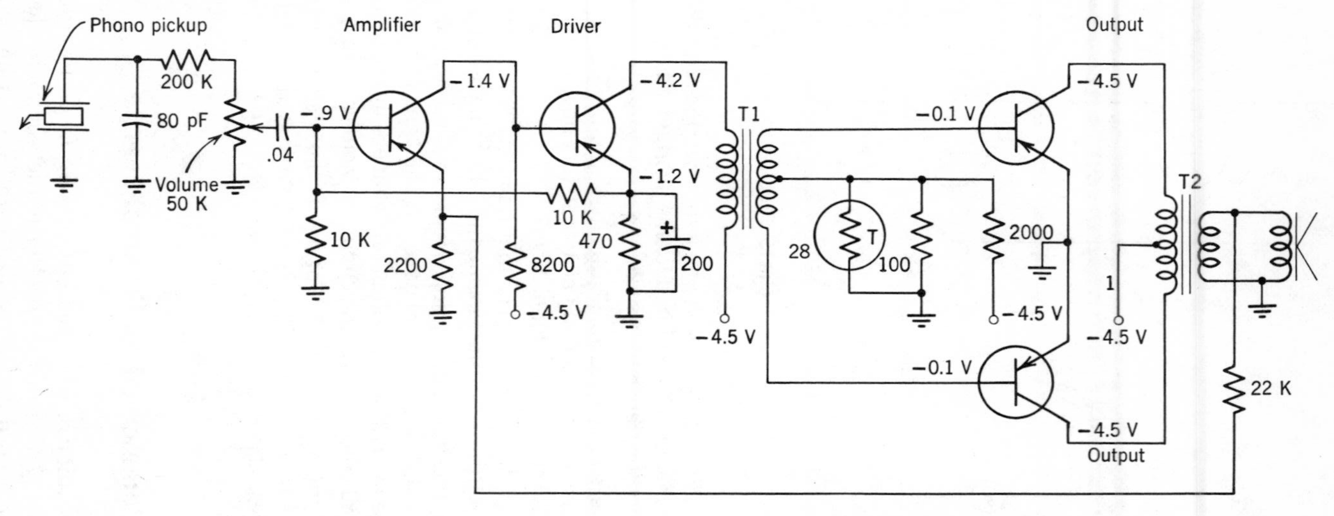
Phono pickup
200 K
80 pF
Volume
50 K
.04
Amplifier
−.9 V
−1.4 V
10 K
2200
Driver
−4.2 V
−1.2 V
10 K
8200
−4.5 V
470
200
T1
−4.5 V
28
T
100
2000
−4.5 V
Output
−0.1 V
−4.5 V
−0.1 V
−4.5 V
Output
T2
1
−4.5 V
22 K
Voltage feedback loop

(a)

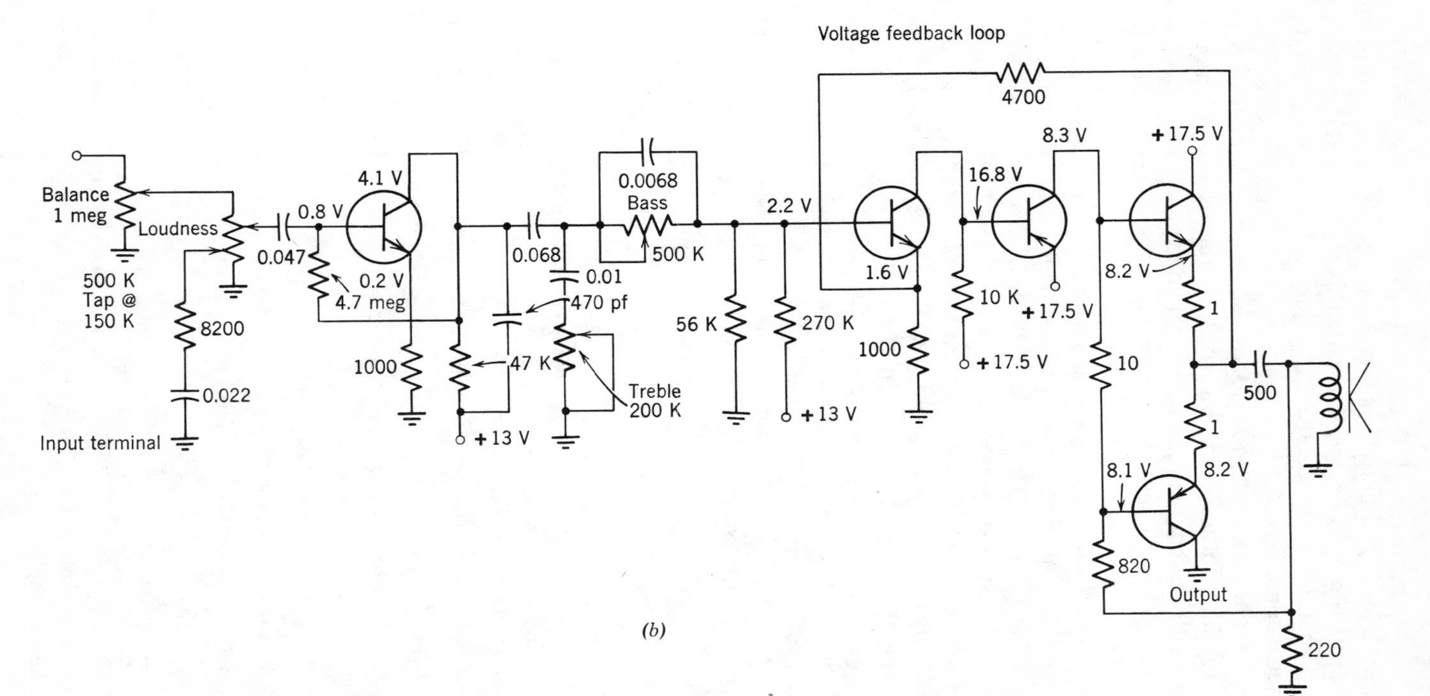

(b)

Figure 18-19 Typical complete power amplifiers. (*a*) Amplifier for a small portable phonograph. (*b*) One channel of a high-fidelity stereo amplifier.

used in the push-pull output amplifier. A thermistor is used to temperature compensate the power stage. The circuit is simple, and the only control is to regulate volume. The expected d-c voltage values are given on the circuit for a no-signal condition.

The amplifier shown in Fig. 18-19*b* is one channel of a two-channel stereo amplifier for a record player. The output is a complementary-symmetry push-pull arrangement, in which only one power source is required. The complementary-symmetry arrangement in the driver stages improve the low-frequency response of the system by avoiding coupling capacitors. The bass control effectively removes and inserts a small coupling capacitor. The treble control controls the amount of the high frequencies that are shorted to ground. The loudness control is not a true loudness compensation, but it controls the response to yield the curves shown in Fig. 15-16. The balance control sets the level of this channel in relation to the other channel. Like channel controls are ganged together so that the same effect is simultaneously introduced in both channels. On the other hand, the balance control is ganged with one channel connection reversed. In this way the center position produces exact balance if both channels are identical.

The power amplifier section of a high-fidelity radio receiver is given in Fig. 18-20. This circuit is representative of the maximum quality audio amplifiers that are available.

Questions

1. Define or explain each of the following terms: (*a*) balance control, (*b*) composite, (*c*) thermistor, (*d*) cascade, (*e*) class A, (*f*) class AB_1, (*g*) class AB_2, (*h*) class B, (*i*) class C, (*j*) crossover distortion.
2. Describe the process for checking the operation of a push-pull amplifier.
3. What does a d-c core saturation do to an a-c waveform in a transformer?
4. What is the effect of a push-pull connection on even harmonics? On odd harmonics?
5. How is a composite load line developed?
6. Give five advantages for push-pull operation.
7. Give three disadvantages for push-pull operation.
8. What is a split-load phase inverter?
9. What is the maximum theoretical efficiency of a class-A amplifier?
10. What is the maximum theoretical efficiency of a class-B amplifier?
11. Does maximum collector dissipation occur at maximum output power conditions in a class-B amplifier?

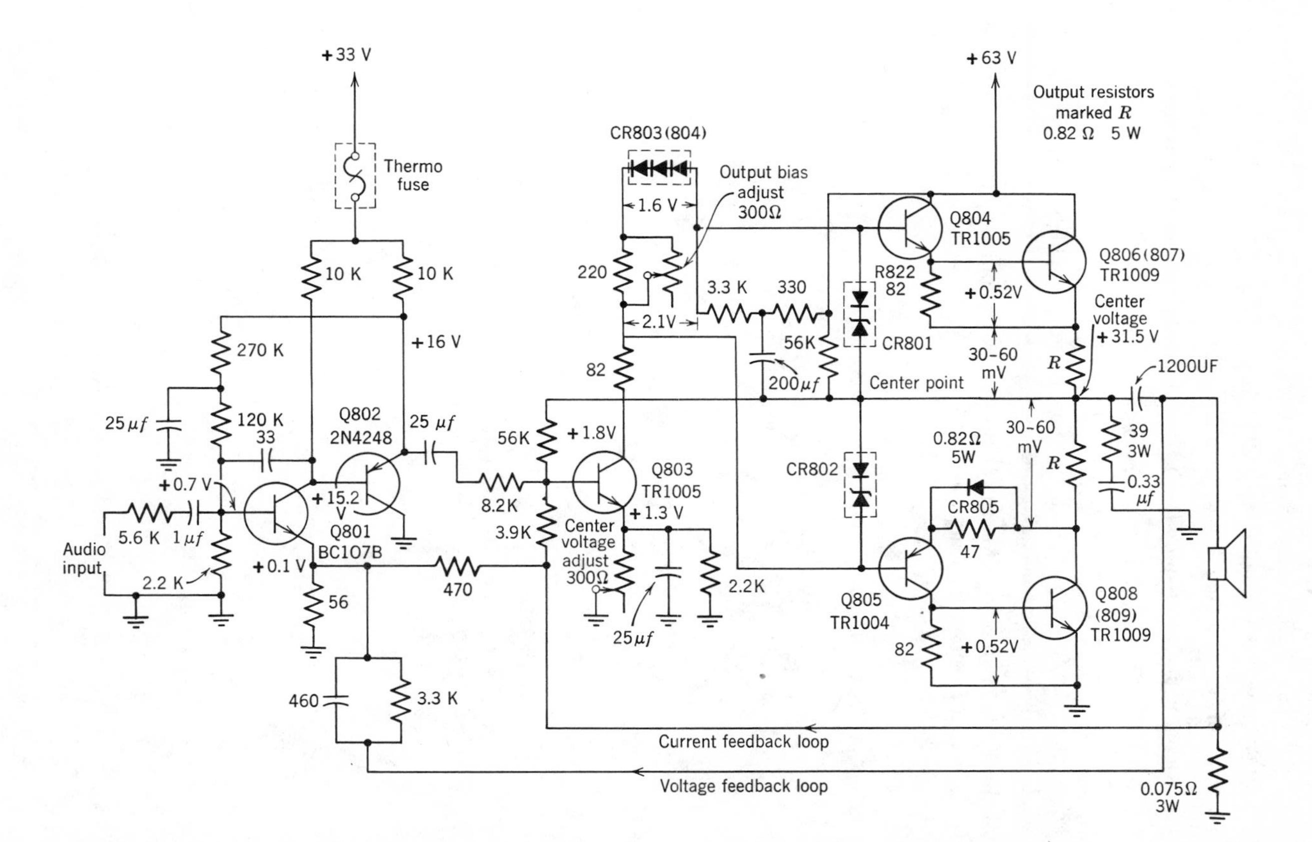

Figure 18-20 Fisher Radio Corporation 45-W audio control amplifier TX-1000 Dual Channel Amplifier. (*Courtesy Fisher Radio Corporation.*)

12. As the input signal is reduced from maximum toward zero, what happens to the efficiency of a class-B amplifier?
13. As the input signal is reduced from maximum toward zero, what happens to the collector dissipation in a class-B amplifier?
14. Explain how to eliminate crossover distortion.
15. Is a phase inverter required with a complementary-symmetry push-pull amplifier?
16. What is the advantage of the complementary-symmetry push-pull amplifier compared to a conventional push-pull amplifier?
17. Explain the function of each component in the amplifier shown in Fig. 18-19*a*. Ignore the voltage feedback loop.
18. Explain the function of each component in the amplifier shown in Fig. 18-19*b*. Ignore the voltage feedback loop.

Chapter Nineteen

CLASS-C AMPLIFIERS

In a class-C amplifier the bias is so adjusted that output current flows for less than one half the a-c cycle. A fixed-bias source is generally associated with class-C amplifiers, but very often in low-power applications the grid-leak or bias-clamp method provides the operating bias (Section 19-1). The loads in class-C amplifiers are tuned circuits (Section 19-2). A load line may be established for this amplifier (Section 19-3), and from this load line an analysis can be made of the plate circuit (Section 19-4) and of the grid circuit (Section 19-5). The adjustment and the method of tuning of a class-C amplifier require special considerations (Section 19-6). A class-C amplifier may also be used as a harmonic generator and amplifier (Section 19-7). A simplified approach for analyzing a class-C amplifier is presented (Section 19-8). When triodes are used as class-C amplifiers, it is necessary to follow the procedure of neutralization (Section 19-9).

The purpose of this chapter is to present a careful analysis of how the class-C amplifier functions and of the meaning of the various problems encountered in its adjustment. It is not the intent of this chapter to develop a design procedure.

Section 19-1 Grid-Leak Bias and the Bias Clamp

A grid-leak bias is a means of developing a negative direct bias voltage between the grid and the cathode of a tube by rectifying the incoming signal for this purpose. The applications of the grid-leak bias go beyond the class-C amplifier and, when these applications are discussed, we shall refer to this section. The principles of operation of the circuit (Fig. 19-1) extend to the solid-state diode and to the semiconductor as well. This concept was used

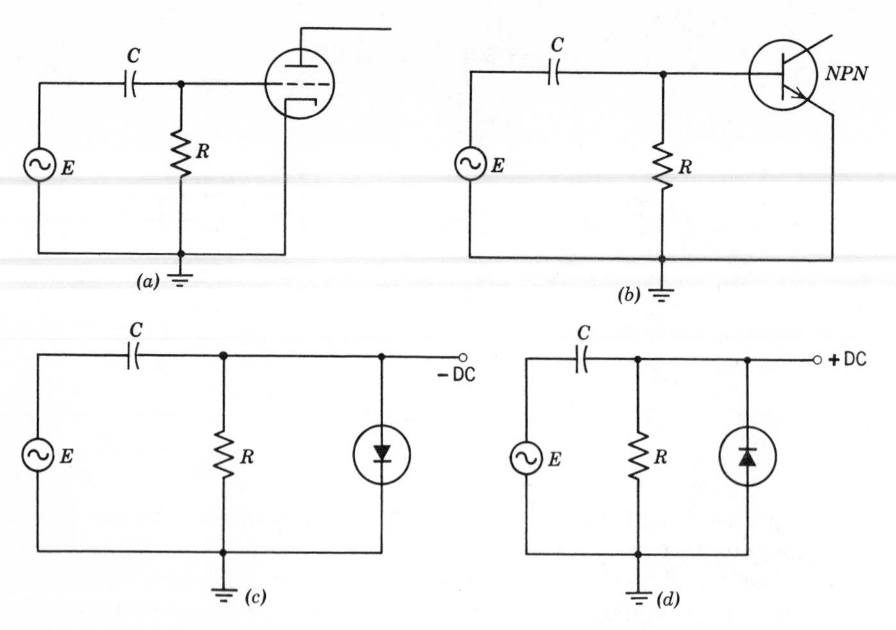

Figure 19-1 Grid-leak bias and bias clamps. (*a*) Vacuum-tube circuit. (*b*) Transistor circuit. (*c*) and (*d*) Diode circuits.

early in the development of the vacuum tube and is carried into solid-state electronics in many applications as the bias clamp. The bias clamp is a form of the shunt rectifier (Section 6-10).

A grid-leak bias or a bias clamp can be recognized in a circuit as it differs from the usual bias arrangement in two respects:

1. The usual biasing networks of resistors and capacitors are missing from the circuits. The emitter or the cathode is usually connected directly to the common return or ground.
2. A resistor R and a capacitor C are placed between the signal source E and the input to the tube, semiconductor, or diode.

A restriction on the circuit is that there must be a d-c path through the source E. We assume for the purpose of this discussion that the d-c resistance of the source is zero. A typical source of voltage that has a negligible d-c resistance is the secondary winding of a transformer.

When the a-c polarity of the signal is such that the grid is positive and the ground is negative (Fig. 19-2), there is a current flow in the grid. The current flow is limited by the grid resistor R, and a direct voltage is developed across this grid resistor with the polarity shown on the circuit diagram. Since the d-c resistance of the source is zero, the grid-leak capacitor C is effectively in

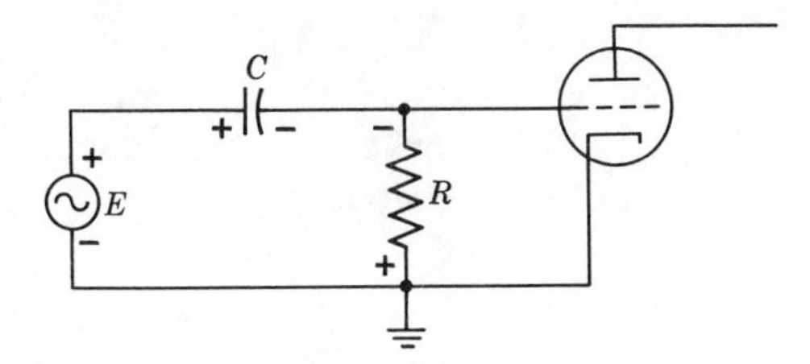

Figure 19-2 Operation of grid-leak bias.

parallel with the grid-leak resistor R, as far as the d-c circuit is concerned. The grid-leak capacitor C charges up to the peak value of the source voltage E.

The capacitor discharges through the grid-leak resistor when the applied signal is less than the peak value. The source E supplies just enough energy to recharge the capacitor each cycle. The action of this circuit is similar to the action of the capacitor filter in a rectifier circuit (Section 6-6). If R is large, if C is small, and if the time constant RC is long with respect to the time of one cycle, $1/f$, the energy absorbed by the grid-leak circuit is small. The negative direct voltage at the grid almost equals the magnitude of the peak of the signal. If the signal changes in level, the bias changes proportionally. Because of this, a grid-leak bias is very often termed an "automatic bias." The waveforms illustrating this discussion are shown in Fig. 19-3.

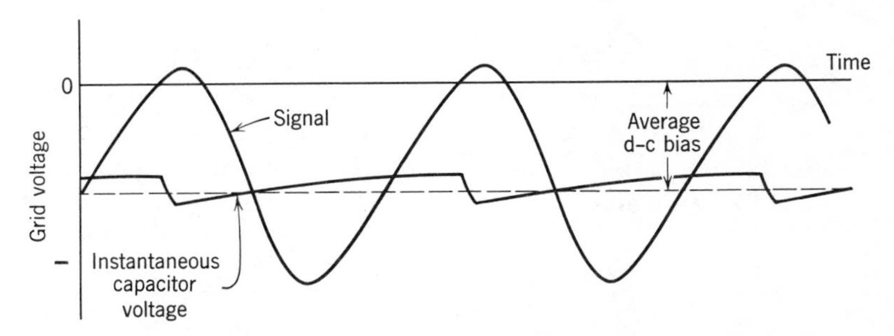

Figure 19-3 Grid-leak bias waveforms.

The bias-clamp circuit shown in Fig. 19-1*b* is directly similar to the vacuum-tube circuit of Fig. 19-2. The diode arrangement of Fig. 19-1*c* also develops a negative direct voltage. If a *PNP* transistor were used in the circuit of Fig. 19-1*b*, the direct voltage at the base would be positive. Similarly, the direct voltage across the diode in Fig. 19-1*d* is also positive.

Since the d-c resistance of the source E is negligible when compared to the clamp resistance, it is possible to rearrange the circuit of Fig. 19-1 as long as the clamp resistor and the clamp capacitor are effectively in parallel in the d-c circuit. Several different circuit arrangements are shown in Fig. 19-4. These circuits are applicable to any device: vacuum tube, transistor, FET, or diode.

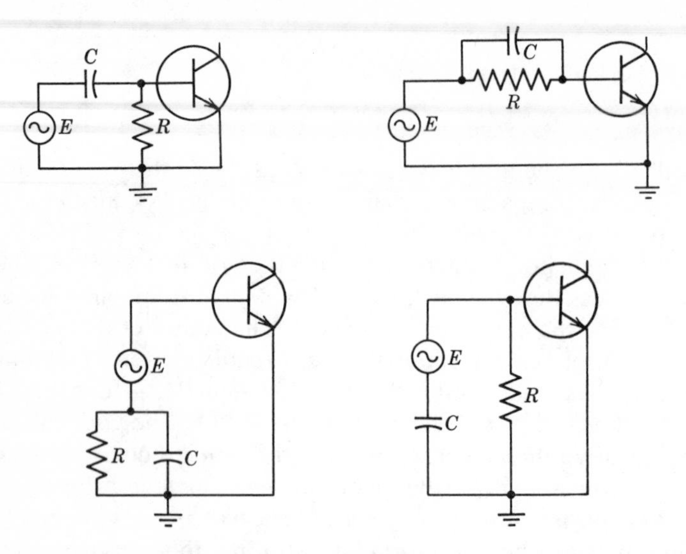

Figure 19-4 Bias-clamp bias circuit arrangements.

A variation of the circuit (Fig. 19-5) is often found in solid-state transmitter design. R_E and C_E have the function of establishing a d-c bias by the concept of the bias clamp. Also R_E can limit the transistor current to a safe value if the signal input falls to zero. In a vacuum-tube circuit in a transmitter, a fuse or a lamp may be placed in series with the cathode to limit the current.

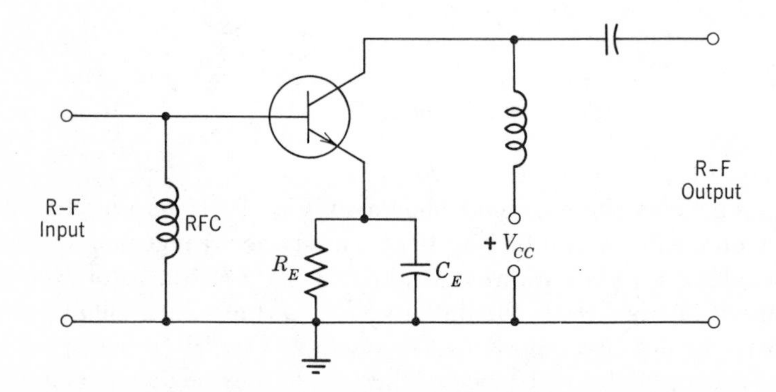

Figure 19-5 Transistor r-f amplifier.

ection 19-2 The Tank Circuit

If a charged capacitor is connected to the coil by closing a switch (Fig. 19-6), energy is transferred to the coil and then back to the capacitor in an oscillatory fashion. The sinusoidal voltage across the capacitor will gradually

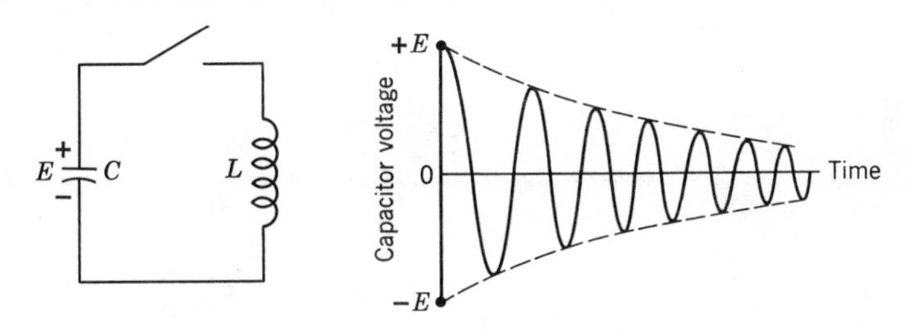

Figure 19-6 The tank circuit and damped wave.

diminish to zero with time. The rate of decay of this transient osciallation is termed the *damping* or the *damping factor* of the circuit and is caused by the effective series resistance of the tank circuit which produces an I^2R heat loss. When the Q of the circuit is at least five, the frequency of the oscillation is determined by:

$$f_0 = \frac{1}{2\pi\sqrt{LC}}$$

If a pulse of energy can be fed into this oscillating system once each cycle to replace the energy lost over the cycle, the *flywheel effect* of the tank circuit carries the action through the complete cycle to the next pulse. It is possible to feed in energy every second, third, or fourth cycle. Practically, an extreme limit is about a pulse of energy every ten cycles. By using this principle, a 1-MHz source of energy can keep a 10-MHz tank in oscillation. This action is called *frequency multiplication.*

At this point, a digression from the general pattern of the text is made. Currently, very high-power class-C amplifiers of the order of kilowatts are primarily the function of the vacuum tube. Transistors are not particularly adapted to the high voltage requirements both forward and reverse of several kilovolts. Solid-state designs are used for medium power amplifiers where the advantage of the size and the weight of solid-state equipment takes priority over cost. An example is the transmitter used on a spacecraft. The principles and the approach taken here in examining the details of class-C operation are directly transferable to the analysis of a semiconductor class-C power amplifier.

When the tank is connected into the plate circuit of an amplifier stage the plate voltage is also oscillatory. Energy must be fed into the tank onl between points x and z of Fig. 19-7. To provide this energy, a grid signa

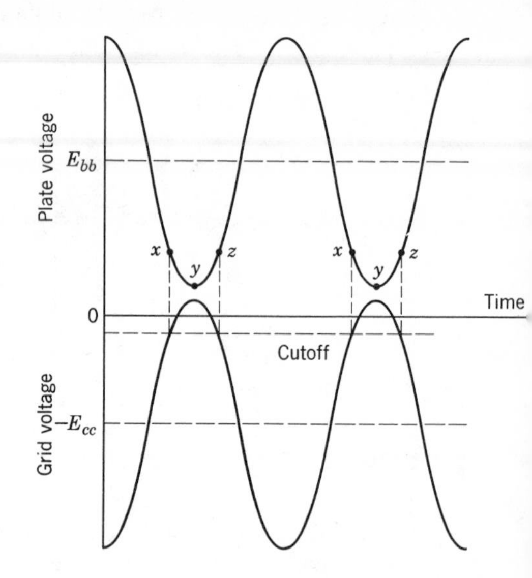

Figure 19-7 Waveforms of class-C grid and plate voltages.

which is biased at $-E_{cc}$ is kept below cutoff at all times except for the in terval xz. During the time xz, plate current flows. If this time xz is expresse in degrees of the full cycle, then xz represents the *angle of plate-curren flow*. If the bias $-E_{cc}$ is increased, and if the peak value of the a-c grid signa is increased, the time interval xz decreases. When the total energy content o each charging pulse is constant and when the charging time interval x decreases, the rate of charge must increase. This charging rate is the plat current.

In Fig. 19-7, the distance between y and the time axis, between x an the time axis, and between z and the time axis are particular values of e during the time of plate-current flow. The instantaneous product of e_b an i_b is the plate dissipation. If the tube were ideal and without losses, the plat dissipation would necessarily have to be zero. Under this ideal conditio point y would lie on the time axis, and the angle of plate current flow woul have to be zero. This pulse of plate current which has zero width would b infinite in amplitude. Thus it is theoretically possible to obtain a figure o 100% for the plate efficiency of the class-C amplifier. Practically, the tub drop at y must be finite and not zero, and the amplitude of the plate current i limited by the saturation value for the tube. Class-C amplifiers using vacuu tubes have typical efficiencies in the plate of the order of 70 to 80%. B the use of transistors, the efficiency can be raised to values well over 90%.

ection 19-3 The Class-C Load Line

When there is a fixed bias on the class-C amplifier, the plate voltage may be expressed when there is a signal as

$$e_b = E_{bb} - E_{pm} \cos \omega t$$

and the grid voltage by

$$e_c = -E_{cc} + E_{gm} \cos \omega t$$

If we solve each of them for $\cos \omega t$ and equate the results, we have

$$\cos \omega t = \frac{e_c + E_{cc}}{E_{gm}} = \frac{E_{bb} - e_b}{E_{pm}}$$

If E_{cc}, E_{bb}, E_{gm}, and E_{pm} are fixed quantities, then

$$\frac{e_c + E_{cc}}{E_{gm}} = \frac{E_{bb} - e_b}{E_{pm}}$$

will plot as a straight line on the constant-current tube characteristic (e_b, e_c curves, Fig. 19-8).

For our example (Fig. 19-8) the following values are assigned to the class-C amplifier:

Plate-supply voltage, E_{bb}	12,000 V
Grid-bias supply, E_{cc}	−1,600 V
Grid-signal drive, $\sqrt{2}\, E_g$	2,800 V peak
or E_g	1,980 V rms
Plate voltage, $\sqrt{2}\, E_p$	10,300 V peak
or E_p	7,280 V rms

These given values establish the operating point M and one extreme point Q on the load line. The other end of the load line is in the direction toward L at a distance so that ML equals MQ. We do not need to show this end of the load line because the tube is cut off and the flywheel effect of the tank circuit is maintaining the oscillation during this cutoff time. Below N on the load line, there is no plate current and the tube is cut off. Above P on the load line, the grid is positive and grid current flows.

The grid-voltage and the plate-voltage signals are taken from this load line for a sinusoidal signal and are plotted in Fig. 19-9. Plate current flows between A and B for a total of 117° out of the full cycle of 360°. There is a grid-current flow between C and D which is 110° of the full cycle. During the rest of the cycle, the tube is cut off.

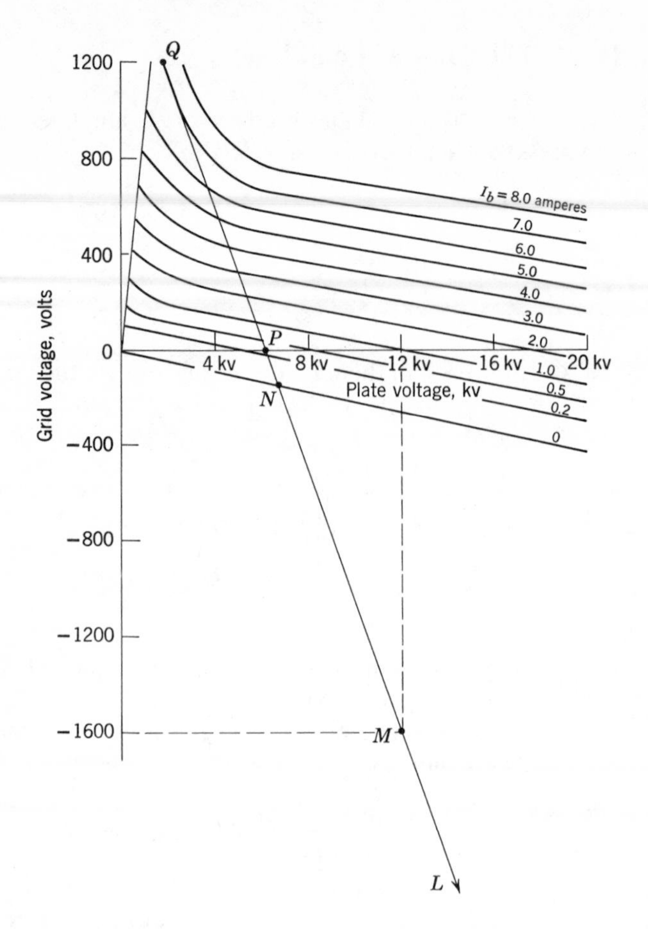

Figure 19-8 Constant plate-current curves with load line.

Section 19-4 Plate-Circuit Analysis

From the load line of Fig. 19-8 sufficient data are taken in the region of plate-current flow (Table A) to develop a detailed graph.

Steps of current are taken at even intervals. The corresponding values of instantaneous plate voltage e_b are read from the load line, and the values of e_p are determined by substituting in

$$e_p = E_{bb} - e_b = 12{,}000 - e_b$$

The ratio of the value of e_p at any point to its peak value, 10,300 V, gives the

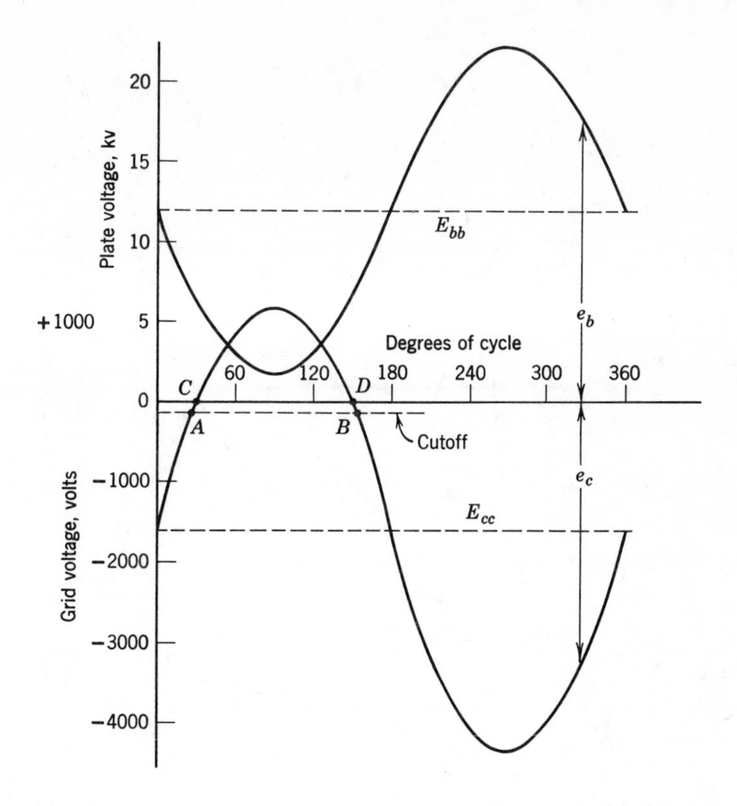

Figure 19-9 Grid and plate waves in a class-C amplifier.

TABLE A

i_b amps	e_b volts	e_p volts	$\sin\theta$	θ degrees	θ degrees	$e_b i_b$ watts
0	6600	5,400	0.524	31.5	148.5	0
0.2	6050	5,950	0.577	35.4	144.6	1,210
0.5	5950	6,050	0.587	36.0	144.0	2,975
1	5600	6,400	0.621	38.5	141.5	5,600
2	5200	6,800	0.660	41.3	138.7	10,400
3	4850	7,150	0.700	44.5	135.5	14,550
4	4500	7,500	0.728	46.6	133.4	18,000
5	4050	7,950	0.771	50.5	129.5	20,250
6	3600	8,400	0.815	54.5	125.5	21,600
7	2400	9,600	0.931	68.5	111.5	16,800
8	1700	10,300	1.000	90	90	11,900

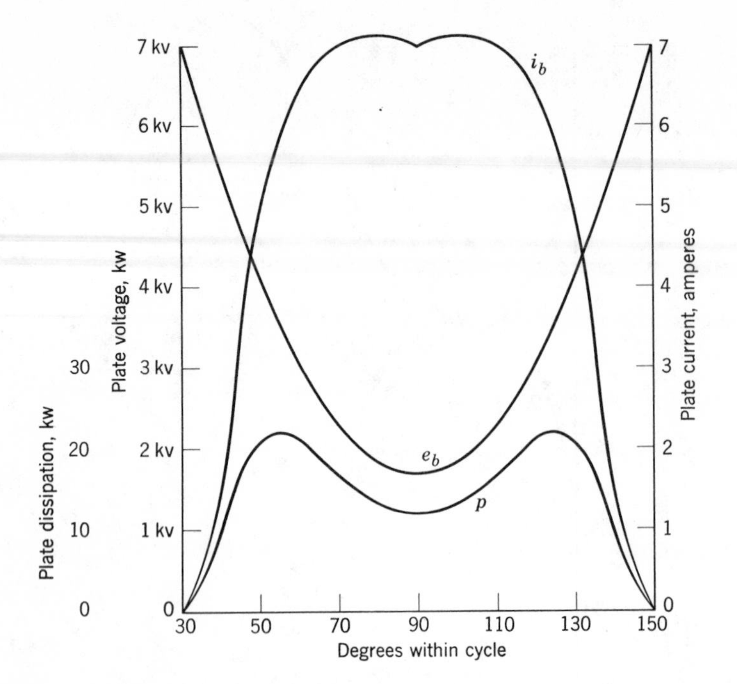

Figure 19-10 Plate analysis of a class-C amplifier.

value of the sine of the angle. The product of e_b and i_b gives the instantaneou plate dissipation. This detailed region is plotted in Fig. 19-10.

The area under the curve of the plate current is determined by countin squares. Dividing this area by the *full length* of 360° gives the average valu of the plate current I_b. From the data used in this example I_b is 1.7 amperes Using the same procedure for the curve of the plate dissipation $e_b i_b$ we find

$$P_{\mathrm{pl}} = 4.68 \text{ kW}$$

The total input power from the plate power supply is

$$\begin{aligned} P_{\mathrm{in}} = E_{bb} I_b &= 12{,}000 \times 1.7 \\ &= 20{,}400 \text{ W} \\ &= 20.4 \text{ kW} \end{aligned}$$

The plate efficiency is

$$\eta_{\mathrm{pl}} = \frac{P_{\mathrm{out}}}{P_{\mathrm{in}}} = \frac{15.72}{20.40} = 0.771 = 77.1\%$$

The a-c power output developed in the plate tank is

$$\begin{aligned} P_{\mathrm{out}} &= P_{\mathrm{in}} - P_{\mathrm{pl}} \\ &= 20.40 - 4.68 = 15.72 \text{ kW} \end{aligned}$$

Since the rms load voltage was given as 7280 V, the rms plate current is

$$I_p = \frac{P_{\text{out}}}{E_p} = \frac{15{,}720}{7280} = 2.16 \text{ A}$$

The a-c load impedance of the tank circuit is

$$Z_L = \frac{E_p}{I_p} = \frac{7280}{2.16} = 3370 \ \Omega$$

PROBLEMS

Problem Data Table

Amplifier	A	B	C	D
E_{bb}, volts	13,000	8,000	1000	10,000
E_{cc}, volts	−1500	−1200	−500	−450
$E_{p,\max}$, volts	11,800	7,000	750	8,000
$E_{g,\max}$, volts	1900	2000	700	600
$I_{b,\max}$, amps	6.00	6.00	0.58	5.00
$I_{c,\max}$, ma	1200	1200	160	1000
e_c for $I_b = 0$, volts	−150	−100	−15	−100

1. Sketch the plate waveform analysis for the data for Amplifier A in the table. Estimate I_b, plate dissipation, and output power.
2. Sketch the plate waveform analysis for the data for Amplifier B in the table. Estimate I_b, plate dissipation, and output power.
3. Sketch the plate waveform analysis for the data for Amplifier C in the table. Estimate I_b, plate dissipation, and output power.
4. Sketch the plate waveform analysis for the data for Amplifier D in the table. Estimate I_b, plate dissipation, and output power.

Section 19-5 Grid-Circuit Analysis

The constant-current characteristic shows not only lines of constant plate currents but also lines of the constant values of the grid current. For this tube and for the set of operating conditions given in Section 19-3, the load line is shown on the grid family in Fig. 19-11. For even increments of grid current, corresponding values of grid voltage are taken from this load line. The values of the grid signal e_g are found from

$$e_g = -E_{cc} + e_c = -1600 + e_c$$

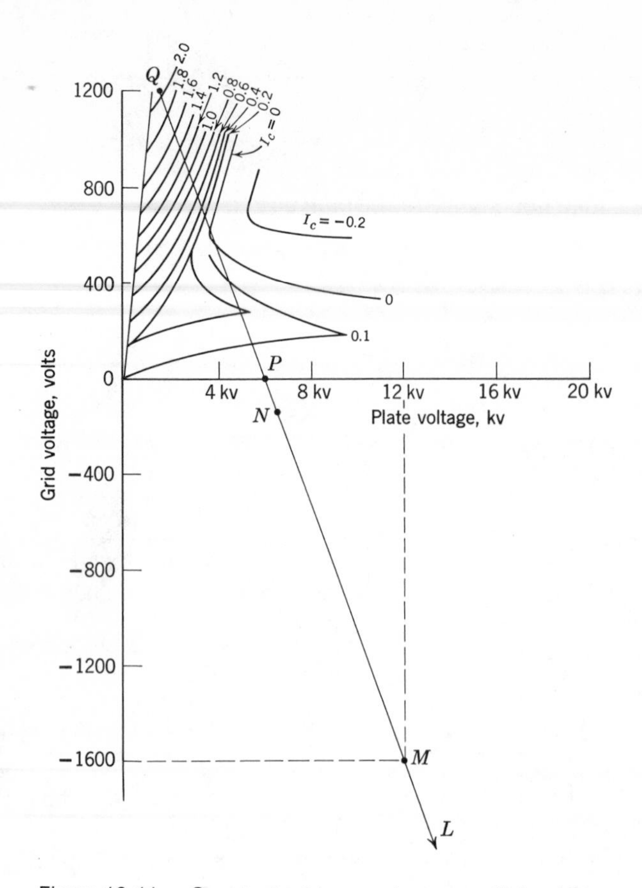

Figure 19-11 Constant grid-current curves with load line.

Using the peak value of grid signal of 2800 V, the value of the sine of the angle is $e_g/2800$. The product of e_c and i_c gives the instantaneous grid dissipation (Table B).

This information is plotted to a time scale in Fig. 19-12. The areas of the grid-current curve and the grid-dissipation curve are determined by the method of counting squares; by dividing these areas by the length of the full 360° cycle, we obtain

Average grid current, I_c	0.312 A
Average grid dissipation, P_g	321 W

This grid current is opposed to the direction of normal current direction through the bias battery and thus represents a charging effect on the bias battery. In other words, power is delivered to the bias battery and not taken

TABLE B

i_c amps	e_c volts	e_g volts	$\sin\theta$	θ degrees	θ degrees	$e_c i_c$ watts
0	0	1600	0.571	34.9	145.1	0
0.1	+120	1720	0.615	38.0	143.0	12
0.2	+290	1890	0.675	42.5	137.5	58
0.1	+420	2040	0.729	46.6	133.3	42
0	+560	2160	0.771	50.5	129.5	0
0	+620	2220	0.793	51.4	127.6	0
0.4	+680	2280	0.815	54.5	125.5	272
0.6	+760	2360	0.844	57.5	122.5	456
0.8	+820	2420	0.865	60.0	120.0	656
1.0	+860	2460	0.879	61.5	118.5	860
1.2	+920	2520	0.900	64.0	116.0	1100
1.4	+960	2560	0.915	66.1	113.9	1340
1.6	+1040	2640	0.944	70.8	109.2	1660
1.8	+1110	2710	0.967	75.0	105.0	2000
2.0	+1180	2780	0.988	81.0	99.0	2360
2.1	+1200	2800	1.000	90	90	2520

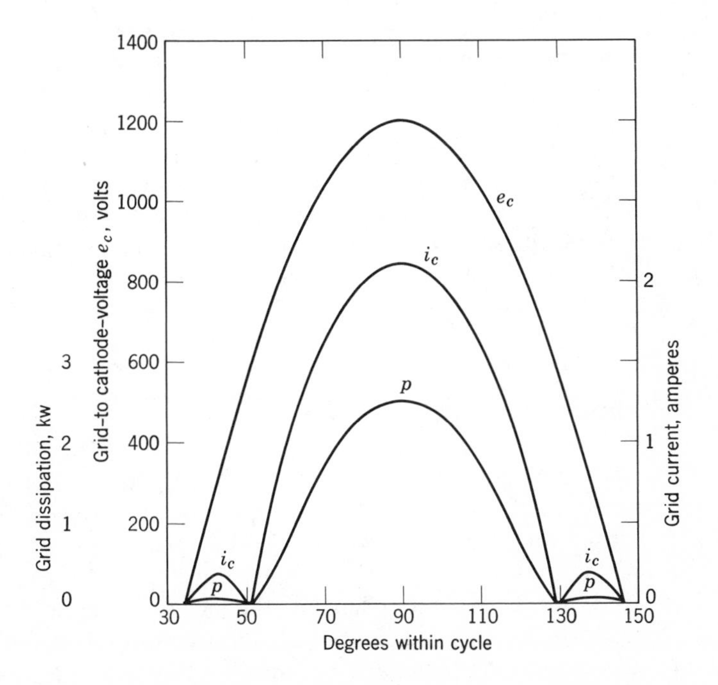

Figure 19-12 Grid analysis of a class-C amplifier.

from the bias battery. This energy amounts to

$$E_{cc}I_c = 1600 \times 0.312 = 495 \text{ W}$$

These two grid powers represent a total of 321 + 495 or 816 W. The outp power of the driving or preceding stage must therefore be 816 W plus t heating loss in its own tank circuit. The alternating grid current in the tube

$$I_g = \frac{P_g}{E_g} = \frac{816}{1980} = 0.412 \text{ A}$$

The a-c grid impedance of the tube is

$$Z_g = \frac{E_g}{I_g} = \frac{1980}{0.412} = 4800\ \Omega$$

The power gain of this class-C amplifier is

$$A_p = \frac{P_{\text{out}}}{P_g} = \frac{15{,}720}{816} = 19.2 \text{ or} + 12.8 \text{ db}$$

PROBLEMS

Refer to the Problem Data Table given in Section 19-4.

1. Sketch the grid waveform analysis for the data for Amplifier A in the tab Estimate I_c, grid dissipation, and driving power requirement.
2. Sketch the grid waveform analysis for the data for Amplifier B in the tab Estimate I_c, grid dissipation, and driving power requirement.
3. Sketch the grid waveform analysis for the data for Amplifier C in the tab Estimate I_c, grid dissipation, and driving power requirement.
4. Sketch the grid waveform analysis for the data for Amplifier D in the tab Estimate I_c, grid dissipation, and driving power requirement.

Section 19-6 Summary and Adjustment

The class-C amplifier, because of the short duration of the plate-curre flow, is not suited for audio applications without producing extreme disto tion. It is used for radio-frequency applications where a fixed-signal level available and a fixed output level is required. Reference to Figs. 19-8, 19- and 19-11 shows that a small change in the location of point Q which is th end of the load line can radically affect the over-all results. Among oth factors, the limitations that must be closely observed in a tube used as class-C amplifier are the peak plate current, the peak grid current, th maximum allowable plate dissipation, and the maximum allowable grid dissipation. This assumes that the voltages used for the plate supply and for the grid supply are suitable values which do not exceed the ratings of the tube. The variables which are normally available for the purposes of tuning and adjustment are:

1. The plate-tank tuning capacitance,
2. The grid-tank tuning capacitance,
3. A variable grid drive, and
4. A variable-output load coupling.

Two meters are needed for adjusting the class-C amplifier of Fig. 19-13. Both meters are d-c movements. One meter reads the plate current, and the

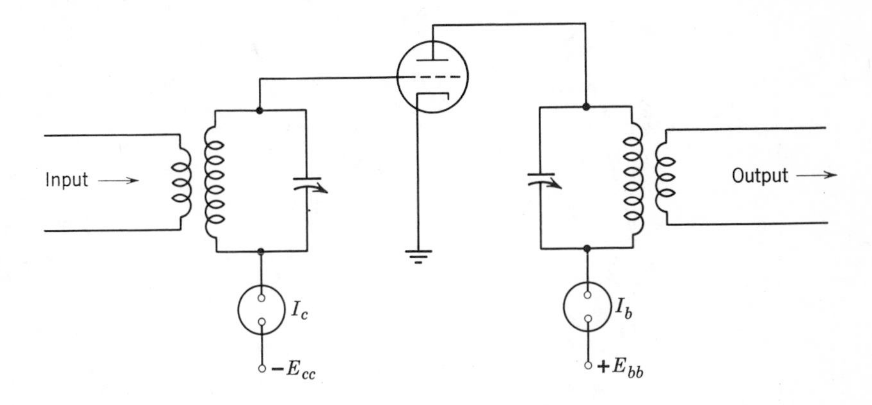

Figure 19-13 Class-C amplifier tuning.

other reads the grid current. When the plate tank is properly tuned, its losses are at a minimum, and the direct plate-current meter indicates a minimum value. When the grid tank is properly tuned, it delivers the largest signal to the grid, and the direct grid-current meter reads a maximum value. When these two adjustments are made, if it is seen that the tube does not deliver sufficient power, the grid drive is increased. If an increase in grid drive to the point of maximum allowable grid dissipation fails to produce the rated or desired output power, the next step is to increase the output coupling into the load. It is quite possible that optimum conditions may be obtained only by making an adjustment in the fixed value of the grid bias $-E_{cc}$. At each step in the adjustment of the class-C amplifier, the grid tank and the plate tank must be carefully retuned to maintain resonance.

Section 19-7 Harmonic Operation

If the plate tank is set to a multiple of the frequency of the grid drive, th class-C amplifier is termed a *harmonic generator* or a *frequency multiplie* Figure 19-14 shows the plate wave of a frequency doubler superimposed c

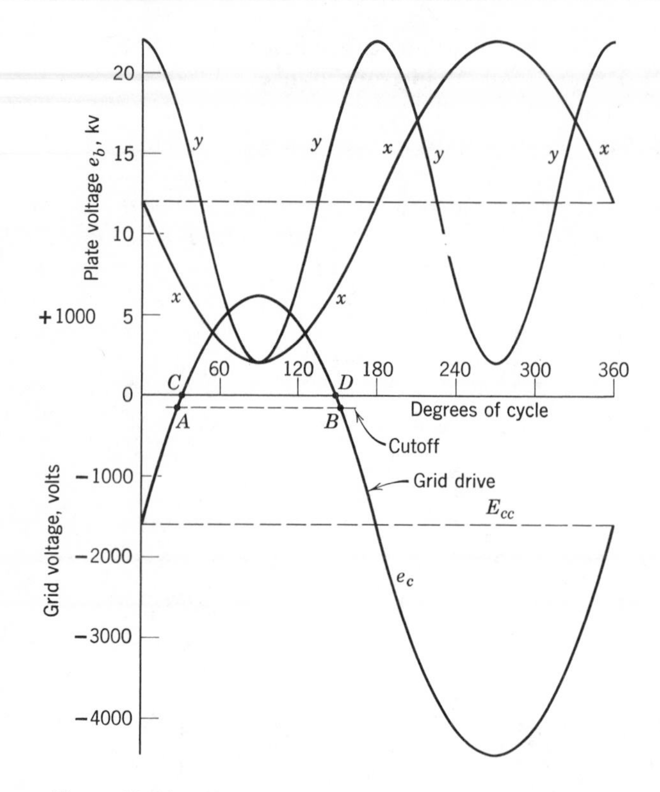

Figure 19-14 Harmonic operation of a class-C amplifier.

the curves for a straight single-frequency class-C amplifier. In both case the duration of the plate-current flow AB and of the grid current flow CD ar the same. However, the corresponding value of the second-harmonic plat voltage, curve y, during current flow is much higher than for the plate-voltag wave at fundamental frequency operation, curve x. This also produces muc larger values of instantaneous plate currents and grid currents. Correspond ingly, the curves for the grid dissipation and for the plate dissipation ar much higher than the curves for the dissipation at the operation at th fundamental frequency. In order to keep the plate circuit and the grid circu at or below the rated dissipation values, it is necessary to reduce materiall the value of the grid drive. Thus, in frequency-multiplying operations, the a-c power output of the tube is lowered considerably. In commercial practice, in order to obtain the greatest overall efficiency, frequency multiplying is done at low-power levels, and the high-power stages are reserved for amplification at one frequency.

PROBLEMS

Refer to the Problem Data Table given in Section 19-4.

1. Sketch the plate waveform analysis and the grid waveform analysis for a doubler, using the data for Amplifier A in the table. Estimate I_b, plate dissipation, output power, I_c, grid dissipation, and driving power requirement.
2. Sketch the plate waveform analysis and the grid waveform analysis for a tripler, using the data for Amplifier B in the table. Estimate I_b, plate dissipation, output power, I_c, grid dissipation, and driving power requirement.

Section 19-8 An Alternative Approach to Class-C Analysis

The current in the plate and in the grid is assumed to be represented by the top part of a sinusoidal waveform (Fig. 19-15). Therefore, the actual plate or grid current is shown by the shaded area and the duration of the current flow

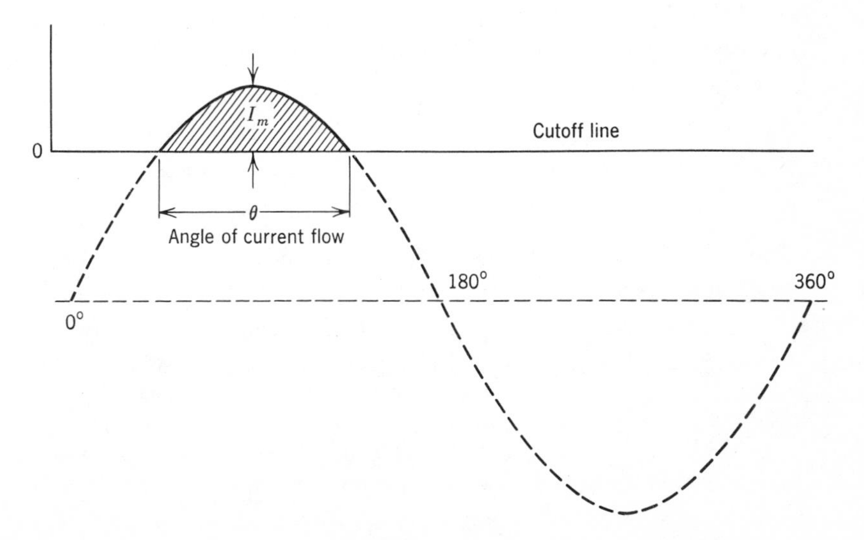

Figure 19-15 Waveform for analyzing currents as a section of a sinusoidal waveform.

is the angle θ. The only section of this wave that is of concern is that part o the waveform that is shaded. The duration of the shaded area is θ degrees At all other angles the waveform is zero. The shaded pulse is the curren waveform that is examined by the techniques of Fourier analysis. Th harmonic content of this shaded waveform is given by a mathematica analysis as

$$A_0 = \frac{\sin \frac{1}{2}\theta - (\pi\theta/360) \cos \frac{1}{2}\theta}{\pi(1 - \cos \frac{1}{2}\theta)} I_m$$

$$A_1 = \frac{\pi\theta/360 - \sin \frac{1}{2}\theta \cos \frac{1}{2}\theta}{\pi(1 - \cos \frac{1}{2}\theta)} I_m$$

$$A_n = \frac{\dfrac{1}{1+n} \sin \dfrac{n+1}{2}\theta + \dfrac{1}{n-1} \sin \dfrac{n-1}{2}\theta - \dfrac{2}{n} \sin \dfrac{n}{2}\theta \cos \frac{1}{2}\theta}{\pi(1 - \cos \frac{1}{2}\theta)} I_m$$

The value of θ for these coefficients is in degrees, not radians. If th coefficient I_m is omitted from these three equations, the resulting values o A_0, A_1, and A_n are decimal fractions which, when multiplied by 100, yiel percentages. The values of A_1 and A_n are peak values. When they ar multiplied by 0.707, rms values result.

These coefficients were evaluated in a computer program, and the result are plotted in a curve shown in Fig. 19-16. The significance of a negative sig is merely a 180° phase inversion. The values for the a-c terms, fundamenta to fifth harmonic, are peak values. A multiplication by 0.707 gives rm values.

The necessary information from the example of the class-C amplifie analyzed by the graphical approach, is

$$E_{bb} = 12{,}000 \text{ V} \qquad E_{cc} = -1600 \text{ V}$$
$$E_p = 7280 \text{ V, rms} \qquad E_g = 1980 \text{ V, rms}$$
$$I_b, \max = 7.3 \text{ A} \qquad I_c \max = 2.1\, A$$
$$\theta_p = 117° \qquad \theta_g = 77°$$

The two small side loops for grid current are ignored in determining θ_g. Th coefficients as determined from Fig. 19-16, are as follows:

Plate circuit	0.216 d-c and 0.387 fundamental
Grid circuit	0.142 d-c and 0.270 fundamental

Then

$$I_b = 0.216 \times 7.3 = 1.57 \text{ A}$$
$$I_p = (0.387 \times 7.3)0.707 = 2.00 \text{ A, rms}$$
$$I_c = 0.142 \times 2.1 = 0.296 \text{ A}$$
$$I_g = (0.270 \times 2.1)0.707 = 0.410 \text{ A, rms}$$

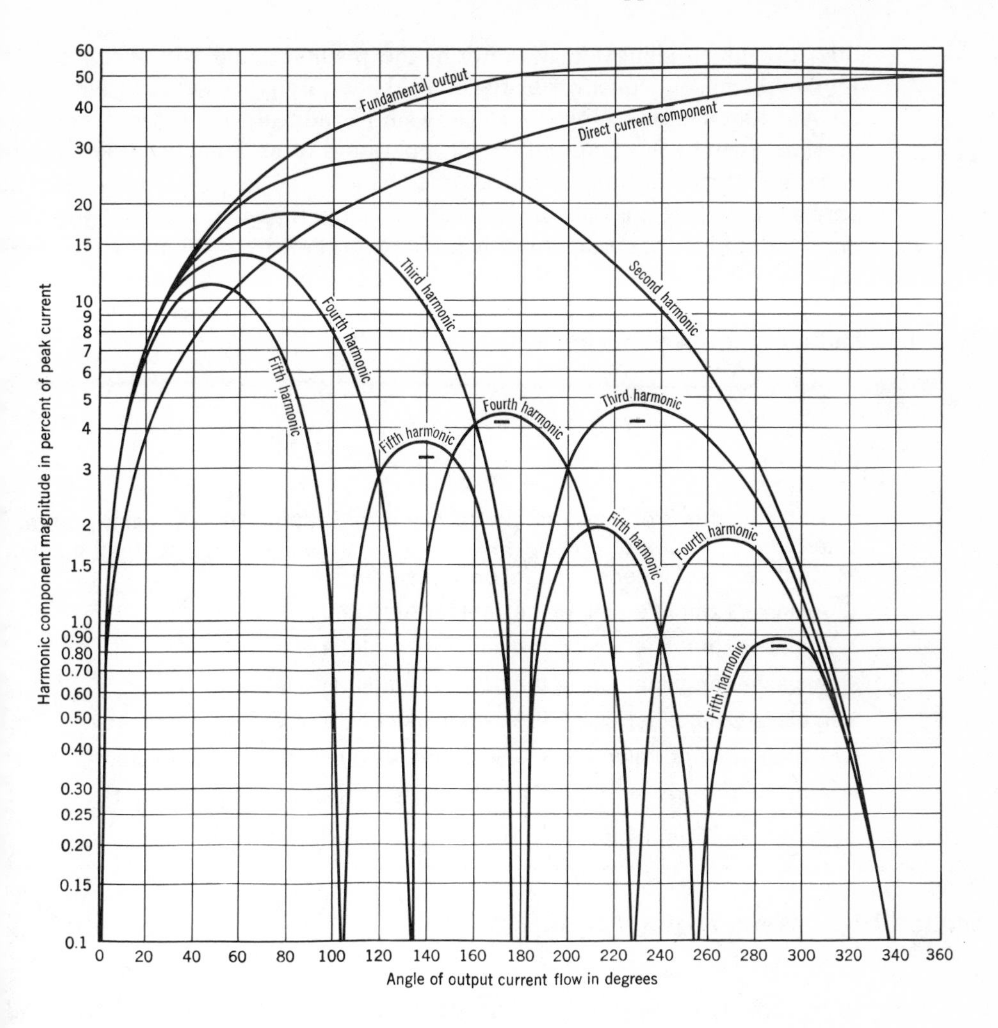

Figure 19-16 Harmonic content in terms of angle of current flow.

The graphical results were, for direct comparison,

$$I_b = 1.7 \text{ A} \qquad I_p = 2.16 \text{ A, rms}$$
$$I_c = 0.312 \text{ A} \qquad I_g = 0.410 \text{ A, rms}$$

These results are obtained quickly and, from them, all of the circuit calculations can be made for powers, efficiencies, and impedances. The values are within a few percent of the graphical approach and a great deal of time is saved. Reference to Fig. 19-10 and to Fig. 19-8 shows that the plate current waveform is flattened. If an allowance had been made to round off this top into a sinusoidal waveform, the results would have been closer However, it is only on a very large drive that we find this flattening effect.

An examination of Fig. 19-16 shows how radically a few degrees change in plate current flow can affect the operation of an amplifier operating to generate a fifth harmonic. On the other hand, a few degrees change in plate current flow does not radically affect a stage operating as a straight amplifier or as a doubler. Accordingly, it is not a major task to tune and to adjust the class-C amplifier at the fundamental frequency or as a doubler or even as a tripler. The adjustment of a stage operating at high orders of frequency multiplication becomes quite "touchy."

PROBLEMS

Refer to the Problem Data Table given in Section 19-4. In each problem determine I_b, I_c, output power, drive power requirement, grid dissipation, plate dissipation, load impedance, and power gain in decibels.

1. A class-C amplifier, using the data for Amplifier A.
2. A class-C amplifier, using the data for Amplifier B.
3. A class-C amplifier, using the data for Amplifier C.
4. A class-C amplifier, using the data for Amplifier D.
5. A frequency-doubler, class-C amplifier, using the data for Amplifier A.
6. A frequency-tripler, class-C amplifier, using the data for Amplifier B.

Section 19-9 Neutralizing Circuits

In Section 15-5, we developed the Miller effect in detail for a resistive load In Section 21-6 we shall show that, when the load on an amplifier becomes inductive, we have the necessary condition that initiates an oscillation in the circuit. In a class-C amplifier, the normal tuning procedure allows the plate load to swing from inductive to capacitive to resistive values freely. We cannot permit the reaction produced by the Miller effect of a changing load impedance to affect adversely the tuning procedure. The process of making the amplifier independent of the Miller effect is termed *neutralization*. The use of tetrodes and pentodes avoids the necessity of neutralization in many cases. However, neutralization is necessary when triodes are used as a conventional radio-frequency amplifier.

Energy is fed back from the plate to the grid within the tube through the grid-to-plate capacitance. In a neutralizing circuit, energy is fed from the plate back into the grid through a circuit parallel to the grid-to-plate capacitance. The energy which is fed back through this external circuit is 180° out of phase to produce a net value of zero by cancelation. The 180° phase difference is obtained by tapping either the grid coil, (Fig. 19-17*a*) or the plate coil (Fig. 19-17*b*). When the tap on the coil is exactly at the center, the value of the neutralizing capacitance C_N equals the value of the grid-to-plate capacitance. In a push-pull circuit (Fig. 19-17*c*), the neutralization is effectively a combination of the two basic methods. In the transistor circuit (Fig. 19-17*d*), the required 180° phase shift in voltage is obtained by using the voltage from the secondary of the transformer in the collector.

The first step in the procedure to neutralize an amplifier circuit is to remove the plate-supply voltage and the screen-supply voltage if the circuit uses a tetrode. The grid-driving voltage is left on, and the filament supply is left on. When the neutralization is correctly established by a proper setting of the neutralizing capacitor C_N, there should be no effect on the reading of the direct grid current meter when the plate tank tuning is varied. If this procedure is not sufficiently sensitive for a critical adjustment, a different approach is used. There should be no energy transferred into the plate-tank circuit. A neon lamp, a wave meter, a grid dip meter, or even a radio receiver may be used to detect energy in the plate tank. The grid tank is adjusted to resonance by setting its tuning for a maximum energy in the plate-tank circuit. The plate tank is adjusted to set this energy at a maximum. Now the neutralizing capacitor C_N is adjusted to reduce this energy to zero. It may be necessary to run through this cycle of the three steps of tuning several times to bring this energy exactly to zero.

When the sufficient degree of neutralization is obtained, the plate (and screen) direct supply voltages are turned back on. It is very important not to disturb any of the tuning adjustments, particularly the setting of C_N, while other class-C adjustments are made.

Usually there is no need to neutralize a circuit using a tetrode. Also, there is no need to neutralize a stage in which frequency multiplication takes place because the energies in the two tank circuits are at different frequencies and normally do not interfere with each other.

The pentode is used exclusively at high frequencies in electron-tube receiving circuits and in any application where the power requirement is less than a few watts. The shielding action of the screen grid and the suppressor grid usually are sufficient to do away with the necessity of neutralization. On the other hand, in solid-state electronics, we do not have a pentode concept as such. Silicon transistors have been developed that can be used in the MHz and GHz region, but very often these circuits must be neutralized. The Hazeltine circuit is readily adaptable to transistors. One of the advantages of the MOSFET is that it can be used in many circuits in place of a transistor that would require neutralization.

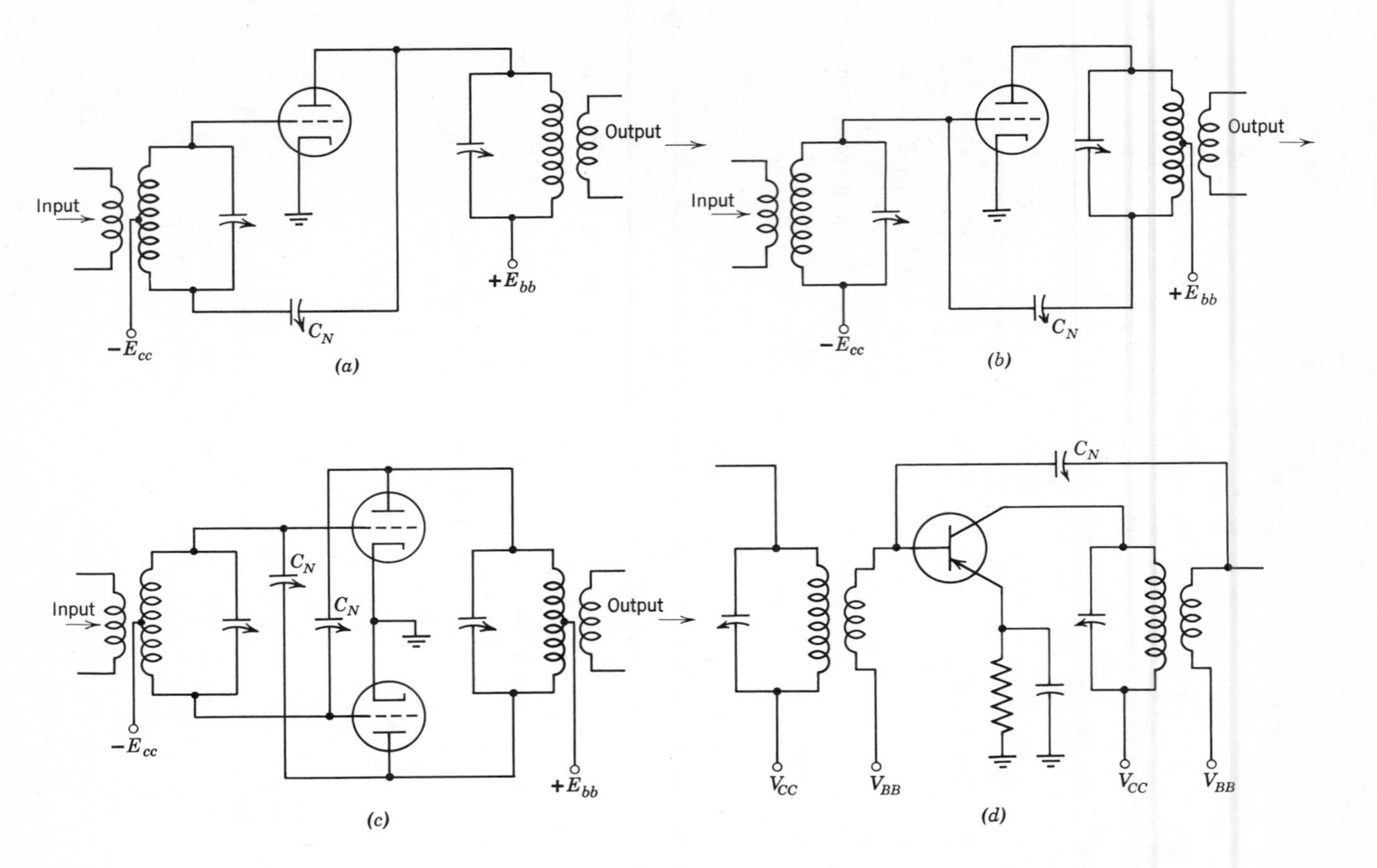

Figure 19-17 Simple neutralizing circuits. (*a*) Grid or Rice neutralization. (*b*) Plate or Hazeline neutralization. (*c*) Push-pull or cross neutralization. (*d*) Neutralized transistor circuit.

Questions

1. Define or explain each of the following terms: (*a*) automatic bias, (*b*) damping factor, (*c*) flywheel effect, (*d*) frequency multiplication, (*e*) grid dip meter, (*f*) driving power, (*g*) grid dissipation, (*h*) harmonic operation, (*i*) wave meter, (*j*) angle of plate-current flow.
2. Compare grid-leak bias with shunt rectification.
3. What different circuit arrangements can be used for grid-leak bias?
4. What is the principle of frequency multiplication?
5. How does the grid drive affect the angle of plate-current flow?
6. How is plate dissipation obtained in the plate analysis?
7. How is average current obtained in the plate analysis?
8. How is power output obtained in the plate analysis?
9. Why is the product of $E_{cc}I_c$ considered a driving power and not a power *from* a battery.
10. Carefully explain the adjustments of the grid and plate tanks for optimum operation.
11. Why is harmonic operation necessary?
12. Why is operation as a frequency multiplier less efficient than straight single-frequency operation?
13. What is the theoretical maximum class-C efficiency? What are typical values?
14. Name three methods of neutralization.
15. Why is neutralization required?
16. Why is neutralization not required with a tetrode power amplifier?
17. Why is neutralization not required in a frequency multiplier?
18. Describe the circuit adjustments necessary to complete the process of neutralization.
19. Describe the process which combines tuning *and* neutralizing.

Chapter Twenty

FEEDBACK

The basic principles of feedback lead to the development of a general formula (Section 20-1). Both positive feedback (Section 20-2) and negative feedback (Section 20-3) are considered along with the advantages and disadvantages of the feedback circuit. Operational amplifiers are adaptable to the basic feedback concepts from a "black-box" circuit (Section 20-4). Negative voltage-feedback circuits (Section 20-5), negative current-feedback circuits (Section 20-6), and shunt-feedback circuits (Section 20-7) are the fundamental basic feedback circuits used in complex equipment. The function of an emitter bypass capacitor is a fundamental feedback concept (Section 20-8). Vacuum-tube circuits are examined from the viewpoint of current feedback (Section 20-9) and as a cathode follower (Section 20-10). To prevent an undesired action between different circuits within the same piece of equipment, decoupling circuits (Section 20-11) are required. A special application of negative feedback is the electronic voltage regulator used in power supplies (Section 20-12).

Section 20-1 The Fundamental Feedback Equation

The most often encountered symbol for feedback is β. In a text dealing with semiconductors, β recurs constantly as the value of the current gain. Consequently, transistor circuits that have feedback require β being used twice—once for feedback and once for current gain. Our approach will be to use the subscript f on β to indicate feedback as β_f.

For an ordinary amplifier (Fig. 20-1) the voltage gain is the output voltage divided by the input signal voltage. The signal E_s is amplified by the factor of A_e to the value E_{out} of output voltage. The gain A_e is often called the *open-loop* gain. If a feedback loop is added to this amplifier (Fig. 20-2), a fractional part β_f of the output voltage is fed back into the input. The total input signal is the original signal plus the feedback voltage. The amplifier amplifies this total signal by the same factor A_e as in Fig. 20-1, producing the

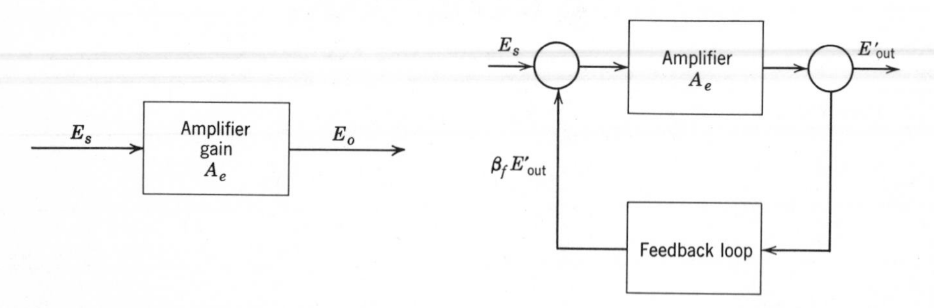

Figure 20-1 Block diagram of amplifier without feedback.

Figure 20-2 Block diagram of amplifier with feedback.

output voltage E'_{out}. We notice that the signal E_s is the same in each case, but that the output voltages E_{out} and E'_{out} are different. The term β_f is the *feedback*, and it is used as a decimal value in the equations, but, in a discussion, β_f is considered to be a percentage. For instance, 15% feedback is 0.15 when used in calculations.

The voltage fed from the output back into the input is $\beta_f E'_{out}$. The total input voltage to the amplifier is $E_s + \beta_f E'_{out}$. Since the input voltage times the gain is the output voltage, we may write

$$(E_s + \beta_f E'_{out})A_e = E'_{out}$$

Expanding gives

$$E_s A_e + \beta_f A_e E'_{out} = E'_{out}$$

Rearranging, we have

$$E_s A_e = E'_{out} - \beta_f A_e E'_{out} = E'_{out}(1 - \beta_f A_e)$$

Then

$$\frac{A_e}{1 - \beta_f A_e} = \frac{E'_{out}}{E_s}$$

But E'_{out}/E_s is the net gain of the circuit with feedback. Calling this gain A'_e, we have

$$A'_e = \frac{A_e}{1 - \beta_f A_e} \tag{20-1}$$

where A_e is the amplifier gain without feedback, β_f is the feedback, and A'_e is the amplifier gain with feedback. The term $\beta_f A_e$ is defined as the *feedback factor.* A'_e is also referred to as the *closed-loop gain.*

Let us take Eq. 20-1 and perform a long division by the method of elementary algebra. The resulting equation is an infinite series

$$\frac{E_{\text{out}}}{E_{\text{in}}} = A_e[1 + \beta_f A_e + (\beta_f A_e)^2 + (\beta_f A_e)^3 + (\beta_f A_e)^4 + \cdots]$$

Rearranging

$$E_{\text{out}} = A_e E_{\text{in}} + \beta_f(A_e E_{\text{in}})A_e + \beta_f[\beta_f(A_e E_{\text{in}})A_e]A_e$$

$$+ \beta_f\{\beta_f[\beta_f(A_e E_{\text{in}})A_e]A_e\}A_e + \cdots$$

The first term of this rearranged expression is the signal times the gain of the amplifier. The second term is the first term output fed through the feedback loop (multiplied by β_f) and then through the amplifier (multiplied again by A_e). The third term is the output represented by the second term, fed through the feedback loop (multiplied by β_f) and then amplified again by A_e. The fourth term is the third term output fed back through the feedback loop (multiplied by β_f) and back into the amplifier to be multiplied again by A_e. If we use the algebraic formula for the sum of an infinite series, we can arrive at Eq. 20-1 by this process of "logic."

Section 20-2 Positive Feedback

In the analysis of the block diagram, we used $(E_s + \beta_f E'_{\text{out}})$ as the total input voltage. Purposely, no reference was made to the algebraic sign of β_f. If β_f is taken as a positive number, the feedback voltage is in phase with and adds to the incoming signal. This circuit condition is termed *positive feedback.*

An understanding of positive feedback may be obtained from a simple numerical example. Let us assume that an amplifier has a gain of 10 without feedback and substitute various values (Table A) of positive feedback into the general equation:

$$A'_e = \frac{A_e}{1 - \beta_f A_e} = \frac{10}{1 - 10\beta_f}$$

TABLE A

β_f	$\beta_f A_e$	$1-\beta_f A_e$	A'_e
0	0	1	10
2%	0.20	0.80	12.5
4%	0.40	0.60	16.7
6%	0.60	0.40	25.0
8%	0.80	0.20	50.0
9%	0.90	0.10	100.0
9.9%	0.99	0.01	1000
9.99%	0.999	0.001	10,000
9.999%	0.9999	0.0001	100,000
10%	1.00	0	∞

The immediate conclusion that can be drawn from the results of this table is that positive feedback increases the gain of an amplifier. For this reason positive feedback is often called *regenerative* feedback. We will show in the next section of this chapter that positive feedback increases the distortion content of the output of an amplifier. Thus, the advantage of an increased gain must be carefully weighed against the disadvantage of an increased distortion level. As a result, we do not find positive feedback used to any great extent in amplifier design. In one application in which the feedback circuit is frequency-selective, a positive feedback is used as either a bass or a treble boost.

As the feedback factor $\beta_f A_e$ approaches unity, we notice from the table that the gain becomes infinite. Mathematically the equation shows that the gain is infinite, but electrically this does not happen. What does happen is that the circuit *oscillates*. Since the gain is infinite, the oscillator supplies its own signal for self-sustained operation. We now can state the very important and necessary conditions that must exist if a circuit is to oscillate:

1. The feedback must be positive, and
2. The feedback factor must be $+1$.

Alternatively, these conditions may be expressed in this form:

> In order to have an oscillator, the feedback must be positive and must be strong enough to sustain the oscillation.

H. Nyquist, in his famous paper, "Regeneration Theory" (*Bell System Technical Journal*, January 1932), originated and extended this theory at length. Both A_e and β_f are complex numbers having magnitude and phase angle. Then, the feedback factor $\beta_f A_e$ has both magnitude and phase angle. Nyquist's conditions for oscillation require that $\beta_f A_e$ contain the point

$(1+j0)$. Thus, it is possible for an amplifier circuit to be stable at one frequency and to oscillate at another. These principles form the basic theory for the operation of the feedback oscillators discussed in Chapter 21.

PROBLEMS

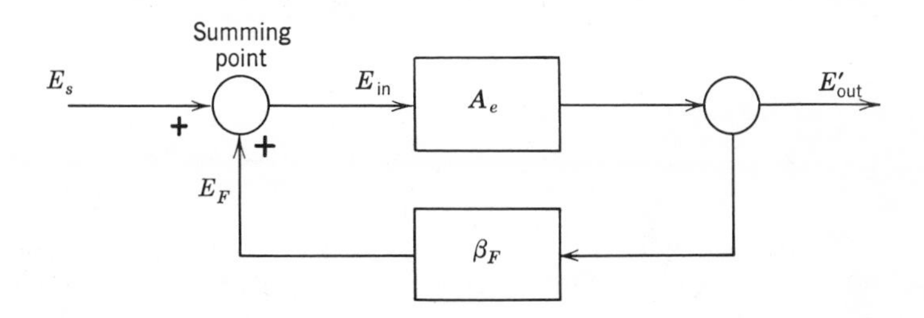

Block diagram for Problems 1 to 5.

Using positive feedback for the block diagram, complete the table:

Problem	E_s	A_e	β_f	$\beta_f A_e$	E_f	E_{in}	E'_{out}
1	20 mv	20	2%				
2	20 mv	50					3 V
3		15	4%				10 V
4	20 mv	100	2%				
5	1 V	100		0.20			

Section 20-3 Negative Feedback

With negative feedback, the voltage β_f/E'_{out} which is fed from the output back to the input is 180° out of phase with the input. The algebraic sign of β_f for negative feedback is minus when used in the feedback equations. To illustrate negative feedback, we consider the effect of negative feedback on the amplifier which was used to illustrate positive feedback in Section 20-2. The amplifier without feedback has a gain of 10, and, substituting the minus sign for β_f in the feedback equation, we have

$$A'_e = \frac{A_e}{1-\beta_f A_e} = \frac{10}{1+10\beta_f}$$

TABLE B

β_f	$\beta_f A_e$	$1-\beta_f A_e$	A_e'
0	0	1	10
−1%	−0.10	1.10	9.09
−2%	−0.20	1.20	8.32
−10%	−1.00	2.00	5.00
−30%	−3.00	4.00	2.50
−40%	−4.00	5.00	2.00
−70%	−7.00	8.00	1.25
−100%	−10.00	11.00	0.909

The results of Table B show that negative feedback reduces the overal gain of an amplifier. Since negative feedback reduces the gain, it is ofte called *degenerative* feedback.

Lest any misconceptions of the magnitude of the effect of a negativ feedback arise, let us determine the gain under conditions of a negativ feedback of 1% for an amplifier that has a gain of 400 without feedback:

$$A_e' = \frac{A_e}{1-\beta_f A_e}$$

$$= \frac{400}{1+0.01\times 400} = \frac{400}{1+4} = \frac{400}{5} = 80$$

A 1% negative feedback on this amplifier reduces the gain by a factor of five A 1% feedback on the amplifier with a gain of 10 used in the table reduce the gain from 10 to 9.09.

In Fig. 20-3, a signal is amplified by the factor A_e. At the same time, th amplifier creates a distortion D in the output. With a feedback loop (Fig 20-4) not only is the output fed back into the input, but also the fractiona part of the distortion which is $\beta_f D'$ appears in the input. The total distortio in the output D' must comprise not only the amplified value of $\beta_f D'$ but als the original distortion of E_s which is produced by the amplifier. The inpu signal is so arranged that E_{out} equals E'_{out}. This may be expressed as

$$D' = D + (\beta_f D')A_e$$

$$D' - \beta_f A_e D' = D$$

$$(1-\beta_f A_e)D' = D$$

$$D' = \frac{D}{1-\beta_f A_e} \qquad (20\text{-}2a)$$

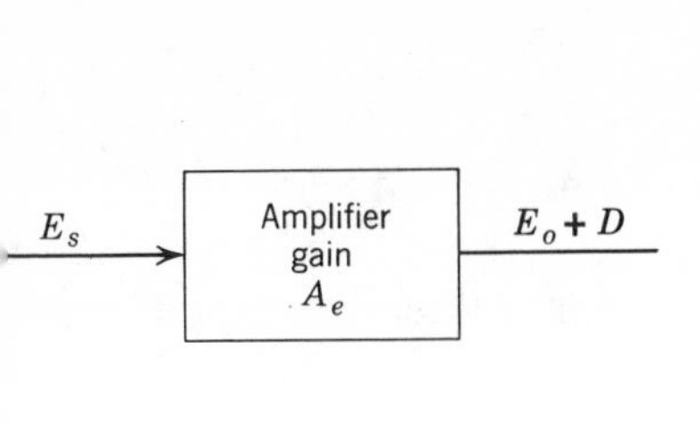

igure 20-3 Block diagram of amplifier with istortion.

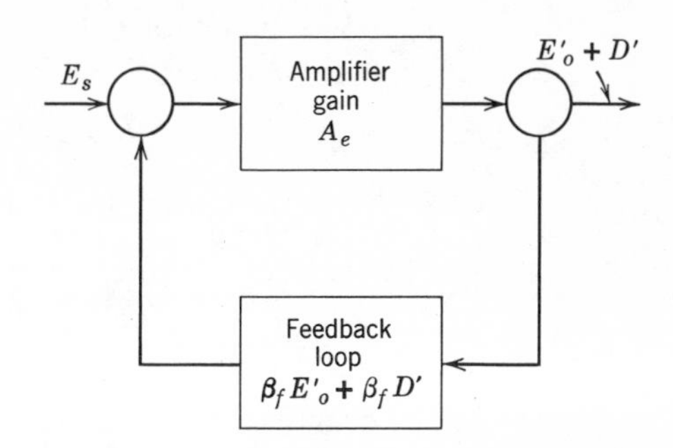

Figure 20-4 Block diagram of amplifier with feedback and distortion.

When the feedback is positive, the distortion with feedback becomes greater than the distortion without feedback. When the feedback is negative, D' is less than D. In other words, regenerative feedback increases distortion whereas degenerative feedback reduces distortion in the same proportion that it reduces gain.

As an example, taking the amplifier having a gain of 10 without feedback, assume that the inherent distortion of the amplifier is 20%. With a 10% feedback, the gain is 5, and the distortion is 10%. With a 40% negative feedback, the gain is 2, and the distortion is 4%. If two amplifiers are connected in cascade, one with the 10% feedback and the other with a 40% feedback, the overall gain is the product of the two gains which is 5 × 2 or 10. The overall distortion is approximately

$$1.10 \times 1.04 - 1 = 0.144 \text{ or } 14.4\%$$

If we take two amplifiers, each with a gain of 10 without feedback and with a distortion of 20%, when these amplifiers are in cascade, we have similarly:

$$\text{Overall gain } A_e = 10 \times 10 = 100$$
$$\text{Overall distortion } D = 1.20 \times 1.20 - 1 = 0.44 \text{ or } 44\%$$

When 9% negative feedback is used over both stages, the gain reduction and distortion reduction factors $(1 - \beta_f A_e)$ are both 10. Now the overall gain is 10, and the overall distortion is 4.4%. By this example, we can show that feedback over a number of stages, instead of feedback on each individual stage alone, produces the same gain but produces a much lower distortion. It is standard practice to use feedback in a single loop over several stages rather than feedback on each stage.

In the feedback formula

$$A_e' = \frac{A_e}{1-\beta_f A_e}$$

If we divide the numerator and denominator by A_e, we find that

$$A_e' = \frac{A_e/A_e}{1/A_e - \beta_f A_e/A_e} = \frac{1}{1/A_e - \beta_f}$$

When the value of the feedback β_f is large compared to $1/A_e$ (i.e., when a heavy negative feedback is used on a high-gain amplifier), the term $1/A_e$ may be neglected, and

$$A_e' = -\frac{1}{\beta_f} \tag{20-2b}$$

We consider only negative feedback in this expression as we have shown that positive feedback would cause this circuit to oscillate. When a 10% negative-feedback loop is applied to an amplifier with a gain of 4000, the gain with feedback is 1/0.1 or 10. This value of gain is independent of transistor, FET, or tube variations, component changes (except the feedback loop), and power-supply variations. The overall gain with feedback is determined by the feedback network alone provided that the feedback does not depend on the parameters of the transistor, FET, or tube. This circuit is very important in its application as a *decade amplifier* used in servo amplifiers, in computers, and as instrument multipliers.

Consider an amplifier that is direct-coupled and has a 3-db break-point at f_2 (Fig. 20-5). The response of the circuit at high frequencies is given by

$$A_{e,\text{HF}} = \frac{A_{e,\text{MF}}}{1+jf/f_2} \tag{15-15a}$$

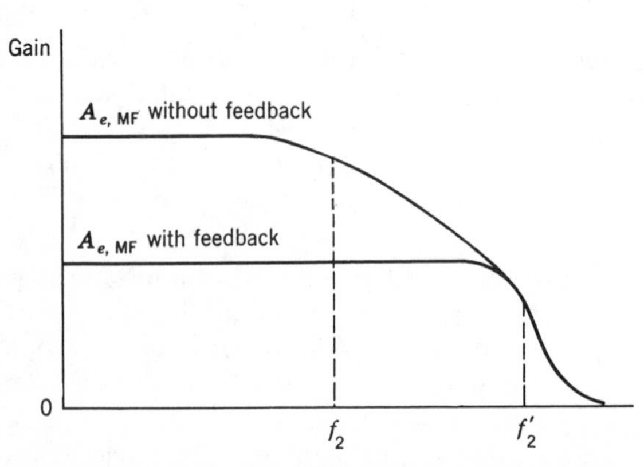

Figure 20-5 Effect of feedback on frequency response.

Substitute this expression directly into Eq. 20-1:

$$A'_e = \frac{A_{e,\text{MF}}/(1+jf/f_2)}{1-\beta_f A_{e,\text{MF}}/(1+jf/f_2)} = \frac{A_{e,\text{MF}}}{1+jf/f_2-\beta_f A_{e,\text{MF}}}$$

Assume that the feedback is negative. Then

$$A'_e = \frac{A_{e,\text{MF}}}{[1+\beta_f A_{e,\text{MF}}]+jf/f_2}$$

Factoring out $[1+\beta_f A_{e,\text{MF}}]$

$$A'_e = \left[\frac{A_{e,\text{MF}}}{1+\beta_f A_{e,\text{MF}}}\right]\left[\frac{1}{1+jf/f_2(1+\beta_f A_{e,\text{MF}})}\right] \qquad (20\text{-}3a)$$

The term contained in the first set of brackets represents the reduced gain of the amplifier at midband that results from the feedback β_f applied to the circuit. This reduced gain is shown in Fig. 20-5. The term within the second set of brackets represents the high-frequency roll-off. Consequently, this term has a new 3-db frequency f'_2 which is determined when this term has the value

$$\frac{1}{1+j1}$$

This means that

$$f'_2 = (1+\beta_f A_{e,\text{MF}})f_2 \qquad (20\text{-}3b)$$

For example, assume that an amplifier has a gain of 40 and a high-frequency breakpoint at 8000 Hz. A 5% negative feedback is used. The new gain is 40/(1+2) or 13.3, but the breakpoint is now at 8000(1+2) or 24,000 Hz. The bandwidth is increased by a factor of three.

A major advantage of negative feedback is that the frequency response is improved. Consequently, negative feedback is a requirement in a high fidelity amplifier. The stereo circuit shown in Fig. 18-19b has a feedback from the load through a 4700-Ω resistor back to the second amplifier. However, when high values of feedback are used, the effect of the phase shifts, discussed in Section 15-3 and 15-4, at the extreme low- and high-frequencies may tend to make the feedback regenerative (Fig. 20-6). There may be a tendency for the circuit to oscillate at a very high ultrasonic frequency. This oscillation absorbs power, causing the circuit to deliver a reduced output power with high distortion when it is used as an audio amplifier. To cure this condition, a low-pass L filter (a resistor and a mica bypass capacitor) is placed in the feedback loop. Then, the feedback at this very high frequency is reduced sufficiently to prevent the circuit from oscillating. A possible oscillation at the low-frequency end is discussed in Section 20-11.

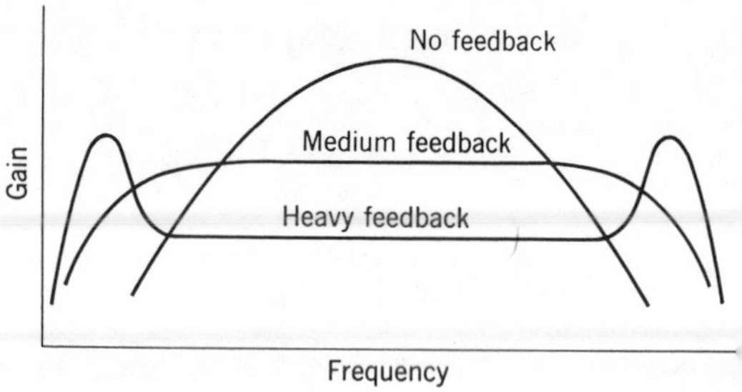

Figure 20-6 Response curves of an amplifier with different values of feedback.

A very important concept in electronics can be developed at this poin[t] from an examination of Eqs. 20-3a and 20-3b. The midband gain of th[e] amplifier with feedback is

$$A'_{e,\mathrm{MF}} = \frac{A_{e,\mathrm{MF}}}{1+\beta_f A_{e,\mathrm{MF}}}$$

Now multiply this expression by Eq. 20-3b. Then

$$A'_e f'_2 = \frac{A_{e,\mathrm{MF}}}{1+\beta_f A_{e,\mathrm{MF}}}(1+\beta_f A_{e,\mathrm{MF}})f_2$$

or

$$A'_e f'_2 = A_e f_2 \qquad (20\text{-}3c)$$

The gain-bandwidth product of an amplifier is a constant.

This statement applies not only to negative feedback amplifiers but als[o] to any electronic circuit. If the gain of an amplifier is reduced by lowerin[g] R_c or R_L or by lowering the Q of the load, the bandwidth of the circuit i[s] widened.

In an audio-frequency amplifier the usual maximum response is that th[e] amplifier is flat to 20,000 Hz. In a video amplifier for use in a televisio[n] receiver, the amplifier must be flat to 4.5 MHz. As a result, the audi[o] amplifier can use a high-stage gain whereas the voltage gain expected in th[e] video amplifier is very low, usually less than 10. The gain-bandwidt[h] product is very important in the design of a high-frequency amplifie[r] (Chapter 23).

If the fundamental feedback gain equation (Eq. 20-1) is differentiate[d] with respect to the open-loop gain A_e, the result simplifies to

$$dA'_e = \frac{1}{(1-\beta_f A_e)^2}dA_e \qquad (20\text{-}3d)$$

If Eq. 20-3*d* is divided by Eq. 20-1, we have

$$\frac{dA'_e}{A'_e} = \frac{1}{(1-\beta_f A_e)} \frac{dA_e}{A_e} \qquad (20\text{-}3e)$$

Assume an amplifier has an open-loop gain of 100 and a negative feedback of 5%. $(1-\beta_f A_e)$ is 6. If the open-loop gain changes by as much as 72%, the closed-loop gain will only change by 72/6 or 12%. Consequently, the use of negative feedback greatly improves the stability of an amplifier. In order to secure this advantage, it should be noted that the network that controls the percent feedback must be stable.

PROBLEMS

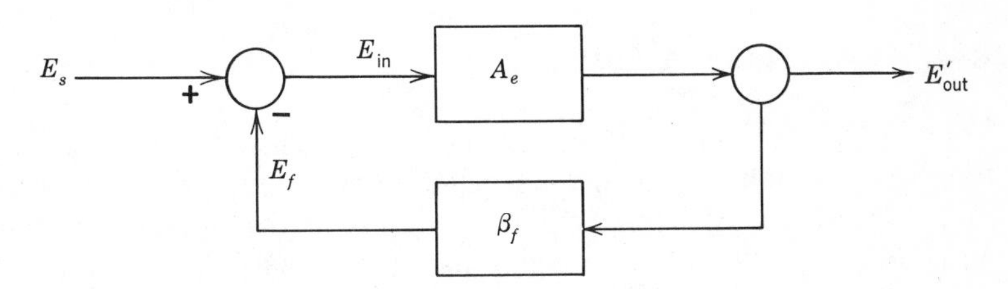

Block diagram for Problems 1 to 7.

Using negative feedback for the block diagram, complete the following table:

Problem	E_s	A_e	β_f	$\beta_f A_e$	E_f	E_{in}	E'_{out}
1	200 mv	20					1 V
2	200 mv	1000					2 V
3		50	3%				5 V
4	1 V	50	8%				
5	0.5 V		20%				2 V
6	5 V	20	100%				
7	1 V	100		−3.0			

8. An audio amplifier consists of three stages:

Stage 1	$A_e = 50$	4% distortion
Stage 2	$A_e = 10$	4% distortion
Stage 3	$A_e = 20$	10% distortion

Each stage has individual feedback that reduces the gain to 10, 2, and 4 respectively. What is the overall gain and distortion with feedback? What percentage feedback is used on each stage?

9. Overall negative feedback is used to reduce the gain of the audio amplifier of Problem 8 to 80. What percentage feedback is required, and what is the overall distortion?
10. An amplifier without feedback has a mid-band gain of 200, and the 3-db high-frequency break point is 50 kHz. What is the mid-band gain and what is the high-frequency break point when 10% negative feedback is used?
11. An amplifier has a mid-band gain without feedback of 200. The 3-db frequency is 200 kHz. This amplifier is to be used as a video amplifier that requires a 5-MHz bandwidth. What gain can be obtained, and what feedback must be used? What bandwidth could be obtained if the feedback were 100%?

Section 20-4 Operational Amplifiers

An operational amplifier is a packaged circuit, usually an integrated circuit that has a very high gain and is treated in circuit analysis as if it were a black box. There are two fundamental versions of the operational amplifier as shown in Fig. 20-7. The conventional operational amplifier (Fig. 20-7*a*) has two inputs and one output. If terminal 2 is grounded and the signal is fed into terminal 1, a phase reversal occurs. Therefore, terminal 1 is the *inverting input.* If terminal 1 is grounded and the signal connected to terminal 2, no phase inversion occurs and terminal 2 is called the *noninverting input.*

A very important feature of the operational amplifier is that neither terminal 1 nor terminal 2 is grounded. Consequently, the input is floating and a ground reference point is determined by the external circuitry. Obviously, the operational amplifier can accept a differential input signal. The use of the operational amplifier of Fig. 20-7*b* provides output signals suitable for difference applications.

The specifications for the typical commercially available operational amplifiers are given in Table C. The open-loop gains of an operational amplifier can go as high as 200 db for special applications. The high gain at open loop does not have a large bandwidth, as shown in Fig. 20-8. The use of negative feedback secures the wide bandwidth that may be required for a circuit application.

The circuit shown in Fig. 20-9 is called a *voltage follower.* The voltage relation is simply

$$E_1 + E_{in} = E_{out}$$

but

$$E_{out} = -AE_{in}$$

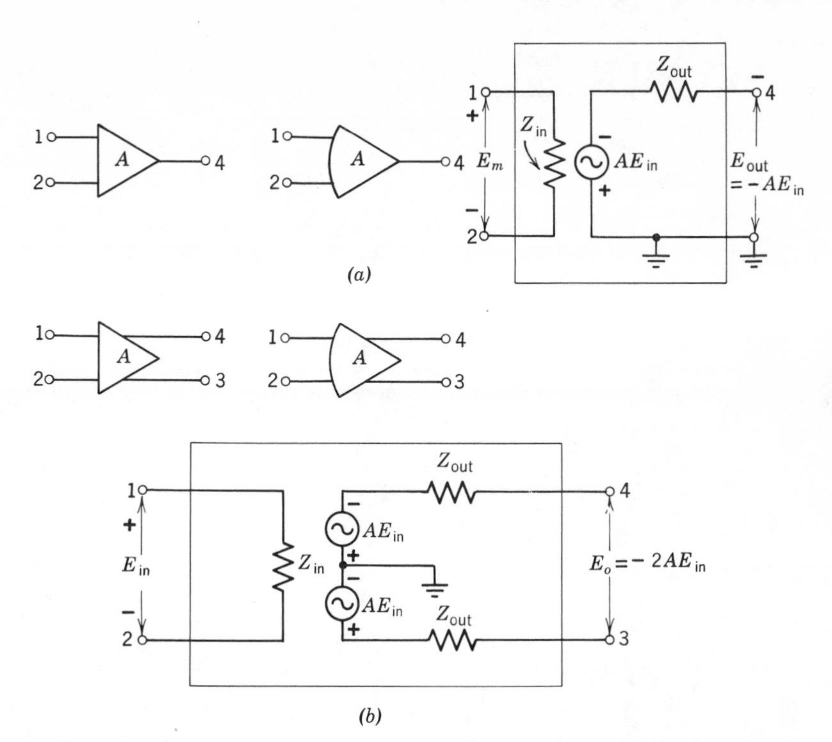

Figure 20-7 The operational amplifier. (*a*) Symbol and model for the operational amplifier. (*b*) Symbol and model for the differential output operational amplifier.

TABLE C Typical Operational Amplifier Specifications

	Micro-electronics	Micro-electronics	Instrumentation Grade General Purpose
Size W″ × L″ × H″	$\frac{1}{8} \times \frac{1}{2} \times 0.035$	$\frac{1}{4} \times \frac{3}{4} \times 0.170$	$1\frac{1}{4} \times 4 \times 2$
Open loop gain-db	62	90	72
Output voltage max, peak	±2.5	±11	±11
Load resistance, ohms	200	500	5000
Input resistance	12K	1.0M	10^6 MΩ, min.
Common mode input volts, max	±4	±5	±200
Common mode rejection ratio, decibels	60	100	limited by precision of external circuitry
Input current, amperes	2×10^{-6}	5×10^{-9}	5×10^{-11}
3-db frequency break point	50 kHz	60 db gain at 230 kHz	—
Unity gain, f_H	10 MHz	—	75 kHz
Frequency roll-off db per decade	—	—	12

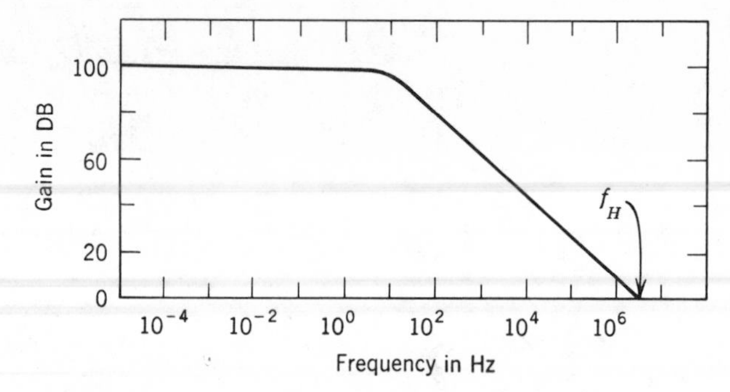

Figure 20-8 Frequency response of operational amplifier, open-loop gain.

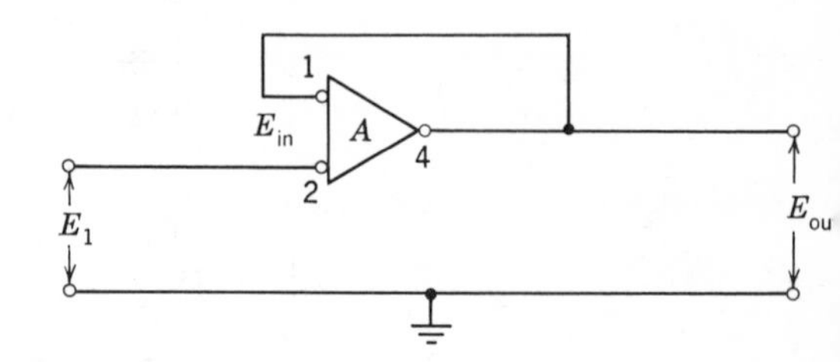

Figure 20-9 The voltage follower.

Substituting

$$E_1 - \frac{E_{out}}{A} = E_{out}$$

or

$$E_1 = \left[1 + \frac{1}{A}\right] E_{out}$$

If the gain of an operational amplifier is at the very least 40 db, A is 100 and $1/A$ is negligible and

$$E_{out} = E_1 \tag{20-4}$$

The typical specifications shown in Table C give an input resistance of a megohm for one of the microamplifiers. Consequently, the circuit can serve as an isolating amplifier. The highest input resistance of the units in the table is 10^6 MΩ. A voltage follower using this unit can be used in electrometer applications where the measuring instrument has no loading effect at all. The output load can be as low as that specified in the table.

Equation 20-4 can be derived in a simplified manner by considering the operational amplifier as ideal and by having a voltage gain of infinity. Thus for any finite output voltage, the input to the amplifier E_{in} is zero. Then Eq. 20-4 can be written directly by inspection of the circuit.

Since the input resistance into the operational amplifier is finite, and since we are taking E_{in} as zero, the input current into the operational amplifier is zero. Therefore, any current flow in R_1 and R_2 of Fig. 20-10*a* is a series circuit with the current in R_2 equal to the current in R_1. In Fig. 20-10*a*,

$$E_{out} = I(R_1 + R_2)$$

and, since terminal 2 is the noninverting input,

$$E_1 = IR_1$$

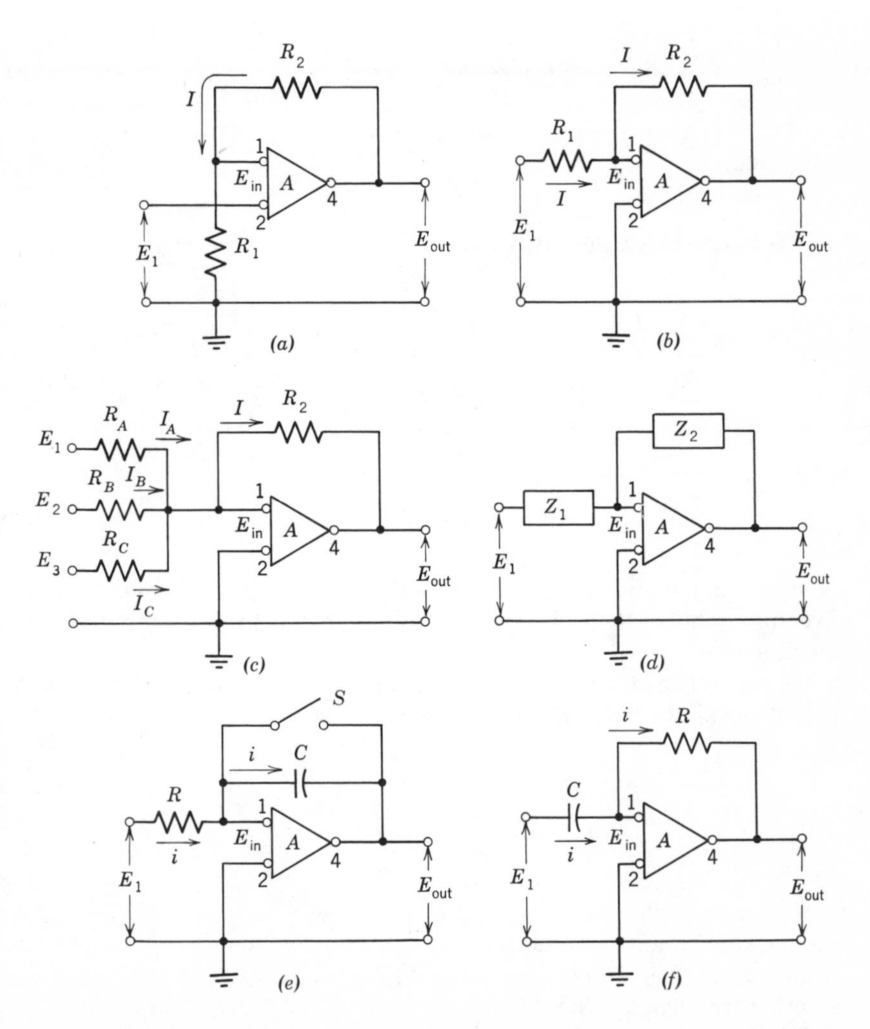

Figure 20-10 Basic operational-amplifier circuits. (*a*) Noninverting amplifier. (*b*) Inverting amplifier. (*c*) The voltage summer. (*d*) General case. (*e*) Integrator. (*f*) Differentiator.

Dividing, we have the gain of this *noninverting amplifier* as

$$A_v = \frac{E_{out}}{E_1} = \frac{R_1 + R_2}{R_1} \quad \text{(20-5}$$

The gain of this circuit is a function of R_1 and R_2 and can be controlle exactly by using precision resistors.

The *inverting amplifier* is shown in Fig. 20-10*b*. Since E_{in} is zero,

$$E_1 = IR_1 \qquad \text{and} \qquad E_{out} = -IR_2$$

Dividing

$$A_v = \frac{E_{out}}{E_1} = -\frac{R_2}{R_1} \quad \text{(20-6}$$

The voltage *summer* or *adder* is shown in Fig. 20-10*c*. Since E_{in} is zer and since

$$I_A + I_B + I_C = I$$

a substitution by Ohm's law yields

$$\frac{E_1}{R_A} + \frac{E_2}{R_B} + \frac{E_3}{R_C} = -\frac{E_{out}}{R_2}$$

$$E_{out} = -R_0\left[\frac{E_1}{R_1} + \frac{E_2}{R_2} + \frac{E_3}{R_3} + \cdots\right] \quad \text{(20-7}a$$

If $R_1 = R_2 = R_3 = \cdots = R_0$

$$E_{out} = -[E_1 + E_2 + E_3 + \cdots] \quad \text{(20-7}b$$

A simple application is the concept of *audio mixing* in which the outputs o five microphones used in a band are combined to one output for powe amplification.

A general case can be made (Fig. 20-10*d*) in which Z_1 and Z_2 are genera impedances. The result is the same as Eq. 20-6 or

$$A_v = \frac{E_{out}}{E_1} = -\frac{Z_2}{Z_1} \quad \text{(20-8}a$$

and

$$E_{out} = -\frac{Z_2}{Z_1}E_{in} \quad \text{(20-8}b$$

As an application, assume that Z_1 is a resistor and Z_2 is a tuned circui resonant at f_0. The impedance ratio, Z_{out}/Z_1, is a maximum at f_0 and lower a all other frequencies. Since the output voltage is directly proportional to th ratio of Z_{out}/Z_1, the output peaks at f_0. A circuit of this nature is called a *active filter* because it uses an amplifying circuit.

The circuit shown in Fig. 20-10*e* is an *integrator*. The voltage across the capacitor is given by

$$e_{out} = -\frac{1}{C}\int i\,dt$$

the current in the circuit is given by

$$e_1 = Ri$$

or

$$i = e_1/R$$

Substituting

$$e_{out} = -\frac{1}{C}\int \frac{e_1}{R}\,dt = -\frac{1}{RC}\int e_1\,dt = -\frac{1}{RC}\int e_1(t)\,dt \qquad (20\text{-}9)$$

Assume that the value of RC is 100 sec and that the capacitor is completely discharged by momentarily closing the switch S. A d-c signal input of 100 mv is applied to the input. The value of $\int e_1 dt$ for the 100 mv signal for one hour is

$$(100\times 10^{-3})(1\times 60\times 60) = 360\text{ V-sec}$$

Substituting this into Eq. 20-9

$$e_{out} = -\frac{1}{100}[360] = -3.6\text{ V}$$

The voltage E_{out} has increased linearly from zero and at the end of one hour is -3.6 V. Now assume that the input is shorted to ground. The output remains steady at -3.6 V either until the input signal is restored or until the capacitor is shorted by the switch. This application concept is a *memory circuit*. If the circuit is to be as stable as this example requires, the specification for offset and drift in the operational amplifier must be very rigid. Any offset voltage in the operational amplifier could destroy the memory as could any leakage in the integrating capacitor.

The derivation of the relation in the *differentiator*, Fig. 20-10*f* is quite similar:

$$i = C\frac{de_1}{dt} \qquad \text{and} \qquad iR = -e_{out}$$

substituting

$$e_{out} = -RC\frac{de_1}{dt} = -RC\frac{d}{dt}e_1(t) \qquad (20\text{-}10)$$

A *differential amplifier* circuit is given in Fig. 20-11. Resistors R_1 must be very carefully selected and matched. Resistors R_2 must be very carefully selected and matched. E_C is the voltage from either terminal 1 or terminal 2 to ground and is the *common-mode voltage*. Since we are considering the gain A to be infinite, the voltage E_{in} between terminals 1 and 2 is zero. The

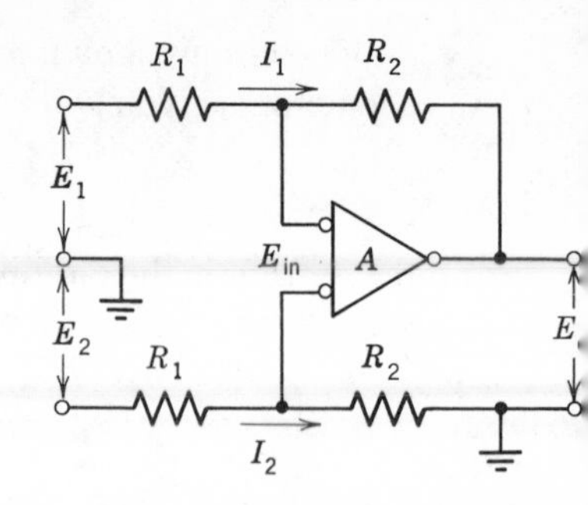

Figure 20-11 Differential amplifier.

current in the top leg of the circuit is

$$I_1 = \frac{E_1 - E_C}{R_1} = \frac{E_C - E_{out}}{R_2}$$

expanding

$$\frac{E_1}{R_1} - \frac{E_C}{R_1} = \frac{E_C}{R_2} - \frac{E_{out}}{R_2}$$

and solving for E_{out}

$$E_{out} = \left[1 + \frac{R_2}{R_1}\right] E_C - \frac{R_2}{R_1} E_1$$

From the bottom leg

$$E_C = \frac{R_2}{R_1 + R_2} E_2$$

and substituting

$$E_{out} = \frac{R_2}{R_1} E_2 - \frac{R_2}{R_1} E_1 = \frac{R_2}{R_1}(E_2 - E_1) \qquad \text{(20-1}$$

PROBLEMS

1. In Fig. 20-8, the break frequency for an operational amplifier occurs at 5 Hz, and f_H occurs at 5 MHz. What feedback is required to reduce th gain to 60 db? Show that the break frequency at a gain of 60 db occurs a 5 kHz.
2. An operational amplifier has an open-loop d-c gain of 80 db, and the roll-o is 20 db per decade. Unity gain occurs at 2 MHz (f_H). In closed-loop opera tion, the feedback is 4%. What is the gain and what is the break frequenc with feedback. Sketch the open-loop and the closed-loop frequency re sponse.
3. In Fig. 20-10*a*, a gain of 15 is required with R_2 established at 120 kΩ. What the value of R_1?

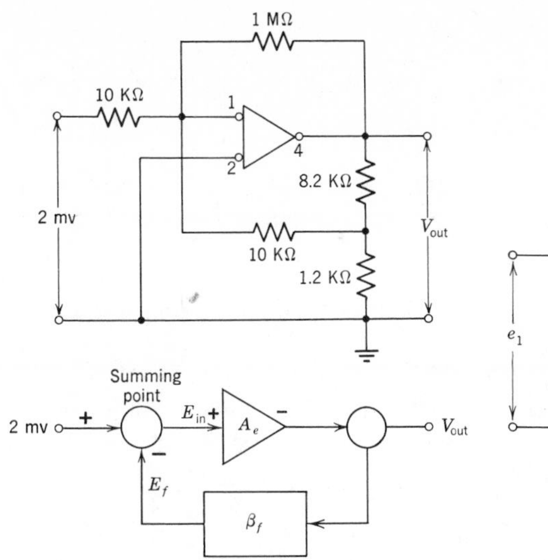

Circuit and Block Diagram for Problem 4.

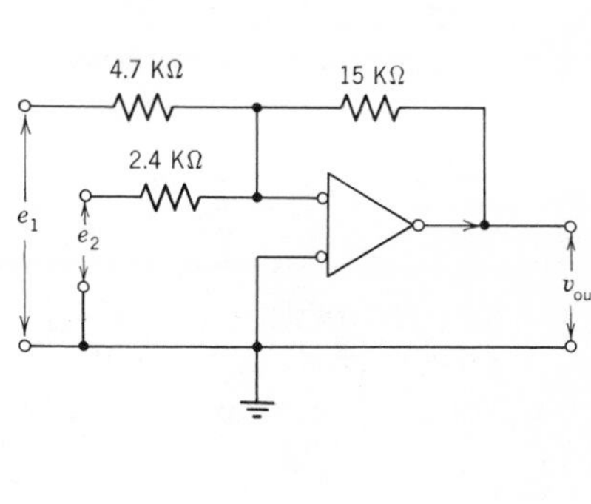

Circuit for Problem 6.

4. A circuit using an operational amplifier has a feedback loop. Determine the numerical values for the block diagram.

5. Using Fig. 20-10*c* as the basic circuit, show the connections and the values for a circuit using operational amplifiers to give

$$e_{\text{out}} = 3e_1 + 6e_2 - 5e_3$$

R_2 is 300,000 Ω for any operational amplifier to be used.

6. Determine v_{out} if $e_1 = 1.8 \cos \omega t$ V, if $e_2 = 2.4 \sin \omega t$ V, and if $v_{\text{out}} = e_1 + e_2$. Draw a connection diagram to show how this addition is accomplished by using operational amplifiers.

7. A 1000-Hz square wave signal has a peak-to-peak amplitude of 2 V. The circuit of Fig. 20-10*e* is used with 100 kΩ for *R*. The time constant *RC* must be 10 times the period of the signal. What value of *C* is used and what is the peak-to-peak amplitude of the output?

8. Show how Problem 5 can be solved by using the differential amplifier (Fig. 20-11) as the basic circuit. R_2 is 300,000 Ω.

Section 20-5 Negative Voltage Feedback Circuits

The circuit shown in Fig. 20-12*a* shows the current, voltage, and impedance relationships in the circuit under open-loop conditions without feedback.

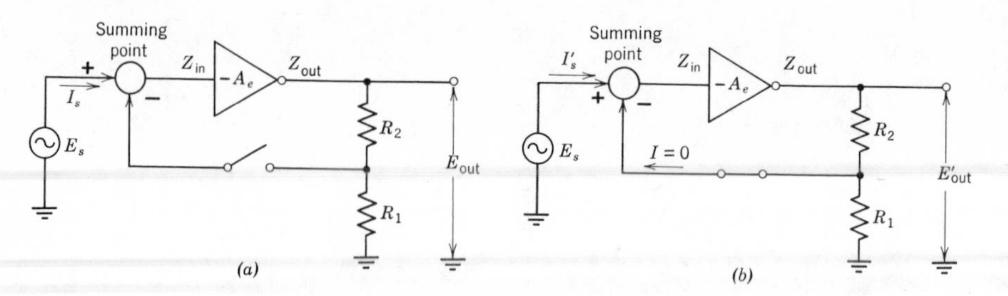

Figure 20-12 Negative voltage feedback. (*a*) Without feedback. (*b*) With feedback.

The feedback is derived from the resistors R_1 and R_2 which are placed in series across the output. The negative feedback is given directly from the voltage divider as

$$\beta_f = \frac{R_1}{R_1 + R_2} \tag{20-12}$$

When the negative feedback loop is closed, the circuit relations are shown in Fig. 20-12*b*. The total voltage to the input of the amplifier is E'_{out}/A_e. Then

$$E_s + \beta_f E'_{out} = E'_{out}/A_e$$

or

$$E_s = E'_{out}/A_e - \beta_f E'_{out}$$

The impedance E_s sees is

$$Z'_{in} = \frac{E_s}{I'_s} = \frac{E'_{out}/A_e - \beta_f E'_{out}}{I'_s} = \frac{E'_{out}}{A_e I'_s} - \beta_f A_e \frac{E'_{out}}{A_e I'_s}$$

$$= (1 - \beta_f A_e)\frac{E'_{out}}{A_e I'_s}$$

But $(E'_{out}/A_e)/I'_s$ is the input impedance to the amplifier, Z_{in}. Therefore

$$Z'_{in} = (1 - \beta_f A_e) Z_{in} \tag{20-13}$$

The significance of Eq. 20-13 is that a negative voltage feedback in an amplifier increases the input impedance to the circuit by the amount $(1 - \beta_f A_e)$.

Figure 20-13*a* is the equivalent circuit of the amplifier output by using Thévenin's theorem. The impedance Z_{out} is the conventional output impedance of the amplifier itself without feedback. When the signal E_s is turned off and a voltage E_{out} is applied to the circuit to measure the back impedance (Fig. 20-13*b*), the internal voltage of the equivalent circuit is not zero but it is $-\beta_f A_e E'_{out}$. This situation results because there is an input voltage pro-

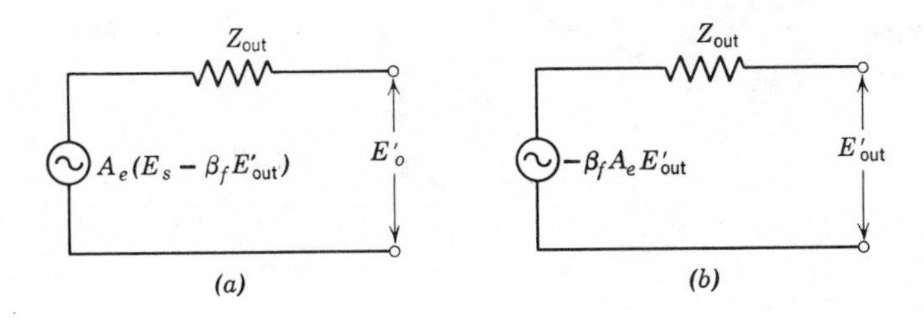

igure 20-13 Equivalent output circuit by Thevenin's theorem. (a) With input signal. (b) Without nput signal.

duced by E'_{out} through the feedback loop. By the voltage equation, the current I'_L in Z_{out} is

$$I'_L = \frac{E'_{out} - \beta_f A_e E'_{out}}{Z_{out}}$$

$$I'_L = \frac{E'_{out}(1 - \beta_f A_e)}{Z_{out}}$$

Solving for E'_{out}/I'_L which is Z'_{out}, the effective output impedance with feedback, we have

$$\frac{E'_{out}}{I'_L} = Z'_{out} = \frac{Z_{out}}{1 - \beta_f A_e} \tag{20-14}$$

The significance of Eq. 20-14 is that the (internal) output impedance of an amplifier with negative voltage feedback is reduced directly as the gain and the distortion are reduced. One advantage of a reduction in output impedance is that the output voltage has better regulation and is more nearly constant when the external load-impedance changes. In transistor circuits, the combined effect of an increased input impedance and a decreased output impedance simplifies the matching problem with RC coupling.

The vacuum-tube amplifier circuits shown in Fig. 20-14 are straight forward applications of the voltage feedback principle. In each case, the feedback is formed by a voltage-divider action between R_1 and R_2. The feedback is

$$\beta_f = \frac{R_1}{R_1 + R_2}$$

When the value of the gain without feedback is determined, notice that the a-c load on the plate is the parallel combination of R_c and $(R_1 + R_2)$.

The operational amplifier circuits of Section 20-4 are voltage-feedback circuits. The gain of the operational amplifier is sufficiently high that the circuit gain with feedback is the reciprocal of β_f.

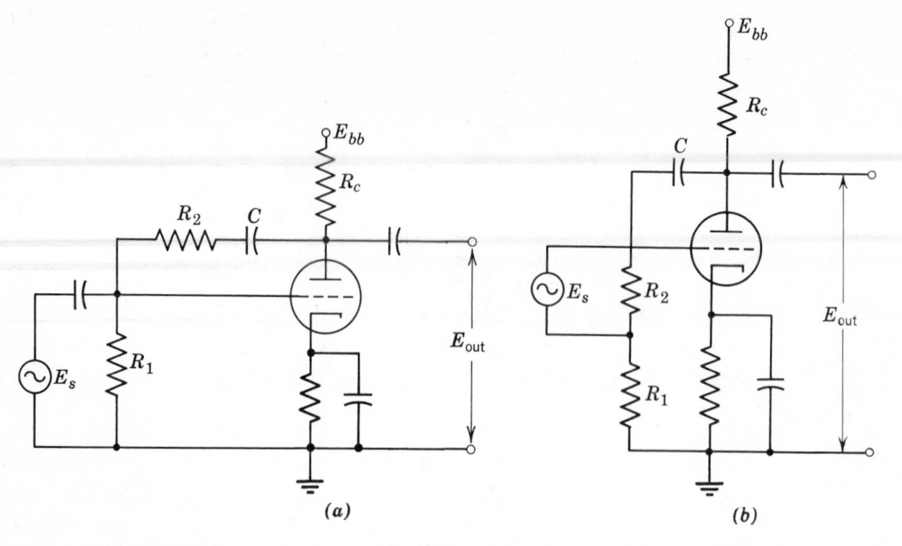

Figure 20-14 Vacuum-tube amplifier circuits using voltage feedback.

PROBLEMS

1. The base bias resistors are adjusted to make V_{CE} 5 V on each transistor. Use 25 mv/I_E for r'_e. The β for the transistors is 100. E_s is 2 mv. Determine V_{out} for S

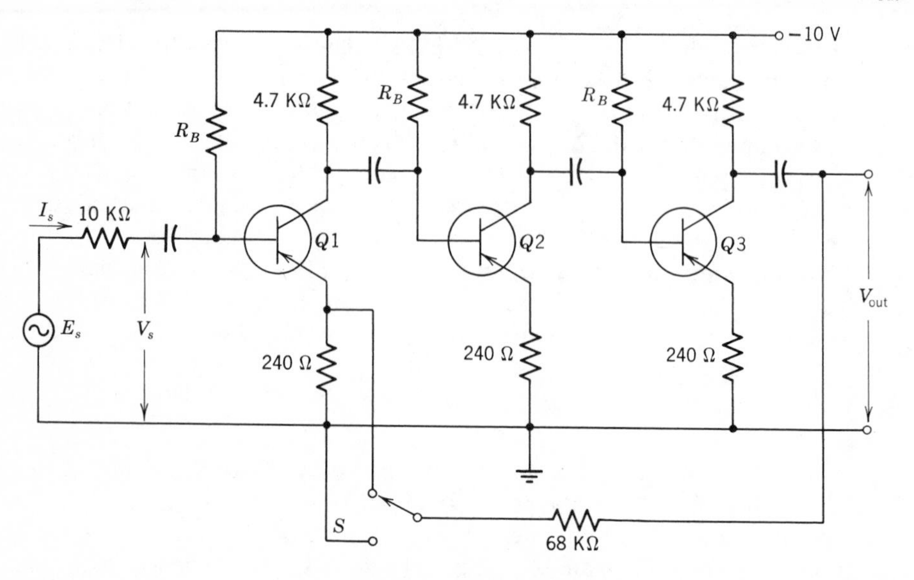

Circuit for Problem 1.

open and for S closed. Neglect the loading effect of the base bias resistors in the calculation. The transistors are germanium.

2. In Fig. 20-12, R_1 is 1000 Ω, R_2 is 16 kΩ, and A_e is $-1000/(1+j\omega/10^5)$. Determine A_e' and the bandwidth.
3. Using the amplifier shown in Fig. 20-14, the plate load is 47,000 Ω, μ is 100, and g_m is 2800 μmhos. The sum of R_1 and R_2 is 100,000 Ω. Determine R_1 and R_2 when A_e' is 20.
4. Repeat Problem 3 for A_e' equal to 10.

Section 20-6 Negative Current Feedback

In a circuit with current feedback, a resistor R_f which is not part of the load is used to develop the feedback. The circuit at open loop without feedback includes R_f and is shown in Fig. 20-15*a*. An equivalent circuit is formed by Thévenin's theorem (Fig. 20-15*b*), and from this circuit the output circuit is a voltage divider. By the voltage divider rule, the output voltage is

$$E_{\text{out}} = \frac{R_L}{Z_{\text{out}} + R_f + R_L} E_s A_e$$

and the open-loop gain of the circuit is

$$A_{OL} = \frac{E_{\text{out}}}{E_s} = \frac{R_L A_e}{Z_{\text{out}} + R_f + R_L} \tag{20-15}$$

When the feedback loop is closed (Fig. 20-15*c*), the feedback voltage to the input of the amplifier is the voltage drop across R_f. The total input signal is

$$E_{\text{in}} = E_s - I'R_f$$

We have in the load, by Ohm's law,

$$E'_{\text{out}} = I'R_L \qquad \text{or} \qquad I' = E'_{\text{out}}/R_L$$

and making the substitution

$$E_{\text{in}} = E_s - \frac{R_f}{R_L} E'_{\text{out}}$$

or

$$E_s = E_{\text{in}} + \frac{R_f}{R_L} E'_{\text{out}}$$

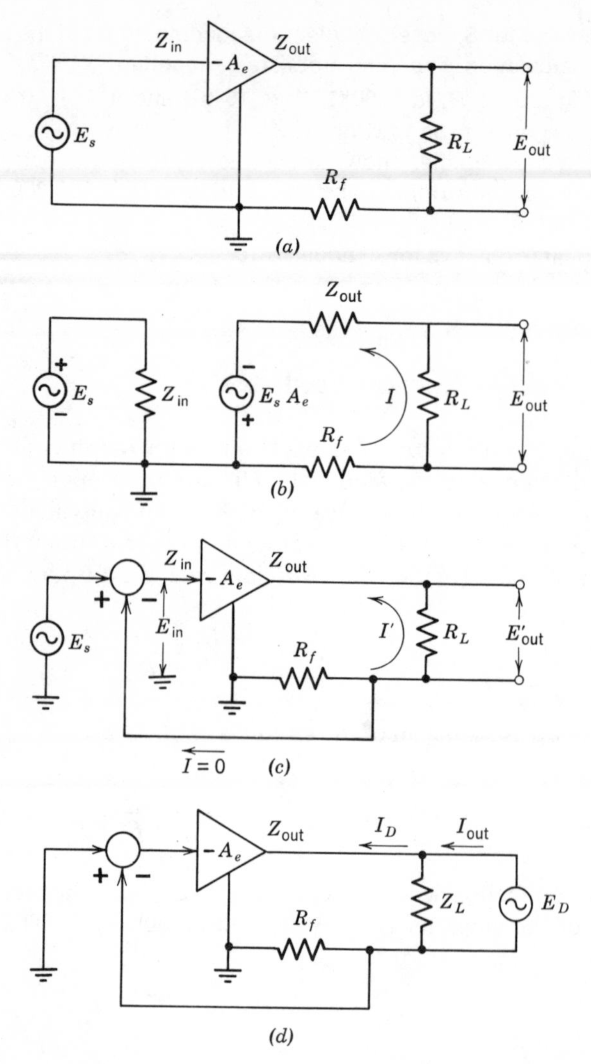

Figure 20-15 Current feedback circuit. (*a*) and (*b*) open loop. (*c*) Closed loop. (*d*) Circuit to determine output impedance.

From Eq. 20-15,

$$E_{in} = \frac{Z_{out} + R_f + R_L}{R_L A_e} E'_{out}$$

and substituting

$$E_s = \frac{Z_{out} + R_f + R_L}{R_L A_e} E'_{out} + \frac{R_f}{R_L} E'_{out} = \frac{Z_{out} + R_L + (1 + A_e) R_f}{R_L A_e} E'_{out}$$

Then the closed-loop gain of the circuit is

$$A_{CL} = \frac{E'_{\text{out}}}{E_s} = \frac{R_L A_e}{Z_{\text{out}} + R_L + (1 + A_e) R_f} \tag{20-16}$$

Equation 20-16 can be rearranged to

$$A_{CL} = \frac{R_L A_e}{Z_{\text{out}} + R_L + R_f + A_e R_f} = \frac{1}{(Z_{\text{out}} + R_L + R_f)/R_L A_e + R_f/R_L} \tag{20-17}$$

Now Eq. 20-15 can be substituted:

$$A_{CL} = \frac{1}{1/A_{OL} + R_f/R_L} = \frac{A_{OL}}{1 + (R_f/R_L) A_{OL}} \tag{20-18}$$

Equation 20-17 is in the same form of the general expression of the basic feedback equation (Eq. 20-1), provided that the feedback is now defined for this circuit as

$$\beta_f = -\frac{R_f}{R_L} \tag{20-19}$$

In this form of current feedback to the input, the input impedance is increased as given by Eq. 20-13. Notice carefully that, while the feedback is derived as a current feedback in the output circuit, the feedback as presented to the summing point is actually a voltage consideration.

To determine the output impedance, a driving voltage E_D is placed across the output, and the source E_s is turned off (Fig. 20-15*d*). The amplified feedback signal now opposes E_D. Then I_D is

$$I_D = \frac{E_D - R_f I_D A_{OL}}{Z_{\text{out}} + R_f}$$

$$E_D = I_D Z_{\text{out}} + R_f I_D + R_f A_{OL} I_D$$

and dividing by I_D

$$Z'_{\text{out}} = Z_{\text{out}} + (1 + A_{OL}) R_f \tag{20-20}$$

Accordingly, with this feedback arrangement, the output impedance is raised.

In Section 12-4, we analyzed the circuit shown in Fig. 20-16 from the viewpoint of an equivalent circuit approach and derived the gain equation

$$A_v = \frac{\beta R_c}{(1 + \beta)(r'_e + R_E)} \tag{12-11a}$$

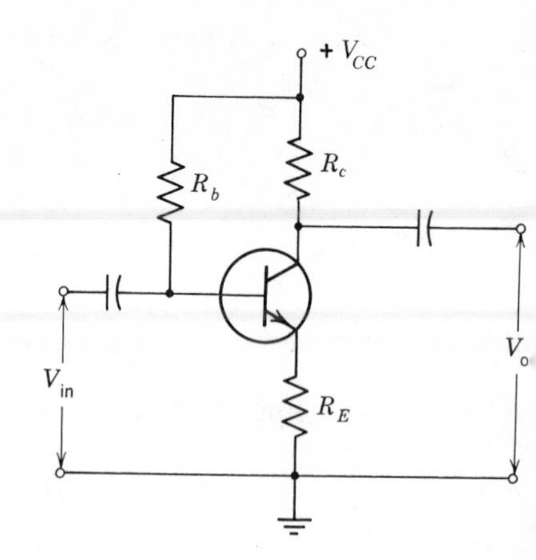

Figure 20-16 Common-emitter amplifier with an external emitter resistor.

Without the feedback resistor R_E the gain of the circuit is

$$A_v = \frac{\beta R_c}{(1+\beta)r'_e} \tag{12-7a}$$

This circuit (Fig. 20-16) has a current feedback, since i_c is the current in R
and i_e is the current in R_E, the feedback resistor. It is necessary to reconci
the fact that i_c and i_e are different, and this can be done by multiplying eac
resistor by the factor necessary to refer each to i_b. Then Eq. 20-19 becomes

$$\beta_f = -\frac{(1+\beta)R_E}{(\beta)R_c} \tag{20-21}$$

Equation 20-21 and Eq. 20-7a are used in the general feedback expression Eq. 20-1

$$A'_v = \frac{A_v}{1-\beta_f A_v} = \frac{\beta R_c/(1+\beta)r'_e}{1+\dfrac{(1+\beta)R_E}{\beta R_c}\left[\dfrac{\beta R_c}{(1+\beta)r'_e}\right]}$$

$$A'_v = \frac{\beta R_c/(1+\beta)r'_e}{1+R_E/r'_e} = \frac{\beta R_c}{(1+\beta)(r'_e+R_E)} \tag{20-22}$$

which is identical to Eq. 12-11a. Consequently, either approach—a genera
equivalent circuit analysis or a feedback analysis—yields the same end
result, as they should.

PROBLEMS

1. R_d is 10,000 Ω and I_D is 1.0 ma. Determine (R_1+R_2). R_1 is varied from zero to (R_1+R_2). As R_1 increases, R_2 decreases to keep the sum of R_1 and R_2 constant

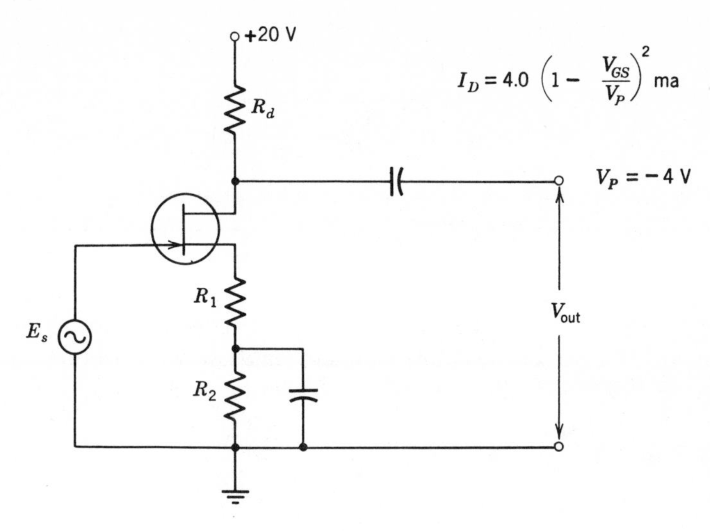

Circuit for Problems 1 and 2.

Obtain data to plot a curve of circuit gain against the value of R_1.

2. Repeat Problem 1 if R_d is 20 kΩ and I_D is 0.5 ma.

Section 20-7 Shunt Feedback

The principles of shunt feedback are shown in the block diagrams of Fig. 20-17. The circuit without feedback (Fig. 20-17*a*) shows the input resistance

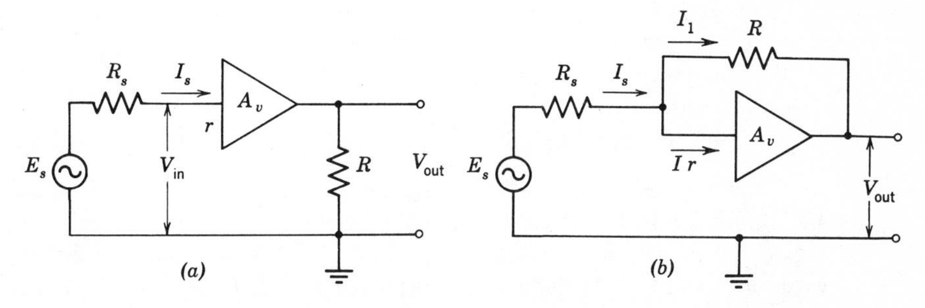

Figure 20-17 Shunt feedback. (*a*) Without feedback. (*b*) With feedback.

to the amplifier as r. R_s and r form a voltage divider input to A_v. Consequently

$$V_{out} = \left(\frac{r}{R_s + r} E_s\right) A_v$$

where

$$A_v = V_{out}/V_{in}$$

Then

$$A_e = \frac{V_{out}}{E_s} = \frac{r}{R_s + r} A_v$$

A_v is determined from the circuit in which the resistor R is definitely a par of the circuit for the calculation of gain, as was done at the end of Sectio 12-5.

Now for shunt feedback, R is connected from the output back into the input (Fig. 20-17*b*). The current I_s from the generator now divides between and I_1. The current I_1 is the same current that was observed in the explana tion of the Miller effect. We are considering only negative feedback. There fore, the voltage across R is $(V_{in} + V_{out})$ which is $(1 + A_v)V_{in}$. By Ohm's law

$$(1 + A_v)V_{in} = RI_1$$

or

$$\frac{V_{in}}{I_1} = \frac{R}{1 + A_v}$$

Consequently, as far as the source is concerned, I_s divides into two resistive branches. One branch is r and the other is $R/(1 + A_v)$. Therefore, a voltage divider action takes place across R_s in series with r in parallel with $R/(1 + A_v)$

The process of the calculation of gain in a circuit with shunt feedback then is performed as a series of steps:

1. A_v is calculated with the feedback resistor R being considered as par of the amplifier in which R is returned to ground.
2. The value of R is divided by $(1 + A_v)$.
3. $R/(1 + A_v)$ is placed in parallel with r and the parallel combinatio evaluated.
4. The gain reduction factor involving R_s and this parallel combination is determined.
5. The gain with feedback is the product of the results of steps 1 and 4.

Consider the basic circuit (Fig. 20-18) that was analyzed in Sections 12-5 and 12-6 from the viewpoint of a circuit analysis of the model. According to the procedure of step 1, the load resistance for the purposes of the evaluation of A_v is R_c in parallel with R_B or $R_cR_B/(R_c + R_B)$. The effect of R_E is to introduce the gain reduction factor $r'_e/(r'_e + R_E)$. Therefore, the gain A_v is

$$A_v = \left(\frac{r'_e}{r'_e + R_E}\right) \frac{\beta[R_cR_B/(R_c + R_B)]}{(1 + \beta)\, r'_e}$$

$$A_v = \left(\frac{r'_e}{r'_e + R_E}\right)\left(\frac{R_B}{R_c + R_B}\right)\frac{\beta R_c}{(1 + \beta) r'_e}$$

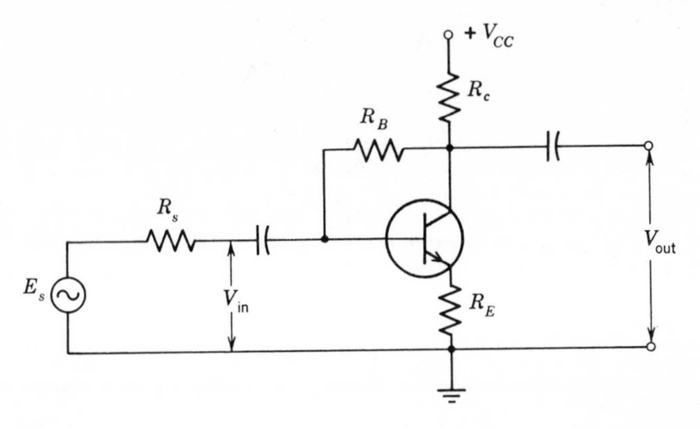

Figure 20-18 Common-emitter circuit using combined emitter feedback and collector feedback.

and if $(1+\beta)$ is approximated as β
or

$$A_v \approx \left(\frac{R_B}{R_c + R_B}\right)\left(\frac{R_c}{r_e' + R_E}\right) \tag{20-23a}$$

Eq. 20-23*a* is identical to Eq. 12-17*c*.

In this circuit $R_B/(1+A_v)$ is in parallel with $(1+\beta)(r_e'+R_E)$. In Chapter 12, we called this parallel combination r_{in}. Then, the overall gain is

$$A_e = \left(\frac{r_{in}}{R_s + r_{in}}\right)A_v$$

$$= \left(\frac{r_{in}}{R_s + r_{in}}\right)\left(\frac{R_B}{R_c + R_B}\right)\left(\frac{R_c}{r_e' + R_E}\right) \tag{20-23b}$$

This last equation is identical to Eq. 12-17*d*.

Now, however, by using the method outlined in this section, we are not restricted to the concept of a single stage with feedback. The principles can easily be adapted to a feedback loop over several stages in cascade.

PROBLEMS

1. The gain of the amplifier A_e is -4500. The input impedance of the amplifier is 50 Ω and the output impedance is zero. Calculate V_{out}/E_s and the impedance the source generator, E_s, sees.
2. For both transistors, β is 100 and r_e' is 40 Ω. R_s is 500 Ω, R_1 and R_2 are each 7500 Ω, and R_3 and R_4 are each 100 Ω. Determine the overall gain of the circuit. Neglect the loadings effect of R_B.
3. For both transistors, β is 80 and r_e' is 20 Ω. R_s is 1 kΩ, R_1 and R_2 are each 10 kΩ,

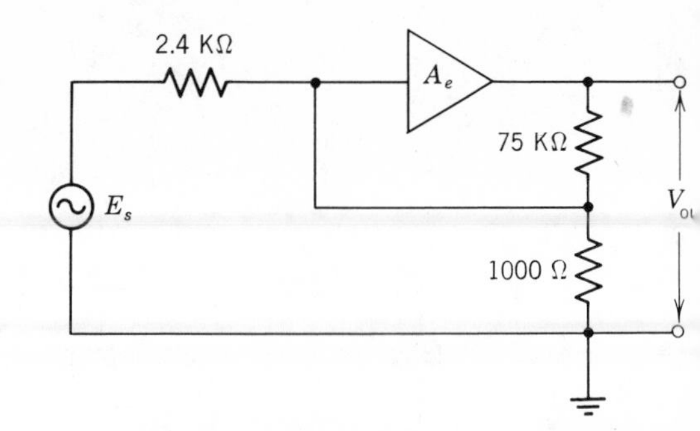

Circuit for Problem 1.

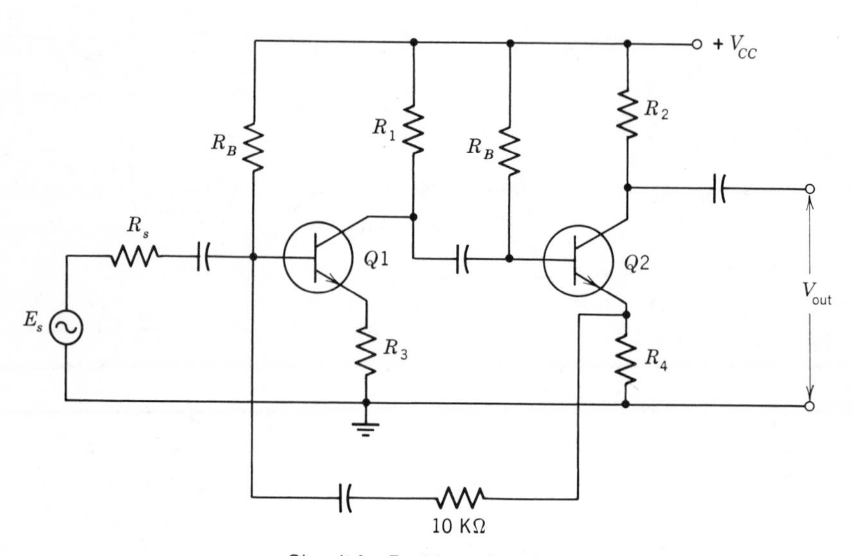

Circuit for Problems 2 and 3.

R_3 is 100 Ω, and R_4 is 240 Ω. Determine the overall gain of the circuit. Negle the loading effect of R_B.

4. Refer to Problem 1, Section 20-5. The 10-kΩ feedback resistor, R_f, is replace by a 2-MΩ resistor and is returned to the base of Q1 instead of being cor nected to the emitter. Determine the circuit gain.

Section 20-8 The Emitter Bypass Capacitor

The circuit shown in Fig. 20-19*a* has an emitter resistor that is adequate bypassed at f_3. By this, we mean that the reactance of C_E is less than 1/10

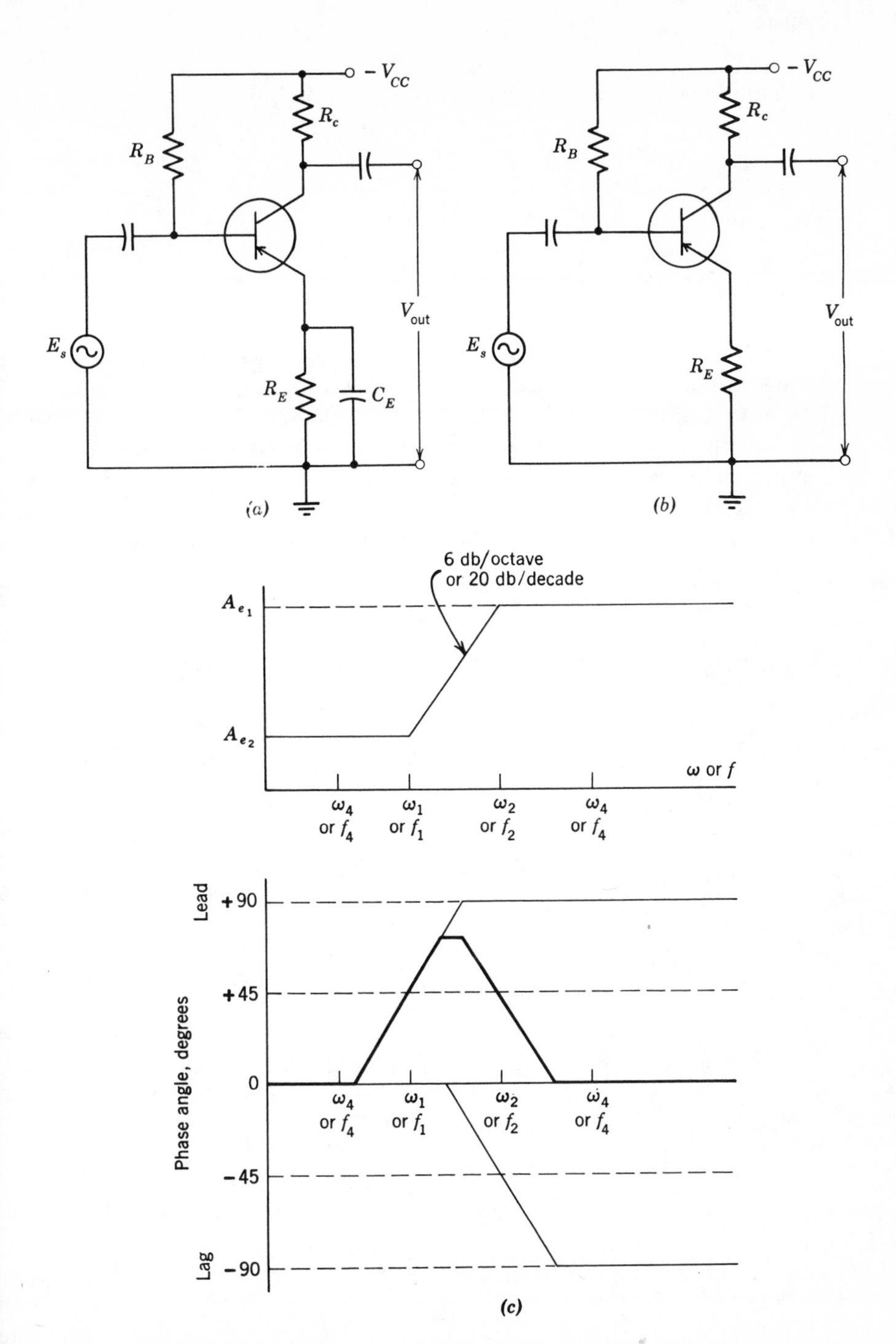

gure 20-19 Emitter bypass capacitor. (*a*) Circuit with capacitor. (*b*) Circuit without capacitor. :) Bode plot.

at that frequency.* Consequently, the gain of the circuit is A_{e_1} which is

$$A_{e_1} = \frac{\beta R_c}{(1+\beta) r_e'}$$

The circuit shown in Fig. 20-19*b* has no C_E, and its gain is A_{e_2}

$$A_{e_2} = \frac{\beta R_c}{(1+\beta)(r_e' + R_E)}$$

When the frequency of operation of the first circuit is so low that the reactance of C_E is much greater than $10R_E$, as at f_4, C_E can be neglected and the circuit gain is A_{e_2}. In a Bode plot (Fig. 20-19*c*), at all frequencies greater than f_2, the gain of the circuit is A_{e_1} and, at all frequencies below f_1, the gain of the circuit is A_{e_2}. Therefore, the values of the break frequencies f_1 and f_2 are required in terms of circuit components. If the reactance of the emitter circuit is Z_e at any frequency, the gain of the circuit at any frequency is

$$A_e = \frac{\beta R_c}{(1+\beta)(r_e' + Z_e)} = \frac{\beta R_c}{1+\beta}\left(\frac{1}{r_e' + Z_e}\right)$$

Z_e is the parallel combination of R_E and C_E, or

$$Z_e = \frac{R_E(-j(1/\omega C_E))}{R_E - j(1/\omega C_E)} = \frac{-jR_E}{\omega R_E C_E - j1} = \frac{R_E}{1 + j\omega R_E C_E}$$

Substitute this into the gain equation:

$$A_e = \left(\frac{\beta R_c}{1+\beta}\right)\frac{1}{r_e' + R_E/(1+j\omega R_E C_E)} = \left(\frac{\beta R_c}{1+\beta}\right)\frac{1+j\omega R_E C_E}{r_e' + j\omega r_e' R_E C_E + R_E}$$

$$= \left(\frac{\beta R_c}{1+\beta}\right)\frac{1+j\omega R_E C_E}{r_e' + R_E + j\omega r_e' R_E C_E} = \left[\frac{\beta R_c}{(1+\beta)(r_e' + R_E)}\right]\frac{1+j\omega R_E C_E}{1+j\omega \dfrac{r_e' R_E}{r_e' + R_E} C_E}$$

The term in brackets is the gain of the circuit of Fig. 20-19 that was given as A_{e_2}.

Then

$$A_e = A_{e_2}\frac{1+j\omega R_E C_E}{1+j\omega \dfrac{r_e' R_E}{r_e' + R_E} C_E} \qquad (20\text{-}24a)$$

*In an FET amplifier and in a vacuum-tube amplifier, r_e' is zero. Consequently, the response of the amplifier is flat down to the frequencies where the capacitor does not properly bypass R_s or R_K. At that point, X_C becomes more than 1/10 R_s or 1/10 R_K. As a result, an emitter bypass capacitor is much larger than a corresponding source-resistor bypass capacitor or a corresponding cathode-resistor bypass capacitor.

In Section 15-6, we used the concept of replacing the reciprocal of a time constant with a specific ω that was the breakpoint in the Bode plot. We can do the same in Eq. 20-24*a* by letting

$$\frac{1}{\omega_1} = R_E C_E \qquad \text{where } f_1 = \omega_1/2\pi \tag{20-24b}$$

and

$$\frac{1}{\omega_2} = \frac{r'_e R_E}{r'_e + R_E} C_E \qquad \text{where } f_2 = \omega_2/2\pi \tag{20-24c}$$

Then

$$A_e = A_{e_2} \frac{1 + j\omega/\omega_1}{1 + j\omega/\omega_2} \tag{20-24d}$$

Since R_E in parallel with r'_e is smaller than R_E, ω_2 must be larger than ω_1.

With these results, the Bode plot (Fig. 20-19*c*) can be formally constructed. At frequencies less than f_1, the gain is a horizontal line of value A_{e_2}. The first break frequency occurs at f_1 and the gain curve rises at a slope of +20 db per decade or at +6 db per octave to the second break frequency f_2. As the second break frequency is in the denominator, the new slope of −20 db per decade cancels the existing slope of +20 db per decade and the response is again flat. At very high frequencies, Eq. 20-24*a* becomes

$$A_e = A_{e_2} \frac{j\omega R_E C_E}{j\omega \left(r'_e R_E C_E / (r'_e + R_E)\right)} = A_{e_2} \frac{r'_e + R_E}{r'_e}$$

Substituting for A_{e_2},

$$A_e = \frac{\beta R_c}{(1+\beta)(r'_e + R_E)} \frac{r'_e + R_E}{r'_e} = \frac{\beta R_c}{(1+\beta) r'_e}$$

which is the expected gain A_{e_1} for which R_E is adequately bypassed.

If a source resistance is considered, the first break frequency remains the same but the second break frequency becomes $rC_E = 1/\omega_2$ where $f_2 = \omega_2/2\pi$ and where r is the parallel combination of R_E in parallel with $[r'_e + R_s/(1+\beta)]$. This derivation is left as an exercise.

The Bode gain plot between A_{e_2} at f_1 and A_{e_1} at f_2 is a straight line with a slope of 20 db per decade. Therefore the change in db from A_{e_2} to A_{e_1} equals the change in db from f_1 to f_2. As a result, the gain ratio is the frequency ratio:

$$\frac{A_{e_1}}{A_{e_2}} = \frac{f_2}{f_1} \tag{20-24e}$$

Based on this relation, f_1 can be found easily by having the values of A_{e_1} and A_{e_2} together with the break frequency of R_E and C_E instead of using the procedure required in evaluating Eq. 20-24*c*. This is especially true when R_s is a factor. The Bode plots for an FET amplifier and a vacuum-tube amplifier can be constructed readily by the use of Eq. 20-24*e*.

PROBLEMS

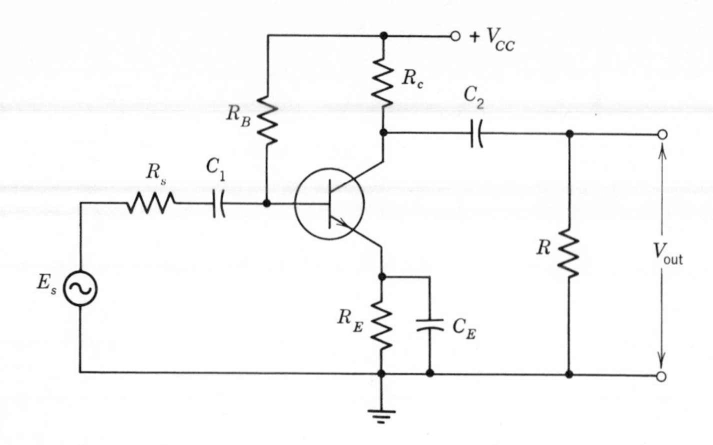

Circuit for Problems 1, 2, 4, and 5.

The coupling capacitors C_1 and C_2 are sufficiently large to be neglected.

1. R_s is zero, R_c is 7.5 kΩ, R is 10 kΩ, and R_E is 1 kΩ. For the transistor, β is 100 and r'_e is 10 Ω. The frequency response is to be flat down to 20 Hz, that is, the break frequency is to occur at 20 Hz. Determine the required value for C_E and draw the Bode plot showing gain and phase response. Neglect R_B.
2. R_s is zero, R_c is 4.7 kΩ, R is 4.7 kΩ, R_E is 470 Ω, and C_E is 80 μF. For the transistor, β is 80 and r'_e is 40 Ω. Draw the Bode plot showing gain and phase response. Neglect R_B.
3. Show how the relations given in the last paragraph of this section are obtained.
4. Repeat Problem 1 if R_s is 5100 Ω.
5. Repeat Problem 2 if R_s is 10 kΩ.

Section 20-9 Current Feedback in Vacuum-Tube Circuits

Figure 20-20*a* shows the circuit of a vacuum-tube amplifier in which the cathode resistor is not bypassed. The gain without feedback is

$$A_e = \frac{\mu R_c}{r_p + R_c + R_K}$$

The feedback β_f is

$$\beta_f = -\frac{R_K}{R_c}$$

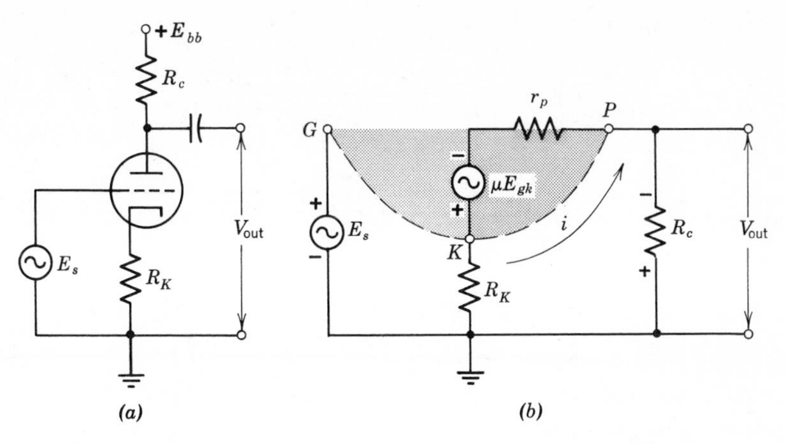

Figure 20-20 Current feedback in vacuum-tube circuits. (*a*) Unbypassed cathode resistor. (*b*) Model.

By substituting in the general gain equation for feedback

$$A_e' = \frac{A_e}{1-\beta_f A_e} = \frac{(\mu R_c/(r_p+R_c+R_K))}{1+(R_K/R_c)(\mu R_c/(r_p+R_c+R_K))}$$

$$A_e' = \frac{\mu R_c}{r_p+R_c+(1+\mu)R_K} \tag{20-25}$$

The model for the circuit is shown in Fig. 20-20*b*. The analysis of the circuit, using the model, can evaluate the gain in a manner similar to the methods of Chapter 12 without using the feedback concept:

$$E_{gk} = E_s - R_K I$$

and

$$V_{\text{out}} = -R_c I$$

Substituting

$$E_{gk} = E_s + \frac{R_K}{R_c} V_{\text{out}}$$

By Ohm's law

$$I = \frac{\mu E_{gk}}{r_p + R_c + R_K}$$

Substituting this value of I in $-R_c I$

$$V_{\text{out}} = -R_c I = -\frac{\mu R_c E_{gk}}{r_p + R_c + R_K}$$

Then substituting for E_{gk}, $(E_s + R_K I)$

$$V_{\text{out}} = -\frac{\mu R_c E_s}{r_p + R_c + R_K} - \frac{\mu R_c R_K I}{r_p + R_c + R_K}$$

But

$$V_{\text{out}} = R_c I$$

Then

$$V_{\text{out}} = -\frac{\mu R_c E_s}{r_p + R_c + R_K} - \frac{\mu R_K V_{\text{out}}}{r_p + R_c + R_K}$$

Cross-multiplying

$$(r_p + R_c + R_K) V_{\text{out}} = -\mu R_c E_s - \mu R_K V_{\text{out}}$$

And

$$(r_p + R_c + R_K) V_{\text{out}} + \mu R_K V_{\text{out}} = -\mu R_c E_s$$

Then

$$A_e' = \frac{V_{\text{out}}}{E_s} = -\frac{\mu R_c}{r_p + R_c + (1 + \mu) R_K} \tag{20-25}$$

The negative sign merely indicates a 180° phase inversion.

PROBLEMS

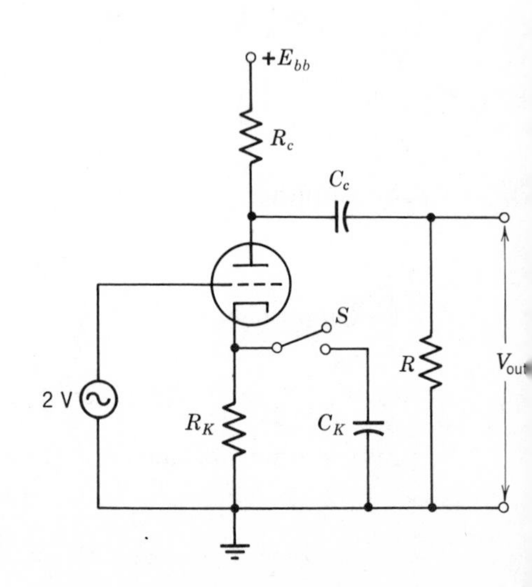

Circuit for Problems 1 to 4.

Table for Problems 1 to 4

Problem	μ	r_p kilohms	g_m μmhos	R_c kilohms	R_K ohms	R kilohms
1	100	70		50	150	120
2		60	2400	50	240	68
3		10	2000	30	700	47
4	20	20		10	1000	10

For each of the Problems 1 to 4, determine V_{out} with S open and with S closed. Assume that C_c and C_K are very large so that the reactances are very small.

Section 20-10 The Cathode Follower

A cathode follower is a circuit (Fig. 20-21) in which the feedback is 100% and is negative. If there were no feedback, the gain would be

$$A_e = \mu \frac{R_K}{r_p + R_K}$$

But, as the feedback is negative and unity,

$$A'_e = \frac{A_e}{1 - \beta_f A_e} = \frac{\left(\mu \dfrac{R_K}{r_p + R_K}\right)}{1 + \left(\mu \dfrac{R_K}{r_p + R_K}\right)}$$

$$= \frac{\mu R_K}{r_p + R_K + \mu R_K} = \frac{\mu R_K}{(1+\mu)R_K + r_p} \tag{20-26a}$$

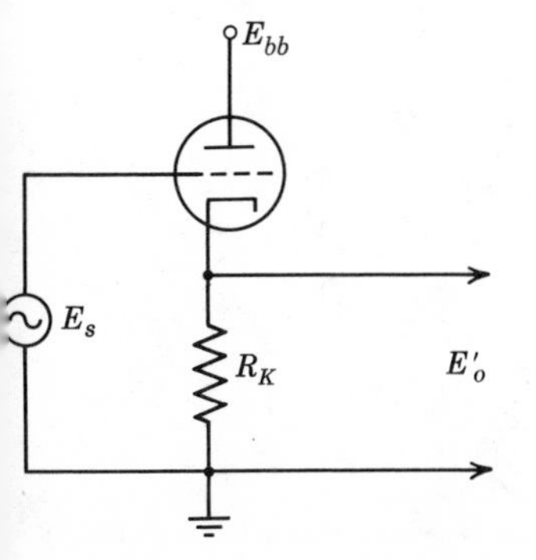

Figure 20-21 The basic cathode follower.

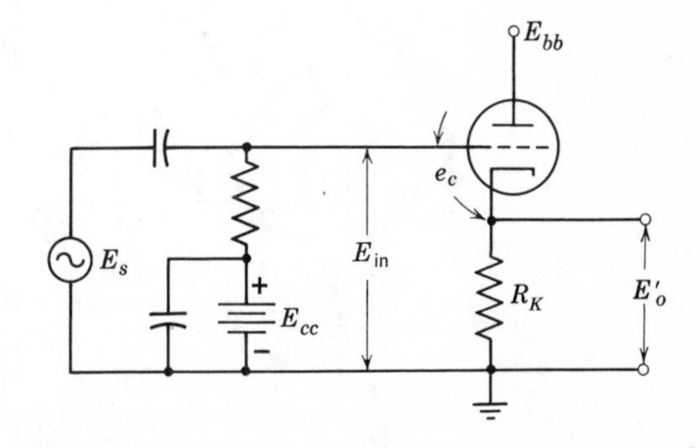

Figure 20-22 Voltage relations in a cathode follower.

This may be rearranged to

$$A'_e = \frac{R_K}{\left(\frac{\mu+1}{\mu}\right)R_K + \frac{1}{g_m}} \tag{20-26b}$$

If a pentode is employed in the circuit, the amplification factor is so much greater than unity that the equation may be written as

$$A'_e = \frac{R_K}{R_K + 1/g_m} = \frac{g_m R_K}{1 + g_m R_K} \tag{20-26c}$$

In any case, the gain of the cathode follower must be less than one.

To determine a load line for a cathode follower (Fig. 20-22) it is evident that

$$E_{bb} = e_b + i_b R_K$$

If R_K is assumed to be 10,000 Ω, using a 6S4 with a supply voltage of 400 V, selection of assumed values of plate current enables the corresponding values of plate voltage to be determined by subtraction. Then the corresponding values of grid voltage e_c are picked off the plate characteristics for the tube. The input voltage e_{in} is obtained from

$$e_{\text{in}} = e_c + i_b R_K$$

The necessary data and calculations for the example are given in Table D.

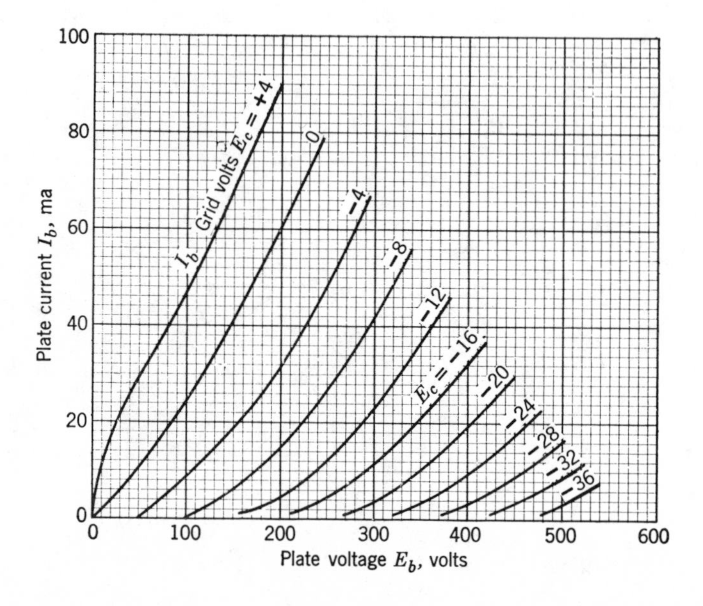

TABLE D Average Plate Characteristics

i_b	e'_{out} i_bR_k	e_b	e_c	e_{in}
0	0	400	−32	−32
5	50	350	−23	+27
10	100	300	−17	+83
15	150	250	−11	+139
20	200	200	−6	+194
25	250	150	−3	+247
28	280	120	−0.5	+279.5
29	290	110	0	+290
30	300	100	+1	+301

The results from this table are plotted as output voltage against input voltage in Fig. 20-23. This circuit can accept a very large and wide-range input signal in contrast to the conventional vacuum-tube amplifier. The optimum bias E_{cc} should be located at the mid-point between cutoff (−32 V) and grid-current flow (+290 V) or at +129 V. Under these conditions, this circuit can accept without a cutoff or a grid current all values of sinusoidal signal from zero to a peak value of 161 V. A peak value of 161 V corresponds to an rms value of 114 V. The slope of the load line in Fig. 20-23 determines the gain of the circuit:

$$A'_e = \frac{290-0}{290-(-32)} = \frac{290}{322} = 0.900$$

In many vacuum-tube voltmeter circuits, the normal full-scale input is

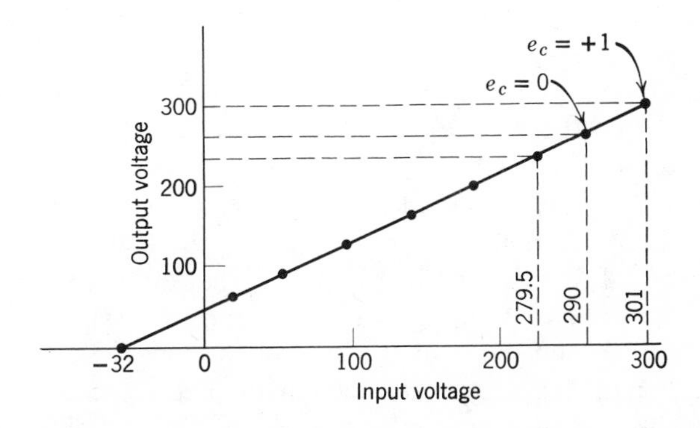

Figure 20-23 Dynamic-response curve for a cathode follower.

5 V rms. If this cathode-follower circuit is used as an input stage to the meter it can withstand an overload of 114/5 or almost 23 times the rated value Another valuable feature of the cathode follower is its ability to handle all signals from zero frequency (d-c) up. In the circuit of Fig. 20-22, the value of the grid resistance R_g is independent of the value of the cathode resistance R_K. This property enables the cathode follower to be used as an ideal device for impedance matching in which R_g can be set to match the source and R_K can either match the load or even be the load itself. Another feature of the cathode-follower circuit is its high impedance which makes it very useful as a high-impedance probe for metering circuits. Because of the low output impedance, cathode followers are often used as drivers for class-AB and class-B amplifiers.

PROBLEMS

1. Using the characteristics of the 6S4, plot the dynamic response of a cathode follower (Fig. 20-23), using a supply voltage of 100 V and a cathode load resistance of 2000 Ω.
2. Repeat Problem 1 for a supply voltage of 250 V and a cathode load resistance of 5 kΩ.
3. Draw the a-c model for a cathode follower. A voltage v_D is placed across the load resistor R_K for the same purpose, as was done in Fig. 12-14 in determmining the output resistance of the emitter follower. Show that the output resistance of the cathode follower is given as the equivalent of R_K in parallel with $r_p/(1+\mu)$.
4. Show that if μ is much larger than one, the output resistance of the cathode follower reduces to $R_K/(1+g_m R_L)$.

Section 20-11 Decoupling

When a d-c power supply has an internal resistance R_{in} (Fig. 20-24), the instantaneous changing current demand on the load creates an a-c drop across this internal resistance of the supply. The output voltage of the supply is no longer a pure direct voltage, but it contains a variation which is a function of the signal. When the instantaneous load-current demand increases, E falls and, when the instantaneous load current decreases, E increases.

In the circuit of Fig. 20-25, a three-stage amplifier is operated from a common positive power supply. Each stage introduces a 180° phase inver-

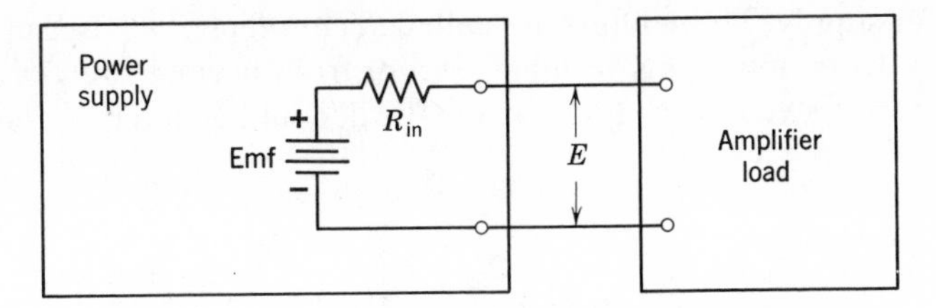

Figure 20-24 Effect of power-supply internal impedance.

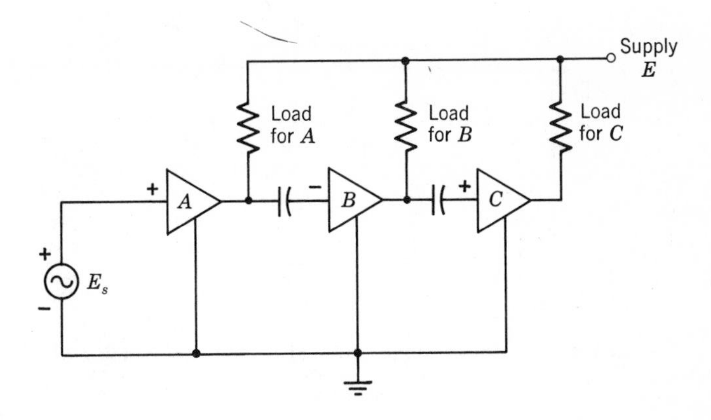

Figure 20-25 Three-stage amplifier operating from a common power supply.

sion. If the instantaneous polarity of the signal, e_s is as shown on the circuit, the signal at the input of stage B is negative, and the signal at the input of stage C is positive. Assume that stages A and B are low-level amplifiers that require low values of supply current and that stage C is a power amplifier, drawing a large value of current from the supply. If the polarity of the signal at the input of stage C is positive, assume that the stage demands a large increase in power supply current. The effect of the internal impedance of the supply causes E to decrease. This decrease of E on stage A causes an unwanted negative signal to appear at the input of stage B, and this decrease of E also produces an unwanted negative signal at the input of stage C. The unwanted signal on the input of stage B is regenerative, whereas the unwanted signal on the input of stage C is degenerative.

This regenerative effect, if severe, can produce a low-frequency oscillation known as *motorboating* which sounds like a "putt putt putt" in a loudspeaker. When an extra filter is placed in the power supply, this condition of motorboating may be cured. However, the best solution to the problem is to make the sensitive stages independent of this interstage coupling through the power supply. These filters are called *decoupling filters* because they prever signal from appearing on tubes except from normal sources. This same ci cuit is redrawn in Fig. 20-26, using the decoupling filters, AB and CD.

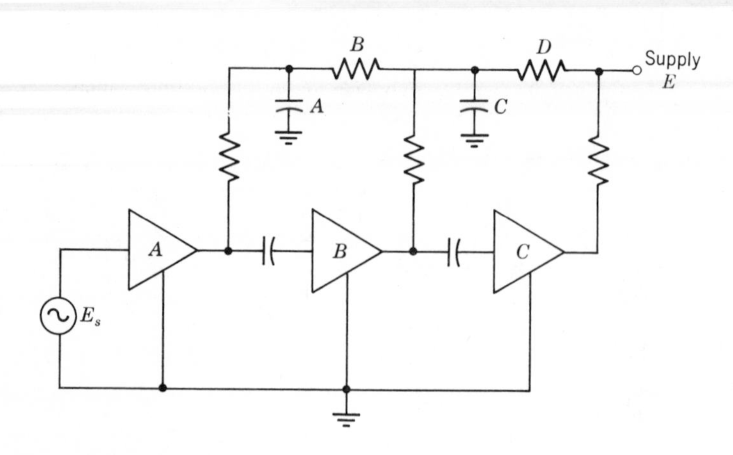

Figure 20-26 Three-stage amplifier using decoupling filters.

The time constant of the decoupling filter must be longer than the perio of the lowest frequency that the circuit is to handle. A rule of thumb state that, in order to have sufficient decoupling, the value of capacitive reactanc should be one-tenth the value of the resistance at the lowest frequency This rule of thumb was used for calculating the values of emitter and cathod bypass capacitors which are actually decoupling filters.

In radio-frequency applications, particularly in transmitters, all leads ar usually decoupled. A radio-frequency choke (RFC) is often substituted fo the resistor. This decoupling must be done to all metering circuits which ar used to tune a class-C amplifier. A typical circuit showing RF decoupling i shown in Fig. 20-27.

Section 20-12 Electronic Regulated Power Supplies

The Zener diode can be used to regulate the output voltage of a powe supply (Fig. 20-28). The d-c voltage regulation was considered as a application for the Zener diode in Section 2-4. In Section 2-4, a 48-V 50-W Zener was used on a 60-V source. The value of Z_{ZT} for the Zener wa

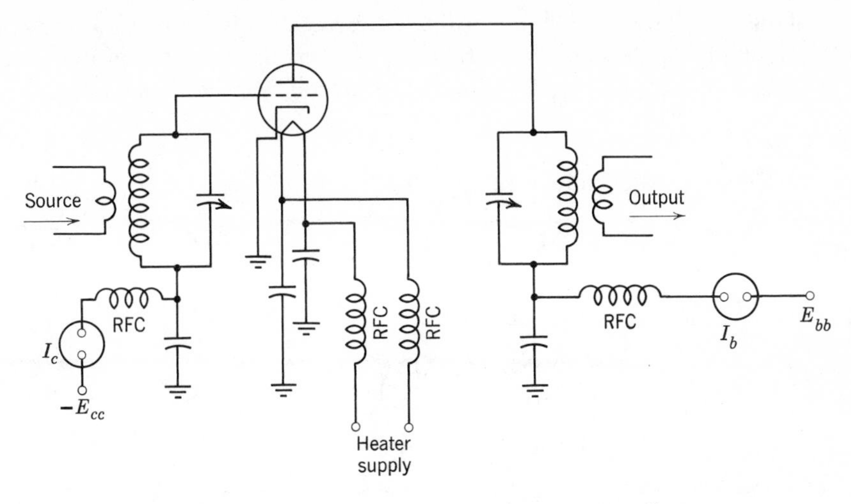

Figure 20-27 Class-C amplifier with decoupling filters.

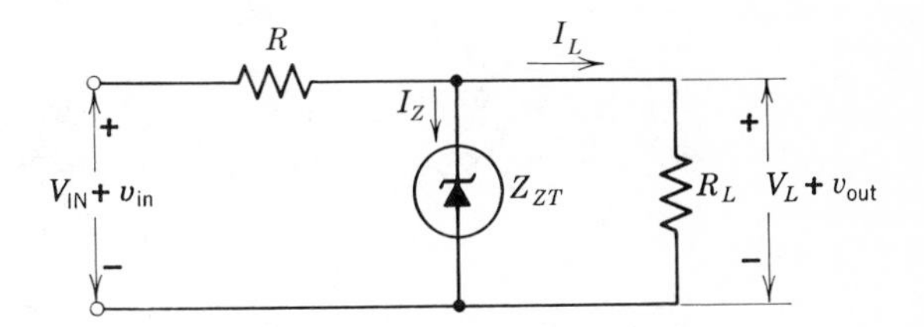

Figure 20-28 Zener diode regulator.

1.2 Ω. The nominal current for the Zener was 800 ma with a 2-A load current.

In this section, we are concerned with the effect of this circuit from an a-c consideration. The current in R is 2.8 A, and the voltage drop across R is 12 V, requiring R to be 12/2.8 or 4.28 Ω. The load resistance is 48/2 or 24 Ω. As far as the ripple is concerned the regulator circuit is a voltage divider circuit in which R is in series with R_L and Z_{ZT} in parallel. Then

$$v_{\text{out}} = \frac{\left(\dfrac{R_L Z_{ZT}}{R_L + Z_{ZT}}\right)}{R + \left(\dfrac{R_L Z_{ZT}}{R_L + Z_{ZT}}\right)} v_{\text{in}} \tag{20-27}$$

Using numerical values

$$v_{\text{out}} = \frac{(24 \times 1.2)/(24 + 1.2)}{2.8 + (24 \times 1.2)/(24 + 1.2)} v_{\text{in}} = \frac{1.14}{3.94} v_{\text{in}} = 0.29 \, v_{\text{in}}$$

This regulator circuit accordingly decreases the ripple to 29% of th incoming ripple value. The regulator circuit has an internal output imped ance that is Z_{ZT} in parallel with R. If R is much greater than Z_{ZT},

$$r_{out} = Z_{ZT} \quad (20\text{-}28$$

The regulator shown in Fig. 20-29 uses a transistor $Q1$ in parallel witl

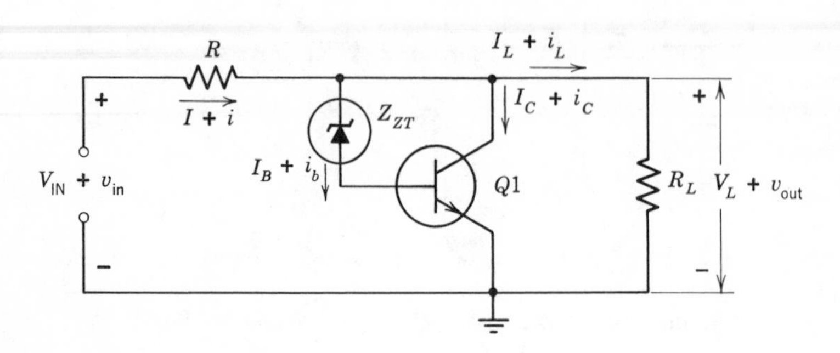

Figure 20-29 Shunt voltage regulator.

the Zener diode. The load voltage across the load is the Zener voltag V_Z plus the base to emitter voltage drop of $Q1$, V_{BE}

$$V_L = V_Z + V_{BE} \quad (20\text{-}29$$

As long as the Zener is maintained in a reverse conduction condition, the circuit regulates.

When the temperature increases, V_Z increases whereas V_{BE} decreases For small values of V_Z, it is possible to secure almost perfect temperatur compensation in the circuit. When V_Z is large, the increase in V_Z is too mucl to be compensated by the fall in V_{BE}.

For the a-c analysis, consider only the a-c current and voltage values By Kirchhoff's current law

$$i = i_b + i_c + i_L$$

and by Kirchhoff's voltage law

$$v_{in} = Ri + v_{out}$$

Substituting

$$v_{in} = R(i_b + i_c + i_L) + v_{out}$$

Since Z_{ZT} is in series with h_{ie},

$$i_b = \frac{v_{out}}{Z_{ZT} + h_{ie}}$$

and

$$i_c = \beta i_b = \frac{\beta v_{\text{out}}}{Z_{ZT} + h_{ie}}$$

From Ohm's law

$$i_L = \frac{v_{\text{out}}}{R_L}$$

Substituting

$$v_{\text{in}} = R\left(\frac{1}{Z_{ZT} + h_{ie}} + \frac{\beta}{Z_{ZT} + h_{ie}} + \frac{1}{R_L}\right) v_{\text{out}} + v_{\text{out}}$$

$$v_{\text{in}} = \left(1 + \frac{R}{R_L} + \frac{(1+\beta)R}{Z_{ZT} + h_{ie}}\right) v_{\text{out}}$$

Then the reduction in ripple is given by

$$\frac{v_{\text{out}}}{v_{\text{in}}} = \frac{1}{1 + (R/R_L) + (1+\beta)R/(Z_{ZT} + h_{ie})} \tag{20-30}$$

The output impedance looking back into the circuit is R in parallel with $[Z_{ZT} + h_{ie}/(1+\beta)]$. When R is large,

$$r_{\text{out}} = Z_{ZT} + \frac{h_{ie}}{1+\beta} = Z_{ZT} + r'_e \tag{20-31}$$

A series voltage regulator is shown in Fig. 20-30. The load voltage is the

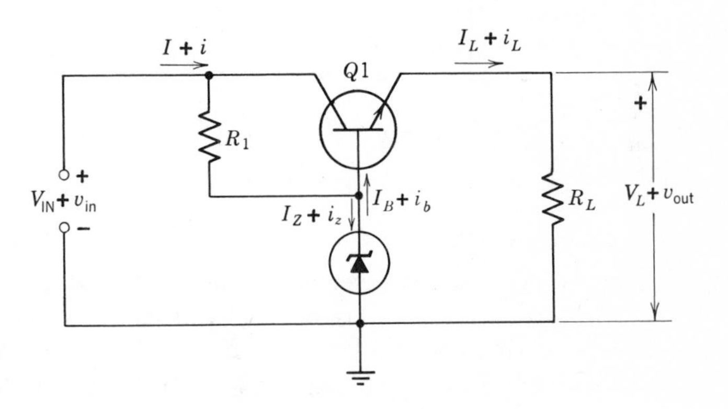

Figure 20-30 Series voltage regulator.

Zener voltage V_Z less the voltage drop across the transistor from base to emitter, V_{BE},

$$V_L = V_Z - V_{BE} \tag{20-32}$$

The attenuation of the ripple is approximately given by the voltage dividing action of R_1, and the Zener

$$\frac{v_{\text{out}}}{v_{\text{in}}} = \frac{Z_{ZT}}{R_1 + Z_{ZT}} \tag{20-33}$$

The control transistor $Q1$ is effectively connected as an emitter follower and, as a result, the impedance looking back into the control circuit is

$$r_{\text{out}} = r'_e + \frac{R_1 Z_{ZT}}{R_1 + Z_{ZT}} \bigg/ (1+\beta) \tag{20-34}$$

The regulated power supply circuit shown in Fig. 20-31 has the additional feature of an output voltage control. The output voltage is controlled by the

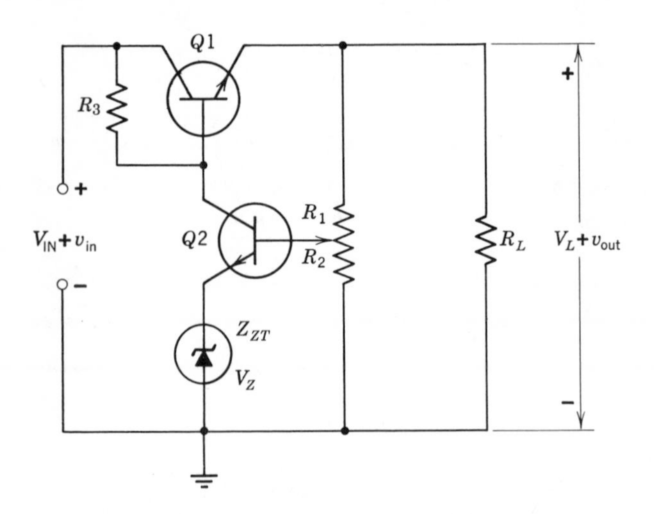

Figure 20-31 Regulated power supply.

setting of the potentiometer arrangement $(R_1 + R_2)$. Assume that $Q1$ functions as a resistor that is controlled by its base current which, in turn, is the collector current of $Q2$. Consider a case in which the load current in R_L increases and V_L falls. This falling V_L reduces the forward bias on $Q2$. The value of V_{CE_2} rises. This rise of V_{CE_2} increases the forward bias on $Q1$. The resistance of $Q1$ to current flow from collector to emitter decreases, and the load current increases. The increase in load current restores V_L to its original value.

An alternative approach is to state that a change in load voltage ΔV_L which is due to either loading or ripple is amplified by $Q2$ with a phase reversal and applied to the control transistor $Q1$. This 180° phase inversion is a negative feedback and the circuit gain is adjusted so the action of $Q1$ cancels out the change in the load. In this circuit the load voltage is monitored. Circuits can be made that monitor the load current or both load current and load voltage.

PROBLEMS

For Problems 1 to 8 use the following data:

$$\text{Zener:}\quad V_Z = 6.40\text{ V at 3 ma} \qquad V_Z = 6.52\text{ V at 33 ma}$$
$$Q1\text{ and }Q2:\quad V_{BE} = 300\text{ mv} \qquad \beta = 50 \qquad r'_e = 50\text{ mv}/I_E$$

1. Refer to Fig. 20-28. R_L is 60 Ω and V_{IN} is 12 V. Find R to set the Zener at the center of its range (18 ma). What is the output impedance of the supply? What range can V_{IN} have and still maintain regulation?
2. In Problem 1, with V_{IN} at 12 V, over what range of variation in R_L and I_L is regulation maintained? What is the short-circuit load current?
3. Refer to Fig. 20-29. I_L is twice I_C when V_{IN} is 12 V and I_Z is at the center of its range (18 ma). Find I_L, R_L, and R. What is the output impedance of the supply? What range can V_{IN} have and still maintain regulation?
4. In Problem 3, with V_{IN} at 12 V, over what range of variation in R_L and I_L is regulation maintained? What is the short-circuit load current when V_{IN} is 12 V?
5. Refer to Fig. 20-30. At the center operation for the Zener (18 ma) the base current in $Q1$ equals the current in the Zener. V_{IN} is 12 V. Find R_1, R_L, and I_L. What is the output impedance of the supply? What range can V_{IN} have and still maintain regulation?
6. In Problem 5, with V_{IN} at 12 V, over what range of variation in R_L and I_L is regulation maintained? What is the short-circuit load current?
7. Refer to Fig. 20-31. V_L is $2V_Z$ and V_N is $3V_Z$. At center operation for the Zener (18 ma), the current in R_2 is 20 times the base current in Q2. I_{B_1} equals I_{C_2}. Find R_1, R_2, R_3, and I_L. V_{IN} varies and find V_{IN}, V_L, and I_L for the limits of Zener current.
8. Repeat Problem 7 if the current in R_2 is fifty times the base current in Q2.

Problems 9 to 14. Repeat Problems 1 to 6 using a nominal V_{IN} of 24 volts where appropriate.

15. Repeat Problem 7 if V_{IN} is 5 V_Z.
16. Repeat Problem 7 if V_{IN} is 4 V_Z.

uestions

1. Define or explain each of the following terms: (*a*) decade amplifier, (*b*) feedback factor, (*c*) regeneration, (*d*) degeneration, (*e*) voltage feedback, (*f*) current

feedback, (*g*) feedback loop, (*h*) open loop, (*i*) closed loop, (*j*) operational amplifier, (*k*) voltage follower, (*l*) integrator, (*m*) differentiator, (*n*) shunt feedback, (*o*) cathode follower, (*p*) ultrasonic, (*q*) motorboating, (*r*) parasitic suppressor.

2. Under what conditions does feedback increase the gain? Decrease the gain?
3. Under what conditions does feedback increase distortion? Decrease distortion?
4. How can a negative feedback loop produce an increase in gain?
5. What forms of feedback increase the input resistance to a circuit?
6. What form of feedback produces a decrease in input resistance?
7. By what amount does feedback increase the bandwidth of an amplifier?
8. How does feedback contribute to the stability of an amplifier?
9. What is the ideal input impedance of an operational amplifier?
10. What is the ideal gain of an operational amplifier?
11. What is an inverting amplifier?
12. What is an adder?
13. Explain how operational amplifiers can be used to obtain the operation of subtraction.
14. What does negative voltage feedback do to the output impedance of an amplifier
15. Explain how the feedback resistor is not part of the load in current feedback. Give an example.
16. What is the output impedance of an amplifier using current feedback?
17. Describe the process of calculating the gain of an amplifier that uses shunt feedback.
18. Why is the analysis of the emitter bypass capacitor considered under the topic feedback?
19. Can the development used for the emitter bypass capacitor be used for a source bypass capacitor? A cathode bypass capacitor?
20. How is current feedback obtained in a vacuum-tube circuit?
21. Why is decoupling necessary in high-gain amplifiers?
22. Explain the operation of the regulator circuit shown in Fig. 20-29.
23. Explain the operation of the regulator circuit shown in Fig. 20-30.
24. Explain the operation of the regulator circuit shown in Fig. 20-31.
25. Trace out the feedback loop in Fig. 18-19*a*.
26. Trace out the feedback loop in Fig. 18-19*b*.

Chapter Twenty-one

OSCILLATORS

Before studying feedback oscillators, it is particularly important to review two topics previously studied: (1) the mechanism of the operation of the bias clamp and the grid-leak circuit (Section 19-1), and (2) the function of the tank circuit which produces a sinusoidal oscillation when pulsed (Section 19-2). A phase-shift oscillator (Section 21-1) ties in directly with the concept of positive feedback, and the bridge oscillator (Section 21-2) uses similar principles. The Armstrong oscillator (Section 21-3) obtains a positive feedback by magnetic coupling. The Hartley oscillator (Section 21-4) and the Colpitts oscillator (Section 21-5) are analyzed from the positive feedback approach. A slightly different approach is taken for the crystal oscillator (Section 21-6) to obtain an understanding of the problems of tuning and adjustment. The electron-coupled oscillator (Section 21-7) is an electron-tube power oscillator. The limitations of oscillators and the methods of obtaining an output from an oscillator are discussed (Section 21-8).

ection 21-1 Phase-Shift Oscillators

In Section 20-2 of the previous chapter, in discussing positive feedback, we set forth two requirements that must be met in order to maintain a sustained oscillation:

1. The feedback must be positive, and
2. The amount of feedback or gain must be large enough to make the feedback factor unity.

In a phase-shift oscillator (Fig. 21-1) the amplifier usually consists of a single-stage circuit in which the output is 180° out of phase with the input.

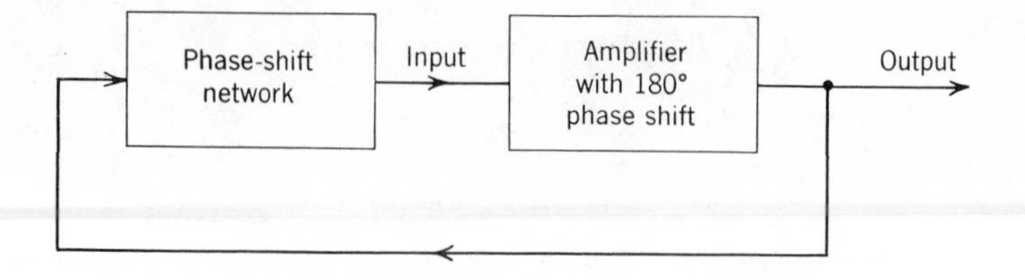

Figure 21-1 Block diagram of a phase-shift oscillator.

The output is fed back into the input through a phase-shift network. Th phase-shift network must produce a 180° phase shift at one particular fre quency to develop the required positive feedback. When the gain of th amplifier is A_e, the loss through the phase-shift network can be no greate than $1/A_e$ in order to maintain the oscillation.

Consider the network shown in Fig. 21-2. The applied voltage V_1 pro duces the currents I_1, I_2, and I_3. The output voltage V_2 is the voltage dro

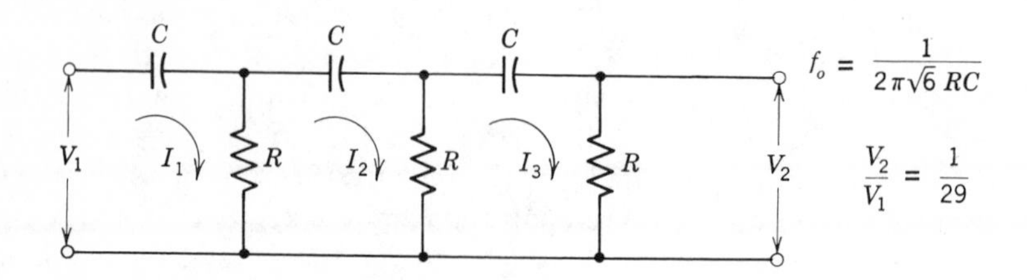

Figure 21-2 A 180° phase-shift network.

RI_3. The network is solved for I_3. The mesh equations for the network are:

$$
\begin{aligned}
(R-j1/\omega C)I_1 - RI_2 \qquad\qquad\qquad\qquad &= V_1 \\
-RI_1 + (2R-j1/\omega C)I_2 - RI_3 \qquad &= 0 \\
-RI_2 + (2R-j1/\omega C)I_3 &= 0
\end{aligned}
$$

Solving these equations for I_3 by determinants yields

$$I_3 = \frac{R^2V_1}{(R-j1/\omega C)(2R-j1/\omega C)^2 - R^2(2R-j1/\omega C) - R^2(R-j1/\omega C)}$$

Then

$$\frac{V_1}{V_2} = \frac{(R-j1/\omega C)(2R-j1/\omega C)^2 - R^2(2R-j1/\omega C) - R^2(R-j1/\omega C)}{R^3}$$

Reducing

$$\frac{V_1}{V_2} = (1 - j1/\omega RC)(2 - j1/\omega RC)^2 - 3 + j2/\omega RC$$

Expanding the first product and then collecting terms

$$\frac{V_1}{V_2} = -\left[\frac{5}{\omega^2 R^2 C^2} - 1\right] - j\left[\frac{6}{\omega RC} - \frac{1}{\omega^3 R^3 C^3}\right]$$

In order to have a 180° phase shift through the network, the first term must be a negative number and the value of the j term must be zero at this frequency, f_0 and ω_0:

$$\frac{6}{\omega_0 RC} - \frac{1}{\omega_0{}^3 R^3 C^3} = 0$$

Solving for ω_0,

$$\omega_0 = \frac{1}{\sqrt{6}RC} \tag{21-1a}$$

or in terms of frequency

$$f_0 = \frac{1}{2\pi\sqrt{6}RC} \tag{21-1b}$$

The first term of the expression for E_1/E_2 is

$$\frac{V_1}{V_2} = -\left[\frac{5}{\omega_0{}^2 R^2 C^2} - 1\right] = -\left[\frac{5}{\left(\frac{1}{\sqrt{6}RC}\right)^2 R^2 C^2} - 1\right] = -[30 - 1] = -29 \tag{21-1c}$$

Accordingly, the gain of the amplifier must be at least 29 for this circuit to oscillate.

Other typical networks using R and C that produce a phase shift of 180° are shown in Fig. 21-3.

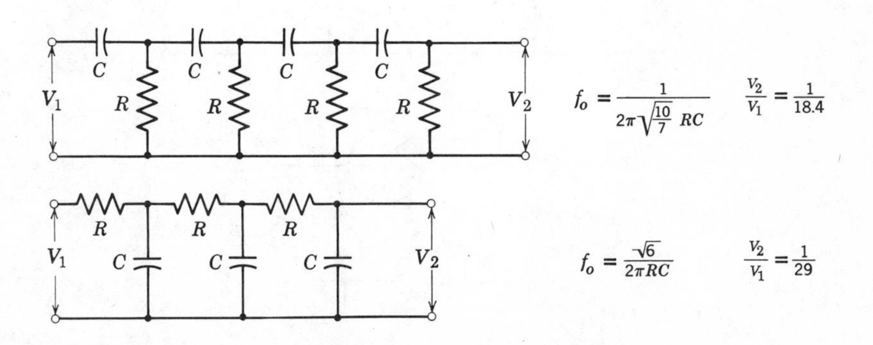

Figure 21-3 Other 180° phase-shift networks.

A phase-shift oscillator is shown in Fig. 21-4. The last resistor R_3 in th phase-shift network is shunted by the other bias resistor R_4 and the inpu resistance h_{ie} to the transistor. Consequently, R_3 is larger than the othe resistors R of the network in order that the parallel combination of R_3, R and h_{ie} is equal to R. If the input resistance to the base is too small, a serie

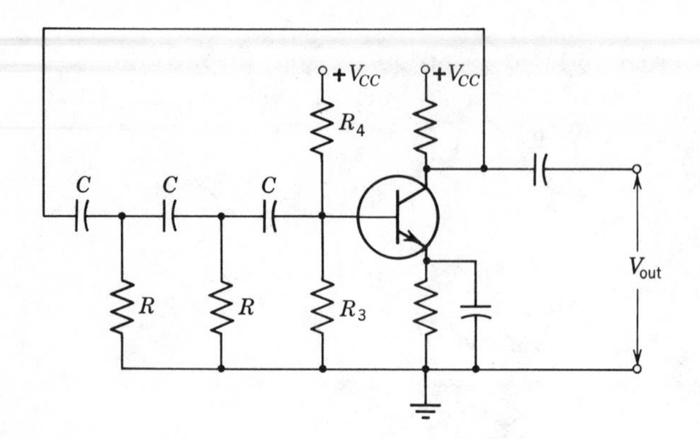

Figure 21-4 RC phase-shift oscillator.

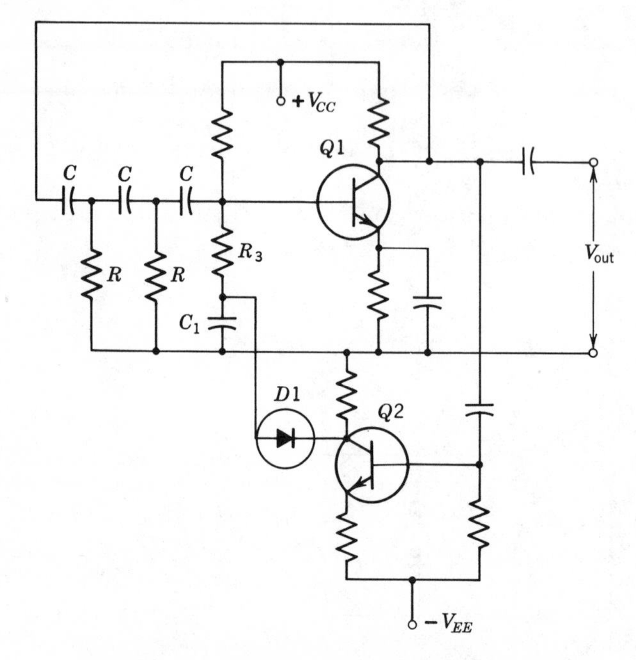

Figure 21-5 Stabilized RC phase-object oscillator.

base-spreading resistor may be used. Alternatively, the bypass capacitor across the emitter resistor can be left off, provided that the gain of the stage is still in excess of 29.

In all oscillator circuits, an excess gain over the one that is required to maintain oscillation produces overdriving which yields a clipped or square waveform. If a sinusoidal waveform is desired, some form of automatic biasing such as the bias-clamp or grid-leak concept must be used. Thus the gain of the stage is reduced as the signal level increases. The gain can be reduced by operating at a higher class-C bias, which reduces the d-c collector current to a lower value. As the d-c collector current decreases to lower values, the β of the transistor falls. An alternative arrangement is to apply an external bias that is derived from the output signal level itself to the amplifier. This arrangement is shown in Fig. 21-5. The signal is rectified in diode $D1$ and filtered in C_1. The d-c voltage on C_1 is proportional to the sinusoidal output voltage, V_{out}.

PROBLEMS

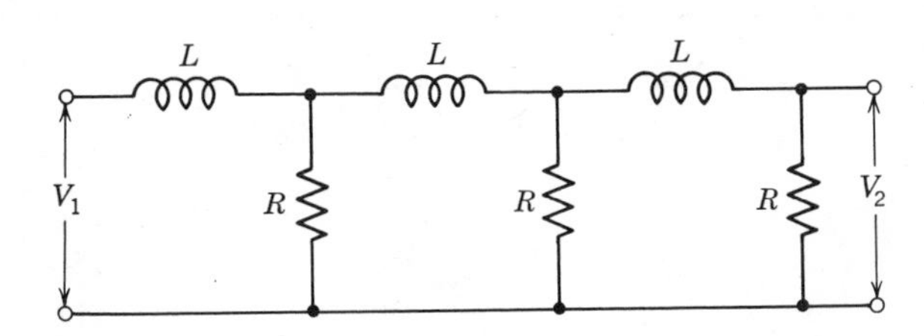

Circuit for Problem 2.

1. Derive the expression for resonant frequency and the expression for minimum gain for the network shown in Fig. 21-3 when used in a phase-shift oscillator.
2. Derive the expression for resonant frequency and the expression for minimum gain for the L-R network when used in a phase-shift oscillator. Assume that the coils have zero resistance.
3. The networks shown in Fig. 21-2 and Fig. 21-3 are fabricated from 0.05 μF capacitors and 10-kΩ resistors. Compare the resonant frequencies obtainable from the network.
4. Find the range of frequencies obtainable for the network by using varactor diodes. Obtain data to plot a curve of resonant frequency against control

voltage. The control voltage is V_R and the effective capacitance of the var- tor diodes can be obtained from

$$C = \frac{600}{\sqrt[2]{0.7 + V_R}} \text{pF}$$

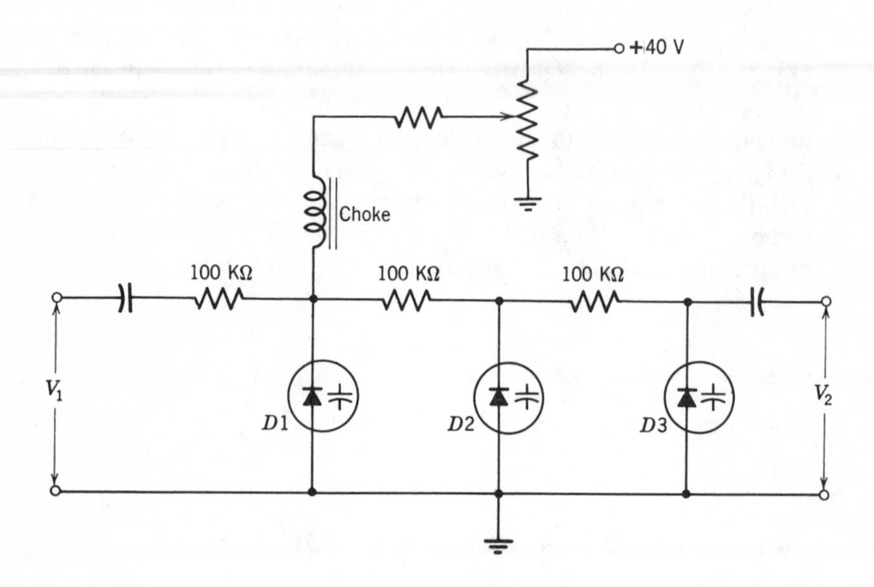

Circuit for Problem 4.

Section 21-2 Bridge Oscillators

Consider a Wheatstone bridge (Fig. 21-6*a*) in which the resistances R_1 ar R_2 are equal. The fixed arm of the bridge is R_3, and R_4 is the variable ar When R_3 equals R_4, the bridge is balanced. Figure 21-6*b* shows the phas diagram for the bridge at balance. Point A which is the junction of R_3 and must be at the same potential as point B because a condition of bridge ba ance requires that there is zero potential across the output terminals of th bridge, A and B.

When R_4 is not equal to R_3, the bridge is unbalanced and a voltage exis across the output. When R_4 is less than R_3, point A is at a lower potenti than point B as seen in the phasor diagram for the unbalance (Fig. 21-6*c* The bridge output voltage, when measured *from A to B* is in phase with th applied voltage V_{OC} since it acts in the same direction as the total applie bridge voltage, V_{OC}. When R_4 is greater than R_3, the potential of point A

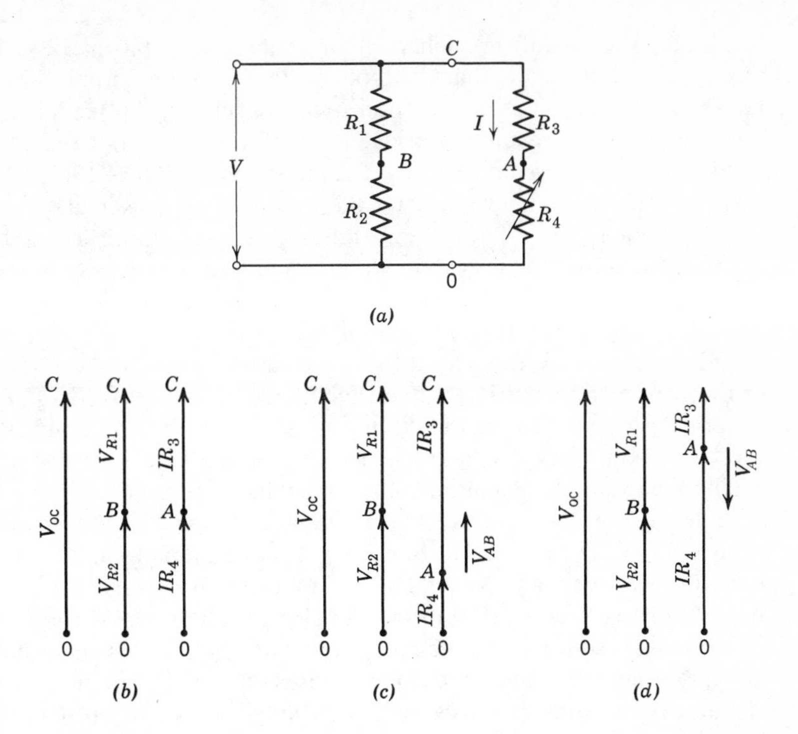

ure 21-6 The Wheatstone bridge. (*a*) Circuit. (*b*) Vector diagram at balance. (*c*) Vector diagram en $R_4 < R_3$. (*d*) Vector diagram when $R_4 > R_3$.

higher than the potential of point B, (Fig. 21-6*d*), and the voltage *from A to B* is 180° out of phase with V_{oc} since it acts in the opposite direction to V_{oc}.

Let us connect this bridge into a two stage-amplifier (Fig. 21-7) which

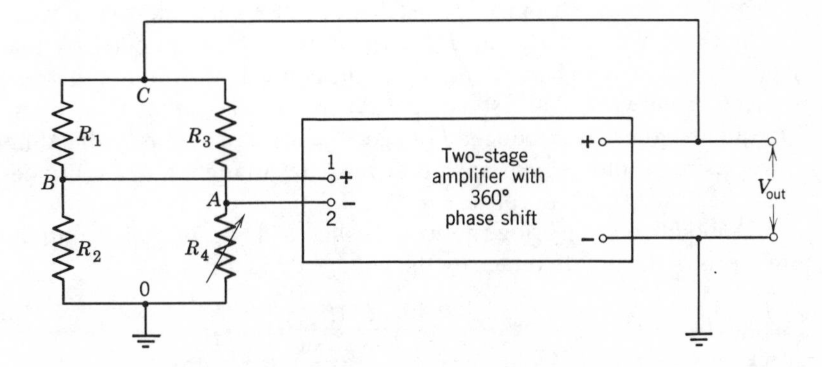

Figure 21-7 Block diagram of bridge oscillator.

has a 180° phase shift in each stage or a total phase shift of 360°. This 3
total phase shift means that the input voltage and the output voltage are
phase. If the circuit is to function as an oscillator, the feedback must
positive, and, as a result, the bridge network cannot introduce a phase sh
Thus, when R_4 is greater than R_3, the input voltage V_{AB} is 180° out of ph
with the output voltage V_{OC}, and the circuit *cannot* oscillate. At balan
when R_4 equals R_3, there is zero output, and the circuit certainly can
oscillate. When R_4 is less than R_3, the input voltage V_{AB} is in phase w
the output voltage V_{OC}, providing positive feedback. The circuit now os
lates if the signal is strong enough, that is, if there is a sufficient amount
bridge unbalance. Reference to Fig. 21-6 shows that V_{AB} is a certain fracti
of V_{OC}. For example, if V_{AB} is one-seventh the magnitude of V_{OC}, an amplif
gain of 7 is needed to set the circuit in oscillation. If the amplifier gain is
the circuit does not oscillate since V_{AB} is only $\frac{1}{7}$ V_{OC}. In order to have
oscillation exist, the magnitude of V_{AB} must increase from $\frac{1}{7}$ to $\frac{1}{5}$ V_{OC}.

Even if the unbalance of the bridge and the gain of the amplifier we
proper for oscillation, we do not, as yet, have the automatic bias arrang
ment that is necessary to establish and to maintain a sinusoidal waveform
we use a resistance panel lamp as the element R_4 in the bridge, we find t
its nonlinear volt-ampere characteristic can provide this required varial
amplitude control. When the voltage across a panel lamp increases, the
sistance of the lamp increases. When the amplitude of the output of the os
lator increases, the voltage across the panel lamp increases, and its resista
increases accordingly. This increase in resistance brings the bridge closer
balance. The bridge output voltage decreases, and this causes the amplitu
of the oscillation to decrease. When the output voltage of the oscillator
creases, the resistance of the lamp decreases, and this provides strong
oscillations to restore the output voltage to a fixed and stable value. By usi
this lamp, the amplitude of the oscillation is constant over a very wide ra
of frequencies.

If the circuit is to provide oscillations at a particular frequency, the
sistance network of R_1 and R_2 must be replaced by a network that is f
quency-selective. Then R_1 and R_2 are the equivalent circuit of this frequenc
selective network "at resonance." At frequencies other than the resona
frequency, there is a reduced voltage V_{AB} and a phase angle that differs fr
the desired inphase relationship. There are a number of a-c bridges whi
may be used for this purpose.

A modified Wien bridge network is shown in Fig. 21-8. The output volta
in terms of the input voltage by the voltage divider rule is

$$V_2 = \frac{\left[\dfrac{R(-j1/\omega C)}{R-j1/\omega C}\right]}{(R-j1/\omega C)+\left[\dfrac{R(-j1/\omega C)}{R-j1/\omega C}\right]} V_{\text{in}}$$

$$\frac{V_2}{V_1} = \frac{-j\dfrac{R}{\omega C}}{\left(R - j\dfrac{1}{\omega C}\right)^2 - j\dfrac{R}{\omega C}} = \frac{\dfrac{R}{\omega C}}{\dfrac{R}{\omega C} + j\left(R - j\dfrac{1}{\omega C}\right)^2}$$

$$\frac{V_2}{V_1} = \frac{R/\omega C}{\dfrac{R}{\omega C} + jR^2 + \dfrac{2R}{\omega C} - j\dfrac{1}{\omega^2 C^2}} = \frac{R/\omega C}{\dfrac{3R}{\omega C} + j\left(R^2 - \dfrac{1}{\omega^2 C^2}\right)}$$

In order to meet the requirements of zero phase shift in the network, the j term must be zero at ω_0,

$$R^2 - \frac{1}{\omega_0^2 C^2} = 0$$

or

$$\omega_0 = \frac{1}{RC} \tag{21-2a}$$

and

$$f_0 = \frac{1}{2\pi RC} \tag{21-2b}$$

Now the ratio of V_2/V_1 becomes

$$\frac{V_2}{V_1} = \frac{R/\omega_0 C}{3R/\omega_0 C} = \frac{1}{3} \tag{21-2c}$$

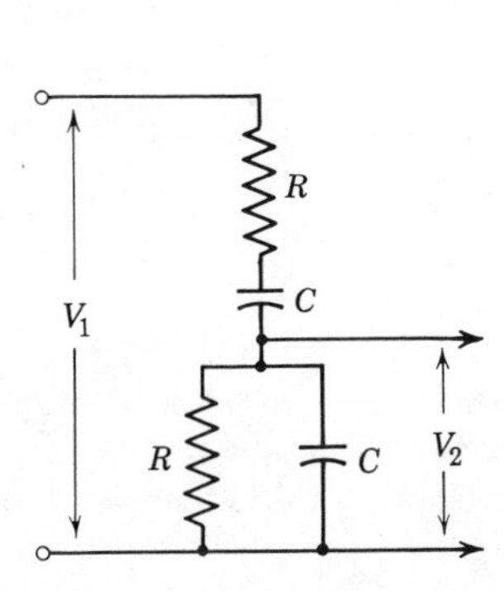

Figure 21-8 The modified Wien bridge circuit.

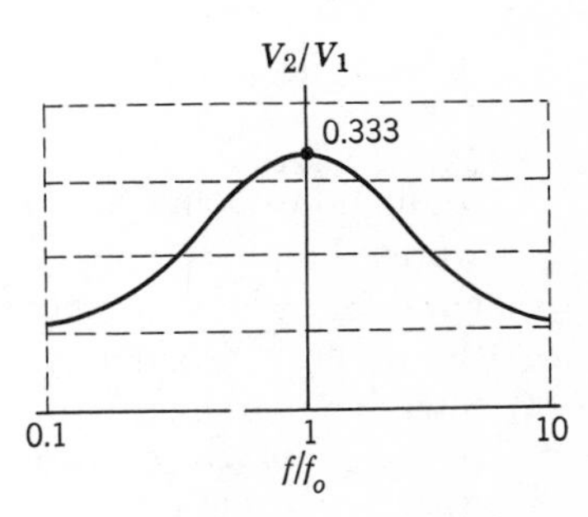

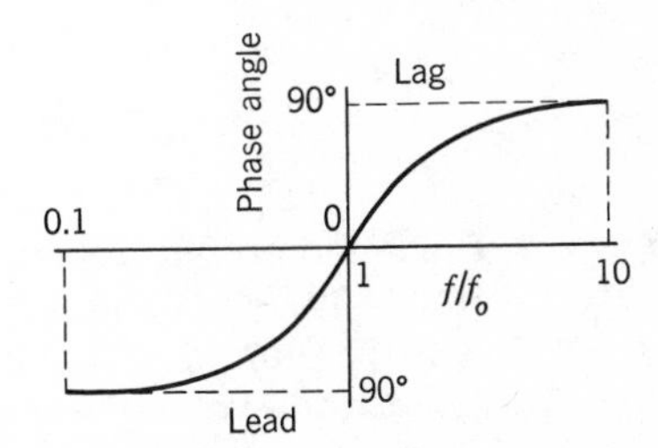

Figure 21-9 Response of the network of Figure 21-8.

If we plot the output voltage for a fixed input voltage over a range
frequencies and obtain the corresponding phase angles, we find that at reso
ance the output voltage is a maximum ($\frac{1}{3}V_{1_{in}}$) and the phase angle is zer
These results are shown in Fig. 21-9.

When this circuit is used in an oscillator (Fig. 21-10), the input signal
the output voltage of Fig. 21-8. The resistance R_2 is a panel lamp. Th

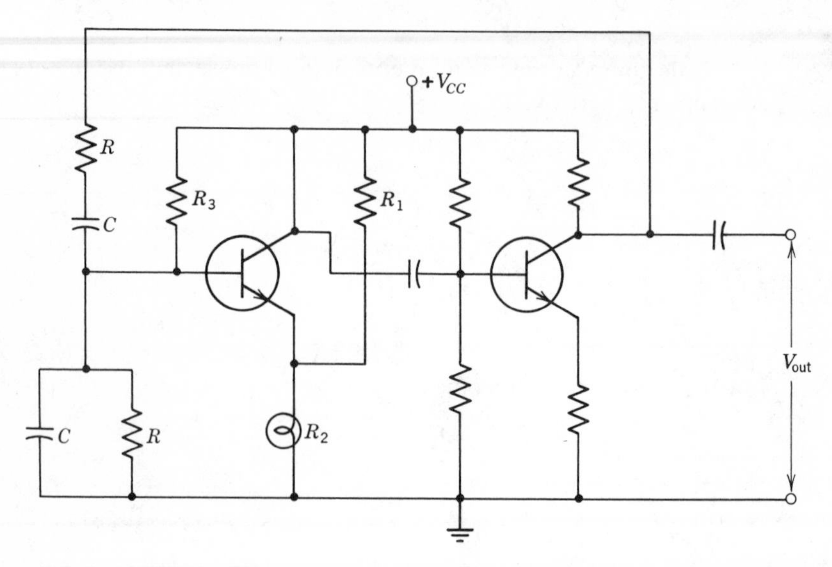

Figure 21-10 Wien bridge oscillator.

oscillation occurs at that frequency at which V_2/V_1 is a maximum and
which the phase angle is zero in the bridge arm. When the frequency of th
circuit is to be continuously variable, switched values of the resistors R giv
the different frequency ranges and the capacitors, C are ganged as a co
tinuously variable tuning capacitor. If the two-stage amplifier is wide-ban
this oscillator can cover, for example, frequencies from sub-audio values t
frequencies of MHz.

PROBLEMS

1. Place the proper polarity markings on the amplifier in order to have oscilla
tion. A two-section ganged capacitor is used to provide C and C for the ci
cuit. The range of tuning of C is from 30 to 330 pF. R is 1 MΩ. What is th
tuning range of the oscillator.
2. In Problem 1, 100-kΩ resistors are switched in place of the 1-MΩ resistor
Now, what is the tuning range of the oscillator?

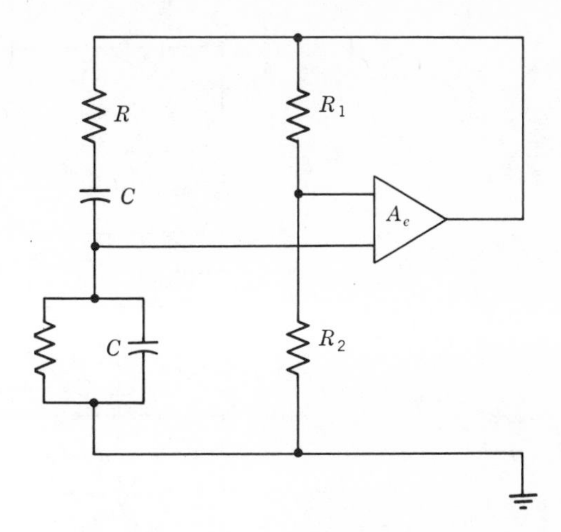

Circuit for Problems 1 to 5.

3. If R is 10 kΩ, C is 0.05 μF, R_1 is 4.7 kΩ, and R_2 is 2.4 kΩ, what is the frequency of oscillation and what minimum gain must the amplifier have to sustain oscillations?
4. In Problem 3, if $R_1 + R_2$ is 10 kΩ and if A_e is 30, what are the values of R_1 and R_2 for sustained oscillation?
5. In Problem 3, if R_1 is fixed at 10 kΩ, find a relation between R_2 amd A_e. Sketch a graph of this relation.

ection 21-3 The Armstrong Oscillator

When an air-core transformer is connected to a transistor as an oscillator (Fig. 21-11), a tuning capacitor C must be placed across either coil L_2, as shown in the diagram, or across coil L_1, in order to have a single stable frequency of operation. The positive feedback is determined by the relative polarity of the coils. If an oscillator is constructed and fails to oscillate, the first check is a reversal of one of two windings to make sure that the feedback actually is positive and not negative. The amount of the feedback is controlled by the mutual inductance M between L_1 and L_2. An increase in the coefficient of coupling k increases the feedback and is accomplished by bringing the coils closer together physically. It should be remembered that most small coils are so wound that the coupling is determined and fixed during manufacture.

When the circuit is first turned on, a rush of collector current develops an increasing flux in the collector coil. This increasing flux produces a secondary voltage in the base winding. The polarity of the base voltage

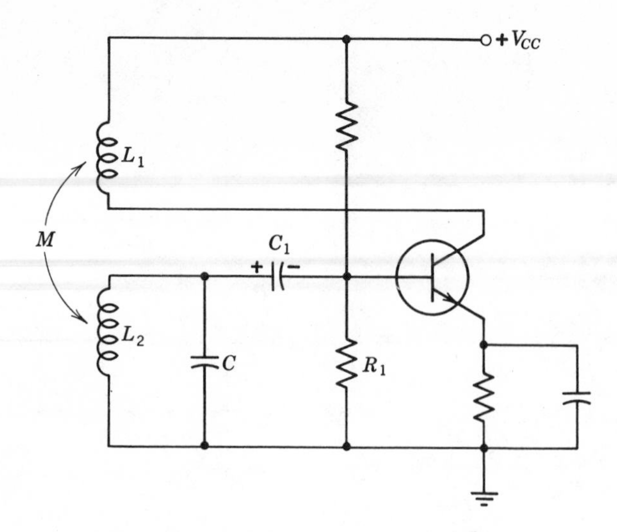

Figure 21-11 The basic Armstrong oscillator.

drives the transistor positive and at the same time charges the tunin capacitor C. The tank starts to oscillate and, when the capacitor voltag becomes negative, this negative voltage on the base reduces the collectc current. The reversed direction of the changing collector current through th collector winding induces a voltage in the base winding which makes th

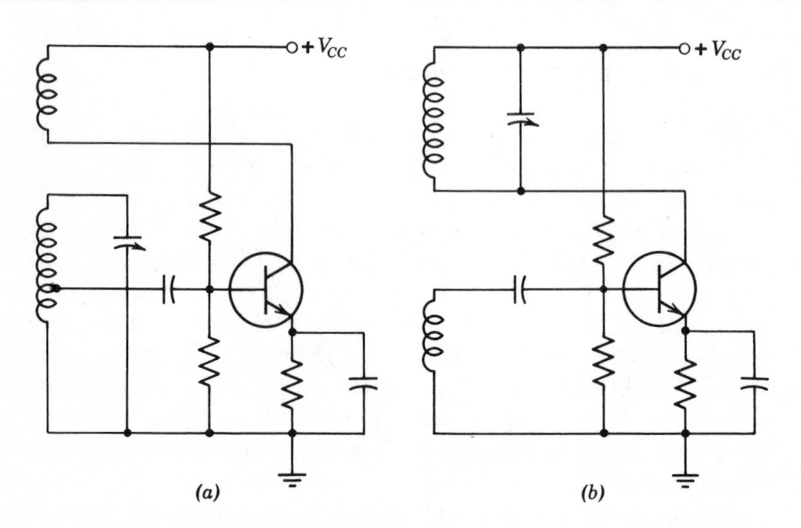

Figure 21-12 Versions of the Armstrong oscillator. (*a*) Untuned-collector-tuned base with tappe coil. (*b*) Tuned-collector—untuned base series feed.

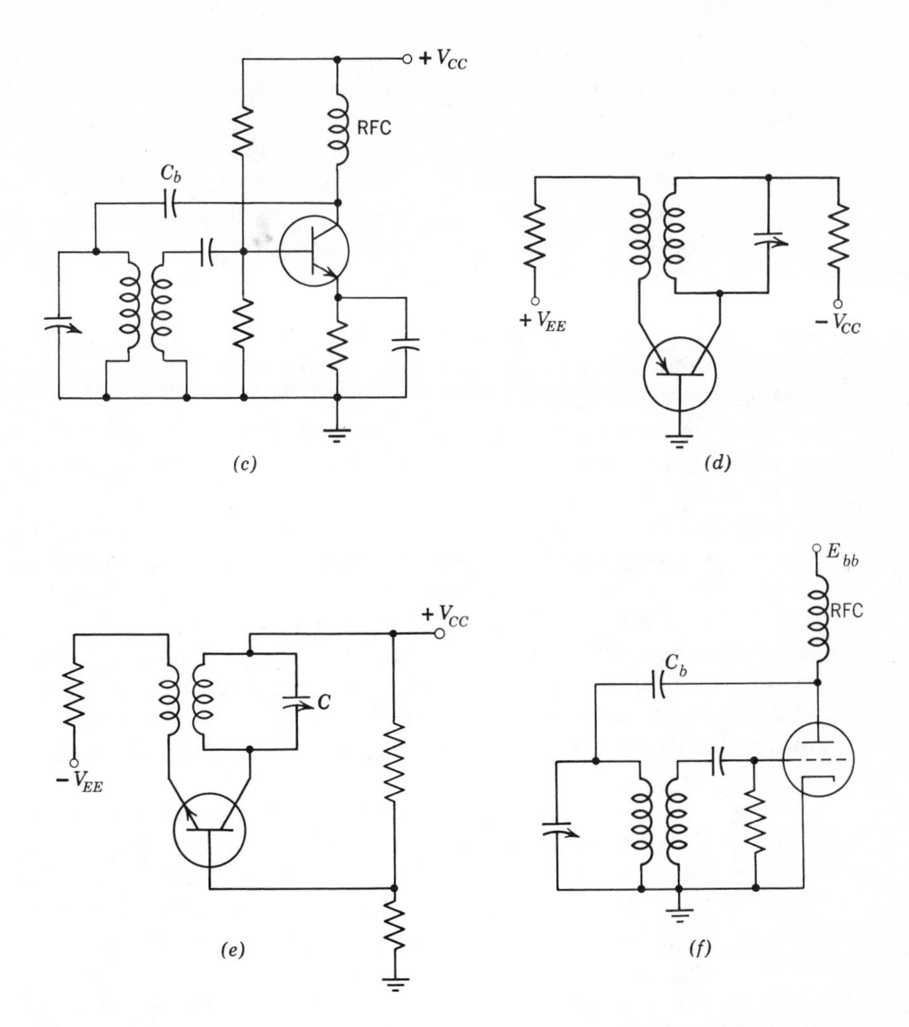

Figure 21-12 (*c*) Tuned-collector–untuned base shunt feed. (*d*) Grounded-base oscillator. (*e*) Grounded-collector oscillator. (*f*) Vacuum-tube version using shunt feed.

current reduction a cumulative process. It takes a few cycles to accomplish the building up of the amplitude of the oscillation to a final value. The limits of the amplitude of the oscillation must be examined. The collector current in the transistor is limited by cutoff at one extreme and by the saturation of the transistor at the other. We expect, then, to have a square-current waveform. Excessive base drive and base current not only prevent a sinusoidal waveform in the tank circuit but also, under normal conditions, cause an overload within the transistor.

To maintain a sinusoidal waveform and to prevent excessive currents from destroying the transistor, a class-C bias arrangement is necessary. If the class-C bias is obtained by the use of a fixed-bias battery, the transistor is permanently cut off at the onset, and oscillations cannot possibly develop from self-excitation. A bias clamp comprising a resistor R_1 and a capacitor C_1 must be used. Now, as the amplitude of the oscillation increases, the d-c voltage across C_1 increases from zero and reduces the angle of collector current flow during each cycle. The oscillator stabilizes at that point where the losses of the circuit equal the developed a-c power output of the transistor. The losses in this circuit, beside the losses in the transistor itself, include the losses in the collector winding, the losses in the base tank, and the losses in the bias resistors.

The resistor in the resistor-capacitor combination in the emitter is necessary to limit the initial current surge to a safe value when the circuit is first turned on.

The Armstrong oscillator may be quickly recognized as the oscillator coil has two windings with four terminals (Fig. 21-12). If the tuning capacitor is placed in the base circuit, the collector winding is referred to as the *feedback* or *tickler winding*. The tuning capacitor may be placed in the collector circuit and the base winding becomes the feedback winding. This latter connection has the material disadvantage that the tuning capacitor is "hot." Not only must the tuning capacitor be insulated from the chassis, but also its shaft must be insulated to prevent a short circuit on, or an electric shock from, the supply voltage.

In a *series-feed* circuit, d-c collector current flows *through* at least part of the oscillator coil. In a *shunt feed* or *parallel feed*, d-c collector current is kept out of the coil by means of a blocking capacitor C_b (Fig. 21-12*c* and *f*). An RF choke must be placed in series with the collector to maintain the collector at a high a-c impedance with respect to ground. The basic versions of the Armstrong oscillator are shown in Fig. 21-12.

In order to prevent a transistor from loading down the oscillator tank, it is quite common to use a tap on the tank coil for the transistor connection as shown in Fig. 21-12*a*. The oscillator coil of this circuit, although it has not four but five terminals, is used in a true Armstrong-type oscillator circuit.

PROBLEMS

1. L_1 and L_2 in a drain-tuned JFET Armstrong oscillator are each 500 μh. The M between the coils is 40 μh. What value of C_1 is required to tune the circuit to 100 kHz? What gain is required in the JFET for sustained oscillation? *Hint.* E_2 is $j\omega MI_1$ and E_1 is $j\omega L_1I_1$. β_f is E_2/E_1.

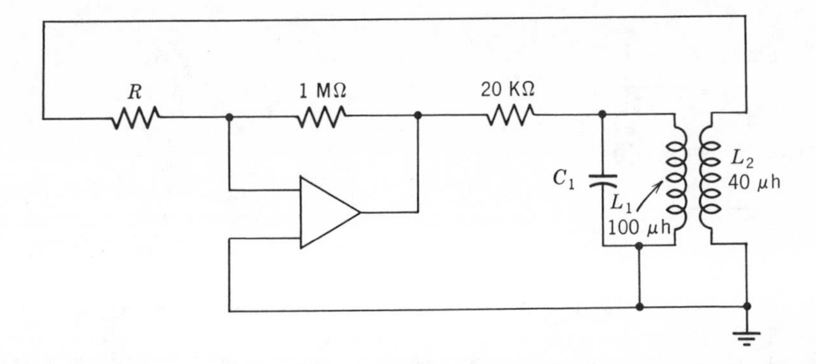

Circuit for Problem 2.

2. The circuit using the operational amplifier oscillates at 1.2 MHz. What is required value of C_1. If the coefficient of coupling k between the coils is 10% and if each coil has a Q of 40, what is the value of R that is required to sustain oscillations?
3. Assume that the oscillator in Problem 2 operates with a 100 peak-to-peak voltage across the tank when power is taken from the circuit from a tertiary winding. The load reduces the effective Q of the circuit to 10. How much power is taken from the circuit?
4. What value of R is required to sustain oscillation in Problem 3?
5. The oscillator circuit of Problem 1 is used with a JFET that has the drain current given by the equation

$$I_D = 4\left(1 - \frac{V_{GS}}{V_P}\right)^2 \text{ ma} \qquad \text{where} \qquad V_P = -4 \text{ V}$$

What is the minimum drain current required to sustain oscillations and what is the value of V_{GS}? Assume r_d is 20 kΩ.

Section 21-4 The Hartley Oscillator

A Hartley oscillator (Fig. 21-13*a*) is characterized by a tapped coil L_T in the oscillator tuned circuit. The Hartley coil, then, has three terminals. If a second tap is taken to prevent transistor loading, the Hartley coil is still one continuous winding with taps. The model (Fig. 21-13*b*) shows only the a-c circuit components. In order for this circuit to oscillate, we must meet the condition that the network of $L_1 + M$, $L_2 + M$, and C provides the requisite 180° phase shift along with a sufficient gain in the amplifier. R_1, R_2, and C_1 form the bias clamp. The voltage v_a is

$$v_a = i_b h_{ie}$$

where h_{ie} is $(1+\beta)r'_e$

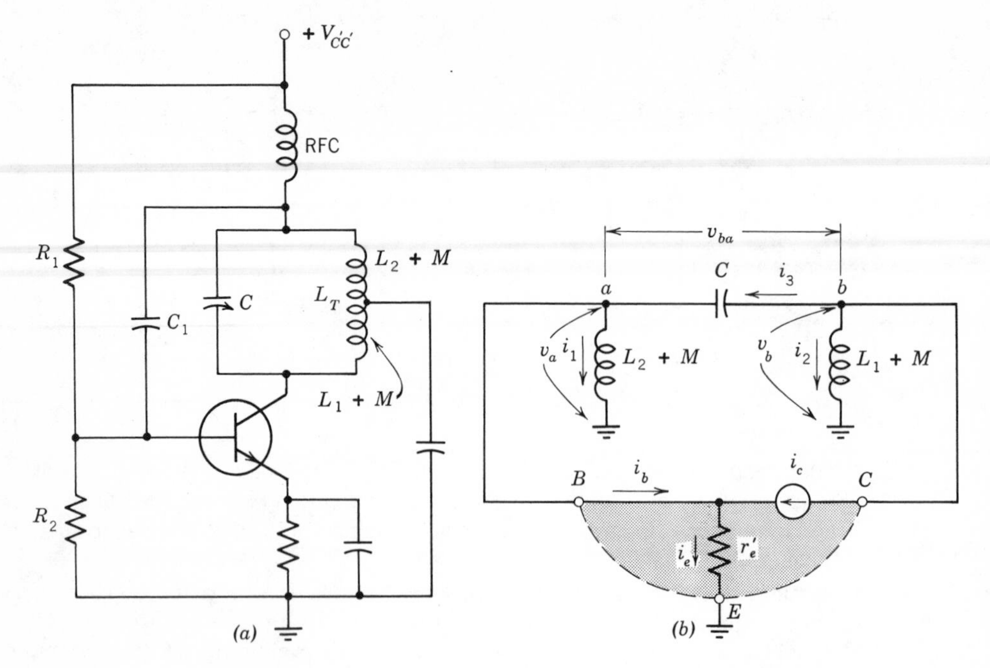

Figure 21-13 The Hartley oscillator. (*a*) Actual circuit. (*b*) Model.

The current i_1 is the voltage v_a divided by the reactance of the branch

$$i_1 = \frac{v_a}{j\omega(L_2+M)} = \frac{h_{ie}i_b}{j\omega(L_2+M)} = -j\frac{h_{ie}i_b}{\omega(L_2+M)}$$

Then

$$i_3 = i_1 + i_b = -j\frac{h_{ie}i_b}{\omega(L_2+M)} + i_b = \frac{\omega(L_2+M) - jh_{ie}}{\omega(L_2+M)}\, i_b$$

v_{ba} is the voltage drop across the capacitor C:

$$v_{ba} = i_3\left(-j\frac{1}{\omega C}\right) = \frac{-h_{ie} - j\omega(L_2+M)}{\omega^2 C(L_2+M)}\, i_b$$

Then v_b is the sum of v_b and v_{ba}:

$$v_b = v_a + v_{ba} = h_{ie}i_b - \frac{h_{ie} + j\omega(L_2+M)}{\omega^2 C(L_2+M)}\, i_b$$

$$= \frac{\omega^2 h_{ie}C(L_2+M) - h_{ie} - j\omega(L_2+M)}{\omega^2 C(L_2+M)}\, i_b$$

Now the current i_2 by Ohm's law is

$$i_2 = \frac{v_b}{j\omega(L_1+M)} = \frac{-(L_2+M) - j[\omega^2 h_{ie}C(L_2+M) - h_{ie}]}{\omega^3 C(L_1+M)(L_2+M)}\, i_b$$

From an inspection of the model, Kirchhoff's current law shows that

$$i_c = -(i_2 + i_3)$$

Substituting

$$i_c = \left\{\frac{(L_2+M) + j[\omega^2 h_{ie} C(L_2+M) - h_{ie}]}{\omega^3 C(L_1+M)(L_2+M)} - \frac{\omega(L_2+M) - jh_{ie}}{\omega(L_2+M)}\right\} i_b$$

Then

$$\frac{i_c}{i_b} = \frac{(L_2+M) - \omega^3 C(L_1+M)(L_2+M) + j[\omega^2 h_{ie} C(L_2+M) - h_{ie} + \omega^2 h_{ie} C(L_1+M)]}{\omega^3 C(L_1+M)(L_2+M)}$$

The j term must be zero at ω_0

$$\omega_0{}^2 h_{ie} C(L_2+M) - h_{ie} + \omega_0{}^2 h_{ic} C(L_1+M) = 0$$

or

$$\omega_0{}^2 C[(L_1+M) + (L_2+M)] = 1$$

But $(L_1+M) + (L_2+M)$ is L_T

Then

$$\omega_0 = \frac{1}{\sqrt{L_T C}} \qquad \text{or} \qquad f_0 = \frac{1}{2\pi\sqrt{L_T C}} \tag{21-3a}$$

Equation 21-3a shows that the resonant frequency is, as expected, not a function of the tap on the coil. Also, we have assumed an ideal coil, and the h_{ie} term is not loading the tank circuit. The analysis in this chapter is developed to show the general concepts of the operation of the circuits. It is beyond the scope of this text to develop an analysis that shows the slight detuning effect of h_{ie} and of the actual tank circuit Q.

When the j term is zero at ω_0, the ratio of i_c/i_b becomes

$$\frac{i_c}{i_b} = \frac{\omega_0(L_2+M) - \omega_0{}^3 C(L_1+M)(L_2+M)}{\omega_0{}^3 C(L_1+M)(L_2+M)} = \frac{1}{\omega_0{}^2 C(L_1+M)} - 1$$

Since

$$\frac{1}{\omega_0{}^2} = C(L_1 + L_2 + 2M)$$

$$\frac{i_c}{i_b} = \frac{L_1 + L_2 + 2M}{L_1 + M} - 1 = \frac{L_1 + L_2 + 2M - L_1 - M}{L_1 + M}$$

Then

$$\frac{i_c}{i_b} = \beta = \frac{L_2 + M}{L_1 + M} \tag{21-3b}$$

Equation 21-3b shows that the tap on the coil is not at the center but at fraction of the whole coil. Usually, the tap is located at about 10% on th coil. The value of β is much larger than 10, but the automatic action of class-C bias clamp will lower the effective beta of the transistor until stable point of sinusoidal oscillation is reached. Experimentation with th tap position can often optimize the performance of the oscillator.

The oscillator circuit shown in Fig. 21-13a is series fed, but it is neces sary to insulate the tuning capacitor C completely, including the shaft. A form of series feed that permits the grounding of the tuning capacitor i

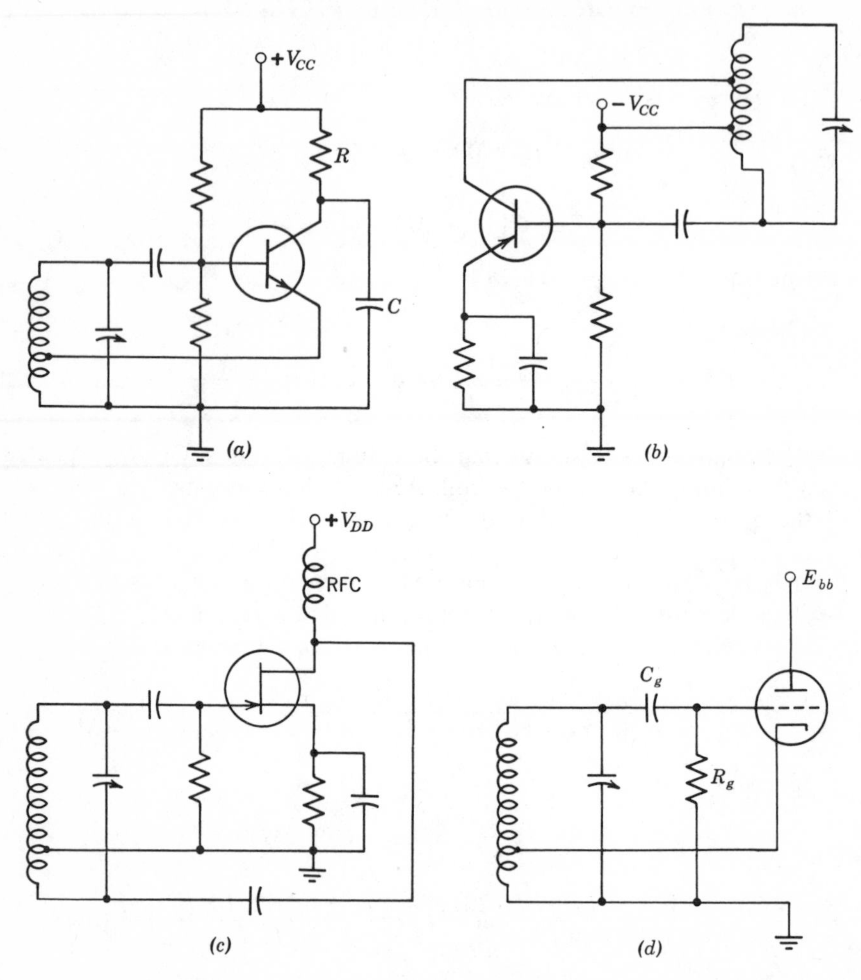

Figure 21-14 Hartley oscillator circuits. (a) Series feed with a grounded-tuning capacitor. (b) Hartley oscillator using a tapped tank to maintain high Q. (c) Shunt fed JFET oscillator. (d) Vacuum-tube circuit.

shown in Fig. 21-14*a*. The resistor in decoupling filter of R and C limits the collector current to a safe maximum initial value. Other versions of Hartley oscillators are shown in Fig. 21-14*b*, *c* and *d*.

PROBLEMS

1. Draw the model for a Hartley oscillator using a JFET. The model should be similar to the transistor model of Fig. 21-13. The input resistance to the JFET is infinite. The model must include r_d. Show that the frequency of oscillation is given by Eq. 21-3*a* and that the condition for oscillation is given by

$$g_m r_d = \frac{L_1 + M}{L_2 + M}$$

2. The drain current in an N-channel JFET is given by

$$I_D = 3.0\left(1 - \frac{V_{GS}}{V_P}\right)^2 \text{ ma} \qquad \text{where} \qquad V_P \text{ is } -4 \text{ V}$$

   A Hartley coil used in an oscillator circuit with this JFET has a tap at 10% of the winding. If r_d is 25,000 Ω, what is the drain current, assuming that the oscillator coil and tuning capacitor are lossless?

3. Draw the complete model for the vacuum-tube Hartley oscillator circuit (Fig. 21-12*f*). Include all resistance values in the model. Using the result of Problem 1, what is the approximate condition for oscillation for this vacuum-tube circuit?

4. The alternating voltage across the whole winding of a Hartley oscillator coil is 100 V rms. The effective Q of the tank circuit is 25 when 2 W are taken from the circuit by means of a secondary winding on the oscillator coil. Find the required value of the tuning capacitor and the inductance of the coil if the circuit operates at 2500 kHz. Assume that the coil and the tuning capacitor are lossless.

Section 21-5 The Colpitts Oscillator

The Colpitts oscillator is shown in Fig. 21-15*a*. Dual tuning capacitors, $C1$ and $C2$, are used, and the characteristic of a Colpitts oscillator is that there is no tap on the oscillator coil. The oscillator coil has two terminals only. The model (Fig. 21-15*b*) shows the a-c circuit components only. In the model, the input resistance to the base is $(1+\beta)r_e'$ or h_{ie}. By the current divider rule

$$i_b = i_L \frac{(-j(1/\omega C_1))}{h_{ie} - j(1/\omega C_1)} = \frac{1}{1 + j\omega C_1 h_{ie}} i_L$$

or

$$i_L = (1 + j\omega C_1 h_{ie})\, i_b$$

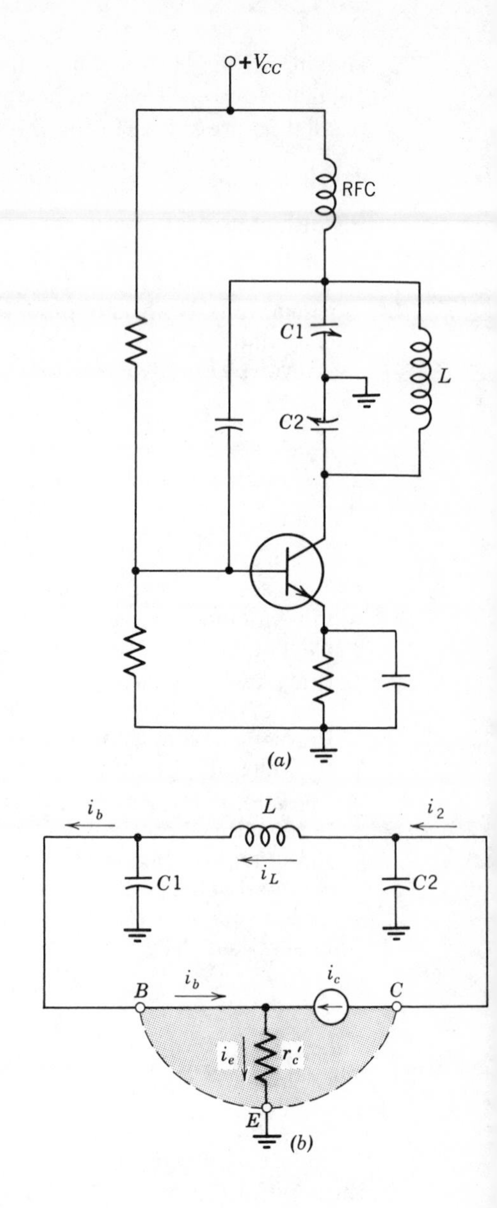

Figure 21-15 Colpitts oscillator. (*a*) Actual circuit. (*b*) Model.

Again, by the current divider rule,

$$i_L = (1+j\omega C_1 h_{ie})_i b = i_2 \frac{\left(-j\dfrac{1}{\omega C_2}\right)}{-j\dfrac{1}{\omega C_2}+j\omega L+\dfrac{h_{ie}\left(-j\dfrac{1}{\omega C_1}\right)}{h_{ie}-j\dfrac{1}{\omega C_1}}}$$

$$(1+j\omega C_1h_{ie})i_b=\frac{1}{1-\omega^2LC_2+j\dfrac{\omega C_2h_{ie}}{1+j\omega C_1h_{ie}}}i_2$$

of

$$i_2=[(1-\omega^2LC_2)(1+j\omega C_1h_{ie})+j\omega C_2h_{ie}]i_b$$

Expanding, collecting, and noticing that $i_2=-i_c$,

$$\frac{i_c}{i_b}=[\omega^2LC_2-1]+j[\omega^3LC_1C_2h_{ie}-\omega C_1h_{ie}-\omega C_2h_{ie}]$$

Now at ω_0 the j term is zero

$$\omega_0{}^3LC_1C_2h_{ie}-\omega_0C_1h_{ie}-\omega_0C_2h_{ie}=0$$

$$\omega_0{}^2LC_1C_2\equiv C_1+C_2$$

Then

$$\omega_0=\frac{1}{\sqrt{L\left(\dfrac{C_1C_2}{C_1+C_2}\right)}} \tag{21-4a}$$

But the capacitance term is merely the result of two capacitors in series

$$C_T=\frac{C_1C_2}{C_1+C_2} \tag{21-4b}$$

Then, an oscillation in this circuit again occurs at the fundamental frequency of the tank circuit alone. The minimum gain relationship is found when the j-term is zero

$$\frac{i_c}{i_b}=\omega_0{}^2LC_2-1$$

and, substituting Eq. 21-4a,

$$\frac{i_c}{i_b}=\beta=\frac{LC_2}{L\left(\dfrac{C_1C_2}{C_1+C_2}\right)}-1=\frac{C_1+C_2}{C_1}-1=\frac{C_2}{C_1} \tag{21-4c}$$

Usually, C_1 and C_2 are equal capacitance values. Consequently, by Eq. 21-4c, there is a least value of unity for beta to maintain oscillations. Therefore, the circuit will operate well into a class-C condition for transistors that have normal values of beta.

In the Hartley oscillator, a tap was used on the oscillator coil. In the Colpitts oscillator a tap is used on the capacitor by using C_1 and C_2 as a capacitor divider in Fig. 21-15a.

Very often, a special *split-stator capacitor* (Fig. 21-16a) is used with the form of the Colpitts oscillator shown in Fig. 12-15a. The capacitor has two stators and a grounded rotor. When the rotor is turned, there is a simultaneous increase or decrease in both C_A and C_B. This capacitor used in the Colpitts oscillator provides both the required tuning and the voltage division.

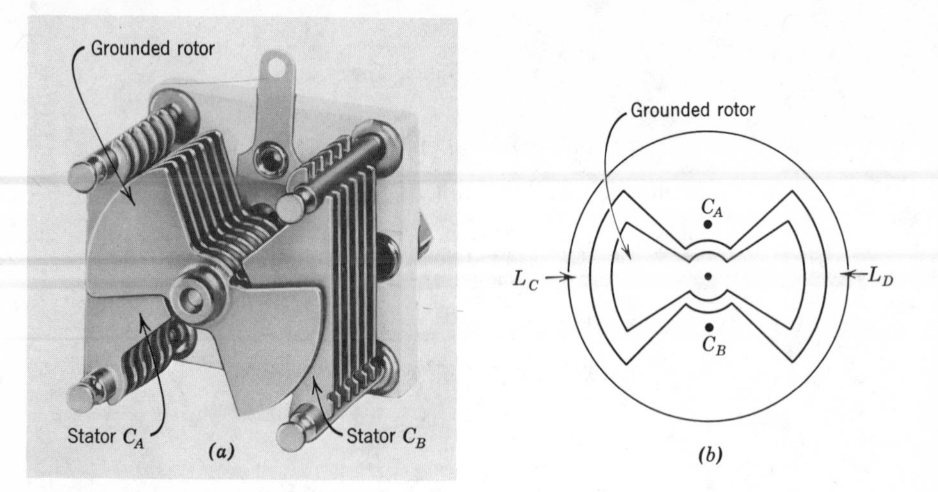

Figure 21-16 Special capacitors used for high-frequency oscillators. (*Courtesy Hammarlund Mfg. Co.*)

Another special tank circuit which is suited to the Colpitts oscillator is the *butterfly* (Fig. 21-16*b*), which is used at very high frequencies. When the rotor plates are meshed with the stator plates, the capacitors C_A and C_B are a maximum. The inductances of the supporting frames, L_C and L_D, are also maximum. When unmeshed, the rotor plates tend to short-circuit the inductance of L_C and L_D by a close capacitance effect. In this tuning circuit both capacitance and inductance vary. Since the resonant frequency of a tuned circuit is

$$f_0 = \frac{1}{2\pi\sqrt{LC}}$$

a simultaneous change of L and C give a much greater range of frequency variation than if C alone is variable.

Several versions of Colpitts oscillator circuits are shown in Fig. 21-17. The first circuit (*a*) uses a single tuning capacitor C. Small mica capacitors C_A and C_B are added to give the required capacitor voltage-divider action. At very high frequencies, C_A and C_B become small enough to be the interelectrode capacitances of the amplifying device. This version is called an *ultra-audion oscillator*, and the vacuum-tube version is given in Fig. 21-17*d*. A shunt-fed circuit is illustrated in Fig. 21-17*b*. The capacitor C_A is not essential for the electrical operation of the circuits. However, it may be required for mechanical reasons of symmetry.

The Clapp oscillator shown in Fig. 21-18 is a modified Colpitts oscillator. C_1 and C_2 serve as the capacitor voltage divider, and C_3 is in series with the coil for tuning.

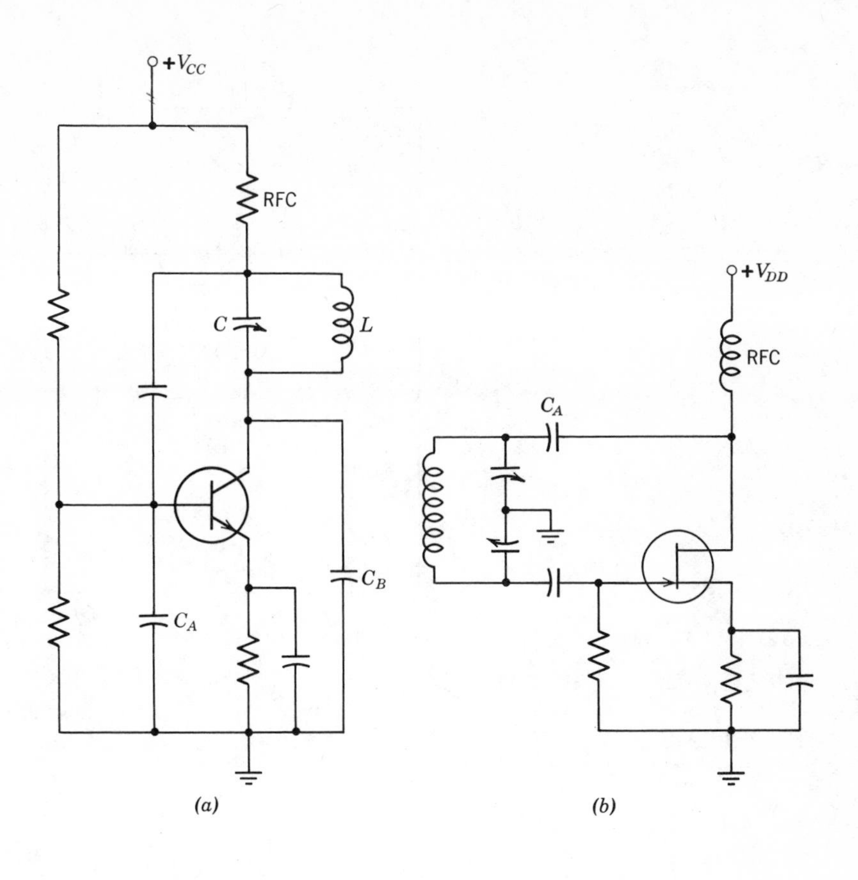

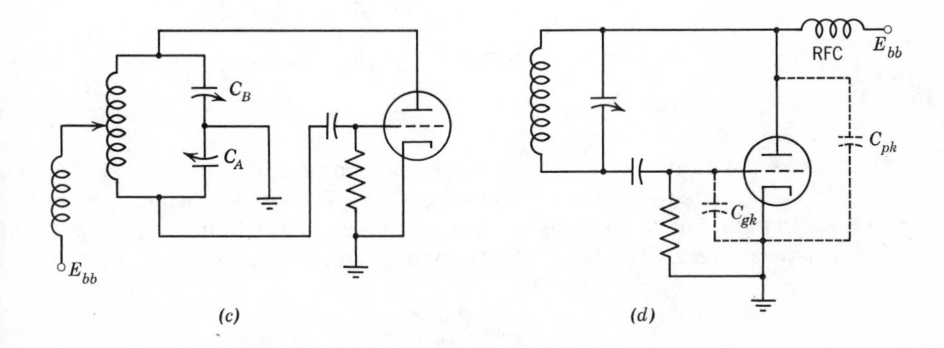

igure 21-17 Colpitts oscillator circuits. (*a*) Oscillator using a simple tuning capacitor. (*b*) JFET rcuit. (*c*) Vacuum-tube oscillator. (*d*) Ultra-audion oscillator.

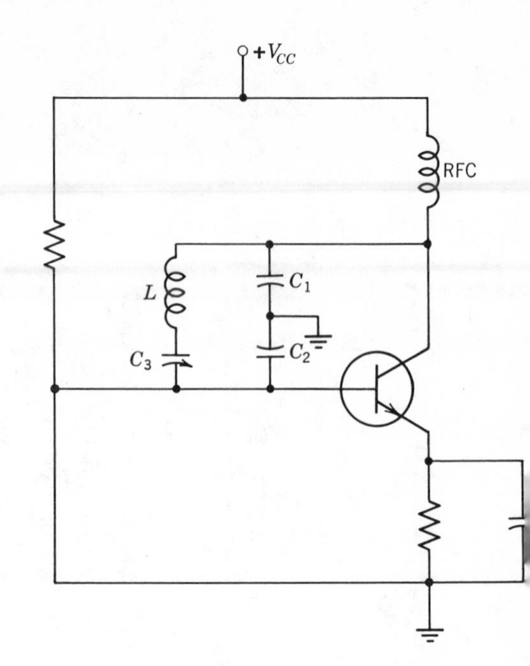

Figure 21-18 The Clapp oscillator.

PROBLEMS

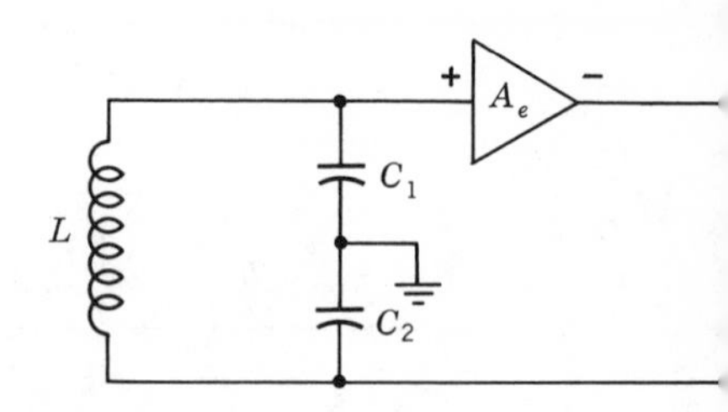

Circuit for Problem 4.

1. Draw the model for a Colpitts oscillator using a JFET. The model should be similar to the transistor model of Fig. 21-15. The input resistance to the JFET is infinite. The model must include r_d. Show that the frequency of oscillation is given by Eq. 21-4*a* and that the condition for oscillation is

$$g_m r_d = \frac{C_1}{C_2}$$

2. The drain current in an N-channel JFET is given by

$$I_D = 3.0\left(1 - \frac{V_{GS}}{V_P}\right)^2 \text{ ma} \qquad \text{where } V_P \text{ is} -4 \text{ V}$$

If C_1 and C_2 in a Colpitts oscillator circuit are each 80 pF, and if r_d is 25,000 Ω, what is the drain current assuming that the coil and the capacitors are lossless?

3. In a Colpitts oscillator circuit, L is 500 μh, C_1 is 100 pF, and C_2 is 250 pF. What is the frequency of oscillation?
4. L is 600 μh with a Q of 30. C_1 is 750 pF and C_2 is 300 pF. Assume that the input impedance to the amplifier is infinite. (a) What is the frequency of oscillation? (b) What is the load impedance on the amplifier? (c) What is the minimum gain A_e of the amplifier in order to have oscillation?

Section 21-6 Crystal Oscillators

Before discussing crystals and crystal oscillators, it is necessary to develop a new approach to the functioning of the oscillator. An approximate block diagram (Fig. 21-19*a*), which is sufficient to serve the needs of this analysis,

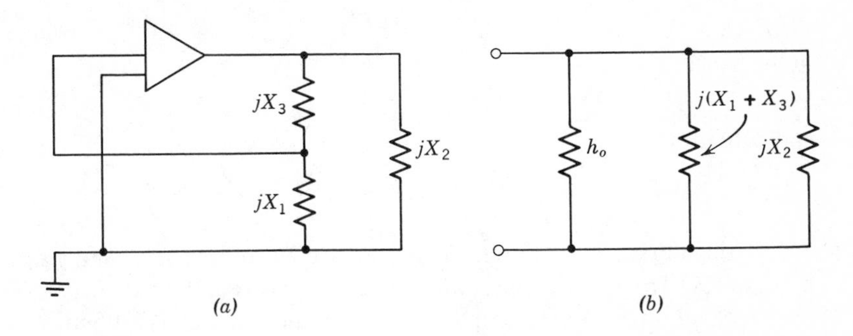

Figure 21-19 Feedback amplifier. (*a*) Block diagram. (*b*) Load circuit.

retains only those elements that contribute to the a-c aspect of the oscillator. The a-c load on the amplifier Z_L (Fig. 21-19*b*) is a parallel circuit and includes h_0 the output conductance of the transistor that is used as the amplifier. The load impedance is determined from

$$\frac{1}{Z_L} = h_0 + \frac{1}{j(X_1+X_3)} + \frac{1}{jX_2}$$

Cross-multiplying and inverting

$$Z_L = \frac{j(X_1+X_3)X_2}{(X_1+X_2+X_3)+jh_0(X_1+X_3)X_2} = -\frac{(X_1+X_3)X_2}{h_0(X_1+X_3)X_2 - j(X_1+X_2+X_3)}$$

The fundamental gain equation for a transistor stage is

$$A_e = \frac{Z_L}{r'_e} = -\frac{(X_1+X_3)X_2}{h_0 r'_e (X_1+X_3)X_2 - j(X_1+X_2+X_3)r'_e}$$

The feedback from the inspection of Fig. 21-19*a* is

$$\beta_f = \frac{jX_1}{jX_1+jX_3} = \frac{X_1}{X_1+X_3}$$

The feedback factor $\beta_f A_e$ is the product of β_f and A_e:

$$\beta_f A_e = -\frac{X_1 X_2}{h_0 r'_e (X_1+X_3)X_2 - j(X_1+X_2+X_3)r'_e}$$

In order to have sustained oscillations, the feedback factor $\beta_f A_e$ must be unity, and the j term must be zero. Since r'_e is finite and not zero, then

$$X_1+X_2+X_3=0$$

This condition reduces the expression for the feedback factor to

$$\beta_f A_e = -\frac{X_1 X_2}{h_0 r'_e (X_1+X_3)X_2} = -\frac{X_1}{h_0 r'_e (X_1+X_3)}$$

Since

$$X_1+X_2+X_3=0$$

$$-X_2 = X_1+X_3 \tag{21-5a}$$

and substituting

$$\beta_f A_e = \frac{1}{h_0 r'_e}\frac{X_1}{X_2} \tag{21-5b}$$

In the derivation of the equations, we used jX_1, jX_2, and jX_3 for the reactances. In the final equation an inductive reactance ωL is used as a positive number, and a capacitive reactance $1/\omega C$ is substituted as a negative number $(-1/\omega C)$. In order to make the feedback factor +1, in Eq. 21-5*b*, X_1 and X_2 must have the same sign; that is, both are either capacitive reactances or inductive reactances. From Eq. 21-5*a* it is evident then that X_2 and X_3 are opposite in sign. These relations are listed in Table A.

The tapped coil of the Hartley circuit provides X_1 and X_2 in Table A, and the tuning capacitor across the whole coil is X_3. The Colpitts oscillator uses the coil as X_3, and the split capacitors as X_1 and X_2.

A similar derivation can be formed for an FET or a vacuum tube. The basic gain equation is used, and the same results are obtained as in Table A and Eq. 21-5*a* and Eq. 21-5*b*. Since Table A applies equally well to all three of these amplifying devices, a Hartley oscillator, for instance, is a function of oscillator tank arrangement.

The concept of "inductive" and "capacitive" as listed in Table A is certainly not restricted to the single component elements of inductors and

TABLE A

X_1	X_2	X_3	Name of Basic Oscillator
Inductance	Inductance	Capacitance	Hartley
Capacitance	Capacitance	Inductance	Colpitts
Inductive crystal	Inductively tuned tank	C_1	Miller crystal oscillator
C_2	C_3	Inductive crystal	Pierce crystal oscillator

C_1 can be external or input-to-output capacitance (C_{bc}, C_{gd}, C_{gp}).
C_2 can be external or input-to-ground capacitance (C_{be}, C_{gs}, C_{gk}).
C_3 can be external or output-to-ground capacitance (C_{ce}, C_{ds}, C_{pk}).

capacitors. When a "black box" is inductive or capacitive, it can be used as an oscillator component for X_1, X_2, or X_3. When a tank circuit operates slightly below its natural frequency, it is inductive. The terms "inductively tuned tank" and "inductive crystal" describe this condition of off-resonance operation. This concept forms the basis for the operation and for the understanding of the tuning of a crystal oscillator.

Among a number of crystalline substances, quartz, Rochelle salt, barium titanate, and tourmaline are the most important that exhibit *piezoelectric* properties. If a slab of one of these crystalline substances is properly cut, a mechanical stress produces an emf across the slab. Conversely, an applied emf produces a mechanical stress. Rochelle salt produces the greatest piezoelectric reaction and is widely used in audio components such as microphones and phonograph pickups. Quartz has the best properties for radio-frequency work. It has a good mechanical strength, a high Q, a low-temperature drift, and a high degree of electrical stability.

A quartz crystal (Fig. 21-20), when properly cut and ground, acts like a

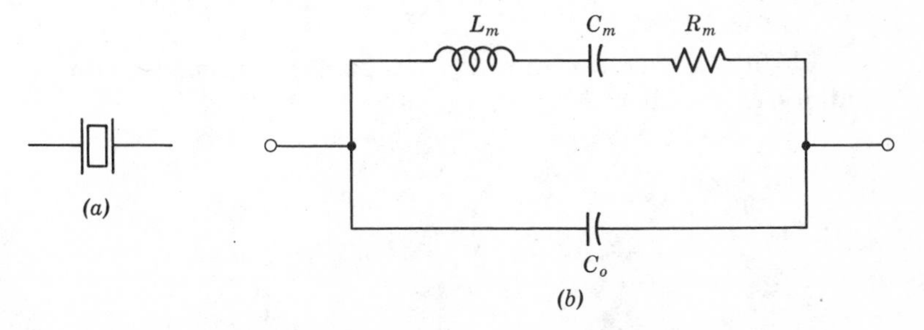

Figure 21-20 Quartz Crystal. (*a*) Circuit symbol. (*b*) Model.

parallel-resonant circuit of high Q and can be used as a tank circuit in man applications. A quartz crystal also has a series resonance which is frequentl used in crystal filters. The principal uses of quartz plates are in oscillator and in filter circuits. Electrical connections are provided by depositing metallic plating on the opposite faces of the crystal. The crystal must b mounted in a special holder and, since it is quite brittle, it must be treated i the same careful manner as a glass microscope slide.

A typical 428-kHz crystal is $2.75 \times 3.33 \times 0.636$ cm in size. The value for the equivalent circuit for this crystal are

C_0 5.8 pF	C_m 0.042 pF
L_m 3.3 H	Q 23,000

The resonant frequency of a crystal is inversely proportional to its size an thickness.

Crystals are commercially available ground for frequencies from 2 kH to about 35 MHz on fundamental and to 150 MHz on overtones. The Q of crystal can be as high as 100,000, but much higher values have been attained

There are many ways of cutting a quartz plate from a crystal. Th orientation of the cutting to the optical (Z) axis, the crystal faces (the Y o mechanical axis), and the corners (the X or electrical axis) determine th electrical properties of the crystal. Some of the characteristics of th different cuts are:

1. Suitability for different frequency ranges.
2. Suitability for filters.
3. Suitability for oscillators.
4. Different or zero temperature coefficients.

Certain cuts have a great deal of waste and, as a result, are more expensive The different cuts from a natural quartz crystal are illustrated in Fig. 21-21.

A plot of the reactance of a crystal is shown in Fig. 21-22. The crossove at f_0 is a series resonance of L_m and C_m (Fig. 21-20b).

$$f_0 = \frac{1}{2\pi\sqrt{L_m C_m}} \tag{21-6a}$$

At a slightly higher frequency, f_a, a parallel resonant or antiresonant condition exists where

$$X_{L_m} - X_{C_m} = X_{C_0}$$

or

$$2\pi f_a L_m - \frac{1}{2\pi f_a C_m} = \frac{1}{2\pi f_a C_0}$$

$$2\pi f_a L_m = \frac{1}{2\pi f_a C_0} + \frac{1}{2\pi f_a C_m} = \frac{1}{2\pi f_a}\left[\frac{1}{C_0} + \frac{1}{C_m}\right]$$

$$2\pi f_a L_m = \frac{1}{2\pi f_a}\left[\frac{C_0 + C_m}{C_0 C_m}\right]$$

solving for f_a

$$f_a = \frac{1}{2\pi\sqrt{L_m C}} \tag{21-6b}$$

where

$$C = \frac{C_0 C_m}{C_0 + C_m}$$

Then

$$f_a - f_0 = \frac{1}{2\pi\sqrt{L_m C_m}\sqrt{\dfrac{C_0}{C_0 + C_m}}} - \frac{1}{2\pi\sqrt{L_m C_m}}$$

Simplifying

$$f_a - f_0 = \frac{1}{2\pi\sqrt{L_m C_m}}\left[\sqrt{\frac{C_0 + C_m}{C_0}} - 1\right]$$

$$f_a - f_0 = f_0\left[\sqrt{1 + \frac{C_m}{C_0}} - 1\right]$$

Since $C_m/C_0 \ll 1$, we can use the expansion

$$\sqrt{1+a} \approx 1 + \frac{1}{2}a$$

Then the bandwidth is

$$f_a - f_0 = \frac{1}{2} f_0 \frac{C_m}{C_0} \tag{21-6c}$$

Using the typical values given for a crystal, we have

$$f_0 = \frac{1}{2\pi\sqrt{3.3 \times 0.042 \times 10^{-6}}} = 428\ \text{kHz}$$

and

$$f_a - f_0 = \frac{1}{2}(428)\,\frac{0.042}{5.8} = 1.350\ \text{kHz}$$

A Miller crystal oscillator using a JFET is shown in Fig. 21-23. The feedback occurs solely through the interelectrode capacitance ($C_{dg} + C_{\text{wiring}}$) of the JFET. According to Table A, when X_3 is capacitive, the elements X_1 and X_2 must be inductive. The crystal X_1 operates in the parallel mode which is f_a on Fig. 21-22. Since the crystal has an extremely high Q, the shift from the ideal parallel resonant frequency f_a is very small. There is a tank circuit in the drain only because a tank circuit is very convenient to adjust at radio frequencies. This drain tank must be so adjusted that it is inductive with respect to the operating frequency of the crystal.

Assume that an alternating voltage of fixed frequency is applied to the parallel combination of a fixed inductance and a variable capacitor. The

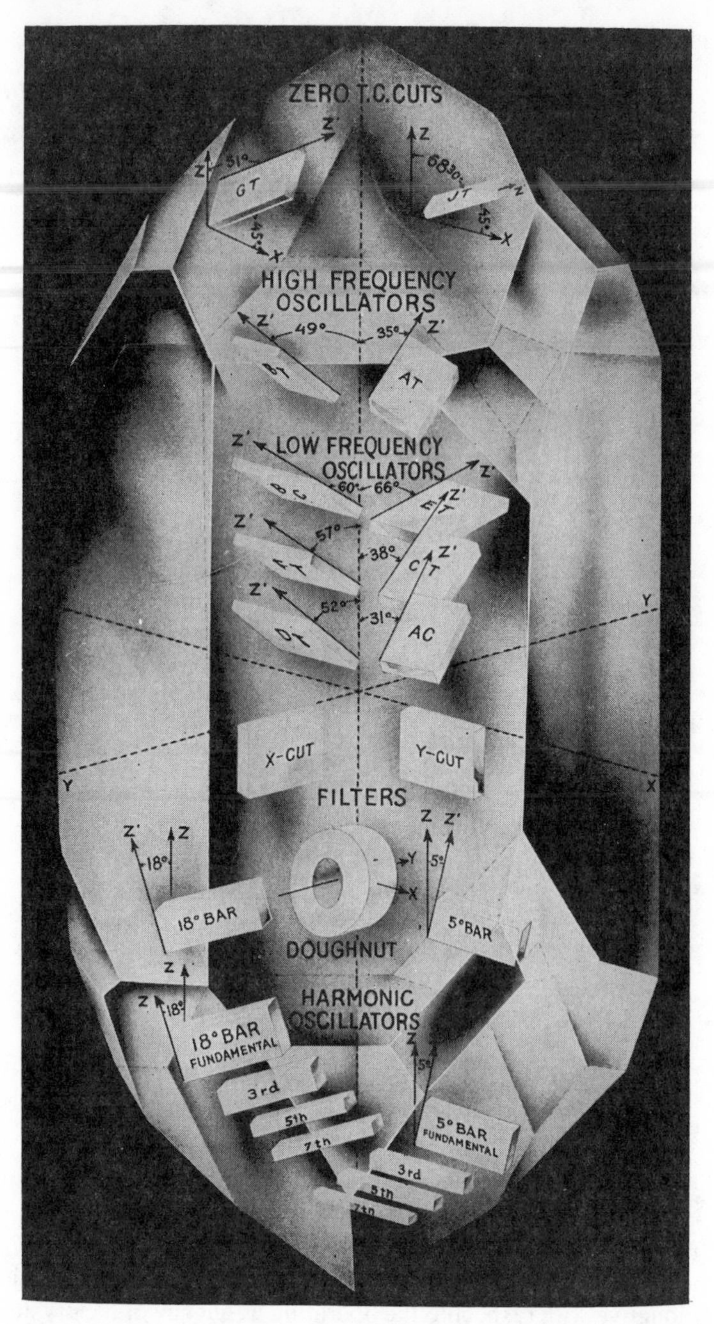

Figure 21-21 Different cuts from a quartz crystal. (*Courtesy Bendz and Scarlott, Electronics for Industry.*)

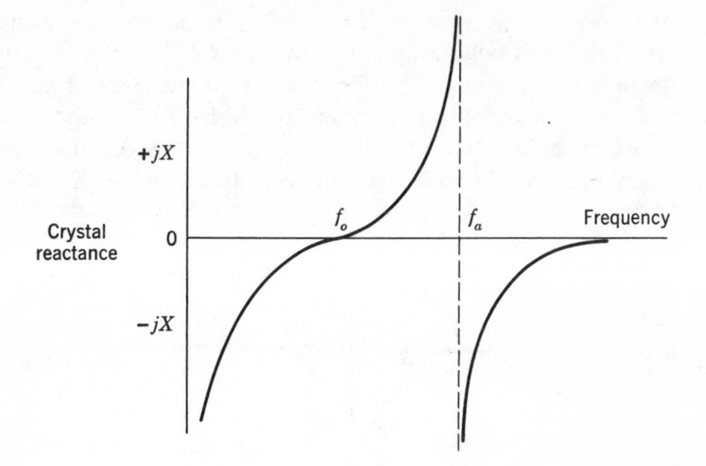

Figure 21-22 Reactance of a crystal.

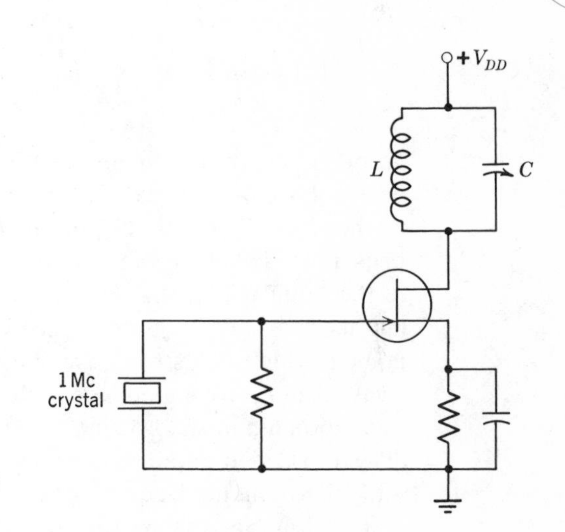

Figure 21-23 Miller crystal oscillator using a JFET.

line current is the sum of the two currents in the branches. If the inductive current is 5 ma and the capacitive current is 7 ma, the line current is 2 ma and it is capacitive. Since the current in the capacitive branch of the circuit is proportional to the value of the capacitance, when the value of the tuning capacitor is greater than that required for resonance, the line current is capacitive, and the overall drain load (Fig. 21-23) is capacitive. Similarly, in order to have the drain load inductive, the setting of the tuning capacitor

C must be less than that value of C that is required to resonante the tank circuit at the operating frequency of the crystal in the gate circuit.

Assume that the operating frequency of the crystal, (Fig. 21-23) is f_0 and that, if the tank in the drain is set to resonate at f_0, the value of C is C_0. When we plot the drain current obtained for this oscillator when C is the only variable, we obtain the solid curve shown in Fig. 21-24. In the region

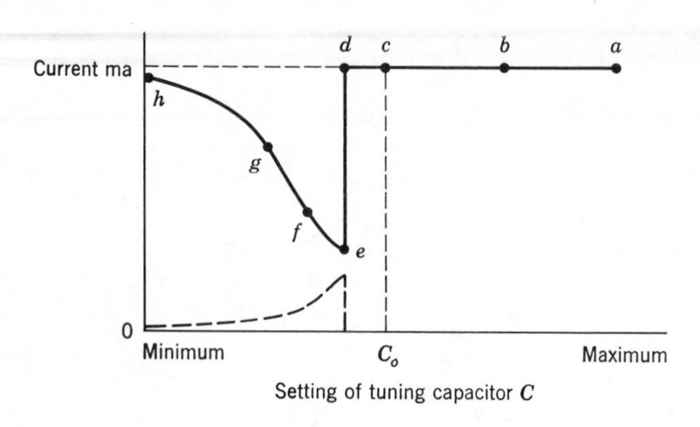

Figure 21-24 Variation in d-c current with tuning.

ab of the curve, the load (the drain tank) is capacitive at f_0, and the circuit does not oscillate. The drain current is a maximum. Where the tank capacitor in the drain is set at C_0, the load is purely resistive and the circuit still does not oscillate (point *c*). It is only when the drain circuit is detuned to the point *d* that the drain tank is sufficiently inductive to produce oscillation. The current jumps sharply from point *d* which is no oscillation to point *e* which is a strong oscillation. When the tank is detuned still further away from f_0, its impedance with respect to f_0 falls, and the strength of the oscillation diminishes, as indicated by the path from *e* to *h*. It is often stated that, in order to obtain an oscillation in the crystal oscillator, a tank circuit is tuned to a higher frequency than the crystal frequency.

The drain tank is set for normal operation, not at point *e*, but at point *f*. If there is any small circuit change or disturbance at *f*, the operation shifts slightly toward *e* or *g*. If the point of operation is at *e*, a small disturbance could easily cause the circuit to go out of oscillation. The small power loss accepted by operating at *f* instead of at *e* is well worth the assurance of continued operation. The d-c current in the bias circuit for the crystal is maximum when the oscillation is strongest. This is shown by point *e* in Fig. 21-24. In a vacuum-tube circuit, the d-c grid current is usually monitored in preference to the plate current. A slight change in the operation of the oscillator gives a greater percentage change in grid current than in plate current. When d-c current meters are used in oscillator circuits, they must be very carefully decoupled with R-F Chokes and bypass capacitors.

A bypassed resistor (in the emitter, or in the source, or in the cathode) may be used as a safety element, should the crystal fail to function. Too large an alternating voltage across a crystal can cause the crystal to shatter from excessive mechanical vibrations. The crystal itself normally serves as the capacitor in the bias clamp or grid-leak.

A small variable capacitor is sometimes placed in parallel with the crystal. It is possible to adjust the frequency of a crystal oscillator up to 50 or 100 cycles per megacycle by varying this capacitor. If a particular frequency must be exact, the grinding process becomes too critical, and this method of fine adjustment is often used.

Very often the temperature variation of a crystal causes the crystal to drift beyond the allowable limits of the required tolerance. Then it is necessary to maintain the crystal in a temperature-controlled oven.

Again referring to Table A, the Pierce crystal oscillator (Fig. 21-25*a*) is a version of the Colpitts oscillator. The crystal operating in the parallel mode (f_a of Fig. 21-22) is offset slightly to function as the inductance X_3. The two capacitors, C_1 and C_2, are X_1 and X_2. External capacitors are used only if the interelectrode capacitances are insufficient or are too far out of balance. In the JFET circuit (Fig. 21-25*b*) a very small capacitor of the

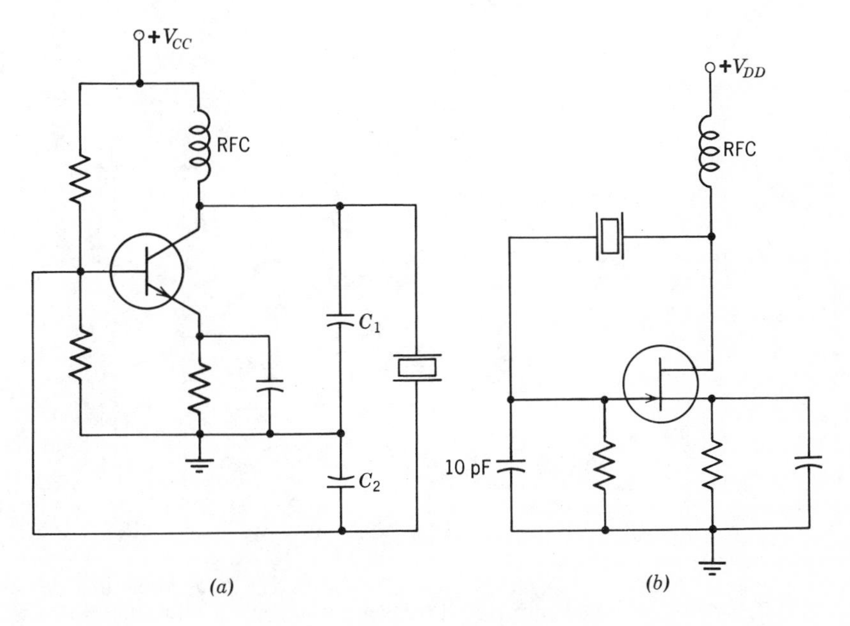

Figure 21-25 The Pierce oscillator. (*a*) Transistor circuit. (*b*) JFET circuit.

order of a few picofarads is used to eliminate parasitic or spurious oscill tions. In this circuit, if the crystal is removed, the drain current is, sa 10 ma. With the crystal in place and without the small capacitor, the curre might be of the order of 0.5 ma. With the oscillator working properly at th fundamental crystal frequency, the drain current is of the order of 25 μ These figures are used to illustrate comparable data values.

The Pierce oscillator is very convenient for transmitters that opera crystal-controlled on several channels. The different crystals can be switche in without having to retune the oscillator circuit.

Some other crystal oscillator circuits are shown in Fig. 21-26.

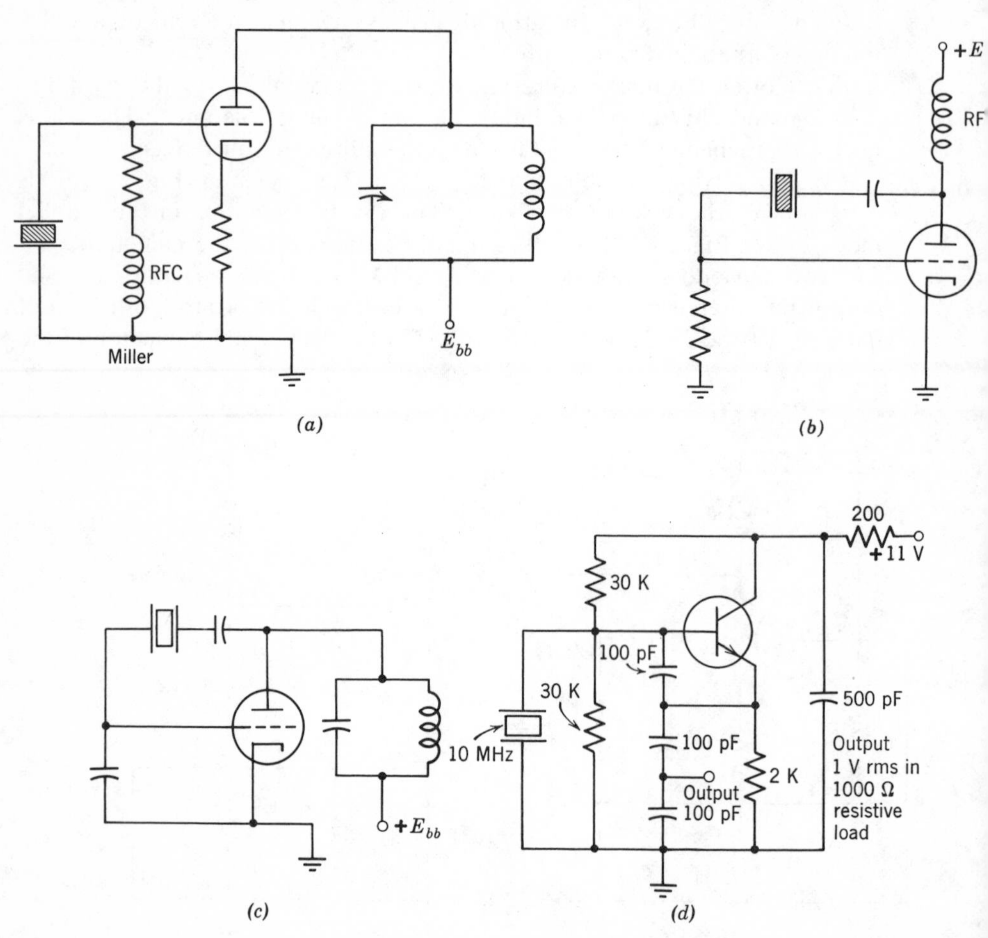

Figure 21-26 Other crystal oscillators. (*a*) Vacuum-tube Miller circuit. (*b*) Vacuum-tube Pierce cryst oscillator. (*c*) Series-mode operation of crystal. (*d*) Microcircuit Colpitts oscillator complete withi a TO5 case.

PROBLEMS

1. Using the values for the equivalent circuit of the crystal given in the text, determine the impedance of the crystal at resonance in the series-resonant mode and in the parallel resonant mode.
2. The values for the equivalent circuit of a crystal are: $L_m = 0.6$ H, $C_m = 0.018$ pF, $C_0 = 4.6$ pF, and $Q = 25{,}000$. Determine f_0 and f_a.
3. Determine the impedance of the crystal given in Problem 2 at resonance in series-resonant mode and in the parallel-resonant mode.
4. When a particular crystal is connected in a circuit in the series-resonant mode, the circuit oscillates at 4.000 MHz. When connected in a circuit in the parallel-resonant mode, the oscillator frequency is 4.004 MHz. A bridge measurement yields 2.0 pF for C_0. Determine the equivalent circuit of the crystal.
5. The crystal given in Problem 2 is placed in an oscillator circuit operating in the parallel-resonant mode. A trimmer varying from 1.8 to 6.4 pF is placed in parallel with the crystal. What range of variation does the trimmer give to the oscillator frequency? What is the effect of the trimmer if it is used in series-mode operation?

Section 21-7 The Electron-Coupled Oscillator (ECO)

The electron-coupled oscillator (Fig. 21-27) is primarily a vacuum-tube circuit that operates at a relatively high power level. A series-fed Hartley oscillator is the basic circuit for the electron-coupled oscillator. The triode is replaced by a tetrode or a pentode, and the screen is used as the plate of the oscillator section of the tube. The electron stream from the cathode, by the time it passes through the screen mesh, is varying in accordance with the fundamental frequency of the oscillator tank circuit. The screen-to-plate region of the tube is used for power amplification, and isolation of the load

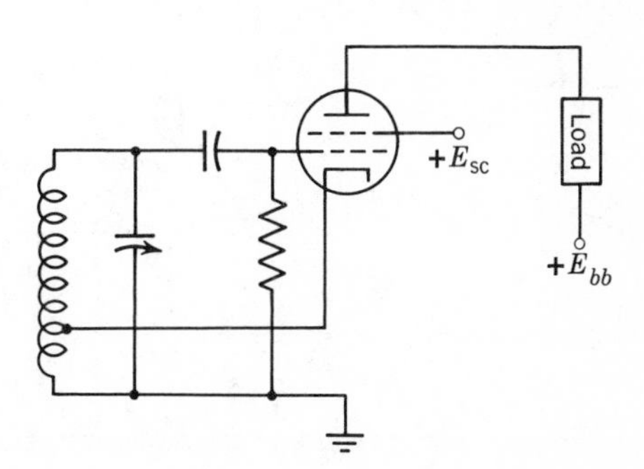

Figure 21-27 The electron-coupled oscillator.

from the actual oscillator is obtained. Because of the shielding action of th screen, a change in the load does not affect the oscillator. In the circu diagram for the electron-coupled oscillator (Fig. 21-27), we show the loa as a block rather than as a specific circuit component. This load may be tuned circuit which is adjusted either to the fundamental frequency or to harmonic frequency. In the latter case, the electron-coupled oscillator serve the dual purpose of an oscillator and a frequency multiplier. In many applica tions, the load is resistive or reactive, depending on the nature of the us of the circuit. Several variations of this circuit are used, but they all have th same basic principle of operation.

The electron-coupled oscillator is so named because energy is transferre from the oscillator to the load through the electron stream of the tube. Th shielding action of the screen (and also of the suppressor if a pentode i used) prevents interaction between the oscillator and the load. The load ma be adjusted or varied without upsetting the electrical conditions within th oscillator section of the circuit.

Section 21-8 Limitations of Oscillators

With the exception of the electron-coupled oscillator, we have not discusse any of the methods for taking power or voltage from an oscillator circuit Very often, the power drain by the load on an oscillator is very small, and a "gimmick" loop (*A* in Fig. 21-28) may be used. Two wires twisted togethe is another variety of "gimmick" which also uses a capacitive coupling be tween the two wires (*B* in Fig. 21-28). A coupling capacitor (*C* in Fig

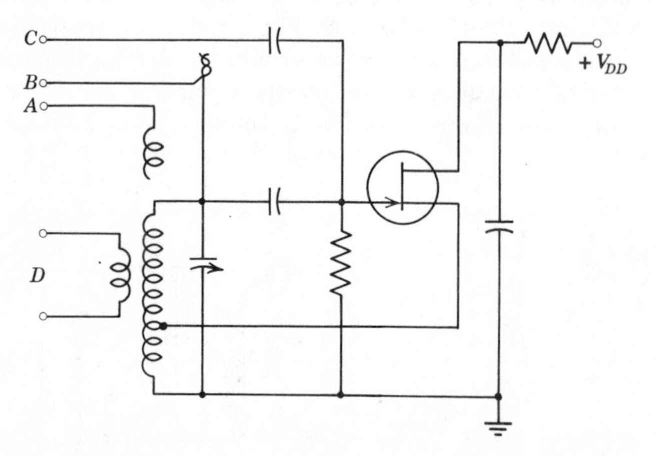

Figure 21-28 Methods of obtaining an output from an oscillator.

21-28) which is not large enough to interfere with the action of the oscillator bias-clamp capacitor may be used. A secondary winding (*D* in Fig. 21-28) is added to the oscillator coil when larger amounts of power are required.

If too much power is demanded from the oscillator, the over-all circuit Q will fall to the point where the flywheel effect of the tank circuit will not be able to maintain a sinusoidal waveform. Certain design factors have been established limiting the output that may be taken from an oscillator. One consideration states that the ratio of stored energy to output energy per cycle must be at least two. Another approach requires that the effective Q of the loaded oscillator tank circuit must be at least 4π. These two factors are mentioned to give the reader a rule of thumb. For the development of these relations, reference should be made to textbooks on circuit design.

When an oscillator is operated at a high power and if the circuit should fail and oscillations cease, the circuit immediately draws excessive supply current. In order to protect the amplifier, a fuse, a resistor, or a panel lamp is often placed in the d-c supply circuit. The panel lamp offers an increased resistance in the circuit when the lamp current increases.

Very often, a rapid means of checking is desired on the operation of an oscillator that does not have permanent metering circuits. One may use an RF vacuum-tube voltmeter which reads the alternating voltage directly. Radio-frequency voltmeters are often unavailable because of their cost and limited usefulness. When an ordinary d-c voltmeter is used to measure the value of the direct class-C bias voltage, the capacitance of the meter and of meter leads either detunes the oscillator so far that the reading is worthless or it loads the circuit to the point where the oscillation ceases. When a resistor of the order of a megohm is placed in series with the "hot" meter lead *at* the point of measurement, the meter will not detune the circuit appreciably, and an accurate reading can be taken of the bias. In order to eliminate the need for correcting the meter reading for the probe resistance drop, commercial vacuum-tube voltmeters have this resistor built in as an integral part of the d-c lead and are calibrated with the probe in place. A neon lamp is often used to check a radio frequency oscillator for operation.

The frequency range of the oscillator is determined by the ratio of the maximum to the minimum capacitance of the tuning capacitor by calculation from the resonant-frequency equation:

$$f_0 = \frac{1}{2\pi\sqrt{LC}}$$

When the inductance is constant, the frequency ratio is inverse to the square root of the capacitance ratio. As an example, assume that a tuning capacitor has a range from 30 to 450 $\mu\mu$f. Assume that the total stray capacitance of the coil, the amplifier, and the wiring is 20 $\mu\mu$f. The total capacitance ratio is (450 + 20) to (30 + 20) or 470/50, and the frequency coverage ratio i $\sqrt{470/50}$ or 3.06 to 1. In a butterfly circuit, if the inductance change is th same as the change in capacitance, say 470/50, the frequency range woul be 470/50 or 9.36 to 1 which is much greater than the frequency ratio in th conventional tank circuit.

The output voltage of one of these oscillator circuits is not constar over the entire range of tuning. The design of the coil and the L/C ratio ar the primary factors. In many applications, a variation in the amplitud of the oscillator output within 2:1 over the band of tuning is generall considered satisfactory.

The factors that involve the stability of an oscillator warrant specia consideration. A variation in supply voltage can produce variations in fre quency and in output level. Very often a regulator is used to control th supply voltage within close limits. At the very least, a decoupling filte should be used. The circuit components should be ruggedly mounted, les mechanical vibrations produce a varying capacitance or inductance. Prope values of R and C in the bias circuit aid in oscillator stability. Changes i temperature produce mechanical contractions and expansions within th coil and capacitor which cause frequency and output variations. Man oscillators have temperature-compensating capacitors placed in paralle with the tank circuit. These capacitors are available with different tempera ture coefficients and are rated in picofarad change per picofarad of capa citance per degree centigrade (pF/pF/°C). Humidity variations can als affect the stability of an oscillator.

In addition to these factors, it must be remembered that a high effectiv Q, a low L/C ratio, and a light loading are the three pre-requisites for os cillator stability. With great care the frequency of a feedback oscillato (other than crystal) can be held to a few thousandths of one per cent. A crystal oscillator can be stabilized to one part in one hundred million wit care.

Many times in electronic circuits undesired negative-resistance pheno mena establish conditions where unwanted, spurious oscillations occur These oscillations are termed *parasitics*. Typical sources of negativ resistances that are encountered are gas discharges and arcs, interaction between the screen and the plate when tetrodes and pentodes are connecte as triodes, and interactions between components. The inductance an capacitance of the leads and of the amplifiers themselves provide a ver high-frequency tank circuit which can oscillate in conjunction with thi negative resistance.

A parasitic oscillation consumes power which otherwise might b converted into useful a-c power output. A parasitic oscillation which occurs within the amplifier envelope increases the heating of the amplifier. Ver often the parasitic oscillations radiate energy into space. This radiation ca severely hamper the operation of commercial services which are properly allocated to that part of the frequency spectrum which suffers the interference. In most cases, the elimination of the parasitic is insured by inserting in the offending lead a small resistance which is sufficient to overcome the negative-resistance effect. These parasitic suppressors are usually of the order of 10 or 100 Ω.

Questions

1. Define or explain each of the following terms: (*a*) phase shift oscillator, (*b*) bridge oscillator, (*c*) series feed, (*d*) shunt feed, (*e*) butterfly tank, (*f*) split stator, (*g*) series tuning, (*h*) bias clamp, (*i*) electron-coupled oscillator, (*j*) tickler winding, (*k*) link, (*l*) parasitics.
2. What gain must an amplifier have to be used with a phase-shift oscillator?
3. Why should a bias clamp or an equivalent be used with oscillators?
4. What is the principle of operation of a bridge oscillator?
5. Explain the build-up process of an oscillation in an Armstrong oscillator.
6. Are phase-shift oscillators suitable for use at high frequencies? Explain.
7. Compare shunt feed with series feed.
8. In a Hartley oscillator, what is the minimum value of beta that will obtain suitable oscillation?
9. What is the minimum value of beta that will obtain oscillation in a Colpitts oscillator.
10. Is an RF oscillator operated class-A, class-B, or class-C? Explain.
11. What determines the approximate resonance frequency in an Armstrong oscillator? In a Hartley oscillator? In a Colpitts oscillator?
12. Why is an oscillator detuned when a person's hand is placed close to the oscillator tank?
13. What is the advantage of a butterfly circuit?
14. How is a Clapp oscillator tuned?
15. What determines the resonant frequency of a Clapp oscillator?
16. Why is a crystal resonant at two different frequencies?
17. How is a crystal oscillator circuit tuned?
18. How is the temperature stability of a crystal specified?
19. What is the advantage of the ECO?

20. How can an ECO be used as a frequency tripler? What are the steps in oscillator adjustment?
21. Why is a panel lamp preferred to a resistor in the emitter, or source, or cathode circuit of a high-power oscillator?
22. Why does an ordinary d-c voltmeter detune an oscillator when it is used to measure circuit voltages?
23. Name three requisites to good oscillator design.
24. How do parasitic suppressors function?

Chapter Twenty-two

NONSINUSOIDAL OSCILLATORS

When an R-C circuit is placed across a d-c voltage, the voltage on the capacitor rises with time. The time interval between two points on this rise is evaluated in terms of the network parameters (Section 22-1). Also the use of this R-C circuit to develop a pulse for triggering is explained. The Schmitt trigger (Section 22-2) brings out some of the fundamental concepts in a bistable circuit. The flip-flop (Section 22-3) is another bistable circuit. The monostable multivibrator (Section 22-4) is an extension of the bistable circuit. A free-running multivibrator (Section 22-5) is an application of an astable condition of operation. A synchronizing pulse (Section 22-6) is used to lock an astable osciilator to an external source. The blocking oscillator (Section 22-7) is another form of nonsinusoidal oscillator that is easily synchronized. A discharge circuit (Section 22-8) is used primarily to convert square waveforms into a waveform that increases linearly with time for use as a sweep voltage for an oscilloscope. Negative resistance concepts (Section 22-9) are most convenient to explain the operation of tunnel-diode oscillators and unijunction oscillators. The use of an astable multivibrator in a power supply (Section 22-10) is examined.

Section 22-1 Basic Concepts

Before starting a discussion of actual circuits for nonsinusoidal oscillators, there are several fundamental concepts that must be treated. The first topic concerns the establishment of conducting and nonconducting times for 1 cycle of the oscillator. In a sinusoidal oscillator, the whole time, or period, of the cycle is usually determined by the resonant frequency of a tuned circuit. The conducting time and the nonconducting time of the amplifier is

established by the class-C bias conditions. In nonsinusoidal oscillators we make use of the charge or discharge of a capacitor in an R-C circuit.
The capacitor (Fig. 22-1a) begins to charge when the switch is closed a

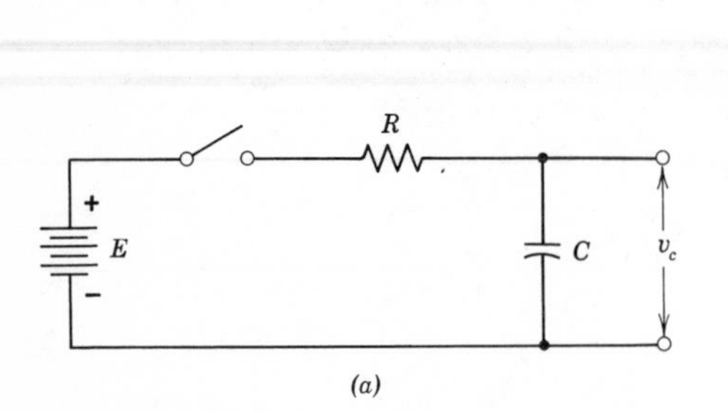

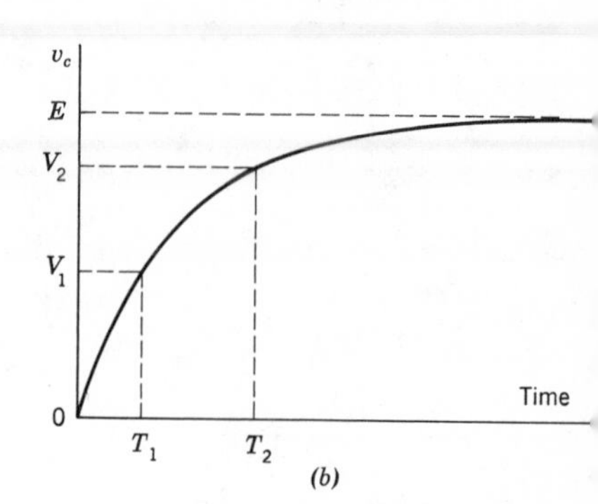

Figure 22-1 R-C charging circuit. (a) Circuit. (b) Response.

zero time. The voltage across the capacitor builds up with time as a function of the time constant

$$\tau = RC \text{ seconds}$$

and eventually charges to the asymptotic value E as shown in Fig. 22-1b. We are interested primarily in the time difference $(T_2 - T_1)$ between two specific voltage levels V_1 and V_2. The resistance-capacitance charging curve for the exponential rise is represented by

$$v_c = E[1 - \epsilon^{-t/RC}] = E - E\epsilon^{-t/RC}$$

The potential V_1 at T_1 is

$$V_1 = E - E\epsilon^{-T_1/RC}$$

and the potential V_2 at T_2 is

$$V_2 = E - E\epsilon^{-T_2/RC}$$

Solving each for T_1 and T_2, we have

$$\epsilon^{-T_2/RC} = \frac{E - V_2}{E} \quad \text{and} \quad \epsilon^{-T_1/RC} = \frac{E - V_1}{E}$$

Then

$$-\frac{T_2}{RC} = \ln \frac{E - V_2}{E} = \ln (E - V_2) - \ln E$$

and

$$-\frac{T_1}{RC} = \ln \frac{E - V_1}{E} = \ln (E - V_1) - \ln E$$

Subtracting the first from the second gives

$$\frac{T_2}{RC} - \frac{T_1}{RC} = \ln(E - V_1) - \ln(E - V_2)$$

Then

$$T_2 - T_1 = RC \ln \frac{E - V_1}{E - V_2} \tag{22-1}$$

If V_1 and V_2 are high on the curve, there is a marked curvature between V_1 and V_2. If V_1 and V_2 are low on the curve or if E is very high in relation to V_1 and V_2, the path between V_1 and V_2 approximates a straight line. In other words, the rise of voltage from V_1 and V_2 would be linear with time.

By using McLaurin's series, the expansion of ϵ^x is

$$\epsilon^x = 1 + \frac{x}{1} + \frac{x^2}{2!} + \frac{x^3}{3!} + \cdots$$

When this expansion is used for

$$v_c = E(1 - \epsilon^{-t/\tau})$$

we have

$$v_c = V\frac{t}{\tau}\left[1 - \frac{t}{2\tau} + \frac{t^2}{6\tau^2} - \frac{t^3}{24\tau^3} + \cdots\right]$$

Let $t/\tau = 1/10$
Then

$$v_c = V\frac{t}{\tau}\left[1 - \frac{1}{20} + \frac{1}{600} - \frac{1}{2400} + \cdots\right] = V\frac{t}{\tau}[0.95]$$

In other words if the time between two intervals, 0 and T, is less than $\frac{1}{10}$ the time constant, RC or τ, the deviation from ideal linearity is less than 5%.

An example of the application of this concept is the thyratron relaxation oscillator (Fig. 22-2). The capacitor C charges to V_i, the ionization potential of the thyratron. The thyratron fires and the voltage falls to V_e, the extinction potential. The thyratron goes out because the resistor R limits the current to such a low value that the recombination of electrons and ions takes place faster than ionization is produced. Then the capacitor starts to charge up again. A change in C or R changes the frequency of the oscillation. A change in grid bias not only changes frequency but also changes the amplitudes.

One other concept is in general use in nonsinusoidal oscillator circuits. Two resistors and a capacitor are connected in series across a positive supply (Fig. 22-3*a*). The capacitor charges to 300 V with the polarity shown. At time T_1 the switch is closed. The voltage across the capacitor cannot change instantly and, since the + side of the capacitor is grounded through

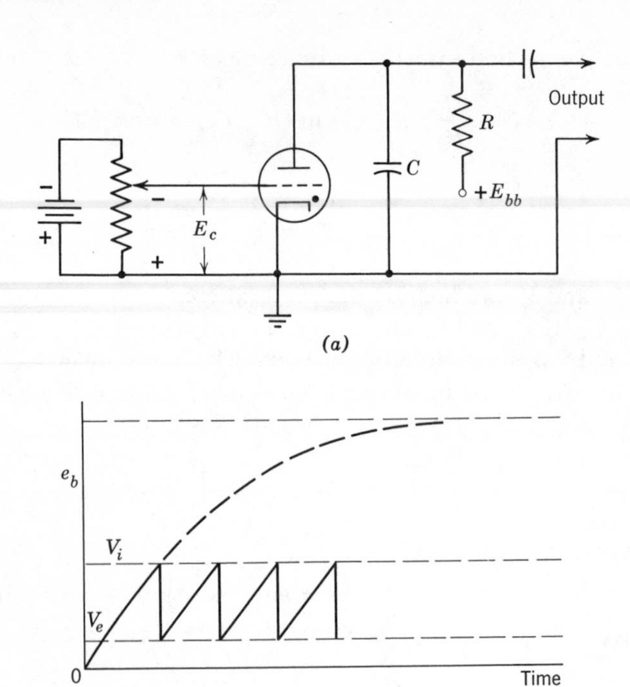

Figure 22-2 The thyration relaxation oscillator. (*a*) Circuit. (*b*) Output waveform.

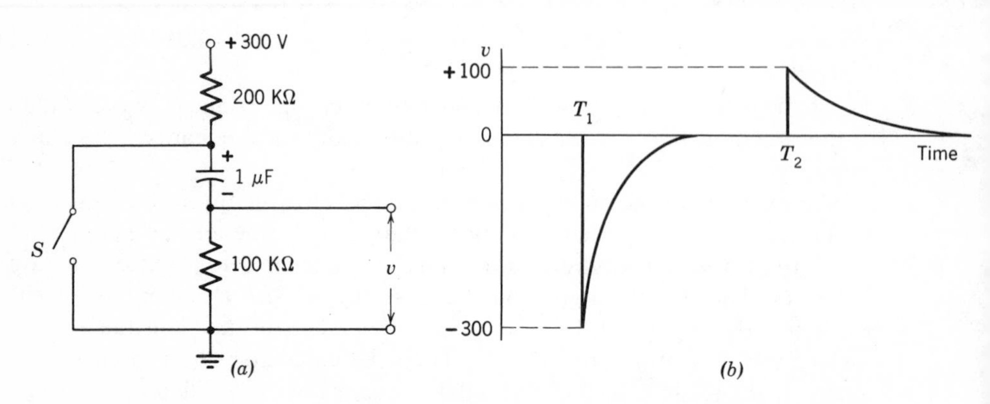

Figure 22-3 Circuit designed to produce pulses. (*a*) Circuit. (*b*) Output waveform.

the switch, v must be -300 V. This voltage falls toward zero with a tim constant of $10^5 \times 10^{-6}$ or 0.1 sec. When the switch is opened at T_2, th capacitor begins to recharge to its initial condition and by voltage divisio

v is now momentarily $+100$ V. This voltage falls toward zero now with a time constant of $3 \times 10^5 \times 10^{-6}$ or 0.3 sec. By means of this circuit, a negative pulse can be obtained even though the only supply voltage is positive.

PROBLEMS

1. A 2-μF capacitor is charged from zero through a 30-kΩ resistor across a 30 V supply. Determine the time interval from $v_C = 3$ V to $v_C = 6$ V. Determine the time interval from $v_C = 6$ V to $v_C = 9$ V. Use both the exact and the approximate methods.
2. A capacitor is charged from zero through a resistor across a d-c supply. The time interval required between $v_C = 4$ V and $v_C = 7$ V is required to be 240 μsec. What is the time constant of the required circuit? The supply is 60 V.
3. In Fig. 22-2, E_{bb} is 200 V, R is 150 kΩ, and C is 0.024 μF. V_i is 70 V and V_e is 16 V. Determine the frequency of the sawtooth.
4. In Fig. 22-3, for how long a time is the output voltage less than -10 V? For how long a time is the output voltage greater than $+10$ V?
5. In Fig. 22-3, both resistors are 100 kΩ. Sketch the output voltage waveform.
6. In Fig. 22-3, what value of capacitor is required that will keep the output voltage below -30 V for one second?

ction 22-2 The Schmitt Trigger

A two-stage d-c amplifier has a positive feedback loop (Fig. 22-4*a*). Also the amplifier is biased below cutoff. When the d-c signal level is raised from zero, the signal must be, at least, V_1 before the circuit begins to amplify. When the feedback factor $\beta_f A_e$ is less than unity, the amplifier output increases linearly from 0 to V as the input level is raised from V_1 to V_2 (Fig. 22-4*b*). At V_2 the amplifier output is saturated and cannot exceed V. The unsaturated gain of the amplifier with the feedback is

$$A'_e = \frac{V - 0}{V_2 - V_1}$$

This expression is the slope of the curve which is positive.

Now, when the feedback factor is raised to unity, the expression for gain

$$A'_e = \frac{A_e}{1 - \beta_f A_e}$$

is infinite, implying a vertical slope for the path from V_1 to V_2. This is shown in Fig. 22-4*c*. Actually, where the signal is raised slowly from zero, when the signal reaches threshhold, the ouptut jumps from zero to saturation as quickly as time constants and transit times of carrier movements will allow.

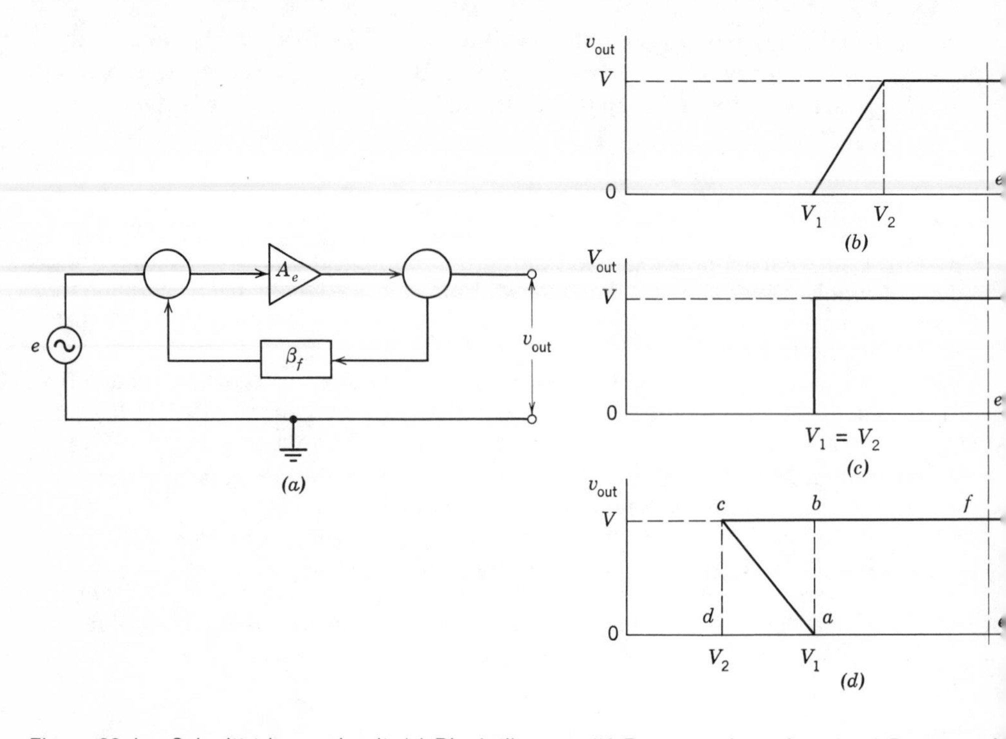

Figure 22-4 Schmitt trigger circuit. (*a*) Block diagram. (*b*) Response for $\beta_f A_e < 1$. (*c*) Response fo $\beta_f A_e = 1$. (*d*) Response for $\beta_f A_e > 1$.

When the feedback factor exceeds unity, the mathematical equatic indicates a negative slope (Fig. 22-4*d*). Actually, the transition from cutc to saturation cannot follow the solid path; that would require that the sign decrease. The transition is actually an abrupt switching from *a* to *b* in ze time. As the signal level is increased from V_1, the output remains at i saturated value V. When the signal is reduced from a high value toward zer the output remains saturated until the input falls below V_2. Again, there an abrupt drop from *c* to *d* in zero time, since there is no way for the inp signal level to go back up to V_1.

The full cycle is described by starting at 0, increasing to *a*, a jump to and a saturation to *f*, a decrease to *c*, a jump to *d*, and a decrease back to 0 complete the cycle. This path is very similar to the type of path that describ a hysteresis loop in steel. Consequently, we define the hysteresis volta V_H as

$$V_H = V_1 - V_2 \tag{22-}$$

The Schmitt trigger has a wide application to reshaping waveforms. For instance, data in space communications is transmitted in the digital form of pulses. The original pulses are square, but the noise added to the signal degrades the waveform. A Schmitt trigger circuit can be used to restore the signals back to clean pulses, as shown in Fig. 22-5.

A circuit used in the laboratory to determine the characteristics of the Schmitt trigger is given in Fig. 22-6. The setting of R_E determines the amount

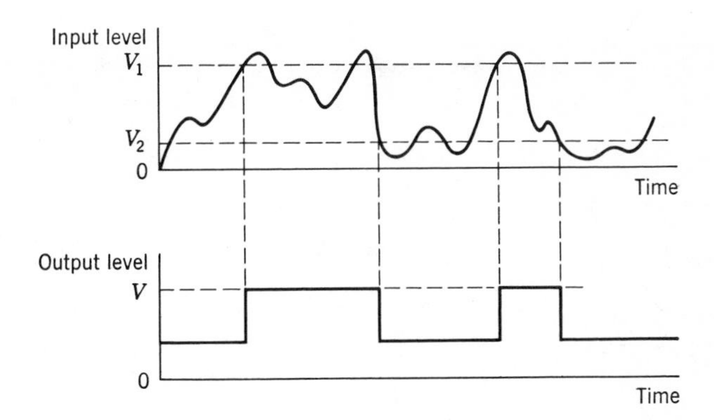

Figure 22-5 Schmitt trigger input and output waveforms.

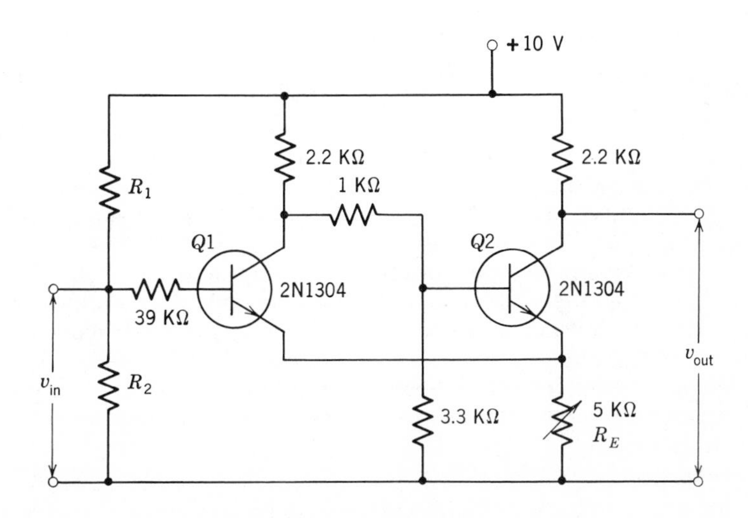

Figure 22-6 Schmitt trigger circuit.

of hysteresis in the response curve. R_1 and R_2 are the two halves of a 10-kΩ potentiometer, used to change the input signal level. When the circuit is used to reshape signals, R_1 is omitted from the circuit. Also, a small capacitor is placed in parallel with the 1-kΩ coupling resistor to increase the switching speed for high frequency incoming signal data. The voltage across this *commutator capacitor* serves to force a higher speed switching by placing an additional pulse on the base of the second transistor, as explained by Fig. 22-3.

PROBLEMS

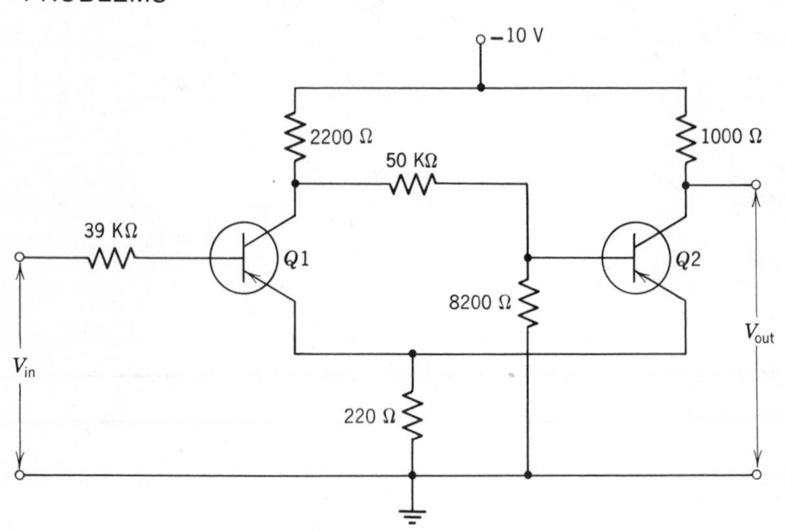

Circuit for Problems 2 and 3.

1. The input voltage to a Schmitt trigger circuit is raised from zero, and when the input voltage reaches +2.4 V, the output collector voltage switches from +2 to +10 V. The output voltage remains at +10 V for higher values of input signal. When the input voltage is lowered from a high voltage back toward zero, when the input voltage decreases to +1 V, the output voltage switches back to +2 V and remains at +2 V for all input levels more negative than +1 V. The circuit will only switch if the input level is raised to at least +2 V again. A 500-Hz signal is applied to the input. The input signal has a peak-to-peak amplitude of 8 V. Draw the output waveform and determine the times of switching.
2. The transistors in the Schmitt trigger circuit have a value of 100 for β, $V_{CE,sat}$ is 0.2 V and V_{BE} is 0.3 V. Determine all the circuit voltages and currents when Q2 is saturated.

3. For Problem 2, determine all circuit voltages and currents when $Q1$ is saturated. Between what limits does V_{out} function?
4. For Problem 2, at what input signal level does $Q2$ switch from a saturated state to a nonconducting state?
5. For Problem 2, if $Q1$ is saturated, to what voltage must the input be reduced in order to switch the circuit back to have $Q2$ saturated?
6. Refer to the Schmitt trigger circuit shown in Fig. 22-6. What is the maximum value that R_E may have in order that $Q2$ can exist in a saturated state? Assume $V_{CE,\text{sat}}$ is 0.2 V, V_{BE} is 0.3 V, and β is 100.

Section 22-3 Bistable Multivibrators – The Flip Flop

The bistable multivibrator commonly called a flip flop (Fig. 22-7*a*) is a two-stage amplifier in which the output is fed back into the input to create a positive feedback. Assume that $Q2$ carries current and is saturated. Assume also that V_{BE_2} and V_{CE_2} are each zero. A series circuit exists from V_{BB} through R_6, R_5, and $Q2$ to ground. The base bias on $Q1$ is

$$V_{BE_1} = \frac{R_5}{R_5 + R_6} V_{BB} = -\frac{360}{360 + 150} 5 = -3.54 \text{ V}$$

Consequently, $Q1$ is reverse biased and must be cut off. The bias network on $Q2$ is the series circuit of R_1, R_2, and R_3. The open-circuit voltage at the junction of R_2 and R_3 for an equivalent circuit by Thévenin's theorem

$$V_{oc} = \frac{R_3}{R_1 + R_2 + R_3}(V_{CC} - V_{BB}) + V_{BB}$$

$$= \frac{150}{75 + 360 + 150} 25 - 5 = 6.4 - 5 = 1.4 \text{ V}$$

and the series resistance by Thévenin's theorem is

$$\frac{(R_1 + R_2)R_3}{R_1 + R_2 + R_3} = \frac{(75 + 360)\,150}{75 + 360 + 150} = 112 \text{ k}\Omega$$

Assuming zero for V_{BE_2}, the base current $Q2$ is

$$I_{B_2} = \frac{1.4}{112} = 0.0126 \text{ ma} = 12.6\ \mu\text{a}$$

Assuming zero for V_{CE_2}, the collector current in $Q2$ is

$$I_{C_2} = \frac{V_{CC}}{R_4} = \frac{20}{75} = 0.268 \text{ ma}$$

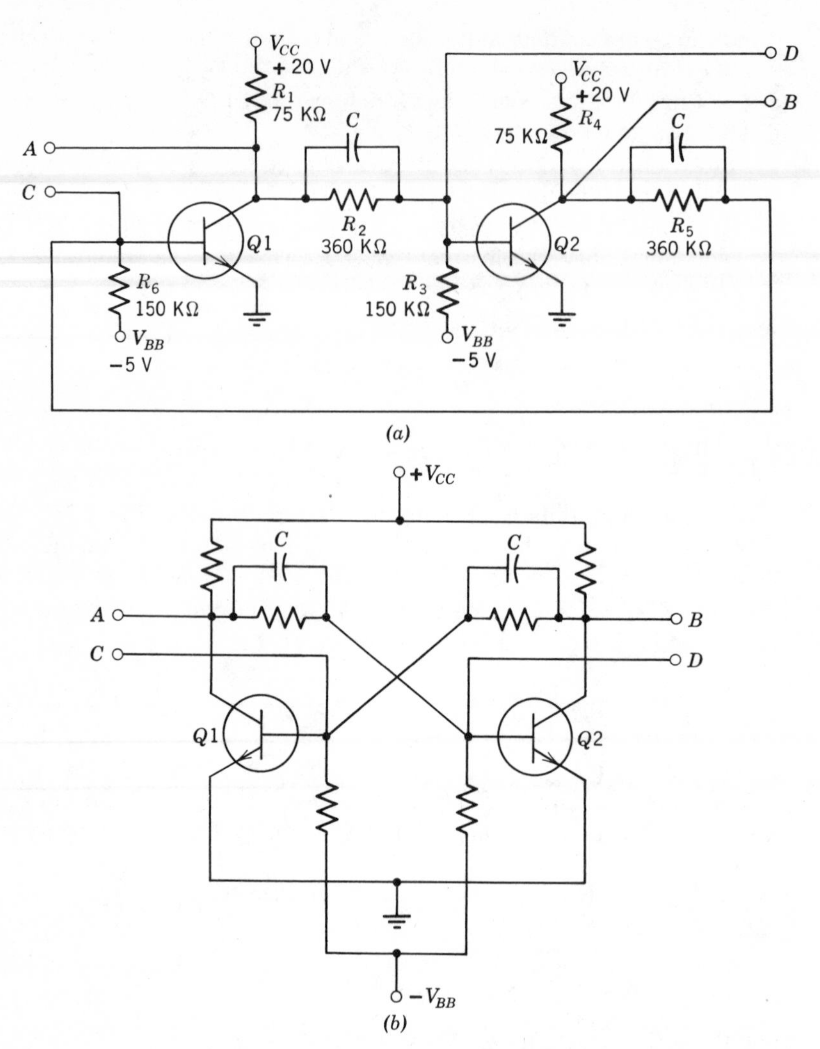

Figure 22-7 Flip-flop. (*a*) Circuit. (*b*) Conventional form of circuit.

Using these values of I_{B_2} and I_{C_2}, the minimum value of β_{dc} that $Q2$ can have is

$$\beta_{\text{dc}} = \frac{I_{C_2}}{I_{B_2}} = \frac{268}{12.6} = 21.2$$

Now, if the value of V_{BE} is 0.7 V, a recalculation for I_B gives

$$I_B = \frac{1.4 - 0.7}{112} = 0.0063 \text{ ma} = 6.3\ \mu\text{a}$$

and since I_C is not affected

$$\beta_{dc} = \frac{268}{6.3} = 42.4$$

The result is that, if we had used silicon transistors having a β_{dc} of 30, the circuit would fail to work. Usually, a design requires a safety factor of 3 or 4 to insure positive operation.

We described the circuit showing that if $Q2$ is saturated, $Q1$ is not conducting. The output at B (Fig. 22-7*a*), is zero, and the output at A is

$$\frac{360}{75+360}20 = +16.6 \text{ V}$$

assuming that V_{BE_2} is zero.

The numerical values we have established for the circuit are listed in Table A. The voltage on the base determines whether that particular transistor is ON or OFF and, consequently, establishes its value of V_{CE}. The

TABLE A Flip-Flop Operating Conditions

	Transistor $Q1$	Transistor $Q2$
	cutoff	on
I_C	0 ma	0.268 ma
V_{CE}	+16.6 V, Terminal A	0 V, Terminal B
V_{BE}	−3.54 V, Terminal C	+0.7 V, Terminal D

arrangement of the circuit is such that the value of V_{CE} on one transistor determines the value of V_{BE} on the other transistor because of the voltage dividing arrangements. Now if one of these voltages is forced to change because of an injected signal, an OFF transistor can be turned ON or an ON transistor can be turned OFF. For example, if a negative pulse is injected into Terminal D, the current in $Q2$ must drop to zero. The rise of the collector voltage in $Q2$ causes V_{BE} on $Q1$ to become positive, and current flows in $Q1$. V_{CE} in $Q1$ drops to zero, and a cutoff bias is now placed on $Q2$. The negative pulse on Terminal D can be removed, and the circuit will remain in the new switched condition. Pulses of sufficient duration can be injected into any one of the four terminals (A, B, C, or D) to force the circuit to switch. Typical switching pulses are shown in Fig. 22-8.

There are a number of fundamental variations to this circuit for use as a basic element in a digital computer. They properly belong to a digital computer course and are not covered in this text.

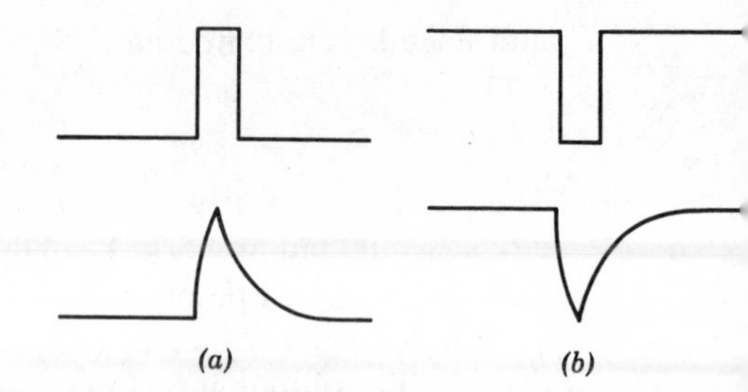

Figure 22-8 Waveforms used to trigger a flip-flop. (*a*) Positive pulses. (*b*) Negative pulses.

This circuit is usually drawn in the manner shown in Fig. 22-7*b*. There is a definite reason for this compact representation. The circuit is a basic building block in a computer, and it has access terminals at *A* and *B* and *C* and *D*. Usually, the circuit is reduced to microcircuits, and many hundreds or thousands of these circuits are used in one computer. Consequently, a method of representation where all the terminals are external with short leads is the most convenient.

PROBLEMS

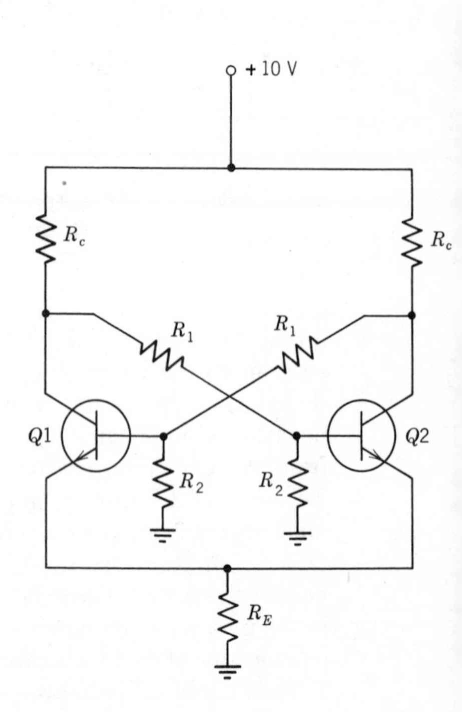

Circuit for Problems 1 to 3.

In the following problems assume that $V_{BE} = 0.7$ V, $V_{CE,\text{sat}} = 0.2$ V, and $\beta = 100$.

1. R_c is 20 kΩ, R_E is 1 kΩ, R_1 is 150 kΩ, and R_2 is 50 kΩ. Determine the currents and voltages for the transistors assuming one is ON and the other is OFF.
2. Solve Problem 1 if R_c is 4 kΩ, R_E is 500 Ω, R_1 is 60 kΩ, and R_2 is 20 kΩ.
3. Solve Problem 1 if R_c is 12 kΩ, R_E is 3000 Ω, R_1 is 300 kΩ, and R_2 is 300 kΩ.

Section 22-4 The Monostable Multivibrator

The block diagram for a monostable multivibrator is shown in Fig. 22-9. The two-stage amplifier provides positive feedback through R and C. C is the usual commutating capacitor to insure rapid switching. In this circuit the

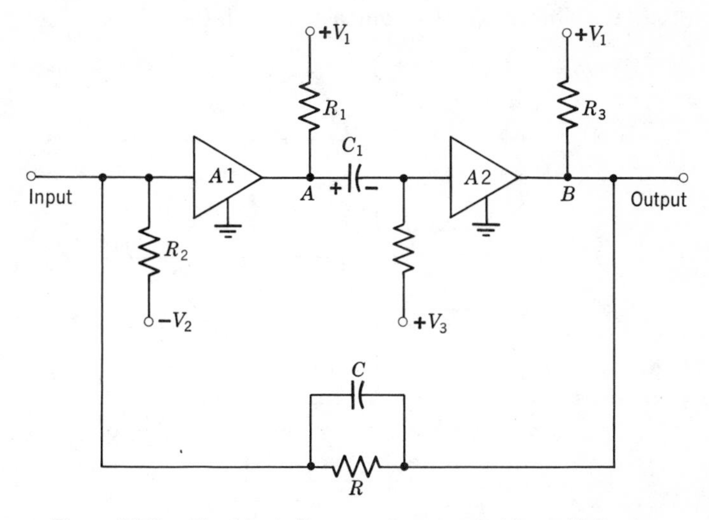

Figure 22-9 The block diagram of a monostable multivibrator.

amplifier $A1$ is biased below cutoff, and the amplifier $A2$ is forward biased into saturation. The voltage at A is V_1, and the voltage at B is the saturated value of $A2$ which is a minimum. The supply polarities are shown for either NPN transistors or vacuum tubes. PNP transistors require a reversal of all polarity markings.

As the circuit stands the output voltage is low, and the voltage across C_1 is a maximum, since there is no IR drop in $R1$. The polarity of this voltage is marked on C_1. Now assume that a positive pulse is applied to the input which is sufficient to overcome the cutoff bias on $A1$. The output voltage on $A1$ falls to the minimum value, but the voltage on $C1$ remains fixed and drives the amplifier $A2$ into cutoff by the concept shown in Fig.

22-3. When $A2$ is thus jammed into cutoff, the voltage at point B increases positively toward V_1. By a voltage divider action in the circuit of R_3 and R_2, amplifier $A1$ is now forward biased, and the initiating pulse is no longer required. The commutator capacitor C effectively removes R from the switching consideration. Now amplifier $A1$ is saturated and amplifier $A2$ is cut off. This condition remains until the capacitor $C1$ discharges to the point where $A1$ is cutoff. Then the voltage at A rises toward $V1$ and $C1$ recharges, placing a positive pulse on the input to $A2$ that causes $A2$ to return to the saturated state. The length of time $A2$ is cutoff is a function of the time constant that involves $C1$. This time is called the *gate time duration*.

Waveforms that illustrate this discussion are shown in Fig. 22-10. The monostable multivibrator is stable in one state only: $A1$ OFF and $A2$ ON. A switched state has a finite time duration only, and then the circuit reverts to the original stable state. In contrast, the flip-flop is bistable. The flip-flop

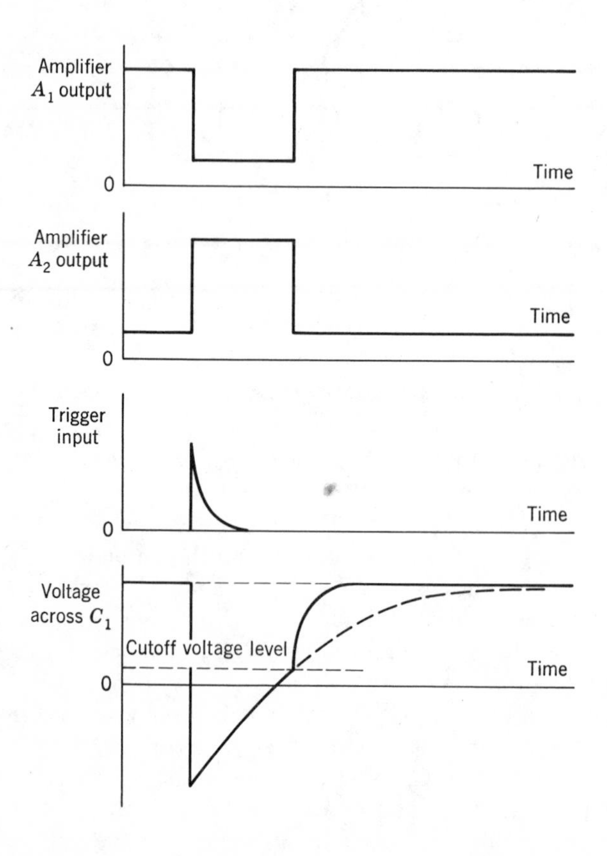

Figure 22-10 Waveforms of a monostable multivibrator.

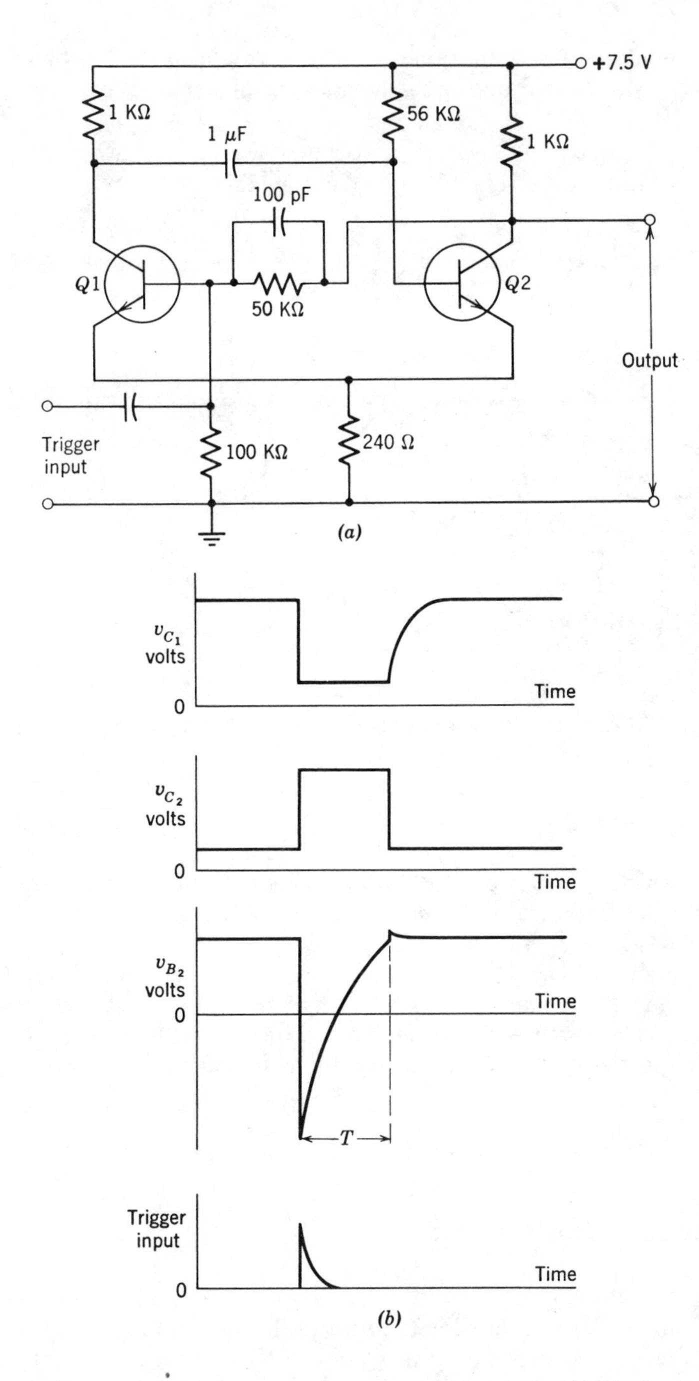

Figure 22-11 Transistor monostable multivibrator. (*a*) Circuit. (*b*) Waveforms.

remains indefinitely in either one of the two conditions, $A1$ ON and $A2$ OFF, or $A1$ OFF and $A2$ ON, until an external control signal inverts the ONN-OFF states.

A transistor circuit and the associated waveforms are shown in Fig. 22-11. A vacuum-tube version is given in Fig. 22-12.

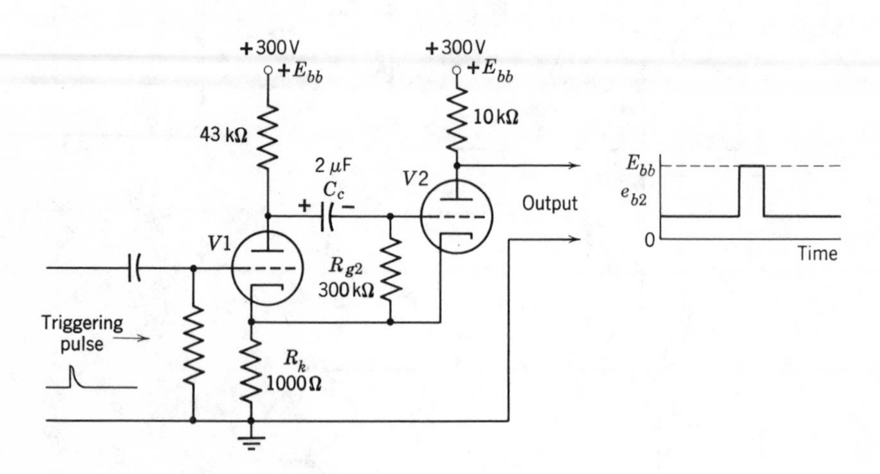

Figure 22-12 The single-shot multivibrator.

PROBLEMS

1. Label waveforms of Fig. 22-11 with numerical values by determining the exponential response of the circuit. Use 0.2 V for $V_{CE\text{-sat}}$, 300 mv for V_{BE} and 100 for β.
2. In the circuit of Fig. 22-12, if $V1$ is removed from the tube socket, the plate current in $V2$ is 15 ma. If $V2$ is removed from the tube socket, the plate current in $V1$ is 5 ma. The cutoff voltage of either tube is −10 V. A triggering pulse is applied to the circuit. Draw all waveforms to scale and indicate numerical values for time.

Section 22-5 Astable Multivibrators

The astable multivibrator, devised by H. Abraham and E. Block in 1918, was one of the earliest developed forms of the vacuum-tube oscillator. This circuit (Fig. 22-13) consists of a two-stage RC-coupled amplifier, which is fed back on itself. Since each stage provides a 180° phase shift, the overall

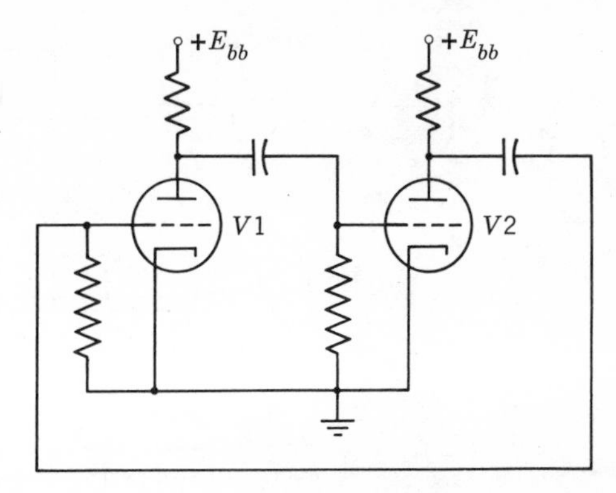

Figure 22-13 The vacuum-tube multivibrator.

feedback is positive. The circuit arrangement does not limit the resultant oscillation to a sine wave, but allows the amplitude to rise to the limits of the tubes. One limit is tube cutoff, and the other is the effect of driving the grids positive. Both these limits clip the sine wave and produce a square wave. The plate voltage of tube $V1$ is 180° out of phase with tube $V2$. The term astable signifies that the circuit does not remain in one state indefinitely until it is switched to the second state by an external triggering signal. The astable condition signifies that there is a continuous switching between the two amplifiers. The solid state version of the multivibrator is shown in Fig. 22-14. Again, it is a two-stage R-C-coupled amplifier fed back into itself. Accordingly, square waveforms (Fig. 22-15) are also characteristic of this circuit.

Unfortunately, this simple explanation does not give us an understanding of the base waveforms, which is most necessary for this circuit. We shall use

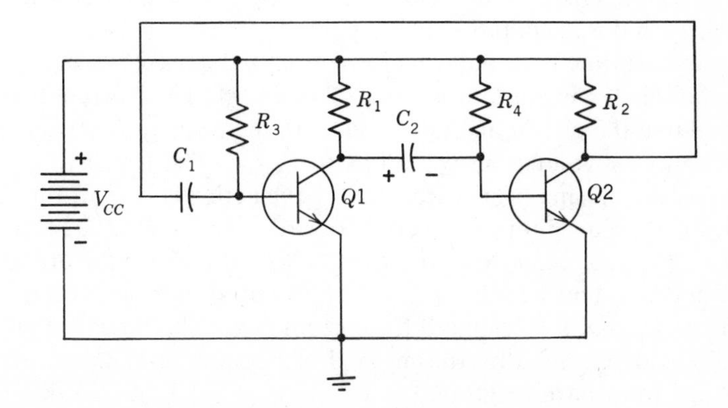

Figure 22-14 The transistor multivibrator.

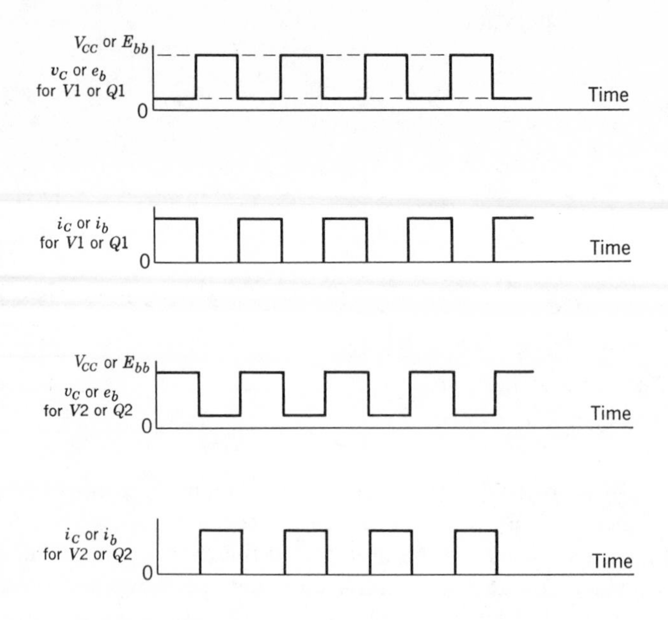

Figure 22-15 Basic waveforms of a multivibrator.

a process of reasoning that is accepted in mathematics: we assume that the circuit does work, and then we show that it must so function.

Assume that $Q1$ is cut off in the circuit of Fig. 22-14 and that $Q2$ is carrying current and is saturated. This state is shown in Fig. 22-16 by the long vertical dashed line $A-A-A-A-A$. The collector voltage of $Q2$ is at the saturation value $V_{CE,\text{sat}}$. The base voltage, when this saturation occurs in the collector, is V_{BE}. There is no current flowing in R_1, and the IR drop in R_1 is zero. There is a d-c voltage across the capacitor with the marked polarity and a magnitude $(V_{CC}-V_{BE})$.

A short time later the waveform of $v_{B,Q1}$ decreases to V_{BE}. V_{BE} is the voltage at which current first starts to flow in $Q1$. This current flow produces a signal that the circuit amplifies and, with the positive feedback in the circuit, the collector voltage on $Q1$ drops to $V_{CE,\text{sat}}$. The transistor $Q1$ effectively acts as the closing of a switch and by the action shown in Fig. 22-3, the left side of the capacitor is lowered toward zero by $(R_1 I_{C,\text{sat.}})$ volts, which is $(V_{CC}-V_{CE,\text{sat.}})$. Therefore, the right side of the capacitor drives the base of $Q2$ negative and cuts off $Q2$. $Q2$ stays cutoff until the voltage on the base falls to V_{BE}—the threshhold for current flow. The time T_2 that $Q2$ remains cutoff is determined by the method of Section 22-1. The magnitude of the voltage from base to ground is $(V_{CC}-V_{CE,\text{sat.}}+V_a)$, and this voltage must charge toward V_{CC}. However, when the voltage falls to V_{BE}, the circuit

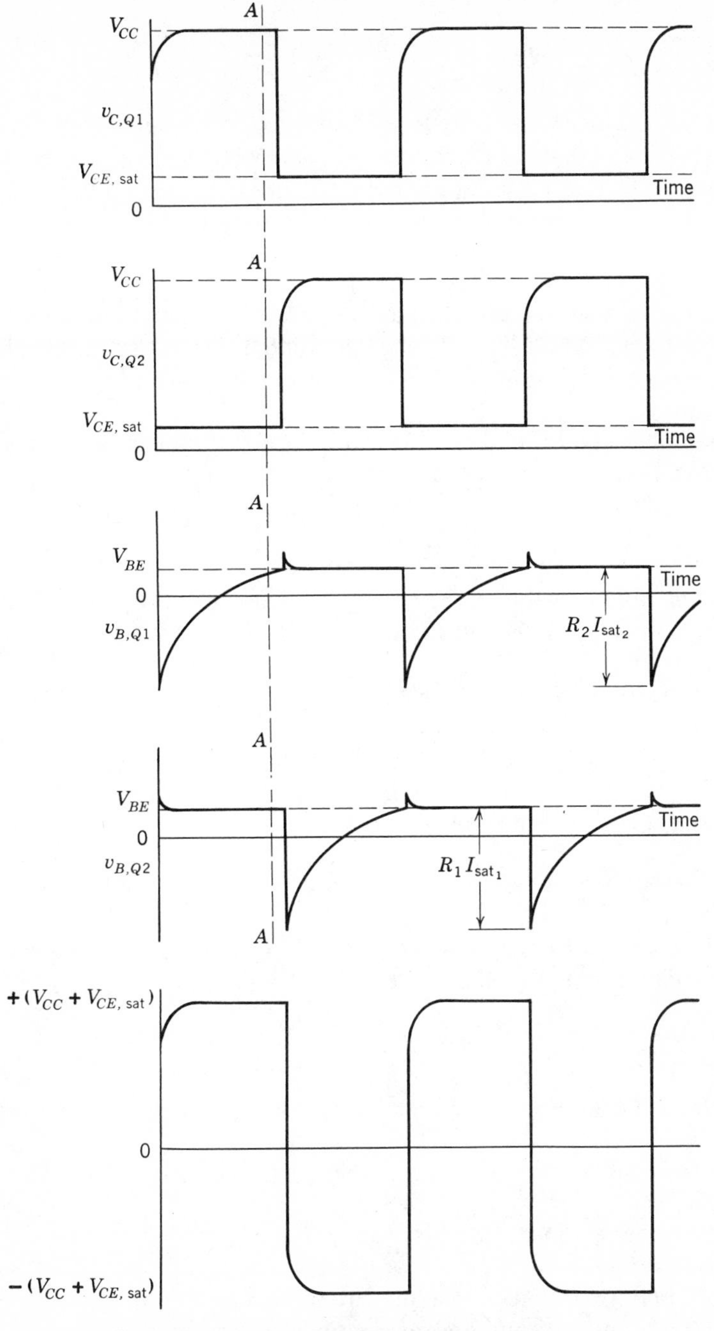

Figure 22-16 Details of multivibrator waveforms.

switches. The time interval T_2 for which $Q2$ is cut off is

$$T_2 = R_4 C_2 \ln \frac{2V_{CC} - V_{CE,\text{sat}} - V_{BE}}{V_{CC} - V_{BE}} \quad \text{(22-3}a$$

A resistance value (the d-c input resistance to the saturated transistor $Q1$ i parallel with R_1) should be added to R_4. However, if R_1 is very small com pared to R_4, this correction need not be made. When the coupling capacito charges back up again in the next half cycle, the charging resistor is R_1 plu the very small d-c input resistance into the base of a saturated transistor Since R_1 is usually much less than R_4, this time is much shorter than T_2 Consequently, the collector-voltage rise shows only a short time delay i reaching its final value V_{CC}. The small pointed tip on the v_B waveforms i caused by the the action of the capacitor in overdriving the transistor int saturation.

The time T_1 that the other transistor is off is a similar equation with th same limitations:

$$T_1 = R_3 C_1 \ln \frac{2V_{CC} - V_{CE,\text{sat}} - V_{BE}}{V_{CC} - V_{BE}} \quad \text{(22-3}b$$

If all like components in the circuit are identical, a complete symmetry o waveforms results, and T_1 and T_2 are equal. This circuit is a *balanced multi vibrator.*

Then the total time T of a cycle is

$$T = T_1 + T_2 = 2R_B C \ln \frac{2V_{CC} - V_{CE,\text{sat}} - V_{BE}}{V_{CC} - V_{BE}} \quad \text{(22-3}c$$

and the frequency f of the multivibrator is

$$f = \frac{1}{T} \quad \text{(22-3}d$$

Similar equations can be developed for the vacuum-tube multivibrator The time either tube is cutoff is given by

$$T_1 \text{ or } T_2 = R_g C \ln \frac{2E_{bb} - E_b}{E_{bb} - E_{co}} \quad \text{(22-3}e$$

where R_g and C are the grid circuit values for the tube that has a cutoff valu of $-E_{co}$ volts.

The push-pull output waveform is shown also on Fig. 22-16. This outpu can yield an excellent square wave if it is fed through a linear circuit (suc as a double Zener diode) that slices off both the rounded top and the rounde bottom.

When the two time constants are different, the waveforms are no symmetrical, and an *unbalanced multivibrator* results. The vacuum tub version of an unbalanced multivibrator is shown in Fig. 22-17 together with the waveforms.

The multivibrator is a circuit of many uses. It is the fundamental square

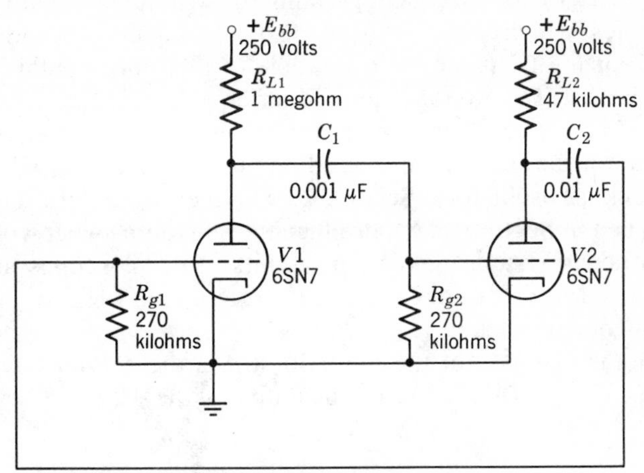

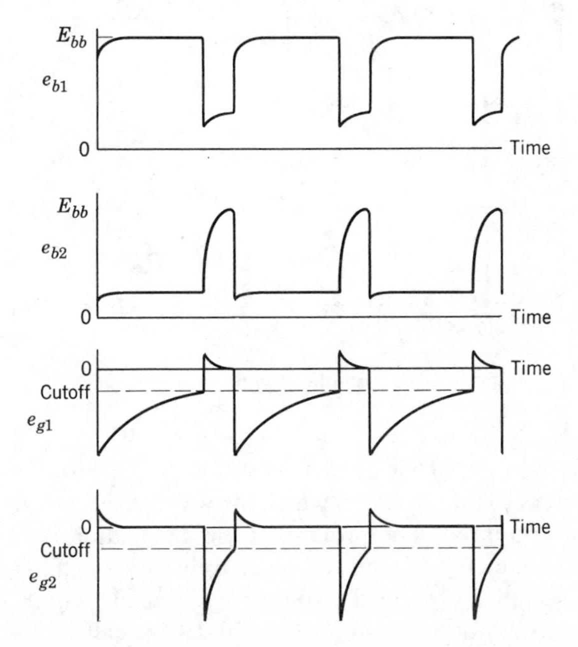

Figure 22-17 The unbalanced multivibrator.

wave or pulse generator for many direct applications. Although the free running frequency is unstable, it is very rapidly and easily synchronized t an outside source. Since the square-wave pattern of the multivibrator con tains many harmonics (useful up to the 150th), when it is synchronize (Section 22-6) to a frequency standard, it produces many high-frequenc harmonics which may be used also as frequency standards. When th synchronizing frequency is a multiple of the fundamental frequency of th multivibrator, the circuit is termed a frequency divider. Frequency divider are used in frequency-modulation transmitters and in "frequency standards" to obtain accurate time measured on an electric clock. Multivibrators (an the blocking oscillators, Section 22-7), when used with a discharge circui (explained in Section 22-8), produce the sawtooth sweep voltages needed i the cathode-ray-tube deflection circuits in oscilloscopes and in televisio receivers. Multivibrators are useful from about 1 Hz to about 500 kHz a fundamental frequency.

Another version of the multivibrator is the *emitter-coupled multivibra tor* (Fig. 22-18). In this circuit, the output of the $Q2$ stage is fed back to $Q1$ i

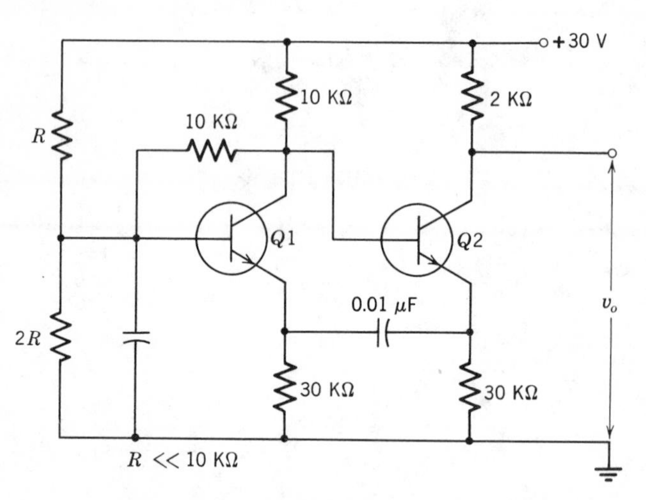

Figure 22-18 Emitter-coupled multivibrator.

the emitter circuit. $Q1$ is direct-coupled to $Q2$. A problem at the end of thi section refers to this circuit in which the waveforms and voltage levels abov ground are required. The process of the calculation is to assume that on transistor conducts and that the other transistor is off. The d-c current cal culations will then give the d-c voltage levels at all points.

The vacuum-tube version (Fig. 22-19) is called the *cathode-couple multivibrator*. The circuit and waveforms are given. To understand the

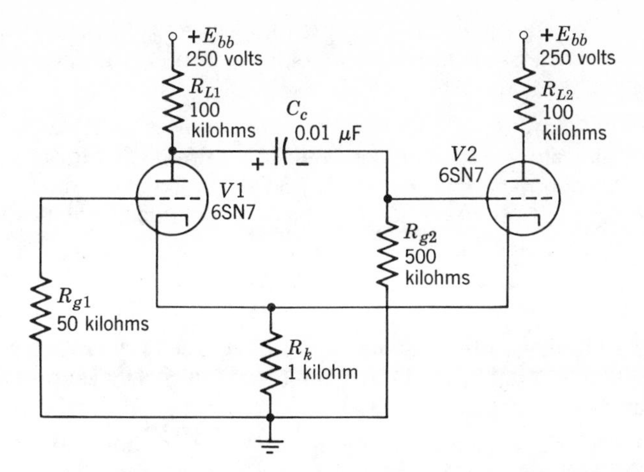

Figure 22-19 The cathode-coupled multivibrator.

operation of the cathode-coupled multivibrator, a different approach will be taken than was used for the standard multivibrator. Assume that the heaters are turned off and that the supply voltage is applied to the circuit. The coupling capacitor C_c is charged to 250 V with the indicated polarity. When the heaters are turned on, both tubes show plate current. The plate current in $V1$ causes its plate voltage to decrease. Since the 250-V charge on C_c cannot change instantly, the grid of $V2$ is driven negative by the amount that the plate voltage of $V1$ falls from 250 V.

The second tube $V2$ stays cut off until C_c discharges to cutoff through the circuit of R_{g2}, R_k and the equivalent d-c resistance of $V1$. When plate current starts to flow in $V2$, there is an increased IR drop in the cathode resistor R_k. This increasing voltage across R_k decreases the plate current in $V1$. The plate voltage of $V1$ rises consequently, and now the low-voltage charge on C_c creates a positive grid voltage on $V2$. A now large current flows in $V2$ and causes a sufficient IR drop in R_k to cut off $V1$. C_c charges rapidly to 250 V since R_{g2} is in the discharge circuit but not in the charge circuit. As soon as C_c recharges, grid current in $V2$ ceases, and $V1$ starts to have plate current. Now the cycle repeats as we have described.

PROBLEMS

1. In Fig. 22-13, E_{bb} is 300 V, the plate load resistors are 10 kΩ, the grid resistors are 240 kΩ, and the coupling capacitors are 0.01 μF. The cutoff voltage of the tubes is −15 V and the maximum plate current is 25 ma. Determine the frequency of oscillation. Draw the waveforms for the circuit to scale.

2. The multivibrator of Problem 1 is to operate at 20 kHz. If the smallest capac tor that can be used for the circuit is 100 pF, determine capacitor and gri resistor values.
3. In Fig. 22-14, V_{CC} is 14 V and the collector load resistors are 20 kΩ. The β the transistors is 80. The base resistors are selected to saturate the transisto during conduction. V_{BE} is 0.7 V and $V_{CE.sat}$ is 0.2 V. The coupling capacitor are 100 pF. Determine the frequency of operation and draw the waveforms t scale.
4. Determine the values of the coupling capacitors in Problem 3 if the frequenc is to be 100 Hz.
5. Draw the waveforms to scale for Fig. 22-17. Determine the frequency of opera tion and the duration of each half of the full cycle. Assume that the least plat voltage is 50 V. Cutoff is −10 V.
6. Draw the waveforms to scale for Fig. 22-18 and show the zero-reference leve What is the frequency of operation? Assume that V_{BE} is 0.7 V and $V_{CE.sat}$ i 0.2 V.
7. Draw the waveforms for Fig. 22-19 to scale. Assume that the least plate vol age is 50 V and cutoff is −10 V.

Section 22-6 Synchronization

To *synchronize* or *sync* a free-running oscillator means to lock it into a external signal source. If the outside signal frequency changes slightly, properly synchronized circuit changes its frequency accordingly. Th synchronized circuit now operates at a *forced frequency* instead of at *free-running frequency*.

The waveform at the transistor base or at the FET gate or at the vacuur tube grid of an astable multivibrator is shown in Fig. 22-20*a*. This wavefor has a designed slight unbalance. The waveform changes exponentially fro point A at V_c volts to point B and to point C at V_b volts. V_b is the voltag value at which this amplifier stage switches into a conducting state. Co sequently, when the voltage reaches V_b, the waveform abruptly switche from point C to point D, which is slightly above V_a volts. The stage remair in conduction until time E, when the circuit again switches back and th voltage at E switches back to point F.

Now a series of triggering pulses (Fig. 22-20*b*) is added to the waveforn As the voltage is changing from along the curve from A toward C, a triggerir pulse appears at B. The pulse has a sufficient amplitude to force the wavefor at B up to and above V_b. The circuit immediately switches without waiting t discharge to C. Pulse 1 has triggered the circuit, and the waveform now take the form shown in Fig. 22-20*c*. The waveform discharges normally from A

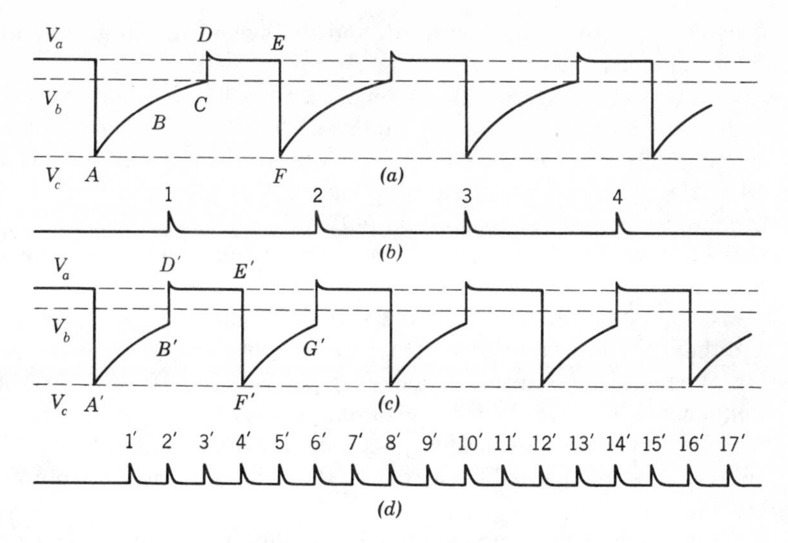

Figure 22-20 Basic synchronization.

B and is forced to switch to D'. This stage is now on from D' to E'. The time from D' to E' is the same as from D to E. We have changed the time from A to D to the new time from A' to D' by an *injected sync pulse*, but we have done nothing to disturb the other stage of the circuit. Therefore, the time from D' to E' is identical to the time from D to E.

The circuit by itself switches from E' to F'. A discharge occurs from F' to G'. At G' sync pulse 2 is added to the waveform, and the circuit is forced to switch. A train of such pulses (1, 2, 3, 4, . . .) is required, and each pulse triggers this stage from nonconduction into conduction. We have drawn Fig. 22-20*a* unbalanced to show how a balanced waveform results only when these particular triggering pulses are applied.

If these synchronizing pulses were less than the voltage level from B to V_b, they would not trigger the circuit each time. Thus, one requirement for synchronization is that *the synchronizing voltage must be sufficiently large to synchronize.* If the synchronizing pulses arrive after the stage has already switched, they are too late. In other words, the time interval from pulse 1 to pulse 2 must be less than the time interval from A to E. Or *the synchronizing frequency must be greater than the free-running frequency of the oscillator.* A third requirement for the synchronizing pulse is that the pulse must have a finite time duration to insure switching.

We have indicated that the sync voltage is a series of short-duration sharp pulses. This waveform is the ideal. The action of synchronization is clearly and sharply defined. It should be understood that a sine wave or an

irregular waveform may be used, and the synchronization functions in th
same manner.

Very often it is desirable to synchronize a low-frequency oscillator to
high-frequency signal, for example, a 1000-Hz oscillator to a 4000-H
signal. Thus, in this example, every fourth cycle is to synchronize. Th
4000-Hz signal is converted into sharp pulses shown in Fig. 22-20*d*. No
the multi-vibrator is triggered on pulses 2′, 6′, 10′, 14′. Actually, synchro
nization could be obtained by using a sinusoidal signal, but the switchin
point would not be especially stable.

Assume that the frequency of the train of pulses (Fig. 22-20*d*) is 2400 H
and that the free-running frequency of the multivibrator (Fig. 22-20*a*) is 48
Hz. When the multivibrator is locked to pulses 2′, 6′, 10′, 14′, etc., the locke
frequency is 600 Hz. When the amplitude of the pulse train is raised suffic
ently, the multivibrator suddenly becomes locked to pulses 1′, 4′, 7′, 10′, 13
16′, etc., and the locked frequency of the multivibrator suddenly switches t
800 Hz.

It is not desirable to have too wide a difference between the free-runnin
and the lock-in frequency. Not only is the synchronization touchy, but ther
is also the possibility of a bad distortion of the waveform. Thus a general ru
should be observed: *use only enough synchronizing voltage to obtain goo
synchronization*.

Figure 22-21 shows a sync control circuit. When the sync voltage-contro
potentiometer is set at the center, the synchronizing voltage to the mult
vibrator is zero, and the oscillator is free-running. When the incoming syn
pulses are +, an upward shift of the potentiometer produces the positive out
put pulses required for synchronization. If the incoming sync pulse is nega
tive, a downward shift of the potentiometer produces a positive output puls
to the oscillator. If both positive and negative pulses are available, eithe
polarity can be selected for synchronization. This is a standard arrangemen
used on oscilloscopes.

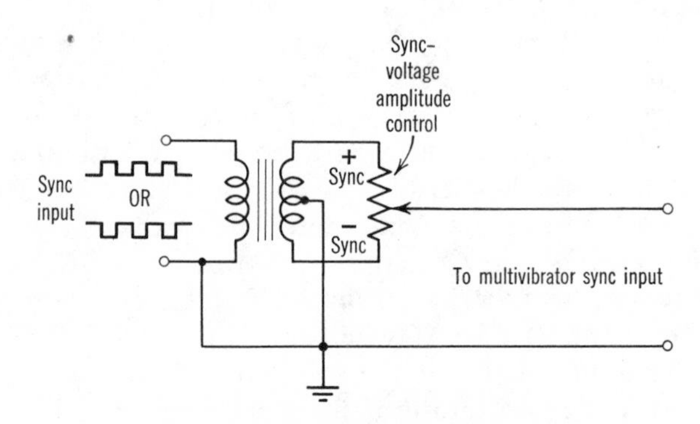

Figure 22-21 Sync control circuit.

ection 22-7 The Blocking Oscillator

The circuit of the blocking oscillator (Fig. 22-22) is a modified Armstrong oscillator. The Armstrong oscillator uses an air-core transformer which has a low coefficient of coupling of the order of a few per cent. A large alternating

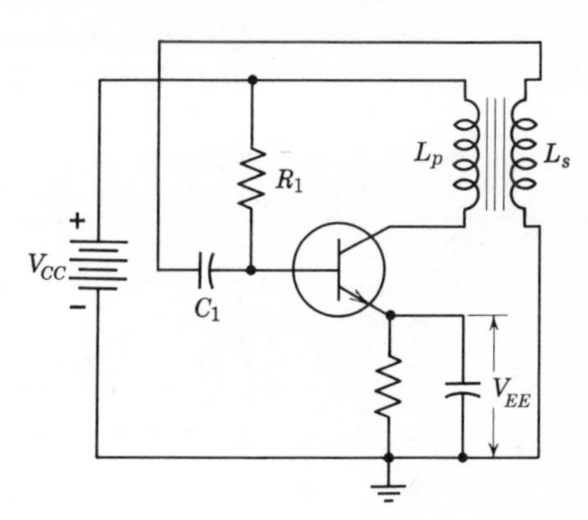

Figure 22-22 The transistor blocking oscillator.

voltage in L_p induces only a few volts in L_s, the grid winding. The bias-clamp action establishes a stable sinusoidal condition where just enough voltage is fed back into the base circuit to maintain the oscillation. In the blocking oscillator, the coefficient of coupling is raised by means of a carefully designed core to almost unity. If the turns ratio of the transformer, N_s/N_p, is three, a 10-V collector signal produces 30 V in the base winding. This 30-V signal violently drives the base into conduction and establishes a high direct voltage on the capacitor. This d-c bias cuts off the transistor completely as soon as the base voltage falls below V_{EE}. This cutoff abruptly ends and "blocks" further current flow. The transistor remains "blocked" until the capacitor bias voltage discharges to the cutoff point of the transistor. Now the oscillator tries to oscillate and is, again, immediately "blocked."

There is a stray winding capacitance within the turns of the blocking transformer which determines a particular resonant frequency for the transformer. When the oscillator is cut off, the collector voltage is the collector supply voltage V_{CC}. When the base discharges to cutoff, collector current starts to flow, inducing a voltage on the base that aids current flow. The circuit shuts off at the peak value of the cycle. Thus, the normal oscillator action of the circuit as an Armstrong oscillator is limited to a quarter of one cycle. This "on" time of the blocking oscillator is established by the transformer design. The "off" time of the blocking oscillator is a function of the time constant of the bias clamp circuit, R and C.

With these ON–OFF circuits, a specific reference to *frequency* coul prove confusing as there are different times involved. To avoid this, a ne term is introduced, *pulse-repetition rate*. With the blocking oscillator, pulse-repetition rate of 400 means that there are 400 ON and 400 OFF condi tions per second. The blocking oscillator and the multivibrator are standar integral parts of television receivers and, in this application, have reache many people whose electronic knowledge is centered on radio. As a result the term *pulse-repetition rate* for these two circuits is dropped and the terr *frequency* is used indescriminately. In equipment where these circuits ar used as pulse generators, the term *pulse-repetition rate* is retained.

The waveforms for the blocking oscillator are shown in Fig. 22-23. Th

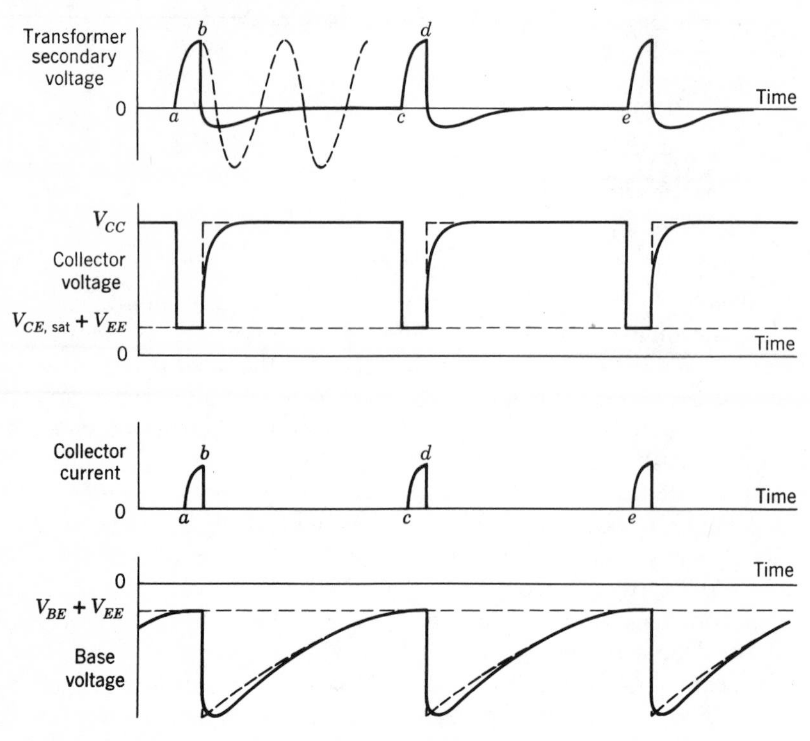

Figure 22-23 Waveforms of the blocking oscillator.

dashed lines are for ideal conditions which are considered first. The circui is ON during the intervals *ab* and *cd* and is OFF during *bc* and *de*. The trans former secondary voltage has a high amplitude. The transformer secondar voltage is purely alternating, and, by definition of an a-c wave, the are under the curve above the zero axis must be equal to the area below the zer axis. Another approach to this waveform is to consider that from a to b energy is stored up in the magnetic field within the transformer. When the circuit is cut off at b, this energy must be dissipated and produces a high "back voltage." This release of stored energy in the transformer causes an oscillation within the secondary winding of the transformer which dies out rapidly. This "die-away" is seen on the waveform. The amount of damping in the transformer and in the circuit controls the amplitude of this transient oscillation. If the circuit did not block, the transformer secondary waveform would be the dotted sine wave. The negative loop on the transformer secondary voltage waveform adds to the base voltage waveform and causes the rounding effect.

Normally, there is no control over the time intervals, ab and cd. The natural resonant frequency of the blocking-oscillator transformer determines this time. The time intervals bc and de are easily controlled by making part of the bias-clamp resistance R_1 a variable resistor.

A vacuum-tube blocking oscillator with provision for synchronization is shown in Fig. 22-24. Vacuum-tube blocking oscillators are readily synchron-

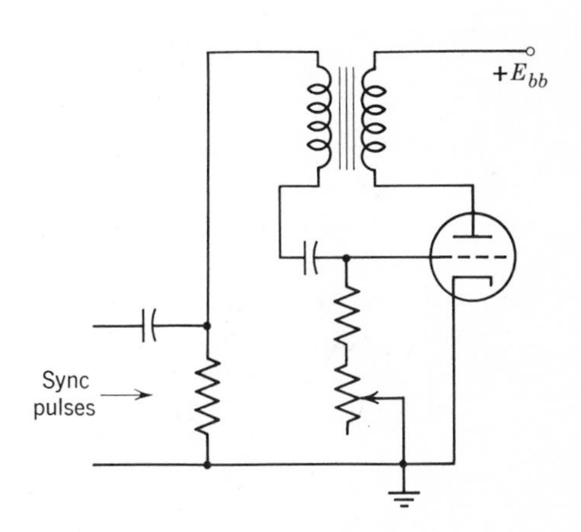

igure 22-24 Blocking oscillator with provision
or synchronization.

ized by positive pulses of the order of 10 to 20 V. The positive pulse brings the grid to and above cutoff earlier than cutoff would be reached for the RC grid-leak discharge action alone. For transistor circuits, the polarity of the sync pulse depends on whether a PNP or an NPN transistor is used. In a television receiver the variable resistance in the R-C circuit adjusts the free-running pulse-repetition rate, so that it is slightly slower than the pulse-repetition rate of the incoming sync signals from the program transmitter. This control in the television receiver is labeled "hold."

Section 22-8 Discharge Circuits

A transistor, an FET, or a vacuum tube may be used in place of the thyratro in the RC sawtooth generating circuit (Fig. 22-2). However, this new circu is not a self-sustaining oscillator; it must be driven as is shown in Fig. 22-2.

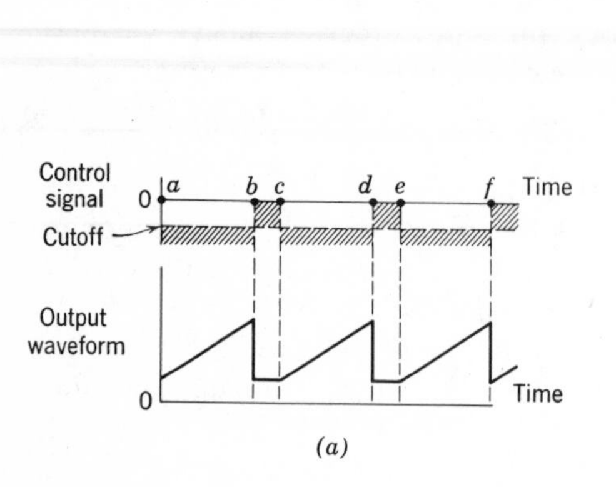

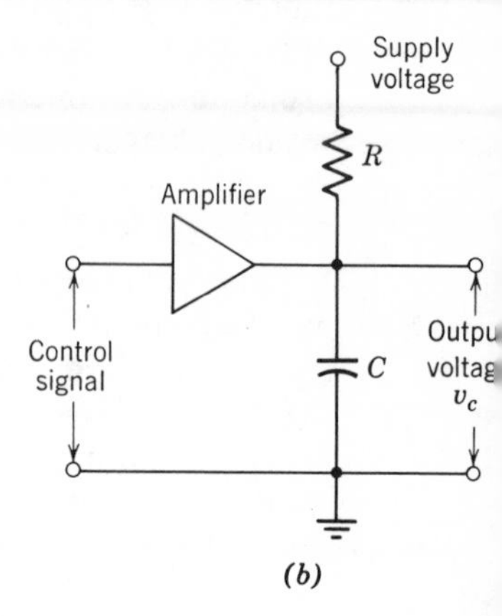

Figure 22-25 The discharge circuit.

During the time of the capacitor charge, a to b, c to d, and e to f, the ampli fier must be kept below cutoff. When the capacitor is being discharged, b to c and d to e, the amplifier must be saturated. The base or grid waveform present in the blocking oscillator and unbalanced multivibrator are ideall suited for this use. In these applications, the frequency is determined by the repetition rate of the blocking oscillator or multivibrator. Then, the R and C of the discharge circuit serve to establish the amplitude of the sawtoot output. A variation in R or C does not affect the frequency but only the amplitude of the waveform.

If the value of R is high enough, the charging current in C is constant By this we mean that the time constant RC is much larger than the time the capacitor is to charge. When the charging current in a capacitor is constant the voltage across the capacitor is

$$v_c = \frac{I}{C} t \qquad (22\text{-}4$$

and it is linear with time. The factor I/C is called the *sweep speed.*

Representative composite circuits are shown in Figs. 22-26 and 22-27 It is possible to combine the discharge circuit with the oscillator circuit This combination simplifies the overall circuit by eliminating an amplifier The Schmitt trigger circuit (Fig. 22-27) is formed by $Q1$ and $Q2$. The inpu

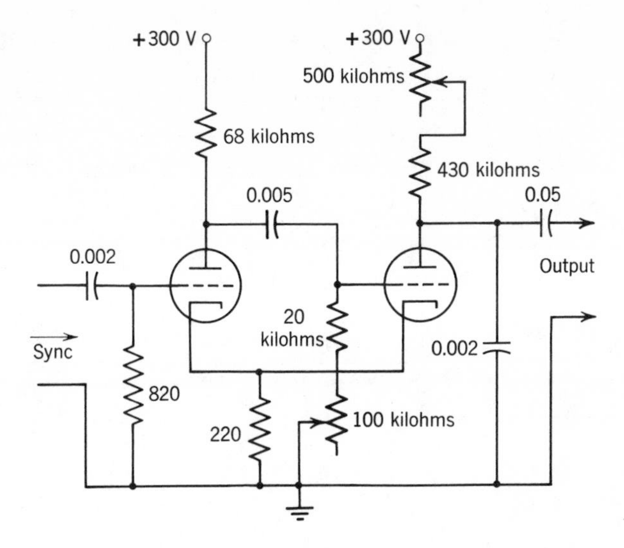

Figure 22-26 Combined multivibrator and discharge circuit.

voltage waveform and the Schmitt trigger output waveform at V are both conventional. The third stage $Q3$ is a discharge circuit driven by the output of the Schmitt trigger. Now the output voltage waveform yields a final voltage that is linearly proportional to the *time* the Schmitt trigger operates. Consequently, this circuit is a means of converting time duration to a voltage signal, and as such it has many applications.

Section 22-9 Negative Resistance Concepts

A conventional resistor is by definition a circuit component that *absorbs* power and produces heat. A *negative resistance* conversely *delivers* power to the rest of the circuit. A negative resistance is not something that can be bought from an electronic supply house; it is merely the equivalent effect between the two terminals of a black box. A tank circuit at resonance is equivalent to an ideal turned circuit in parallel with a resistor of QX ohms (Section 10-2). Now if a negative resistance of value $(-QX)$ ohms were connected in parallel with the tank, the whole circuit is ideal and any oscillation induced in the circuit continues to infinity.

All oscillators can be analyzed by investigating the negative resistance effect of the amplifier stage that is used with the tank. We did not use this approach in Chapter 21, nor have we used it for the nonsinusoidal oscillators

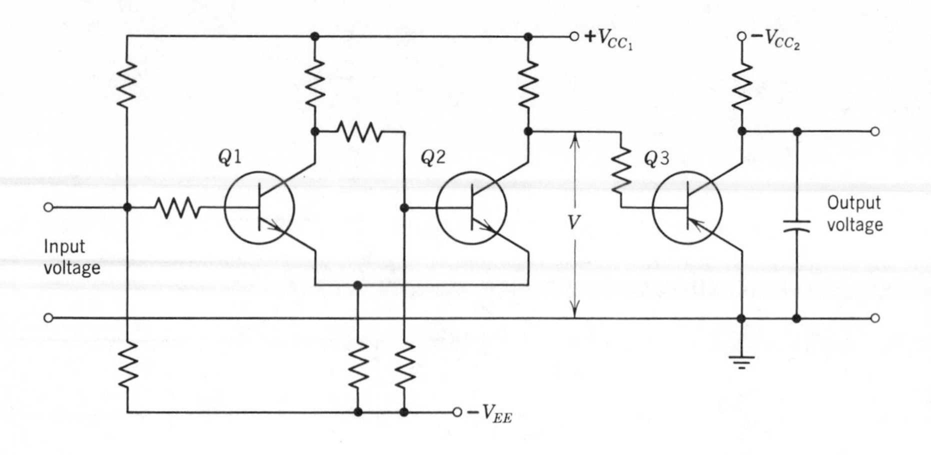

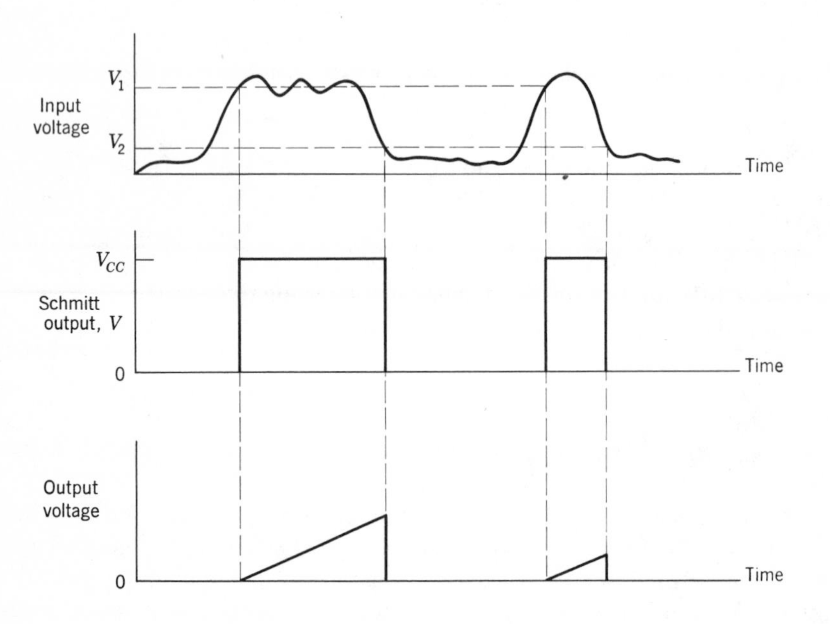

Figure 22-27 Schmitt trigger circuit with discharge circuit.

to this point in this chapter. However, the use of the negative resistance concept is the simplest approach to certain oscillator circuits.

A negative resistance occurs in electronics when an increase in voltage results in a decrease in current (a voltage-controlled negative resistance) or where an increase in current results in a decrease in voltage (a current controlled negative resistance). In either case, any a-c resistance is defined as

$$r = dv/di$$

and if this equation shows a negative slope

$$r = -dv/di$$

r is a negative resistance. Volt-ampere curves that have negative resistances in the vicinity of point Q are shown in Fig. 22-28. A typical voltage-controlled negative-resistance characteristic (Fig. 22-28*a*) is the property of

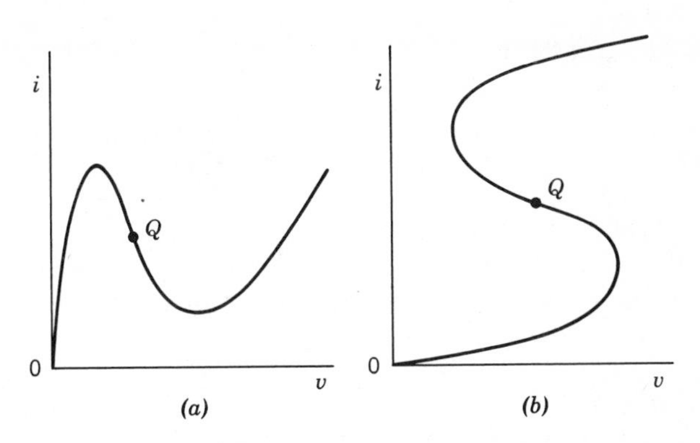

Figure 22-28 Volt-ampere curves showing a negative resistance in the portion of the curves near Q. (*a*) Voltage controlled. (*b*) Current controlled.

a tunnel diode. Unijunction transistors, silicon-controlled rectifiers and thyristors have a current-controlled negative-resistance characteristic (Fig. 22-28*b*).

Four different load lines are superimposed on a tunnel-diode characteristic in Fig. 22-29. Load line 1, using a supply voltage V_A, intersects the characteristic curve at A. The slope of the diode curve at A is positive and, therefore, operation at A is unconditionally stable. Point A can be the normal operating point for a monostable circuit. Similarly, load line 2, drawn from V_D, is also stable. Load line 3, using supply voltage V_E, intersects the characteristic at three points—D, E, and F. There are two stable points, D and F, and one unstable point, E. A bistable circuit, such as a Schmitt trigger and a flip-flop, has two stable points, D and F. The bistable circuit is externally switched back and forth between D and F. Load line 4, using V_B as the supply, intersects the characteristic only at one point, B, which is a point of negative resistance. If a tuned circuit is placed in series with the diode, the whole circuit will oscillate. In this manner the tunnel diode is used as an oscillator at microwave frequencies.

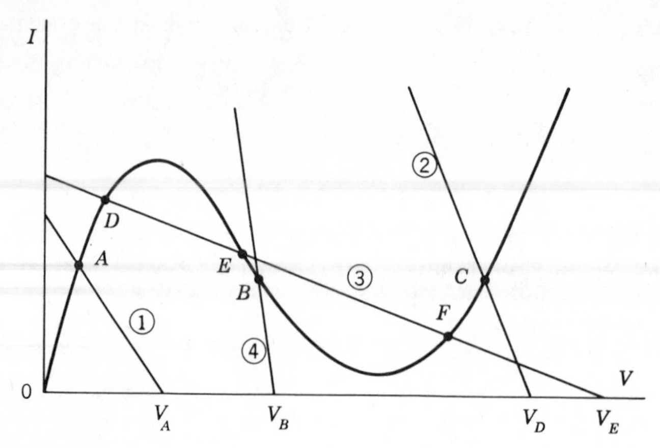

Figure 22-29 Load lines superimposed on a tunnel-diode characteristic.

The astable free-running oscillators have the operating point at B.

Consider a circuit operating at Q-point 1 which is stable (Fig. 22-30 The supply voltage is V_A. Now an additional voltage is added to V_A th

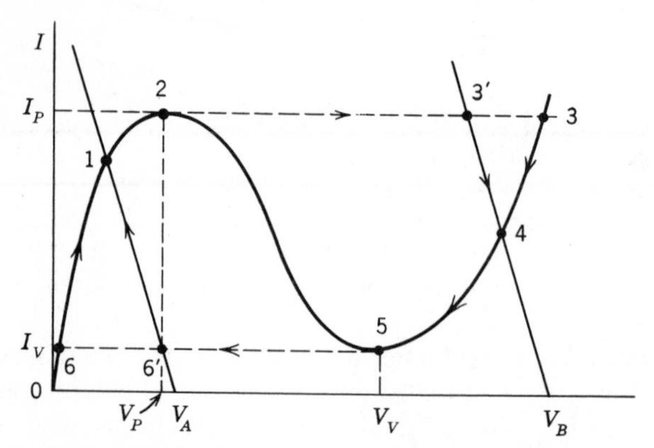

Figure 22-30 Bistable operation of a tunnel diode.

causes the load line to transfer to V_B. Again, we have a stable point of opera tion, Q-point 4. The mechanism of switching is to be examined to show transfer from Q-point 1 to Q-point 4.

From the diagram an obvious path is 1–2–5–4. However, the path fro 2 to 5 has a negative slope and dV/dt is negative. But if V_A is increased t V_B, the value of dV/dt is positive, and operation from 2 to 5 is impossible Likewise operation along a path from 3 to 4 is prohibited. When the circui voltage is switched from V_B back to V_A, 5 to 2 and 6′ to 1 become prohibite paths.

When the supply is changed from V_A to V_B, the current increases to the peak current I_P at 2. A jump occurs from 2 to 3 that is an abrupt change in voltage at the peak current value. The current then falls along the characteristic curve to the final stable value at point 4. When the voltage across the circuit is switched back from V_B to V_A, the current falls from 4 to 5—the valley point I_V. Again, there is an abrupt jump in voltage from 5 to 6. Then the current follows the characteristic curve up to Q-point 1. The effect of the hysteresis concept is clear. There are inductive and capacitive time constants which, together with internal effects in the semiconductor, establish a finite switching time.

A tunnel-diode monostable oscillator is shown in Fig. 22-31, together

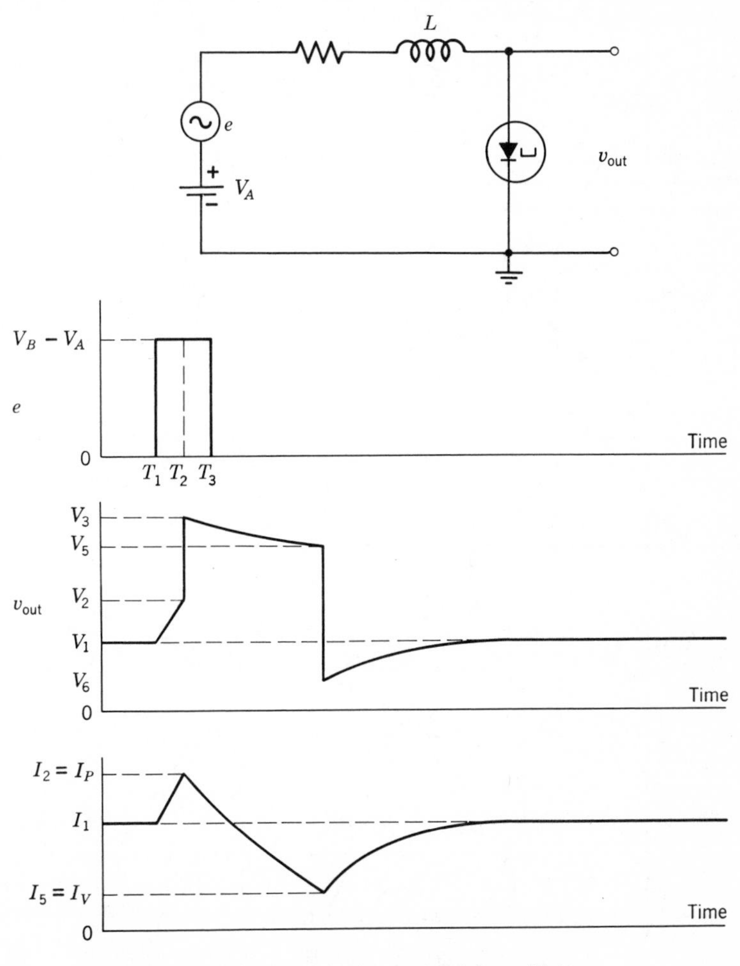

Figure 22-31 A tunnel-diode monostable oscillator.

with the waveforms. The operation is switched from Q-point 1 to Q-point 4 on Fig. 22-30 by means of a triggering pulse e. The magnitude of this triggering pulse is $V_B - V_A$ and must last until the circuit switches. The pulse dura-

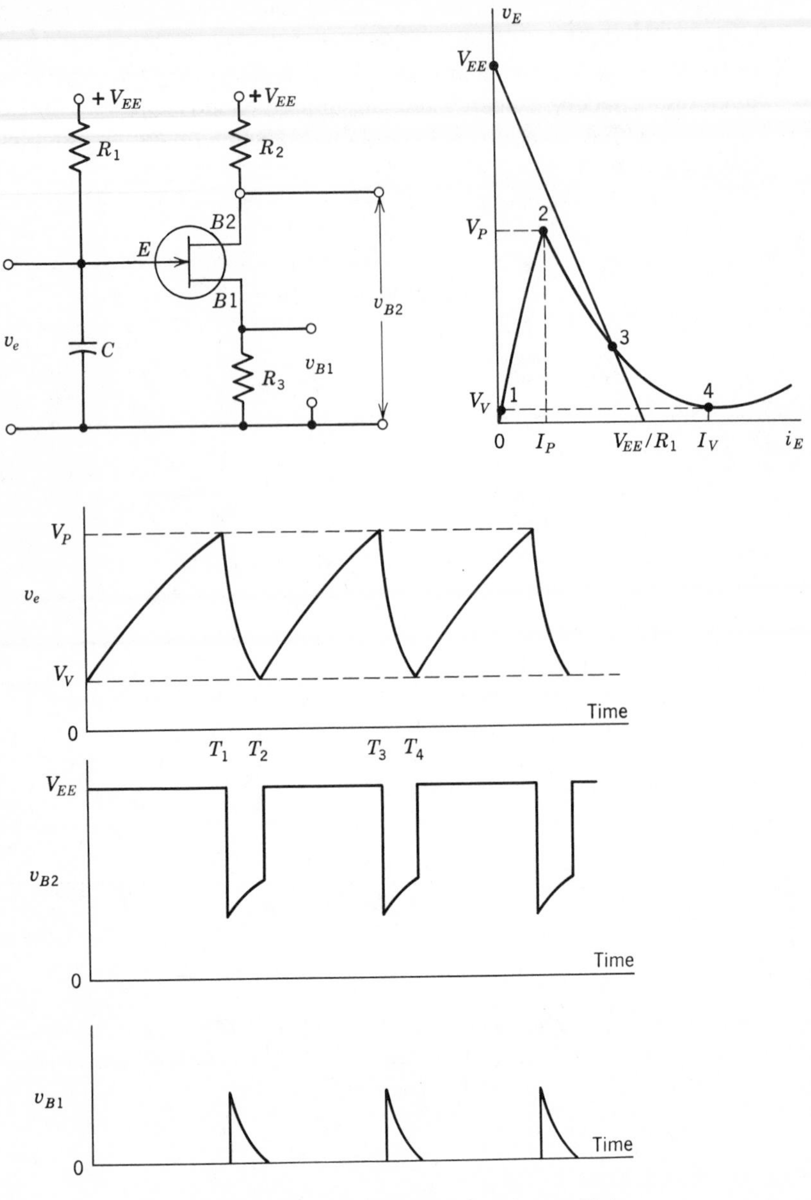

Figure 22-32 Unijunction relaxation oscillator.

tion time is from T_1 to T_3 but the minimum required duration time for the circuit to function is from T_1 to T_2. The duration of the output pulse is a function of the time constant of the inductance L in the circuit.

The circuit and the waveforms for a unijunction relaxation oscillator are shown in Fig. 22-32. In order for the circuit to function, the load line drawn between V_{EE} and V_{EE}/R_1 must intersect the characteristic in the negative resistance region at point 3, which is between the peak point, 2, and the valley point, 4. The capacitor C charges up to the peak point voltage V_P. The unijunction transistor "fires," and the voltage drops to the valley point, V_V. The path is from 1 to 2 to 3 to 4, the valley point. The circuit switches to point 1, and the process repeats.

The emitter waveform is a sawtooth and can be used for time delay up to a minute. The v_{B1} output could be used to fire an SCR. The OFF time for the UJT is given by

$$T_3 - T_2 = R_1 C \ln \frac{1}{1-\eta} \tag{22-4}$$

where η is the intrinsic standoff ratio. A typical value of η is 0.65. The derivation of Eq. 22-4 is left as an exercise at the end of the section.

Typical values used in a unijunction circuit that is used to fire an SCR rated at 10 A in an automobile circuit (12 V) are 100 Ω for $R3$ and 510 Ω for $R2$. $R1$ is a 51-kΩ resistor in series with a 500-kΩ potentiometer. When C is 50 μF, the cycling of the circuit varies from 3 to 30 sec. The circuit is used to control a windshield wiper. With this circuit a single-stroke cycle of the blade occurs at intervals separated by the time delay of the circuit. A cam-operated switch operates at the end of the cleaning cycle and restores the SCR to a non-conducting stage and the UJT to the start of a new timing cycle.

PROBLEMS

Refer to the tunnel-diode characteristic given in Fig. 4-31*c*.

1. Specify supply voltages and load lines that will produce the characteristics shown in Fig. 22-29.
2. If the circuit shown in Fig. 22-31 is operated at the inflection point of Fig. 4-31 it will oscillate, provided that the Q of the coil is sufficiently high. The inductance L is 7.6 μh and resonates with the tunnel-diode capacitance at 100 MHz. What is the least Q that the coil may have in order to oscillate? What is the input power to the oscillator?
3. Using Fig. 4-31, specify a circuit for Fig. 22-31. Draw the waveforms to scale.
4. Derive Eq. 22-4.
5. Calculate the exact cycling rate of the windshield washer.

Section 22-10 Transistor Power Supplies

Two power transistors can be arranged as a multivibrator to produce a square-wave power source (Fig. 22-33). The feedback is obtained from the ends of a center-tapped transformer winding. The same winding can be used

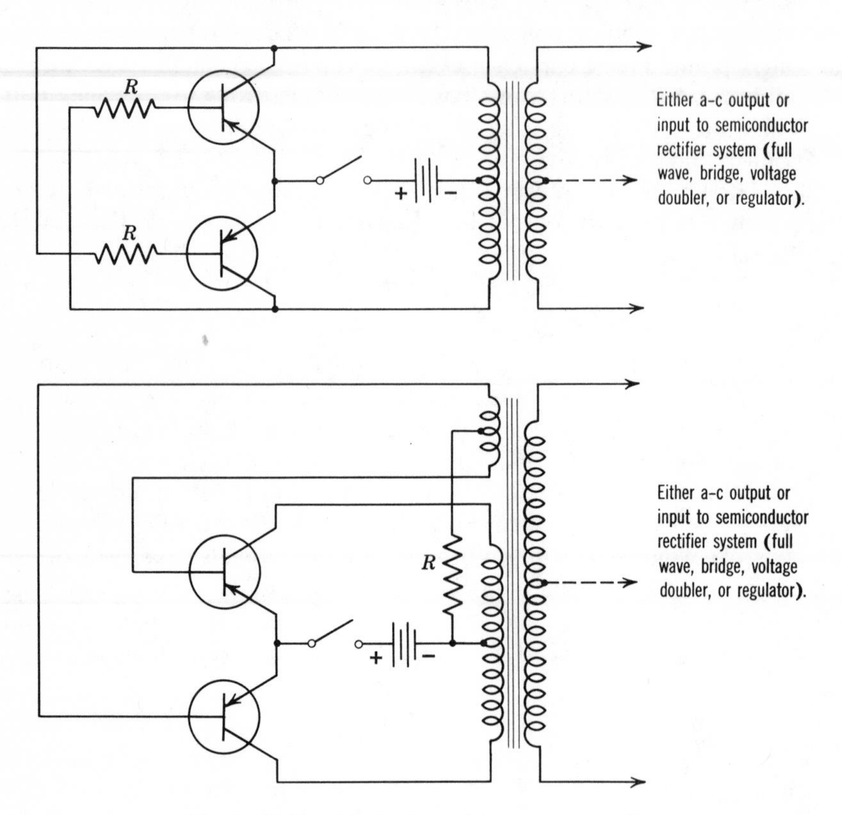

Figure 22-33 Typical transistor power supplies.

for both the bases and the collectors by cross-connecting the base leads, or a separate winding can be used as a feedback winding. There is a resistor in series with the base leads to limit the current to safe values. If the circuit were adjusted to establish sinusoidal oscillation, we could call these circuits push-pull oscillators of the Armstrong type. Often a capacitor is placed across one of the windings to "tune" the transformer to generate a particular frequency for use as an a-c source of power.

The circuit is often used to convert a low battery voltage into a high d-c voltage for such applications as a portable or mobile radio transmitter. The rectifier system used to convert a high alternating voltage into a direct voltage depends on the application. Conventional rectifier circuits (Chapter 6) or voltage regulators (Section 20-12) may be used with either circuit arrangement.

In a typical application that is used for changing the low voltage obtained from solar paddles on a satellite to the high voltage necessary for the equipment, an 85% efficiency can be obtained in a regulated system that operates on a 12-V input. When the input voltage increases, a regulator circuit maintains the output to ±1%. The whole unit is packaged in an enclosure $3'' \times 3\frac{1}{2}'' \times \frac{3}{4}''$ and has a life expectancy of three years operating 10 to 20 minutes twice daily.

PROBLEM

1. Assume that a full-wave bridge rectifier is connected to the secondary of the power supply shown in Fig. 22-33. The load on the bridge rectifier is pure resistance. Show current and voltage waveforms throughout the circuit. Assume that there are a total of 100 turns on the primary of the transformer and 400 turns on the secondary of the transformer. The battery voltage is 24 V.

Questions

1. Define or explain each of the following terms: (*a*) time constant, (*b*) trigger, (*c*) the Schmitt trigger, (*d*) commutator capacitor, (*e*) the flip-flop, (*f*) bistable oscillators, (*g*) monostable oscillators, (*h*) astable oscillators, (*i*) gate time duration, (*j*) forced frequency, (*k*) free running frequency, (*l*) pulse repetition rate, (*m*) damping, (*n*) sweep speed, (*o*) negative resistance.
2. What is an application of the Schmitt trigger circuit?
3. What is an application of the flip-flop?
4. What is an application of a monostable multivibrator?
5. What is an application of the astable multivibrator?
6. How is an astable miltivibrator unbalanced?
7. Explain carefully how synchronization is obtained?
8. In a synchronized oscillator, is the free running frequency higher or lower than the forced frequency? Explain.
9. Can the pulse width be varied in a blocking oscillator?
10. What is the purpose of a discharge circuit?

11. What provisions can be made to improve the linearity of the output of a discharge circuit?
12. What controls the frequency of the output of a discharge circuit?
13. If the driving oscillator fails, what is the output of the discharge circuit?
14. Explain what a voltage controlled negative resistance is. Give examples.
15. Explain what a current controlled negative resistance is. Give examples.
16. Show how bistable operation is possible on a tunnel-diode characteristic.
17. Show how astable operation is possible on a tunnel-diode characteristic.
18. If two tunnel diodes are connected in series, what does the composite characteristic look like?

Chapter Twenty-three

HIGH-FREQUENCY AMPLIFIERS

This chapter is concerned with small-signal amplifiers at high frequencies. The hybrid-π model for a transistor is examined in detail (Section 23-1). The high-frequency model for a FET (Section 23-2) is quite similar to the model for the vacuum tube. Tuned amplifiers of various types (Section 23-3) are direct applications of these high-frequency considerations.

Section 23-1 The Hybrid-π Model for a Transistor

The hybrid-π model is used primarily to analyze a transistor circuit at high frequencies and is valid for radio frequencies up to, at least, 100 MHz or 200 MHz. Models that are used for microwave frequencies are beyond the scope of this text. Also, another restriction is placed on our discussion in this section. The common emitter circuit configuration is assumed, and any resistor placed in the emitter circuit is adequately bypassed.

The basic model for a transistor that we have used in the transistor circuit analysis is shown in Fig. 23-1*a*. We specified r_e' by evaluating

$$\frac{25mv}{I_E} \leqslant r_e' \leqslant \frac{50mv}{I_E} \tag{12-1}$$

In a common emitter circuit, r_e' reflects into the input as $(1+\beta)r_e'$. If we consider only common-emitter circuits with any external R_E properly bypassed, an alternative model is shown in Fig. 23-1*b*. We never did use this particular model because of the problem of the unbypassed R_E which would require a modification in the model. This model is adequate for low frequencies, and we could have treated the low-frequency and high-frequency response in R-C-coupled amplifiers with this modified model.

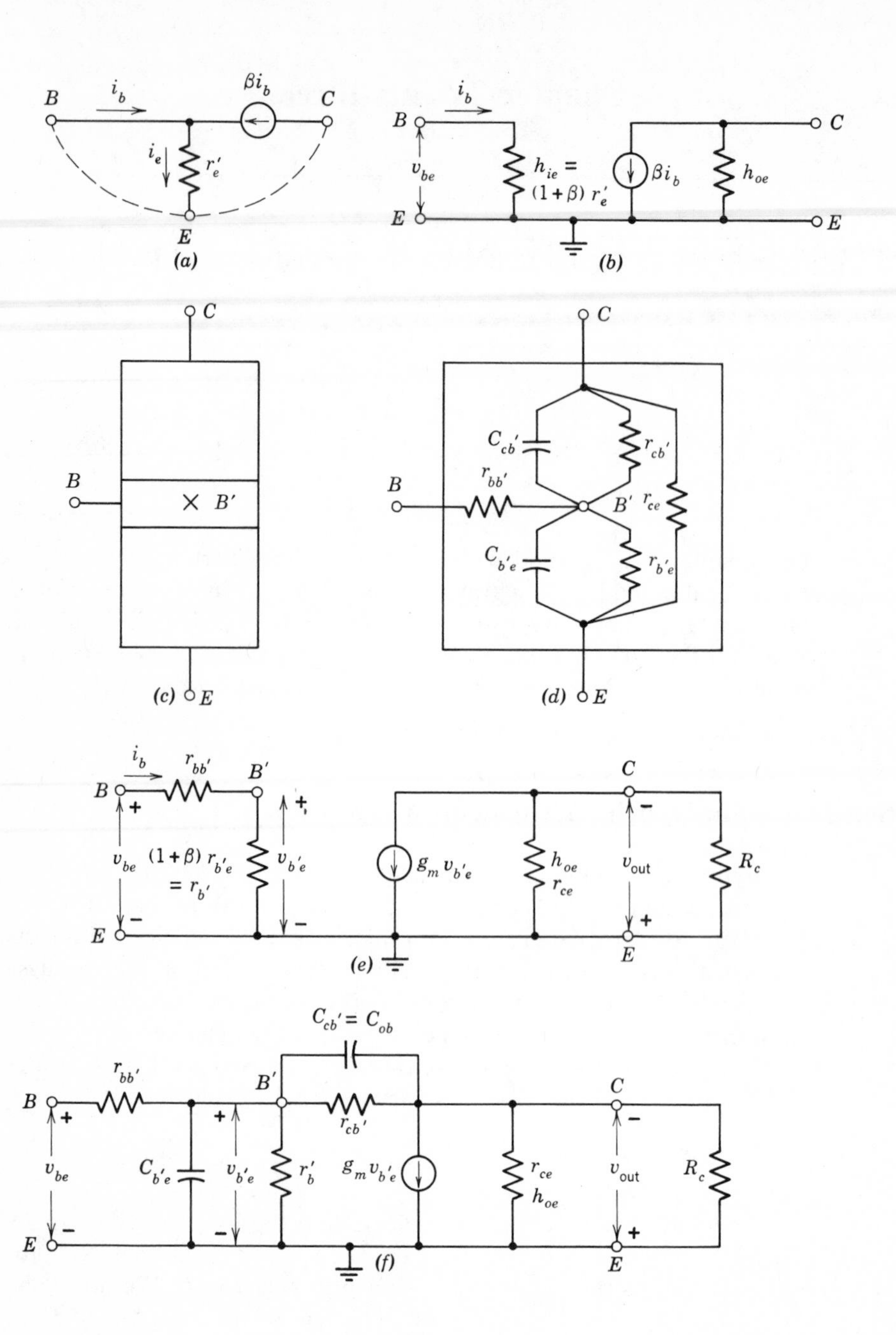

Figure 23-1 Transistor models. (*a*) Low-frequency a-c model. (*b*) Low-frequency a-c model for a common-emitter amplifier. (*c*) A geometric form of a transistor. (*d*) Internal electrical effects. (*e*) High-frequency resistance model. (*f*) The hybrid-π model.

At high frequencies, however, we must examine the internal conditions of the transistor (Fig. 23-1*c*). We considered all the action in a transistor that takes place at a point B' inside the transistor and, with the exception of r'_e, all internal resistive effects were disregarded.

Now we must account for the fact that the action in a transistor is spread out. Point B' is not infinitesimally close to the base lead B, and there are several resistance paths and capacitive effects (Fig. 23-1*d*) that must be included. A model that shows resistance effects only is given in Fig. 23-1*e*. h_{ie} is now broken up into two parts r_{bb}, and $r_{b'}$. The internal diagram (Fig. 23-1*e*) shows that, as we go from the base lead to the emitter lead, we encounter two resistances $r_{bb'}$ and $r_{b'e}$. $r_{bb'}$ is a low resistance of the order of 50 or 100 Ω. The use of either of these numberical values is an accepted design procedure if the exact value is not given in the transistor specification sheet. $r_{b'e}$, when it reflects into the base input circuit, is multiplied by $(1+\beta)$ in the same concept that was used to find h_{ie} from r'_e. The reflected value of $r_{b'e}$ is given a new symbol $r_{b'}$:

$$r_{b'} \equiv (1+\beta) r_{b'e} \tag{23-1}$$

The literature also uses the notation r_π for $r_{b'}$.

Then $$h_{ie} = r_{bb'} + (1+\beta) r_{b'e} = r_{bb'} + r_{b'} \tag{23-2}$$

The input signal to the transistor is v_{be}, but the signal is reduced by an IR drop in $r_{bb'}$ to $v_{b'e}$.

$$v_{b'e} = (1+\beta) r_{b'e} i_b = r_{b'} i_b$$

Then

$$i_b = \frac{v_{b'e}}{(1+\beta) r_{b'e}}$$

The current source in the collector, βi_b, is

$$\beta i_b = \frac{\beta v_{b'e}}{(1+\beta) r_{b'e}} = \frac{\beta}{1+\beta}\left(\frac{v_{b'e}}{r_{b'e}}\right)$$

Since $\beta/(1+\beta)$ is approximately unity

$$\beta i_b = \frac{1}{r_{b'e}} v_{b'e}$$

In Fig. 23-1*b*, r'_e reflects into the base when it is multiplied by $(1+\beta)$. Consequently, $r_{bb'}$ reflects into the emitter divided by $(1+\beta)$, and the numerical value of such a reflection in a transistor that has a β of 100 is only 1 Ω. Therefore, only a small error is made if the collector current source is written as

$$\beta i_b = \frac{1}{r'_e} v_{b'e}$$

A new term is now defined. The *transconductance* g_m of the transistor is the reciprocal of r'_e and can be evaluated from

$$\frac{I_E}{50 \text{ mv}} \leqslant g_m \equiv \frac{1}{r'_e} \leqslant \frac{I_E}{25 \text{ mv}} \tag{23-3}$$

The transistor specification sheets often give a curve that shows the variation of g_m with operating frequency. The current source is now

$$\beta i_b = g_m v_{b'e} \tag{23-4}$$

and this current source $g_m v_{b'e}$ is shown as the current generator in the hybrid-π models (Fig. 23-1*e* and *f*). The capacitances within the transistor are included in the complete hybrid-π model along with a leakage resistance $r_{cb'}$. Usually $r_{cb'}$ is sufficiently large that it can be omitted.

If we ignore $r_{cb'}$, the current in C_{ob} (Fig. 23-1*f*) is

$$i_{b'c} = \frac{v_{b'e} - v_{\text{out}}}{-j(1/\omega C_{ob})} = j\omega C_{ob}(v_{b'e} - v_{\text{out}})$$

But
$$v_{\text{out}} = -(g_m v_{b'e}) R_c$$

This last equation ignores h_{oe} and assumes that the parallel combination of h_{oe} and R_c is effectively R_c. If h_{oe} is a sufficiently high conductance, the value of R_c can be reduced to the value of the parallel combination.

Now
$$i_{b'c} = j\omega C_{ob}(v_{b'e} + g_m R_c v_{b'e}) = j\omega C_{ob}(1 + g_m R_c) v_{b'e}$$

dividing
$$\frac{v_{b'e}}{i_{b'c}} = -j \frac{1}{\omega C_{ob}(1 + g_m R_c)}$$

Therefore, the equivalent capacitance effect of C_{ob} is $C_{ob}(1 + g_m R_c)$. This is the Miller effect previously developed in Section 15-5 and can be used in a modified hybrid-π model (Fig. 23-2*a*). $C_{b'e}$ is combined with C_{ob} into C_T, as shown in Fig. 23-2*b* where

$$C_T = C_{b'e} + (1 + g_m R_c) C_{ob} \tag{23-5}$$

In an actual transistor the order of magnitude of $C_{b'e}$ is about 100 pF and the value of C_{ob} ranges from 0.22 to 2.8 pF. Now apply a short circuit across the output of the circuit of Fig. 23-1*f*. If a short circuit is applied across the output of the circuit of Fig. 23-1*f*, C_{ob} will be in parallel with $C_{b'e}$ and since C_{ob} is small, it can be neglected. The circuit is now the one shown in Fig. 23-3. The *beta cutoff frequency* f_β is that frequency at which the short-circuit current in the output i_{out} is down 3 db. At this frequency the current, i_1, in $C_{b'e}$ equals the current, i_2, in $r_{b'}$. Accordingly, the impedances of the two branches are equal

$$\frac{1}{\omega_\beta C_{b'e}} = r_{b'}$$

or

$$\omega_\beta = \frac{1}{r_{b'} C_{b'e}} \quad \text{and} \quad f_\beta = \frac{1}{2\pi r_{b'} C_{b'e}} \tag{23-6}$$

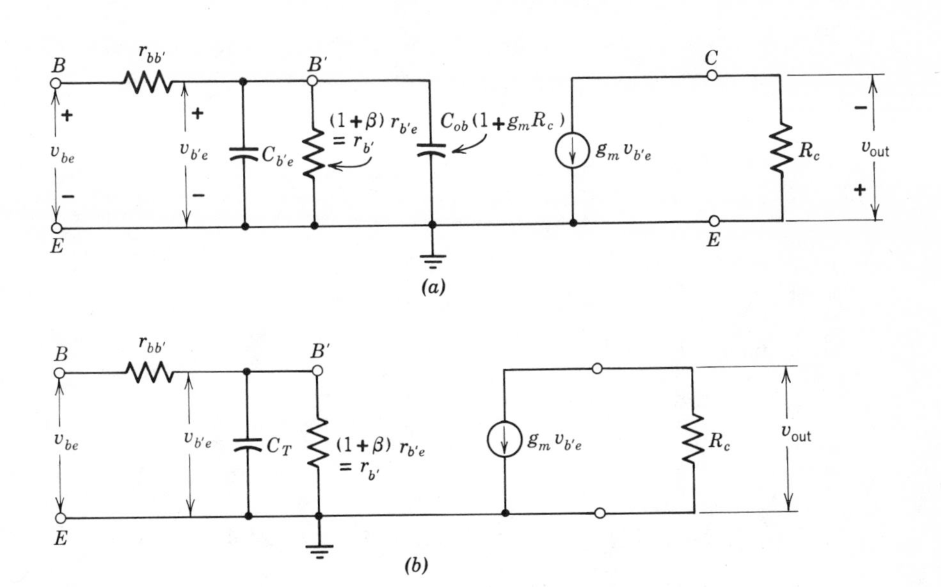

Figure 23-2 Hybrid-π models. (*a*) Uses the Miller effect value of *Cob*. (*b*) Combines all input capacitances into C_T.

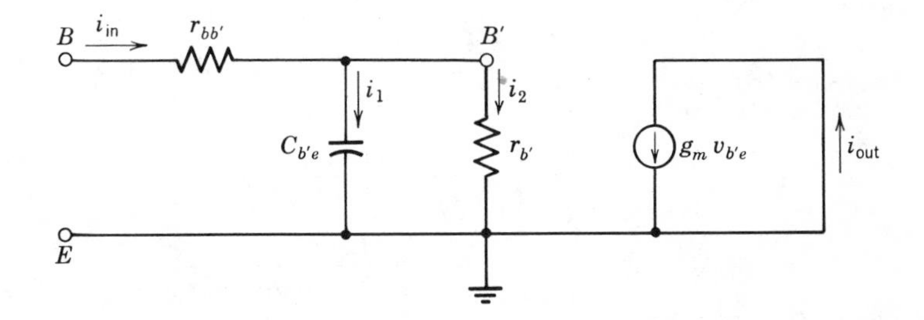

Figure 23-3 Circuit used to determine f_β.

The *short-circuit unity gain frequency*, f_T, requires that the collector current $g_m v_{b'e}$ equals i_{in} in the circuit of Fig. 23-3. By the current divider rule

$$i_2 = i_{in} \frac{-j(1/\omega_T C_{b'e})}{r_{b'} - j(1/\omega_T C_{b'e})}$$

At this very high frequency, the reactance of $C_{b'e}$ is very low, and this equation can be approximated to

$$i_2 = -j\frac{i_{\text{in}}}{\omega_T r_{b'} C_{b'e}}$$

But

$$g_m v_{b'e} = \beta i_2$$

Then, by the definition of f_T,

$$\beta i_2 = i_{\text{in}}$$

and using magnitudes

$$i_{\text{in}} = \omega_T r_{b'} C_{b'e} i_2 = \beta i_2$$

Then

$$\omega_T = \frac{\beta}{r_{b'} C_{b'e}}$$

By using Eq. 23-6,

$$\omega_T = \frac{\beta}{r_{b'} C_{b'e}} = \beta\omega_\beta \tag{23-7a}$$

and

$$f_T = \frac{\beta}{2\pi r_{b'} C_{b'e}} = \beta f_\beta \tag{23-7b}$$

A generator whose emf is e_s and whose internal resistance is R_s is connected to the input of the circuit of Fig. 23-2b. An expression for voltage gain A_e is found by evaluating v_{out}/e_s. The parallel combination of C_T and r_b is

$$\frac{r_{b'}(-j(1/\omega C_T))}{r_{b'} - j(1/\omega C_T)} = \frac{r_{b'}}{1 + j\omega r_{b'} C_T}$$

By the voltage divider rule

$$v_{b'e} = \frac{r_{b'}/(1 + j\omega r_{b'} C_T)}{R_s + r_{bb'} + r_{b'}/(1 + j\omega r_{b'} C_T)} e_s$$

$$v_{b'e} = \frac{r_{b'} e_s}{R_s + r_{bb'} + r_{b'} + j\omega r_{b'}(R_s + r_{bb'}) C_T}$$

The output voltage is

$$v_{\text{out}} = -g_m R_c v_{b'e}$$

Then

$$A_e = \frac{v_{\text{out}}}{e_s} = -\frac{g_m R_c r_{b'}}{R_s + r_{bb'} + r_{b'} + j\omega r_{b'}(R_s + r_{bb'}) C_T} \tag{23-8}$$

The 3-db frequency f_2 occurs when the j term in the denominator increases to equal the real term in magnitude:

$$\omega_2 r_{b'}(R_s + r_{bb'}) C_T = R_s + r_{bb'} + r_{b'}$$

or

$$\omega_2 = \frac{R_s + r_{bb'} + r_{b'}}{r_{b'}(R_s + r_{bb'})C_T} \tag{23-9a}$$

and

$$f_2 = \frac{\omega_2}{2\pi} = \frac{R_s + r_{bb'} + r_{b'}}{2\pi r_{b'}(R_s + r_{bb'})C_T} \tag{23-9b}$$

The equation for the midband gain of this amplifier is

$$A_{e,MF} = \frac{h_{ie}}{R_s + h_{ie}} \frac{\beta R_c}{h_{ie}} = \frac{\beta R_c}{R_s + h_{ie}} = \frac{\beta R_c}{R_s + r_{bb'} + r_{b'}} \tag{23-9c}$$

The *Gain-Bandwidth product (G-BW)* is defined as

$$\text{G-BW} \equiv (A_{e,MF})\,\omega_2$$

and substituting Eq. 23-9c for $A_{e,MF}$ and Eq. 23-9a for ω_2

$$\text{G-BW} = \frac{\beta R_c}{r_{b'}(R_s + r_{bb'})C_T} \tag{23-10a}$$

The units of Eq. 23-10a are in radians per second, since $A_{e,MF}$ is dimensionless.
If the units of G-BW are in hertz, the gain-bandwidth product is

$$\text{G-BW} = \frac{\beta R_c}{2\pi r_{b'}(R_s + r_{bb'})C_T} \text{ hertz} \tag{23-10b}$$

The gain-bandwidth product of typical RF transistors ranges from 300 MHz to 1000 MHz.

Assume that a circuit has the following values

$$\begin{aligned} r_{bb'} &= 50\ \Omega & C_{b'e} &= 100\text{ pF} \\ r_{b'} &= 5000\ \Omega & C_{ob} &= 1.5\text{ pF} \\ R_c &= 20\text{ k}\Omega & \beta &= 100 \\ R_s &= 0 & h_{oe} &= 0 \end{aligned}$$

The following information is required:

1. The short-circuit beta cutoff frequency
2. The short-circuit unity gain frequency, and
3. The circuit 3 db break frequency.

r'_e is determined by

$$r'_e = (r_{bb'} + r_{b'})/(1+\beta) \approx r_{b'}/\beta = 5000/100 = 50\ \Omega$$

Then $g_m = 1/r'_e = 1000/50 = 20$ millimhos
The low-frequency gain is

$$A_{e,MF} = -g_m R_c = -20 \times 20 = -400$$

By Eq. 23-5, $C_T = C_{b'e} + (1 + g_m R_c) C_{ob} = 100 + (401)1.5 = 700 \text{ pF}$

By Eq. 23-6,

$$f_\beta = \frac{1}{2\pi r_{b'} C_{b'e}} = \frac{1}{2\pi 5000 \times 100 \times 10^{-12}} = 0.318 \text{ MHz}$$

By Eq. 23-7*b*, $f_T = \beta f_\beta = 100(0.318) = 31.8 \text{ MHz}$

By Eq. 23-9*b*,

$$f_2 = \frac{R_s + r_{bb'} + r_{b'}}{2\pi r_{b'}(R_s + r_{bb'})C_T} = \frac{0 + 50 + 5000}{2\pi 5000(0 + 50)700 \times 10^{-12}}$$

$$= 4.54 \text{ MHz}$$

The gain-bandwidth product is

$$\text{G-BW} = (A_{e,MF})(f_2) = 400 \times 4.54 = 1816 \text{ MHz.}$$

As a check on the results, the gain-bandwidth product is

$$\text{G-BW} = \frac{\beta R_c}{2\pi r_{b'}(R_s + r_{bb'})C_T} = \frac{100 \times 20000}{2\pi 5000(0 + 50)700 \times 10^{-12}} = 1820 \text{ MHz}$$

PROBLEMS

TABLE A High-Frequency Transistor Data

Transistor	A	B	C
I_E ma	4	4	2
h_{fe} at 1 kHz	20	80	80
C_{ob} pF	1.7	0.75	3.5
$C_{b'e}$ pF	50	50	80
$r_{bb'}$ ohms	100	50	50

For r_e' use 50 mv/I_E.

1. Determine f_β for Transistor A.
2. Determine f_β for Transistor B.
3. Determine f_β for Transistor C.
4. Determine f_T for Transistor A.
5. Determine f_T for Transistor B.
6. Determine f_T for Transistor C.
7. Determine the mid-frequency gain for a circuit using Transistor A, a 50-Ω source, and a 3,000-Ω load. Determine the 3-db frequency. What is gain-bandwidth product?
8. Repeat Problem 7 for a 500-Ω load.

9. Determine the mid-frequency gain for a circuit using Transistor B, a 50-Ω source, and a 3000-Ω load. Determine the 3-db frequency. What is the gain-bandwidth product?
10. Repeat Problem 9 for a 500-Ω load.
11. Determine the mid-frequency gain for a circuit using Transistor C, a 50-Ω source, and a 3000-Ω load. Determine the 3-db frequency. What is the gain-bandwidth product?
12. Repeat Problem 11 for a 500-Ω load.

Section 23-2 The High-Frequency Model for the FET

In Section 12-8, the FET was considered in the middle-frequency range. This analysis was concerned with the gain of the FET that is used as an amplifier without a consideration of the capacitive effects within the FET. The model that was used is shown in Fig. 23-4a. At high frequencies, the

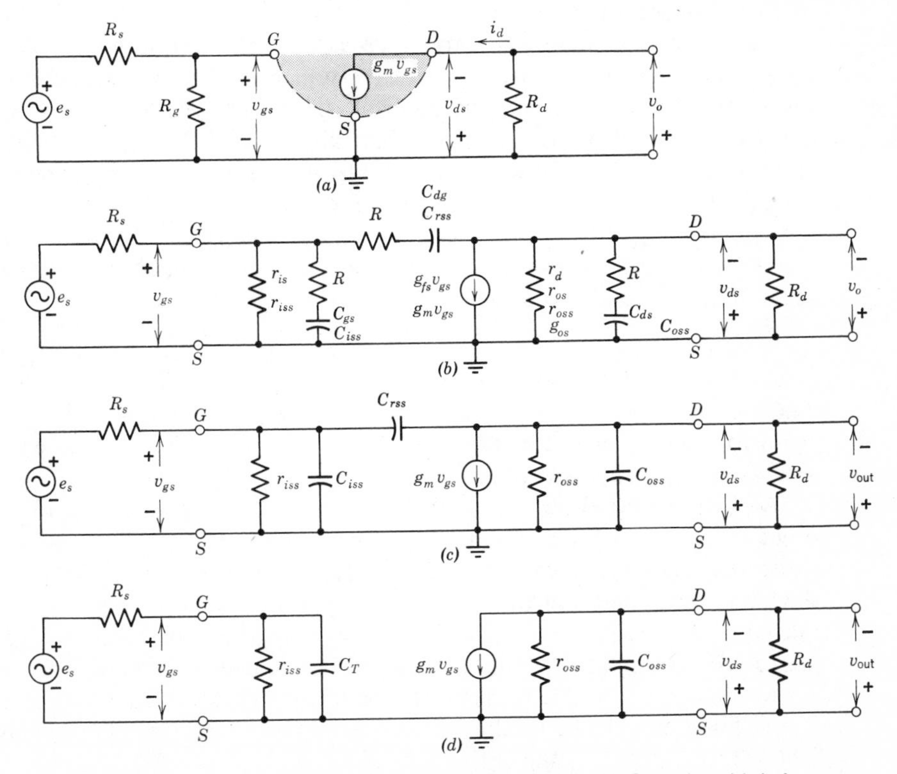

Figure 23-4 FET models. (a) Midband low-frequency model. (b) Complete high-frequency model. (c) Modified high-frequency model. (d) Rearranged high-frequency model.

interelectrode capacitances must be included. They are similar to the one of the transistor and are added to the mid-frequency model to give the mode shown in Fig. 23-4*b*. This model is labeled to show the various terminologie that exist in the literature.

Some new terms that are used in this section require a formal definition All of these definitions require that the FET or MOSFET be connected i the common-source mode for the definition and for the actual measuremen of the parameter.

r_{iss}	or	r_{is}	is the small-signal, short-circuit input resistance
r_{oss}	or	r_{os}	is the small-signal, short-circuit output resistance
C_{iss}	or	C_{gss}	is the small-signal, short-circuit input capacitance
C_{oss}			is the small-signal, short-circuit output capacitance
C_{rss}	or	C_{dg}	is the small-signal, short-circuit reverse or feedbac capacitance

These definitions are made in the same sense that h_{11}, h_{22}, and h_{12} are define in the hybrid parameter system.

The interelectrode capacitors are not ideal; consequently, each i represented with a series equivalent resistance R. The numerical value o these series resistances is in the order of 100 Ω. However, these resistance can be neglected at frequencies below about 200 MHz. The model nov simplifies to the one shown in Fig. 23-4*c*. Typical values for C_{rss} are fron 0.1 to 0.3 pF. A special type is available that has the extremely low value o 0.02 pF for C_{rss}. C_{oss} ranges from 1 to 2.5 pF, and C_{iss} is about 4 to 6 pF The Miller effect causes C_{rss} to reflect back into the input circuit the same a in the transistor and in the vacuum tube

$$C_T = C_{iss} + (1 + g_m r_L)C_{rss} \tag{23-11}$$

where r_L is the parallel combination of r_{oss} and R_d. Now the model simplifie again to the one shown in Fig. 23-4*d*. R_d is the equivalent of any externa resistance in parallel with a tuned circuit of which C_{oss} is a part.

At 100 or 200 MHz, r_{oss} is of the order of 3 or 4 kΩ, and r_{iss} is of the order of 1.5 to 3.5 kΩ. As the frequency drops toward zero, the outpu resistance increases toward about 20 kΩ, and the input resistance approache the very high values of thousands of megohms, which is characteristic o the gate. When we have an amplifier operating at 100 MHz, we are no interested in calculating a voltage gain; the power gain is the sole concern.

Assume r_{oss} is 4 kΩ, r_{iss} is 3 kΩ and R_d is 10 kΩ for a FET that has a g_m of 7500 μmhos. The total load is the parallel combination of r_{oss} and R_d o 2.86 kΩ. If v_{gs} is 0.5 V, the output voltage v_{ds} is

$$v_{ds} = g_m r_L = 7.5 \times 2.86 = 21.5 \text{ V}$$

The power gain is

$$A_p = \frac{21.5^2/2.86}{0.5^2/3} = 1930 \text{ or } 32.8 \text{ db}$$

This example ignores the effect of R_s, which cannot be zero at 200 MHz because of skin effect and the shunting effect of C_T. Although C_{oss} is small (2 pF at 200 MHz), its reactance at 200 MHz is only 400 Ω. Consequently, to prevent C_{oss} from reducing the gain to a low value, C_{oss} must be part of a parallel tuned circuit.

In any event, the calculation for the FET does not have the complexity of the calculation for a transistor. The vacuum-tube calculation is still less complex since "r_{iss}" for a vacuum tube is infinite at all but microwave frequencies. "r_{oss}" is the a-c plate resistance of the tube and, since pentodes are used, the plate resistance is so high that it, too, is neglected. Therefore, in a vacuum-tube circuit, a capacitance

$$C_{\text{in}} = C_{gk} + C_{gp}(1 + A_v) \tag{15-18a}$$

loads the input circuit, and the voltage gain of the stage is simply

$$A_e = g_m Z_L \tag{15-18b}$$

PROBLEMS

TABLE B High-Frequency FET Data

FET	A	B	C	D
g_{fs}	6000	4500	4000	4500
g_{os}	50	50	50	35
At 1 MHz				
C_{iss}	4	4.5	8	6
C_{rss}	0.8	1	1.4	2
C_{oss}	2	2.5	4	3.0
At 100 MHz				
$g_{is} = Re\,(y_{is})$	100	100		
$b_{is} = Im\,(y_{is})$	2500	3000		
$g_{os} = Re\,(y_{os})$	75	75		
$b_{os} = Im\,(y_{os})$	1000	1000		
At 400 MHz				
g_{is}	1000	1000		
b_{is}	10,000	12,000		
g_{fs}	4000	3000		
g_{os}	100	100		
b_{os}	4000	4000		

All conductances and susceptances are in micromhos and all capacitances are in pF.

1. An RF amplifier with a load tuned to 1 MHz uses FET A. The load impedance at resonance is 10,000 Ω and the source resistance is 1000 Ω. Determine the circuit gain.
2. Repeat Problem 1 using FET C.
3. Repeat Problem 1 using FET D.
4. A 100-MHz amplifier using FET A has a source resistance of 500 Ω and a tuned load impedance of 5000 Ω. Determine the circuit gain.
5. Repeat Problem 4 using FET B.
6. A 400-MHz amplifier has a source impedance of 50 Ω and an effective load impedance of 500 Ω. Using FET A, determine the circuit gain.
7. Repeat Problem 5 using FET B.
8. What is the gain of the amplifier of Problem 4 if a coil with a Q of 30 is placed across the input and is adjusted to tune the input to resonance?
9. Repeat Problem 8 for Problem 5.
10. Repeat Problem 8 for Problem 6.
11. Repeat Problem 8 for Problem 7.

Section 23-3 Tuned Amplifier

Single-tuned amplifiers, using a parallel tuned circuit and a coupling capacitor, are shown in Fig. 23-5. The tank circuit is loaded in each case by various resistances. In all three circuits the output resistance of the amplifier is in parallel with the tank. The bias circuit and the input resistance of the second stage also load down the tuned circuit in the transistor amplifier.

The parallel circuit is treated in Section 10-2. The impedance of a tuned circuit is maximum at resonance and is a pure resistance. This resistance value is lowered by the shunting resistances to yield an equivalent resistance that can be used in the gain equations we have developed. Then the maximum gain, which is the gain at resonance, can be determined for any one of the three circuit configurations.

The gain falls off either side of resonance, as shown in Fig. 23-6*a*. This fall-off is discussed in Section 10-3 which is devoted to bandwidth. Having the gain at center frequency, the fall-off can be evaluated by considering the reduction factor of the tuned circuit response.

The gains of two amplifiers, A and B, are plotted on the basis of unity gain at resonance in Fig. 23-6*b*. Amplifier A has a bandwidth of $f_2 - f_1$, and amplifier B has a bandwidth of $f'_2 - f'_1$. When these two amplifiers are connected in cascade, the overall gain is the product of the gains of the two stages (curve C). When the gain of stage A is 0.40 and the gain of stage B is 0.30, the combined gain is 0.12. The bandwidth of the composite curve is $f''_2 - f''_1$.

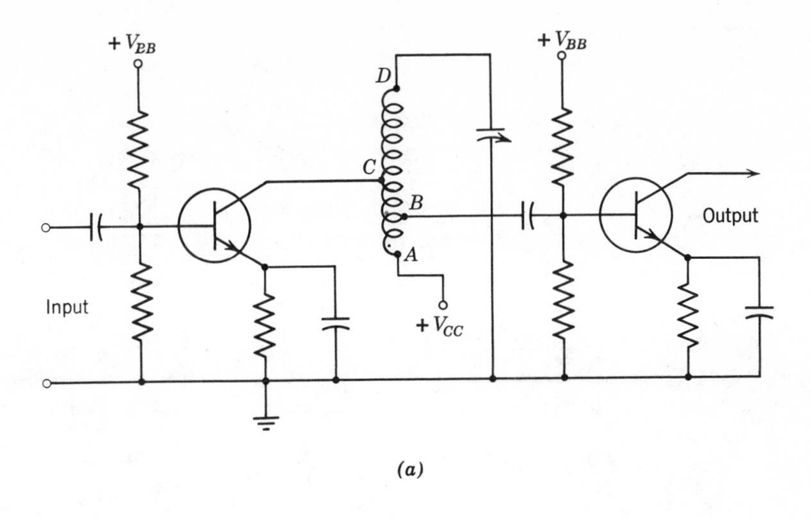

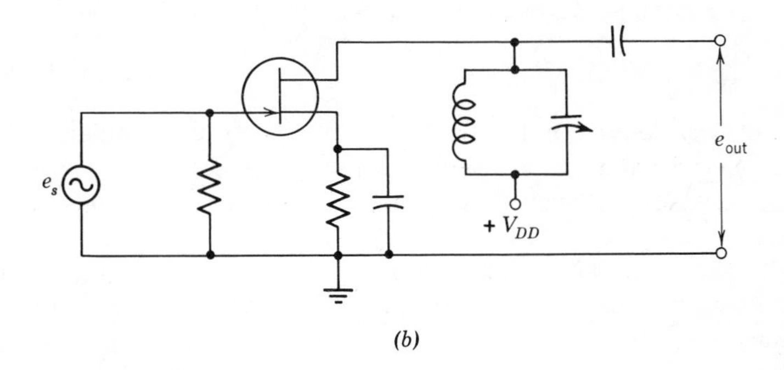

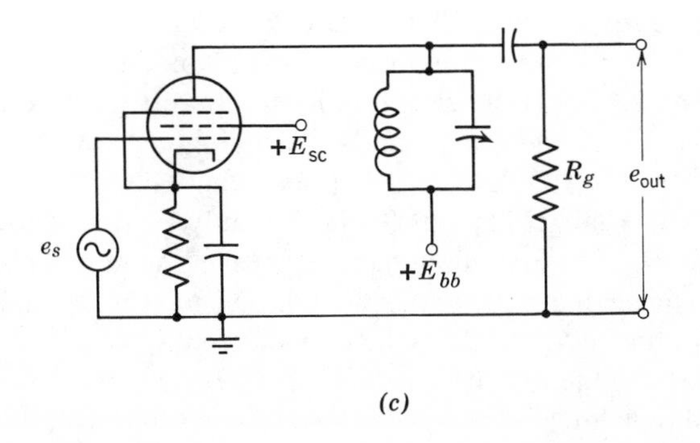

Figure 23-5 Single-tuned amplifiers. (*a*) Transistor circuit. (*b*) JFET circuit. (*c*) Vacuum-tube circuit.

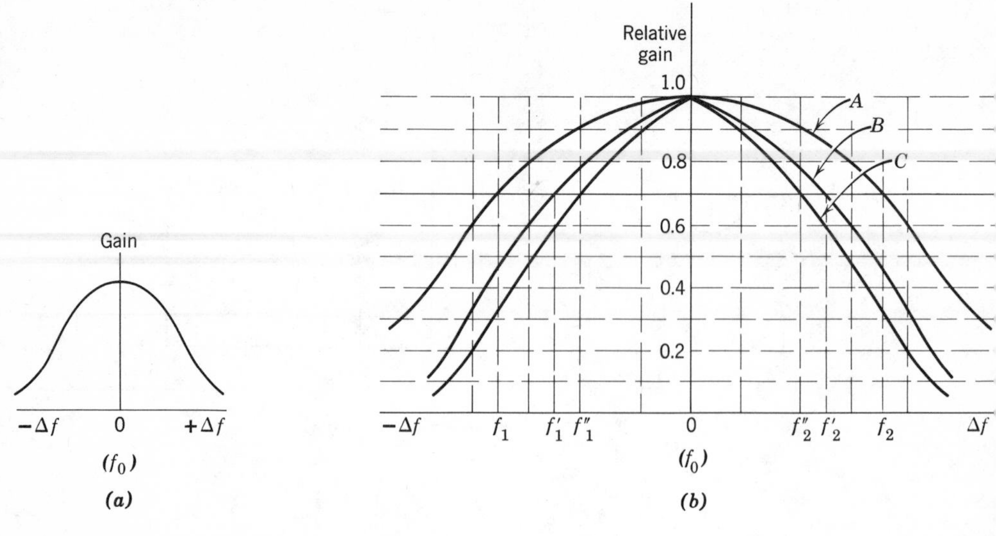

Figure 23-6 Response curves. (*a*) Single-tuned circuit. (*b*) Relative gain for a two-stage amplifier.

It is evident from this example that, when tuned amplifiers are cascaded the overall bandwidth depends on the bandwidths of the several individual stages. Also, the overall bandwidth is less than the bandwidth of any one stage.

Where there are several stages, n with the restriction that each stage has an identical bandwidth response, BW, the overall bandwidth, BW_n can be found from

$$BW_n = BW\sqrt{2^{1/n} - 1} \tag{23-12}$$

In Fig. 23-7, the amplifiers use as the coupling device between stages the single-tuned air core transformer treated in Section 10-5. The transistor circuit (Fig. 23-7*a*) is the circuit most commonly used in radio receivers. The primary is tuned, and the secondary winding is a coupling method to extract energy for the next stage. The response curves are shown in Fig. 23-7*c*. The coupling in an air-core transformer is only of the order of a few percent. Consequently, k is low enough that the reflected impedance of the secondary load into the primary does not overcompensate for the increase in the voltage induced in the secondary when k is increased. Therefore, as k increases, the maximum voltage in the secondary increases with k. The vacuum-tube circuit, shown in Fig. 23-7*b*, is used as the RF amplifier stage in a radio receiver or in a television receiver. Here the secondary is tuned. Also this is the circuit that is usually used with the FET. The circuit that couples the

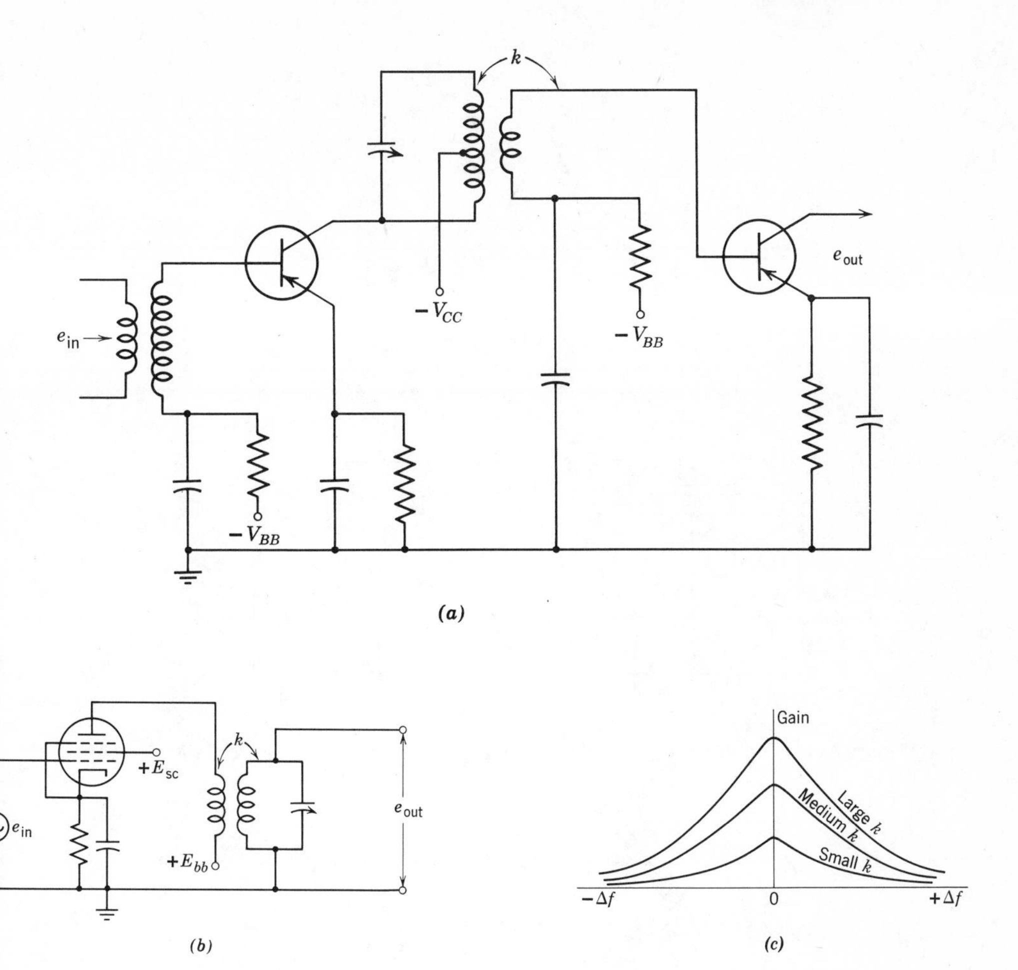

Figure 23-7 Amplifiers using single-tuned air-core transformers. (*a*) Transistor circuit. (*b*) Vacuum-tube circuit. (*c*) Response.

antenna to the first transistor in a radio receiver usually has the tuning capacitor placed across the secondary of the transformer.

A circuit using a double-tuned air-core transformer is shown in Fig. 23-8. Each of the two tuned circuits is adjusted to the same resonant frequency. The double-tuned air-core transformer analysis is given in Section 10-6. A FET can be used directly in place of the vacuum tube, and the only change in the equations would be to replace r_p with r_{ds}. The equivalent circuit for this amplifier is given in Fig. 23-9*a*. The source $g_m e_g$ and the circuit elements r_p and C_1 may be replaced by Thévenin's theorem to yield

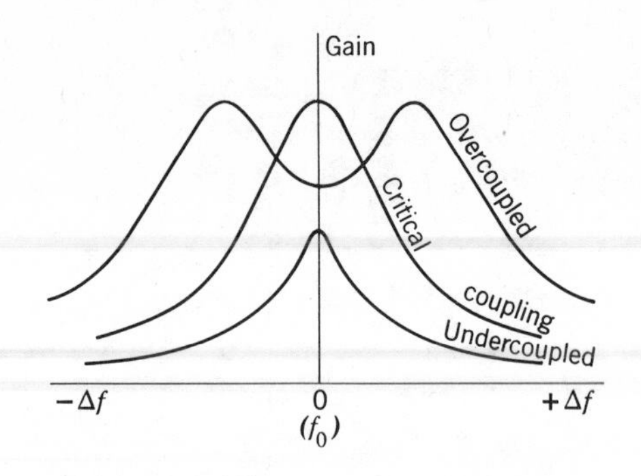

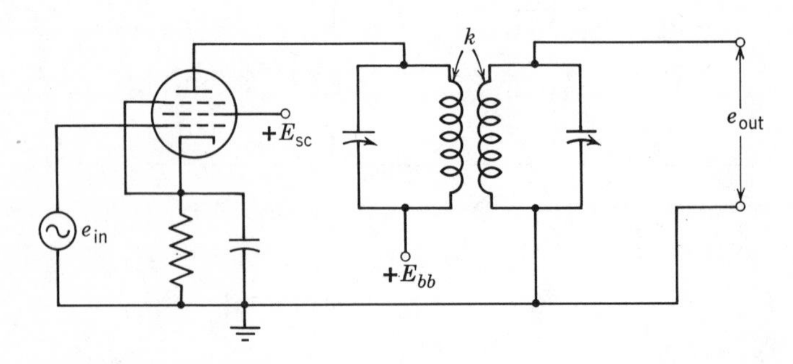

Figure 23-8 Amplifier using a tuned-primary-tuned-secondary air-core transformer.

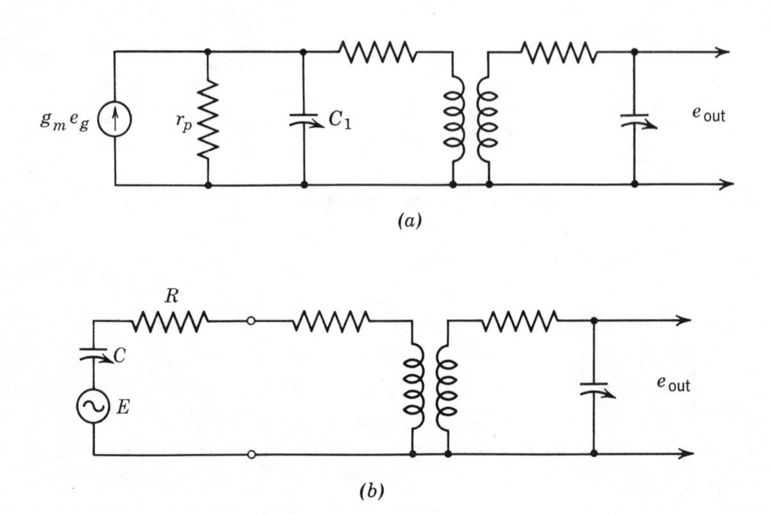

Figure 23-9 Equivalent circuits for the double-tuned amplifier. (*a*) Equivalent circuit. (*b*) Equivalent transformed by Thevenin's theorem.

a series circuit of E, C, and R. This series circuit is combined into a new equivalent circuit (Fig. 23-9b). If the plate resistance is high, it may be neglected in relation to the rest of the circuit. The plate resistance r_p converts into the series circuit as the resistance R. In a series circuit the value of R, in order to be negligible, must be very small, and it can be combined with the resistance of the primary for the purpose of calculations.

When the primary coil of the circuit is arranged in series, and the secondary is in parallel, we can use the analysis that was developed in Section 10-6. In this analysis we showed that, as the coefficient of coupling increases from zero, the gain increases and the bandwidth decreases. The gain increases until the critical coupling is reached. At this point the gain is for high Q coils

$$A_e = \frac{g_m \omega_0 M}{k_c^2} = g_m \frac{\omega_0 \sqrt{L_p L_s}}{k_c} \tag{23-13}$$

When the coefficient of coupling is increased beyond the value of critical coupling, the response curves expand to produce the double humped characteristic. The gain of the amplifier at the two peaks is given by Eq.

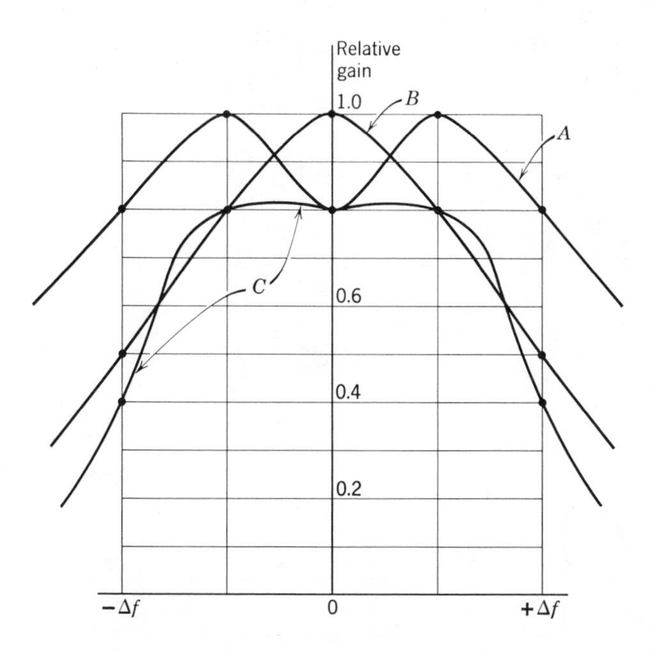

Figure 23-10 Response of a combined two-stage amplifier.

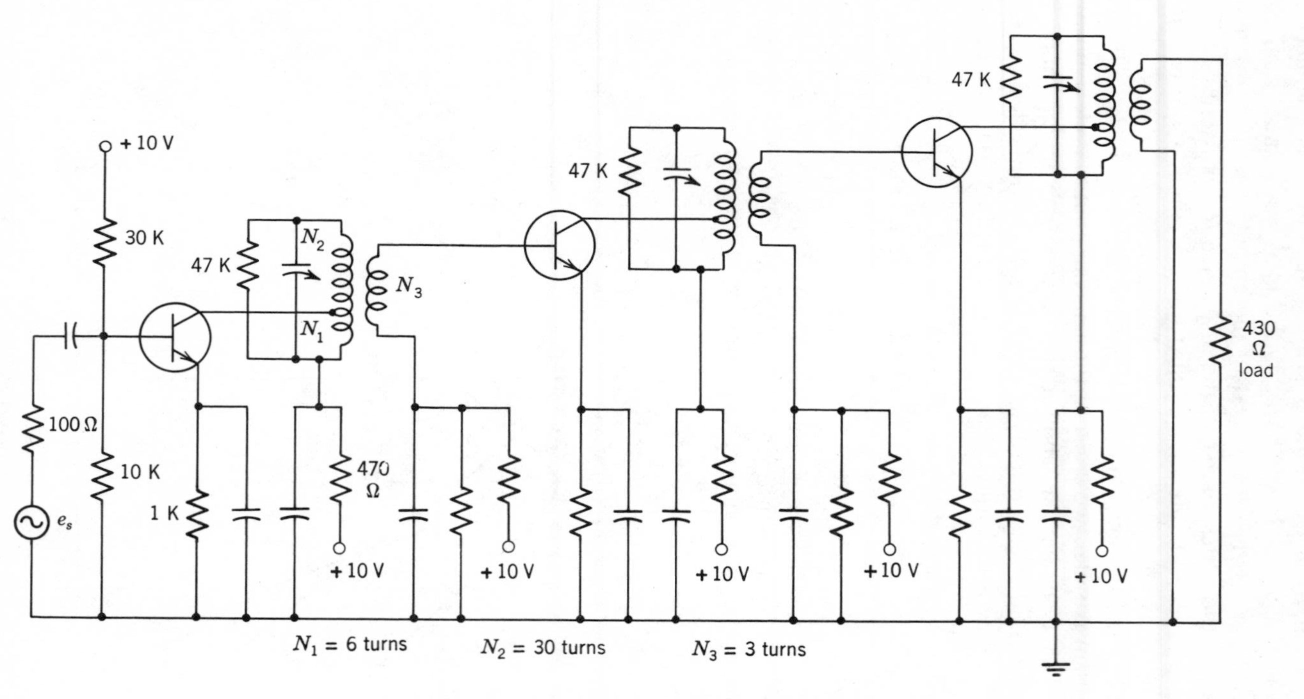

Figure 23-11 Three-stage 5-MHz *IF* amplifier.

23-13, and the separation of the two peaks is given by

$$\frac{f_a}{f_0} = 1 - \frac{k}{2}, \quad \frac{f_b}{f_0} = 1 + \frac{k}{2} \tag{10-13a}$$

In Fig. 23-10 the response curve for a single-tuned circuit is shown as B, and a double-tuned amplifier response curve is shown as A. The product of the two gain curves is curve C. Curve C has a wide bandwidth, and it has almost constant gain over this region. In this manner of combining two such circuit arrangements, a very steep slope is obtained on the sides of a flat response curve.

A three-stage 5-MHz IF amplifier is shown in Fig. 23-11. The power gain is 60 db, and the bandwidth is 200 kHz. Loading is required across the coils to broadband the response of each stage. All stages are identical, and the capacitors provide adequate bypassing. This strip is formed on a small subassembly. A common alternative arrangement is to use fixed capacitors in the tuned circuits and to employ slug tuning in the coils.

PROBLEMS

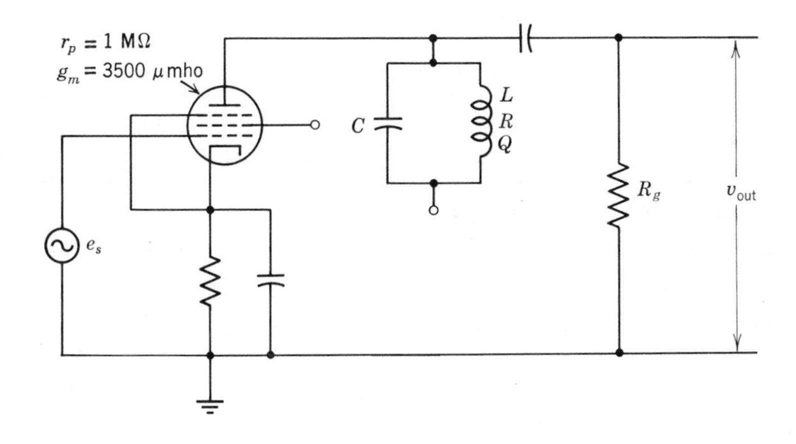

Circuit for Problems 1 to 3.

1. In the given circuit, C is 60 pF and L is 800 μh. The Q of the coil itself is 170. R_g is 100,000 Ω. Determine the voltage gain of the circuit at the resonant frequency, and determine the bandwidth of the response curve.

2. Solve Problem 1 for the following circuit values:

$C = 125$ pF, $Q = 80$
$L = 1.2$ mh, $R_g = 240{,}000\ \Omega$

3. Solve Problem 1 for the following circuit values:

$C = 30$ pF, $Q = 35$
$L = 180\ \mu$h, $R_g = 27{,}000\ \Omega$

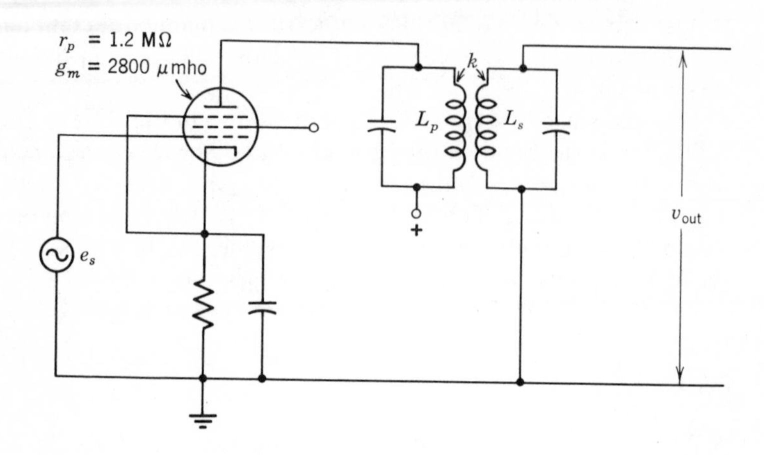

Circuit for Problems 4 and 5.

4. In the circuit diagram, L_p and L_s are each 130 μh with a Q of 45 at the resonan frequency, 455 kHz. The coefficient of coupling is critical. Determine the value of the capacitors needed to resonate the coils. Determine the gain a the resonant frequency. Determine k_c.
5. Repeat Problem 4 if L_p and L_s are each 55 μh with a Q of 80 in a circuit tune to 1 MHz.

Questions

1. Define or explain each of the following terms: (*a*) $r_{bb'}$, (*b*) $r_{b'e}$, (*c*) $r_{cb'}$, (*d*) $C_{cb'}$ (*e*) $C_{cb'}$, (*f*) C_{ob}, (*g*) C_T, (*h*) ω_β, (*i*) ω_T, (*j*) r_{iss}, (*k*) r_{oss}, (*l*) C_{iss}, (*m*) C_{oss}, (*n*) C_{rss} (*o*) Miller effect, (*p*) over-coupled.
2. Why is a hybrid-π model so called?
3. What factors are introduced in the hybrid-π model that were not considered in earlier chapters?

4. How is transconductance related to r'_e?
5. Define beta cutoff frequency.
6. Define the short-circuit unity gain frequency.
7. How is f_T related to f_β?
8. How is the gain-bandwidth product defined?
9. What is the Miller effect in a transistor amplifier?
10. What is the Miller effect in a FET amplifier?
11. What is the Miller effect in a vacuum-tube amplifier?
12. What restrictions, if any, apply toward using $g_m Z_L$ as the voltage gain for an FET amplifier?
13. Why is a MOSFET more suited than a JFET as a high-frequency amplifier?
14. Explain the effect of Eq. 23-12 on the performance of an amplifier.
15. A transistor is used in an amplifier that has a flat response from zero to 15 MHz. An identical transistor is used in an amplifier that has a tank circuit tuned to 15 MHz. Which circuit would be expected to have the higher gain? Explain.

Chapter Twenty-four

MODULATION

The fundamental purpose of modulation is to superimpose the desired intelligence signals on a high-frequency carrier for transmission at that high frequency (Section 24-1). Amplitude modulation (Section 24-2) and amplitude-modulation circuits (Section 24-3) are discussed. The general considerations of frequency modulation (Section 24-4) are followed by a study of two methods of producing frequency modulation, the reactance circuit (Section 24-5) and the balanced modulator (Section 24-6). Preemphasis and deemphasis are considered (Section 24-7). The fundamentals of suppressed-carrier and single side-band transmissions are examined (Section 24-8).

Section 24-1 The General Problem of Modulation

A signal is transmitted from one point to another for a variety of purposes. The most common example is telephone communication within a geographic area. A remote-metering problem is another form of signal transmission that is in the same classification. For a telephone circuit existing between two distant cities, the physical equipment involves an enormous quantity of poles, cross-arms, insulators, and wires. When the demand on the circuit becomes large, it is necessary to add additional facilities. Now the problem resolves into the method of providing the additional circuit. The situation is similar in the remote-metering problem when it becomes necessary to add further metering circuits.

One method of increasing the facility of a circuit is to use a method of modulation. Let us assume that a voice-frequency band from 200 to 3000 Hz is required for telephone communications. If we take bands of frequencies, 0 to 3 kHz, 3 to 6 kHz, 6 to 9 kHz, 9 to 12 kHz, and 12 to 15 kHz a total band of 0 to 15 kHz apparently could provide five separate *channels* for five separate telephone circuits over one pair of wires, provided that the original band of 200 to 3000 Hz can be transferred to each of the high-frequency bands. The process of superimposing the information contained within a frequency band onto another frequency band is called *modulation*. The process of decoding or converting the signal back to its original form is called *demodulation* or *detection*. The problems of detection are considered in Chapter 25.

The energy medium by which the signal is to be transferred is called the *carrier*. The signal is often termed the *modulating frequency*. If we consider a single-frequency carrier, we may write

$$e_{out} = E_m \cos(\omega_0 t + \theta)$$

or

$$e_{out} = E_m \cos(2\pi f_0 t + \theta)$$

When the amplitude of the carrier E_m is varied in accordance with the signal information, we have *amplitude modulation*. When the frequency of the carrier f_0 is varied in accordance with the signal, we have *frequency modulation*. When the phase angle θ is varied in accordance with the signal, we have *phase modulation*. There are also a number of special methods of modulation which are in use but are beyond the scope of this textbook. For example, if a series of short-duration pulses are transmitted, the signal information can vary the width of the pulse (*pulse-width modulation*), or the signal can vary the height of the pulses (*pulse-amplitude modulation*), or the signal can vary the exact starting point of each pulse (*pulse-position modulation* or *pulse-time modulation*).

Pulse-code modulation is a widely used commercial communications system in which the instantaneous value of the signal is measured in volts at fixed close intervals. The voltage readings are converted into a set of pulses which represent the value of the measured voltage in a binary code. At the receiver the binary code is converted back to voltage values. The reconstruction of the original signal is effectively a series of dots. When the dots are close enough, the original waveform is reconstructed. PCM can be used to transmit both black-and-white and color television.

Section 24-2 Amplitude Modulation

A carrier signal may be represented by

$$e_0 = E_m \cos 2\pi f_0 t \tag{24-1}$$

where f_0 is the frequency of the carrier wave (Fig. 24-1*a*). The phase angle is taken as zero because it does not contribute to amplitude modulation, and

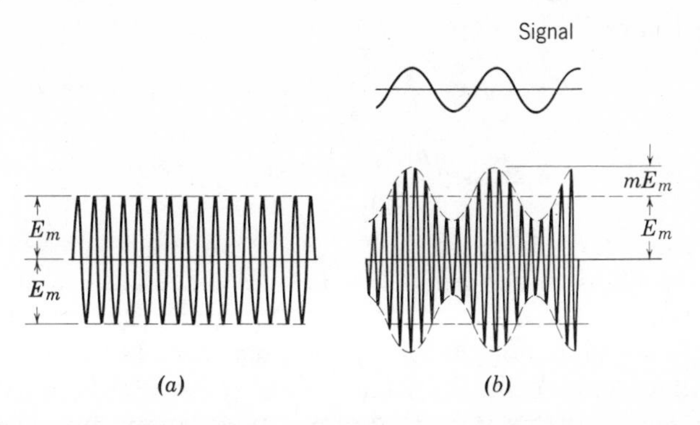

Figure 24-1 Amplitude modulation. (*a*) Unmodulated wave. (*b*) Modulated wave.

its retention complicates the algebra unnecessarily. In amplitude modulation, the magnitude of the carrier E_m is varied in accordance with the modulating signal (Fig. 24-1*b*). Amplitude modulation produces a modulation envelope on the carrier waveform. This modulation envelope follows the signal waveform, and it should be noted quite carefully that, when the envelope increases in the positive direction, it also increases in the negative direction. The amplitude of the modulation envelope is a fraction m of the amplitude of unmodulated wave. This fraction m is called the *modulation* and is usually given in per cent and, in this form, is termed *per cent modulation*. Based on this definition for modulation, we may write the equation for the information signal e_s as

$$e_s = mE_m \cos \omega_s t$$

or

$$e_s = mE_m \cos 2\pi f_s t$$

When a carrier is amplitude-modulated by a sine-wave signal, the amplitude of the carrier contains the sinusoidal variations as expressed by

$$(1 + m \cos \omega_s t)E_m$$

The instantaneous voltage of the resultant wave is

$$e = (1 + m \cos \omega_s t)E_m \cos \omega_0 t$$

Expanding this expression yields

$$e = E_m \cos \omega_0 t + mE_m \cos \omega_0 t \cos \omega_s t$$

From trigonometry, we have the expansion formula

$$\cos x \cos y = \tfrac{1}{2} \cos (x+y) + \tfrac{1}{2} \cos (x-y)$$

Substituting

$$e = E_m \cos \omega_0 t + \frac{mE_m}{2} \cos (\omega_0 + \omega_s)t + \frac{mE_m}{2} \cos (\omega_0 - \omega_s)t \qquad (24\text{-}2)$$

or $$e = E_m \cos 2\pi f_0 t + \frac{mE_m}{2} \cos 2\pi(f_0 + f_s)t + \frac{mE_m}{2} \cos 2\pi(f_0 - f_s)t$$

By this derivation, we show that the equation of an amplitude-modulated wave contains three terms. The first term is identical with Eq. 24-1 which is the unmodulated wave. Thus, it is apparent that the process of amplitude modulation does not change the original wave but adds to it by producing two additional terms. The frequency of the second term is $(f_0 + f_s)$, and the frequency of the third term is $(f_0 - f_s)$. As an example, when the carrier is 5000 Hz and the signal is 100 Hz, the frequencies of the three terms are 5000 Hz, 5100 Hz and 4900 Hz. The term that is at 5100 Hz $(f_0 + f_s)$ is called the *upper side-band* the term that is at 4900 Hz $(f_0 - f_s)$ is called the *lower side-band.* In this example, the signal is 100 Hz, but the total band width required is from 4900 Hz to 5100 Hz, or 200 Hz. The very important conclusion to be made at this point in the discussion is that *the bandwidth required in amplitude modulation is twice the frequency of the modulating signal.* In standard broadcast transmission, the carriers of the stations are allocated at intervals of 10 kHz: for example, 960 kHz, 970 kHz, 980 kHz. Thus each station would appear to have a bandwidth allocation of 10 kHz and a maximum permissible modulating frequency of 5000 Hz. Actually the Federal Communications Commission (FCC) imposes a limit of 7500 Hz only when justifiable complaints of interference are received and proved.

From Eq. 24-2, we can see that the side bands go to zero when the modulation m is zero and the equation is that of the carrier alone, (Eq. 24-1). When the modulation is 100% (Fig. 24-2a), the maximum instantaneous

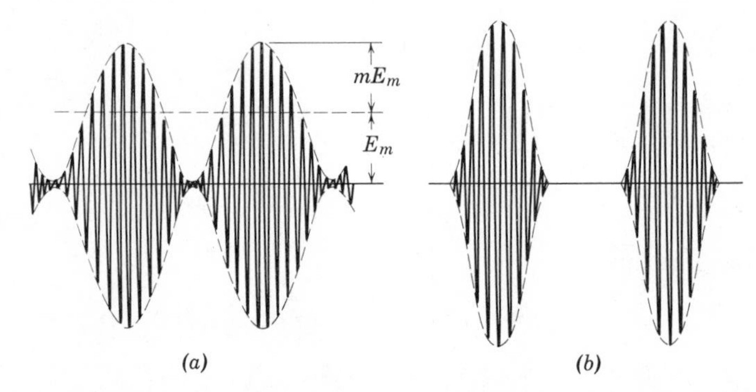

Figure 24-2 100% modulation and overmodulation. (*a*) 100% modulation ($m = 1$). (*b*) Overmodulation ($m > 1$).

voltage is $2E_m$ and the minimum instantaneous voltage of the envelope is zero. The condition of overmodulation is shown in Fig. 24-2*b*. The waveform is clipped and, since the envelope is discontinuous, it cannot be represented by Eq. 24-2. Under conditions of overmodulation, the envelope is no longer sinusoidal, but is represented by a fundamental and many harmonics. These harmonics also produce side bands. When the modulation is 98%, there are only two side bands. When the modulation is 105%, there are many side bands. This condition of overmodulation produces a side-band *splattering* by requiring a bandwidth much greater than the normal bandwidth for modulations not exceeding 100%. This splattering creates interference for the stations in the adjacent channel assignments.

The modulation patterns of Fig. 24-1*b* and Fig. 24-2*a* are called *modulated continuous waves (MCW)*. The simplest method of radio transmission is accomplished by turning a radio transmitter on and off by means of a telegraph key. The modulation pattern (Fig. 24-3) is a sequence of square dots

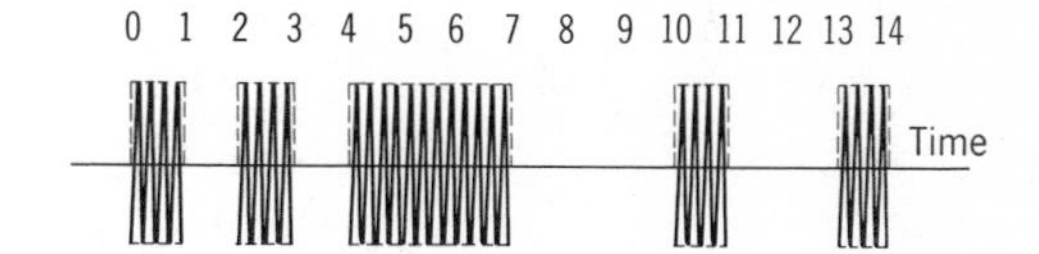

Figure 24-3 Waveform of radio telegraphy (ICW).

and dashes. This method of transmission is called *interrupted continuous waves (ICW)*. The time duration of a dot is called a *baud*. The length of a dash is three bauds, and the space between the dots and the dashes of the code for a letter is one baud. The space between successive letters is three bauds. If a word is taken as five letters and each letter is taken as 10 bauds including the spacing interval, the word is 50 bauds long. If the spacing between words is nine bauds, ten words require 10(50+9) or 590 bauds. Since a baud corresponds to either a positive half-cycle or a negative half-cycle, 590 bauds represents 295 Hz. When the speed of transmission of signals is ten words per minute, this figure becomes 295/60 Hz of bauds per second. In order to produce a dot or a dash, a square-wave modulation is needed and requires a harmonic content of all harmonics up to the eleventh to obtain a clean waveform. With this modification, ten words per minute requires a modulation containing frequencies up to

$$(295/60) \times 11 = 54.1 \text{ Hz}$$

Since amplitude modulation requires both an upper side band and a lower side band, the total bandwidth necessary to transmit code at ten words per minute is 108.2 Hz.

Very often it is necessary to determine the modulation from the waveform (Fig. 24-4). The maximum peak-to-peak amplitude is A, and the spread between the minimum points is B. These values may be determined easily

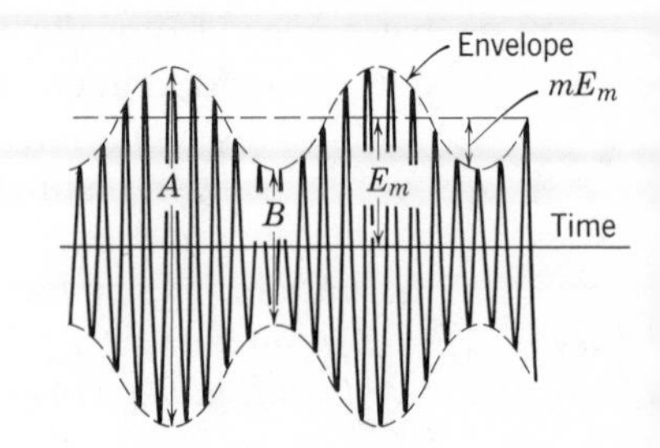

Figure 24-4 Details of modulated wave.

from an oscilloscope pattern. From this information, the peak of the unmodulated waveform is

$$E_m = \frac{A/2 + B/2}{2} = \frac{A+B}{4}$$

The peak of the modulating signal mE_m is

$$mE_m = \frac{A/2 - B/2}{2} = \frac{A-B}{4}$$

When these two expressions are taken as a ratio

$$\frac{mE_m}{E_m} = \frac{(A-B)/4}{(A+B)/4}$$

$$m = \frac{A-B}{A+B} \tag{24-3}$$

When the total modulation is fed into the vertical or the Y-deflection terminals of an oscilloscope and the modulating signal is fed into the horizontal or X-deflection terminals, a trapezoidal pattern results (Fig. 24-5). Several different patterns are in the diagram, including one that shows the pattern f of a developed second harmonic in the modulating circuit. The values of the per cent modulation may be determined from these figures by use of Eq. 24-3. The pattern is quite useful in checking the operation of a modulator and is often used in monitoring the operation of a transmitter.

The coefficients of the terms of Eq. 24-2 are

$$E_m \qquad \frac{mE_m}{2} \qquad \frac{mE_m}{2}$$

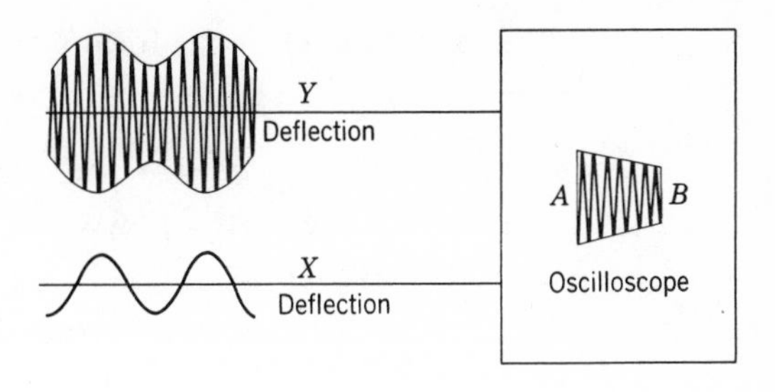

(a) Test equipment

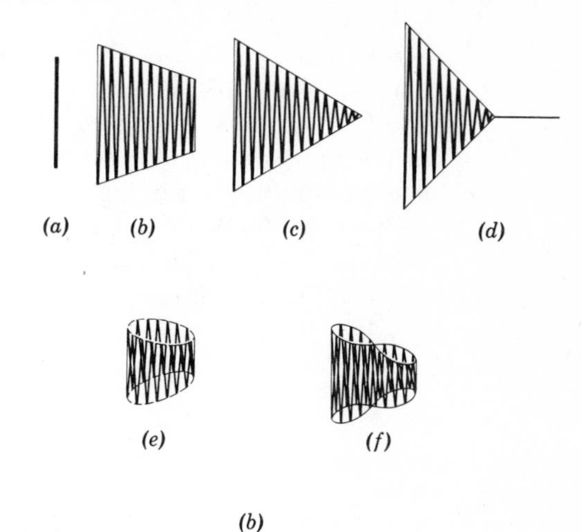

(b)

Figure 24-5 Visual observation of modulation. Trapezoidal modulation patterns. (a) $m = 0\%$. (b) $m = 25\%$. (c) $m = 100\%$. (d) $m > 100\%$. (e) Phase shift not proper to produce trapezoid. (f) Second harmonic distortion.

and are in the ratio

$$1 \quad m/2 \quad m/2$$

Since power may be expressed as E^2/R, these terms can be converted to a power ratio:

$$1 \quad \frac{m^2}{4} \quad \frac{m^2}{4}$$

The total power is in the ratio

$$1 + \frac{m^2}{4} + \frac{m^2}{4} = 1 + \frac{m^2}{2}$$

We may state this in the form of an equation:

$$\frac{P_T}{P_o} = 1 + \frac{m^2}{2} \qquad (24\text{-}4a)$$

where P_o is the carrier power, and P_T is the total power for a modulation m. If R is the resistance,

and
$$\frac{P_T}{P_o} = \frac{I_T^2 R}{I_0^2 R} = 1 + \frac{m^2}{2}$$

$$I_T = I_0\sqrt{1 + m^2/2} \qquad (24\text{-}4b)$$

When the modulation is 100%, m is unity and the total power ratio becomes 1.5.

As an example, when the carrier power is 500 W, the total power, under conditions of 100% modulation, becomes 1.5 × 500 or 750 W. The additional 250 W represents the energy content of the side-bands. There are 125 W in the upper side-band and 125 W in the lower side-band. The side-band energy represents the signal content, and the unchanged carrier-energy content of 500 W is that energy which is required as the means of transmission. As a further example, if the antenna or load current for an unmodulated transmitter is 8 A, when the modulation is 40%, the current rises to

$$\begin{aligned} I_T &= I_0\left(1 + \frac{m^2}{2}\right)^{1/2} \\ &= 8\left(1 + \frac{0.40^2}{2}\right)^{1/2} \\ &= 8(1.08)^{1/2} \\ &= 8.32 \text{ A} \end{aligned}$$

When the average power of an unmodulated transmitter is 1000 W, the peak instantaneous power is 2000 W. Since the peak voltage of a wave at 100% modulation is twice the peak of the unmodulated wave, the peak power at 100% modulation is 4000 W.

We found from Eq. 24-2 that an amplitude-modulated wave may be expressed as

$$e = E_m \cos \omega_0 t + \frac{mE_m}{2} \cos (\omega_0 + \omega_s) t + \frac{mE_m}{2} \cos (\omega_0 - \omega_s) t$$

This wave may be represented graphically in three forms (Fig. 24-6). The waveform (Fig. 24-6a) showing the instantaneous total value of the carrier and the sidebands is the conventional form of representing the modulated wave. The horizontal axis in this case is time. When the horizontal axis is frequency (Fig. 24-6b), energy appears only at three places, the lower side

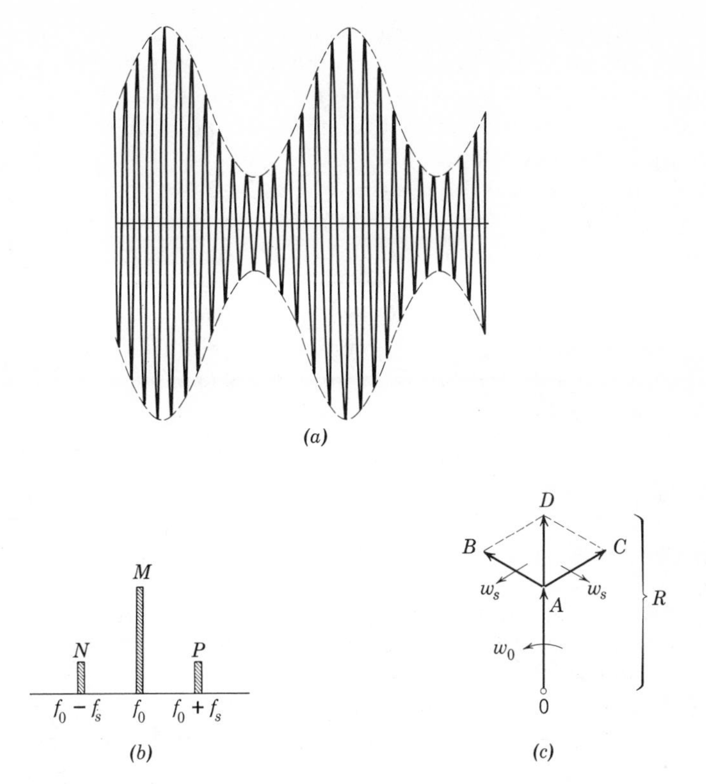

Figure 24-6 Three representations of an AM wave. (*a*) Time axis. (*b*) Frequency axis. (*c*) Phasor diagram.

band, the carrier, and the upper side band. The amplitudes of the representations N, M, and P are proportional either to the voltage or to the power content of the three frequencies. The separation between N and P is the bandwidth. When the modulation changes, the amplitude of M is fixed, but the amplitudes of N and P vary. When the modulating frequency changes, N and P are nearer to or further from M, depending on whether the modulating frequency decreases or increases.

In the phasor diagram (Fig. 24-6*c*), OA represents the carrier and has a length proportional to E_m. This carrier rotates about point O counterclockwise at the angular velocity of the carrier ω_0. Two phasors, AB and AC, are added to the carrier. The length of these phasors is proportional to $mE_m/2$, and they represent the side bands. The phasor AC rotates about point A clockwise, and the phasor AB rotates about point A counterclockwise, both at the angular velocity of the signal ω_s. The relative speed of AC about point O is $(\omega_0 - \omega_s)$, and the relative speed of AB about point O is $(\omega_0 + \omega_s)$. Thus,

AB is the upper side band, and AC is the lower side band. The phase relation of the phasors must be such that the sum of the three phasors is R, and R must at all times be in phase with the carrier phasor OA. When the projection of R is traced out as it rotates about point O, the modulation pattern of Fig. 24-6*a* is developed.

In discussing the three terms of Eq. 24-2, we pointed out that the signal content of the total was the side band energy alone. It is possible to have communications systems that do not transmit the energies of all the three terms, but suppress all or part of the carrier term and/or one of the side bands. These methods have been developed into systems of great importance in modern communications equipment which fall into a general classification called *single side-band transmission*. Single side-band transmission will be covered in Section 24-8.

PROBLEMS

1. A high-speed radio-telegraph system operates at 350 words per minute, transmitting five-letter code words. Determine the bandwidth required for AM transmission of the signals.
2. Solve Problem 1 for a transmission speed of 500 words per minute.
3. The carrier strength of an AM signal is 1000 W. When the modulation is 25% determine the average sideband power.
4. Repeat Problem 3 for a modulation of 45%.
5. Repeat Problem 3 for a modulation of 85%.
6. Repeat Problem 3 for a modulation of 100%.
7. A transmitter deliver a carrier power of 500 W into a 300-Ω antenna. If the modulation is 30%, draw the diagrams shown in Fig. 24-5*b* and Fig. 24-6 using numerical values. The carrier frequency is 1 MHz and the modulating frequency is 5 kHz.
8. Repeat Problem 7 for a modulation of 60%.
9. Repeat Problem 7 for a modulation of 80%.
10. Determine the antenna current and the sideband power for Problem 7.
11. Determine the antenna current and the sideband power for Problem 8.
12. Determine the antenna current and the sideband power for Problem 9.
13. At 80% modulation, the antenna current is 12 A. Determine the antenna current at 30% and at 100% modulation.
14. If the antenna current is 6.4 A at 40% modulation, what is the antenna current at 75% modulation?
15. If the antenna current is 7.2 A at 45% modulation, what is the modulation when the antenna current rises to 7.9 A?

Section 24-3 AM Circuits

In order to generate an amplitude-modulated wave, the fundamental requirement of the circuit, the *modulator*, is that it must be nonlinear. This requirement is in direct contrast to the design objective of conventional amplifiers in which a great deal of trouble is taken to insure linearity. The output of a nonlinear device is represented by the usual power series,

$$i = a_0 + a_1 e + a_2 e^2 + a_3 e^3 + a_4 e^4 + \cdots$$

Now let us use the input signal

$$e = E_1 \sin \omega_0 t + E_2 \sin \omega_s t$$

where ω_0 represents the carrier and ω_s represents the information signal. We consider only the first four terms of the power series, and for e we substitute the expression for the input signal.

$$\begin{aligned} i = a_0 &+ a_1 (E_1 \sin \omega_0 t + E_2 \sin \omega_s t) \\ &+ a_2 (E_1 \sin \omega_0 t + E_2 \sin \omega_s t)^2 \\ &+ a_3 (E_1 \sin \omega_0 t + E_2 \sin \omega_s t)^3 \end{aligned}$$

$$\begin{aligned} i = a_0 &+ a_1 E_1 \sin \omega_0 t + a_1 E_2 \sin \omega_s t \\ &+ a_2 E_1^2 \sin^2 \omega_0 t + 2a_2 E_1 E_2 \sin \omega_0 t \sin \omega_s t + a_2 E_2^2 \sin^2 \omega_s t \\ &+ a_3 E_1^3 \sin^3 \omega_0 t + 3a_3 E_1^2 E_2 \sin^2 \omega_0 t \sin \omega_s t \\ &\qquad + 3a_3 E_1 E_2^2 \sin \omega_0 t \sin^2 \omega_s t + a_3 E_2^3 \sin^3 \omega_s t \end{aligned}$$

To reduce these equations the following identities from trigonometry are useful

$$2 \cos x \cos y = \cos (x+y) + \cos (x-y)$$

$$2 \sin x \sin y = -\cos (x+y) + \cos (x-y)$$

$$2 \sin x \cos y = \sin (x+y) + \sin (x-y)$$

$$2 \cos x \sin y = \sin (x+y) - \sin (x-y)$$

$$\sin^2 x = \tfrac{1}{2} - \tfrac{1}{2} \cos 2x$$

$$\sin^3 x = \sin x (\sin^2 x) = \sin x - \tfrac{1}{4} \sin 3x$$

$$\sin^4 x = (\sin^2 x)^2 = \tfrac{3}{8} - \tfrac{1}{2} \cos 2x + \tfrac{1}{8} \cos 4x$$

The expression for i reduces to

$$\begin{aligned} i = k_0 &+ k_1 \sin \omega_0 t + k_2 \cos 2\omega_0 t + k_3 \sin 3\omega_0 t \\ &+ k_4 \sin \omega_s t + k_2 \cos 2\omega_s t + k_5 \sin 3\omega_s t \\ &- k_6 \cos (\omega_0 + \omega_s) t + k_6 \cos (\omega_0 - \omega_s) t \\ &+ k_7 \sin (2\omega_0 + \omega_s) t + k_7 \sin (2\omega_0 - \omega_s) t \\ &- k_8 \sin (\omega_0 + 2\omega_s) t - k_8 \sin (\omega_0 - 2\omega_s) t \end{aligned}$$

The a-c terms in the first line represent the fundamental and the harmonics of the carrier. The a-c terms in the second line represent the fundamental and the harmonics of the information signal. Assume that the load in this nonlinear circuit is a tuned circuit tuned to the carrier frequency f_0. This circuit rejects all components except the k_1 term. Likewise this circuit will pass energy from the k_6 terms which are the upper and lower side bands. The k_7 terms, fourth line, show energy at a harmonic which is rejected. Energy from the k_8 terms is present in the output. The k_8 terms show that there is a second harmonic distortion in the side bands. Consequently, the only energy components that pass the tuned circuit are

$$\begin{aligned} i_0 = k_1 \sin \omega_0 t - k_6 \cos (\omega_0 + \omega_s)t + k_6 \cos (\omega_0 - \omega_s)t \\ - k_8 \sin (\omega_0 + 2\omega_s)t - k_8 \sin (\omega_0 - 2\omega_s)t \end{aligned} \tag{24-5b}$$

The diode modulator (Fig. 24-7) can accomplish amplitude modulation

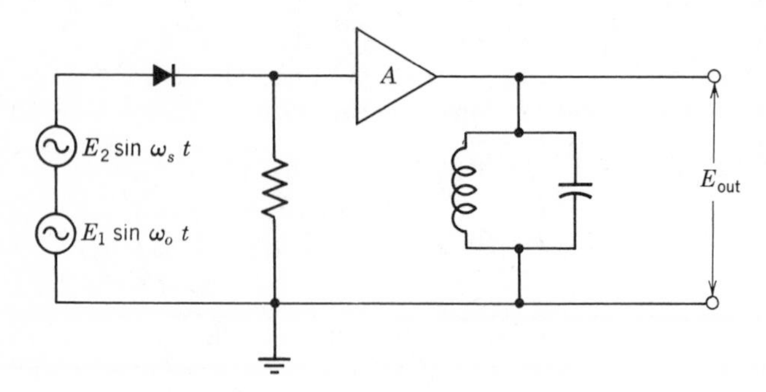

Figure 24-7 Diode modulator.

of a carrier. The circuit uses an isolating amplifier A and a tuned circuit to remove the unwanted modulation components. This circuit is useful, but it does not readily yield high values of undistorted modulation.

To obtain high power levels in the order of kilowatts, vacuum-tube circuits are used in preference to a semiconductor for the same reasons that are outlined in Chapter 19. Typical modulation circuits are shown in Fig. 24-8. In these circuits, all RF decoupling networks are omitted for simplicity of illustration. Amplitude modulation is obtained by increasing the gain of an RF amplifier when the audio signal is positive and by decreasing the gain when the audio signal is negative. Amplitude modulation may be accomplished by any of several forms: *grid modulation*, *screen modulation, plate modulation*, or, if the modulator tube is a pentode, *suppressor grid modulation*. When these circuits are arranged for push-pull operation, the bias is class AB or class B to improve the overall operating efficiency. At high-power levels, class-B operation is standard practice in order to obtain the greatest

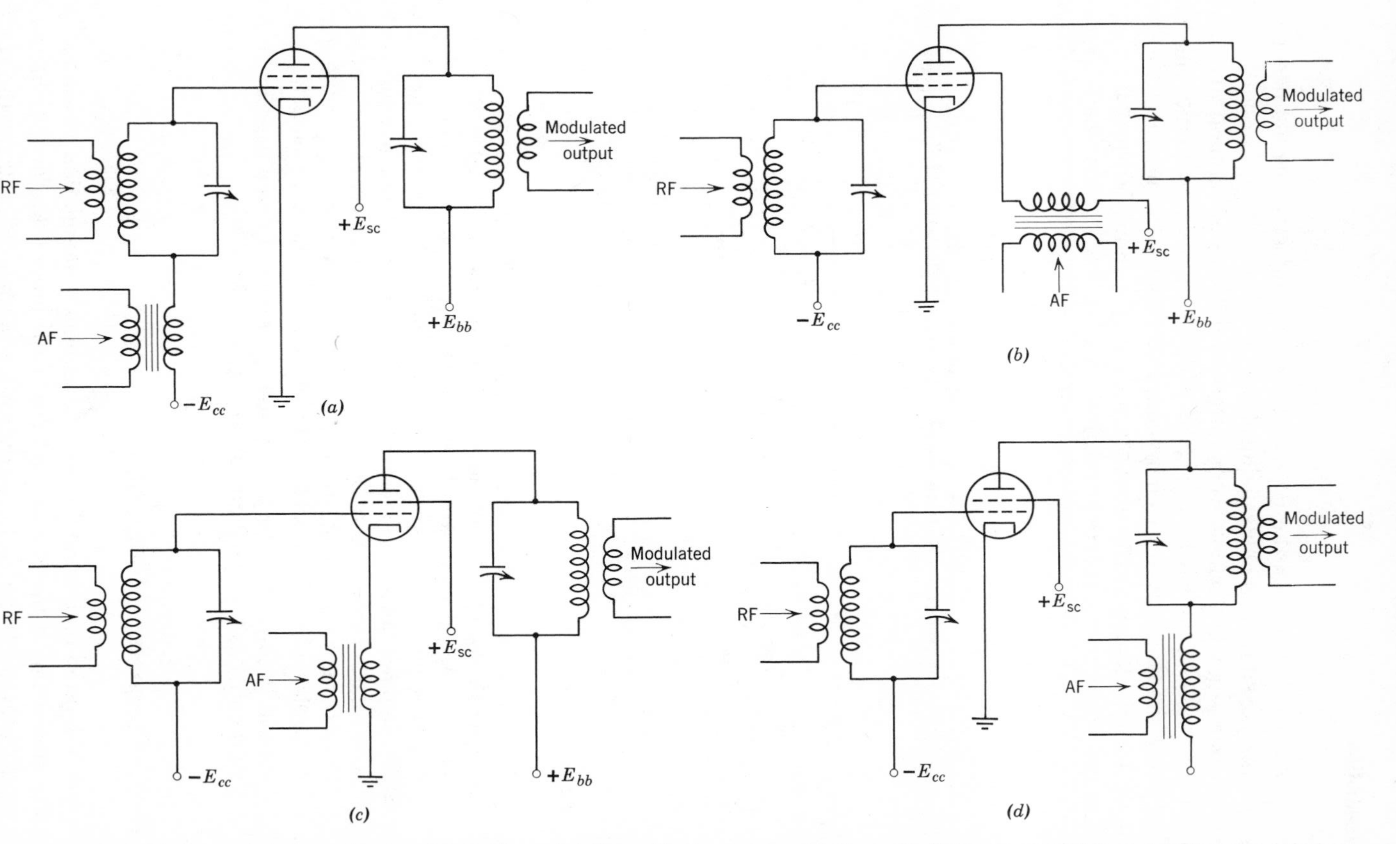

Figure 24-8 Basic modulation circuits. (*a*) Grid modulation. (*b*) Screen modulation. (*c*) Cathode modulation. (*d*) Plate modulation.

efficiency. Very often the RF amplifiers in the modulator are triodes which require an additional neutralizing circuit.

In low-power portable transmitters, transistors are completely adaptable to the circuits shown in Fig. 24-8. Collector, base, or emitter modulation circuit arrangements are used in these same configurations.

The block diagram for an AM transmitter is shown in Fig. 24-9.

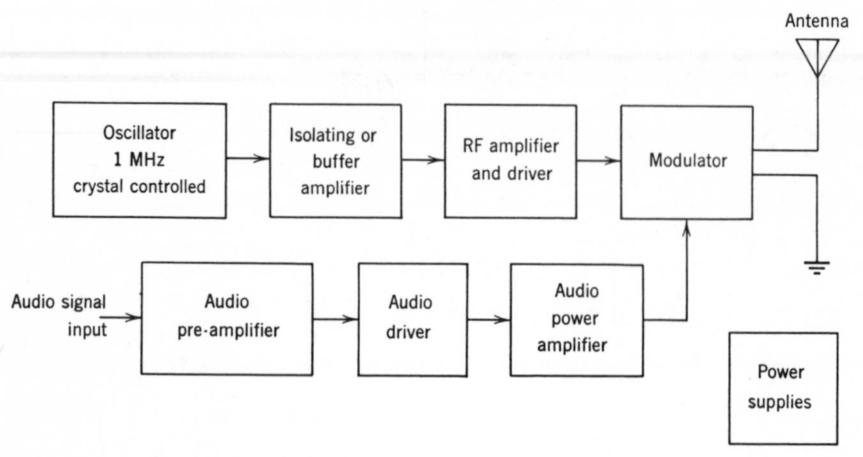

Figure 24-9 AM transmitter block diagram.

PROBLEMS

1. An AM class-B plate-modulated amplifier (Fig. 24-8*d*) operates at 70% efficiency. Without modulation, the plate current is 2.5 A from a 2000-V supply. The AF is derived from a class-B push-pull amplifier. What audio power is required to produce 90% modulation? What impedance does the RF amplifier present to the secondary of the modulating transformer? What is the carrier power and what is the sideband power at 90% modulation? Sketch the waveform at the bottom of the RF tank and sketch the waveform at the plate of the RF amplifier. The efficiency of the audio amplifier is 90%.
2. Repeat Problem 1 if the supply voltage is 1000 V and if the current without modulation is 1 A. The modulation requirement is 60%.

Section 24-4 Frequency Modulation

The invention and development of frequency modulation by Major Edwin H. Armstrong (*Proceedings of the Institute of Radio Engineers*, May 1936) was the result of his search for a method of reducing the static and noise present in home reception of the standard AM broadcasts. Since most natural and manmade electrical noise is in the form of amplitude-modulated signals, a method of keeping the amplitude E_m constant while incorporating the signal into variations of the carrier frequency f_0 accomplishes the initial objective.

The terms and definitions used in frequency modulation and the principles of frequency modulation can be shown best by a numerical example. Let the carrier frequency f_0 be 1000 kHz, the audio-signal frequency f_s be 1 kHz, and the amplitude of the audio signal E_s be 1 V. At the instant the audio is zero, the FM wave is 1000 kHz. When the audio increases in a positive direction, let us assume that the output wave increases its frequency, and, when the audio signal cycle is negative, the output wave decreases in frequency. Assume that at the instant the signal is + 1 V, the instantaneous frequency of the output is 1010 kHz, and that, at the instant the audio cycle is − 1 V, the output frequency is 990 kHz. This concept is shown in Fig. 24-10.

For each complete cycle of audio, the instantaneous frequency of the output follows:

Signal	0	+1	0	−1	0
Output frequency, kHz	1000	1010	1000	990	1000

If this relation is linear, a 2-V signal changes these figures to

Signal	0	+2	0	−2	0
Output frequency, kHz	1000	1020	1000	980	1000

When the audio signal is reduced to 0.5 V, we find that

Signal	0	+0.5	0	−0.5	0
Output frequency, kHz	1000	1005	1000	995	1000

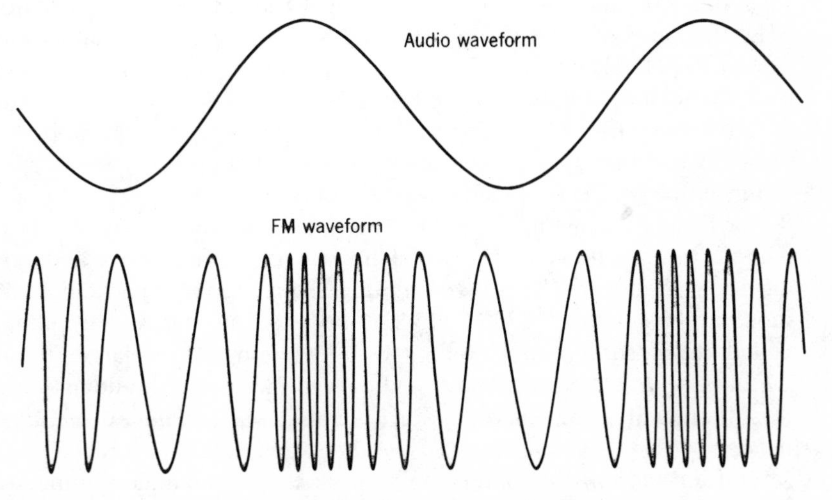

Figure 24-10 Frequency modulation.

It is evident that the amplitude of the audio signal determines the *frequency deviation* f_d from the carrier. For a 0.5-V signal, the deviation is 5 kHz. For a 1-V signal, f_d is 10 kHz, and, for a 2-V signal, f_d is 20 kHz. Notice that f_d is measured *one way* from the carrier. It is not a total spread of frequency. The limit on f_d is established by the application. For FM broadcasting, the Federal Communications Commission limits f_d to a maximum of 75 kHz and, in television broadcasting, it limits f_d to 25 kHz for the sound portion of the program.

We may summarize by stating that the deviation frequency f_d contains the information on the amplitude or volume of the signal f_s.

If we take the original conditions where f_s is 1 kHz and E_s is 1 V, the deviation is 10 kHz. This indicates that the output is changing between 1010 kHz and 990 kHz at the rate of 1000 times a second. If the audio is kept at 1 V, and if the audio frequency is changed from 1 kHz to 2 kHz, the deviation stays the same at 10 kHz, but the output frequency changes between 1010 kHz and 990 kHz 2000 times a second instead of 1000 times a second. From this, we observe that the frequency of the signal is the *rate of change* of the output frequency.

In order to correlate these two concepts, we define a new term, *index of modulation*, m_f, as

$$m_f \equiv \frac{f_d}{f_s} \tag{24-6}$$

In the original example, m_f is 10 kHz/1 kHz, or 10. When the signal is increased to 2 V, the index of modulation changes from 10 to 20 since the deviation is doubled. When the signal is 0.5 V, the index of modulation is 5. An infinite number of combinations of signal and amplitude can produce the same index of modulation. If a 1-V signal at 1 kHz produces a deviation of 10 kHz, the index of modulation is 10. A 2-V signal at 2 kHz produces a deviation of 20 kHz. The index of modulation is still 10. A 1-V signal at 2 kHz produces a deviation of 10 kHz, and the index of modulation is 5. By using this assumed data, Fig. 24-11 shows the relation among f_s, f_d, and m_f. It should be remembered that f_d is proportional to the signal and that the vertical axis of this graph could be labeled E_s instead of f_d.

In FM broadcasting, 15 kHz is the maximum allowed audio frequency. The maximum allowed deviation is a swing of 75 kHz above and below the carrier frequency. These two limiting figures give a particular index of modulation which is 75 kHz/15 kHz, or 5. This index of modulation is called the *deviation ratio*. When the deviation ratio is larger than unity, we have *wide-band* frequency modulation and, when the deviation ratio is less than unity, the classification *narrow-band frequency modulation* is used.

Our discussion up to this point uses the instantaneous time wave of

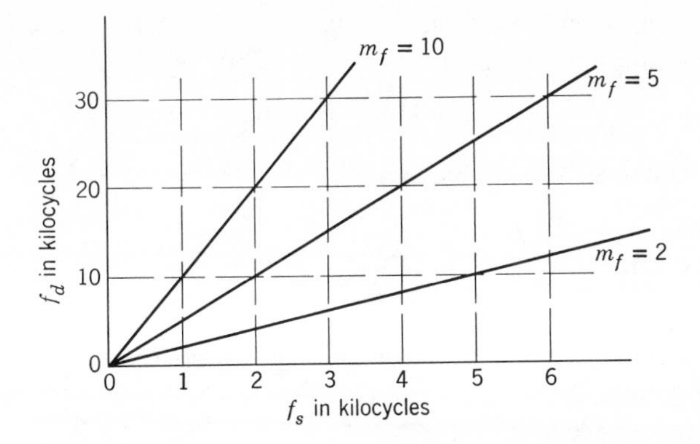

Figure 24-11 Plot showing relationship of f_d, f_s, and m_f.

an FM signal (Fig. 24-10). In order to illustrate frequency modulation by means of a frequency axis and by a phasor diagram, a mathematical development must be formed in order to present the FM output wave as a carrier with sidebands. The general equation of a sine wave, neglecting the phase angle, is

$$e = E_m \cos \omega t$$

In frequency modulation, as we have shown, the instantaneous frequency is a function of f_0, f_d, f_s, and E_s. Since the index of modulation m_f joins together, f_d, f_s, and E_s, we can reduce the variables to f_0, f_s, and m_f. The equation for the instantaneous frequency f_i of the FM wave may be expressed as

$$f_i = f_0 + f_d \cos 2\pi f_s t$$
$$2\pi f_i = 2\pi f_0 + 2\pi f_d \cos 2\pi f_s t$$
$$\omega_i = \omega_0 + \omega_d \cos \omega_s t$$

The expression for ω_i may be converted by means of calculus* to an expression for instantaneous voltage:

$$e = E_m \cos (\omega_0 t + m_f \sin \omega_s t)$$

or

$$\frac{e}{E_m} = \cos (\omega_0 t + m_f \sin \omega_s t) \qquad (24\text{-}7)$$

$$\begin{aligned} {}^*e(t) &= E_m \cos [\textstyle\int \omega_i(t)\, dt] \\ &= E_m \cos [\textstyle\int (\omega_0 + \omega_d \cos \omega_s t)\, dt] \\ &= E_m \cos \left(\omega_0 t + \frac{\omega_d}{\omega_s} \cos \omega_s t\right) \\ &= E_m \cos (\omega_0 t + m_f \sin \omega_s t) \end{aligned}$$

This expression is similar to cos $(x+y)$ and, from the expansion formula of trigonometry, we have

$$\cos (x+y) = \cos x \cos y + \sin x \sin y$$

which enables us to write

$$\frac{e}{E_m} = \cos \omega_0 t \cos (m_f \sin \omega_s t) + \sin \omega_0 t \sin (m_f \sin \omega_s t)$$

The expressions cos (sin x) and sin (sin x), although they appear to be quite simple, are, in fact, very complex and require advanced methods of mathematical analysis for their evaluation.* When the last equation for e/E_m is expanded, we find that

$$\begin{aligned}\frac{e}{E_m} &= J_0(m_f) \cos \omega_0 t \\ &\quad + J_1(m_f) \cos (\omega_0 + \omega_s)t - J_1(m_f) \cos (\omega_0 - \omega_s)t \\ &\quad + J_2(m_f) \cos (\omega_0 + 2\omega_s)t + J_2(m_f) \cos (\omega_0 - 2\omega_s)t \\ &\quad + J_3(m_f) \cos (\omega_0 + 3\omega_s)t - J_3(m_f) \cos (\omega_0 - 3\omega_s)t \\ &\quad + J_4(m_f) \cos (\omega_0 + 4\omega_s)t + J_4(m_f) \cos (\omega_0 - 4\omega_s)t \\ &\quad + \cdots \end{aligned} \qquad (24\text{-}8)$$

From Eq. 24-8, it is evident that in frequency modulation there are many side bands whereas in amplitude modulation there were only two side bands. The side bands in frequency modulation occur in pairs. There are an upper side band and a lower side band for the signal frequency, for the second harmonic of the signal frequency, for the third harmonic of the signal frequency, for the fourth harmonic of the signal frequency, and so on. As in amplitude modulation there is a term that represents energy at the carrier frequency—the J_0 term. Since the equation is for e/E_m, the vector sum of the coefficients of the carrier and side band terms must add to unity. The coefficients [$J_0(m_f)$, $J_1(m_f)$, $J_2(m_f)$, etc.] of the terms are called *Bessel functions of the first kind.* The subscript of the J is called the *order*. Thus the fourth-order Bessel function is the coefficient of the fourth side bands which are located at (f_0+4f_s) and at (f_0-4f_s). These Bessel functions may be evaluated numerically and are plotted as a function of the index of modulation in Fig. 24-12. Some representative values are given in Table A.

It is apparent from Fig. 24-12 that the magnitude of the side band coefficients varies with the index of modulation m_f. When the index of modulation is zero, all the energy is contained within the carrier. The graph shows that $J_0(m_f)$ is unity and all higher-order coefficients are zero. When modulation is applied, the index of modulation increases to finite values.

*E. Jahnke and F. Emde, *Tables of Functions* (Leipzig: Teubner, 1938); (New York: Dover Publications, 1943).

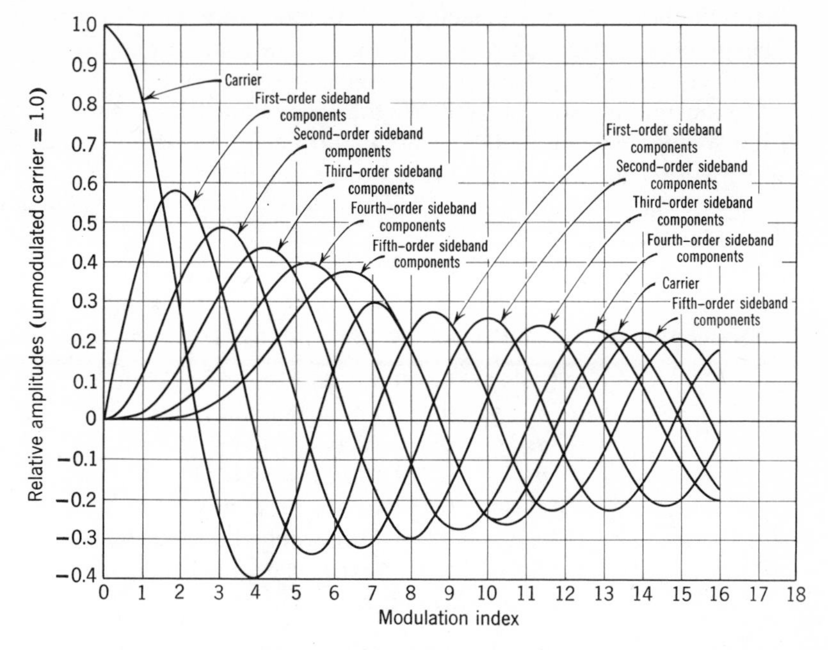

Figure 24-12 Bessel functions.

TABLE A Bessel Function Values

m_f	$J_0(m_f)$	$J_1(m_f)$	$J_2(m_f)$	$J_3(m_f)$	$J_4(m_f)$	$J_5(m_f)$
0.0	1.0000	0.0000	0.0000	0.0000	0.0000	0.0000
0.5	0.9385	0.2423	0.0306	0.0026	0.0002	0.0000
1.0	0.7652	0.4401	0.1149	0.0196	0.0025	0.0002
1.5	0.5118	0.5579	0.2321	0.0610	0.0118	0.0018
2.0	0.2239	0.5767	0.3528	0.1289	0.0340	0.0070
2.5	−0.0484	0.4971	0.4461	0.2166	0.0738	0.0195
3.0	−0.2601	0.3391	0.4861	0.3091	0.1320	0.0430
3.5	−0.3801	0.1374	0.4586	0.3868	0.2044	0.0804
4.0	−0.3971	−0.0660	0.3641	0.4302	0.2811	0.1321
5.0	−0.1776	−0.3276	0.0466	0.3648	0.3912	0.2611
6.0	0.1506	−0.2767	−0.2429	0.1148	0.3576	0.3621

The coefficient of the J_0 term decreases, and other side band coefficients appear. Thus, it is evident that the energy content of the carrier decreases and energy shifts into the side bands. As we stated in the original presentation of the material on frequency modulation, the amplitude of the final output wave is constant. One of the main advantages of the FM transmitter is that, unlike the AM transmitter, the output power is at all times constant. The process of frequency modulation reduces the carrier power and puts this decreased energy into useful signal carrying side-band energy. In an AM transmitter rated at 1000 W, the circuit must be capable of handling 1500 W average and 4000 W peak at 100% modulation. On the other hand, an FM transmitter has an average power, for the same conditions, of 1000 W at any modulation level. As a result, smaller equipment can be used in the FM transmitter.

We notice from Fig. 24-12 that, as the order of modulation m_f increases, there appear a greater number of side bands. As an example, for an index of modulation of 3, we find from the graph that:

$J_0(m_f) = -0.260$ carrier
$J_1(m_f) = +0.339$ first-order side band
$J_2(m_f) = +0.486$ second-order side band
$J_3(m_f) = +0.309$ third-order side band
$J_4(m_f) = +0.132$ fourth-order side band
$J_5(m_f) = +0.043$ fifth-order side band
$J_6(m_f) = +0.011$ sixth-order side band
Higher-order side bands are negligible.

An index of modulation of 3 can be produced by a 4 kHz signal at a sufficient level to cause a deviation frequency of 12 kHz. The fifth-order side band means that energies exist at 5×4 or 20 kHz, above and below the carrier. This energy distribution can be represented on a frequency axis (Fig. 24-13*a*) in the same fashion as was done with the AM signal (Fig. 24-6*b*). From this energy spectrum, we observe that, for this index of modulation of three, the second-order side band contains the greatest amount of energy.

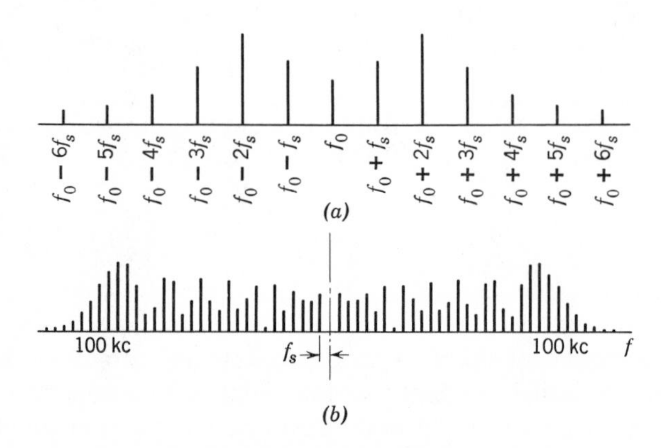

Figure 24-13 Side-band energy distribution. (*a*) Frequency distribution for $m_f = 3$. (*b*) Frequency distribution for $m_f = 25$. (*Courtesy Gray, Applied Electronics.*)

At high modulation levels where the index of modulation is large, the energy spectrum contains a great many side bands. If the deviation is 25 kHz and the signal is 1 kHz, the index of modulation is 25. The energy distribution is shown in Fig. 24-13*b*. Side bands are present up to the 30th order. In this case a total bandwidth of 60 kHz is required to transmit the complete signal.

A phasor diagram may be established for frequency modulation as was done for amplitude modulation, and we can show this by using the coefficients determined for the index of modulation of three. The resultant phasor is the sum of the carrier and the side-band terms, which are added to each other at right angle. Each phasor is at a right angle to the preceding phasor and the phasor is shifted 90° clockwise from the previous one. By the term phasor, we mean the sum of the coefficients for each side-band pair. When a coefficient is negative, the phasor is reversed from the normal positive direction. Each coefficient is represented twice since there are two side bands for each coefficient. The resulting phasor diagram is shown in Fig. 24-14. The resultant phasor R, which is the sum of the carrier and

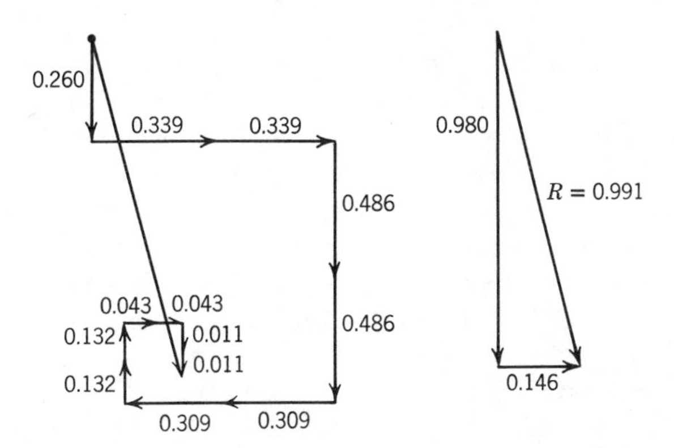

Figure 24-14 Addition of side-band phasors to show total energy.

sidebands up to the sixth order, is 0.991, whereas it should be 1.000. This small error is brought about by neglecting the higher-order terms. When the sixth-order side band is also neglected, R is 0.979.

It is possible to transmit frequency modulation without some of the higher-order side band pairs. The resultant phasor is not exactly unity, but varies slightly in length. A variation in length is an amplitude modulation. In a receiver, a circuit that restores the constant amplitude by clipping or limiting the amplitude to a fixed level is called a *limiter* (Section 25-4), and the process is called *limiting*. In this sense, the action of the limiter is to restore the side bands which are needed to produce a pure FM wave without amplitude variation.

We have shown in developing the results of Eq. 24-8 that the resultant phasor R of the carrier and side bands in frequency modulation is a phasor of constant amplitude. This brings us back to the initial premise illustrated in the time wave of frequency modulation (Fig. 24-10), that the signal is independent of a variation in the amplitude of the output.

If we show the resultant phasor R in reference to the unmodulated carrier OY in Fig. 24-15, the locus of the tip of R is a circle. R advances and falls

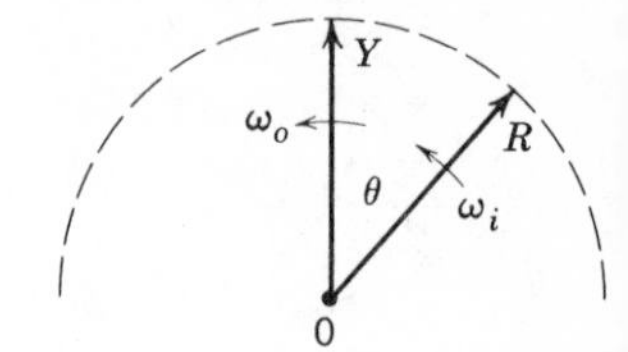

Figure 24-15 Phasor diagram showing frequency and phase modulation.

behind the carrier phasor OY. OY rotates at a speed ω_0, whereas R has a varying speed ω_i. When ω_i is greater than ω_0, R advances ahead of OY, and, when ω_i is less than ω_0, R falls behind OY. The rate of change of ω_i in frequency modulation is the signal ω_s. At all times there exists a phase angle θ between R and OY. If the method of modulation provides that θ is proportional to the amplitude of the signal e_s, and if it provides that the rate of change of θ is proportional to the frequency of the signal f_s, we have *phase modulation.* Phase modulation and frequency modulation are necessarily simultaneous. Whether we call the system frequency or phase modulation is determined by whether the modulation is proportional to frequency or to phase angle.

It is quite possible and reasonable to bring the discussion of frequency modulation to this point without mention of the complex concept of the Bessel function. However, an electronic technician should know how to check out an FM transmitter and, without an appreciation of the side band concepts of frequency modulation, the technique is meaningless.

We notice that in Fig. 24-12 the carrier energy, the J_0 term, goes to zero at successive values of the index of modulation. These null points occur at

Number 1 $m_f = 2.4048$
Number 2 $m_f = 5.5201$
Number 3 $m_f = 8.6537$
Number 4 $m_f = 11.7915$
Number 5 $m_f = 14.9309$
Number 6 $m_f = 18.0711$ etc.

As an example, assume that a 0.1-V signal at 1000 Hz produces a deviation of 2404.8 Hz. The index of modulation is 2.4048, and the carrier energy is zero. If the signal frequency is raised to 2000 Hz, if the amplitude of the audio signal is increased, when the carrier energy is again zero, the index of modulation does not change, and the deviation must be doubled, or 4809.6 Hz. If the modulation is linear, the signal level must be 0.2 V. If the voltage of the signal is held fixed at 0.1 V, and if the frequency of the signal is varied, we can, by observing the audio frequencies at which the carrier energy disappears, determine if the modulation is linear. Now, if the initial conditions produce a null at 0.1 V and 1000 Hz, and this null is the first null as the signal level is raised from zero, the second null should occur at an audio frequency of $(2.4048/5.5201) \times 1000$ Hz, the third null occurs at $(2.4048/8.6537) \times 1000$ Hz, the fourth null occurs at $(2.4048/11.7915) \times 1000$ Hz, etc.

If the original condition calls for the first null at 1000 Hz and 0.1 volt for the audio signal, maintaining the frequency and increasing the signal level increases the deviation. At $(5.5201/2.4048) \times 0.1$ V, we obatain the second null. At $(8.6537/2.4048) \times 0.1$ V we find the third null. Any deviation from these values of signal levels indicates that there is a nonlinearity in the modulation.

Unfortunately, there is no other method of checking the linearity of a source of frequency modulation. This indirect procedure must be followed and the results must be analyzed to determine whether they correspond to the proper cross over or null points of the carrier term. There are several methods in use for determining the null of the carrier term. The laboratory technique for the procedure is involved. If the principles are understood, a demonstration of the actual procedure on test equipment becomes meaningful.

PROBLEMS

1. In an FM system, when the audio frequency is 400 Hz and the audio voltage is 2.2 V, the deviation is 4.7 kHz. When the audio is changed to 6.3 V, what is the deviation? What is the index of modulation in each case?
2. In an FM system, when the audio is 15,000 Hz and the audio voltage is 50 V, the deviation is 75,000 Hz. When the audio frequency is reduced to 20 Hz, what is the deviation for the same index of modulation? What is the audio voltage?
3. In an FM system, if the carrier vanishes when the audio is 5 kHz and 6 V, what voltage at 300 Hz causes the carrier to go to zero?
4. In an FM system, as the voltage is increased from zero at 400 Hz, the carrier first goes to zero when the audio level is 0.30 V. At what successive audio

levels does the carrier energy go to zero? When the audio frequency i
changed to 1000 Hz, at what audio voltages will the carrier go to zer
Assume that the modulation is linear.

5. A 5-MHz signal with a peak amplitude of 100 V is frequency modulated a
5 kHz with a deviation of 30 kHz. Determine the amplitude of the carrier an
the first five sideband pairs. Sketch and dimension the line spectra diagram
Show the phasor diagram. (Use Fig. 24-12.)
6. A 100-V 500-kHz carrier is frequency modulated at 10 kHz with a deviation c
50 kHz. Determine the carrier level and levels of the first five sideband pairs
Determine the line spectra diagram and the phasor diagram. (Use Fig. 24-12.
7. Repeat Problem 6 if the only change is a doubling of the audio modulatin
frequency.

Section 24-5 Reactance Circuits

A varactor diode (Section 4-6) can be used to generate an FM signal directl
(Fig. 24-16). The capacitance across a varactor diode varies inversely wit

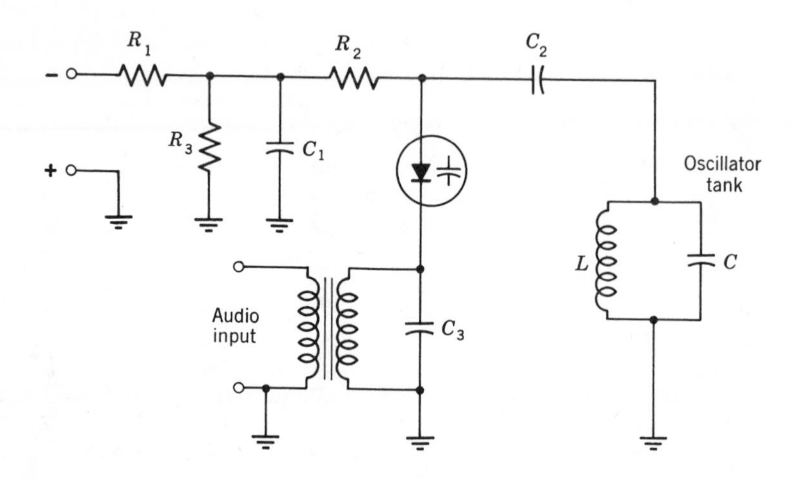

Figure 24-16 Varactor diode modulator.

the reverse voltage across the diode. A d-c reverse voltage is placed on the
diode by the divider network $R1$ and $R3$ and is bypassed by $C1$. $R2$ isolates
the RF and, since the leakage current of the diode is very small, $R2$ can be
large. The instantaneous voltage across the varactor diode is the sum of the

d-c bias plus the audio input. Consequently, a variable capacitance is placed in parallel with the oscillator tank L and C and, since the ΔC of the varactor diode varies with the amplitude of the signal, deviation can be produced in the oscillator that is proportional to the amplitude of the signal. The transmitter circuit, shown in Fig. 24-23, uses a varactor diode to produce the FM.

The basic reactance circuits are shown in Fig. 24-17*a* and *c*. For simplicity, all blocking capacitors and biasing arrangements are omitted. The

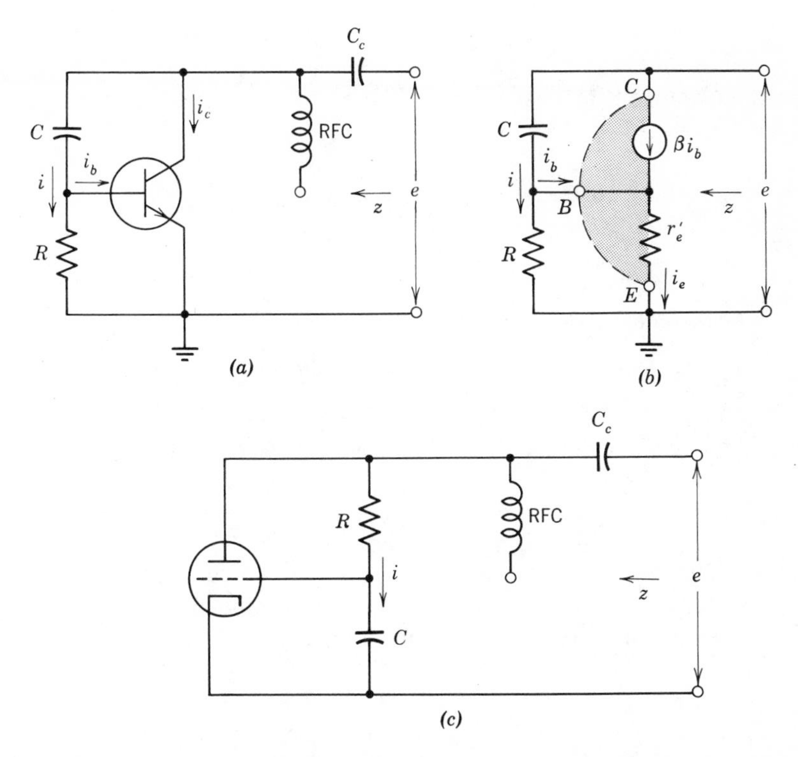

Figure 24-17 Basic reactance circuits. (*a*) Transistor circuit. (*b*) Transistor model. (*c*) Vacuum-tube circuit.

reactance circuit itself is first analyzed, and then it is incorporated in a complete oscillator circuit. The reactance network is the capacitor C and the resistor R. This circuit for an amplifier is not conventional because in these circuits, an external voltage is applied to the collector or to the plate circuit, and the reaction of the circuit to the phase angle of the resulting current is evaluated. The model for the transistor circuit is given in Fig. 24-17*b*. The capacitor C_c is merely a blocking capacitor, and the reactance of the RF choke is sufficiently high that its effect can be neglected. The applied voltage e produces a current i in the reactance network. The component values used in the reactance network are such that

$$X_c \gg R$$

The input resistance to the base is h_{ie} or $(1+\beta)\,r'_e$ and is much larger than R and can be neglected in determining i.

Then the current i is determined solely by X_c:

$$i = \frac{e}{-jX_c} = j\frac{e}{X_c} = j2\pi fCe$$

The voltage from base to ground is

$$Ri = j2\pi fRCe$$

and the base current is this voltage divided by $(1+\beta)r'_e$

$$i_b = \frac{j2\pi fRCe}{(1+\beta)r'_e}$$

The collector current is

$$i_c = \beta i_b = j\frac{\beta}{1+\beta}2\pi f\frac{RC}{r'_e}e \approx j2\pi f\frac{RC}{r'_e}e$$

The impedance presented by the transistor to the external voltage is

$$z = \frac{e}{i_c} = \frac{e}{j2\pi f(RC/r'_e)e} = -j\frac{1}{2\pi f(RC/r'_e)}$$

The final form of this equation is that of a capacitive reactance in which the term in parentheses is an equivalent capacitance. In this manner, the circuit arrangement acts as a capacitance C_{eq}, which is given by

$$C_{eq} = \frac{RC}{r'_e}$$

This equivalent capacitance is often described as an *injected capacitance* since the circuit places across the source of e a parallel capacitance.

This equivalent capacitance is inversely proportional to r'_e. r'_e is determined from

$$\frac{25 \text{ mv}}{I_E} \leqslant r'_e \leqslant \frac{50 \text{ mv}}{I_E} \qquad (12\text{-}1)$$

As far as C_{eq} is concerned, the use for C_{eq} is at radio frequencies in the oscillator. Although Eq. 12-1 is based on I_E, a d-c value, r'_e can be varied at a rate determined by the audio information signal.

The vacuum-tube version (Fig. 24-17c) shows a different arrangement of the reactance network. In this case

$$R \gg X_c$$

and

$$i = e/R$$

The grid voltage is

$$e_g = iX_c = -j\frac{e}{\omega RC}$$

The plate current is

$$i_p = g_m e_g = -j\frac{g_m}{\omega RC}e$$

The impedance seen by e is

$$z = \frac{e}{i_p} = \frac{e}{-j(g_m/\omega RC)e} = j\omega\frac{RC}{g_m}$$

Now the injected impedance is an inductance

$$L_{\text{eq}} = \frac{RC}{g_m}$$

The four basic circuit arrangements are shown in Fig. 24-18. The amplifiers can be transistors, FET's, or vacuum tubes. The calculations for the equivalent injected values of L or C are left to the exercises at the end of the

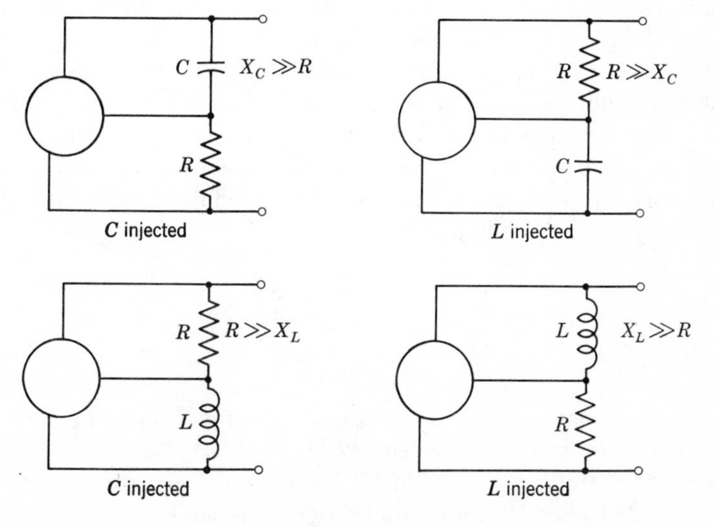

Figure 24-18 Basic forms of a reactance circuits.

section. These basic circuits are not restricted to applications of FM. They are also found in many industrial circuits where an electronically controlled L or C is desired.

The full circuit of a reactance-tube modulator is given in Fig. 24-19. The

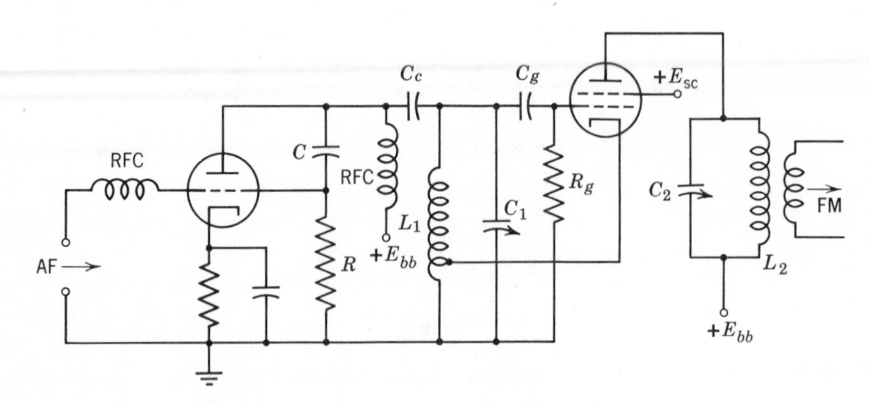

Figure 24-19 The reactance-tube modulator.

equivalent capacitance of the reactance tube is in parallel with the oscillator tank, L_1 and C_1. When there is an audio signal on the grid of the reactance tube, the instantaneous g_m varies, and C_{eq} changes. Since the total tuning capacitance of L_1 includes C_{eq}, the frequency of the electron-coupled oscillator changes. In this manner, the output of the electron-coupled oscillator is an FM signal. Usually, the reactance tube is a remote-cutoff multigrid tube because these tubes have very large variations in the transconductance with changing bias. When a multigrid tube is used, the junction of the R and C of the reactance network is connected to either the grid or the cathode, and the signal is fed into one of the electrodes which is not used by the reactance network. In practice, a reactance tube modulator must be compensated to produce a linear modulation.

PROBLEMS

1. Derive equations for the values of C_{injected} or L_{injected} for the basic forms of the reactance networks shown in Fig. 24-18.
2. An oscillator has a value of 100 μh for L and 30 pF for C in its tuned circuit. The capacitor C is paralleled by a varactor diode that is biased to add 6 pF to the tuned circuit. Draw the circuit and determine the carrier frequency. An

audio signal varies the capacitance of the varactor diode ±1 pF. What deviation results in the modulator?

3. A varactor diode has a capacitance of 10 pF for a given bias. The variation of capacitance is ±1.5 pF. It is desired to obtain a deviation of ±40 kHz at 2.9 MHz. What are the value of *L* and *C* for the oscillator tuned circuit?

Section 24-6 The Balanced Modulator

The original FM transmitter developed by Major Armstrong was not the direct FM system using the reactance-tube modulator, but was an indirect method using a *ring* or *balanced modulator*. First, we develop the relations between the input and the output of the widely used balanced modulator, and then we apply this circuit to the generation of an FM signal.

The balanced modulator (Fig. 24-20) is essentially a push-pull circuit

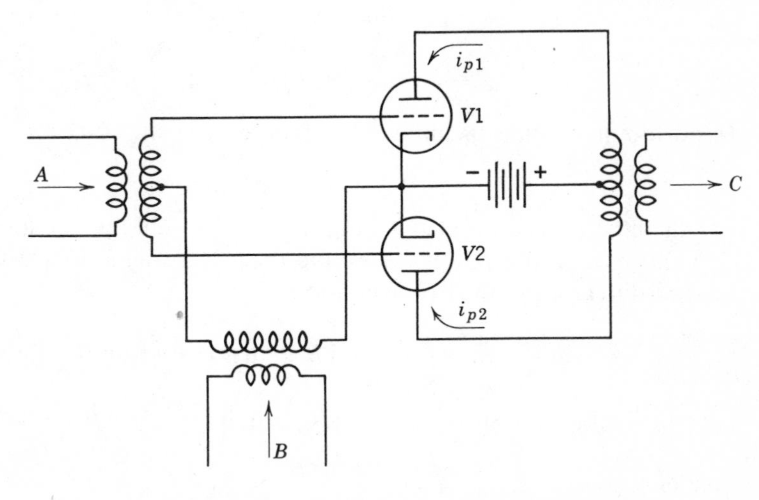

Figure 24-20 The balanced or ring modulator.

with two inputs, *A* and *B*, and one output, *C*. If input *B* did not exist, the circuit would be an ordinary push-pull amplifier. Since the signal *B* simultaneously appears on both grids at the same phase angle, the input *B* is called a *push-push* input.

The circuit shown in Fig. 24-20 uses vacuum tubes. Any nonlinear device can be used in place of the vacuum tubes, diodes, transistors, FET's, or nonlinear magnetic materials. The only change that would occur in the development of the analysis would be the symbols used in place of i_{p1} and i_{p2}.

Let us assume for the derivation that the frequency of signal B is greater than the frequency of signal A. In a specific application, the input B is radio frequency and the input A is audio frequency. The signal A produces a voltage on the grid of tube $V1$ which may be given as $\sin A$. Since the circuit is in push-pull, the signal from A on the grid of $V2$ is $-\sin A$. The signal from the push-push input, $\sin B$, is simultaneous on both grids. The total signals on the grids are

$$e_{g1} = \sin B + \sin A$$

and

$$e_{g2} = \sin B - \sin A$$

A power series is used to represent the plate currents of the tubes, and, since the tubes are identical, the coefficients of corresponding terms are the same for the two tubes:

$$i_{p1} = a_0 + a_1 e_{g1} + a_2 e_{g1}^2 + a_3 e_{g1}^3 + \cdots$$

$$i_{p2} = a_0 + a_1 e_{g2} + a_2 e_{g2}^2 + a_3 e_{g2}^3 + \cdots$$

The push-pull output is the difference of these two plate currents

$$C = i_{p1} - i_{p2}$$

The algebraic procedure at this point is to substitute the equations for the grid voltages into the power series and then to examine the results of the push-pull difference of the two expansions.

$$i_{p1} = a_0 + a_1(\sin B + \sin A) + a_2(\sin B + \sin A)^2 + a_3(\sin B + \sin A)^3 + \cdots$$

$$i_{p2} = a_0 + a_1(\sin B - \sin A) + a_2(\sin B - \sin A)^2 + a_3(\sin B - \sin A)^3 + \cdots$$

Expanding gives

$$i_{p1} = a_0 + a_1 \sin B + a_1 \sin A + a_2 \sin^2 B + 2a_2 \sin B \sin A + a_2 \sin^2 A + a_3 \sin^3 B + 3a_3 \sin^2 B \sin A + 3a_3 \sin B \sin^2 A + a_3 \sin^3 A + \cdots$$

$$i_{p2} = a_0 + a_1 \sin B - a_1 \sin A + a_2 \sin^2 B - 2a_2 \sin B \sin A + a_2 \sin^2 A + a_3 \sin^3 B - 3a_3 \sin^2 B \sin A + 3a_3 \sin B \sin^2 A - a_3 \sin^3 A + \cdots$$

On subtraction, we find that the difference is

$$i_{p1} - i_{p2} = 2a_1 \sin A + 4a_2 \sin B \sin A + 6a_3 \sin^2 B \sin A + 2a_3 \sin^3 A$$

This last equation contains a number of terms. We are not interested in the magnitude of these terms but only in their frequency content. For example, if the term is $2a_1 \sin A$, the energy content is at f_A.

By using the trigonometric expansion formula

$$2 \sin x \sin y = \cos(x+y) - \cos(x-y)$$

it is evident that the energy content of the term $4a_2 \sin B \sin A$ is at frequencies (f_B+f_A) and (f_B-f_A).

In considering the term $6a_3 \sin^2 B \sin A$, we neglect the coefficient and write the term as

$$(\tfrac{1}{2}-\tfrac{1}{2}\cos 2B)\sin A$$

Expanding, it becomes

$$\tfrac{1}{2}\sin A - \tfrac{1}{2}\cos 2B \sin A$$

But

$$2 \cos x \sin y = \sin(x+y) - \sin(x-y)$$

Substituting, we have

$$\tfrac{1}{2}\sin A - \tfrac{1}{4}\sin(2B+A) + \tfrac{1}{4}\sin(2B-A)$$

This means that the term $\sin^2 B \sin A$ contains energies at

$$f_A,\ (2f_B+f_A),\ \text{and}\ (2f_B-f_A)$$

The remaining term, $2a_3 \sin^3 A$, may be treated by considering $\sin^3 x$:

$$\sin^3 x = \tfrac{3}{4}\sin x - \tfrac{1}{4}\sin 3x$$

Thus, the term contains energy at f_A and at $3f_A$.

When the energy terms are collected and duplications are ignored, we find that the push-pull output C contains energies at

$$f_A,\ (f_B+f_A),\ (f_B-f_A),\ (2f_B+f_A),\ (2f_B-f_A),\ 3f_A$$

If the output transformer is an RF transformer tuned to f_B, the audio frequencies, f_A and $3f_A$, are attenuated completely. Since $(2f_B+f_A)$ and $(2f_B-f_A)$ are very close to the second harmonic, $2f_B$, of f_B, these frequencies are also not present in the output. The output, then, from this tuned circuit consists only of energies at (f_B+f_A) and at (f_B-f_A). The output *does not* contain the carrier f_B. Hence, *the dual function of the ring modulator is to develop amplitude modulation and to separate the sidebands from the carrier.*

In the phasor diagram for amplitude modulation (Fig. 24-21*a*), the resultant phasor R is at all times in phase with the carrier. If the side bands are rotated 90° with respect to the carrier, the resultant phasor is not in phase with the carrier phasor and represents phase modulation (Fig. 24-21*b*). This resultant phasor does vary in length, but, if the resultant wave is

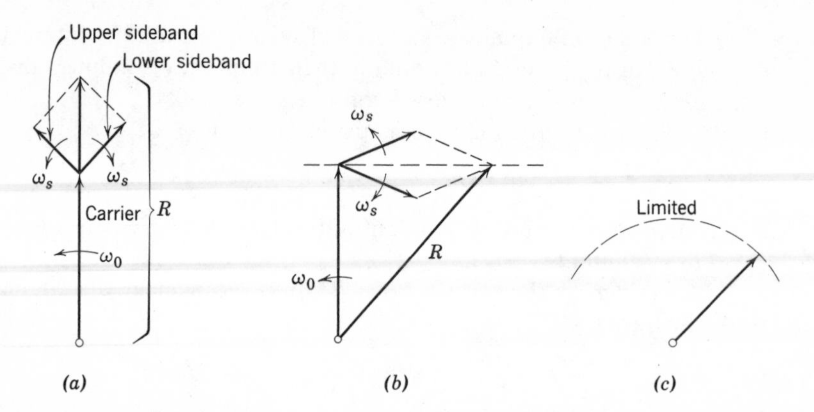

Figure 24-21 Phasor diagrams for the Armstrong method of producing frequency modulation. (*a*) Phasor diagram of amplitude modulation. (*b*) Side-bands rotated 90°. (*c*) Pure frequency modulation.

amplified in a limiter circuit, the output of the limiter has a constant amplitude (Fig. 24-21*c*), and, together with frequency-response correction networks in the audio input stages, produces a pure FM wave.

The block diagram of the Armstrong transmitter (Fig. 24-22) shows how

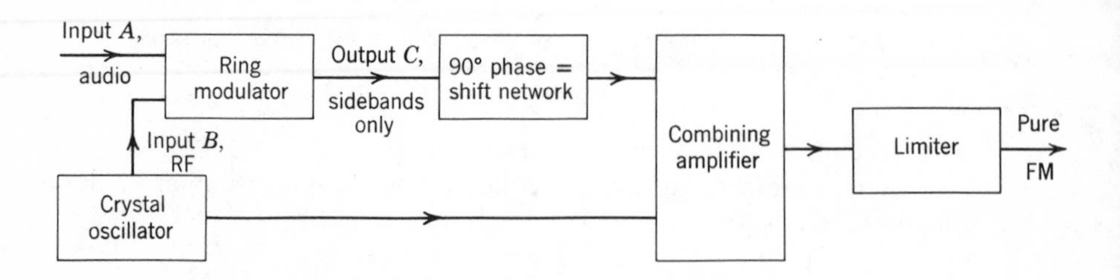

Figure 24-22 Block diagram of the Armstrong transmitter.

these principles are joined together in the complete circuit. The audio and the radio frequency from the crystal drive a balanced modulator which produces the sidebands only in the output. The output coupling circuit gives the required 90° phase shift to the side bands, and this side band signal is fed into the combining amplifier. A signal from the crystal oscillator is also fed into the combining amplifier. The amplitude modulation present in the output of the combining amplifier is removed in the limiter stage to produce a pure FM output.

The reactance-tube modulator cannot be crystal-controlled, and various complicated methods are used in later amplifier stages to provide the necessary close frequency stability. The Armstrong method provides crystal control of the output signal as a part of the balanced-modulator circuit itself.

There are other methods used to develop frequency modulation, but, since their application is limited to FM transmitters, they are not considered here. These circuits can be found in the specialized textbooks on frequency modulation that are in print. The circuit diagram for a small FM transmitter is shown in Fig. 24-23.

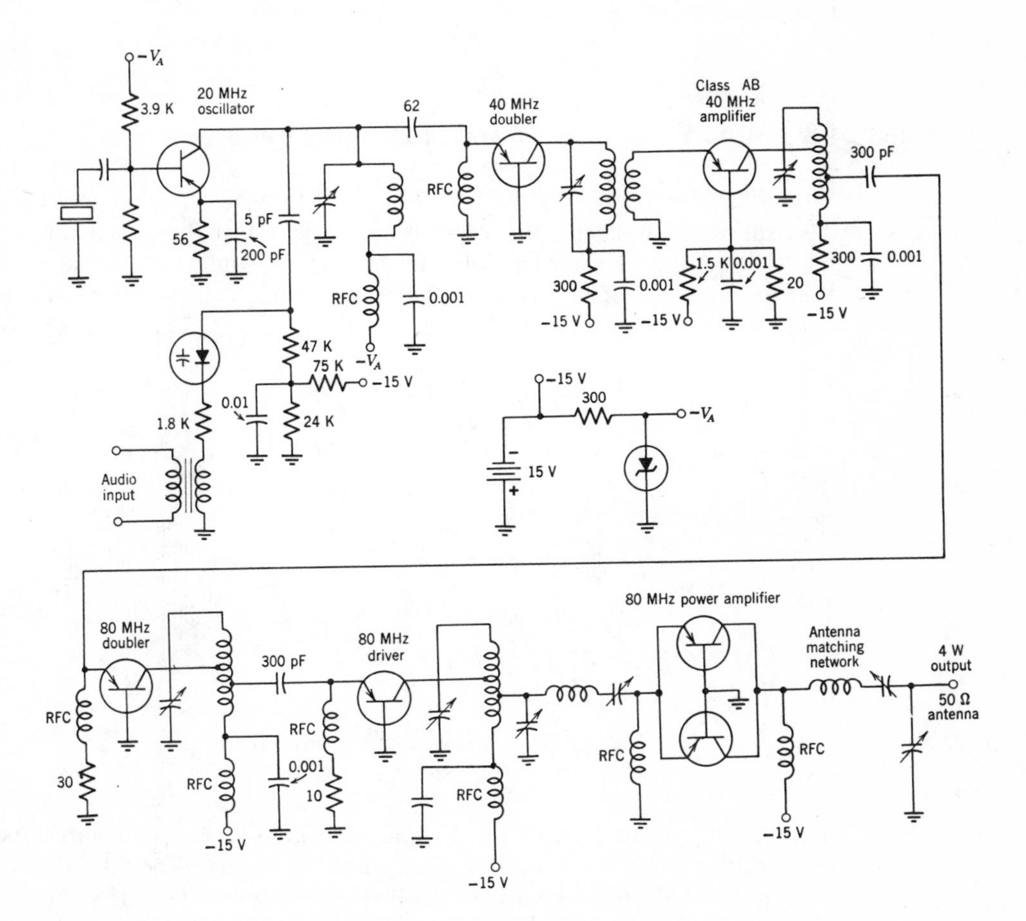

Figure 24-23 Portable 4-W FM transmitter.

Section 24-7 Preemphasis and Deemphasis

In FM broadcasting and receiving systems, the volume of the high-frequency audio signals is increased for transmission and is corrected within the audio stages of the receiver. The increase of the level of the high frequencies which is termed *preemphasis* increases the energy content of the high-order side bands, with the result that reception can be made with lower signal-to-noise values than without preemphasis. In the audio amplifiers of the transmitter, a simple high-pass L filter made up of a resistor and an inductor is used. The time constant of this filter is 75 μsec.

It is necessary to use in the receiver the inverse, or *deemphasis*, characteristic. The deemphasis must have the same time constant, 75 μsec, as the preemphasis.

Section 24-8 Suppressed Carrier and Single Side Band

The discussion on energy content of an AM transmission centered on Eq. 24-4 showed that most of the transmitted energy was in the carrier and not in the side bands. The carrier does not convey any information but wastes transmitter power. The carrier can be eliminated by using a balanced modulator. The resulting waveform is shown in Fig. 24-24. The two side bands

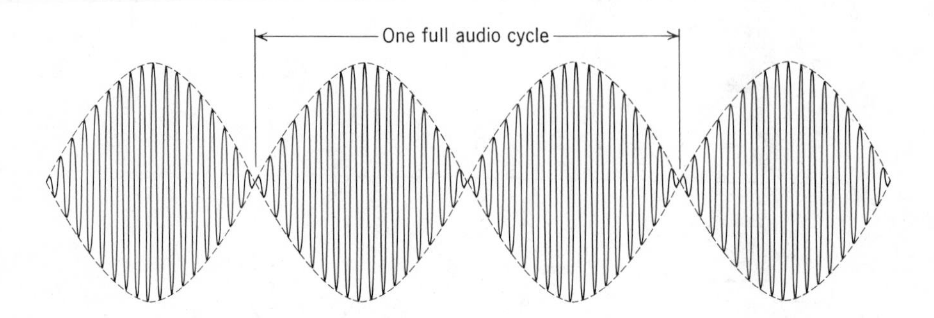

Figure 24-24 Suppressed carrier waveform.

that make up this waveform are $2f_s$ apart and, consequently, the envelope is the result of a beat between two equal signal $2f_s$ apart. When the signals of a suppressed carrier signal are received, the carrier must be restored to convert this waveform to the conventional AM waveform in order that it can be detected and converted back to the original information signal. The phase angle of the restored carrier must correspond exactly to the phase angle of the original carrier. This becomes a rather difficult problem, especially at a distance of several thousand miles. As a result, enough of the carrier is transmitted so that the received carrier itself can be used. This system is described as having a *suppressed carrier*, since the carrier is not completely eliminated.

Another system of transmission that is used extensively is single side-band (SSB) transmission. One side band carries all the necessary signal information. The other side band only duplicates the signal information contained in the first side band. SSB transmission has the major advantage of requiring only one half the bandwidth taken by conventional AM transmission. In SSB equipment, the carrier may or may not be partially suppressed.

Fig. 24-25 shows a block diagram of a method of securing a single side

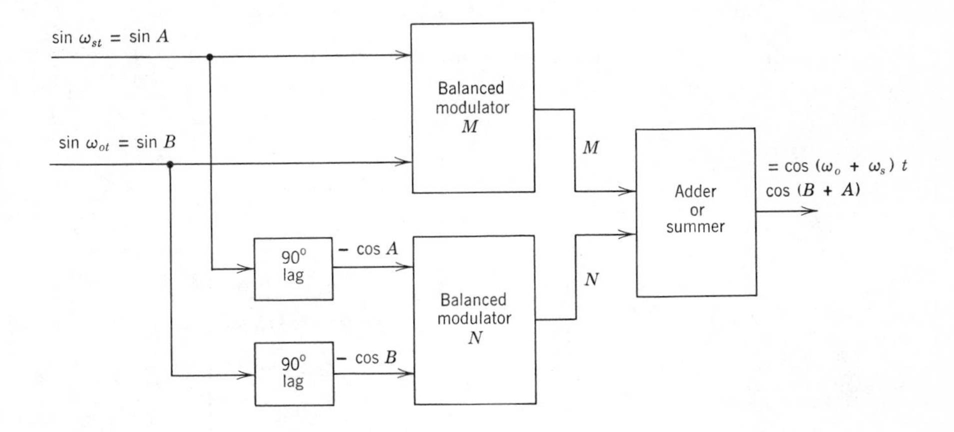

Figure 24-25 Block diagram of a single side-band generator.

band without carrier content. This block diagram is stripped of amplifiers, isolators, filters, and tuned circuits. The sideband output of the balanced modulator M was determined in Section 24-6 as

$$M = 2a_2 \cos (B+A) - 2a_2 \cos (B-A)$$
$$= 2a_2 \cos (\omega_0+\omega_s)t - 2a_2 \cos (\omega_0+\omega_s)t$$

The 90° lag networks change $\sin A$ and $\sin B$ to $-\cos A$ and $-\cos B$, respectively, and then the squared term $a_2 e_g^2$ in the output of the balanced modulator N is

$$i_1 = a_2 e_g^2 = a_2(-\cos B - \cos A)^2 = a_2 \cos^2 B + 2a_2 \cos B \cos A + a_2 \cos^2 A$$

and

$$i_2 = a_2 e_g^2 = a_2(-\cos B + \cos A)^2 = a_2 \cos^2 B - 2a_2 \cos B \cos A + a_2 \cos^2 A$$

and the output of the modulator is the difference

$$i_1 - i_2 = 4a_2 \cos B \cos A$$

By trigonometric expansion, this output is

$$N = 2a_2 \cos (B + A) + 2a_2 \cos (B - A)$$

$$= 2a_2 \cos (\omega_0 + \omega_s)t + 2a_2 \cos (\omega_0 - \omega_s)t$$

The output of the adder is simply

$$M + N = 4a_a \cos (\omega_0 + \omega_s)t$$

which is the upper side band without the carrier. A portion of the carrier can be reinserted in a second adder.

PROBLEMS

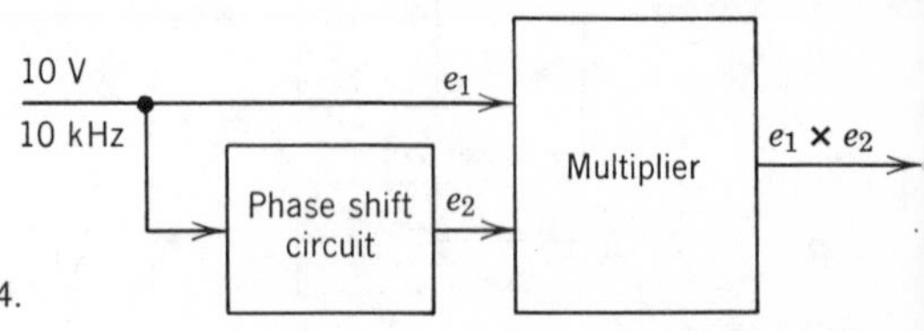

Block diagram for Problems 3 and 4.

1. A 10-V 1-MHz carrier is modulated by a 5-V, 5-kHz signal. Draw to scale the time function, the phasor diagram, and the line spectra diagram for (*a*) normal AM transmission, (*b*) suppressed carrier transmission, and (*c*) single sideband transmission.
2. The current in a square-law modulator has the equation

$$i = 20 + 15e + 10e^2 \text{ ma}$$

The voltage impressed across the modulator is

$$e = 10 + 6 \cos 100{,}000t + 4 \cos 1000t$$

Calculate the amplitude and frequency of each of the current components in the modulator.

3. Determine the amplitude and frequency of the output. The phase-shift circuit introduces a 90° lead without attenuation.
4. Repeat Problem 3 of the phase-shift circuit introduces a 90° lag without attenuation. If the output of the circuit is added to the output obtained from Problem 3, what is the combined output?
5. The inputs to a multiplier are $e_1 = A \sin \omega_0 t$ and $e_2 = B \sin \omega_0 t$ in which ω_0 is RF and ω_s is audio. The output of the multiplier is $e_1 \times e_2$. What is the output? Sketch the time waveform of the output. Show a circuit other than a multiplier that accomplishes this result.

Questions

1. Define or explain each of the following terms used in the development of the theory of amplitude modulation: (*a*) upper sideband, (*b*) per cent modulation, (*c*) overmodulation, (*d*) splattering, (*e*) peak power, (*f*) MCW, (*g*) ICW, (*h*) baud, (*i*) SSB, (*j*) trapezoidal pattern.
2. Define or explain each of the following terms used in the development of the theory of frequency modulation: (*a*) index of modulation, (*b*) deviation frequency, (*c*) deviation ratio, (*d*) wide-band FM, (*e*) narrow-band FM, (*f*) multiple sidebands, (*g*) reactance tube, (*h*) push-push, (*i*) ring modulator, (*j*) signal-to-noise ratio.
3. What is the purpose of a modulation?
4. Distinguish among AM, FM and PM.
5. How can the per cent modulation be determined from a screen pattern on an oscilloscope?
6. Describe four methods of producing amplitude modulation.
7. How does the sideband energy content vary with audio signal in AM?
8. Compare AM and FM based on bandwidth considerations.
9. Compare AM and FM based on power considerations.
10. Explain in detail the method of checking an FM transmitter or oscillator for linearity.
11. What is meant by "linearity" in an AM modulator and in an FM modulator?
12. What are the advantages and disadvantages of a reactance circuit?
13. What is the advantage of a balanced modulator circuit in producing FM?
14. Describe the operation of the balanced-modulator circuit.
15. Explain pre-emphasis and de-emphasis.
16. Using Fig. 24-12, explain how the first-order side-bands could be used to check out the linearity of a frequency modulator.

17. Explain the function of each component shown in the portable FM transmitter circuit given in Fig. 24-23.
18. What is an advantage of suppressed carrier transmission? A disadvantage?
19. What are the advantages of single side-band transmission?
20. Using the block diagram of Fig. 24-25, show how single side-band transmission can be generated.

Chapter Twenty-five

DETECTION

In a receiving device, it is necessary to convert a modulated wave, AM or FM, back to the audio signal. At the same time it is necessary to remove the high-frequency carrier from the composite wave. The *detector* is the electronic device that restores the audio, and the *filter section* separates the signal from the carrier.

A mathematical approach to the problem of detection is very complex, and a quantitative treatment is sufficient for the needs of the technician. The diode (Section 25-1) and the other AM detectors (Section 25-2) are examined. Automatic gain control (Section 25-3) is a fundamental circuit used in the radio receiver. The limiter-discriminator circuit (Section 25-4) and the ratio detector (Section 25-5) are alternative methods of FM detection. Frequency converters (Section 25-6) are also forms of the detector. An automatic frequency-control circuit (Section 25-7) combines several basic circuits to provide a stabilized frequency source. The basic principles of stereophonic FM transmission and reception are discussed from a block diagram viewpoint (Section 25-8).

Section 25-1 AM Detectors Using Diodes

A modulated carrier wave is shown in Fig. 25-1*a*. When this signal is fed into a diode rectifier, only one half the composite signal is passed to the load (Fig. 25-1*b*). When this rectifier output is filtered, the high frequency is removed and the audio is retained. The resultant wave corresponds to the envelope of the original input signal (Fig. 25-1*c*). The action of rectification produces a simultaneous direct voltage in the filtered output, E_{dc}. A blocking capacitor removes the direct voltage leaving the pure a-c audio signal shown in Fig. 25-1*d*.

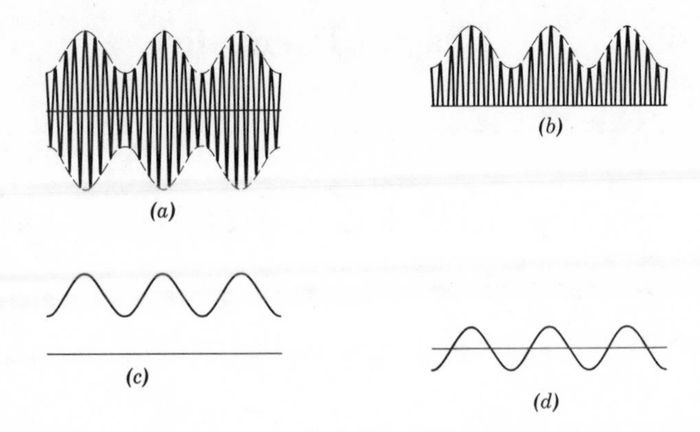

Figure 25-1 The mechanism of diode detection. (*a*) Modulated carrier wave. (*b*) Waveform resulting from half-wave rectification. (*c*) Envelope of the output of half-wave rectification. (*d*) D-c value removed by blocking capacitor.

A circuit that accomplishes this detection is shown in Fig. 25-2. The incoming modulated AM signal is fed into the detector from a tuned circuit, L_1 and C_1. This tuned circuit is resonant at the frequency of the carrier, and

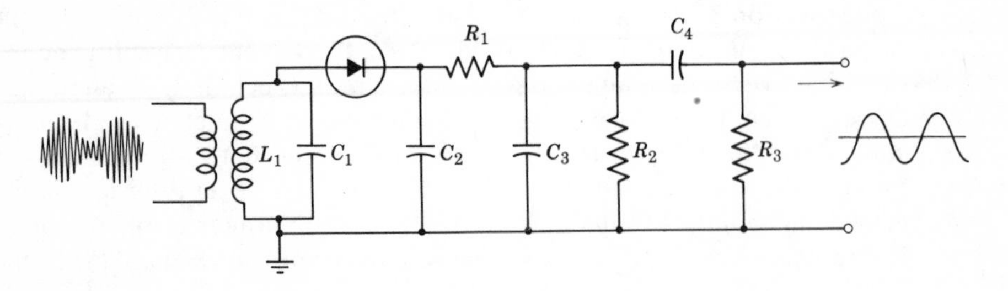

Figure 25-2 The simple diode detector and filter.

its response is broad enough to pass the sideband energy content of the modulated wave. Since a diode is a unilateral device, it rectifies this modulated wave and allows one half, in this case the positive half, to pass on to the filter circuit. The filter section comprising C_2, R_1, and C_3 is a low-pass π filter. Its cutoff frequency is adjusted by component selection to attenuate the high carrier and sideband frequencies. The audio frequencies and the direct current pass through the π filter and appear across the load resistor R_2. The coupling capacitor C_4 blocks the direct voltage and a pure a-c audio signal appears across R_3.

There are several sources of distortion in this circuit. The d-c load on the diode, neglecting the resistance of the transformer secondary and R_1, is R_2. The a-c load on the diode is the parallel combination of R_2 and R_3. This difference can cause the diode to operate in a nonlinear fashion.

The rectification characteristic of a typical diode under dynamic conditions is shown in Fig. 25-3. If we assume in Fig. 25-2 that R_1 is zero and

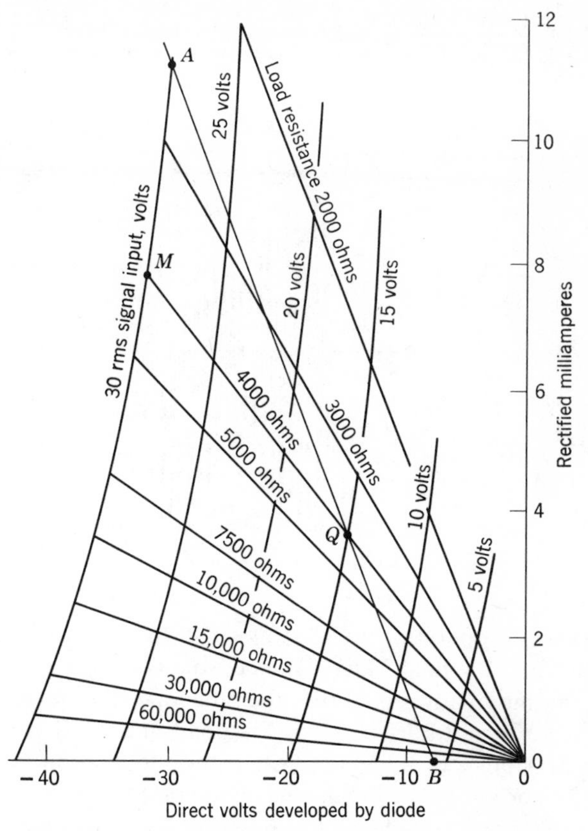

Figure 25-3 Rectification characteristics of a diode.

that R_3 is infinite resistance, the action of the circuit for different values of the load resistance R_2 is shown in Fig. 25-3. This load line is somewhat different from the concepts of the amplifier load line. When R_2 is 4000 Ω, the load line is $0M$. The operating point is not fixed, but it is determined by the level of the incoming carrier signal. When the carrier signal is 15 V rms, the operating point is Q. If the carrier-signal strength increases, the operating point shifts toward M and, if the carrier drops to zero, the operating point moves along the load line from Q to 0.

When the coupling network (R_2, C_4, and R_3) exists in Fig. 25-2, both a d-c and an a-c load line must be considered. If we assume that both R_2 and R_3 are 4000 Ω, the d-c load line for 4000 Ω is used, and an a-c load line for 2000 Ω, the parallel combination, must be drawn through the operating point in the same manner as was done with the voltage amplifier a-c load line. In this circuit, the operating point Q is valid only for a specific input rms signal. In Fig. 25-3, the load line AQB is drawn through point Q with a slope of 2000 Ω. A carrier signal of 15 V rms or 21 V peak establishes the operation at Q. When the carrier signal is modulated, the modulation causes the

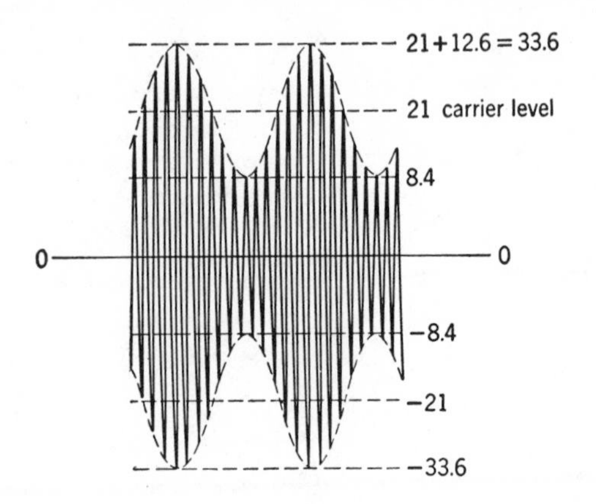

instantaneous operation to shift along AQB. The distance from Q to A is 15 V rms or 21 V peak. The distance from Q to B is 9 V rms or 12.6 V peak. When the modulation is 100%, the envelope rises from the carrier level of 21 V to 42 V and falls to zero. The load line AQB can accommodate the modulation peak of 42 V but, when the modulation envelope falls below (21 − 12.6) or 8.4 V, the diode cuts off at point B.

Using Eq. 24-3, we have, for this condition, a modulation of

$$m = \frac{A-B}{A+B} \times 100 = \frac{2\times 33.6 - 2\times 8.4}{2\times 33.6 + 2\times 8.4} \times 100$$

$$= \frac{25.2}{42.0} \times 100 = 60\%$$

Thus, for this carrier input and these specific values of R_2 and R_3, the action of the detector clips the lower part of the modulation envelope when the modulation exceeds 60%.

To minimize this distortion, it is necessary to make R_3 much greater than R_2. Since R_3 usually serves as the bias resistor for the first audio stage, the value of R_3 that can be used has a practical upper limit of the order of one half, or 1 MΩ.

A detailed examination of the input to the filter (Fig. 25-4*a*) shows that

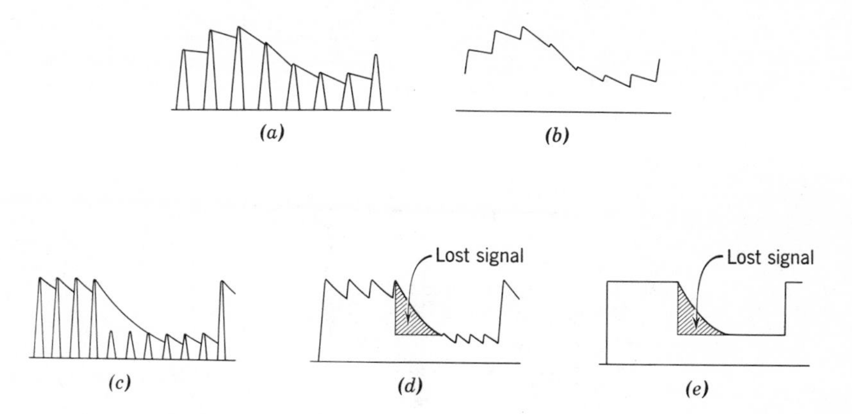

Figure 25-4 Effect of filter time constant on diode output. (*a*) Input to filter. (*b*) Output of filter. (*c*) Input to filter for square-wave signal. (*d*) Output of filter for square-wave signal. (*e*) Signal output.

the input capacitor charges to the peak of the cycle and then discharges until the next positive half-cycle. As a result, the output of the filter (Fig. 25-4*b*) is not exactly smooth, but is serrated. If the signal abruptly drops from a high level to a low level (Fig. 25-4*c*), the time constant of the filter prevents the filter output from following this change exactly (Fig. 25-4*d*) and introduces a distortion. This distortion, called *diagonal clipping*, cannot occur when the abrupt change is from a low level to a high level because the capacitor voltage rises with the first positive half-cycle to the peak value. If the time constant of the filter is decreased, this distortion decreases, but the magnitude of the serrations or sawteeth increases in the output. Since these serrations occur at the frequency of the carrier, an increase in the magnitude indicates that the filtering action is reduced. An actual circuit design is a compromise between these two factors.

Another source of distortion arises from the fact that the a-c load for the audio frequency is the parallel combination of R_2 and R_3 (Fig. 25-2). This loading is not the same as the a-c load at the carrier frequency. The load impedance on the transformer is essentially the low impedance of the filter at the carrier frequency.

As far as the alternating signal output voltage is concerned (Fig. 25-5), a reversed diode has no effect on the magnitude of the output wave. One diode connection rectifies the top envelope, and the other diode connection

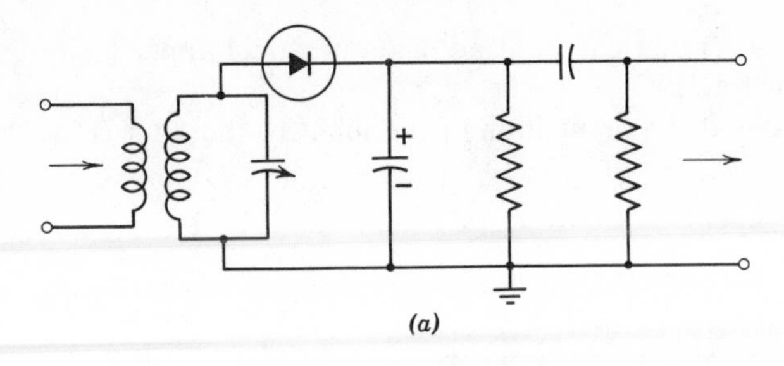
(a)

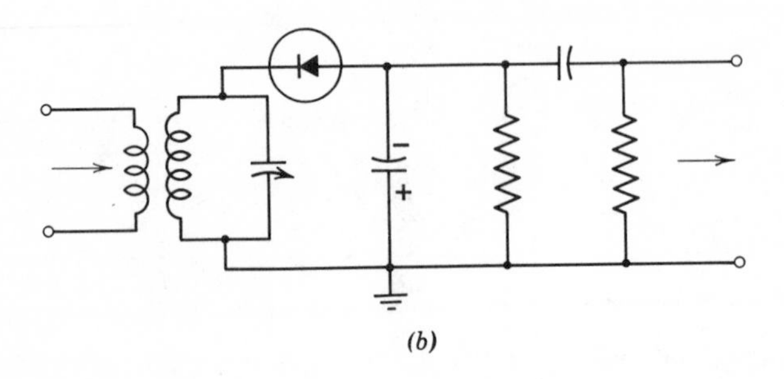
(b)

Figure 25-5 Polarity of detector load voltage. (a) Positive direct voltage. (b) Negative direct voltage.

rectifies the bottom envelope. These two envelopes are equal in magnitude but opposite in phase. This means that the only difference between the two circuits is a phase difference which is usually not at all important. The two diode connections produce different direct voltage polarities across the load. This direct voltage is proportional to the magnitude of the incoming carrier and is not a function of the modulation. The direct voltage can be used to measure the carrier level and is employed in many applications for this purpose. The specific application of the circuit determines whether a positive or a negative voltage is required. The direct voltage is used to provide the control voltage for automatic gain control (Section 25-3).

The general equation for an AM waveform is, as given in the previous chapter,

$$e = E_m \cos \omega_0 t + \frac{mE_m}{2} \cos (\omega_0 + \omega_s) t + \frac{mE_m}{2} \cos (\omega_0 - \omega_s) t \qquad (24\text{-}2)$$

This voltage is the signal voltage that produces a current in a nonlinear device. The power series for the current is, in the nonlinear detector,

$$i = a_0 + a_1 e + a_2 e^2 + a_3 e^3 + a_4 e^4 + \cdots$$

The only term of interest for the process of detection is the second term a_2e^2. Making the substitution for this term

$$a_2e^2 = a_2\left[E_m \cos \omega_0 t + \frac{mE_m}{2} \cos (\omega_0 + \omega_s)t + \frac{mE_m}{2} \cos (\omega_0 - \omega_s)t\right]^2$$

$$= a_2E_m{}^2 \cos^2 \omega_0 t + a_2\frac{m^2E_m{}^2}{4} \cos^2 (\omega_0 + \omega_s)t + a_2\frac{m^2E_m{}^2}{4} \cos^2 (\omega_0 + \omega_s)t$$

$$+ a_2mE_m{}^2 \cos^2 (\omega_0 + \omega_s)t \cos \omega_0 t + a_2mE_m{}^2 \cos \omega_0 t \cos (\omega_0 - \omega_s)t$$

$$+ a_2\frac{m^2E_m{}^2}{2} \cos (\omega_0 + \omega_s)t \cos (\omega_0 - \omega_s)t$$

Remembering that all frequencies except ω_s are rejected by the filter, the first three terms produce the rejected frequencies $2f_0$, $2f_0 + 2f_s$, and $2f_0 - 2f_s$, respectively. The fourth term yields a sum at $2f_0 + f_s$, which is rejected, and a difference component

$$\tfrac{1}{2}a_2mE_m{}^2 \cos \omega_s t$$

The sum component of the fifth term is $2f_0 - f_s$, which is rejected by the filter, and the difference component is

$$\tfrac{1}{2}a_2mE_m{}^2 \cos (-\omega_s)t = \tfrac{1}{2}a_2mE_m{}^2 \cos \omega_s t$$

The sum component of the last term, $2f_0$, is also rejected and the difference term is

$$\tfrac{1}{4}a_2m^2E_m{}^2 \cos 2\omega_s t$$

The sum of the three passed components is

$$e_{\text{out}} = a_2mE_m{}^2 \cos \omega_s t + \tfrac{1}{4}m^2E_m{}^2 \cos 2\omega_s t$$

The first term is the output signal. It is properly proportional to m, and it is the proper frequency f_s. The second term represents inherent distortion in *square-law detection*. When the modulation is 100%, there is a 25% distortion. Fortunately, in entertainment programming, most of the information is at low modulation levels and the distortion is much less.

As a result of this concomitant distortion, if extreme high fidelity reception is desired, the method of detection must be changed. One approach is to use FM and to change the whole concept of program transmission. A *linear detector* can be used for AM, however, to reduce this inherent distortion of diode detection. One linear detection method is the grid-leak detector concept of the next section.

PROBLEMS

1. In the circuit of Fig. 25-2, R_1 is zero, R_2 is 7500 Ω, and R_3 is 10 kΩ. The carrie level is 20 V. Determine by use of the rectifier characteristic shown in Fig 25-3, at what modulation level distortion occurs.
2. Solve Problem 1 for R_2 15,000 Ω and R_3 20,000 Ω.

Section 25-2 Other AM Detectors

The dynamic transfer characteristic for either an FET or a triode is shown in Fig. 25-6. If the circuit employs a transistor, the transfer curve and this discussion are both valid without modification if the horizontal axis of the

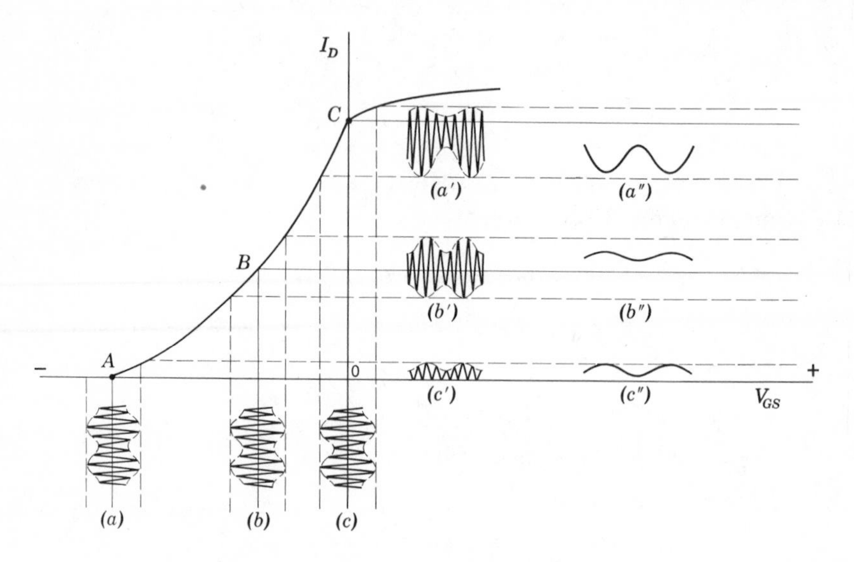

Figure 25-6 Waveforms for output circuit detection.

graph is taken as the input current. It is possible to obtain detection along this characteristic as long as curvature exists. Waveforms *a*, *b*, and *c* represent an incoming modulated carrier. Waveforms *a′*, *b′*, and *c′* are the output-current waves before filtering, and waveforms *a″*, *b″*, and *c″* show the results of the smoothing effect of the filter.

When the bias is held at cutoff *A*, the action is very similar to the diode but, in this circuit, a voltage gain is obtained. A cutoff bias is not obtained from a fixed external battery, but from a self-bias (Fig. 25-7). The source re-

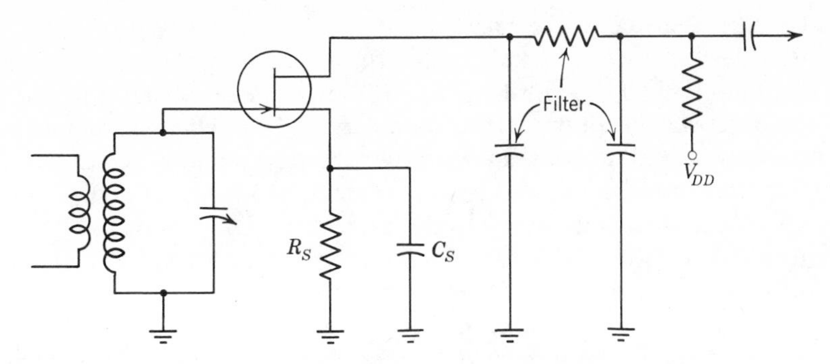

Figure 25-7 Circuit for output circuit detection.

sistor R_S is very high, of the order of 100,000 Ω or 1 MΩ. This very high bias resistance does not quite cut off the FET, but it is near enough to cutoff that the operation is effectively at cutoff.

When the circuit is biased at B, the values of R_S and C_S are those of a normal amplifier. Since the action of the detection depends on the degree of nonlinearity of the transfer characteristic at B, the difference between the top half of the waveform b' and the bottom half cannot be large. The output b'', which is the average of the top envelope and the bottom envelope, is not a large signal. Normally, operating near point B is not used for detection purposes.

Operation at point C yields a somewhat larger output signal, but the high d-c output current is enough of a disadvantage to overcome the slight increase in output in comparison to operation near cutoff (Point A).

The circuit for grid-leak detection is shown in Fig. 25-8. This circuit was one of the fundamental circuits used in the early days of radio. An FET can

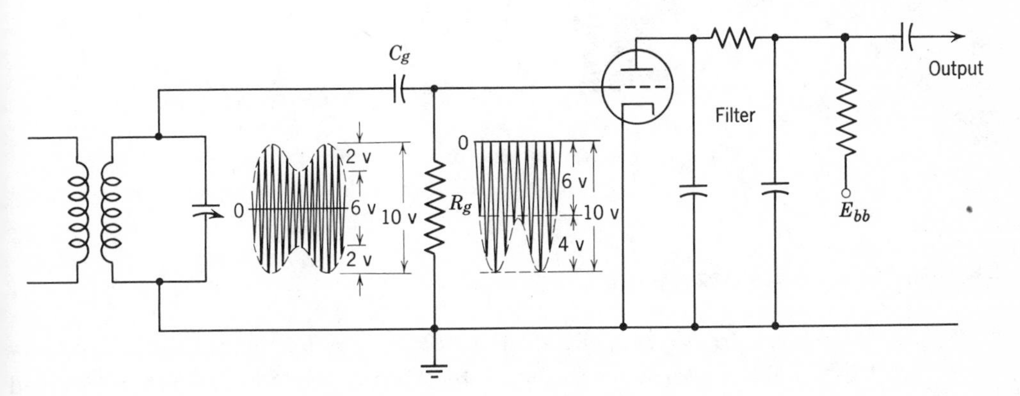

Figure 25-8 Circuit for grid-leak detection.

be used in place of the vacuum tube without any circuit change. The grid leak combination, C_g and R_g, develops a d-c bias in the manner explained in Section 19-1. The time constant of the grid leak is such that the d-c bias follows the envelope of the modulation. For example, assume that the maximum peak-to-peak value of the incoming modulated wave is 10 V and that the minimum peak-to-peak value is 6 V. The d-c bias follows the envelope. Each cycle drives the grid very slightly positive to develop a direct voltage on the capacitor C_g equal to the peak value of the voltage of the cycle. In this example, the d-c bias varies between −3 and −5 V. Except at the time when the capacitor is recharging, the wave is at all times negative measured *down from zero*. As can be seen from the grid waveform on Fig. 25-8, the relocated peak-to-peak amplitude of the envelope is 4 V instead of 2 volts. This action of the grid leak provides a gain of two in addition to the amplification within the triode, which serves as the detector amplifier. The waveforms within the tube (Fig. 25-9) are very similar to those obtained from the plate detector. Now, however, the signal is being obtained from the linear portion of the transfer curve and, consequently, this circuit is a *linear detector* without the severe second harmonic distortion of the diode detector.

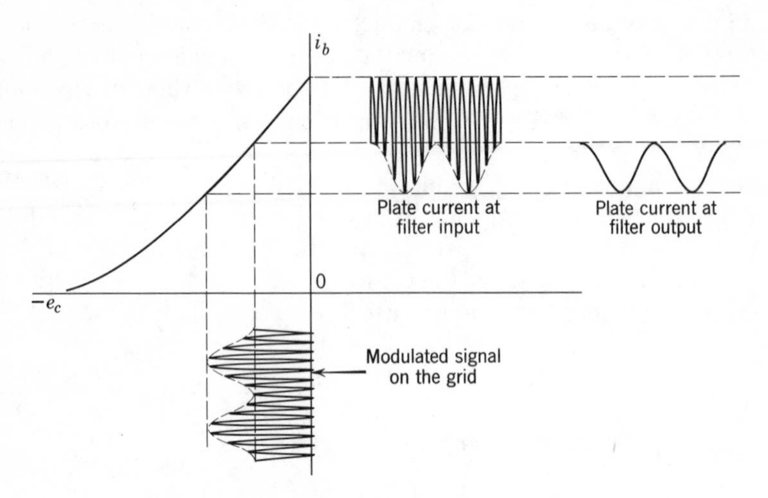

Figure 25-9 Waveforms of grid-leak detection.

Section 25-3 Automatic Gain Control

If we have a radio receiver tuned to a station which induces a signal of 50 μv in the antenna, a proper setting of the volume control produces a pleasing output sound level. When the receiver is tuned in to a strong station which induces a signal of 50,000 μv in the antenna, the input signal has increased by a factor of 1000. If the sound level in the speaker is increased by a factor of 1000 also, the listener would be extremely discomforted. It is necessary to provide an *automatic volume-control* (*AVC*) circuit which reduces the gain of the receiver on strong signals. The term AVC is applied solely to entertainment-type radio receivers. The concept of maintaining a constant output-signal level with a varying input signal has many applications in other systems, for example, space communications receivers, television receivers, and even signal generators. In all of these fields, the terminology that is used is *automatic gain control* or *AGC*.

In a radio receiver, the direct voltage at the detector is proportional to the strength of the carrier level that is being detected. If this direct voltage is filtered to remove all signal components, it can be used to bias the transistors or the remote cutoff tubes which amplify the signal. Fig. 3-11 shows that the beta of a transistor is reduced sharply when the emitter current is reduced. Consequently, a stronger input signal produces a larger d-c voltage at the detector which is used to reduce the forward bias on the previous stages of the receiver. When *PNP* transistors are used in the common-emitter connection, an increasing negative voltage applied to the emitter circuit reduces the forward bias on the transistors. When *NPN* transistors are used, the AGC voltage must be negative. In vacuum-tube circuits, the AGC voltage is negative in order to reduce the transconductance of the remote cutoff tubes used (Fig. 5-9).

A negative direct voltage proportional to the carrier level is developed across the load resistor R_L from A to ground in Fig. 25-10. This direct voltage is well filtered in the L filter comprised of R and C. This L filter is

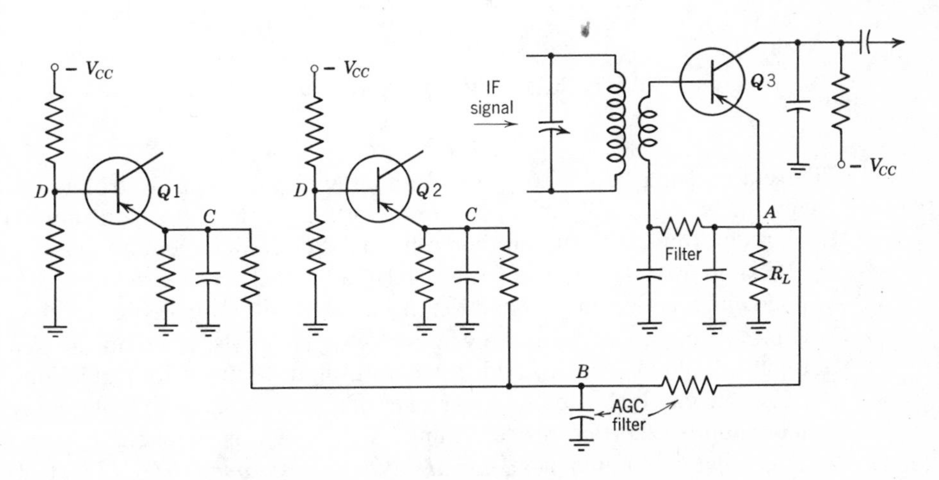

Figure 25-10 Automatic gain control (AGC) circuit.

called the AGC filter, and its time constant is longer than the period of the lowest audio frequency. If the time constant of this filter were too short, the AGC action would tend to suppress the audio signal itself. The AGC voltage is applied to each bias point of the preceding stages. A decoupling filter is usually placed in each AGC lead to prevent motorboating.

In the circuit shown in Fig. 25-10, the forward bias on $Q1$ and $Q2$ is determined by the difference in voltage between C and D, both of which are negative. Point D is more negative than point C, giving a forward bias on the transistor. When the carrier level increases, the negative voltage at A increases. This negative voltage is filtered in the AGC filter, and it causes the negative voltage at point C to increase. An increased negative voltage on the emitters of $Q1$ and $Q2$ reduces the net forward bias and reduces I_E in these transistors. The value of beta decreases, and the stage gains decrease sufficiently to compensate for the increase in signal strength.

In Fig. 25-11, the effect of the automatic gain control is shown for

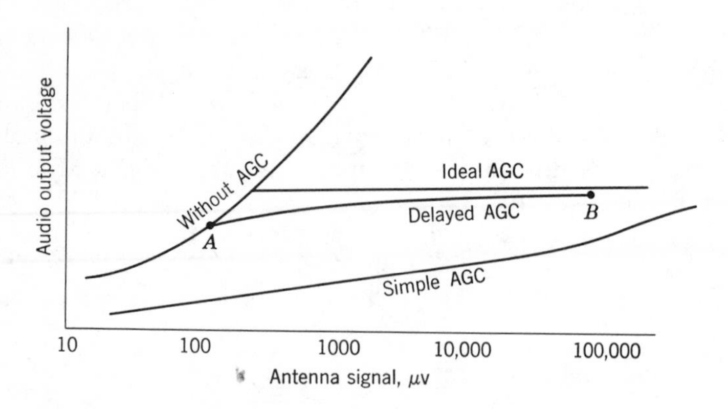

Figure 25-11 Effects of AGC action.

different antenna signal strengths. Without automatic gain control, the audio signal increases rapidly to a point of distortion. Ideally, the gain should rise to a predetermined point and then remain fixed for all larger antenna signals. In the simple AGC circuit of Fig. 24-10, all signal levels produce an AGC voltage and a consequent reduction in gain. This situation decreases the gain when it is most needed—at the very weak signal levels. If a diode is used for automatic volume control and a separate diode is used for detection, the AGC diode can be biased in the manner of a gate or clamp so that it does not function until a certain predetermined signal level is reached. This circuit arrangement is known as *delayed automatic gain control, DAGC* (Fig. 25-12). A d-c amplifier can be placed in series with the AGC line at point A

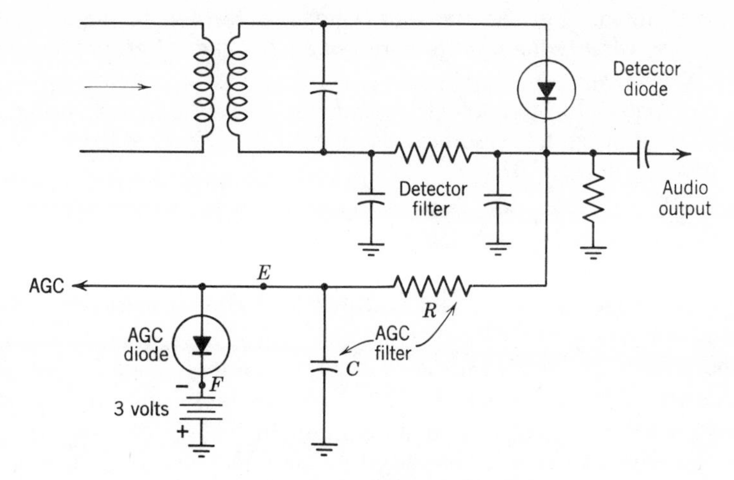

Figure 25-12 Delayed AGC circuit.

in Fig. 25-12. This *AGC amplifier* has a differential input connection. One lead is connected to point *E* and the other input lead is connected to point *F* in Fig. 25-12. The circuit is also arranged so that the amplifier only amplifies when the magnitude of the voltage at *E* exceeds the magnitude of the voltage at *F*. The input to the amplifier is effectively the difference between points *A* and *B* on Fig. 25-11. The result of amplifying the AGC voltage yields an overall receiver response that approaches the ideal curve very closely.

In an AGC system, the AGC voltage is a d-c voltage that has been carefully filtered to remove all signal components. The AGC voltage is applied to stages that amplify RF and IF frequencies. There can be no positive feedback which could produce an oscillation at RF, at IF, or at information frequencies. An AGC system is theoretically unconditionally stable.

PROBLEMS

1. Ideally the voltage level across a detector is 7.07 V for all incoming signals ranging from 10 μv to 100,000 μv at the antenna of a receiver. The impedance levels are 50 Ω at all points in the receiver. What is the input range of signals in db (the dynamic range of the receiver)? If there are five equal stages in the receiver, what is the range of gain in db of each stage? Assume that AGC is applied equally to each stage and that 1 V of AGC voltage will reduce the gain of the stage 2 db. How much AGC voltage is required to control the receiver gain?

2. In Problem 1, if the detector voltage is allowed to increase +3 db from 7.07 V, what is the gain required of an AGC amplifier that feeds all stages? The delay voltage is 7.07 V.
3. In Problem 1, if the detector voltage is allowed to increase +1 db from 7.07 V, what is the gain required of an AGC amplifier that feeds all stages? The delay voltage is 7.07 V.

Section 25-4 FM Detectors – The Limiter and Discriminator

The limiter (Fig. 25-13*a*) is a form of clipping circuit which, although widely used in the field of applied electronics, is usually associated with the Foster–Seeley discriminator circuit in FM detection. The purpose of the limiter is to remove amplitude variations from an incoming FM signal before it is detected. A relatively low collector-voltage operating point is used at *A* (Fig. 25-13*b*), and as soon as the incoming signal exceeds the limits of *B* and *C*, a further increase in signal cannot increase the output beyond the limits *B* and *C*. The least incoming signal level that clips on both sides is termed *threshold limiting*. Signals larger than threshold, point *T* (Fig. 25-13*c*) have clipped tops and bottoms, but there is a tuned circuit in the collector that restores the sinusoidal waveform.

Obviously, the concept of conventional gain cannot be used for a limiter. If limiting occurs at point *T* (Fig. 25-13*c*) for a 1-V signal and if the limiter output is 10 V, the voltage gain is 10. When the input voltage is increased to 30 V (point *W*) the output is still 10 V. If the usual concept of gain is used, the gain at point *W* is 10/30 or 0.33. In order to eliminate the difficulty of this gain concept, the term threshold limiting is introduced. Threshold limiting occurs at point *T* on the response curve. The action of a limiter is usually specified by stating the minimum signal that is required to produce this limiting action.

The limiter action restores the FM signal to a constant level and removes any amplitude variations which may have been introduced between the transmitting antenna and this point in the receiver. Another way of looking at the action of a limiter is to state that the limiting action restores any lost side bands that were in the original signal provided the signal is at or above threshold. This can be seen by referring to Fig. 24-14 and to Fig. 24-15.

The limiter circuit shown in Fig. 25-14*a* can be used with an FET or a pentode that has a sharp cutoff characteristic. This circuit uses a bias clamp or a grid-leak bias. A weak signal (*A* in Fig. 25-14*b*) is clipped on one side by the action of the grid leak, but amplitude variations are still present in the plate circuit. When the signal level reaches the level shown at *B* in the

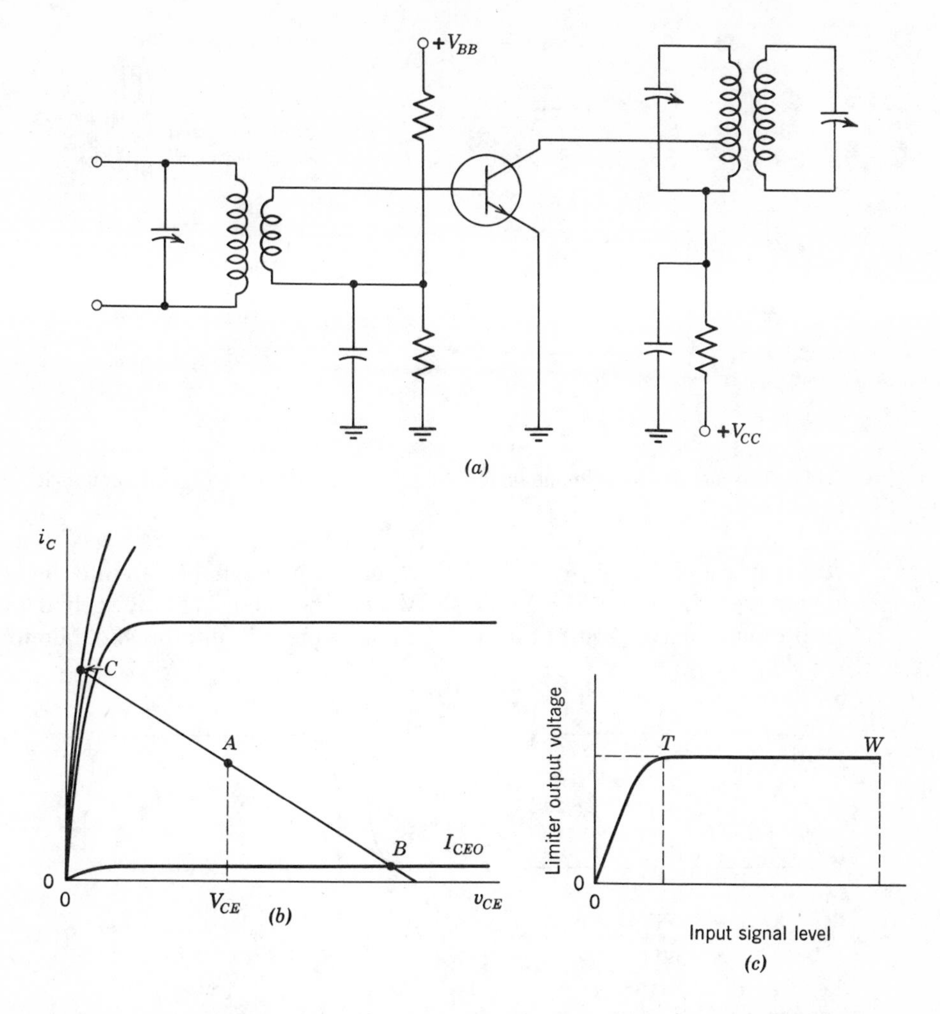

Figure 25-13 The limiter. (*a*) Circuit. (*b*) Range of operation on the collector characteristic. (*c*) Response characteristic.

diagram, clipping occurs on both sides of the signal, and all amplitude variations are removed. A still larger signal *C* does not produce a greater output, and the limiting is complete. The sharp characteristic of the FET or the tube used as a limiter is necessary to provide proper clipping, and a high-gain amplifier insures saturation at low-input signal levels. It is possible to improve the action of the limiting by using a *double limiter*. In a double-limiter circuit, two limiters are connected in cascade. The coupling between the two stages is usually *RC* coupling. Let us assume that threshold limiting in each stage

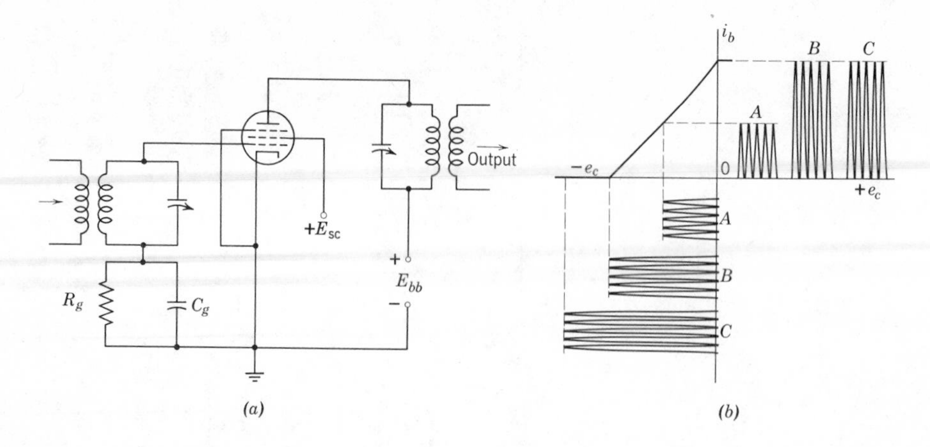

Figure 25-14 The vacuum-tube limiter. (*a*) Circuit. (*b*) Waveforms on transfer characteristic.

occurs at a 1-V signal (Fig. 25-15*a*). When a 0.2 V signal is fed into the first limiter grid, the output is 1 V. This 1-V signal on the grid of the second limiter produces a 10-V output. In this double-limiter circuit, threshold limiting

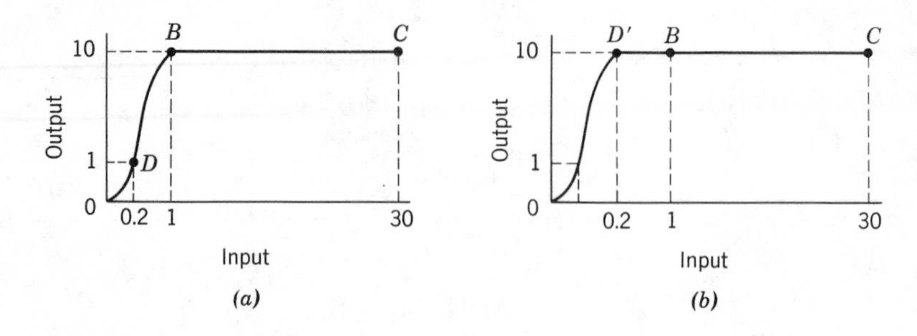

Figure 25-15 Action of a double limiter. (*a*) Single limiter. (*b*) Double limiter.

is accomplished at an input level of 0.2 V, whereas a single limiter requires 1 V. In other words, in this example, the use of a double limiter increases the limiting sensitivity of the circuit by a factor of five. The characteristic of the double limiter is shown in Fig. 24-15*b*.

In a *discriminator circuit* (Fig. 25-16*a*), a constant level signal from a limiter is impressed across the primary winding of a double-tuned transformer. Both the primary L_1 and the secondary L_2 are carefully tuned to the same resonant frequency. The key to the operation of this circuit is the phase relation between the primary and the secondary voltages. The angle between

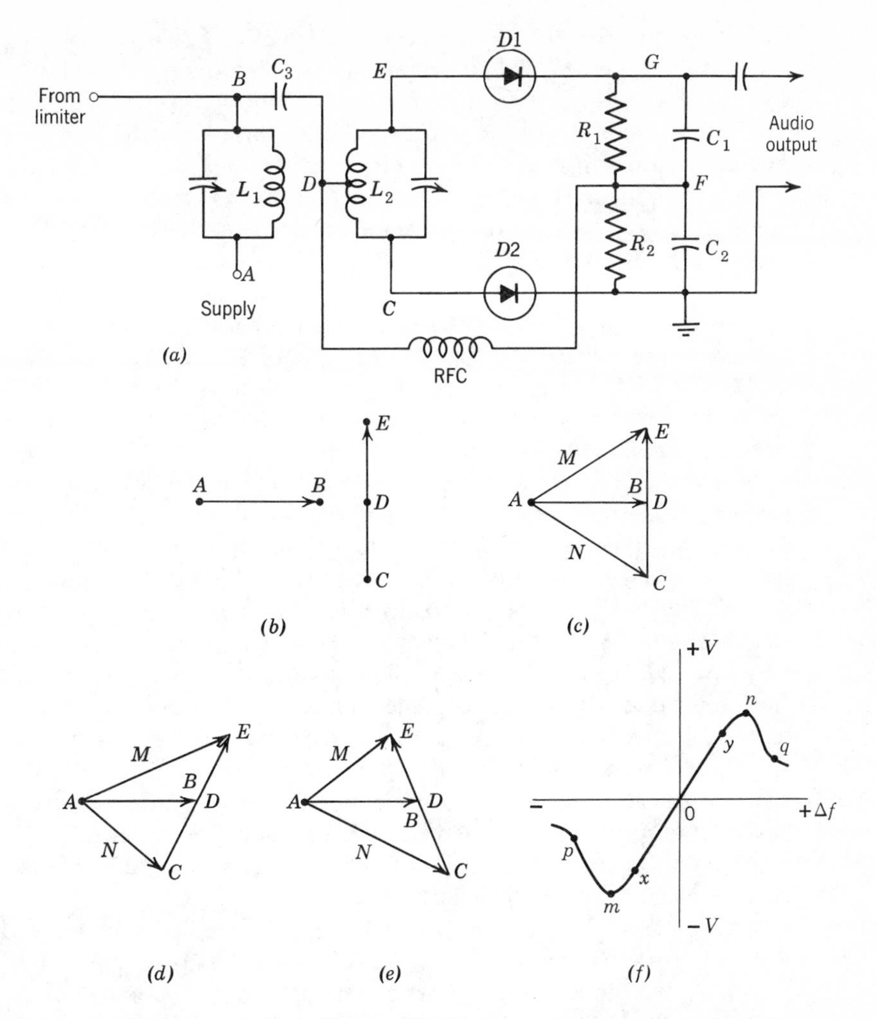

Figure 25-16 The discriminator. (*a*) Circuit. (*b*) Primary and secondary-voltage-phasors at resonance. (*c*) Phasor diagram at $f = f_0$. (*d*) Phasor diagram at $f > f_0$. (*e*) Phasor diagram at $f < f_0$. (*f*) Discriminator characteristic.

the primary and secondary voltages is 90° at resonance (Fig. 25-16*b*). When the incoming frequency goes above the resonant frequency, the angle between these phasors increases. When the frequency shifts below the resonant frequency, the angle decreases. Over a range of frequencies near resonance, this phase-angle change is linear with a change in frequency.

The supply voltage (point *A* in the circuit) is at a-c ground potential because the supply voltage is properly bypassed. A coupling capacitor

C_3 electrically ties the tip of the primary voltage phasor (point B) to the center tap of the transformer secondary winding (point D). The alternating voltage on the top diode $D1$ is the sum of the primary voltage AB plus one half the secondary voltage DE, producing the phasor M in Fig. 25-16*c*. The alternating voltage on the bottom diode $D2$ is the sum of the primary voltage AB plus the other half of the secondary voltage DC. This is shown as the phasor N. The load resistors, R_1 and R_2, are equal. Since the rectifier anode voltages, M and N, are equal, the direct output voltages across R_1 and R_2 are equal. The direction of rectification of the diodes produces direct voltages across R_1 and R_2 which are opposing in polarity. The net output voltage across the total load is zero. This value is represented as point O in Fig. 25-16*f*. The RF choke provides a return path for the direct currents. This choke is often replaced by a resistor in order to reduce the cost of the circuit components.

When the incoming frequency is greater than the resonant frequency, the magnitudes of AB and CE remain constant because of limiter action, and only the phase angle changes. The voltage M to the top diode is greater than the voltage N on the bottom diode (Fig. 25-16*d*), and the net output across both load resistors is positive, locating point y on the discriminator characteristic. When the frequency of the signal is less than the resonant frequency, the secondary phasor shifts in a leading direction (Fig. 25-16*e*), and a negative output voltage is developed which is point x on the characteristic. There are maximum output values that can be obtained, and they are shown as points m and n on the characteristic. When the frequency changes beyond the deviation corresponding to m and n, we are working beyond the range of the tuned transformer, and the output falls off toward zero (points p and q). The overall characteristic (Fig. 25-16*f*) is the so-called S curve of the discriminator.

It is necessary that complete and exact symmetry be maintained in the S curve. An examination of Fig. 25-16*c* shows that the magnitude of the phasor N is a direct function of the length of AB. The length of N can be monitored by measuring the direct voltage across R_2. When the transformer primary winding is properly tuned to resonance, the voltage across R_2 is a maximum. When the direct voltage across the full discriminator load is metered, it should be zero at the resonant frequency. When L_2 is properly tuned to resonance, this meter reading is zero. It is necessary to work back and forth between the primary and the secondary windings several times since the tuning of one affects the other slightly.

It can be seen from Fig. 25-16*f* that a varying frequency produces a changing voltage across the full diode load. Thus a frequency modulated wave can be detected into audio. An increase in deviation which is proportional to volume produces greater swings on the S curve, giving the required greater amplitudes of output signal. The rate of deviation or the rate of changing frequency from above resonance to below resonance is the frequency of the signal and produces the frequency of the alternating current in the output of the discriminator.

PROBLEMS

1. In Fig. 25-16*b*, the two phasors are each 10 V. As the frequency changes, the angle changes from 90° between the phasors. Determine the detector output for phase changes of ±30°, ±45°, ±60°, and ±90° changes from 90°. Plot the response characteristic.
2. Repeat Problem 1 is the phasor AB is 10 V and phasor CE is 4 V.

Section 25-5 FM Detectors—The Ratio Detector

The basic ratio detector (Fig. 25-17) is similar in form to the discriminator. The diodes are connected in series to produce an additive d-c polarity across the load resistors, R_1 and R_2. The primary voltage is added to the secondary

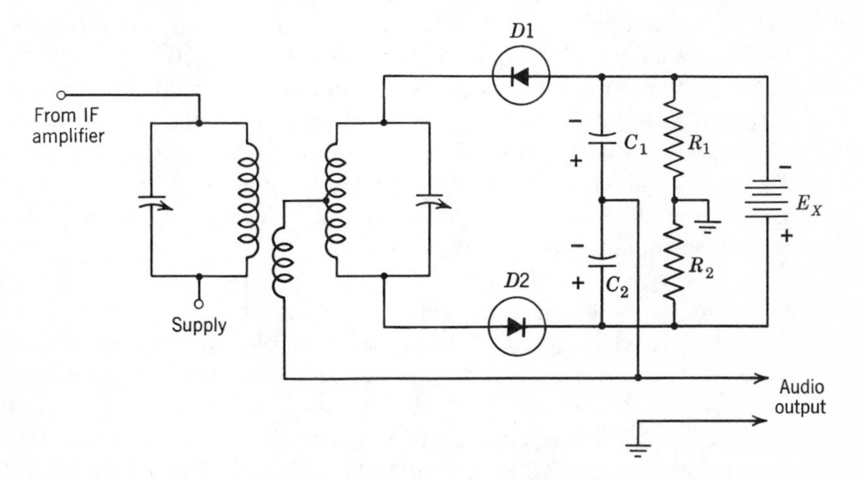

Figure 25-17 The basic ratio detector.

voltage by means of a tertiary winding L_3. When the peak-to-peak incoming signal across the secondary is less than the battery voltage E_x, the diodes cannot conduct and a signal does not appear across the load. A load voltage exists only when the incoming peak-to-peak signal is greater than the battery voltage. In any case, the load voltage cannot be greater than E_x. This limiting action of the battery serves the purpose of a limiter in the circuit. Threshold limiting is that point at which the diodes begin to rectify.

The sum of the voltages across C_1 and C_2 must equal the limiting voltage E_x for amplitude-limited conditions:

$$e_a + e_b = E_x$$

The voltage across C_1 is e_a and is the peak of the incoming signal on $D1$. The voltage across C_2 is e_b and is the peak of the incoming signal on $D2$. When the incoming signal is at center frequency, the two voltages e_a and e_b are equal, and the output voltage taken between the junction of the resistors R_1 and R_2 and the junction of the capacitors is zero. When the incoming frequency is above the center frequency of the transformer, the alternating voltage on $D1$ exceeds the alternating voltage on $D2$, and e_a, is larger than e_b, producing an instantaneous positive voltage. When the incoming signal is below the center frequency, e_b is larger than e_a, and an instantaneous negative output voltage results. The total voltage of $(e_a + e_b)$ is a constant, and a change in voltage between e_a and e_b must divide proportionally; hence, the name ratio detector.

Assume that E_x is 10 V. At the center frequency, e_a and e_b are each 5 V. When the deviation above the center frequency is sufficient, e_a becomes 7 V and e_b is 3 V. The output voltage for this condition is +2 V. The reference point of measurement is the center of the 10 V that is established by the fixed voltages across R_1 and R_2. If we consider these voltages in regard to the discriminator circuit of Fig. 25-16, the voltage across R_1 is 7 V and the voltage across R_2 is 3 V. In the discriminator, the output is the *difference* voltage between the two load voltages, or +4 V. From these figures, we find that the output of the ratio detector is one half the output of the discriminator for the same conditions.

It is very inconvenient to use a fixed battery across the ratio detector to provide limiting. When the battery is replaced by an electrolytic capacitor C in Fig. 25-18, the long time constant of $(R_1 + R_2)$ and C has the same energy storage effect as the battery to eliminate the variations in the amplitude of the incoming signal. The use of a battery to provide limiting produces an S curve similar to the characteristic of the discriminator circuit. When a capacitor is used to provide the limiting, the magnitude of the S curve depends on the amplitude of the incoming signal, since the equivalent E_x depends on the direct voltage on the capacitor, which is the peak-to-peak carrier level. In order to prevent a distortion which may occur when changing from a small S curve to a large S curve, an AGC circuit is usually used with the ratio detector. The direct voltage from the limiting capacitor C is used for the automatic gain control, since the voltage across the capacitor is produced by and is proportional to the incoming carrier-signal strength.

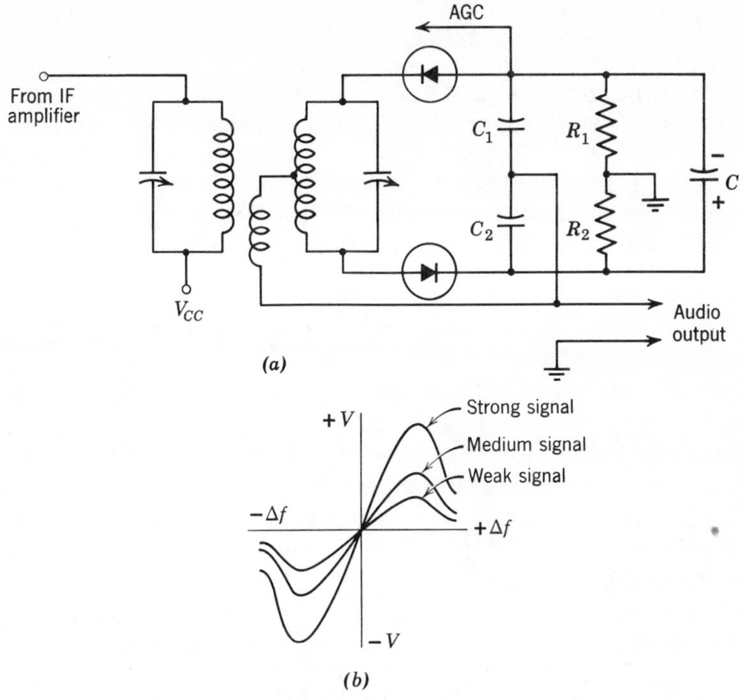

Figure 25-18 The ratio detector. (*a*) Circuit. (*b*) Frequency response.

The polarity of the AGC lead in Fig. 25-18*a* is negative. If the AGC lead were taken from the other side of the capacitor C, the AGC polarity would be positive.

The ratio-detector transformer may be tuned to the carrier frequency by using a visual alignment procedure in which the test equipment is an FM sweep signal generator and an oscilloscope. If it is desired to align the circuit with a d-c voltmeter, resistors are temporarily placed in parallel with C_1 and C_2. These resistors provide a d-c circuit for the voltmeter and should be much higher in value than R_1 and R_2. When the incoming signal is set to the carrier frequency, the primary winding of the ratio detector transformer is tuned for a maximum voltage across the limiting capacitor C. The secondary is tuned for a zero reading on the d-c voltmeter connected between the junction of the temporary resistors and the junction of R_1 and R_2. As in the case of discriminator alignment, this sequence of tuning must be repeated several times in order to eliminate detuning from the coupling of one winding of the transformer into the other.

A number of simple versions of this circuit have been developed for use in television receivers wherein a reduced cost is the primary consideration.

PROBLEMS

1. The phasor diagrams shown in Fig. 25-16*c*, *d*, and *e* apply to the ratio detector if the sum of *AE* and *AC* is a constant voltage. When frequency changes, the angle at *D* changes linearly from 90°. Assume the total voltage output of the ratio detector is 10 V. What is the audio output for ±30°, ±45°, ±60° and ±90° changes from 90°? Plot the response characteristic?
2. If the signal level causes the total output voltage of the ratio detector of Problem 1 to fall to 4 V, what is the audio output using the previous data?

Section 25-6 Frequency Conversion

Frequency conversion or *heterodyning* is used to produce a *beat* or difference frequency and can be accomplished by using any nonlinear device. In a nonlinear device, the current is represented by a power series in terms of the voltage:

$$i = a_0 + a_1 e + a_2 e^2 + a_3 e^3 + \cdots$$

The input signal is the sum of two sinusoidal signals:

$$e = A \sin \omega_1 t + B \sin \omega_2 t$$

It is only necessary to use the first three terms of the power series for the expansion and, by substitution, we have

$$\begin{aligned} i &= a_0 + a_1(A \sin \omega_1 t + B \sin \omega_2 t) + a_2(A \sin \omega_1 t + B \sin \omega_2 t)^2 \\ &= a_0 + a_1 A \sin \omega_1 t + a_1 B \sin \omega_2 t + a_2 A^2 \sin^2 \omega_1 t \\ &\quad + 2a_2 AB \sin \omega_1 t \sin \omega_2 t + a_2 B^2 \sin^2 \omega_2 t \end{aligned}$$

This expression is reduced by using the relations

$$\sin^2 x = \tfrac{1}{2} - \tfrac{1}{2} \cos 2x$$

and

$$\sin x \sin y = \tfrac{1}{2} \cos (x + y) - \tfrac{1}{2} \cos (x - y)$$

The substitutions are made, and the terms that correspond are collected. The exact coefficients are not needed for the discussion, and they are simplified to give

$$\begin{aligned} i &= k_0 + k_1 \sin \omega_1 t + k_2 \sin \omega_2 t + k_3 \cos 2\omega_1 t + k_4 \cos 2\omega_2 t \\ &\quad + k_5 \cos (\omega_1 + \omega_2)t + k_6 \cos (\omega_1 - \omega_2)t \end{aligned}$$

The k_1 and k_2 terms are the signal frequencies present in the output, and the k_3 and k_4 terms represent the second harmonics. The energy at the sum of the two frequencies $(f_1 + f_2)$ is the k_5 term, and the energy at the difference frequency $(f_1 - f_2)$ is the k_6 term. The k_1, the k_5, and the k_6 terms taken together represent amplitude modulation.

The objective of frequency conversion is to separate from the composite output the difference or beat frequency, the $(f_1 - f_2)$ term. The simplest circuit to accomplish this is called a *diode mixer* (Fig. 25-19). The output

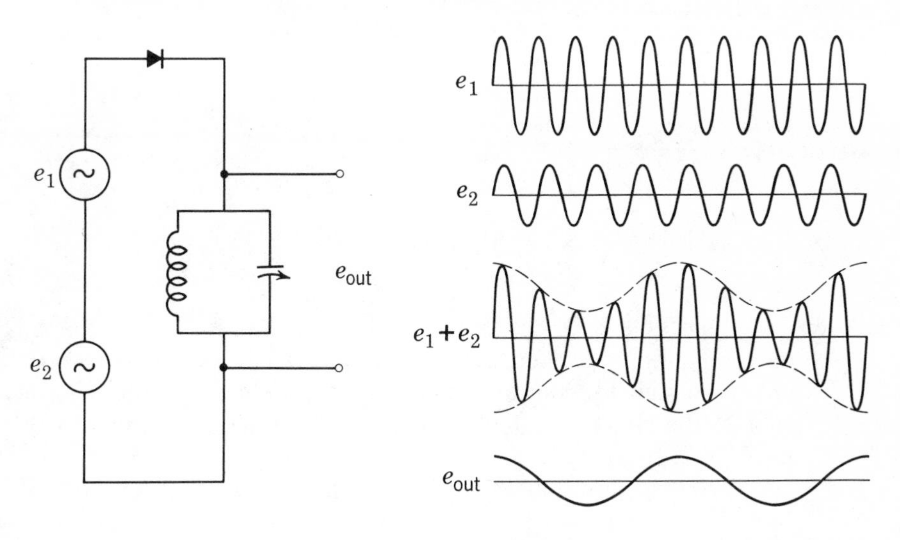

Figure 25-19 Diode mixing.

circuit is tuned to the difference frequency $(f_1 - f_2)$ in order to eliminate all the undesired products of mixing. If the input waveforms are added, point by point, a waveform results that varies at the difference frequency. This difference frequency is shown in the diagram as the envelope of the sum. This composite envelope is rectified by the diode and, by filtering action of the tuned circuit, the output is a pure sine wave whose frequency is $(f_1 - f_2)$.

A transistor can be used as a converter for heterodyning (Fig. 25-20). In this circuit, the varying transistor current from the oscillator at frequency f_2 beats with a signal at frequency f_1 to produce the difference frequency $(f_1 - f_2)$ in the output. If both f_1 and f_2 are radio frequencies, f_2 can be greater than f_1. In this case, the frequency of the output is $(f_2 - f_1)$. The transistor has a circuit advantage in comparison to the diode, because a voltage gain is obtained in the amplifying action of the transistor. In a mixing circuit the gain is less than the gain obtained from a straight amplifier. The gain of the stage is called *conversion gain*.

To be technically exact, when frequency conversion is obtained from a

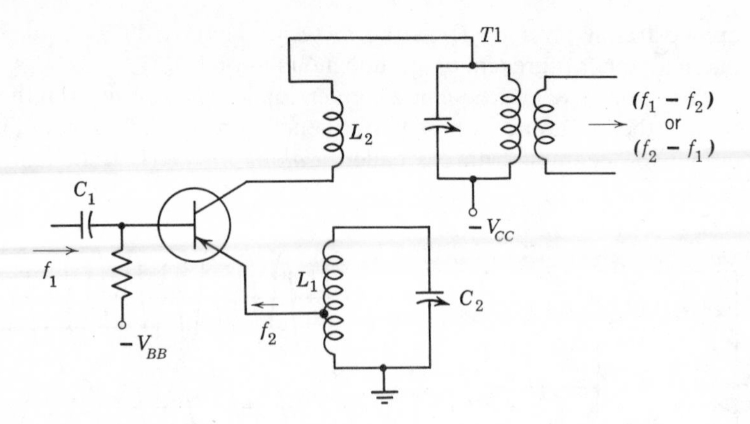

Figure 25-20 Heterodyning in a transistor circuit.

circuit such as Fig. 25-20 that uses a single transistor, FET, or tube for both oscillation and heterodyning, the circuit is called a *converter*. If the oscillator circuit uses an amplifier that is separate from the heterodyning amplifier, the circuit is called a *mixer*. Such a separate oscillator is used in Fig. 25-21.

A FET is ideally suited for use as a mixer. The current in the FET has a square-law characteristic that does not contain higher order terms. Consequently there is no interference with spurious harmonics. Also the isolated characteristic of the gate minimizes detracting energy transfers in a reverse direction through the FET. A circuit using a JFET in an FM tuner at 88 to 108 MHz is shown in Fig. 15-21.

A dual-gate MOSFET can be used in a radio receiver as a converter. The source and the first gate is used in the Hartley-oscillator configuration, and the information signal is fed into the second gate. An IF transformer serves as the tuned circuit in the drain. At very high frequencies, the dual-gate MOSFET can be used as a mixer to isolate the oscillator from the incoming signal. In this use, the information signal is fed into the first gate and a portion of the oscillator output is fed into the second gate.

A block diagram of a typical radio receiver is given in Fig. 25-22.

When interrupted continuous-wave (ICW) telegraph signals are picked up on an ordinary receiver, the dots and dashes cannot be interpreted since the original transmitted signal does not have an audio modulation. In order to produce an audible signal a *beat-frequency oscillator* (*BFO*) is used. Assume that the incoming code-signal carrier frequency is f_x. Energy from the beat-frequency oscillator at a frequency f_y is also injected into the final detector circuit. The detector acts as a mixer, and its output is $(f_x - f_y)$ or $(f_y - f_x)$. The frequency of the beat-frequency oscillator is such that the

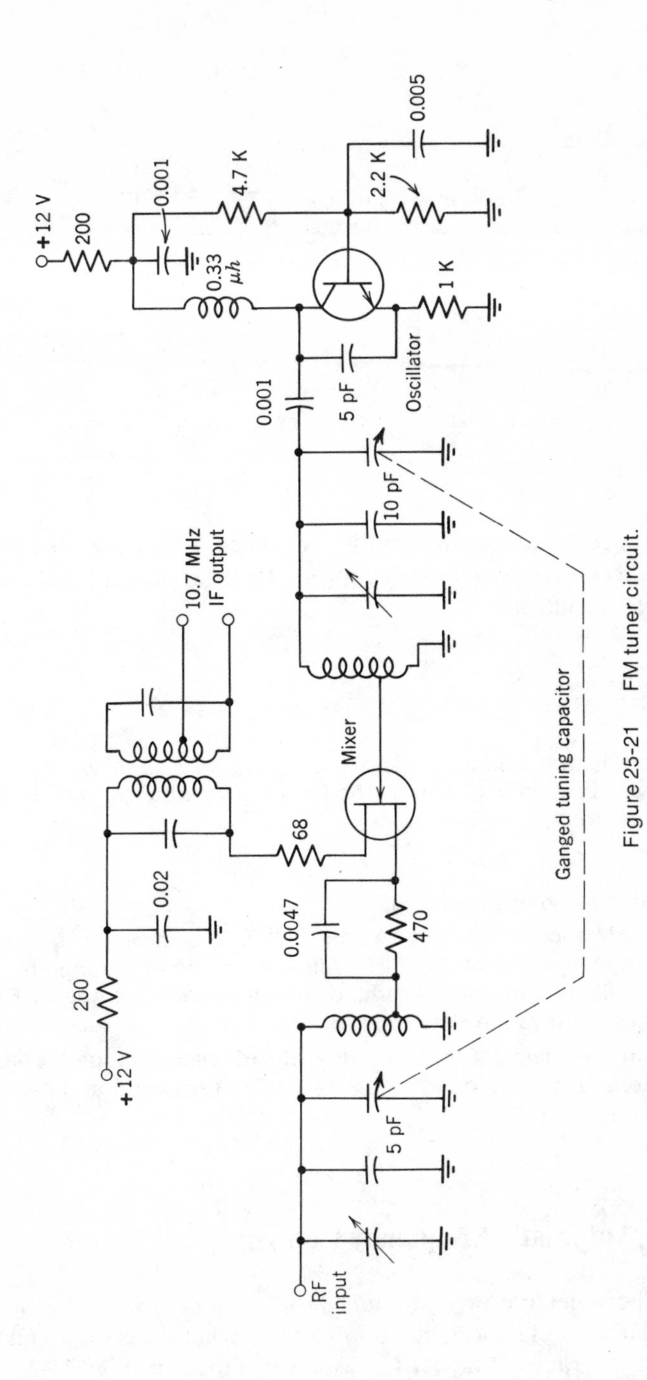

Figure 25-21 FM tuner circuit.

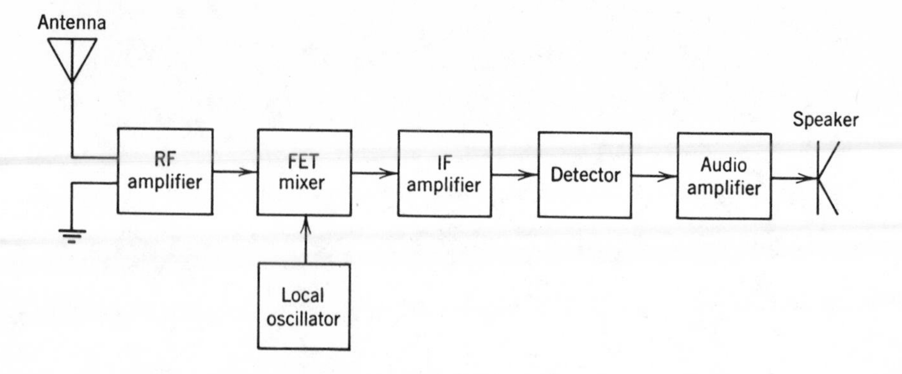

Figure 25-22 Block diagram for a radio receiver.

difference frequency is audible. A control that varies the frequency of the beat-frequency oscillator f_y controls the beat output frequency and is called the *pitch* control.

PROBLEMS

1. Modulation products result when a signal f_s is mixed with an oscillator f_0 to produce an IF output f_1. The frequency of the modulation products can be found from

$$|nf_s \pm mf_0|$$

where n and m are integers.

An FM receiver is tuned to 96.3 MHz. The oscillator operates at 107 MHz producing an IF of 10.7 MHz. Allowing n and m to range from 0 to 10, what modulation products produce interference on a nearby FM receiver that tunes to the range 88 to 106 MHz.

2. Using the data of Problem 1, does the FM receiver tuned to 96.3 MHz produce interference in a nearby space tracking receiver that tunes from 134 to 138 MHz?

Section 25-7 Automatic Frequency Control

The objective of *automatic frequency control* (*AFC*) is to stabilize an oscillator to close tolerance in circuits where a crystal cannot be used. A block diagram of the AFC system is shown in Fig. 25-23. The oscillator

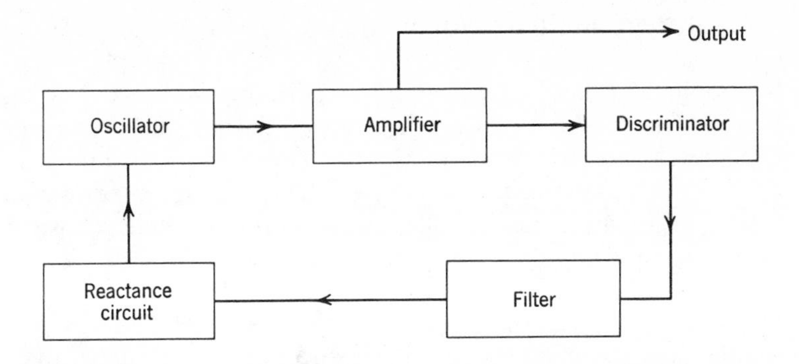

Figure 25-23 Block diagram for automatic frequency control (AFC).

is one of the conventional forms. An amplifier between the oscillator and the discriminator may or may not be necessary, according to the needs of the application. When the frequency of the oscillator is correct, the output from the discriminator is zero. When the oscillator is off the frequency at which the control is made, an output is developed in the discriminator load. This output voltage is negative or positive, according to which way from the center frequency the oscillator has changed. This direct output voltage is filtered carefully and adds or subtracts to a d-c bias on the reactance injected filtered carefully and adds or subtracts to a d-c bias on the reactance circuit. A changing bias on the reactance circuit changes the injected L or C of the circuit and, since the reactance circuit is connected in parallel with the oscillator, the oscillator returns to the proper frequency.

The reactance circuit must be left in the circuit during an alignment procedure since it places a "static" capacitance across the oscillator tank circuit. The connection between the filter and the reactance circuit is disabled, and the normal no-signal bias is maintained on the reactance circuit. Now there can be no correcting action by the reactance circuit. The oscillator is set to the desired frequency, and the discriminator circuit is carefully aligned to give zero output. When the d-c circuit from the filter to the reactance circuit is restored, there is no additional bias on the reactance circuit, and the oscillator will maintain the stable frequency. When one of these circuits is constructed, it should be remembered that, if the phasing of the detector is reversed, the tendency of the AFC action is to shift the oscillator as far as possible away from the desired frequency.

When the oscillator is replaced by a converter or mixer-oscillator circuit, the automatic frequency control can be used to provide automatic tuning for a radio or a television receiver.

Section 25-8 Stereophonic Broadcasting – FM Multiplex

Stereophonic broadcasting, *stereo*, is currently authorized by the FCC for the FM band. The concept of stereo is a means of having a dimensional effect or "presence" to sound reproduction. The original program is picked up by two microphones, one at the left, L, and the other on the right, R, of the stage where the broadcast or recording is made. Thus two separate channels are required, the L channel and the R channel.

If only one microphone had been used in the first place at the center of the stage, a monophonic signal which is equal to $L+R$, is produced. A simple *summer* or *adder* circuit produces the $L+R$ signal that is used to modulate the FM transmitter in the normal fashion. Thus a person with a monophonic receiver hears the $(L+R)$ signal that is the equivalent of a single microphone at the center.

The stereo signal which is broadcast as stereo information is an $(L-R)$ signal obtained from a *subtractor*. The L signal is fed into an adder. The R signal is fed into an inverting amplifier with a gain of one. The signal that is now $-R$ is the second signal in the adder. The output of this second adder is now $(L-R)$.

The problem of stereo broadcasting is to transmit simultaneously $(L+R)$ and $(L-R)$ in such a way that the $(L-R)$ signal is not detected by a monophonic receiver. In a stereo receiver the $(L+R)$ and $(L-R)$ signals are separated. When they are fed into an adder matrix the output is $2L$. When the two separated signals are fed into a subtractor matrix, the output is $2R$. Thus the audio signal $2L$ is available for the left channel and the audio signal $2R$ is available for the right channel.

The allocated FM bandwidth is ± 75 kHz and, in stereo broadcasting, the bandwidth is assigned in a specific manner to the transmitter (Fig. 25-24). The audio program of $(L+R)$ is assigned to 0 to 15 kHz providing a full 15 kHz audio spectrum. The band of frequencies from 23 kHz to 53 kHz, a total bandwidth of 30 kHz, is used for the $(L-R)$ channel. The $(L-R)$

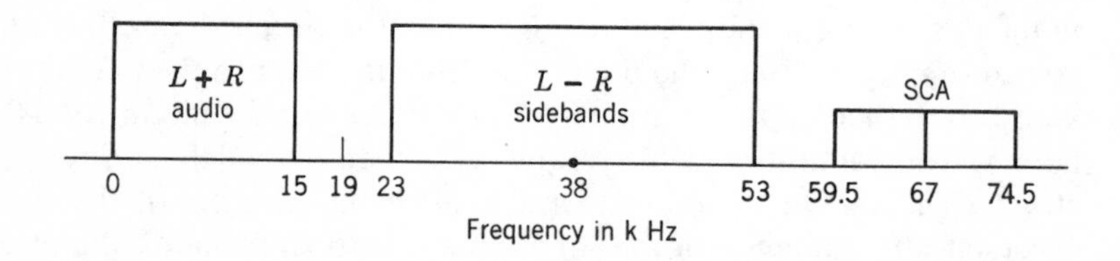

Figure 25-24 FM stereo bandwidth assignment.

signal is fed into a balanced modulator at the transmitter. The other input is a 2×19 kHz or 38 kHz oscillator. The balanced-modulator output is an AM double-sideband suppressed carrier output that has a 15 kHz audio signal capability. If the 38 kHz carrier were also transmitted, an interference problem would result with the SCA band. Consequently, the carrier for the $(L-R)$ information is transmitted at the *submultiple* 19 kHz.

The SCA band is reserved for *S*ubcarrier *C*ommunication *A*uthorization for those transmitters having FCC clearance to employ SCA. The SCA signal usually is information signals that turn off the $(L+R)$ and $(L-R)$ audio systems in a receiver, so that station announcements and commercials are cut out when the program is used for "storecast" purposes in public places.

The carriers at 19 kHz and at 67 kHz are called *subcarriers*. The use of subcarriers is quite common in commercial, military and aerospace applications. The detection of a subcarrier signal has the formal name of *subcarrier demodulation*. The block diagram of the stereo receiver illustrates a subcarrier demodulation.

The normal FM detection process is included in a stereo receiver Fig. 25-25, that produces the $(L+R)$ output. The low pass filter in the $(L+R)$ detector passes only those audio frequencies below 15 kHz, which is the $(L+R)$ signal on the spectrum (Fig. 25-24). Simultaneously the detector output is fed to two other band-pass filters. One passes frequencies between 23 to 53 kHz and the other is a narrow-band tuned circuit that selects the subcarrier 19 kHz. The subcarrier is doubled to 38 kHz and is reinserted with the side bands in an adder matrix. The output of this adder is a true AM signal that is detected and filtered to yield the $(L-R)$ output. The output matrices yield the $2L$ and the $2R$ signals for a pair of stereo speakers. An additional feature is the inclusion of a relay that may be electronic or mechanical. When stereo is transmitted, the 38 kHz signal energizes the relay that is used to operate a stereo indicating lamp. When monophonic programs are transmitted, the 38 kHz signal is absent. The light is off and the relay removes the $(L-R)$ signal line from the adders. Now the output in each speaker is $(L+R)$.

The SCA signal requires a similar channel for detection.

In a stereo receiver, the alignment is very critical as the phase relationships in the subcarrier detector circuits must be maintained exactly. Any slight mistuning in any of these subcarrier circuits results in a degradation of the *separation* between the two outputs, $(2R)$ and $(2L)$. The amount of R signal that is in the L channel can be expressed in db when compared to the L signal. Consequently, stereo separation is usually given in decibels in the sense of rejection.

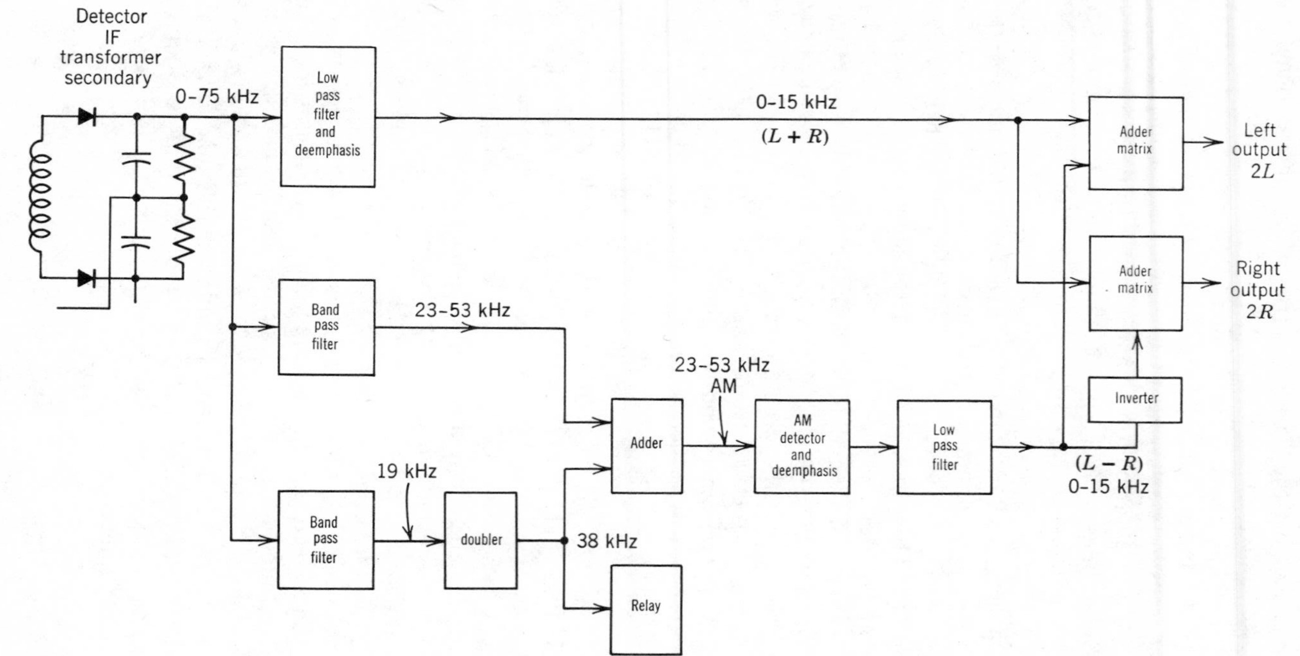

Figure 25-25 Block diagram of a stereo receiver.

Questions

1. Define or explain each of the following terms: (*a*) diode filter, (*b*) diagonal clipping, (*c*) AGC, (*d*) DAGC, (*e*) limiter, (*f*) threshold limiting, (*g*) S curve, (*h*) mixer, (*i*) converter, (*j*) conversion transconductance.
2. Explain how a diode filter separates the audio and the RF.
3. Show how a diode detector is nonlinear, owing to a shifting of the operating point on Fig. 25-3.
4. Explain how diode clipping produces distortion.
5. Explain how a linear AM detector functions.
6. Why is AGC required in a radio receiver?
7. What is ideal AGC action in a radio receiver?
8. What determines the time constant for an AGC filter?
9. When might an AGC amplifier be required in a radio receiver?
10. What characteristic must an amplifying device have if it is to be controlled by AGC?
11. Compare simple AGC to delayed AGC and to ideal AGC.
12. Explain the operation of a limiter circuit.
13. Compare a single limiter with a double limiter.
14. What is the essential characteristic of an amplifier if it is to be used as a limiter?
15. Explain how a discriminator functions.
16. Explain how a ratio detector functions.
17. Point out the ways in which the discriminator and the ratio detector are similar.
18. Point out the ways in which the discriminator and the ratio detector are not similar.
19. Give the alignment procedure in detail for the limiter-discriminator circuit.
20. Give the alignment procedure in detail for the ratio-detector circuit.
21. What is the difference between a mixer and a converter?
22. Compare conversion gain with normal gain.
23. Show how a BFO detects ICW telegraph signals.
24. Explain the principles of an AFC circuit.
25. Explain in detail the steps that must be taken to align an AFC circuit.
26. What is SCA and what is its purpose?
27. What is stereo and what is its purpose?
28. Explain how a subcarrier is used in FM multiplex.
29. How is subcarrier demodulation accomplished in an FM multiplex receiver circuit.
30. What is $L+R$ and $L-R$ in stereo broadcasting?
31. Explain the function of each block in Fig. 25-25.

ANSWERS TO SELECTED PROBLEMS

Chapter 2

Section 2-2. **1.** 141 Ω; 0.383 A. **3.** 1N1095, 1.6 Ω, 100 M.

Section 2-3. **1.** 0, 21, 33, 44, 53, 62 V. **3.** 150 ma; 40 V.

Section 2-4. **1.** 80 Ω; 50.6 Ω. **3.** 19.00 V; 13.15 V.

Section 2-5. **1.** *a* Sinusoid −10 V loops only. *c* Sinusoid + loop clipped at 2 V, −8 V loop. **3.** Forward (0 A, 0.7 V), (1.18 A, 4.24 V); reverse (0.298 ma, 29.8 V), (167 ma, 30 V). **5.** 0.72 mv. **7.** 3.88 mv. **9.** 220 μv.

Section 2-6. **1.** 281 V; 193.7 to 206.3 V; 127.5 to 172.5 ma. **3.** 635 Ω; 155.7 to 184.3 V; 52.5 to 97.5 ma.

Chapter 3

Section 3-3. **1.** 0.994; 160. **3.** 2.265 to 3.615 μa. **5.** 0.980, 0.990, 0.992, 0.993, 0.995. **7.** 199, 99, 79, 41.6.

Section 3-4. **1.** 0.99; 100. **3.** 60.45 mw. **5.** 0.275 mw; 60.3 mw.

Section 3-10. **1.** −35; 10 kΩ; 3500 μmhos. **3.** 9760 Ω; 4670 μmhos; −70; 7140 Ω; 5650 μmhos; −48.

Chapter 4

Section 4-3. **1.** 8.40, 5.85, 3.75, 2.10, 0.90, 0 ma. **4.** 3.33, 2.22 mmhos. **5.** 0.9, 1.8, 3.1, 4.8, 6.7, 9.0 ma. **9.** 0.73, 1.64, 2.55, 3.46 mmhos.

Section 4-6. **1.** 167; 64.5; 21.9 pF. **3.** 7.3 V reverse; 0.175 V forward.

Section 4-8. **1.** 40.25 Ω; 17.1 V; 1.11 Ω. **3.** Breakdown at 10.4°. **5.** Breakdown at 10.4°; 190.4°.

Chapter 6

Section 6-1. **1.** 2.63/1; 246 W; 349 VA. **3.** (V_{in}, V_{out}), (−100, 0), (0, 0), (100, 100).

Section 6-2. **1.** 246.5 V; 24.65 ma; 64.5 ma; 7.5 W; 8.8 VA. **3.** 117/(22.2-0-22.2) V; 123 W; 144 VA. **5.** (V_{in}, V_{out}), (−60, 50), (0, 0), (60, 50) V.

Section 6-3. **1.** 225 V; 22.5 ma; 53.4 ma; 6.25 W. **3.** 117/22.2 V; 494 VA; 4.21/22.2 A.

Section 6-4. **1*a*.** 117/26.7 V; 590 W; 838 VA. **1*c*.** 117/13.35 V; 296 W; 296 VA.

Section 6.5. **3.** 141 V; 139.3 V; 0%; 0.68%.

Section 6-7. **1.** 245 V; 7.5 H; 11 μF. **3.** 134 V; 40.2 ma; 0.2%.

Section 6-8. **1.** 6.9 H; 289 V; 96.3 ma.

Section 6-10. **1.** 10 μF, peak output −200 V; 10 μF, peak output 0 V.

Chapter 7

Section 7-1. **1.** 63.6 ma to 0; 54.4 ma; 31.8 ma; 9.4 ma. **3.** 82° to 117.5°. **5.** 41 Ω; 0.56 to 1.275 A; 0° to 97°.

Section 7-2. **3.** 1885 Ω. **5.** 106 kΩ.

Section 7-3. **1.** 64.3°; 3.80 A; 38 V; 0.72 ma. **3.** 14°; 0.868 A; 52 V; 1.61 ma. **5.** 1.09, 10.9, 109, 1090, 10900 Hz.

Section 7-4. **1.** 6.9°; 7.1 kΩ 14.5 kΩ; 19.0 kΩ; 21.8 Ω. **3.** 24°; 90°; 156°. **5.** 1.17, 0.974, 0.850, 0.826, 0.802, 0.645, 0 A; 137, 95, 72.4, 68.5, 64.5, 41.9, 0 W.

Section 7-5. **1.** (E_c, I_b), (−20, 0), (−15.6, 0), (−15.6, 10.7), (−10.3, 18.7), (0, 21.5 ma). **3.** 44.8°.

Chapter 8

Section 8-3. **1*a*.** 3.423. ***c*.** 5.879. ***e*.** 18.468. ***g*.** −0.0584. ***i*.** −1.329. **2*a*.** 288. ***c*.** 1.585×10^{14}. ***e*.** 1.0568. ***g*.** 1.000. ***i*.** 1.155.

Section 8-4. **1.** 3.35 V. **3.** 1.13 mv. **5.** 1db/100 ft. **7.** 0.266 W; +16.5 db. **9.** 238 db space loss; 1.58 μv.

Chapter 9

Section 9-1. **1.** 31.6 V, 31.6 ma; 354 V, 354 μa; 369 V, 5.4 ma; 36.8 V, 13.6 ma. **3.** 14.1, 22.4, 31.6, 71, 100 ma. **5.** 15, 20 W. **7.** 30.1 ma.

Section 9-3. **1.** 1.13, 11.3, 113, 1130, 11300 Ω; 1.00, 0.970, 0.405, 0.040, 0.005. **3.** 11.5%. **5.** 1.65 mh. **7.** 16.3 mh.

Section 9-5. **1.** 3920 pF. **3.** 0.0868 μF. **5.** 350 pF. **7.** 0.03%; 7.57 MΩ. **9.** 3.75 Ω; 175 Ω; 8170 Ω.

Chapter 10

Section 10-1. **1.** 1.25 pF; 3.96 mv. **3.** 39.4 pF. **5.** $R = X_L$ at 400 kHz. **6.** $R = X_L$ at 1570 kHz.

Section 10-2. **3.** 25.5 kΩ. **5.** 56.2 kΩ. **7.** 10.4. **9.** 900 Ω; 33.3 ma.

Section 10-3. **1.** 86 kHz. **3.** 60 kHz.

Section 10-4. **1.** 204.4 + j 3535 Ω; 2.82 ma at −86.5°. **3.** 182 − j 40 Ω; 53.5 ma at 12.4°. **5.** 410 + j 840 Ω; 21.5 ma at −63.9°.

Section 10-5. **1.** 0.8 ma at −18.5°. **3.** 0.528 ma at −4.5°.

Section 10-6. **1.** 3.43 ma; 30 V. **3.** 11.1 μh; 2.5 mv. **5.** 27.5 ma at 12°.

Section 10-8. **1.** 1500 Ω; 500 μmhos; $-\frac{1}{2}$; $\frac{1}{2}$. **3.** 10.656 Ω; 48.5 μmhos; −0.0338; 0.0338.

Chapter 11

Section 11-1. **1.** 1.7 A; 69 V; 51 V.

Section 11-2. **1.** (I_C, V_{CB}), (2.8, 13), (1.85, 19), (0.90, 25); 0.71 ma, rms; 0.674 ma, rms; 4.24 V, rms. **3.** (I_C, V_{CB}), (1, 40), (2, 20), (4, 10), (8, 5).

Section 11-3. **1.** (I_C, V_{CE}), (31, 11.5), (21, 17.5), (11, 23.5); 141 μa, rms; 7.1 ma, rms; 4.24 V, rms. **3.** (I_C, V_{CE}), (20, 35), (35, 20), (50, 14), (70, 10).

Section 11-4. **1.** 1.8 ma; 11 V. **3.** 0.5 ma; 10 V. **5.** (I_D, V_{GS}), (3.6, 0), (3.2, −1.0), (2.5, −1.5), (1.8, −2.0), (0.9, −3.0), (0.3, −4.0), (0, −5.0).

Section 11-5. **1.** 2960 Ω; 27 μF; 15. **3.** (I_b, E_c), (4.4, 0), (4, −2), (3, −6.5), (2.7, −8), (2, −12), (1, −18), (0, −28).

Section 11-7. **1.** 8.4, 4.8, 3.3, 10.2%; 68.8 W; 8.32 A; 41.4%.

Section 11-8. **1.** 3.5 kΩ; 326 kΩ. **3.** 2.5 kΩ; 256 kΩ. **5.** 2 kΩ; 247 kΩ. **7.** 10.4 μa; 0.728 ma; 5.61 V. **9.** 512 kΩ. **11.** 258 kΩ. **13.** 213 kΩ; 5.48 kΩ. **15.** 1.00 V. **17.** 600 kΩ. **19.** 4.95 kΩ; 269 kΩ. **21.** $R_B = 62.6\ R_c - 6.2$ kΩ.

23. 12.96 V; 12.1 kΩ. **25.** 30.5 kΩ. **27.** 43 kΩ. **29.** 9010 Ω.

Section 11-9. **1.** 1 ma; 511 kΩ.

Chapter 12

Section 12-1. **1.** 39.4, 26.0, 18.5, 13.2, 5.6 db.

Section 12-2. **1.** 312; 50 mv, peak-peak. **3.** 0.483 ma; 10.34 V; 138.5. **5.** 72.5 μa; 2.55 V; 268 mv; 88 mv.

Section 12-4. **1.** 5570 Ω; 149; 52.6. **3.** 45.7. **5.** 129 kΩ; 13.96 kΩ; 5.81. **7.** 1.30 MΩ; 1.97 V, peak-peak.

Section 12-5. **1.** 5630 Ω; 89. **2.** 7680 Ω; 36. **5.** 1850 Ω; 44.8. **7.** 82 mv, peak-peak. **8.** 374 kΩ; 100 mv, peak-peak.

Section 12-6. **1.** 1370 Ω; 109.8. **2.** 2070 Ω; 56. **3.** 28.5. **6.** 4 V, peak-peak; 21.8; 184 mv, peak-peak.

Section 12-7. **1.** 17.7 μa; 1.98 V. **3.** 0.22 ma; 0.72 V. **5.** 1.19 V.

Section 12-8. **1.** 1000 μmhos; 5 V, V_{DS}; 2 V, peak. **2.** 6.25%. **5.** 2400 μmhos; 8 V, V_{DS}; 24. **6.** 5%. **8.** 1.818 V, V_{GS}; 1970 μmhos; 11.18 V; 19.7.

Section 12-9. **1.** 30. **3.** 33.75. **5.** 3.46 + j 1.52; 5.66 + j 13.76.

Chapter 13

Section 13-1. **1.** 0.95 V; 1.9 ma. **2.** 2.5 mw; 1600 Ω; 1.25 ma; 2.0 V. **3.** 1731 Ω; 1.03 mw; 0.77 ma; 1.34 V. **8.** 126.5 Ω; 834 kΩ; 456; −0.964; −439. **9.** 395 V; 981 kΩ; 1215; −0.774; −940. **12.** 993; −0.905; −899. **13.** 100.6 Ω; 316 kΩ; 1630. **15.** 187 Ω; 376 kΩ; 1685.

Section 13-2. **2.** −16.3; 63; −1025. **3.** 1700 Ω; 42 kΩ; 22,000. **6.** 2620 Ω; 112.5 kΩ; −180; 47.3; −8525. **7.** 1720 Ω; 102.6 kΩ; −1230; 20.5; −25,200.

Section 13-3. **3.** 99.3 kΩ; 1904 Ω; 1.00; −48.2; −48.2. **4.** +17.8 db.

Chapter 14

Section 14-2. **1.** 6.54 A; 1735; 109.3. **3.** 13.75; 57.6 ma; 12.6 V. **5.** 387; tap at 40, 49, 127. **7.** 300; 89; 32; 13.

Section 14-5. **1.** 1.5 W; 5.15; 9850 Ω. **3.** 10 W; 0.5 A; 7940 Ω. **5.** 15.6 V; 112 ma; 875 mw. **7.** 8 W; 400 ma; 5950 Ω; 800 ma; 80 V. **9.** 9650 × 30,000 =

$+84.6$ db; 68.3×10^{-12} W; 0.566 V; 61.35×10^{-6} V; 1 to 2.86; 1 to 6.6; 1 to 47.2.

Section 14-6. **1.** 8.43 W; 23.1 W; 36.5%. **2.** 12%.

Chapter 15

Section 15-1. **3.** 700 kΩ; 6 V, peak-peak; 50 mv, peak-peak. **5.** 1.31 MΩ; 20 V, peak-peak; 110 mv, peak-peak. **7.** 1.48 MΩ; 23.8 V, peak-peak; 72.5 mv, peak-peak. **9.** 1.345 ma; 215.5 V; 161 V, peak-peak.

Section 15-2. **1.** 534. **3.** 11.1 mv; 8.8 V, peak-peak; 180,000.

Section 15-3. **1.** 13.37; 0.265 μF; 0.492 μF; 6.69 A_e at 90° lead. **3.** 0.53 μF; 0.53 μF; 0.40 μF. **5.** 25.3; 20.6 Hz; 156; 201 Hz.

Section 15-4. **1.** 33 kHz; 800 pF. **3.** 3.18 MHz; 39.3 kHz.

Section 15-5. ***1a.*** 38.5. ***c.*** 49 kHz; 50.5 kHz. ***2a.*** 30 kΩ. ***c.*** 0.57 μF. ***e.*** 8.1. ***g.*** 2.25 MHz. ***i.*** 10.5. **3.** $-101.2\,A_e$; $36.1\,A_i$; $-23.7\,A_e$; $39.1\,A_i$; $2400\,A_e$; $1411\,A_i$; $475\,A_{e,\mathrm{LF}}$; $278\,A_{i,\mathrm{LF}}$; $700\,A_{e,\mathrm{HF}}$; $410\,A_{i,\mathrm{HF}}$. **5.** 20.8, 15.6, 325 A_e; 28 Hz; 59 Hz; 81.4 kHz.

Section 15-6. ***1a.*** $\omega_0 = 10^5$. ***c.*** $\omega_0 = 0.5$. ***e.*** $\omega_0 = 0.05$. **2.** $\omega_1 = 0.835 \times 10^4$; $\omega_2 = 5 \times 10^4$.

Chapter 16

Section 16-1. **1.** 2.24 V, peak-peak; 172 μa; 4.3 μa; 52.5 kΩ; 75. **3.** 460 Ω; 1920 Ω; 392 Ω; 2000 Ω; 4550 Ω; 4800. **5.** 9.4 kΩ; 2110 Ω; 300 Ω; 59.8.

Section 16-2. **1.** 1.87 MΩ; 3.6 V, peak; 3.28 μa, peak. **3.** 94.4 kΩ; 0.518; 5250; 1.91 Ω. **5.** 145 to 13.8 kΩ.

Section 16-3. **1.** 1.465 ma; 18.1 μa; 1.447 ma; 16.2 V; 221; 131 mv, peak-peak. **3.** 0.825 ma; 8.15 μa; 0.817 ma; 6.10 V; 9.2 V, peak-peak; 52; 177 mv, peak-peak.

Section 16-4. **1.** 0.352 ma; 4.35 μa; 0.348 ma; 9.0 V; 14.0 V, peak-peak; 5130 Ω; 282. **3.** 0.415 ma; 5.14 μa; 0.410 ma; 6.6 V; 8.2 V, peak-peak; 3280 Ω; 166.

Section 16-5. **1.** 1.38 ma; 13.7 μa; 1.37 ma; 21.3 V; 12 V; 18.6 V, peak-peak; 1690 Ω; 188. **3.** 0.217 ma; 2.15 μa; 0.215 ma; 10.75 V; 6.45 V; 8.6 V, peak-peak; 7430 Ω; 86.

Section 16-6. **1.** 106 db. **3.** 20 V desired, 21.5 V actual peak-peak.

Section 16-7. **1.** 1.0 μa; 18.8. **3.** 0.707 ma; 0.487 ma; 1.29 ma; 2870 Ω; 2500; $+4.42$ to -4.55 V; ± 1.75 V.

Section 16-8. **1.** 29.8 V; 54.7 V, peak.

Chapter 17

Section 17-1. **1.** 3.285, 6.57 ma.

Section 17-2. **1.** 6.57 ma. **3.** 1.76, 3.87 ma; 14.69, 11.66, 8.33 V. **5.** 76 to 127. **7.** 35 to 145. **9.** 1.64 ma. **11.** 4450 Ω; 330 kΩ; 0.414; 0.917 ma; 4.5 V. **13.** 0.998 ma; 308 kΩ; 0.506; 0.306; 0.797 to 1.120 ma.

Section 17-3. ***a.*** 92°C. **c.** 65°C. **e.** 103°C.

Section 17-4. **1.** 114.5°C. **3.** 88.9°C. **4.** 51.4 μa; 44.4 μa. **7.** 244 kΩ; 87°C. **8.** 5.85 μa; 10 μa. **11.** 7.28 MΩ; 49°C. **12.** 1.28 μa; 1.0 μa. **15.** 1.265 μa; 7.35 MΩ; 49.7°C; 0.993 μa. **16a.** 67°C. **c.** 44°C.

Section 17-5. **1.** Q1, 5 ma, 13.8 V, 4 A_e; Q2, 3.2 ma, 17.1 V, 4.8 A_e; Q3, 1.6 ma, 20.6 V, 6.0 A_e. **2.** 14.8, 10, 4.6 V, peak-peak. **3.** Q1, 5.4 ma, 10 V, 2.5 A_e; Q2, 4.7 ma, 12.1 V, 3.2 A_e; Q3, 4.1 ma, 13.8 V, 4.8 A_e. **4.** 11.4, 9.4, 2.0 V, peak-peak.

Section 17-7. **1.** 75°C/W; 10°C/W. **3.** 2.5 mw/°C. **5.** 0.3°C/mw. **7.** NC441 with 5°C reserve. **9.** 8°C, 1.36°C, 0.37°C drop.

Chapter 18

Section 18-5. **1.** 102, A_e; 124 kΩ, R_1; 495 kΩ, R. **3.** 1.97 V; 1.94 V.

Section 18-6. **1.** 75 ma, I_{CQ}; 24 kΩ, R_B; 80.5 mv, peak, E_s; 2.25 W; 4.50 W; 50.4 μw; +46.5 db. **3.** 200 W, P_L; 3.18 A; 40 W, each transistor; 2.70 V, E_s; +39.5 db. **5.** 24.6 mw; 2.46 V, peak, E_s; +27.1 db; 5 mw.

Section 18-7. **1.** ±43.9 V; 22.6 ma, I_b peak; 43.9 V, E_s peak; 500 mw; 8.6 W; 22.6%; 6.6 ma, I_b peak; 12.64 V, E_s peak; 41.6 mw. **2.** 4.14 mw; 0.35 mw. **3.** ±25.3 V; 0.502 A.

Chapter 19

Amplifier A. 89° I_b angle; 76° I_c angle; 0.987 A, I_b; 156 ma, I_c; 10.9 kW, P_0; 85%; 6450 Ω; 538 W, grid drive; +13.1 db.

Amplifier C. 92.6° I_b angle; 89° I_c angle; 98.2 ma, I_b; 26.8 ma, I_c; 69.1 W, P_0; 70.5%; 4080 Ω; 30.6 W grid drive; +3.5 db.

Amplifier A – Doubler. 25.5% I_b; 9.04 kW, P_0; 3.79 kW, dissipation; 70.2%; 7700 Ω; +12.2 db.

Chapter 20

Section 20-2

Problem	E_s	β_f	$\beta_f A_e$	E_f	E_i	E_0'
1.	20 mv	+2%	0.40	13.3 mv	33.3 mv	667 mv
3.	267 mv	+4%	0.60	400 mv	667 mv	10 V
5.	1 V	+0.2%	0.20	0.25 V	1.25 V	125 V

Section 20-3

Problem	E_s	A_e	β_f	$\beta_f A_e$	E_f	E_i	E_0'
1.	200 mv	20	−15%	−3	−150 mv	50 mv	1 V
3.	250 mv	50	−3.0%	−1.5	−150 mv	100 mv	5 V
5.	0.5 V	20	−20%	−4.0	−0.4 V	100 mv	2 V
7.	1 V	100	−3%	−3.0	−750 mv	250 mv	25 V

Section 20-3. **8.** 8, 40, 20%; 80; 3.6%. **10.** 9.51; 1.05 MHz.

Section 20-4. **1.** 01. %. **3.** 8.56 kΩ. **6.** 16.1 sin $(\omega t - 159°)$;

Section 20-5. **1.** 930 kΩ, R_B; 2700 A_e; 5.4 V; 135 A_e'; 0.27 V. **3.** 2.89, 97.1 kΩ.

Section 20-6. **1.** $R_1 + R_2 = 2000\ \Omega$; 10 to 3.33.

Section 20-7. **1.** 5420 Ω; 41.5. **3.** 340.

Section 20-8. **1.** +52.4, +12.8 db; 800 μF; 0.2 Hz. **4.** +37.0, +12.2 db; 141 μF; 1.13 Hz.

Section 20-9. **1.** 33.5; 29.3. **3.** 15.5; 9.7.

Section 20-10. **1.** (I_b, ma; E_0'; V; E_s, V), (2,4,−3.6); (10; 20; + 18); (20; 40; +40); (30; 60; +64.5).

Section 20-12. **1.** 3.66 Ω; 44 Ω; 11.24 V to 12.75 V. **2.** 91.6 ma at 72 Ω; 124.3 ma at 51.5 Ω; 272 ma. **5.** 153.5 Ω; 6.73 Ω; 6.16 V; 0.131 Ω; 14.1 to 9.3 V. **6.** 143 ma at 43.5 Ω; 1710 ma at 3.57 Ω; 3.88 A. **7.** 955 Ω; 830 Ω; 173.5 Ω; 895 ma; (22.4 V, V_{in}; 13.30 V, V_L; 920 ma, I_L); (16.4 V, V_{in}; 12.60 V, V_L; 871 ma, I_L). **9.** 140 Ω; 3.90 Ω; 21.7 to 26.3 V. **10.** 92 ma at 71 Ω; 125 ma at 51.2 Ω; 172 ma. **13.** 918 ma; 6.72 Ω; 487 Ω; 0.135 Ω; 16.4 to 31.5 V. **14.** 148 ma at 42 Ω to 1690 ma at 3.6 Ω; 2.48 A.

Chapter 21

Section 21-1. **3.** 130, 267, 654 Hz.

Section 21-2. **1.** 5300 to 483 Hz. **3.** 318 Hz; 213. **5.** $A_e = \pm(30 + 3R_2)/(2R_2 - 10)$.

Section 21-3. **1.** 5050 pF; 12.5. **3.** 122.5 mv. **5.** 0.392 ma; −2.75 V.

Section 21-4. **2.** 360 μmhos; 0.173 ma; −3.04 V.

Section 21-5. **3.** 843 kHz.

Section 21-6. **1.** 387 Ω; 207,000 MΩ. **3.** 230 Ω; 145,000 MΩ. **5.** Series no effect; 900 Hz.

Chapter 22

Section 22-1. **1.** 7.06, 6 millisec.; 7.86, 6 millisec. **3.** 785 Hz. **5.** 150 V, 0.2 sec; −300 V, 0.1 sec.

Section 22-2. **1.** 862 μsec, 10 V; 1138 μsec, 2 V. **6.** 2080 Ω.

Section 22-3. **1.** 0.466 ma, I_C; 0.494 ma, I_E; 27.9 μa, I_B. **3.** 0.651 ma, I_C; 0.666 ma, I_E; 14.5 μa, I_B.

Section 22-4. **1.** 38.2 millisec.

Section 22-5. **1.** 136 Hz. **3.** 1.54 MΩ; $T_1 = 0.71$ RC; 4680 Hz. **5.** 0.81, 8.1 millisec; 112 Hz **7.** 15, 3 millisec; 55.5 Hz.

Section 22-9. **1.** Answers vary: 140 mv, 100 Ω; 500 mv, 100 Ω; 900 mv, 600 Ω; 160 mv, 20 Ω. **3.** Answers vary: 70 Ω R; 140 mv switched to 540 mv.

Section 22-10. **1.** +192 V output.

Chapter 23

Section 23-1. **1.** 19.5 MHz. **3.** 1 MHz. **4.** 390 MHz. **6.** 81 MHz. **7.** 3.44 MHz; 153; 527. **9.** 7.69 MHz; 217; 1650. **11.** 3.32 MHz; 114.2; 377.

Section 23-2. **1.** 39 at 12.7° lag. **3.** 30.2 at 25° lag. **5.** 2.3 at 81.7° lag. **7.** 1.11 at 42.5° lag. **9.** 12.7 at 0°. **11.** 1.46 at 0°.

Section 23-3. **1.** 277; 33.5 kHz. **3.** 69.4; 269 kHz. **5.** 77.5; 460 pF.

Chapter 24

Section 24.2. **1.** 3784 Hz. **3.** 31.25 W. **5.** 361 W. **7.** 740, 1376 V; 529, 79.3 V. **9.** 220, 1972 V; 548, 219.2 V. **11.** 1.41 A, 90 W. **13.** 10.68 A; 12.8 A. **15.** 76%.

Section 24-3. **1.** 3500, 1414, 1575 W; 1030 Ω.

Section 24-4. **1.** 11.75; 13.45 kHz; 33.6. **3.** 0.36 V. **5.** 15.1, −27.7, −24.3, 11.5, 35.8, 36.2 V; 136 V. **7.** −4.8, 49.7, 44.6, 21.6, 7.4, 2.0 V; 98.4 V.

Section 24-5. **3.** 44.38 pF; 55.5 μh.

Section 24-8. **3.** 50 sin 2 π(20 kHz) t. **5.** Suppressed carrier; from a balanced modulator.

Chapter 25

Sections 25-1. **1.** 59%.

Section 25-3. **1.** 80 db; 22.8 to 6.8 db; 8 V. **3.** 9.15.

Section 24-4. **1.** ±4.63, ±6.59, ±8.37, ±10 V.

Section 24-5. **1.** ±1.05, ±1.65, ±2.01, ±2.50 V.

Section 25-6. **1.** $9f_s - 8f_0 = 10.7$; $9f_0 - 9f_s = 96.3$;
$10f_s - 8f_0 = 10f_0 - 10f_s = 107$.

INDEX

α_{ac}, 62
α_{dc}, 58
Acceptor atom, 17
Active filter, 604
Adder, FM stereo, 804
Adder amplifier, 604
Adjustment class-C amplifier, 578
Adjustment class-C neutralization, 584
AFC, 802
AGC, 786
Air-core transformer, 268
  tuned primary, untuned secondary, 269
  untuned primary, tuned secondary, 268
Air dielectric capacitor, 247
Alpha, 58, 62
Alpha frequency cutoff, 432
AM, *see* Amplitude modulation
Ambient temperature, 526
Amplification curve, universal, 426
Amplification factor, triode, 81
Amplifier, adder, 604
  class-AB push-pull, 547
  class-B push-pull, 547
  class-C, 565
  commercial audio, 559
  common-base, *h* parameters, 371
    load line, 283
    model, 328
  common-collector, *h* parameters, 388
    model, 352
  common-emitter, *h* parameters, 382
    load line, 289
    model, 334
  complementary symmetry, 451
    push-pull, 555
  Darlington pair, 455
  differential, 460
    equation summary, 476
  feedback, 589
  FET, graphical analysis, 513
    load line, 293
    model, 358
  grounded grid, 481
  high frequency, FET, 725
    transistor, 717
  hybrid-π model, 717
  operational, 600
  phase inverter, 543
  push-pull, 535
  resistance-capacitance coupled, 413
  simple transistor, 67
  transformer coupled, 397
    load line, 405
  tuned circuit, 728
  vacuum tube, load line, 296
    model, 365
    transformer coupled, 411
Amplitude distortion, 303
Amplitude modulation, 740
  circuits, 749
  definition, 740
  detectors, 777
Analysis, numerical, class-C, 572, 575, 581
Analytic approach, distortion, 304

load lines, 321
Angstrom unit, 10
Anode, 36
Arc discharge, 21
Armstrong FM transmitter, 770
Armstrong oscillator, 647
Astable multivibrator, 692
Atom, 2
Attenuator, 447
Audio amplifier, commercial, 559
Audio choke, 240
Audio mixing, 604
Automatic bias, 567
Automatic frequency control, 802
Automatic gain control, 786
Automatic volume control, 787
Avalanche, 33
AVC, 787

β, 63
Back current, 30
Balance control, 537
Balanced modulator, 767
Band, semiconductor, 7
Bandwidth, amplifier, 440
feedback, 596
gain-bandwidth product, HF, 723
n stages, 730
tuned circuits, 262
Barrier, 28
Base, 55
Baud, 743
Beam-power tube, 143
Beat detector, 798
Beat-frequency oscillator, 800
Bel, definition, 213
Bessel functions, 757
Beta, 63
Beta frequency cutoff, 433, 720
Beta stability, 485
circuit analysis, 490
Bias, clamp, 181, 565
oscillator, 650
diode, 518
FET, 293, 513
grid-leak, 565
oscillator, 650
transistor circuits, 310
vacuum-tube circuits, 296
Bidirectional dipole thyristor, 127
Bidirectional triode thyristor, 131
Bipolar junction transistor, BJT, 55
Bistable multivibrator, 685
Bleeder resistor, 171
Blocking oscillator, 703
Bode plots, 439
Bohr model, 2
Breakdown diode, 41, 123
Breakdown region, transistor, 72
Breakdown voltage, FET, 105
Break frequency, 441
Breakover voltage, 124
Breakpoint, 441
Bridge oscillator, 642
Bridge rectifier, 157
Bulk resistance, 47
Butterfly capacitor, 658
Bypass capacitor, analysis, 618
FET amplifier, 295
vacuum-tube amplifier, 296

Capacitance, barrier, 34
junction, 32
transistor, 88
Capacitor, 242
butterfly, 658
bypass, analysis, 618
commutator, 684
diode, 117, 442, 762
equivalent circuit, 244
filter, 163
formula, 242
integrated circuit, 246
metallized, 246
parallel resistance, 244
series resistance, 244
split-stator, 657
Carbon potentiometer, 232
Carbon resistor, 228
Carrier, 740
Carrier suppression, 772
Case temperature, 526
Cathode, 9, 36
Cathode-coupled multivibrator, 698
Cathode fall, 22
Cathode follower, 625
Ceramic capacitor, 245
Channel, FET, 100
modulation, 740
Charge-discharge of a capacitor, 678
Child's law, 39
Chip, 93, 96

Choke, 240
Choke filter, 169
Clamp, 179
 SCR circuit, 194
Clapp oscillator, 658
Class-AB push-pull, 547
Class-B amplifier, definition, 478
Class-B push-pull, 547
Class-C amplifier, 565
 adjustment, 578
 grid circuit analysis, 575, 581
 harmonic generation, 580
 load line, 571
 neutralization, 584
 plate circuit analysis, 572, 581
Clipping circuits, 478
Closed-loop gain, 591
Coefficient of coupling, 270, 398
Coil, equivalent, circuit, 238
 integrated circuit, 241
 parallel resistance, 238
 series resistance, 238
 tapped, 259
Cold-cathode diode, 51
Collector, 55
Collector dissipation, 71
 optimum operating point, 515
Collector resistance, 346
Collector-to-base feedback, 342, 348, 616
Colpitts oscillator, 655
Commercial audio amplifier, 559
Common-base amplifier, 59, 67
 h parameters, 371
 load line, 283
 model, 328
Common-collector amplifier, 69
 h parameters, 388
 model, 352
Common-emitter amplifier, 62, 68
 h parameters, 382
 load line, 289
 model, 334
Common-mode rejection, 470
Commutator capacitor, 684
Comparison of basic rectifier circuits, 159
Compensation, diode biasing, 518
Complementary symmetry, 451
 push-pull, 555
Conduction band, 8, 13
Conductivity (light), 18
Conductor, 8
Constant-current curves, triode, 76, 571
Constant-current stabilization, 472
Controlled rectifier, phase shift, 189
 principles, 183
Conversion, frequency, 798
Conversion gain, 799
Conversions, α to β, 65, 66
 transistor current, 311
Converter, 800
Coupled circuit theory, 265
Coupling, critical, 270
Coupling impedance, 266
Critical coupling, 270
Critical grid curve, thyratron, 206
Cross-over distortion, 553
Crystal, quartz, 663
 structure, 12
Crystal filter, 161
Crystal oscillator, 661
Current feedback, vacuum tubes, 622
Cutoff, in transistors, 285
 in triodes, 74
Cutoff current, 30, 59, 65
Cutoff frequency, 161
 alpha, 432
 beta, 433, 720
Cutoff line, triode, 78

DAGC, 788
Damping factor, 569
Dark current, 21
Darlington pair, 455
dbm, 219
Decade, Bode plot, 441
Decade amplifier, 596
Decibels, Bode plots, 439
 calculations, 216
 definition, 213
 reference, 219
Decoupling, 628
Decoupling filter, 630
Deemphasis, 772
Degenerative feedback, 594
Deionization time, 22
Delayed AGC, 788
Depletion, *P-N* junction, 29
Depletion region, 32, 34
Depletion-type MOSFET, 108
Deposited carbon resistor, 229
Derating, resistors, 228

thermal in transistors, 526
Detection, 777
Detector, diode, 777
grid-leak, 785
linear, 783, 786
square law, 783
Deviation frequency, 754
Deviation ratio, 754
DIAC, 127
Diagonal clipping, 781
Dielectric, 243
Dielectric strength, 245
Difference amplifier, *see* Differential amplifier
Differential amplifier, common-mode rejection, 470
constant-current stabilization, 472
dual input, balanced output, 464
unbalanced output, 468
equation summary, 476
operational amplifier, 605
single input, balanced output, 460
Differentiating amplifier, 605
Diffused-alloy transistor, 88
Diode, 27
biasing and compensation, 518
capacitor, 117, 442, 762
characteristics, dynamic, 44
cold cathode, 51
light emitting, 52
mixing, 799
model, 44
vacuum, 36
varactor, 117, 442, 762
Zener, 41
Discharge circuit, 706
Discharge of a capacitor, 678
Discriminator, 792
Dissipation, class B, 550
collector, 71
optimum operating point, 515
Distortion, 273, 303
analysis, 304
feedback, 594
push-pull, 537
Donor atom, 15
Doping, 15
Double-based diode, 120
Double breakdown diode, 44
Double ionization, 20
Double limiter, 791
Doubler, 580
Drain, 99
Drain resistance, 103
Drift transistor, 115
Dual insulated gate MOSFET, 112, 800
Dynamic coefficients, triode, 80
Dynamic diode characteristic, 44
Dynamic load line, R-C coupling, 415
Dynamic plate resistance, 40, 82
Dynamic transfer curve, 288, 301

ECO, 671
Eddy current, 402
Edison effect, 37
Efficiency, class-A, 287, 288
transformer, 407
class-B, 551
class-C, 570
Electron, 2
Electron-coupled oscillator, 671
Electron flow in solids, 6
Electron-hole pair, 14
Electron-volt, 8
Electrostatic shield, pentode, 140
tetrode, 135
transformer, 242, 403
Emission, 5, 9
Emitter, 55
Emitter bypass capacitor analysis, 618
Emitter feedback, 337, 348
Emitter-follower amplifier, 69
h parameters, 388
model, 352
Emitter resistance, 325
Emulsion, 96
Energy gap, 13
Energy level, 5
Enhancement-type MOSFET, 110
Epitaxial mesa transistor, 90
Equalization, 449
Equal loudness contour, 223
Equivalent circuit, *see* Model
Esaki diode, 115, 710
Excess current, tunnel diode, 117
Exciting current, 402
Extrinsic, 15

Face-centered lattice, 12
Feedback, 589
bandwidth, 596
cathode follower, 625
collector to base, 342, 348
distortion, 594

emitter, 337, 348
factor, 591
fundamental equation, 591
negative, 593
negative current, 611
negative voltage, 607
operational amplifier, 600
positive, 591
shunt, 615
vacuum-tube current, 622
winding, 650
Fermi level, 14
Field Effect Transistor, amplifier principles, 105
bias, 293, 513
breakdown voltages, 105
characteristics, JFET, 102
construction and operation, 99
dual insulated gate MOSFET, 112
enhancement type MOSFET, 110
graphical amplifier analysis, 513
high-frequency amplifier, 725
high-frequency model, 725
insulated gate (IGFET), 108
integrated circuit, 107
junction, 99
leakage current, 105
load line, 293, 513
model, 358, 725
MOSFET, 108
operation, 99
Figure of merit, *Q*, 238, 259
Filament, 10
Filter, active, 604
capacitor, 163
choke, 169, 240
L, 169
low-pass, 161
π, 173
Firing, SCR, 192
thyratron, 205
thyristor, 126
triac, 202
Fixed-film resistor, 229
Fletcher-Munson curves, 223
Flip-flop, 685
Fluorescence, 24
FM, *see* Frequency modulation
Forbidden band, 7
Forced frequency, 700
Forward bias, 29
Four-terminal networks, 277
Fourier series for rectifier output, 163
Free-running frequency, 700
Frequency, short-circuit, unity gain, 721
Frequency conversion, 798
Frequency cutoff, alpha, 432
beta, 433, 720
Frequency deviation, 754
Frequency distortion, 303
Frequency modulation, 752
deemphasis, 772
definition, 740
discriminator, 792
limiter, 790
multiplex, 804
narrow band, 754
preemphasis, 772
ratio detector, 795
reactance circuits, 762
stereo, 804
wide band, 754
Frequency multiplication, 569
Frequency multiplier, class-C, 580
Frequency response, transformer, 403
Frequency response plotting, 212
Full-wave bridge rectifier, 157
Full-wave rectifier, 153

Gain-bandwidth product, feedback, 598
HF, 723
Gas discharge, 19
Gate, 100
Gate time duration, 690
Glow discharge, 21
Grain boundaries, 12
Graphical analysis, FET amplifier, 513
R-C coupling, 415
Grid, action of, 73
Grid circuit analysis, class-C, 575, 581
Grid-leak bias, 181, 565
oscillator, 650
Grid-leak detector, 785
Grid-plate transconductance, 82
Ground, 328
Grounded-grid amplifier, 481

h_{fb}, 62
h_{FB}, 58
Half-wave rectifier, 34, 147
Harmonic, 273
analysis, push-pull, 537
Harmonic generator, class-C, 580

Hartley oscillator, 651
Heater, 10
Heat sink, 524
Heterodyning, 798
High field emission, 5
High frequencies, R-C coupling, 420, 428
High frequency amplifiers, 717
Holding current, 126, 129
Hole, 14, 18
h parameters, 371
  summary of equations, 393
Hybrid parameters, 371
  general definitions, 279
  summary of equations, 393
Hybrid-π model, 717
Hysteresis, 402

ICW, 743
IGFET, 108
Impedance matching transformer, 400
Impedance ratio, 399
Index of modulation, 754
Inductance, definition, 234
Inductor, 234
Inflection point, tunnel diode, 117
Input loading, 434
Insulated-gate FET, 108
Insulator, 8
Integrated circuit, capacitor, 246
  coil, 241
  construction and definitions, 92
  JFET, 107
  resistor, 229
Integrating amplifier, 605
Interbase resistance, 121
Interelectrode capacitance, pentode, 140
  tetrode, 135
  triode, 135
Interface wiring, 93
Intermodulation distortion, 304
Interrupted continuous wave, 743
Intrinsic, 14
Intrinsic regions, 113
Intrinsic standoff ratio, 121
Inverting amplifier, 600, 604
Ionization, 5, 19
Ionization time, 22
Island, 113

Junction FET, 99
Junction temperature, 524

Knee, Zener diode, 42

Langmuir-Child law, 39
Large-scale integration, 94
LAS, 124
LASCR, 131
Latching current, 129, 499
Leakage current, FET, 105
  I_{CBO}, 59, 499
  I_{CEO}, 63, 501
Leakage inductance, 398
L filter, 169
Light, 5
  generation, 22
Light-activated SCR, LASCR, 131
Light-activated switch, LAS, 124
Light emitting diode, 52
Light in crystals, 18
Lightning arrestor, 8
Light packet, 23
Limitations, of a transistor, 71
  of a triode, 80
Limiter, 790
Linear detector, 783, 786
Linear gate, 107
Linearity, FM transmitter, 760
Litz wire, 237
Loading, input, 434
Loading resistor, 257
Load line, analytic approach, 321
  class-C, 571
  common-base amplifier, 283
  common-emitter amplifier, 289
  concept, 281
  dc and ac, *R-C* coupling, 413
  FET amplifier, 293, 513
  R-C coupling, 413
  transformer coupling, 405
  vacuum-tube amplifier, 296
Logarithms, 213
Loudness, 223
Loudness control, 225, 447
Lower side band, 742
Low frequencies, R-C coupling, 420, 423
Low-pass filter, 161
LSI, 94

Majority current carrier, 14, 17
Mask, 96
Matching, *h* parameters, 376
Matching transformer, 400
Maximum available gain, 377
MCW, 743
Mean free path, 20

Medium scale integration, 93
Meltback process, 88
Mesa transistor, 90
Metallized capacitor, 246
Metallized contacts, 96
Metal-oxide-semiconductor-FET, 108
Mica capacitor, 245
Microbar, 224
Microelectronics, *see* Integrated circuit
Microetched diffused transistor, 115
Micrometer, 10
Middle frequencies, R-C coupling, 420
Miller crystal oscillator, 665
Miller effect, 435, 720
Minority current carrier, 14, 17, 34
Mixing, audio, 604
  detection, 800
Model, common-base amplifier, 328
  common-collector amplifier, 352
  common-emitter amplifier, 334
  diode, 44
  FET amplifier, 358
  FET high frequency, 725
  hybrid-π, 717
  transformer, 402
  vacuum-tube amplifier, 365
Modulated continuous wave, 743
Modulating frequency, 740
Modulation, 739
  amplitude, 740
    circuits, 749
  balanced modulator, 767
  frequency, 752
  reactance circuits, 762
  suppressed carrier, 772
Molecule, 1
Monolithic, 96
Monostable multivibrator, 689
MOSFET, 108
MOST, 108
Motorboating, 629
MSI, 93
Multiple ionization, 20
Multiplex, FM, 804
Multivibrator, astable, 692
  bistable, 685
  cathode coupled, 698
  monostable, 689
  synchronization, 700
  unbalanced, 696
Mutual characteristics, triode, 76, 82
Mutual impedance, 266
Mutual inductance, 259, 270
Mylar capacitor, 245
Narrow band FM, 754
Negative current feedback, 611
Negative feedback, 593
Negative resistance, arc discharge, 21
  oscillators, 707
  tetrode, 139
  tunnel diode, 117
  unijunction transistor, 121
Negative voltage feedback, 607
Neutralization, 584
Neutron, 3
N material, 15
Nomenclature, 299
Noninverting amplifier, 600, 604
Nonsinusoidal oscillators, 677
NPIN transistor, 114
N-P junction, 27
N+ substrate, 94
Nucleus, 2
Numerical analysis, class-C, 572, 575, 581

Octave, Bode plot , 441
Opaque, 18
Open-loop gain, 590
Operational amplifier, 600
Optical window, 124
Orbit, 2
Oscillator, 637
  Armstrong, 647
  astable multivibrator, 692
  bistable multivibrator, 685
  blocking, 703
  bridge, 642
  cathode-coupled multivibrator, 698
  Clapp, 658
  Colpitts, 655
  conditions for, 592
  crystal, 661
  discharge circuit, 706
  electron-coupled, 671
  flip-flop, 685
  Hartley, 651
  limitations, 672
  Miller crystal, 665
  monostable multivibrator, 689
  negative resistance, 707
  phase shift, 637
  Pierce crystal, 669
  Schmitt trigger, 681
  thyratron relaxation, 679
  tunnel diode, 710
  ultra-audion, 658
  unijunction transistor, 709, 713
Oxide-coated cathode, 10

π filter, 173
Pad, 447
Padder, 247
Paper capacitor, 245
Parallel ac circuits, 255
Parallel resistance, capacitor, 244
  coil, 238
Parasitics, 674
Paschen's law, 20
Passivation, 91
Pauli exclusion principle, 7
Peak inverse voltage, 35
  bridge rectifier, 157
  capacitor filter, 164
  full-wave rectifier, 154
  half-wave rectifier, 35
  voltage doubler, 176
Peak point, tunnel diode, 117
  unijunction transistor, 121
Peak-to-valley ratio, tunnel diode, 117
Pentode, 139
Permeability, 234
Permissive energy level, 5
Phase, common-base amplifier, 68
  common-collector amplifier, 70
  common-emitter amplifier, 69
  FET amplifier, 105
  vacuum-tube amplifier, 80
Phase distortion, 304
Phase inversion, 69
Phase inverter, 543
Phase modulation, 740, 760
Phase-shift circuits, 189
Phase-shift oscillator, 637
Photoemission, 6, 10
Photolithography, 96
Photon, 23
Pierce crystal oscillator, 669
Piezoelectric effect, 663
Pinchoff, voltage, 100
Pitch control, 802
Plasma, 22
Plate, 36
Plate characteristics, triode, 76
Plate circuit analysis, class-C, 572, 581
Plate dissipation, triode, 80
Plate resistance, 40
  triode, 82
Plate saturation, 38
P material, 17
Polycrystalline, 12
Positive feedback, 591
Potentiometer, 231
Power, sideband, AM, 745
  FM, 756
Power supplies, full-wave bridge, 157
  full-wave rectifier, 153
  half-wave rectifier, 147
  regulated, 630
  shunt rectifier, 178
  transistor, 714
  voltage multiplier, 175
PNIP transistor, 114
P-N junction, 27
Preemphasis, 772
Proton, 3
Pulse modulation, 740
Pulse-repetition rate, 704
Punch-through voltage, 73
Push-pull amplifier, 535

Q, coil, parallel and series, 240
  definition, 238, 259
Q gain, 254
Q-point, 283
  R-C coupling, 414
Quanta, 4
Quartz crystal, 663

r'_e, 326
Rate-grown transistor, 87
Ratio detector, 795
Ratio of rectification, 151, 155, 158
R-C coupled amplifiers, 413
Reach-through voltage, 73
Reactance circuits, 762
Rectangular gate, 107
Rectifier, bridge, 157
  comparison of basic circuits, 159
  controlled rectifier principles, 183
  filter, capacitor, 163
    choke, 169
    L, 169
    π, 173
  full-wave, 153
  half-wave, 34, 147
  phase-shift circuits, 189
  shunt, 178
  silicon controlled rectifier, 128, 192
  thyratron, 205
  triac, 131, 202
  voltage doublers, 175
  voltage multipliers, 175
Reference diode, 41
Regenerative feedback, 592
Regulated power supplies, 630

Regulation, definition, 171
Rejection, common-mode, 470
Relaxation oscillator, thyratron, 679
  unijunction, 709, 713
Remote cutoff tube, 143
Resistance collector, 346
Resistance-capacitance coupled amplifiers, 413
Resistance of a square, 230
Resistor, 227
  integrated circuit, 229
Response, R-C coupled amplifier, 423, 428
Reverse bias, 29
Reverse blocking diode thyristor, 124
Reverse blocking triode thyrister, 129
Reverse current, 30, 34
RF amplifier, 717
RF tuned amplifier, 728
Rheostat, 231
Ring, 4
Ring modulator, 767
Ripple, 149, 155, 158
Rochelle salt crystal, 663
Role-off, Bode plot, 444
  low-pass filter, 161
Runaway, thermal, 530

Saturable-core reactor, 191
Saturation, 285
  choke, 240
  current, 30
  gas discharge, 21
  transistor, 71
SCA, 805
Schmitt trigger, 681, 706
Screen decoupling filter, 298
Screen grid, beam power tube, 144
  pentode, 139
  tetrode, 135
Secondary emission, 5, 138, 139
Semiconductor, 8
Separation, stereo, 805
Series, Fourier for rectifier output, 163
Series ac circuits, 251
Series resistance, capacitor, 244
  coil, 238
Sharp cutoff tube, 142
Shell, 4
Shockley diode, 124
Short-circuit, unity-gain frequency, 721
Shunt feedback, 615
Shunt rectifier, 178
Sideband power, AM, 745
  FM, 756
Sidebands, AM, 742
  FM, 756
Silicon controlled rectifier, 128
  rectifier circuits, 192
  specification data, 192
Silicon controlled switch, 131
Silicon planar transistor, 90
Single side band, 773
Sink, heat, 524
Skin effect, 236
Small signal analysis, 325
Solid state switch, 126
Sound, 223
Sound pressure, 224
Source, 99
Space charge, 37
Space-charge equation, 39
Space charge limited, 38
Space charge neutrality, 15
Spectral response, 11
Split-stator capacitor, 657
Square, 230
Square law detector, 783
Stability, beta, 485
  circuit analysis, 490
  temperature, 502
Standard nomenclature, 299
Static load line, R-C coupling, 415
Static resistance, 40
Stereo, 804
Stereo separation, 805
Storage temperature, 524
Storecast, 805
Subcarrier, FM stereo, 805
Subcarrier Communications Authorization, 805
Subshell, 6
Substrate, 94
Subtractor, FM stereo, 804
Summer, FM stereo, 804
Summer amplifier, 604
Supercontrol, 143
Suppressed carrier, 772
Suppressor, 139
  beam power tube, 144
Surface-barrier transistor, 89
Sustaining voltage, 72
Swamping, 330, 338, 546
Sweep speed, 706
Swinging choke, 171
Switch, solid state, 126
Synchronization, 700

Tank circuit, class-C, 569

Tantalum capacitor, 245
Taper, 232
Tapped coil, 259
Temperature, 13, 17
  transistor, 524
Temperature sensitivity, 502
Temperature stability, 502
Tetrode, 135
Tetrode transistor, 91
Thermal breakdown, 32
Thermal lag, 524
Thermal resistance, 525
Thermal runaway, 32, 530
Thermionic emission, 9
Thermistor, 14, 521, 537
Thoriated tungsten, 10
Three-halves power law, 39
Threshold limiting, 790
Threshold voltage, 110
Thyratron, 205, 679
Thyristor, 123
  SCR, 128
  triac, 131
Thyrite, 8
Tickler winding, 650
Tone control, 446
Transconductance, FET, 104, 112
  transistor, 720
  tunnel diode, 117
  vacuum-tube, 82
Transfer characteristics, FET, 102, 301
  triode, 76
Transformer, air-core, 268
  frequency response, 403
  model, 402
  steel-core, 241, 397
  tuned primary, tuned-secondary, 269
  untuned primary, tuned secondary, 268
Transformer-coupled amplifier, 397
  load line, 405
  vacuum-tube, 411
Transformer utilization factor, 151, 155, 158
Transistor, 55
  bias circuits, 310
  construction, 87
  drift transistor, 115
  integrated circuit, 95
  intrinsic regions, 113
  limitations, 71
  microetched diffused, 115
  NPIN, PNIP, 114
  power supply, 714
  types, 87
  unijunction, 120
Transit time, 481
Transmitter, AM, 752
  FM, 770
    FM linearity check, 760
Transparent, 18
Triac, operation and characteristics, 131–133
  rectifier circuits, 202
Trigger, Schmitt, 681, 706
Trigger diac, 127
Triggering, thyristor, 126
Trimmer, 247
Triode, 73
  amplifier principles, 78
  characteristic, 75
  dynamic coefficients, 80
  limitations, 80
Tripler, 580
Tuned amplifier, 728
Tuned primary, tuned secondary coil, 269
Tungsten, 9
Tuning, class-C amplifier, 578
  class-C neutralization, 584
Tuning capacitor, 247
Tuning core, 235
Tunnel diode, 115
Tunnel diode oscillator, 710
Turns ratio, 399

Ultra-audion oscillator, 658
Unbalanced multivibrator, 696
Unijunction transistor, 120
  relaxation oscillator, 709, 713
  SCR firing, 197
Unipolar transistor, 99
Unity-gain short-circuit frequency, 721
Universal amplification curve, 426
Untuned primary, tuned secondary coil, 268
Upper sideband, 742

Vacancy level, 7
Vacuum in a tube, 21
Vacuum tube, beam-power, 143
  diode, 36
  miscellaneous types, 145
  pentode, 139
  remote and sharp cutoff, 140
  tetrode, 135
  triode, 73
Vacuum-tube amplifier, 78
  current feedback, 622
  load line, 296
  model, 365
  transformer-coupled, 411
Valence, 4

Valence band, 8
Valley point, tunnel diode, 117
  unijunction transistor, 121
Varactor diode, 117, 442
Varactor diode modulator, 762
Variable-mu, 143
Variable-voltage capacitor, 117, 442, 762
Variational plate resistance, 40
Varistor diode, 44
Virtual cathode, 38
Virtual suppressor, 144
Voltage-controlled variable-resistance, 107
Voltage doublers, 175
Voltage follower, 600
Voltage gradient, 32
Voltage multiplier, 175
Voltage regulation, definition, 171
Voltage regulator, cold cathode diode, 51
  Zener diode, 42
Volume control, 446
Volume unit, 220

Wafer, 96
Wavelength of light, 10
Weber-Fechner law, 212
Wheatstone bridge, 642
Wide band FM, 754
Wien bridge, 644
Wire-wound potentiometer, 232
Wire-wound resistor, 228
Work function, 9

Yield, 96

Zener, voltage regulator, 42
Zener breakdown, 33
Zener diode, 41
Zener potential, 33
Zero decibel reference, 219

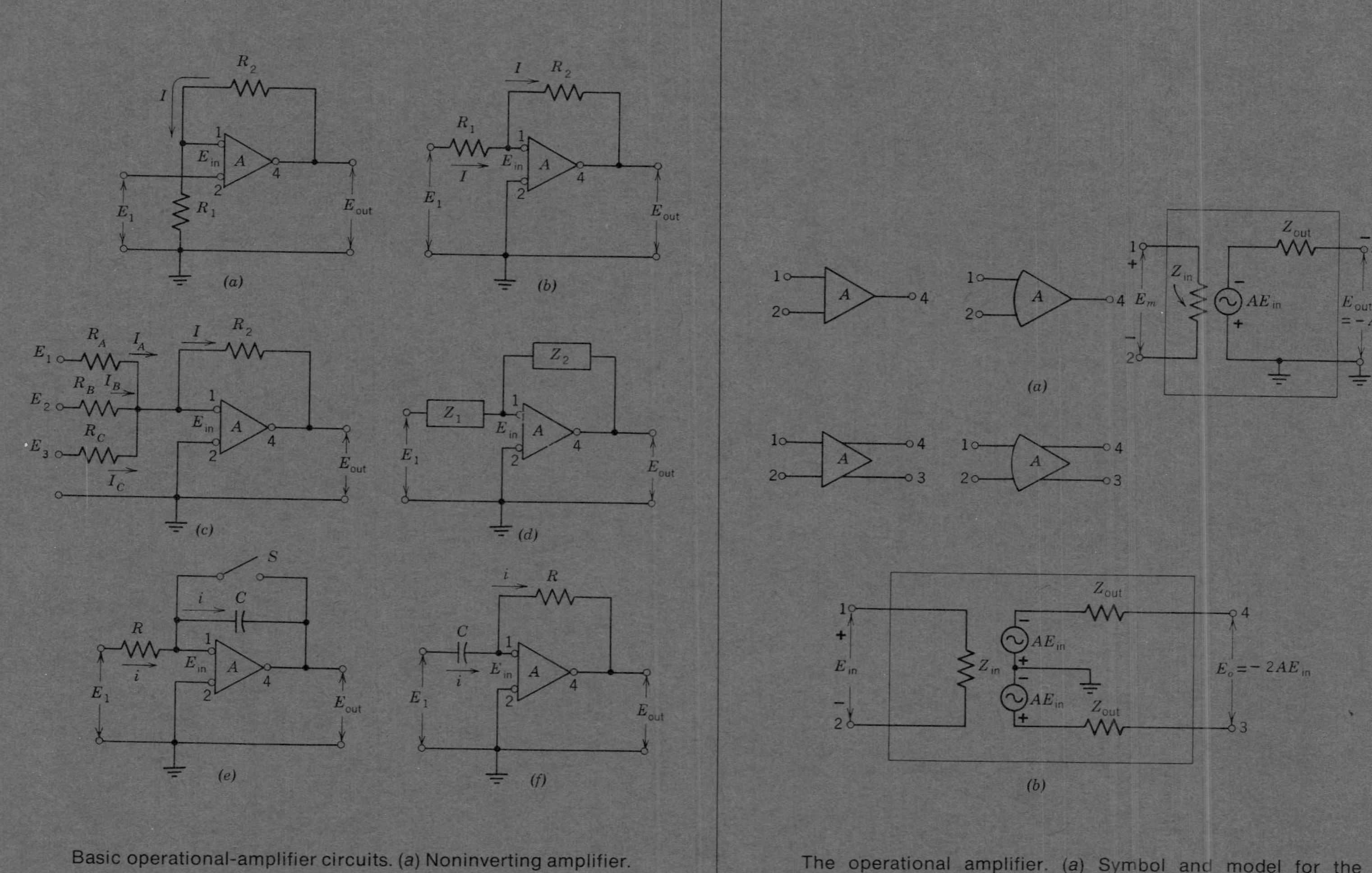

Basic operational-amplifier circuits. (*a*) Noninverting amplifier. (*b*) Inverting amplifier. (*c*) The voltage summer. (*d*) General case. (*e*) Integrator. (*f*) Differentiator.

The operational amplifier. (*a*) Symbol and model for the operational amplifier. (*b*) Symbol and model for the differential output operational amplifier